Continuous Distributions

Beta
$0 < \alpha$
$0 < \beta$

$$f(x) = \frac{\Gamma(\alpha + \beta)}{\Gamma(\alpha)\Gamma(\beta)} x^{\alpha-1}(1-x)^{\beta-1}, \qquad 0 < x < 1$$

$$\mu = \frac{\alpha}{\alpha + \beta}, \qquad \sigma^2 = \frac{\alpha\beta}{(\alpha + \beta + 1)(\alpha + \beta)^2}$$

Chi-square
$\chi^2(r)$
$r = 1, 2, \ldots$

$$f(x) = \frac{1}{\Gamma(r/2)2^{r/2}} x^{r/2-1} e^{-x/2}, \qquad 0 \le x < \infty$$

$$M(t) = \frac{1}{(1 - 2t)^{r/2}}, \qquad t < \frac{1}{2}$$

$$\mu = r, \qquad \sigma^2 = 2r$$

Exponential
$0 < \theta$

$$f(x) = \frac{1}{\theta} e^{-x/\theta}, \qquad 0 \le x < \infty$$

$$M(t) = \frac{1}{1 - \theta t}, \qquad t < \frac{1}{\theta}$$

$$\mu = \theta, \qquad \sigma^2 = \theta^2$$

Gamma
$0 < \alpha$
$0 < \theta$

$$f(x) = \frac{1}{\Gamma(\alpha)\theta^{\alpha}} x^{\alpha-1} e^{-x/\theta}, \qquad 0 \le x < \infty$$

$$M(t) = \frac{1}{(1 - \theta t)^{\alpha}}, \qquad t < \frac{1}{\theta}$$

$$\mu = \alpha\theta, \qquad \sigma^2 = \alpha\theta^2$$

Normal
$N(\mu, \sigma^2)$
$-\infty < \mu < \infty$
$0 < \sigma$

$$f(x) = \frac{1}{\sigma\sqrt{2\pi}} e^{(x-\mu)^2/2\sigma^2}, \qquad -\infty < x < \infty$$

$$M(t) = e^{\mu t + \sigma^2 t^2/2}$$

$$E(X) = \mu, \qquad \mathrm{Var}(X) = \sigma^2$$

Uniform
$U(a, b)$
$-\infty < a < b < \infty$

$$f(x) = \frac{1}{b - a}, \qquad a \le x \le b$$

$$M(t) = \frac{e^{tb} - e^{ta}}{t(b - a)}, \qquad t \ne 0; \qquad M(0) = 1$$

$$\mu = \frac{a + b}{2}, \qquad \sigma^2 = \frac{(b - a)^2}{12}$$

PROBABILITY
AND
STATISTICAL INFERENCE

PROBABILITY
AND
STATISTICAL INFERENCE

EIGHTH EDITION

Robert V. Hogg
University of Iowa

Elliot A. Tanis
Hope College

Prentice Hall
is an imprint of

Upper Saddle River, NJ 07458

Library of Congress Cataloging-in-Publication Data

Hogg, Robert V.
  Probability and statistical inference/Robert V. Hogg, Elliot A. Tanis.–8th ed.
    p. cm.
  Includes bibliographical references and index.
  ISBN 0-321-58475-9 (978-0-321-58475-5)
  1. Probabilities. 2. Mathematical statistics. I. Tanis, Elliot A. II. Title.
  QA273.H694 2010
  519.2–dc22

                                                    2008043025

Editor-in-Chief: *Deirdre Lynch*
Acquisitions Editor: *Christopher Cummings*
Assistant Editor: *Christina Lepre*
Project Manager: *Raegan Keida Heerema*
Associate Managing Editor: *Bayani Mendoza deLeon*
Senior Managing Edior: *Linda Mihatov Behrens*
Senior Operations Supervisor: *Diane Peirano*
Editorial Assistant: *Dana Jones*
Marketing Manager: *Alex Gay*
Marketing Assistant: *Kathleen DeChavez*
Art Director: *Jayne Conte*
Cover Designer: *Bruce Kenselaar*
Art Studio/Formatter: *Laserwords*
Cover Image: *The image on the cover shows the joint p.d.f. of the sum of two U-shaped random variables (see Figure 5.6-4 on page 261) and a single U-shaped random variable*

Prentice Hall
is an imprint of

© 2010, 2006, 2001, 1997 by Pearson Education, Inc.
Pearson Prentice Hall
Pearson Education, Inc.
Upper Saddle River, New Jersey 07458

Printed in the United States of America

10 9 8 7 6 5 4 3 2 1

ISBN-10:   0-321-58475-9
ISBN-13: 978-0-321-58475-5

Pearson Education Ltd., *London*
Pearson Education Singapore, Pte. Ltd.
Pearson Education Canada, Inc.
Pearson Education—Japan
Pearson Education Australia Pty. Limited, *Sydney*
Pearson Education North Asia Ltd., *Hong Kong*
Pearson Educación de Mexico, S.A. de C.V.
Pearson Education Malaysia, Pte. Ltd.
Pearson Education, Inc., Upper Saddle River, New Jersey

Contents

Preface

We are pleased with the reception that was given to the first seven editions of *Probability and Statistical Inference*. We believe, after seven trys, that this is our best edition for students at the junior level who have taken a good standard course in calculus. While the text is designed for a two-semester course, we've provided suggestions on how it can be adapted for a one-semester course around four semester hours. Of course, no previous study of probability or statistics is needed. The material in this text is taught by many departments of mathematics or statistics and is actually required of many students who want to major in one of those departments.

One of the authors has been associated with the excellent program in actuarial science at the University of Iowa for a long time and no doubt has been influenced by the needs of actuarial students. However, we certainly have not designed the text for their needs only, but have written it to reflect what we believe is the best course for all students taking probability and mathematical statistics at this level. We must note, however, that actuarial students who truly understand the first five chapters, including the sections on Bayes's theorem and additional models, and who can work the exercises should be able to pass the first actuarial examination. This exam is now called Exam P by the Society of Actuaries and Exam 1 by the Casualty Actuarial Society. The first four chapters of the book contain material that most statisticians believe provides an excellent course in probability and in probability distributions of the discrete and continuous types, both of one and two random variables. In Chapter 5, these concepts are extended to many random variables, particularly those which are independent. This important chapter includes transformations of random variables and the moment-generating function technique, which leads to the central limit theorem, although its proof is given in Chapter 10, which is titled "Some Theory."

For a two-semester course, we assume that the instructor will cover most, if not all, of the sections included in the first seven chapters. Chapter 6 is a fine one on estimation, both point and interval, with a clear explanation concerning the sample size required to achieve a certain accuracy. Chapter 7 is about the other aspect of statistical inference, namely, tests of statistical hypotheses, and goes through elementary analysis of variance and simple linear regression. The connection between tests of statistical hypotheses and confidence intervals is clearly explained, including that between one-sided tests and confidence intervals.

Then, to complete the course, the instructor can select topics of his or her interests from the chapters titled "Nonparametric Methods," "Bayesian Methods," "Some

Theory," and "Quality Improvement Through Statistical Methods." We believe that we have included those additional topics which will appeal to most instructors' interests.

Chapter 8 on nonparametric methods include the topic of order statistics, leading to distribution-free confidence intervals for percentiles, resampling methods, and many standard methods in this area. Bayesian methods are in the substantial, but brief Chapter 9, which includes an interesting discussion of subjective probability. We decided to collect much of the theory given in a mathematical statistics course in Chapter 10. This includes consideration of sufficient statistics, the power of statistical tests, best critical regions, likelihood ratio tests, and many limiting ideas, including a proof of the central limit theorem. Many instructors might want to include these theoretical concepts in their course. Finally, Chapter 11 gives some basic ideas about quality improvement in the workplace via statistical methods, with a short description of the Six Sigma program.

For the one-semester course, we believe that instructors can select sections from the first seven chapters.

We have tried to make the eighth edition more "user-friendly;" yet we do want to reinforce certain basic concepts of mathematics, particularly calculus. To help the student with methods used in the algebra of sets and in calculus, we include Appendix D, a "Review of Selected Mathematical Techniques," on the companion CD-ROM bound into the back of this book. This review includes a method that makes integration by parts easier. Also, we derive the important "Rule of 72," which provides an approximation to the number of years necessary for money to double.

Some Special Features

- There is a better and more logical flow in this edition, and that is one of the reasons we believe it to be our best. For example, the normal distribution has been moved to the end of Chapter 3. Now all of the usual one-variable distributions which are used by statisticians are in Chapters 2 and 3.

- In the first five chapters, we have included several "real" probability examples and exercises that will appeal to students of actuarial science, finance, economics, and so on. We are less dependent upon problems involving coins, cards, and dice, although we find that students are interested in some of the latter problems.

- Chapter 6 on estimation and Chapter 7 on tests of hypotheses are now back-to-back. Furthermore, confidence intervals and tests of hypotheses are tied together.

- There is a short, but excellent, Bayesian chapter, including real examples and an indication of how Bayesians prove theorems by establishing "Dutch books." The technique was pointed out to us by George Woodworth.

- There is a very good coverage of conditional distributions, correlation, and distributions that are mixtures of the discrete and continuous cases.

- At the end of each chapter, we have added interesting historical comments, such as one that explains how the normal distribution was discovered.

- In the section on bootstrapping, there is an explanation of the origin of this word.

- Examples are given of Simpson's paradox.

- An explanation of the Six Sigma program is given in Chapter 11.

- The figures include the use of color.

- The companion CD-ROM includes not only all of the data sets in several formats—ASCII files, Minitab worksheets, as text files, and a format used by *Maple*—but also many more applications of Minitab and *Maple*, as well as Appendix D, a Review of Selected Mathematical Techniques of the book.
- Illustrations of *Maple* as a computer algebra system are given in the text and on the CD-ROM.

Important Points in This Edition

We recognize the importance of technology in modern statistics by including supplements that actively engage students in using the computer. Some of these supplements illustrate the use of Minitab for calculating probabilities, analyzing data, and applying statistical inference in the form of tests of hypotheses and confidence intervals. The power of a computer algebra system (CAS) for theoretical computations is illustrated with *Maple*. The importance of simulation is also demonstrated. Some of the illustrations are printed in the text. There are also new illustrations on the CD-ROM and you are encouraged to check these out.

There are many exercises in the text illustrating the mathematics of probability and statistics but a great number are concerned with applications. We've included applications in the areas of biology, education, economics, engineering, environmental studies, exercise science, health science, manufacturing, opinion polls, psychology, sociology, and sports.

Throughout the book, figures and real applications will help the student understand probability and statistics and what they can accomplish. For some exercises, it is assumed that calculators or computers are available; thus, the solutions will not always involve "nice" numbers. Data sets for all of the exercises are available on the companion CD-ROM. The data are provided in different formats so that they should be accessible to most computer programs. Finally, in the first part of the book, concerning probability, supplementary comments are inserted about statistics and computation.

Ancillaries

An **Instructor's Solutions Manual** containing worked-out solutions to the even-numbered exercises in the text is available for download from Pearson Education's online catalog. Many of the numerical exercises were solved with *Maple*. For additional exercises that involve simulations, a separate manual, *Probability & Statistics: Explorations with MAPLE*, second edition, by Zaven Karian and Elliot Tanis, is available for download from Pearson Education's online catalog. Several exercises in that manual also make use of the power of *Maple* as a computer algebra system.

The **CD-ROM** contains all of the data sets in various formats. There are also applications of Minitab for drawing figures, calculating probabilities, calculating characteristics of a sample, and statistical inference. One folder contains some supplementary *Maple* programs that are useful in probability and statistics. These programs were used to construct the figures in this text and for other applications. *Maple* was also used to animate some of the figures in the text. All you need is a browser. Simply load in the directory and you can pull up these animations. Appendix D, a "Review of Selected Mathematical Techniques," is also available on the CD-ROM.

If you find any errors in this text, please send them to tanis@hope.edu so that they can be corrected in a future printing. These **errata** will also be posted on http://www.math.hope.edu/tanis/.

Acknowledgments

We wish to thank our colleagues, students, and friends for many suggestions and for their generosity in supplying data for exercises and examples. In particular we would like to acknowledge the excellent suggestions from our copy editor, Brian Baker, and the fine work of our accuracy checkers, Tom Wegleitner and Roxane Barrows. We also thank the University of Iowa and Hope College for providing office space and encouragement. Finally, our families, through eight editions, have been most understanding during the preparation of all of this material. We would especially like to thank our wives, Ann and Elaine. We truly appreciate their patience and needed their love.

Robert V. Hogg
Elliot A. Tanis
tanis@hope.edu

Prologue

The discipline of statistics deals with the collection and analysis of data. Advances in computing technology, particularly in relation to changes in science and business, have increased the need for more statistical scientists to examine the huge amount of data being collected. We know that data are not equivalent to information. Once data (hopefully of high quality) are collected, there is a strong need for statisticians to make sense of them. That is, data must be analyzed, in order to provide information upon which decisions can be made. In light of this great demand, opportunities for the discipline of statistics have never been greater, and there is a special need for more bright young persons to go into statistical science.

If we think of fields in which data play a major part, the list is almost endless: accounting, actuarial science, atmospheric science, biological science, economics, educational measurement, environmental science, epidemiology, finance, genetics, manufacturing, marketing, medicine, pharmaceutical industries, psychology, sociology, sports, and on and on. Because statistics is useful in all of these areas, it really should be taught as an applied science. Nevertheless, to go very far in such an applied science, it is necessary to understand the importance of creating models for each situation under study. Now, no model is ever exactly right, but some are extremely useful as an approximation to the real situation. Most appropriate models in statistics require a certain mathematical background in probability. Accordingly, while alluding to applications in the examples and the exercises, this textbook is really about the mathematics needed for the appreciation of probabilistic models necessary for statistical inferences.

In a sense, statistical techniques are really the heart of the scientific method. Observations are made that suggest conjectures. These conjectures are tested, and data are collected and analyzed, providing information about the truth of the conjectures. Sometimes the conjectures are supported by the data, but often the conjectures need to be modified and more data must be collected to test the modifications, and so on. Clearly, in this iterative process, statistics plays a major role with its emphasis on proper design and analysis of experiments and the resulting inferences upon which decisions can be made. Through statistics, information is provided that is relevant to taking certain actions, including improving manufactured products, providing better services, marketing new products or services, forecasting energy needs, classifying diseases better, and so on.

Statisticians recognize that there are often small errors in their inferences, and they attempt to make the probabilities of those mistakes as small as possible. That these uncertainties even exist is due to the fact that there is variation in the data. Even though experiments are repeated under seemingly the same conditions, the results vary from trial to trial. We try to improve the quality of the data by making them as reliable as possible, but the data simply do not fall on given patterns. In light of this uncertainty, the statistician tries to determine the pattern in the best possible way, always explaining the error structures of the statistical estimates.

This is an important lesson to be learned: Variation is almost everywhere. It is the statistician's job to understand variation. Often, as in manufacturing, the desire is to reduce variation because the products will be more consistent. In other words, car doors will fit better in the manufacturing of automobiles if the variation is decreased by making each door closer to its target values.

Many statisticians in industry have stressed the need for "statistical thinking" in everyday operations. This need is based upon three points (two of which have been mentioned in the preceding paragraph): (1) Variation exists in all processes, (2) understanding and reducing undesirable variation is a key to success, and (3) all work occurs in a system of interconnected processes. W. Edwards Deming, an esteemed statistician and quality improvement "guru," stressed these three points, particularly the third one. He would carefully note that you could not maximize the total operation by maximizing the individual components unless they are independent of each other. However, in most instances, they are highly dependent, and persons in different departments must work together in creating the best products and services. If not, what one unit does to better itself could very well hurt others. He often cited an orchestra as an illustration of the need for the members to work together to create an outcome that is consistent and desirable.

Any student of statistics should understand the nature of variability and the necessity for creating probabilistic models of that variability. We cannot avoid making inferences and decisions in the face of this uncertainty; however, these inferences and decisions are greatly influenced by the probabilistic models selected. Some persons are better model builders than others and accordingly will make better inferences and decisions. The assumptions needed for each statistical model are carefully examined; it is hoped that thereby the reader will become a better model builder.

Finally, we must mention how modern statistical analyses have become dependent upon the computer. Statisticians and computer scientists really should work together in areas of exploratory data analysis and "data mining." Statistical software development is critical today, for the best of it is needed in complicated data analyses. In light of this growing relationship between these two fields, it is good advice for bright students to take substantial offerings in statistics and in computer science.

Students majoring in statistics, computer science, or a joint program are in great demand in the workplace and in graduate programs. Clearly, they can earn advanced degrees in statistics or computer science or both. But, more importantly, they are highly desirable candidates for graduate work in other areas: actuarial science, industrial engineering, finance, marketing, accounting, management science, psychology, economics, law, sociology, medicine, health sciences, etc. So many fields have been "mathematized" that their programs are begging for majors in statistics or computer science. Often, such students become "stars" in these other areas. We truly hope that we can interest students enough that they want to study more statistics. If they do, they will find that the opportunities for very successful careers are numerous.

PROBABILITY
AND
STATISTICAL INFERENCE

Probability

CHAPTER

1

1.1 BASIC CONCEPTS

It is usually difficult to explain to the general public what statisticians do. Many think of us as "math nerds" who seem to enjoy dealing with numbers. And there is some truth to that concept. But if we consider the bigger picture, many recognize that statisticians can be extremely helpful in many investigations.

Consider the following:

1. There is some problem or situation that needs to be considered; so statisticians are often asked to work with investigators or research scientists.

2. Suppose that some measure (or measures) are needed to help us understand the situation better. The measurement problem is often extremely difficult, and creating good measures is a valuable skill. As an illustration, in higher education, how do we measure good teaching? This is a question to which we have not found a satisfactory answer, although several measures, such as student evaluations, have been used in the past.

3. After the measuring instrument has been developed, we must collect data through observation, possibly the results of a survey or an experiment.

4. Using these data, statisticians summarize the results, often with descriptive statistics and graphical methods.

5. These summaries are then used to analyze the situation. Here it is possible that statisticians make what are called statistical inferences.

6. Finally a report is presented, along with some recommendations that are based upon the data and the analysis of them. Frequently such a recommendation might be to perform the survey or experiment again, possibly changing some of

the questions or factors involved. This is how statistics is used in what is referred to as the scientific method, because often the analysis of the data suggests other experiments. Accordingly, the scientist must consider different possibilities in his or her search for an answer and thus performs similar experiments over and over again.

The discipline of statistics deals with the *collection* and *analysis of data*. When measurements are taken, even seemingly under the same conditions, the results usually vary. Despite this variability, a statistician tries to find a pattern; yet due to the "noise," not all of the data fit into the pattern. In the face of the variability, the statistician must still determine the best way to describe the pattern. Accordingly, statisticians know that mistakes will be made in data analysis, and they try to minimize those errors as much as possible and then give bounds on the possible errors. By considering these bounds, decision makers can decide how much confidence they want to place in the data and in their analysis of them. If the bounds are wide, perhaps more data should be collected. If, however, the bounds are narrow, the person involved in the study might want to make a decision and proceed accordingly.

Variability is a fact of life, and proper statistical methods can help us understand data collected under inherent variability. Because of this variability, many decisions have to be made that involve uncertainties. In medical research, interest may center on the effectiveness of a new vaccine for mumps; an agronomist must decide whether an increase in yield can be attributed to a new strain of wheat; a meteorologist is interested in predicting the probability of rain; the state legislature must decide whether decreasing speed limits will result in fewer accidents; the admissions officer of a college must predict the college performance of an incoming freshman; a biologist is interested in estimating the clutch size for a particular type of bird; an economist desires to estimate the unemployment rate; an environmentalist tests whether new controls have resulted in a reduction in pollution.

In reviewing the preceding (relatively short) list of possible areas of applications of statistics, the reader should recognize that good statistics is closely associated with careful thinking in many investigations. As an illustration, students should appreciate how statistics is used in the endless cycle of the scientific method. We observe nature and ask questions, we run experiments and collect data that shed light on these questions, we analyze the data and compare the results of the analysis with what we previously thought, we raise new questions, and on and on. Or if you like, statistics is clearly part of the important "plan–do–study–act" cycle: Questions are raised and investigations planned and carried out. The resulting data are studied and analyzed and then acted upon, often raising new questions.

There are many aspects of statistics. Some people get interested in the subject by collecting data and trying to make sense out of their observations. In some cases the answers are obvious and little training in statistical methods is necessary. But if a person goes very far in many investigations, he or she soon realizes that there is a need for some theory to help describe the error structure associated with the various estimates of the patterns. That is, at some point, appropriate probability and mathematical models are required to make sense out of complicated data sets. Statistics and the probabilistic foundation on which statistical methods are based can provide the models to help people do this. So in this book, we are more concerned with the mathematical, rather than the applied, aspects of statistics. Still, we give enough real examples so that the reader can get a good sense of a number of important applications of statistical methods.

In the study of statistics, we consider experiments for which the outcome cannot be predicted with certainty. Such experiments are called **random experiments**. Each experiment ends in an outcome that cannot be determined with certainty before the experiment is performed. However, the experiment is such that the collection of every possible outcome can be described and perhaps listed. This collection of all outcomes is called the **outcome space**, the **sample space**, or, more simply, the **space** S. The following examples will help illustrate what we mean by random experiments, outcomes, and their associated spaces.

EXAMPLE 1.1-1 Two dice are cast, and the total number of spots on the sides that are "up" are counted. The outcome space is $S = \{2, 3, 4, 5, 6, 7, 8, 9, 10, 11, 12\}$.   ■

EXAMPLE 1.1-2 Each of six students has a standard deck of playing cards, and each student selects one card randomly from his or her deck. We are interested in whether at least two of these six cards match (M) or whether all are different (D). Thus, $S = \{M, D\}$.   ■

EXAMPLE 1.1-3 A fair coin is flipped successively at random until the first head is observed. If we let x denote the number of flips of the coin that are required, then $S = \{x : x = 1, 2, 3, 4, \ldots\}$, which consists of an infinite, but countable, number of outcomes.   ■

EXAMPLE 1.1-4 A box of breakfast cereal contains one of four different prizes. The purchase of one box of cereal yields one of the prizes as the outcome, and the sample space is the set of four different prizes.   ■

EXAMPLE 1.1-5 In Example 1.1-4, assume that the prizes are put into the boxes randomly. A family continues to buy this cereal until it obtains a complete set of the four different prizes. The number of boxes of cereal that must be purchased is one of the outcomes in $S = \{b : b = 4, 5, 6, \ldots\}$.   ■

EXAMPLE 1.1-6 A fair coin is flipped successively at random until heads is observed on two successive flips. If we let y denote the number of flips of the coin that are required, then $S = \{y : y = 2, 3, 4, \ldots\}$.   ■

EXAMPLE 1.1-7 To determine the percentage of body fat on a person, one measurement that is made is a person's weight under water. If w denotes this weight in kilograms, then the sample space could be $S = \{w : 0 < w < 7\}$, as we know from past experience that this weight does not exceed 7 kilograms.   ■

EXAMPLE 1.1-8 An ornithologist is interested in the clutch size (number of eggs in a nest) for gallinules, a species of bird that lives in a marsh. If we let c equal the clutch size, then a possible sample space would be $S = \{c : c = 0, 1, 2, \ldots, 15\}$, as 15 is the largest known clutch size.   ■

Note that the outcomes of a random experiment can be numerical, as in Examples 1.1-3, 1.1-6, 1.1-7, and 1.1-8, but they do not have to be, as shown by Examples 1.1-2 and 1.1-4. Often we "mathematize" those latter outcomes by assigning numbers to them. For instance, in Example 1.1-2, we could denote the outcome

$\{D\}$ by the number zero and the outcome $\{M\}$ by the number one. In general, measurements on outcomes associated with random experiments are called **random variables**, and these are usually denoted by some capital letter toward the end of the alphabet, such as X, Y, or Z.

Note the numbers of outcomes in the sample spaces in these examples. In Examples 1.1-1, 1.1-4, and 1.1-8, the number of outcomes is finite. In Examples 1.1-3 and 1.1-6, the number of possible outcomes is infinite but countable. That is, there are as many outcomes as there are counting numbers (positive integers). The space for Example 1.1-7 is different from that of the other examples in that the set of possible outcomes is an interval of numbers. Theoretically the weight could be any one of an infinite number of possible weights; here the number of possible outcomes is not countable. However, from a practical point of view, reported weights are selected from a finite number of possibilities because we can read and record the answer only to an accuracy determined by our scale. Many times, however, it is better to conceptualize the space as an interval of outcomes, and Example 1.1-7 is an example of a space of the continuous type.

If we consider a random experiment and its space, we note that under repeated performances of the experiment, some outcomes occur more frequently than others. For instance, in Example 1.1-3, if this coin-flipping experiment is repeated over and over, the first head is observed on the first flip more often than on the second flip. If we can somehow, by theory or observations, determine the fractions of times a random experiment ends in the respective outcomes, we have described a **distribution of the random variable** (sometimes called a **population**). Often we cannot determine this distribution through theoretical reasoning, but must actually perform the random experiment a number of times to obtain guesses or **estimates** of these fractions. The collection of the observations that are obtained from such repeated trials is often called a **sample**. The making of a conjecture about the distribution of a random variable based on the sample is called a **statistical inference**. That is, in statistics, we try to infer from the sample to the population. To understand the background behind statistical inferences that are made from the sample, we need a knowledge of some probability, basic distributions, and sampling distribution theory; these topics are considered in the early part of this book.

Given a sample or set of measurements, we would like to determine methods for describing the data. Suppose that we have some **counting** or **discrete data**. For example, you record the number of children in the families of each of your classmates. Or perhaps your state, on a regular basis, selects 5 integers out of the first 37 positive integers for the state lottery; you could count the number of odd integers among the 5 that were chosen.

In general, suppose we repeat a random experiment a number of times—say, n. If a certain outcome—say, A—has occurred $f = \mathcal{N}(A)$ times in these n trials, then the number $f = \mathcal{N}(A)$ is called the **frequency** of the outcome A. The ratio $f/n = \mathcal{N}(A)/n$ is called the **relative frequency** of the outcome. A relative frequency is usually very unstable for small values of n, but it tends to stabilize about some number—say, p—as n increases. The number p is called the **probability** of the outcome.

To develop an understanding of a particular set of discrete data, we can summarize the data in a frequency table and then construct a histogram of the data. A **frequency table** provides the number of occurrences of each possible outcome. A **histogram** presents the tallied data graphically, as illustrated in the next example.

EXAMPLE 1.1-9 In Table 1.1-1, the number of children in each family of 100 students in two statistics classes is recorded.

TABLE 1.1-1: Number of Children per Family																			
2	2	5	3	4	4	3	3	6	4	3	4	4	4	4	2	5	9	2	3
1	3	5	2	4	4	4	3	3	2	2	4	2	2	6	6	1	3	3	3
3	2	3	4	7	3	3	3	2	2	2	2	3	2	3	2	3	2	5	2
3	2	2	2	4	3	3	2	3	2	4	3	3	3	4	2	4	1	2	2
2	4	3	3	3	5	2	3	3	2	2	3	3	4	2	2	2	7	2	3

We shall construct a frequency table in which we tally the number of times that each outcome was observed. We also add a column that gives the relative frequency of each outcome.

TABLE 1.1-2: Frequency Table of Number of Children per Family			
Number of Children	Tabulation	Frequency	Relative Frequency
1	|||	3	0.03
2	ⅣⅢ ⅣⅢ ⅣⅢ ⅣⅢ ⅣⅢ ⅣⅢ ||||	34	0.34
3	ⅣⅢ ⅣⅢ ⅣⅢ ⅣⅢ ⅣⅢ ⅣⅢ ||||	34	0.34
4	ⅣⅢ ⅣⅢ ⅣⅢ |||	18	0.18
5	ⅣⅢ	5	0.05
6	|||	3	0.03
7	||	2	0.02
8		0	0.00
9			1	0.01
Totals		100	1.00

In a table of numbers like Table 1.1-1, it is often difficult to detect features of the measurements, such as the minimum, the maximum, or the measurement that occurs most often (known as the typical value). Often this typical value is called the **mode**. The frequency table, Table 1.1-2, helps to point out these features. A more visual way to present such data is with a **histogram**. To construct a **frequency histogram**, center a rectangle with a base of length 1 at each observed integer value and make the height equal to the frequency of this integer. To construct a **relative frequency histogram** or a **density histogram**—say, $h(x)$—make the height of each rectangle equal to the relative frequency of the outcome. Note that since each base equals 1, the area of each rectangle is the respective relative frequency of the midpoint of the base and the total area of a relative frequency histogram is equal to 1. This relationship is illustrated in Figure 1.1-1 on next the page. ∎

One reason for constructing histograms is to picture the data more easily. Also, a histogram helps us to use the observed data to say something about the source of the data or the population from which the data were selected, provided that the data were selected randomly. For example, these 100 observations of number of children could perhaps help us say something about the number of children in families having

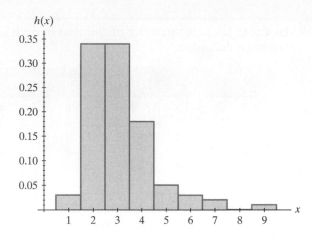

FIGURE 1.1-1: Relative frequency histogram of number of children per family

a student at Hope College. If we would like to say something about the number of children in families in the United States, we should select a representative sample of many families and look at the number of children in those families. Note that 0 is a possible outcome for this latter experiment, while that is not true of families sending at least one of their children to Hope College.

We have noted that the relative frequency of x—say, $f/n = \mathcal{N}(x)/n$—is very unstable for small n but tends to stabilize around some probability p as n increases. Since the relative frequency histogram $h(x)$ represents the relative frequencies of various values of x, we would believe that as n increases, $h(x)$ approaches some ideal function $f(x)$ that represents the probabilities of the respective values of x. That is, if x_0 is one of the possible x values with frequency $f_0 = \mathcal{N}(x_0)$, then, as n increases,

$$h(x_0) = \frac{f_0}{n} = \frac{\mathcal{N}(x_0)}{n} \to p_0 = f(x_0),$$

where p_0 is that probability about which $f_0/n = \mathcal{N}(x_0)/n$ stabilizes. Suppose that we have a certain function $f(x)$ that serves as a model for the probabilities of the outcomes of a random experiment. If we repeat that random experiment a large number of times and construct the corresponding relative frequency histogram $h(x)$, we would anticipate that $h(x)$ would be approximately equal to $f(x)$. We call this function, $f(x)$, the **probability mass function** and abbreviate it **p.m.f.** In order to graphically compare $h(x)$ and $f(x)$, we construct a **probability histogram** of $f(x)$. This is a graphical representation that has a rectangle of height $f(x)$ and a base of length 1, centered at each $x \in S$, where $x \in S$ means "x belongs to the space S." Since each base is equal to 1, the area of each rectangle is equal to the probability $f(x)$.

Let us consider a simple example in which we can intuitively assign a reasonable probability model to the outcomes of a given random experiment.

EXAMPLE 1.1-10 Place 10 chips of the same size in a bowl. The number 1 is placed on one of the chips, the number 2 on two of the chips, the number 3 on three of the chips, and finally the number 4 on four of the chips. The experiment is to select one chip at random and read the number x on the chip. Here $S = \{1, 2, 3, 4\}$. If this experiment is performed in a "fair" manner, most of us would intuitively assign the probabilities 1/10, 2/10,

3/10, and 4/10 to the respective x values $x = 1$, $x = 2$, $x = 3$, $x = 4$. That is, our model is given by the p.m.f.

$$f(1) = \frac{1}{10}, \qquad f(2) = \frac{2}{10}, \qquad f(3) = \frac{3}{10}, \qquad f(4) = \frac{4}{10}$$

or, equivalently, by

$$f(x) = \frac{x}{10}, \qquad x = 1, 2, 3, 4.$$

In Figure 1.1-2, we plot the probability histogram associated with the p.m.f. $f(x)$ on the left. To simulate this experiment, we used the random numbers in Table IX in the appendix. There are 1260 digits in this table. We assigned the digit $\{0\}$ to $x = 1$, the digits $\{1,2\}$ to $x = 2$, the digits $\{3,4,5\}$ to $x = 3$, and the digits $\{6,7,8,9\}$ to $x = 4$. The respective frequencies of $x = 1, 2, 3$, and 4 are 119, 254, 376, and 511. The respective relative frequencies are

$$h(1) = \frac{119}{1260}, \quad h(2) = \frac{254}{1260}, \quad h(3) = \frac{376}{1260}, \quad h(4) = \frac{511}{1260}.$$

In Figure 1.1-2 on the right, we plot the relative frequency histogram $h(x)$ with the p.m.f. superimposed. The top of $f(x)$ is dashed. Note that $h(x)$ approximates $f(x)$ very well—almost too well. ∎

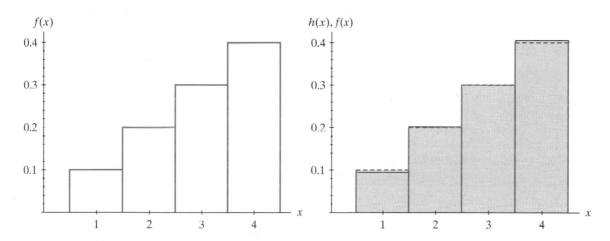

FIGURE 1.1-2: Probability histogram alone and with relative frequency histogram

The value of x that has a greatest probability (or highest relative frequency) is called the **mode** of the distribution (or of the histogram determined by the data). So in the last example, $x = 4$ is the mode.

EXAMPLE 1.1-11   A student in a statistics class was asked to write down a string of one-digit numbers $(0, 1, 2, \ldots, 9)$ that seemed to her to be random. She recorded the following 40 numbers in this order:

8, 2, 6, 0, 3, 9, 1, 6, 5, 8, 7, 4, 9, 5, 0, 5, 2, 7, 5, 2
4, 8, 1, 3, 6, 5, 2, 8, 4, 1, 0, 8, 1, 2, 7, 6, 1, 9, 4, 0

Let us count the number of times she recorded the same number next to itself (like 5, 5), the number of times adjacent numbers were next to each other (like 7, 6 and, say, 9, 0 are next to each other) and the number of times neither of these occurred. If the 40 numbers were truly selected at random, consecutive numbers would be the same about 10% of the time, consecutive numbers would differ by one about 20% of the time (recall that 9 and 0 differ by one in this example), and consecutive numbers would differ by more than one 70% of the time. The student's results were as follows:

	Frequency	Relative Frequency
Same	0	0/39 = 0.00
Differ by one	6	6/39 = 0.15
Differ by more than one	33	33/39 = 0.85
Totals	39	39/39 = 1.00

It seems as if 0.00 differs too much from 0.10, 0.15 is fairly close to 0.20, and 0.85 is somewhat higher than the expected 0.70. Intuitively, her sequence does not seem to be a very good approximation to a sequence of random digits. Later in the text, you will study a statistical method to test for the difference. ■

In many applied problems, not just those associated with probability, we want to find a model that describes the reality of the situation as well as possible. In truth, no model is exactly right, but often models are close enough to being correct that they are extremely useful in practice. However, before they can be used, models must be tested with data collected from the real situation. In this text, we are interested in constructing probability models for different random experiments. In the next few chapters, we study enough probability that allows us to assign certain reasonable probability models to important cases. Often we do not know everything about these models; that is, they frequently have unknown parameters. Thus, we must observe the results of several trials of the experiment to make further inferences about the models; in particular, we must estimate the unknown parameter(s). This process of refining a model by using the data is referred to as making a **statistical inference** and is the focus of the latter part of the text.

STATISTICAL COMMENTS   (**Simpson's Paradox**) While most of the first five chapters are about probability and probability distributions, from time to time we will mention certain statistical concepts. As an illustration, the relative frequency, f/n, is called a **statistic** and is used to **estimate** a probability p, which is usually unknown. For example, if a major league batter gets $f = 152$ hits in $n = 500$ official at bats during the season, then the relative frequency $f/n = 0.304$ is an estimate of his probability of getting a hit and is called his batting average for that season.

Once while speaking to a group of coaches, one of us (Hogg) made the comment that it would be possible for batter A to have a higher average than batter B for each season during their careers and yet B could have a better overall average at the end of their careers. While no coach spoke up, you could tell that they were thinking, "And that guy is supposed to know something about math."

Of course, the following simple example convinced them that the statement was true: Suppose A and B played only two seasons, with these results:

Season	Player A			Player B		
	AB	Hits	Average	AB	Hits	Average
1	500	126	0.252	300	75	0.250
2	300	90	0.300	500	145	0.290
Totals	800	216	0.270	800	220	0.275

Clearly, A beats B in the two individual seasons, but B has a better overall average. Note that during their better season (the second) B had more at bats than did A. This kind of result is often called **Simpson's paradox** and it can happen in real life. (See Exercises 1.1-11, 1.1-12, and 1.1-13.)

EXERCISES

1.1-1. Describe the outcome space for each of the following experiments:

(a) A student is selected at random from a statistics class, and the student's ACT score in mathematics is determined. HINT: ACT test scores in mathematics are integers between 1 and 36, inclusive.

(b) A candy bar with a 20.4-gram label weight is selected at random from a production line and is weighed.

(c) A coin is tossed three times, and the sequence of heads and tails is observed.

1.1-2. Describe the outcome space for each of the following experiments:

(a) Consider families that have three children each, and select one such family at random. Describe the sample space S in terms of their children as 3-tuples, agreeing, for example, that "gbb" would indicate that the youngest is a girl and the two oldest are boys.

(b) A rat is selected at random from a cage, and its sex is determined.

(c) A state selects a three-digit integer at random for one of its lottery games.

1.1-3. The State of Arizona generates a three-digit number at random six days a week for its daily lottery. The numbers are generated one digit at a time. Consider the following set of 50 three-digit numbers as 150 one-digit integers that were generated at random from May 1, 2000 to July 3, 2000:

```
951  728  818  922  850  835  003  406  203  603
011  217  803  776  397  019  785  185  632  245
945  929  508  849  516  729  306  305  278  100
089  860  918  124  675  220  728  751  786  609
076  320  732  911  913  556  367  897  540  979
```

Let X denote the outcome when a single digit is generated.

(a) With true random digits, what probability would you assign to each digit? That is, what is the p.m.f. of X?

(b) For the 150 observations, determine the relative frequencies of 0, 1, 2, 3, 4, 5, 6, 7, 8, and 9, respectively.

(c) Construct a probability histogram and a relative frequency histogram like those in Figure 1.1-2.

(d) A set of 32 three-digit numbers was generated from May 4, 1998 to June 9, 1998. Answer (b) and (c) for these data. What is true about the fit of the probability model to the data? (For a more complete discussion of these numbers, see "The Case of the Missing Lottery Number," by W. D. Kaigh, in *The College Mathematics Journal*, January, 2001, pages 15–19.) The numbers are as follows:

```
287  215  846  873  485  812  432  415  348  383
334  075  655  871  824  704  533  014  822  477
880  316  012  010  557  881  222  256  282  704
111  124
```

1.1-4. Some ornithologists were interested in the clutch size of the common gallinule. They observed the number of eggs in each of 117 nests, yielding the following data:

7	5	13	7	7	8	9	9	9	8	8	9	9	7	7
5	9	7	7	4	9	8	8	10	9	7	8	8	8	7
9	7	7	10	8	7	9	7	10	8	9	7	11	10	9
9	4	8	6	8	9	9	9	8	8	5	8	8	9	9
14	10	8	9	9	9	8	7	9	7	9	10	10	7	6
11	7	7	6	9	7	7	6	8	9	4	6	9	8	9
7	9	9	9	9	8	8	8	9	9	9	8	10	9	9
8	5	7	8	7	6	7	7	7	6	5	9			

(a) Construct a frequency table for these data.

(b) Draw a histogram.

(c) What is the mode (the typical clutch size)?

1.1-5. Noticing that some of the gallinules in the last exercise had a second brood during the summer, the ornithologists became interested in comparing the clutch sizes for the second brood with those for the first brood. They were able to collect the following clutch sizes for the second brood:

4	4	6	5	5	9	6		5	6	9	7	9	4	8	6	5	9
8	8	7	6	8	8	9	10	9	9	7	4	6	8	7	5	

(a) Construct a frequency table for these data.

(b) Draw a histogram.

(c) Is there a typical clutch size for second broods?

1.1-6. Before buying enough cereal to obtain a set of four prizes (Example 1.1-4), a family decided to use simulation to estimate the number of boxes it would have to buy. Each member of the family rolled a four-sided die until he or she observed each face at least once. The family members repeated this exercise 100 times and recorded the numbers of rolls needed as follows:

8	6	6	4	13	11	19	13	4	15	8	13	5	8	9
16	9	5	12	6	4	8	6	6	6	11	6	5	10	5
6	5	8	4	5	14	5	7	5	7	5	8	4	9	8
9	13	8	14	5	24	4	5	6	7	5	4	8	7	6
11	4	5	6	5	10	10	4	5	14	10	15	6	9	8
7	10	14	10	8	8	10	7	9	10	8	10	9	7	5
12	7	16	6	5	11	4	11	5	5					

(a) Construct a frequency table for these data.

(b) Draw a histogram.

1.1-7. For a Texas lottery game, Cash Five, 5 integers are selected randomly out of the first 37 positive integers. The following table lists the numbers of odd integers out of each set of five integers selected in 100 consecutive drawings in 2007:

3	2	2	2	2	3	3	2	4	2	4	2	2	3	2	3	1	3	2	3
4	3	3	4	5	3	2	1	2	0	2	2	2	3	1	2	1	3	3	1
2	5	4	0	3	3	0	1	3	3	3	5	2	2	3	1	4	4	3	3
4	3	3	1	2	3	1	3	2	3	3	2	2	3	3	4	3	2	1	5
3	0	1	2	2	4	3	3	3	1	2	4	4	1	1	4	4	1	1	3

(a) Construct a frequency table for these data.

(b) Draw a histogram.

1.1-8. Consider a bowl containing 10 chips of the same size such that 2 are marked "one," 3 are marked "two," 3 are marked "three," and 2 are marked "four." Select a chip at random and read the number. Here $S = \{1, 2, 3, 4\}$.

(a) Assign a reasonable p.m.f. $f(x)$ to the outcome space.

(b) Simulate this experiment at least $n = 100$ times and find the relative frequency histogram $h(x)$. HINT: Here you can use a computer to perform the simulation; or simply use the table of random numbers (Table IX in the appendix), start at a random spot, and let an integer in the set $\{0,1\} = 1$, in $\{2,3,4\} = 2$, in $\{5,6,7\} = 3$, in $\{8,9\} = 4$.

(c) Plot $f(x)$ and $h(x)$ on the same graph.

1.1-9. Toss two coins at random and count the number of heads that appear "up." Here $S = \{0,1,2\}$. In Chapter 2, we discover that a reasonable probability model is given by the p.m.f. $f(0) = 1/4$, $f(1) = 1/2$, $f(2) = 1/4$. Repeat this experiment at least $n = 100$ times, and plot the resulting relative frequency histogram $h(x)$ on the same graph with $f(x)$.

1.1-10. Let the random variable X be the number of tosses of a coin needed to obtain the first head. Here $S = \{1,2,3,4,\ldots\}$. In Chapter 2, we find that a reasonable probability model is given by the p.m.f. $f(x) = (1/2)^x$, $x \in S$. Do this experiment a large number of times, and compare the resulting relative frequency histogram $h(x)$ with $f(x)$.

1.1-11. In 1985, Al Bumbry of the Baltimore Orioles and Darrell Brown of the Minnesota Twins had the following numbers of hits (H) and official at bats (AB) on grass and artificial turf:

Playing Surface	Bumbry			Brown		
	AB	H	BA	AB	H	BA
Grass	295	77		92	18	
Artificial Turf	49	16		168	53	
Total	344	93		260	71	

(a) Find the batting averages BA (namely, H/AB) of each player on grass.
(b) Find the BA of each player on artificial turf.
(c) Find the season batting averages for the two players.
(d) Interpret your results.

1.1-12. In 1985, Kent Hrbek of the Minnesota Twins and Dion James of the Milwaukee Brewers had the following numbers of hits (H) and official at bats (AB) on grass and artificial turf:

Playing Surface	Hrbek			James		
	AB	H	BA	AB	H	BA
Grass	204	50		329	93	
Artificial Turf	355	124		58	21	
Total	559	174		387	114	

(a) Find the batting averages BA (namely, H/AB) of each player on grass.
(b) Find the BA of each player on artificial turf.
(c) Find the season batting averages for the two players.
(d) Interpret your results.

1.1-13. If we had a choice of two airlines, we would possibly choose the airline with the better "on-time performance." So consider Alaska Airlines and America West, using data reported by Arnold Barnett (see references):

Airline	Alaska Airlines	America West
Destination	Relative Frequency on Time	Relative Frequency on Time
Los Angeles	$\dfrac{497}{559}$	$\dfrac{694}{811}$
Phoenix	$\dfrac{221}{233}$	$\dfrac{4840}{5255}$
San Diego	$\dfrac{212}{232}$	$\dfrac{383}{448}$
San Francisco	$\dfrac{503}{605}$	$\dfrac{320}{449}$
Seattle	$\dfrac{1841}{2146}$	$\dfrac{201}{262}$
Five-City Total	$\dfrac{3274}{3775}$	$\dfrac{6438}{7225}$

(a) For each of the five cities listed, which airline has the better on-time performance?
(b) Combining the results, which airline has the better on-time performance?
(c) Interpret your results.

1.2 PROPERTIES OF PROBABILITY

In Section 1.1, the collection of all possible outcomes (the **universal set**) of a random experiment is denoted by S and is called the **outcome space**. Given an outcome space S, let A be a part of the collection of outcomes in S; that is, $A \subset S$. Then A is called an **event**. When the random experiment is performed and the outcome of the experiment is in A, we say that **event A has occurred**.

Since, in studying probability, the words *set* and *event* are interchangeable, the reader might want to review **algebra of sets**, found in Appendix D on the CD-ROM. For convenience, however, here we remind the reader of some terminology:

- ∅ denotes the **null** or **empty** set;
- $A \subset B$ means A is a **subset** of B;
- $A \cup B$ is the **union** of A and B;
- $A \cap B$ is the **intersection** of A and B;
- A' is the **complement** of A (i.e., all elements in S that are not in A).

Some of these sets are depicted by the shaded region in Figure 1.2-1, in which S is the interior of the rectangles. Such figures are called **Venn diagrams**.

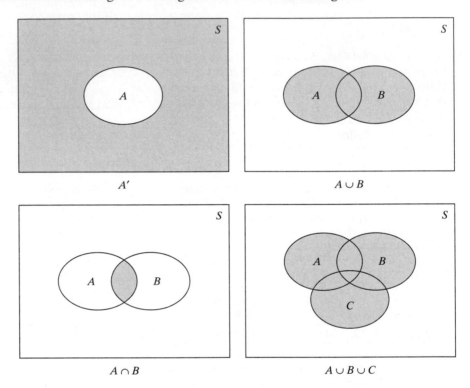

FIGURE 1.2-1: Algebra of sets

Special terminology associated with events that is often used by statisticians includes the following:

1. $A_1, A_2, \ldots, A_k$ are **mutually exclusive events** means that $A_i \cap A_j = \emptyset, i \neq j$, that is, $A_1, A_2, \ldots, A_k$ are disjoint sets;
2. $A_1, A_2, \ldots, A_k$ are **exhaustive events** means that $A_1 \cup A_2 \cup \cdots \cup A_k = S$.

So if $A_1, A_2, \ldots, A_k$ are **mutually exclusive and exhaustive** events, we know that $A_i \cap A_j = \emptyset, i \neq j$, and $A_1 \cup A_2 \cup \cdots \cup A_k = S$.

We are interested in defining what is meant by the probability of event A, denoted by $P(A)$ and often called the chance of A occurring. To help us understand what is meant by the probability of A, consider repeating the experiment a number of times—say, n times. Count the number of times that event A actually occurred throughout these n performances; this number is called the frequency of event A and is denoted by $\mathcal{N}(A)$. The ratio $\mathcal{N}(A)/n$ is called the **relative frequency** of event A

in these n repetitions of the experiment. A relative frequency is usually very unstable for small values of n, but it tends to stabilize as n increases. This suggests that we associate with event A a number—say, p—that is equal to or approximately equal to the number about which the relative frequency tends to stabilize. This number p can then be taken as the number that the relative frequency of event A will be near in future performances of the experiment. Thus, although we cannot predict the outcome of a random experiment with certainty, we can, for a large value of n, predict fairly accurately the relative frequency associated with event A. The number p assigned to event A is called the **probability** of event A and is denoted by $P(A)$. That is, $P(A)$ represents the proportion of outcomes of a random experiment that terminate in the event A as the number of trials of that experiment increases without bound.

The next example will help to illustrate some of the ideas just presented.

EXAMPLE 1.2-1   A fair six-sided die is rolled six times. If the face numbered k is the outcome on roll k for $k = 1, 2, \ldots, 6$, we say that a match has occurred. The experiment is called a success if at least one match occurs during the six trials. Otherwise, the experiment is called a failure. The sample space is $S = \{\text{success, failure}\}$. Let $A = \{\text{success}\}$. We would like to assign a value to $P(A)$. Accordingly, this experiment was simulated 500 times on a computer. Figure 1.2-2 depicts the results of this simulation, and the following table summarizes a few of the results:

n	$\mathcal{N}(A)$	$\mathcal{N}(A)/n$
50	37	0.740
100	69	0.690
250	172	0.688
500	330	0.660

The probability of event A is not intuitively obvious, but it will be shown in Example 1.5-6 that $P(A) = 1 - (1 - 1/6)^6 = 0.665$. This assignment is certainly supported by the simulation (although not proved by it). ■

FIGURE 1.2-2: Fraction of experiments having at least one match

Example 1.2-1 shows that at times intuition cannot be used to assign probabilities, although simulation can perhaps help to assign a probability empirically. The next example illustrates where intuition can help in assigning a probability to an event.

EXAMPLE 1.2-2 A disk 2 inches in diameter is thrown at random on a tiled floor, where each tile is a square with sides 4 inches in length. Let C be the event that the disk will land entirely on one tile. In order to assign a value to $P(C)$, consider the center of the disk. In what region must the center lie to ensure that the disk lies entirely on one tile? If you draw a picture, it should be clear that the center must lie within a square having sides of length 2 and with its center coincident with the center of a tile. Since the area of this square is 4 and the area of a tile is 16, it makes sense to let $P(C) = 4/16$. ∎

Sometimes the nature of an experiment is such that the probability of A can be assigned easily. For example, when a state lottery randomly selects a three-digit integer, we would expect each of the 1000 possible three-digit numbers to have the same chance of being selected, namely, $1/1000$. If we let $A = \{233, 323, 332\}$, then it makes sense to let $P(A) = 3/1000$. Or if we let $B = \{234, 243, 324, 342, 423, 432\}$, then we would let $P(B) = 6/1000$, the probability of the event B. Probabilities of events associated with many random experiments are perhaps not quite as obvious and straightforward as was seen in Example 1.2-1.

So we wish to associate with A a number $P(A)$ about which the relative frequency $\mathcal{N}(A)/n$ of the event A tends to stabilize with large n. A function such as $P(A)$ that is evaluated for a set A is called a **set function**. In this section, we consider the probability set function $P(A)$ and discuss some of its properties. In succeeding sections, we shall describe how the probability set function is defined for particular experiments.

To help decide what properties the probability set function should satisfy, consider properties possessed by the relative frequency $\mathcal{N}(A)/n$. For example, $\mathcal{N}(A)/n$ is always nonnegative. If $A = S$, the sample space, then the outcome of the experiment will always belong to S, and thus $\mathcal{N}(S)/n = 1$. Also, if A and B are two mutually exclusive events, then $\mathcal{N}(A \cup B)/n = \mathcal{N}(A)/n + \mathcal{N}(B)/n$. Hopefully, these remarks will help to motivate the following definition.

DEFINITION 1.2-1 **Probability** is a real-valued set function P that assigns, to each event A in the sample space S, a number $P(A)$, called the probability of the event A, such that the following properties are satisfied:

(a) $P(A) \geq 0$,

(b) $P(S) = 1$,

(c) If $A_1, A_2, A_3, \ldots$ are events and $A_i \cap A_j = \emptyset, i \neq j$, then

$$P(A_1 \cup A_2 \cup \cdots \cup A_k) = P(A_1) + P(A_2) + \cdots + P(A_k)$$

for each positive integer k, and

$$P(A_1 \cup A_2 \cup A_3 \cup \cdots) = P(A_1) + P(A_2) + P(A_3) + \cdots$$

for an infinite, but countable, number of events.

The theorems that follow give some other important properties of the probability set function. When one considers these theorems, it is important to understand the theoretical concepts and proofs. However, if the reader keeps the relative frequency concept in mind, the theorems should also have some intuitive appeal.

THEOREM 1.2-1

For each event A,

$$P(A) = 1 - P(A').$$

Proof. We have

$$S = A \cup A' \qquad \text{and} \qquad A \cap A' = \varnothing.$$

Thus, from properties (b) and (c), it follows that

$$1 = P(A) + P(A').$$

Hence,

$$P(A) = 1 - P(A'). \qquad \square$$

EXAMPLE 1.2-3

A fair coin is flipped successively until the same face is observed on successive flips. Let $A = \{x : x = 3, 4, 5, \ldots\}$; that is, A is the event that it will take three or more flips of the coin to observe the same face on two consecutive flips. To find $P(A)$, we first find the probability of $A' = \{x : x = 2\}$, the complement of A. In two flips of a coin, the possible outcomes are $\{HH, HT, TH, TT\}$, and we assume that each of these four points has the same chance of being observed. Thus,

$$P(A') = P(\{HH, TT\}) = \frac{2}{4}.$$

It follows from Theorem 1.2-1 that

$$P(A) = 1 - P(A') = 1 - \frac{2}{4} = \frac{2}{4}. \qquad \blacksquare$$

THEOREM 1.2-2

$P(\varnothing) = 0.$

Proof. In Theorem 1.2-1, take $A = \varnothing$ so that $A' = S$. Then

$$P(\varnothing) = 1 - P(S) = 1 - 1 = 0. \qquad \square$$

THEOREM 1.2-3

If events A and B are such that $A \subset B$, then $P(A) \le P(B)$.

Proof. We have

$$B = A \cup (B \cap A') \qquad \text{and} \qquad A \cap (B \cap A') = \varnothing.$$

Hence, from property (c),

$$P(B) = P(A) + P(B \cap A') \ge P(A)$$

because, from property (a),

$$P(B \cap A') \geq 0. \qquad \square$$

THEOREM 1.2-4

For each event A, $P(A) \leq 1$.

Proof. Since $A \subset S$, we have, by Theorem 1.2-3 and property (b),

$$P(A) \leq P(S) = 1,$$

which gives the desired result. $\qquad \square$

Property (a), along with Theorem 1.2-4, shows that, for each event A,

$$0 \leq P(A) \leq 1.$$

THEOREM 1.2-5

If A and B are any two events, then

$$P(A \cup B) = P(A) + P(B) - P(A \cap B).$$

Proof. The event $A \cup B$ can be represented as a union of mutually exclusive events, namely,

$$A \cup B = A \cup (A' \cap B).$$

Hence, by property (c),

$$P(A \cup B) = P(A) + P(A' \cap B). \qquad (1.2\text{-}1)$$

However,

$$B = (A \cap B) \cup (A' \cap B),$$

which is a union of mutually exclusive events. Thus,

$$P(B) = P(A \cap B) + P(A' \cap B)$$

and

$$P(A' \cap B) = P(B) - P(A \cap B).$$

If the right-hand side of this equation is substituted into Equation 1.2-1, we obtain

$$P(A \cup B) = P(A) + P(B) - P(A \cap B),$$

which is the desired result. $\qquad \square$

EXAMPLE 1.2-4 A faculty leader was meeting two students in Paris, one arriving by train from Amsterdam and the other arriving by train from Brussels at approximately the same time. Let A and B be the events that the respective trains are on time. Suppose we know from past experience that $P(A) = 0.93$, $P(B) = 0.89$, and $P(A \cap B) = 0.87$. Then

$$P(A \cup B) = P(A) + P(B) - P(A \cap B)$$
$$= 0.93 + 0.89 - 0.87 = 0.95$$

is the probability that at least one train is on time.   ■

THEOREM 1.2-6

If A, B, and C are any three events, then

$$P(A \cup B \cup C) = P(A) + P(B) + P(C) - P(A \cap B)$$
$$- P(A \cap C) - P(B \cap C) + P(A \cap B \cap C).$$

Proof. Write

$$A \cup B \cup C = A \cup (B \cup C)$$

and apply Theorem 1.2-5. The details are left as an exercise.   □

EXAMPLE 1.2-5   A survey was taken of a group's viewing habits of sporting events on TV during the last year. Let $A = \{$watched football$\}$, $B = \{$watched basketball$\}$, $C = \{$watched baseball$\}$. The results indicate that if a person is selected from the group surveyed, then $P(A) = 0.43$, $P(B) = 0.40$, $P(C) = 0.32$, $P(A \cap B) = 0.29$, $P(A \cap C) = 0.22$, $P(B \cap C) = 0.20$, and $P(A \cap B \cap C) = 0.15$. It then follows that

$$P(A \cup B \cup C) = P(A) + P(B) + P(C) - P(A \cap B) - P(A \cap C)$$
$$- P(B \cap C) + P(A \cap B \cap C)$$
$$= 0.43 + 0.40 + 0.32 - 0.29 - 0.22 - 0.20 + 0.15$$
$$= 0.59$$

is the probability that this person watched at least one of these sports.   ■

Let a probability set function be defined on a sample space S. Let $S = \{e_1, e_2, \ldots, e_m\}$, where each e_i is a possible outcome of the experiment. The integer m is called the total number of ways in which the random experiment can terminate. If each of these outcomes has the same probability of occurring, we say that the m outcomes are **equally likely**. That is,

$$P(\{e_i\}) = \frac{1}{m}, \qquad i = 1, 2, \ldots, m.$$

If the number of outcomes in an event A is h, then the integer h is called the number of ways that are favorable to the event A. In this case, $P(A)$ is equal to the number of ways favorable to the event A divided by the total number of ways in which the experiment can terminate. That is, under this assumption of equally likely outcomes, we have

$$P(A) = \frac{h}{m} = \frac{N(A)}{N(S)},$$

where $h = N(A)$ is the number of ways A can occur and $m = N(S)$ is the number of ways S can occur. Exercise 1.2-19 considers this assignment of probability in a more theoretical manner.

It should be emphasized that in order to assign the probability h/m to the event A, we must assume that each of the outcomes $e_1, e_2, \ldots, e_m$ has the same probability $1/m$. This assumption is then an important part of our probability model; if it is not realistic in an application, then the probability of the event A cannot be computed in this way. Actually, we have used this result in the simple case given in Example 1.2-3 because it seemed realistic to assume that each of the possible outcomes in $S = \{HH, HT, TH, TT\}$ had the same chance of being observed.

EXAMPLE 1.2-6   Let a card be drawn at random from an ordinary deck of 52 playing cards. Then the sample space S is the set of $m = 52$ different cards, and it is reasonable to assume that each of these cards has the same probability of selection, $1/52$. Accordingly, if A is the set of outcomes that are kings, then $P(A) = 4/52 = 1/13$ because there are $h = 4$ kings in the deck. That is, $1/13$ is the probability of drawing a card that is a king, provided that each of the 52 cards has the same probability of being drawn.   ∎

In Example 1.2-6, the computations are very easy because there is no difficulty in the determination of the appropriate values of h and m. However, instead of drawing only one card, suppose that 13 are taken at random and without replacement. Then we can think of each possible 13-card hand as being an outcome in a sample space, and it is reasonable to assume that each of these outcomes has the same probability. For example, to use the preceding method to assign the probability of a hand consisting of seven spades and six hearts, we must be able to count the number h of all such hands, as well as the number m of possible 13-card hands. In these more complicated situations, we need better methods of determining h and m. We discuss some of these counting techniques in Section 1.3.

EXERCISES

1.2-1. Of a group of patients having injuries, 28% visit both a physical therapist and a chiropractor and 8% visit neither. Say that the probability of visiting a physical therapist exceeds the probability of visiting a chiropractor by 16%. What is the probability of a randomly selected person from this group visiting a physical therapist?

1.2-2. An insurance company looks at its auto insurance customers and finds that (a) all insure at least one car, (b) 85% insure more than one car, (c) 23% insure a sports car, and (d) 17% insure more than one car, including a sports car. Find the probability that a customer selected at random insures exactly one car and it is not a sports car.

1.2-3. Draw one card at random from a standard deck of cards. The sample space S is the collection of the 52 cards. Assume that the probability set function assigns 1/52 to each of the 52 outcomes. Let

$A = \{x : x \text{ is a jack, queen, or king}\}$,
$B = \{x : x \text{ is a 9, 10, or jack and } x \text{ is red}\}$,
$C = \{x : x \text{ is a club}\}$,
$D = \{x : x \text{ is a diamond, a heart, or a spade}\}$.

Find (a) $P(A)$, (b) $P(A \cap B)$, (c) $P(A \cup B)$, (d) $P(C \cup D)$, and (e) $P(C \cap D)$.

1.2-4. A coin is tossed four times, and the sequence of heads and tails is observed.

(a) List each of the 16 sequences in the sample space S.

(b) Let events A, B, C, and D be given by $A = \{\text{at least 3 heads}\}$, $B = \{\text{at most 2 heads}\}$, $C = \{\text{heads on the third toss}\}$, and $D = \{1 \text{ head and 3 tails}\}$. If the probability set function assigns 1/16 to each outcome in the sample space, find (i) $P(A)$, (ii) $P(A \cap B)$, (iii) $P(B)$,

(iv) $P(A \cap C)$, (v) $P(D)$, (vi) $P(A \cup C)$, and (vii) $P(B \cap D)$.

1.2-5. A field of beans is planted with three seeds per hill. For each hill of beans, let A_i be the event that i seeds germinate, $i = 0, 1, 2, 3$. Suppose that $P(A_0) = 1/64$, $P(A_1) = 9/64$, and $P(A_2) = 27/64$. Give the value of $P(A_3)$.

1.2-6. Consider the trial on which a 3 is first observed in successive rolls of a six-sided die. Let A be the event that 3 is observed on the first trial. Let B be the event that at least two trials are required to observe a 3. Assuming that each side has probability 1/6, find (a) $P(A)$, (b) $P(B)$, and (c) $P(A \cup B)$.

1.2-7. A fair eight-sided die is rolled once. Let $A = \{2, 4, 6, 8\}$, $B = \{3, 6\}$, $C = \{2, 5, 7\}$, and $D = \{1, 3, 5, 7\}$. Assume that each face has the same probability.

(a) Give the values of (i) $P(A)$, (ii) $P(B)$, (iii) $P(C)$, and (iv) $P(D)$.

(b) Give the values of (i) $P(A \cap B)$, (ii) $P(B \cap C)$, and (iii) $P(C \cap D)$.

(c) Give the values of (i) $P(A \cup B)$, (ii) $P(B \cup C)$, and (iii) $P(C \cup D)$, using Theorem 1.2-5.

1.2-8. If $P(A) = 0.4$, $P(B) = 0.5$, and $P(A \cap B) = 0.3$, find (a) $P(A \cup B)$, (b) $P(A \cap B')$, and (c) $P(A' \cup B')$.

1.2-9. Given that $P(A \cup B) = 0.76$ and $P(A \cup B') = 0.87$, find $P(A)$.

1.2-10. During a visit to a primary care physician's office, the probability of having neither lab work nor referral to a specialist is 0.21. Of those coming to that office, the probability of having lab work is 0.41 and the probability of having a referral is 0.53. What is the probability of having both lab work and a referral?

1.2-11. Roll a fair six-sided die three times. Let $A_1 = \{1 \text{ or } 2 \text{ on the first roll}\}$, $A_2 = \{3 \text{ or } 4 \text{ on the second roll}\}$, and $A_3 = \{5 \text{ or } 6 \text{ on the third roll}\}$. It is given that $P(A_i) = 1/3$, $i = 1, 2, 3$; $P(A_i \cap A_j) = (1/3)^2, i \neq j$; and $P(A_1 \cap A_2 \cap A_3) = (1/3)^3$.

(a) Use Theorem 1.2-6 to find $P(A_1 \cup A_2 \cup A_3)$.

(b) Show that $P(A_1 \cup A_2 \cup A_3) = 1 - (1 - 1/3)^3$.

1.2-12. Prove Theorem 1.2-6.

1.2-13. For each positive integer n, let $P(\{n\}) = (1/2)^n$. Consider the events $A = \{n : 1 \leq n \leq 10\}$, $B = \{n : 1 \leq n \leq 20\}$, and $C = \{n : 11 \leq n \leq 20\}$. Find (a) $P(A)$, (b) $P(B)$, (c) $P(A \cup B)$, (d) $P(A \cap B)$, (e) $P(C)$, and (f) $P(B')$.

1.2-14. Let x equal a number that is selected randomly from the closed interval from zero to one, $[0, 1]$. Use your intuition to assign values to

(a) $P(\{x : 0 \leq x \leq 1/3\})$.

(b) $P(\{x : 1/3 \leq x \leq 1\})$.

(c) $P(\{x : x = 1/3\})$.

(d) $P(\{x : 1/2 < x < 5\})$.

1.2-15. A typical roulette wheel used in a casino has 38 slots that are numbered $1, 2, 3, \ldots, 36, 0, 00$, respectively. The 0 and 00 slots are colored green. Half of the remaining slots are red and half are black. Also, half of the integers between 1 and 36 inclusive are odd, half are even, and 0 and 00 are defined to be neither odd nor even. A ball is rolled around the wheel and ends up in one of the slots; we assume that each slot has equal probability of 1/38, and we are interested in the number of the slot into which the ball falls.

(a) Define the sample space S.

(b) Let $A = \{0, 00\}$. Give the value of $P(A)$.

(c) Let $B = \{14, 15, 17, 18\}$. Give the value of $P(B)$.

(d) Let $D = \{x : x \text{ is odd}\}$. Give the value of $P(D)$.

1.2-16. The five numbers 1, 2, 3, 4, and 5 are written respectively on five disks of the same size and placed in a hat. Two disks are drawn without replacement from the hat, and the numbers written on them are observed.

(a) List the 10 possible outcomes of this experiment as unordered pairs of numbers.

(b) If each of the 10 outcomes has probability 1/10, assign a value to the probability that the sum of the two numbers drawn is (i) 3; (ii) between 6 and 8 inclusive.

1.2-17. Divide a line segment into two parts by selecting a point at random. Use your intuition to assign a probability to the event that the longer segment is at least two times longer than the shorter segment.

1.2-18. Let the interval $[-r, r]$ be the base of a semicircle. If a point is selected at random from this interval, assign a probability to the event that the length of the perpendicular segment from the point to the semicircle is less than $r/2$.

1.2-19. Let $S = A_1 \cup A_2 \cup \cdots \cup A_m$, where events $A_1, A_2, \ldots, A_m$ are mutually exclusive and exhaustive.

(a) If $P(A_1) = P(A_2) = \cdots = P(A_m)$, show that $P(A_i) = 1/m$, $i = 1, 2, \ldots, m$.

(b) If $A = A_1 \cup A_2 \cup \cdots \cup A_h$, where $h < m$, and (a) holds, prove that $P(A) = h/m$.

1.2-20. Let p_n, $n = 0, 1, 2, \ldots$, be the probability that an automobile policyholder will file for n claims in a five-year period. The actuary involved makes the assumption that $p_{n+1} = (1/4)p_n$. What is the probability that the holder will file two or more claims during this period?

1.3   METHODS OF ENUMERATION

In this section, we develop counting techniques that are useful in determining the number of outcomes associated with the events of certain random experiments. We begin with a consideration of the multiplication principle.

Multiplication Principle: Suppose that an experiment (or procedure) E_1 has n_1 outcomes and, for each of these possible outcomes, an experiment (procedure) E_2 has n_2 possible outcomes. Then the composite experiment (procedure) $E_1 E_2$ that consists of performing first E_1 and then E_2 has $n_1 n_2$ possible outcomes.

EXAMPLE 1.3-1   Let E_1 denote the selection of a rat from a cage containing one female (F) rat and one male (M) rat. Let E_2 denote the administering of either drug A (A), drug B (B), or a placebo (P) to the selected rat. Then the outcome for the composite experiment can be denoted by an ordered pair, such as (F, P). In fact, the set of all possible outcomes, namely, $(2)(3) = 6$ of them, can be denoted by the following rectangular array:

$$(F, A) \quad (F, B) \quad (F, P)$$
$$(M, A) \quad (M, B) \quad (M, P)$$ ∎

Another way of illustrating the multiplication principle is with a tree diagram like that in Figure 1.3-1. The diagram shows that there are $n_1 = 2$ possibilities (branches) for the sex of the rat and that, for each of these outcomes, there are $n_2 = 3$ possibilities (branches) for the drug.

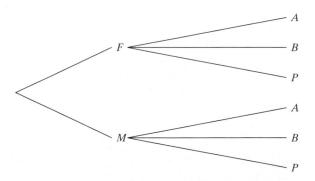

FIGURE 1.3-1: Tree diagram

Clearly, the multiplication principle can be extended to a sequence of more than two experiments or procedures. Suppose that the experiment E_i has n_i ($i = 1, 2, \ldots, m$) possible outcomes after previous experiments have been performed. Then the composite experiment $E_1 E_2 \cdots E_m$ that consists of performing E_1, then $E_2, \ldots$, and finally E_m has $n_1 n_2 \cdots n_m$ possible outcomes.

EXAMPLE 1.3-2   A certain food service gives the following choices for dinner: E_1, soup or tomato juice; E_2, steak or shrimp; E_3, French fried potatoes, mashed potatoes, or a baked potato; E_4, corn or peas; E_5, jello, tossed salad, cottage cheese, or coleslaw; E_6, cake, cookies, pudding, brownie, vanilla ice cream, chocolate ice cream, or orange sherbet; E_7, coffee, tea, milk, or punch. How many different dinner selections are possible if one of the listed choices is made for each of $E_1, E_2, \ldots,$ and E_7? By the multiplication principle, there are

$$(2)(2)(3)(2)(4)(7)(4) = 2688$$

different combinations.   ∎

Although the multiplication principle is fairly simple and easy to understand, it will be extremely useful as we now develop various counting techniques.

Suppose that n positions are to be filled with n different objects. There are n choices for filling the first position, $n - 1$ for the second, ..., and 1 choice for the last position. So, by the multiplication principle, there are

$$n(n - 1)\cdots(2)(1) = n!$$

possible arrangements. The symbol $n!$ is read "n factorial." We define $0! = 1$; that is, we say that zero positions can be filled with zero objects in one way.

DEFINITION 1.3-1   Each of the $n!$ arrangements (in a row) of n different objects is called a **permutation** of the n objects.

EXAMPLE 1.3-3   The number of permutations of the four letters a, b, c, and d is clearly $4! = 24$. However, the number of possible four-letter code words using the four letters a, b, c, and d if letters may be repeated is $4^4 = 256$, because in this case each selection can be performed in four ways.   ∎

If only r positions are to be filled with objects selected from n different objects, $r \leq n$, then the number of possible ordered arrangements is

$$_nP_r = n(n - 1)(n - 2)\cdots(n - r + 1).$$

That is, there are n ways to fill the first position, $(n - 1)$ ways to fill the second, and so on, until there are $[n - (r - 1)] = (n - r + 1)$ ways to fill the rth position.

In terms of factorials, we have

$$_nP_r = \frac{n(n - 1)\cdots(n - r + 1)(n - r)\cdots(3)(2)(1)}{(n - r)\cdots(3)(2)(1)} = \frac{n!}{(n - r)!}.$$

DEFINITION 1.3-2   Each of the $_nP_r$ arrangements is called a **permutation of n objects taken r at a time**.

EXAMPLE 1.3-4   The number of possible four-letter code words, selecting from the 26 letters in the alphabet, in which all four letters are different is

$$_{26}P_4 = (26)(25)(24)(23) = \frac{26!}{22!} = 358{,}800.$$ ∎

EXAMPLE 1.3-5 The number of ways of selecting a president, a vice president, a secretary, and a treasurer in a club consisting of 10 persons is

$$_{10}P_4 = 10 \cdot 9 \cdot 8 \cdot 7 = \frac{10!}{6!} = 5040.$$ ∎

Suppose that a set contains n objects. Consider the problem of drawing r objects from this set. The order in which the objects are drawn may or may not be important. In addition, it is possible that a drawn object is replaced before the next object is drawn. Accordingly, we give some definitions and show how the multiplication principle can be used to count the number of possibilities.

DEFINITION 1.3-3 If r objects are selected from a set of n objects, and if the order of selection is noted, then the selected set of r objects is called an **ordered sample of size r.**

DEFINITION 1.3-4 **Sampling with replacement** occurs when an object is selected and then replaced before the next object is selected.

By the multiplication principle, the number of possible ordered samples of size r taken from a set of n objects is n^r when sampling with replacement.

EXAMPLE 1.3-6 A die is rolled five times. The number of possible ordered samples is $6^5 = 7776$. Note that rolling a die is equivalent to sampling with replacement. ∎

EXAMPLE 1.3-7 An urn contains 10 balls numbered $0, 1, 2, \ldots, 9$. If 4 balls are selected one at a time and with replacement, then the number of possible ordered samples is $10^4 = 10{,}000$. Note that this is the number of four-digit integers between 0000 and 9999, inclusive. ∎

DEFINITION 1.3-5 **Sampling without replacement** occurs when an object is not replaced after it has been selected.

By the multiplication principle, the number of possible ordered samples of size r taken from a set of n objects without replacement is

$$n(n-1)\cdots(n-r+1) = \frac{n!}{(n-r)!},$$

which is equivalent to $_nP_r$, the number of permutations of n objects taken r at a time.

EXAMPLE 1.3-8 The number of ordered samples of 5 cards that can be drawn without replacement from a standard deck of 52 playing cards is

$$(52)(51)(50)(49)(48) = \frac{52!}{47!} = 311{,}875{,}200.$$ ∎

Often the order of selection is not important, and interest centers only on the selected set of r objects. That is, we are interested in the number of subsets of size r that can be selected from a set of n different objects. In order to find the number of (unordered) subsets of size r, we count, in two different ways, the number of ordered subsets of size r that can be taken from the n distinguishable objects. Then, by equating the two answers, we are able to count the number of (unordered) subsets of size r.

Let C denote the number of (unordered) subsets of size r that can be selected from n different objects. We can obtain each of the $_nP_r$ ordered subsets by first selecting one of the C unordered subsets of r objects and then ordering these r objects. Since the latter ordering can be carried out in $r!$ ways, the multiplication principle yields $(C)(r!)$ ordered subsets; so $(C)(r!)$ must equal $_nP_r$. Thus, we have

$$(C)(r!) = \frac{n!}{(n-r)!},$$

or

$$C = \frac{n!}{r!\,(n-r)!}.$$

We denote this answer by either $_nC_r$ or $\binom{n}{r}$; that is,

$$_nC_r = \binom{n}{r} = \frac{n!}{r!\,(n-r)!}.$$

Accordingly, a set of n different objects possesses

$$\binom{n}{r} = \frac{n!}{r!\,(n-r)!}$$

unordered subsets of size $r \leq n$.

We could also say that the number of ways in which r objects can be selected without replacement from n objects when the order of selection is disregarded is $\binom{n}{r} = {}_nC_r$, and the latter expression can be read as "n choose r." This result motivates the next definition.

DEFINITION 1.3-6   Each of the $_nC_r$ unordered subsets is called a **combination of n objects taken r at a time**, where

$$_nC_r = \binom{n}{r} = \frac{n!}{r!\,(n-r)!}.$$

EXAMPLE 1.3-9   The number of possible 5-card hands (in 5-card poker) drawn from a deck of 52 playing cards is

$$_{52}C_5 = \binom{52}{5} = \frac{52!}{5!\,47!} = 2{,}598{,}960.$$ ∎

EXAMPLE 1.3-10 The number of possible 13-card hands (in bridge) that can be selected from a deck of 52 playing cards is

$$_{52}C_{13} = \binom{52}{13} = \frac{52!}{13!\,39!} = 635{,}013{,}559{,}600.$$ ∎

The numbers $\binom{n}{r}$ are frequently called **binomial coefficients**, since they arise in the expansion of a binomial. We illustrate this property by giving a justification of the binomial expansion

$$(a + b)^n = \sum_{r=0}^{n} \binom{n}{r} b^r a^{n-r}. \tag{1.3-1}$$

In the expansion of

$$(a + b)^n = (a + b)(a + b)\cdots(a + b),$$

either an a or a b is selected from each of the n factors. One possible product is then $b^r a^{n-r}$; this occurs when b is selected from each of r factors and a from each of the remaining $n - r$ factors. But the latter operation can be completed in $\binom{n}{r}$ ways, which then must be the coefficient of $b^r a^{n-r}$, as shown in Equation 1.3-1.

The binomial coefficients are given in Table I in the appendix for selected values of n and r. Note that for some combinations of n and r, the table uses the fact that

$$\binom{n}{r} = \frac{n!}{r!\,(n-r)!} = \frac{n!}{(n-r)!\,r!} = \binom{n}{n-r}.$$

That is, the number of ways in which r objects can be selected out of n objects is equal to the number of ways in which $n - r$ objects can be selected out of n objects.

EXAMPLE 1.3-11
Assume that each of the $\binom{52}{5} = 2{,}598{,}960$ five-card hands drawn from a deck of 52 playing cards has the same probability of being selected. Then the number of possible 5-card hands that are all spades (event A) is

$$N(A) = \binom{13}{5}\binom{39}{0},$$

because the 5 spades can be selected from the 13 spades in $\binom{13}{5}$ ways, after which zero nonspades can be selected in $\binom{39}{0} = 1$ way. We have

$$\binom{13}{5} = \frac{13!}{5!\,8!} = 1287$$

from Table I in the appendix. Thus, the probability of an all-spade 5-card hand is

$$P(A) = \frac{N(A)}{N(S)} = \frac{1287}{2{,}598{,}960} = 0.000495.$$

Suppose now that the event B is the set of outcomes in which exactly 3 cards are kings and exactly 2 cards are queens. We can select the three kings in any one of $\binom{4}{3}$ ways and the two queens in any one of $\binom{4}{2}$ ways. By the multiplication principle, the number of outcomes in B is

$$N(B) = \binom{4}{3}\binom{4}{2}\binom{44}{0},$$

where $\binom{44}{0}$ gives the number of ways in which 0 cards are selected out of the nonkings and nonqueens and of course is equal to 1. Thus,

$$P(B) = \frac{N(B)}{N(S)} = \frac{\binom{4}{3}\binom{4}{2}\binom{44}{0}}{\binom{52}{5}} = \frac{24}{2{,}598{,}960} = 0.0000092.$$

Finally, let C be the set of outcomes in which there are exactly two kings, two queens, and one jack. Then

$$P(C) = \frac{N(C)}{N(S)} = \frac{\binom{4}{2}\binom{4}{2}\binom{4}{1}\binom{40}{0}}{\binom{52}{5}} = \frac{144}{2{,}598{,}960} = 0.000055$$

because the numerator of this fraction is the number of outcomes in C.   ■

Now suppose that a set contains n objects of two types: r of one type and $n - r$ of the other type. The number of permutations of n different objects is $n!$. However, in this case, the objects are not all distinguishable. To count the number of distinguishable arrangements, first select r out of the n positions for the objects of the first type. This can be done in $\binom{n}{r}$ ways. Then fill in the remaining positions with the objects of the second type. Thus, the number of distinguishable arrangements is

$$_nC_r = \binom{n}{r} = \frac{n!}{r!\,(n-r)!}.$$

DEFINITION 1.3-7	Each of the $_nC_r$ permutations of n objects, r of one type and $n - r$ of another type, is called a **distinguishable permutation**.

EXAMPLE 1.3-12   A coin is flipped 10 times and the sequence of heads and tails is observed. The number of possible 10-tuplets that result in four heads and six tails is

$$\binom{10}{4} = \frac{10!}{4!\,6!} = \frac{10!}{6!\,4!} = \binom{10}{6} = 210.$$ ■

EXAMPLE 1.3-13 In an orchid show, seven orchids are to be placed along one side of the greenhouse. There are four lavender orchids and three white orchids. Considering only the color of the orchids, we see that the number of lineups of the orchids is

$$\binom{7}{4} = \frac{7!}{4!\,3!} = 35.$$

If the colors of the seven orchids are white, lavender, yellow, mauve, crimson, orange, and pink, the number of different displays is $7! = 5040$. ∎

The foregoing results can be extended. Suppose that in a set of n objects, n_1 are similar, n_2 are similar, $\ldots$, n_s are similar, where $n_1 + n_2 + \cdots + n_s = n$. Then the number of distinguishable permutations of the n objects is (see Exercise 1.3-18)

$$\binom{n}{n_1, n_2, \ldots, n_s} = \frac{n!}{n_1!\, n_2! \cdots n_s!}. \tag{1.3-2}$$

EXAMPLE 1.3-14 Among nine orchids for a line of orchids along one wall, three are white, four lavender, and two yellow. The number of different color displays is then

$$\binom{9}{3, 4, 2} = \frac{9!}{3!\,4!\,2!} = 1260.$$ ∎

The argument used in determining the binomial coefficients in the expansion of $(a + b)^n$ can be extended to find the expansion of $(a_1 + a_2 + \cdots + a_s)^n$. The coefficient of $a_1^{n_1} a_2^{n_2} \cdots a_s^{n_s}$ is

$$\binom{n}{n_1, n_2, \ldots, n_s} = \frac{n!}{n_1!\, n_2! \cdots n_s!}.$$

This is sometimes called a **multinomial coefficient**.

When r objects are selected out of n objects, we are often interested in the number of possible outcomes. We have seen that for ordered samples, there are n^r possible outcomes when sampling with replacement and $_nP_r$ outcomes when sampling without replacement. For unordered samples, there are $_nC_r$ outcomes when sampling without replacement. Each of the preceding outcomes is equally likely, provided that the experiment is performed in a fair manner.

REMARK   Although not needed in the study of probability, it is interesting to count the number of possible samples of size r that can be selected out of n objects when the order is irrelevant and when sampling with replacement. For example, if a six-sided die is rolled 10 times (or 10 six-sided dice are rolled once), how many possible unordered outcomes are there? To count the number of possible outcomes, think of listing r 0's for the r objects that are to be selected. Then insert $(n - 1)$ |'s to partition the r objects into n sets, the first set giving objects of the first kind, and so on. So if $n = 6$ and $r = 10$ in the die illustration, a possible outcome is

$$0\,0\,|\,|\,0\,0\,0\,|\,0\,|\,0\,0\,0\,|\,0,$$

which says there are two 1's, zero 2's, three 3's, one 4, three 5's, and one 6. In general, each outcome is a permutation of r 0's and $(n - 1)$ |'s. Each distinguishable permutation is equivalent to an unordered sample. The number of distinguishable permutations, and hence the number of unordered samples of size r that can be selected out of n objects when sampling with replacement, is

$$_{n-1+r}C_r = \frac{(n-1+r)!}{r!\,(n-1)!}.$$

■

EXERCISES

1.3-1. A boy found a bicycle lock for which the combination was unknown. The correct combination is a four-digit number, $d_1 d_2 d_3 d_4$, where d_i, $i = 1, 2, 3, 4$, is selected from 1, 2, 3, 4, 5, 6, 7, and 8. How many different lock combinations are possible with such a lock?

1.3-2. How many different orchid displays in a line are possible using four orchids of different colors if exactly three orchids are used?

1.3-3. How many different license plates are possible if a state uses

    (a) Two letters followed by a four-digit integer (leading zeros are permissible and the letters and digits can be repeated)?

    (b) Three letters followed by a three-digit integer? (In practice, it is possible that certain "spellings" are ruled out.)

1.3-4. In designing an experiment, the researcher can often choose many different levels of the various factors in order to try to find the best combination at which to operate. As an illustration, suppose the researcher is studying a certain chemical reaction and can choose four levels of temperature, five different pressures, and two different catalysts.

    (a) To consider all possible combinations, how many experiments would need to be conducted?

    (b) Often in preliminary experimentation, each factor is restricted to two levels. With the three factors noted, how many experiments would need to be run to cover all possible combinations with each of the three factors at two levels? (NOTE: This is often called a 2^3 design.)

1.3-5. How many four-letter code words are possible using the letters in HOPE if

    (a) The letters may not be repeated?

    (b) The letters may be repeated?

1.3-6. The "eating club" is hosting a make-your-own sundae at which the following are provided:

Ice Cream Flavors	Toppings
Chocolate	Caramel
Cookies-n-cream	Hot fudge
Strawberry	Marshmallow
Vanilla	M&M's
	Nuts
	Strawberries

    (a) How many sundaes are possible using one flavor of ice cream and three different toppings?

    (b) How many sundaes are possible using one flavor of ice cream and from zero to six toppings?

    (c) How many different combinations of flavors of three scoops of ice cream are possible if it is permissible to make all three scoops the same flavor?

1.3-7. In a state lottery, four digits are drawn at random one at a time with replacement from 0 to 9. Suppose that you win if any permutation of your selected integers is drawn. Give the probability of winning if you select

    (a) 6, 7, 8, 9.

    (b) 6, 7, 8, 8.

    (c) 7, 7, 8, 8.

    (d) 7, 8, 8, 8.

1.3-8. From a collection of nine paintings, four are to be selected to hang side by side on a gallery wall in positions 1, 2, 3, and 4. In how many ways can this be done?

1.3-9. Some albatrosses return to the world's only mainland colony of royal albatrosses, on Otago Peninsula near Dunedin, New Zealand, every two years to nest and raise their young. In order to learn more about the albatross, colored plastic bands are placed on their legs so that they can be identified from a distance. Suppose that three bands are placed on one leg, with the color of the band selected from the colors red, yellow, green, white, and blue. Find the number of different color codes that are possible for banding an albatross if

(a) The three bands are of different colors.

(b) Repeated colors are permissible.

1.3-10. Hope and Calvin play volleyball until one team wins three games. Using **H** for a Hope victory and **C** for a Calvin victory, and considering the possible orderings for the winning team, in how many ways could the volleyball match end?

1.3-11. The World Series in baseball continues until either the American League team or the National League team wins four games. How many different orders are possible (e.g., *ANNAAA* means the American League team wins in six games) if the series goes

(a) Four games?

(b) Five games?

(c) Six games?

(d) Seven games?

1.3-12. How many different varieties of pizza can be made if you have the following choices: small, medium, or large size; thin 'n crispy, hand-tossed, or pan crust; and 12 toppings (cheese is automatic), from which you may select from 0 to 12?

1.3-13. A cafe lets you order a deli sandwich your way. There are six choices for bread, four choices for meat, four choices for cheese, and 12 different garnishes (condiments). How many different sandwich possibilities are there if you choose

(a) One bread, one meat, and one cheese?

(b) One bread, one meat, one cheese, and from 0 to 12 garnishes?

(c) One bread; 0, 1, or 2 meats; 0, 1, or 2 cheeses; and from 0 to 12 garnishes?

1.3-14. Pascal's triangle gives a method for calculating the binomial coefficients; it begins as follows:

$$
\begin{array}{ccccccccccc}
 & & & & & 1 & & & & & \\
 & & & & 1 & & 1 & & & & \\
 & & & 1 & & 2 & & 1 & & & \\
 & & 1 & & 3 & & 3 & & 1 & & \\
 & 1 & & 4 & & 6 & & 4 & & 1 & \\
1 & & 5 & & 10 & & 10 & & 5 & & 1 \\
 & \cdot & & \cdot & & \cdot & & \cdot & & \cdot & \\
 & & \cdot & & \cdot & & \cdot & & \cdot & & \\
\end{array}
$$

The nth row of this triangle gives the coefficients for $(a + b)^{n-1}$. To find an entry in the table other than a 1 on the boundary, add the two nearest numbers in the row directly above. The equation

$$
\binom{n}{r} = \binom{n-1}{r} + \binom{n-1}{r-1},
$$

called **Pascal's equation**, explains why Pascal's triangle works. Prove that this equation is correct.

1.3-15. Three students (S) and six faculty members (F) are on a panel discussing a new college policy.

(a) In how many different ways can the nine participants be lined up at a table in the front of the auditorium?

(b) How many lineups are possible, considering only the labels S and F?

(c) For each of the nine participants, you are to decide whether the participant did a good job or a poor job stating his or her opinion of the new policy; that is, give each of the nine participants a grade of G or P. How many different "scorecards" are possible?

1.3-16. Prove:

$$
\sum_{r=0}^{n} (-1)^r \binom{n}{r} = 0 \qquad \text{and} \qquad \sum_{r=0}^{n} \binom{n}{r} = 2^n.
$$

HINT: Consider $(1 - 1)^n$ and $(1 + 1)^n$, or use Pascal's equation and proof by induction.

1.3-17. A poker hand is defined as drawing 5 cards at random without replacement from a deck of 52 playing cards. Find the probability of each of the following poker hands:

(a) Four of a kind (4 cards of equal face value and 1 card of a different value).

(b) Full house (one pair and one triple of cards with equal face value).

(c) Three of a kind (three equal face values plus 2 cards of different values).

(d) Two pairs (two pairs of equal face value plus 1 card of a different value).

(e) One pair (one pair of equal face value plus 3 cards of different values).

1.3-18. Prove Equation 1.3-2. HINT: First select n_1 positions in $\binom{n}{n_1}$ ways. Then select n_2 from the remaining $n - n_1$ positions in $\binom{n - n_1}{n_2}$ ways, and so on. Finally, use the multiplication rule.

1.3-19. There are three teams in a cross-country race. Team A has five runners, team B has six runners, and team C has seven runners. In how many ways can the runners cross the finish line if we are interested only in the team for which they run? That is, what is the number of distinguishable permutations of five A's, six B's, and seven C's? (Note that, for scoring purposes, only the scores of the first five runners for each team count.)

1.3-20. A box of candy hearts contains 52 hearts, of which 19 are white, 10 are tan, 7 are pink, 3 are purple, 5 are yellow, 2 are orange, and 6 are green. If you select 9 pieces of candy randomly from the box, without replacement, give the probability that

    (a)　Three of the hearts are white.

    (b)　Three are white, 2 are tan, 1 is pink, 1 is yellow, and 2 are green.

1.3-21. An office furniture manufacturer that makes modular storage files offers its customers two choices for the base and four choices for the top, and the modular storage files come in five different heights. The customer may choose any combination of the five different-sized modules so that the finished file has a base, a top, and one, two, three, four, five, or six storage modules.

    (a)　How many choices does the customer have if the completed file has four storage modules, a top, and a base? The order in which the four modules are stacked is irrelevant.

    (b)　In its advertising, the manufacturer would like to use the number of different files that are possible—selecting one of the two bases, one of the four tops, and then either one, two, three, four, five, or six storage modules. The manufacturer may select any combination of the five different sizes, with the order of stacking irrelevant. What is the number of possibilities?

1.3-22. A bag of 36 dum-dum pops (suckers) contains up to 10 flavors. That is, there are from 0 to 36 suckers of each of 10 flavors in the bag. How many different flavor combinations are possible?

1.4 CONDITIONAL PROBABILITY

We introduce the idea of conditional probability by means of an example.

EXAMPLE 1.4-1　Suppose that we are given 20 tulip bulbs that are similar in appearance and told that 8 will bloom early, 12 will bloom late, 13 will be red, and 7 will be yellow, in accordance with the various combinations listed in Table 1.4-1. If one bulb is selected at random, the probability that it will produce a red tulip (R) is given by $P(R) = 13/20$, under the assumption that each bulb is "equally likely." Suppose, however, that close examination of the bulb will reveal whether it will bloom early (E) or late (L). If we consider an outcome only if it results in a tulip bulb that will bloom early, only eight outcomes in the sample space are now of interest. Thus, under this limitation, it is natural to assign the probability 5/8 to R; that is, $P(R \mid E) = 5/8$, where $P(R \mid E)$ is read as the probability of R given that E has occurred. Note that

$$P(R \mid E) = \frac{5}{8} = \frac{N(R \cap E)}{N(E)} = \frac{N(R \cap E)/20}{N(E)/20} = \frac{P(R \cap E)}{P(E)},$$

where $N(R \cap E)$ and $N(E)$ are the numbers of outcomes in events $R \cap E$ and E, respectively.　∎

TABLE 1.4-1: Tulip Combinations

	Early (E)	Late (L)	Totals
Red (R)	5	8	13
Yellow(Y)	3	4	7
Totals	8	12	20

This example illustrates a number of common situations. That is, in some random experiments, we are interested only in those outcomes which are elements of a subset B of the sample space S. This means, for our purposes, that the sample space is effectively the subset B. We are now confronted with the problem of defining a probability set function with B as the "new" sample space. That is, for a given event A, we want to define $P(A \mid B)$, the probability of A, considering only those outcomes of the random experiment that are elements of B. The previous example gives us the clue to that definition. That is, for experiments in which each outcome is equally likely, it makes sense to define $P(A \mid B)$ by

$$P(A \mid B) = \frac{N(A \cap B)}{N(B)},$$

where $N(A \cap B)$ and $N(B)$ are the numbers of outcomes in $A \cap B$ and B, respectively. If we then divide the numerator and the denominator of this fraction by $N(S)$, the number of outcomes in the sample space, we have

$$P(A \mid B) = \frac{N(A \cap B)/N(S)}{N(B)/N(S)} = \frac{P(A \cap B)}{P(B)}.$$

We are thus led to the following definition.

DEFINITION
1.4-1

The **conditional probability** of an event A, given that event B has occurred, is defined by

$$P(A \mid B) = \frac{P(A \cap B)}{P(B)},$$

provided that $P(B) > 0$.

A formal use of the definition is given in the next example.

EXAMPLE 1.4-2 If $P(A) = 0.4$, $P(B) = 0.5$, and $P(A \cap B) = 0.3$, then $P(A \mid B) = 0.3/0.5 = 0.6$; $P(B \mid A) = P(A \cap B)/P(A) = 0.3/0.4 = 0.75$. ∎

We can think of "given B" as specifying the new sample space for which, to determine $P(A \mid B)$, we now want to calculate the probability of that part of A which is contained in B. The next two examples illustrate this idea.

EXAMPLE 1.4-3 Suppose that $P(A) = 0.7$, $P(B) = 0.3$, and $P(A \cap B) = 0.2$. These probabilities are listed on the Venn diagram in Figure 1.4-1. Given that the outcome of the experiment belongs to B, what then is the probability of A? We are effectively restricting the sample space to B; of the probability $P(B) = 0.3$, 0.2 corresponds to $P(A \cap B)$ and hence to A. That is, $0.2/0.3 = 2/3$ of the probability of B corresponds to A. Of course, by the formal definition, we also obtain

$$P(A \mid B) = \frac{P(A \cap B)}{P(B)} = \frac{0.2}{0.3} = \frac{2}{3}.$$ ∎

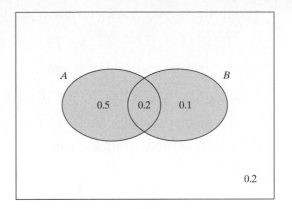

FIGURE 1.4-1: Conditional probability

EXAMPLE 1.4-4 A pair of four-sided dice is rolled and the sum is determined. Let A be the event that a sum of 3 is rolled, and let B be the event that a sum of 3 or a sum of 5 is rolled. In a sequence of rolls, the probability that a sum of 3 is rolled before a sum of 5 is rolled can be thought of as the conditional probability of a sum of 3 given that a sum of 3 or 5 has occurred; that is, the conditional probability of A given B is

$$P(A \mid B) = \frac{P(A \cap B)}{P(B)} = \frac{P(A)}{P(B)} = \frac{2/16}{6/16} = \frac{2}{6}.$$

Note that for this example, the only outcomes of interest are those having a sum of 3 or a sum of 5, and of these six equally likely outcomes, two have a sum of 3. (See Figure 1.4-2 and Exercise 1.4-13.) ■

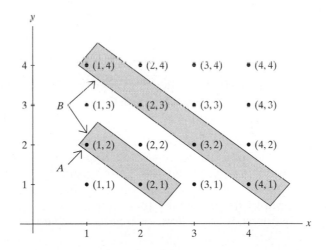

FIGURE 1.4-2: Dice example

It is interesting to note that conditional probability satisfies the axioms for a probability function, namely, with $P(B) > 0$,

(a) $P(A \mid B) \geq 0$.

(b) $P(B \mid B) = 1$.

(c) If $A_1, A_2, A_3, \ldots$ are mutually exclusive events, then

$$P(A_1 \cup A_2 \cup \cdots \cup A_k \mid B) = P(A_1 \mid B) + P(A_2 \mid B) + \cdots + P(A_k \mid B),$$

for each positive integer k, and

$$P(A_1 \cup A_2 \cup \cdots \mid B) = P(A_1 \mid B) + P(A_2 \mid B) + \cdots,$$

for an infinite, but countable, number of events.

Properties (a) and (b) are evident because

$$P(A \mid B) = \frac{P(A \cap B)}{P(B)} \geq 0,$$

since $P(A \cap B) \geq 0, P(B) > 0$, and

$$P(B \mid B) = \frac{P(B \cap B)}{P(B)} = \frac{P(B)}{P(B)} = 1.$$

Property (c) holds because, for the second part of (c),

$$P(A_1 \cup A_2 \cup \cdots \mid B) = \frac{P[(A_1 \cup A_2 \cup \cdots) \cap B]}{P(B)}$$

$$= \frac{P[(A_1 \cap B) \cup (A_2 \cap B) \cup \cdots]}{P(B)}.$$

But $(A_1 \cap B), (A_2 \cap B), \ldots$ are also mutually exclusive events; so

$$P(A_1 \cup A_2 \cup \cdots \mid B) = \frac{P(A_1 \cap B) + P(A_2 \cap B) + \cdots}{P(B)}$$

$$= \frac{P(A_1 \cap B)}{P(B)} + \frac{P(A_2 \cap B)}{P(B)} + \cdots$$

$$= P(A_1 \mid B) + P(A_2 \mid B) + \cdots.$$

The first part of property (c) is proved in a similar manner.

Many times, the conditional probability of an event is clear because of the nature of an experiment. The next example illustrates this.

EXAMPLE 1.4-5 At a county fair carnival game there are 25 balloons on a board, of which 10 balloons are yellow, 8 are red, and 7 are green. A player throws darts at the balloons to win a prize and randomly hits one of them. Given that the first balloon hit is yellow, what is the probability that the next balloon hit is also yellow? Of the 24 remaining balloons, 9 are yellow, so a natural value to assign to this conditional probability is 9/24. ∎

In Example 1.4-5, let A be the event that the first balloon hit is yellow, and let B be the event that the second balloon hit is yellow. Suppose that we are interested in the probability that both balloons hit are yellow. That is, we are interested in finding $P(A \cap B)$. We noted in that example that

$$P(B \mid A) = \frac{P(A \cap B)}{P(A)} = \frac{9}{24}.$$

Thus, multiplying through by $P(A)$, we have

$$P(A \cap B) = P(A)P(B|A) = P(A)\left(\frac{9}{24}\right), \qquad (1.4\text{-}1)$$

or

$$P(A \cap B) = \left(\frac{10}{25}\right)\left(\frac{9}{24}\right).$$

That is, Equation 1.4-1 gives us a general rule for the probability of the intersection of two events once we know the conditional probability $P(B|A)$.

DEFINITION 1.4-2	The probability that two events, A and B, both occur is given by the **multiplication rule**, $$P(A \cap B) = P(A)P(B|A)$$ or by $$P(A \cap B) = P(B)P(A|B).$$

Sometimes, after considering the nature of the random experiment, one can make reasonable assumptions so that it is easier to assign $P(B)$ and $P(A|B)$ rather than $P(A \cap B)$. Then $P(A \cap B)$ can be computed with these assignments. This approach will be illustrated in Examples 1.4-6 and 1.4-7.

EXAMPLE 1.4-6   A bowl contains seven blue chips and three red chips. Two chips are to be drawn successively at random and without replacement. We want to compute the probability that the first draw results in a red chip (A) and the second draw results in a blue chip (B). It is reasonable to assign the following probabilities:

$$P(A) = \frac{3}{10} \text{ and } P(B|A) = \frac{7}{9}.$$

The probability of obtaining red on the first draw and blue on the second draw is

$$P(A \cap B) = \frac{3}{10} \cdot \frac{7}{9} = \frac{7}{30}. \qquad \blacksquare$$

Note that in many instances it is possible to compute a probability by two seemingly different methods. For instance, consider Example 1.4-6, but find the probability of drawing a red chip on each of the two draws. Following that example, it is

$$\frac{3}{10} \cdot \frac{2}{9} = \frac{1}{15}.$$

However, we can also find this probability by using combinations as follows:

$$\frac{\binom{3}{2}\binom{7}{0}}{\binom{10}{2}} = \frac{\frac{(3)(2)}{(1)(2)}}{\frac{(10)(9)}{(1)(2)}} = \frac{1}{15}.$$

Thus, we obtain the same answer, as we should, provided that our reasoning is consistent with the underlying assumptions.

EXAMPLE 1.4-7 From an ordinary deck of playing cards, cards are to be drawn successively at random and without replacement. The probability that the third spade appears on the sixth draw is computed as follows: Let A be the event of two spades in the first five cards drawn, and let B be the event of a spade on the sixth draw. Thus, the probability that we wish to compute is $P(A \cap B)$. It is reasonable to take

$$P(A) = \frac{\binom{13}{2}\binom{39}{3}}{\binom{52}{5}} = 0.274 \qquad \text{and} \qquad P(B \mid A) = \frac{11}{47} = 0.234.$$

The desired probability, $P(A \cap B)$, is the product of those numbers:

$$P(A \cap B) = (0.274)(0.234) = 0.064.$$ ∎

EXAMPLE 1.4-8 Continuing with Example 1.4-4, in which a pair of four-sided dice is rolled, the probability of rolling a sum of 3 on the first roll and then, continuing the sequence of rolls, rolling a sum of 3 before rolling a sum of 5 is

$$\frac{2}{16} \cdot \frac{2}{6} = \frac{4}{96} = \frac{1}{24}.$$ ∎

The multiplication rule can be extended to three or more events. In the case of three events, using the multiplication rule for two events, we have

$$P(A \cap B \cap C) = P[(A \cap B) \cap C]$$
$$= P(A \cap B)P(C \mid A \cap B).$$

But

$$P(A \cap B) = P(A)P(B \mid A).$$

Hence,

$$P(A \cap B \cap C) = P(A)P(B \mid A)P(C \mid A \cap B).$$

This type of argument can be used to extend the multiplication rule to more than three events, and the general formula for k events can be officially proved by mathematical induction.

EXAMPLE 1.4-9 Four cards are to be dealt successively at random and without replacement from an ordinary deck of playing cards. The probability of receiving, in order, a spade, a heart, a diamond, and a club is

$$\frac{13}{52} \cdot \frac{13}{51} \cdot \frac{13}{50} \cdot \frac{13}{49},$$

a result that follows from the extension of the multiplication rule and reasonable assignments to the probabilities involved. ∎

We close this section with a different type of example.

EXAMPLE 1.4-10 A grade school boy has five blue and four white marbles in his left pocket and four blue and five white marbles in his right pocket. If he transfers one marble at random from his left to his right pocket, what is the probability of his then drawing a blue marble from his right pocket? For notation, let BL, BR, and WL denote drawing blue from left pocket, blue from right pocket, and white from left pocket, respectively. Then

$$P(BR) = P(BL \cap BR) + P(WL \cap BR)$$
$$= P(BL)P(BR \mid BL) + P(WL)P(BR \mid WL)$$
$$= \frac{5}{9} \cdot \frac{5}{10} + \frac{4}{9} \cdot \frac{4}{10} = \frac{41}{90}$$

is the desired probability. ∎

EXAMPLE 1.4-11 An insurance company sells a number of different policies, among which 60% are for autos, 40% are for homeowners, and 20% are for both types of policy. Let A_1 be people with only an auto policy, A_2 people with only a homeowner policy, A_3 people with both types of policy, and A_4 those with only other types of policies. If a person is selected at random from the policyholders, then $P(A_1) = 0.4$, $P(A_2) = 0.2$, $P(A_3) = 0.2$, and $P(A_4) = 0.2$, as these four events are mutually exclusive and exhaustive. Further, let B be the event that a policyholder will renew at least one of the auto or homeowner policies. Say from past experience that we can assign the conditional probabilities $P(B \mid A_1) = 0.6$, $P(B \mid A_2) = 0.7$, and $P(B \mid A_3) = 0.8$. Given that the person selected at random has an auto or homeowner policy, what is the conditional probability that the person will renew at least one of those policies? The desired probability is

$$P(B \mid A_1 \cup A_2 \cup A_3) = \frac{P(A_1 \cap B) + P(A_2 \cap B) + P(A_3 \cap B)}{P(A_1) + P(A_2) + P(A_3)}$$
$$= \frac{(0.4)(0.6) + (0.2)(0.7) + (0.2)(0.8)}{0.4 + 0.2 + 0.2}$$
$$= \frac{0.54}{0.80} = \frac{27}{40}$$
$$= 0.675.$$ ∎

EXAMPLE 1.4-12 A device has two components, C_1 and C_2, but it will continue to operate with only one active component for a one-year period. The probability that each component will fail when both are in operation is 0.01 in that one-year period. However, when one component fails, the probability of the other failing is 0.03 in that period, due to added strain. Thus, the probability that the device fails in one year is

$$P(C_1 \text{ fails})P(C_2 \text{ fails} \mid C_1 \text{ fails}) + P(C_2 \text{ fails})P(C_1 \text{ fails} \mid C_2 \text{ fails}) =$$
$$(0.01)(0.03) + (0.01)(0.03) = 0.0006.$$ ∎

EXERCISES

1.4-1. A common test for AIDS is called the ELISA (enzyme-linked immunosorbent assay) test. Among 1 million people who are given the ELISA test, we can expect results similar to those given in the following table:

	B_1: Carry AIDS Virus	B_2: Do Not Carry AIDS Virus	Totals
A_1: Test Positive	4,885	73,630	78,515
A_2: Test Negative	115	921,370	921,485
Totals	5,000	995,000	1,000,000

If one of these 1 million people is selected randomly, find the following probabilities: (a) $P(B_1)$, (b) $P(A_1)$, (c) $P(A_1 | B_2)$, (d) $P(B_1 | A_1)$. (e) In words, what do parts (c) and (d) say?

1.4-2. The following table classifies 1456 people by their gender and by whether or not they favor a gun law.

	Male (S_1)	Female (S_2)	Totals
Favor (A_1)	392	649	1041
Oppose (A_2)	241	174	415
Totals	633	823	1456

Compute the following probabilities if one of these 1456 persons is selected randomly: (a) $P(A_1)$, (b) $P(A_1 | S_1)$, (c) $P(A_1 | S_2)$. (d) Interpret your answers to parts (b) and (c).

1.4-3. Let A_1 and A_2 be the events that a person is left-eye dominant or right-eye dominant, respectively. When a person folds his or her hands, let B_1 and B_2 be the events that the left thumb and right thumb, respectively, are on top. A survey in one statistics class yielded the following table:

	B_1	B_2	Totals
A_1	5	7	12
A_2	14	9	23
Totals	19	16	35

If a student is selected randomly, find the following probabilities: (a) $P(A_1 \cap B_1)$, (b) $P(A_1 \cup B_1)$, (c) $P(A_1 | B_1)$, (d) $P(B_2 | A_2)$. (e) If the students had their hands folded and you hoped to select a right-eye-dominant student, would you select a "right thumb on top" or a "left thumb on top" student? Why?

1.4-4. Two cards are drawn successively and without replacement from an ordinary deck of playing cards. Compute the probability of drawing

(a) Two hearts.

(b) A heart on the first draw and a club on the second draw.

(c) A heart on the first draw and an ace on the second draw.

HINT: In part (c), note that a heart can be drawn by getting the ace of hearts or one of the other 12 hearts.

1.4-5. Suppose that $P(A) = 0.7$, $P(B) = 0.5$, and $P([A \cup B]') = 0.1$.

(a) Find $P(A \cap B)$.

(b) Calculate $P(A | B)$.

(c) Calculate $P(B | A)$.

1.4-6. A hand of 13 cards is to be dealt at random and without replacement from an ordinary deck of playing cards. Find the conditional probability that there are at least three kings in the hand, given that the hand contains at least two kings.

1.4-7. Suppose that the genes for eye color for a certain male fruit fly are (R, W) and the genes for eye color for the mating female fruit fly are (R, W), where R and W represent red and white, respectively. Their offspring receive one gene for eye color from each parent.

(a) Define the sample space of the genes for eye color for the offspring.

(b) Assume that each of the four possible outcomes has equal probability. If an offspring ends up with either two red genes or one red and one white gene for eye color, its eyes will look red. Given that an offspring's eyes look red, what is the conditional probability that it has two red genes for eye color?

1.4-8. A researcher finds that, of 982 men who died in 2002, 221 died from some heart disease. Also, of the 982 men, 334 had at least one parent who had some heart disease. Of the latter 334 men, 111 died from some heart disease. A man is selected from the group of 982. Given that neither of his parents had some heart disease, find the conditional probability that this man died of some heart disease.

1.4-9. An urn contains four colored balls: two orange and two blue. Two balls are selected at random without replacement, and you are told that at least one of them is orange. What is the probability that the other ball is also orange?

1.4-10. An urn contains 17 balls marked LOSE and 3 balls marked WIN. You and an opponent take turns selecting a single ball at random from the urn without replacement. The person who selects the third WIN ball wins the game. It does not matter who selected the first two WIN balls.

   (a) If you draw first, find the probability that you win the game on your second draw.

   (b) If you draw first, find the probability that your opponent wins the game on his second draw.

   (c) If you draw first, what is the probability that you win? HINT: You could win on your second, third, fourth, . . . , or tenth draw, but not on your first.

   (d) Would you prefer to draw first or second? Why?

1.4-11. In a string of 12 Christmas tree light bulbs, 3 are defective. The bulbs are selected at random and tested, one at a time, until the third defective bulb is found. Compute the probability that the third defective bulb is the

   (a) Third bulb tested.

   (b) Fifth bulb tested.

   (c) Tenth bulb tested.

1.4-12. A small grocery store had 10 cartons of milk, 2 of which were sour. If you are going to buy the sixth carton of milk sold that day at random, compute the probability of selecting a carton of sour milk.

1.4-13. In the gambling game "craps," a pair of dice is rolled and the outcome of the experiment is the sum of the points on the up sides of the six-sided dice. The bettor wins on the first roll if the sum is 7 or 11. The bettor loses on the first roll if the sum is 2, 3, or 12. If the sum is 4, 5, 6, 8, 9, or 10, that number is called the bettor's "point." Once the point is established, the rule is as follows: If the bettor rolls a 7 before the point, the bettor loses; but if the point is rolled before a 7, the bettor wins.

   (a) List the 36 outcomes in the sample space for the roll of a pair of dice. Assume that each of them has a probability of 1/36.

   (b) Find the probability that the bettor wins on the first roll. That is, find the probability of rolling a 7 or 11, $P(7 \text{ or } 11)$.

   (c) Given that 8 is the outcome on the first roll, find the probability that the bettor now rolls the point 8 before rolling a 7 and thus wins. Note that at this stage in the game the only outcomes of interest are 7 and 8. Thus, find $P(8 \mid 7 \text{ or } 8)$.

   (d) The probability that a bettor rolls an 8 on the first roll and then wins is given by $P(8)P(8 \mid 7 \text{ or } 8)$. Show that this probability is $(5/36)(5/11)$.

   (e) Show that the total probability that a bettor wins in the game of craps is 0.49293. HINT: Note that the bettor can win in one of several mutually exclusive ways: by rolling a 7 or an 11 on the first roll or by establishing one of the points 4, 5, 6, 8, 9, or 10 on the first roll and then obtaining that point on successive rolls before a 7 comes up.

1.4-14. A single card is drawn at random from each of six well-shuffled decks of playing cards. Let A be the event that all six cards drawn are different.

   (a) Find $P(A)$.

   (b) Find the probability that at least two of the drawn cards match.

1.4-15. Consider the birthdays of the students in a class of size r. Assume that the year consists of 365 days.

   (a) How many different ordered samples of birthdays are possible (r in sample) allowing repetitions (with replacement)?

   (b) The same as part (a), except requiring that all the students have different birthdays (without replacement)?

   (c) If we can assume that each ordered outcome in part (a) has the same probability, what is the probability that at least two students have the same birthday?

   (d) For what value of r is the probability in part (c) about equal to 1/2? Is this number surprisingly small? HINT: Use a calculator or computer to find r.

1.4-16. You are a member of a class of 18 students. A bowl contains 18 chips: 1 blue and 17 red. Each student is to take 1 chip from the bowl without replacement. The student who draws the blue chip is guaranteed an A for the course.

   (a) If you have a choice of drawing first, fifth, or last, which position would you choose? Justify your choice on the basis of probability.

   (b) Suppose the bowl contains 2 blue and 16 red chips. What position would you now choose?

1.4-17. A drawer contains four black, six brown, and eight olive socks. Two socks are selected at random from the drawer.

(a) Compute the probability that both socks are the same color.

(b) Compute the probability that both socks are olive if it is known that they are the same color.

1.4-18. Bowl A contains three red and two white chips, and bowl B contains four red and three white chips. A chip is drawn at random from bowl A and transferred to bowl B. Compute the probability of then drawing a red chip from bowl B.

1.4-19. An urn contains four balls numbered 1 through 4. The balls are selected one at a time without replacement. A match occurs if ball numbered m is the mth ball selected. Let the event A_i denote a match on the ith draw, $i = 1, 2, 3, 4$.

(a) Show that $P(A_i) = \dfrac{3!}{4!}$.

(b) Show that $P(A_i \cap A_j) = \dfrac{2!}{4!}$.

(c) Show that $P(A_i \cap A_j \cap A_k) = \dfrac{1!}{4!}$.

(d) Show that the probability of at least one match is

$$P(A_1 \cup A_2 \cup A_3 \cup A_4)$$
$$= 1 - \frac{1}{2!} + \frac{1}{3!} - \frac{1}{4!}.$$

(e) Extend this exercise so that there are n balls in the urn. Show that the probability of at least one match is

$$P(A_1 \cup A_2 \cup \cdots \cup A_n)$$
$$= 1 - \frac{1}{2!} + \frac{1}{3!} - \frac{1}{4!}$$
$$+ \cdots + \frac{(-1)^{n+1}}{n!}$$

$$= 1 - \left(1 - \frac{1}{1!} + \frac{1}{2!} - \frac{1}{3!}\right.$$
$$\left. + \cdots + \frac{(-1)^n}{n!}\right).$$

(f) What is the limit of this probability as n increases without bound?

1.4-20. Paper is often tested for "burst strength" and "tear strength." Say we classify these strengths as low, middle, and high. Then, after examining 100 pieces of paper, we find the following:

Tear Strength	Burst Strength		
	A_1 (low)	A_2 (middle)	A_3 (high)
B_1 (low)	7	11	13
B_2 (middle)	11	21	9
B_3 (high)	12	9	7

If we select one of the pieces at random, what are the probabilities that it has the following characteristics:

(a) A_1,

(b) $A_3 \cap B_2$,

(c) $A_2 \cup B_3$,

(d) A_1, given that it is B_2,

(e) B_1, given that it is A_3?

1.4-21. An urn contains eight red and seven blue balls. A second urn contains an unknown number of red balls and nine blue balls. A ball is drawn from each urn at random, and the probability of getting two balls of the same color is 151/300. How many red balls are in the second urn?

1.5 INDEPENDENT EVENTS

For certain pairs of events, the occurrence of one of them may or may not change the probability of the occurrence of the other. In the latter case, they are said to be **independent events**. However, before giving the formal definition of independence, let us consider an example.

EXAMPLE 1.5-1 Flip a coin twice and observe the sequence of heads and tails. The sample space is then

$$S = \{HH, HT, TH, TT\}.$$

It is reasonable to assign a probability of 1/4 to each of these four outcomes. Let

$$A = \{\text{heads on the first flip}\} = \{HH, HT\},$$
$$B = \{\text{tails on the second flip}\} = \{HT, TT\},$$
$$C = \{\text{tails on both flips}\} = \{TT\}.$$

Then $P(B) = 2/4 = 1/2$. Now, on the one hand, if we are given that C has occurred, then $P(B \mid C) = 1$, because $C \subset B$. That is, the knowledge of the occurrence of C has changed the probability of B. On the other hand, if we are given that A has occurred, then

$$P(B \mid A) = \frac{P(A \cap B)}{P(A)} = \frac{1/4}{2/4} = \frac{1}{2} = P(B).$$

So the occurrence of A has not changed the probability of B. Hence, the probability of B does not depend upon knowledge about event A, so we say that A and B are independent events. That is, events A and B are independent if the occurrence of one of them does not affect the probability of the occurrence of the other. A more mathematical way of saying this is

$$P(B \mid A) = P(B) \qquad \text{or} \qquad P(A \mid B) = P(A),$$

provided that $P(A) > 0$ or, in the latter case, $P(B) > 0$. With the first of these equalities and the multiplication rule (Definition 1.4-2), we have

$$P(A \cap B) = P(A)P(B \mid A) = P(A)P(B).$$

The second of these equalities, namely, $P(A \mid B) = P(A)$, gives us the same result

$$P(A \cap B) = P(B)P(A \mid B) = P(B)P(A). \qquad \blacksquare$$

This example motivates the following definition of independent events.

DEFINITION 1.5-1	Events A and B are **independent** if and only if $P(A \cap B) = P(A)P(B)$. Otherwise A and B are called **dependent** events.

Events that are independent are sometimes called **statistically independent**, **stochastically independent**, or **independent in a probabilistic sense**, but in most instances we use *independent* without a modifier if there is no possibility of misunderstanding. It is interesting to note that the definition always holds if $P(A) = 0$ or $P(B) = 0$, because then $P(A \cap B) = 0$, since $(A \cap B) \subset A$ and $(A \cap B) \subset B$. Thus, the left-hand and right-hand members of $P(A \cap B) = P(A)P(B)$ are both equal to zero and thus are equal to each other.

EXAMPLE 1.5-2   A red die and a white die are rolled. Let event $A = \{4$ on the red die$\}$ and event $B = \{$sum of dice is odd$\}$. Of the 36 equally likely outcomes, 6 are favorable to A, 18 are favorable to B, and 3 are favorable to $A \cap B$. Thus,

$$P(A)P(B) = \frac{6}{36} \cdot \frac{18}{36} = \frac{3}{36} = P(A \cap B).$$

Hence, A and B are independent by Definition 1.5-1. $\blacksquare$

EXAMPLE 1.5-3   A red die and a white die are rolled. Let event $C = \{5$ on red die$\}$ and event $D = \{$sum of dice is 11$\}$. Of the 36 equally likely outcomes, 6 are favorable to C, 2 are favorable to D, and 1 is favorable to $C \cap D$. Thus,

$$P(C)P(D) = \frac{6}{36} \cdot \frac{2}{36} = \frac{1}{108} \neq \frac{1}{36} = P(C \cap D).$$

Hence, C and D are dependent events by Definition 1.5-1. ∎

THEOREM 1.5-1

If A and B are independent events, then the following pairs of events are also independent:

(a) A and B'.

(b) A' and B.

(c) A' and B'.

Proof. We know that conditional probability satisfies the axioms for a probability function. Hence, if $P(A) > 0$, then $P(B'|A) = 1 - P(B|A)$. Thus,

$$P(A \cap B') = P(A)P(B'|A) = P(A)[1 - P(B|A)]$$
$$= P(A)[1 - P(B)]$$
$$= P(A)P(B'),$$

since $P(B|A) = P(B)$ by hypothesis. Consequently, A and B' are independent events. The proofs of parts (b) and (c) are left as exercises. □

Before extending the definition of independent events to more than two events, we present the following example.

EXAMPLE 1.5-4 An urn contains four balls numbered 1, 2, 3, and 4. One ball is to be drawn at random from the urn. Let the events A, B, and C be defined by $A = \{1, 2\}$, $B = \{1, 3\}$, and $C = \{1, 4\}$. Then $P(A) = P(B) = P(C) = 1/2$. Furthermore,

$$P(A \cap B) = \frac{1}{4} = P(A)P(B),$$

$$P(A \cap C) = \frac{1}{4} = P(A)P(C),$$

$$P(B \cap C) = \frac{1}{4} = P(B)P(C),$$

which implies that A, B, and C are independent in pairs (called **pairwise independence**). However, since $A \cap B \cap C = \{1\}$, we have

$$P(A \cap B \cap C) = \frac{1}{4} \neq \frac{1}{8} = P(A)P(B)P(C).$$

That is, something seems to be lacking for the complete independence of A, B, and C. ∎

This example illustrates the reason for the second condition in the next definition.

DEFINITION 1.5-2	Events A, B, and C are **mutually independent** if and only if the following two conditions hold:

(a) A, B, and C are pairwise independent; that is,

$$P(A \cap B) = P(A)P(B), \qquad P(A \cap C) = P(A)P(C),$$

and

$$P(B \cap C) = P(B)P(C).$$

(b) $P(A \cap B \cap C) = P(A)P(B)P(C)$.

Definition 1.5-2 can be extended to the mutual independence of four or more events. In such an extension, each pair, triple, quartet, and so on, must satisfy this type of multiplication rule. If there is no possibility of misunderstanding, *independent* is often used without the modifier *mutually* when several events are considered.

EXAMPLE 1.5-5   A rocket has a built-in redundant system. In this system, if component K_1 fails, it is bypassed and component K_2 is used. If component K_2 fails, it is bypassed and component K_3 is used. (An example of a system with these kinds of components is a three-computer system.) Suppose that the probability of failure of any one component is 0.15, and assume that the failures of these components are mutually independent events. Let A_i denote the event that component K_i fails for $i = 1, 2, 3$. Because the system fails if K_1 fails and K_2 fails and K_3 fails, the probability that the system does not fail is given by

$$
\begin{aligned}
P[(A_1 \cap A_2 \cap A_3)'] &= 1 - P(A_1 \cap A_2 \cap A_3) \\
&= 1 - P(A_1)P(A_2)P(A_3) \\
&= 1 - (0.15)^3 \\
&= 0.9966.
\end{aligned}
$$

One way to increase the reliability of such a system is to add more components (realizing that this also adds weight and takes up space). For example, if a fourth component K_4 were added to this system, the probability that the system does not fail is

$$P[(A_1 \cap A_2 \cap A_3 \cap A_4)'] = 1 - (0.15)^4 = 0.9995. \qquad \blacksquare$$

If A, B, and C are mutually independent events, then the following events are also independent:

(a) A and $(B \cap C)$;

(b) A and $(B \cup C)$;

(c) A' and $(B \cap C')$.

In addition, A', B', and C' are mutually independent. (The proofs and illustrations of these results are left as exercises.)

Many experiments consist of a sequence of n trials that are mutually independent. If the outcomes of the trials, in fact, do not have anything to do with one another, then events, such that each is associated with a different trial, should be independent in the probability sense. That is, if the event A_i is associated with the ith trial, $i = 1, 2, \ldots, n$, then

$$P(A_1 \cap A_2 \cap \cdots \cap A_n) = P(A_1)P(A_2)\cdots P(A_n).$$

EXAMPLE 1.5-6   A fair six-sided die is rolled six independent times. Let A_i be the event that side i is observed on the ith roll, called a match on the ith trial, $i = 1, 2, \ldots, 6$. Thus, $P(A_i) = 1/6$ and $P(A_i') = 1 - 1/6 = 5/6$. If we let B denote the event that at least one match occurs, then B' is the event that no matches occur. Hence,

$$P(B) = 1 - P(B') = 1 - P(A_1' \cap A_2' \cap \cdots \cap A_6')$$

$$= 1 - \frac{5}{6} \cdot \frac{5}{6} \cdot \frac{5}{6} \cdot \frac{5}{6} \cdot \frac{5}{6} \cdot \frac{5}{6} = 1 - \left(\frac{5}{6}\right)^6$$

is the probability of B.   ∎

The sample space for an experiment of n trials is a set of n-tuples, where the ith component denotes the outcome on the ith trial. For example, if a six-sided die is rolled five times, then

$$S = \{(O_1, O_2, O_3, O_4, O_5) : O_i = 1, 2, 3, 4, 5, \text{ or } 6, \text{ for } i = 1, 2, 3, 4, 5\}.$$

That is, S is a set of five-tuples, where each component is one of the first six positive integers.

If a coin is tossed two times, then

$$S = \{(O_1, O_2) : O_i = H \text{ or } T, i = 1, 2\}.$$

We often drop the commas and parentheses and let, for example, $(H, T) = HT$, as in Example 1.5-1.

EXAMPLE 1.5-7   The probability that a company's workforce has at least one accident during a certain month is $(0.01)k$, where k is the number of days in that month (say, February has 28 days). Assume that the numbers of accidents is independent from month to month. If the company's year starts with January, the probability that the first accident is in April is

$P(\text{none in Jan., none in Feb., none in March, at least one in April}) =$
$(1 - 0.31)(1 - 0.28)(1 - 0.31)(0.30) = (0.69)(0.72)(0.69)(0.30) = 0.103.$ ∎

EXAMPLE 1.5-8   Three inspectors look at a critical component of a product. Their probabilities of detecting a defect are different, namely, $0.99, 0.98$, and 0.96, respectively. If we assume independence, then the probability of at least one detecting the defect is

$$1 - (0.01)(0.02)(0.04) = 0.999992.$$

The probability of *only* one finding the defect is

$$(0.99)(0.02)(0.04) + (0.01)(0.98)(0.04) + (0.01)(0.02)(0.96) = 0.001576.$$

As an exercise, compute the following probability exactly: (a) that two find the defect, (b) that three find the defect.   ∎

EXAMPLE 1.5-9 Suppose that on five consecutive days an "instant winner" lottery ticket is purchased and the probability of winning is 1/5 on each day. Assuming independent trials, we have

$$P(WWLLL) = \left(\frac{1}{5}\right)^2\left(\frac{4}{5}\right)^3,$$

$$P(LWLWL) = \frac{4}{5} \cdot \frac{1}{5} \cdot \frac{4}{5} \cdot \frac{1}{5} \cdot \frac{4}{5} = \left(\frac{1}{5}\right)^2\left(\frac{4}{5}\right)^3.$$

In general, the probability of purchasing two winning tickets and three losing tickets is

$$\binom{5}{2}\left(\frac{1}{5}\right)^2\left(\frac{4}{5}\right)^3 = \frac{5!}{2!3!}\left(\frac{1}{5}\right)^2\left(\frac{4}{5}\right)^3 = 0.2048,$$

because there are $\binom{5}{2}$ ways to select the positions (or the days) for the winning tickets and each of these $\binom{5}{2}$ ways has the probability $(1/5)^2(4/5)^3$. ■

EXERCISES

1.5-1. Let A and B be independent events with $P(A) = 0.7$ and $P(B) = 0.2$. Compute (a) $P(A \cap B)$, (b) $P(A \cup B)$, and (c) $P(A' \cup B')$.

1.5-2. Let $P(A) = 0.3$ and $P(B) = 0.6$.

   (a) Find $P(A \cup B)$ when A and B are independent.

   (b) Find $P(A \mid B)$ when A and B are mutually exclusive.

1.5-3. Let A and B be independent events with $P(A) = 1/4$ and $P(B) = 2/3$. Compute (a) $P(A \cap B)$, (b) $P(A \cap B')$, (c) $P(A' \cap B')$, (d) $P[(A \cup B)']$, and (e) $P(A' \cap B)$.

1.5-4. Prove parts (b) and (c) of Theorem 1.5-1.

1.5-5. If $P(A) = 0.8$, $P(B) = 0.5$, and $P(A \cup B) = 0.9$, are A and B independent events? Why or why not?

1.5-6. Show that if A, B, and C are mutually independent, then the following pairs of events are independent: A and $(B \cap C)$, A and $(B \cup C)$, A' and $(B \cap C')$. Show also that A', B' and C' are mutually independent.

1.5-7. Each of three football players will attempt to kick a field goal from the 25-yard line. Let A_i denote the event that the field goal is made by player i, $i = 1, 2, 3$. Assume that A_1, A_2, A_3 are mutually independent and that $P(A_1) = 0.5, P(A_2) = 0.7$, $P(A_3) = 0.6$.

   (a) Compute the probability that exactly one player is successful.

   (b) Compute the probability that exactly two players make a field goal (i.e., one misses).

1.5-8. Die A has orange on one face and blue on five faces, Die B has orange on two faces and blue on four faces, Die C has orange on three faces and blue on three faces. All are unbiased dice. If the three dice are rolled, find the probability that exactly two of the three dice come up orange.

1.5-9. Suppose that A, B, and C are mutually independent events and that $P(A) = 0.5$, $P(B) = 0.8$, and $P(C) = 0.9$. Find the probabilities that (a) all three events occur, (b) exactly two of the three events occur, and (c) none of the events occur.

1.5-10. Let D_1, D_2, D_3 be three four-sided dice whose sides have been labeled as follows:

$$D_1 : 0\,3\,3\,3 \qquad D_2 : 2\,2\,2\,5 \qquad D_3 : 1\,1\,4\,6$$

The three dice are rolled at random. Let A, B, and C be the events that the outcome on die D_1 is larger than the outcome on D_2, the outcome on D_2 is larger than the outcome on D_3, and the outcome on D_3 is larger than the outcome on D_1, respectively. Show that (a) $P(A) = 9/16$, (b) $P(B) = 9/16$, and (c) $P(C) = 10/16$. Do you find it interesting that each of the probabilities that D_1 "beats" D_2, D_2 "beats" D_3, and D_3 "beats" D_1 is greater than 1/2? Thus, it is difficult to determine the "best" die.

1.5-11. Let A and B be two events.

    (a) If the events A and B are mutually exclusive, are A and B always independent? If the answer is no, can they ever be independent? Explain.

    (b) If $A \subset B$, can A and B ever be independent events? Explain.

1.5-12. Flip an unbiased coin five independent times. Compute the probability of

    (a) *HHTHT*.

    (b) *THHHT*.

    (c) *HTHTH*.

    (d) Three heads occurring in the five trials.

1.5-13. An urn contains two red balls and four white balls. Sample successively five times at random and with replacement, so that the trials are independent. Compute the probability of each of the two sequences *WWRWR* and *RWWWR*.

1.5-14. In Example 1.5-5, suppose that the probability of failure of a component is $p = 0.4$. Find the probability that the system does not fail if the number of redundant components is

    (a) 3.

    (b) 8.

1.5-15. An urn contains 10 red and 10 white balls. The balls are drawn from the urn at random, one at a time. Find the probability that the fourth white ball is the sixth ball drawn if the sampling is done

    (a) With replacement.

    (b) Without replacement.

    (c) In the World Series, the American League (red) and National League (white) teams play until one team wins four games. Do you think that the urn model presented in this exercise could be used to describe the probabilities of a 4-, 5-, 6-, or 7-game series? If your answer is yes, would you choose sampling with or without replacement in your model? (For your information, the numbers of 4-, 5-, 6-, and 7-game series, up to and including 2008, were 20, 23, 21, and 36, respectively. The World Series was canceled in 1994, and in 1903 and 1919–1921 winners had to take five out of nine games. Three of those series went eight games and one went seven.)

1.5-16. An urn contains five balls, one marked WIN and four marked LOSE. You and another player take turns selecting a ball at random from the urn, one at a time. The first person to select the WIN ball is

the winner. If you draw first, find the probability that you will win if the sampling is done

    (a) With replacement.

    (b) Without replacement.

1.5-17. Each of the 12 students in a class is given a fair 12-sided die. In addition, each student is numbered from 1 to 12.

    (a) If the students roll their dice, what is the probability that there is at least one "match" (e.g., student 4 rolls a 4)?

    (b) If you are a member of this class, what is the probability that at least one of the other 11 students rolls the same number as you do?

1.5-18. An eight-team single-elimination tournament is set up as follows:

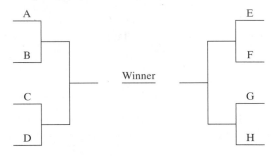

For example, eight students (called A–H) set up a tournament among themselves. The top-listed student in each bracket calls heads or tails when his or her opponent flips a coin. If the call is correct, the student moves on to the next bracket.

    (a) How many coin flips are required to determine the tournament winner?

    (b) What is the probability that you can predict all of the winners?

    (c) In NCAA Division I basketball, after a "play-in" game between the 64th and 65th seeds, 64 teams participate in a single-elimination tournament to determine the national champion. Considering only the remaining 64 teams, how many games are required to determine the national champion?

    (d) Assume that for any given game, either team has an equal chance of winning. (That is probably not true.) On page 43 of the March 22, 1999, issue, *Time*, claimed that the "mathematical odds of predicting all 63 NCAA games correctly is 1 in 75 million." Do you agree with this statement? If not, why not?

1.5-19. Extend Example 1.5-6 to an n-sided die. That is, suppose that a fair n-sided die is rolled n independent times. A match occurs if side i is observed on the ith trial, $i = 1, 2, \ldots, n$.

(a) Show that the probability of at least one match is

$$1 - \left(\frac{n-1}{n}\right)^n = 1 - \left(1 - \frac{1}{n}\right)^n.$$

(b) Find the limit of this probability as n increases without bound.

1.5-20. An urn contains n balls numbered from 1 through n. A sample of n balls is selected at random from the urn one at a time. A match occurs if ball i is selected on the ith draw.

(a) For $n = 1$ to 15, find the probability of at least one match if the sampling is done (i) with replacement (see Exercise 1.5-19), (ii) without replacement (see Exercise 1.4-19).

(b) How much does n affect these probabilities?

(c) How does sampling with and without replacement affect these probabilities?

(d) Illustrate these probabilities empirically, either performing the experiments physically or simulating them on a computer.

1.6 BAYES'S THEOREM

We begin this section by illustrating Bayes's theorem with an example.

EXAMPLE 1.6-1 Bowl B_1 contains two red and four white chips, bowl B_2 contains one red and two white chips, and bowl B_3 contains five red and four white chips. Say that the probabilities for selecting the bowls are not the same but are given by $P(B_1) = 1/3$, $P(B_2) = 1/6$, and $P(B_3) = 1/2$, where B_1, B_2, and B_3 are the events that bowls B_1, B_2, and B_3 are respectively chosen. The experiment consists of selecting a bowl with these probabilities and then drawing a chip at random from that bowl. Let us compute the probability of event R, drawing a red chip—say, $P(R)$. Note that $P(R)$ is dependent first of all on which bowl is selected and then on the probability of drawing a red chip from the selected bowl. That is, the event R is the union of the mutually exclusive events $B_1 \cap R$, $B_2 \cap R$, and $B_3 \cap R$. Thus,

$$P(R) = P(B_1 \cap R) + P(B_2 \cap R) + P(B_3 \cap R)$$
$$= P(B_1)P(R \mid B_1) + P(B_2)P(R \mid B_2) + P(B_3)P(R \mid B_3)$$
$$= \frac{1}{3} \cdot \frac{2}{6} + \frac{1}{6} \cdot \frac{1}{3} + \frac{1}{2} \cdot \frac{5}{9} = \frac{4}{9}.$$

Suppose now that the outcome of the experiment is a red chip, but we do not know from which bowl it was drawn. Accordingly, we compute the conditional probability that the chip was drawn from bowl B_1, namely, $P(B_1 \mid R)$. From the definition of conditional probability and the preceding result, we have

$$P(B_1 \mid R) = \frac{P(B_1 \cap R)}{P(R)}$$

$$= \frac{P(B_1)P(R \mid B_1)}{P(B_1)P(R \mid B_1) + P(B_2)P(R \mid B_2) + P(B_3)P(R \mid B_3)}$$

$$= \frac{(1/3)(2/6)}{(1/3)(2/6) + (1/6)(1/3) + (1/2)(5/9)} = \frac{2}{8}.$$

Similarly,

$$P(B_2 \mid R) = \frac{P(B_2 \cap R)}{P(R)} = \frac{(1/6)(1/3)}{4/9} = \frac{1}{8}$$

and

$$P(B_3 \mid R) = \frac{P(B_3 \cap R)}{P(R)} = \frac{(1/2)(5/9)}{4/9} = \frac{5}{8}.$$

Note that the conditional probabilities $P(B_1 \mid R)$, $P(B_2 \mid R)$, and $P(B_3 \mid R)$ have changed from the original probabilities $P(B_1)$, $P(B_2)$, and $P(B_3)$ in a way that agrees with your intuition. Once the red chip has been observed, the probability concerning B_3 seems more favorable than originally because B_3 has a larger percentage of red chips than do B_1 and B_2. The conditional probabilities of B_1 and B_2 decrease from their original ones once the red chip is observed. Frequently, the original probabilities are called *prior probabilities* and the conditional probabilities are the *posterior probabilities*. ∎

We generalize the result of Example 1.6-1. Let $B_1, B_2, \ldots, B_m$ constitute a *partition* of the sample space S. That is,

$$S = B_1 \cup B_2 \cup \cdots \cup B_m \text{ and } B_i \cap B_j = \emptyset, i \neq j.$$

Of course, the events $B_1, B_2, \ldots, B_m$ are mutually exclusive and exhaustive (since the union of the disjoint sets equals the sample space S). Furthermore, suppose the **prior probability** of the event B_i is positive; that is, $P(B_i) > 0, i = 1, \ldots, m$. If A is an event, then A is the union of m mutually exclusive events, namely,

$$A = (B_1 \cap A) \cup (B_2 \cap A) \cup \cdots \cup (B_m \cap A).$$

Thus,

$$P(A) = \sum_{i=1}^{m} P(B_i \cap A)$$

$$= \sum_{i=1}^{m} P(B_i)P(A \mid B_i). \tag{1.6-1}$$

If $P(A) > 0$, then

$$P(B_k \mid A) = \frac{P(B_k \cap A)}{P(A)}, \qquad k = 1, 2, \ldots, m. \tag{1.6-2}$$

Using Equation 1.6-1 and replacing $P(A)$ in Equation 1.6-2, we have **Bayes's theorem**:

$$P(B_k \mid A) = \frac{P(B_k)P(A \mid B_k)}{\displaystyle\sum_{i=1}^{m} P(B_i)P(A \mid B_i)}, \qquad k = 1, 2, \ldots, m.$$

The conditional probability $P(B_k \mid A)$ is often called the **posterior probability** of B_k. The next example illustrates one application of Bayes's theorem.

EXAMPLE 1.6-2 In a certain factory, machines I, II, and III are all producing springs of the same length. Of their production, machines I, II, and III, respectively produce 2%, 1%, and 3% defective springs. Of the total production of springs in the factory, machine I produces 35%, machine II produces 25%, and machine III produces 40%. If one spring is selected at random from the total springs produced in a day, the probability that it is defective, in an obvious notation, equals

$$P(D) = P(I)P(D \mid I) + P(II)P(D \mid II) + P(III)P(D \mid III)$$
$$= \left(\frac{35}{100}\right)\left(\frac{2}{100}\right) + \left(\frac{25}{100}\right)\left(\frac{1}{100}\right) + \left(\frac{40}{100}\right)\left(\frac{3}{100}\right) = \frac{215}{10,000}.$$

If the selected spring is defective, the conditional probability that it was produced by machine III is, by Bayes's formula,

$$P(III \mid D) = \frac{P(III)P(D \mid III)}{P(D)} = \frac{(40/100)(3/100)}{215/10,000} = \frac{120}{215}.$$

Note how the posterior probability of III increased from the prior probability of III after the defective spring was observed, because III produces a larger percentage of defectives than do I and II. ∎

EXAMPLE 1.6-3 A Pap smear is a screening procedure used to detect cervical cancer. For women with this cancer, there are about 16% *false negatives*; that is,

$$P(T^- = \text{test negative} \mid C = \text{cancer}) = 0.16.$$

Thus,

$$P(T^+ = \text{test positive} \mid C = \text{cancer}) = 0.84.$$

For women without cancer, there are about 19% *false positives*; that is,

$$P(T^+ \mid C' = \text{not cancer}) = 0.19.$$

Hence,

$$P(T^- \mid C' = \text{not cancer}) = 0.81.$$

In the United States, there are about 8 women in 100,000 who have this cancer; that is,

$$P(C) = 0.00008; \quad \text{so } P(C') = 0.99992.$$

By Bayes's theorem,

$$P(C \mid T^+) = \frac{P(C \text{ and } T^+)}{P(T^+)}$$
$$= \frac{(0.00008)(0.84)}{(0.00008)(0.84) + (0.99992)(0.19)}$$
$$= \frac{672}{672 + 1,899,848} = 0.000354.$$

What this means is that for every million positive Pap smears, only 354 represent true cases of cervical cancer. This low ratio makes one question the value of the procedure. The reason that it is ineffective is that the percentage of women having that cancer is so small and the error rates of the procedure—namely, 0.16 and 0.19—are so high. (See Yobs, A .R., Swanson, R. A., and Lamotte, L. C. "Laboratory Reliability of the Papanicolaou Smear." *Obstetrics and Gynecology*, Volume 65, February 1985, pp. 235–244.) ∎

EXERCISES

1.6-1. Bowl B_1 contains two white chips, bowl B_2 contains two red chips, bowl B_3 contains two white and two red chips, and bowl B_4 contains three white chips and one red chip. The probabilities of selecting bowl B_1, B_2, B_3, or B_4 are 1/2, 1/4, 1/8, and 1/8, respectively. A bowl is selected using these probabilities and a chip is then drawn at random. Find

   (a) $P(W)$, the probability of drawing a white chip.

   (b) $P(B_1 \mid W)$, the conditional probability that bowl B_1 had been selected, given that a white chip was drawn.

1.6-2. Bean seeds from supplier A have an 85% germination rate and those from supplier B have a 75% germination rate. A seed-packaging company purchases 40% of its bean seeds from supplier A and 60% from supplier B and mixes these seeds together.

   (a) Find the probability $P(G)$ that a seed selected at random from the mixed seeds will germinate.

   (b) Given that a seed germinates, find the probability that the seed was purchased from supplier A.

1.6-3. A doctor is concerned about the relationship between blood pressure and irregular heartbeats. Among her patients, she classifies blood pressures as high, normal, or low and heartbeats as regular or irregular and finds that (a) 16% have high blood pressure; (b) 19% have low blood pressure; (c) 17% have an irregular heartbeat; (d) of those with an irregular heartbeat, 35% have high blood pressure; and (e) of those with normal blood pressure, 11% have an irregular heartbeat. What percentage of her patients have a regular heartbeat and low blood pressure?

1.6-4. Assume that an insurance company knows the following probabilities relating to automobile accidents:

Age of Driver	Probability of Accident	Fraction of Company's Insured Drivers
16–25	0.05	0.10
26–50	0.02	0.55
51–65	0.03	0.20
66–90	0.04	0.15

A randomly selected driver from the company's insured drivers has an accident. What is the conditional probability that the driver is in the 16–25 age group?

1.6-5. At a hospital's emergency room, patients are classified and 20% of them are critical, 30% are serious, and 50% are stable. Of the critical ones, 30% die; of the serious, 10% die; and of the stable, 1% die. Given that a patient dies, what is the conditional probability that the patient was classified as critical.

1.6-6. A life insurance company issues standard, preferred, and ultrapreferred policies. Of the company's policyholders of a certain age, 60% have standard policies and a probability of 0.01 of dying in the next year, 30% have preferred policies and a probability of 0.008 of dying in the next year, and 10% have ultrapreferred policies and a probability of 0.007 of dying in the next year. A policyholder of that age dies in the next year. What are the conditional probabilities of the deceased having had a standard, a preferred, and an ultrapreferred policy?

1.6-7. Among 60-year-old college professors, 90% are nonsmokers and 10% are smokers. The probability of a nonsmoker dying in the next year is 0.005 and the probability for smokers is 0.05. Given that one of this group of college professors dies in the next year, what is the conditional probability that the professor was a smoker?

1.6-8. A store sells four brands of DVD players. The least expensive brand, B_1, accounts for 40% of the sales. The other brands (in order of their price) have the following percentages of sales: $B_2, 30\%$; $B_3, 20\%$; and $B_4, 10\%$. The respective probabilities of needing repair during warranty are 0.10 for B_1, 0.05 for B_2, 0.03 for B_3, and 0.02 for B_4. A randomly selected purchaser has a DVD player that needs repair under warranty. What are the four conditional probabilities of being brand B_i, $i = 1, 2, 3, 4$?

1.6-9. There is a new diagnostic test for a disease that occurs in about 0.05% of the population. The test is not perfect, but will detect a person with the disease 99% of the time. It will, however, say that a person without the disease has the disease about 3% of the time. A person is selected at random from the population, and the test indicates that this person has the disease. What are the conditional probabilities that

   (a) the person has the disease?

   (b) the person does not have the disease?

Discuss. HINT: Note that the fraction 0.0005 of diseased persons in the population is much smaller than the error probabilities of 0.01 and 0.03.

1.6-10. Suppose we want to investigate the percentage of abused children in a certain population. To do this, doctors examine some of these children taken at random from that population. However, doctors are not perfect: They sometimes classify an abused child (A) as one not abused (ND) or they classify a nonabused child (N) as one that is abused (AD). Suppose these error rates are $P(ND|A) = 0.08$ and $P(AD|N) = 0.05$, respectively; thus, $P(AD|A) = 0.92$ and $P(ND|N) = 0.95$ are the probabilities of the correct decisions. Let us pretend that only 2 percent of all children are abused; that is, $P(A) = 0.02$ and $P(N) = 0.98$.

(a) Select a child at random. What is the probability that the doctor classifies this child as abused. That is, compute

$$P(AD) = P(A)P(AD|A)$$
$$+ P(N)P(AD|N).$$

(b) Given that the child is classified by the doctor as abused, compute $P(N|AD)$ and $P(A|AD)$.

(c) Also, compute $P(N|ND)$ and $P(A|ND)$.

(d) Are the probabilities in (b) and (c) alarming? This happens because the error rates of 0.08 and 0.05 are high relative to the fraction 0.02 of abused children in the population.

1.6-11. At the beginning of a certain study of a group of persons, 15% were classified as heavy smokers, 30% as light smokers, and 55% as nonsmokers. In the five-year study, it was determined that the death rates of the heavy and light smokers were five and three times that of the nonsmokers, respectively. A randomly selected participant died over the five-year period; calculate the probability that the participant was a nonsmoker.

1.6-12. A test indicates the presence of a particular disease 90% of the time when the disease is present and the presence of the disease 2% of the time when the disease is not present. If 0.5% of the population has the disease, calculate the conditional probability that a person selected at random has the disease if the test indicates the presence of the disease.

1.6-13. A hospital receives two-fifths of its flu vaccine from Company A and the remainder from Company B. Each shipment contains a large number of vials of vaccine. From Company A, 3% of the vials are ineffective; from Company B, 2% are ineffective. A hospital tests $n = 25$ randomly selected vials from one shipment and finds that 2 are ineffective. What is the conditional probability that this shipment came from Company A?

1.6-14. Two processes of a company produce rolls of materials: The rolls of Process I are 3% defective and the rolls of Process II are 1% defective. Process I produces 60% of the company's output, Process II 40%. A roll is selected at random from the total output. Given that this roll is defective, what is the conditional probability that it is from Process I?

1.6-15. A chemist wishes to detect an impurity in a certain compound that she is making. There is a test that detects an impurity with probability 0.90; however, this test indicates that an impurity is there when it is not about 5% of the time. The chemist produces compounds with the impurity about 20% of the time, that is, 80% do not have the impurity. A compound is selected at random from the chemist's output. The test indicates that an impurity is present. What is the conditional probability that the compound actually has an impurity?

HISTORICAL COMMENTS Most probabilists would say that the mathematics of probability began when, in 1654, Chevalier de Méré, a French nobleman who liked to gamble, challenged Blaise Pascal to explain a puzzle and a problem created from his observations concerning rolls of dice. Of course, there was gambling well before this, and actually, almost 200 years before this challenge, a Franciscan monk, Luca Paccioli, proposed the same puzzle. Here it is:

> *A* and *B* are playing a fair game of balla. They agree to continue until one has six rounds. However, the game actually stops when *A* has won five and *B* three. How should the stakes be divided?

And over 100 years before de Méré's challenge, a 16th-century doctor, Girolamo Cardano, who was also a gambler, had figured out the answers to many dice problems, but not the one that de Méré proposed. Chevalier de Méré had observed this: If a single die is tossed 4 times, the probability of obtaining at least one six was slightly greater than 1/2. However, keeping the same proportions, if a pair of dice is tossed 24 times, the probability of obtaining at least one double-six seemed to be slightly less than 1/2; at least de Méré was losing money betting on it. This is when he approached Blaise Pascal with the challenge. Not wanting to work on the problems alone, Pascal formed a partnership with Pierre de Fermat, a brilliant young mathematician. It was this 1654 correspondence between Pascal and Fermat that started the theory of probability.

Today an average student in probability could solve both problems easily. For the puzzle, note that B could win with six rounds only by winning the next three rounds, which has probability of $(1/2)^3 = 1/8$ because it was a fair game of balla. Thus, A's probability of winning six rounds is $1 - 1/8 = 7/8$, and stakes should be divided seven units to one. For the dice problem, the probability of at least one six in four rolls of a die is

$$1 - \left(\frac{5}{6}\right)^4 \approx 0.518,$$

while the probability of rolling at least one double-six in 24 rolls of a pair of dice is

$$1 - \left(\frac{35}{36}\right)^{24} \approx 0.491.$$

It seems amazing to us that de Méré could have observed enough trials of those events to detect the slight difference in those probabilities. However, he won betting on the first but lost by betting on the second.

Incidentally, the solution to the balla puzzle led to a generalization—namely, the binomial distribution—and to the famous Pascal triangle. Of course, Fermat was the great mathematician associated with "Fermat's last theorem."

The Reverend Thomas Bayes, who was born in 1701, was a Nonconformist (a Protestant who rejected most of the rituals of the Church of England). While he published nothing in mathematics when he was alive, two works were published after his death, one of which contained the essence of Bayes's theorem and a very original way of using data to modify prior probabilities to create posterior probabilities. It has had such an influence on modern statistics that many modern statisticians are associated with the Neo-Bayesian movement and we devote Chapter 9 to some of these methods.

Discrete Distributions

2.1   RANDOM VARIABLES OF THE DISCRETE TYPE

An outcome space S may be difficult to describe if the elements of S are not numbers. We shall now discuss how we can use a rule by which each outcome of a random experiment, an element s of S, may be associated with a real number x. We begin the discussion with an example.

EXAMPLE 2.1-1   A rat is selected at random from a cage and its sex is determined. The set of possible outcomes is female and male. Thus, the outcome space is $S = \{\text{female, male}\} = \{\text{F, M}\}$. Let X be a function defined on S such that $X(\text{F}) = 0$ and $X(\text{M}) = 1$. X is then a real-valued function that has the outcome space S as its domain and the set of real numbers $\{x : x = 0,1\}$ as its range. We call X a random variable, and in this example, the space associated with X is $\{x : x = 0,1\}$. ∎

We now formulate the definition of a random variable.

DEFINITION 2.1-1   Given a random experiment with an outcome space S, a function X that assigns one and only one real number $X(s) = x$ to each element s in S is called a **random variable**. The **space** of X is the set of real numbers $\{x : X(s) = x, s \in S\}$, where $s \in S$ means that the element s belongs to the set S.

REMARK   As we give examples of random variables and their probability distributions, the reader will soon recognize that, when observing a random experiment, the experimenter must take some type of measurement (or measurements). This measurement can be thought of as the outcome of a random variable. We would simply like to know the probability of a measurement ending in A, a subset of the space of X. If this is known for all subsets A, then we know the probability distribution of the random variable. Obviously, in practice, we do not very often know this distribution exactly. Hence, statisticians make conjectures about these distributions; that is, we construct probabilistic models for random variables. The ability of a statistician to model a real situation appropriately is a valuable trait. In this chapter, we introduce some probability models in which the spaces of the random variables consist of sets of integers. ∎

It may be that the set S has elements that are themselves real numbers. In such an instance, we could write $X(s) = s$, so that X is the identity function and the space of X is also S. This situation is illustrated in Example 2.1-2.

EXAMPLE 2.1-2   Let the random experiment be the cast of a die. Then the outcome space associated with this experiment is $S = \{1, 2, 3, 4, 5, 6\}$, with the elements of S indicating the number of spots on the side facing up. For each $s \in S$, let $X(s) = s$. The space of the random variable X is then $\{1, 2, 3, 4, 5, 6\}$.

If we associate a probability of 1/6 with each outcome, then, for example, $P(X = 5) = 1/6$, $P(2 \le X \le 5) = 4/6$, and $P(X \le 2) = 2/6$ seem to be reasonable assignments, where, in this example, $\{2 \le X \le 5\}$ means $\{X = 2, 3, 4, \text{ or } 5\}$ and $\{X \le 2\}$ means $\{X = 1 \text{ or } 2\}$. ∎

The student will no doubt recognize two major difficulties here:

1. In many practical situations, the probabilities assigned to the events are unknown.

2. Since there are many ways of defining a function X on S, which function do we want to use?

As a matter of fact, the solutions to these problems in particular cases are major concerns in applied statistics. In considering (2), statisticians try to determine what *measurement* (or measurements) should be taken on an outcome; that is, how best do we "mathematize" the outcome? These measurement problems are most difficult and can be answered only by getting involved in a practical project. For (1), we often need to estimate these probabilities or percentages through repeated observations (called sampling). For example, what percentage of newborn girls in the University of Iowa Hospital weigh less than 7 pounds? Here a newborn baby girl is the outcome, and we have measured her one way (by weight), but obviously there are many other ways of measuring her. If we let X be the weight in pounds, we are interested in the probability $P(X < 7)$, and we can estimate this probability only by repeated observations. One obvious way of estimating it is by the use of the relative frequency of $\{X < 7\}$ after a number of observations. If it is reasonable to make additional assumptions, we will study other ways of estimating that probability. It is this latter aspect with which mathematical statistics is concerned. That is, if we assume certain models, we find that the theory of statistics can explain how best to draw conclusions or make predictions.

In many instances, it is clear exactly what function X the experimenter wants to define on the outcome space. For example, the caster in the dice game called craps is concerned about the sum of the spots (say X) that are facing upwards on the pair of dice. Hence, we go directly to the space of X, which we shall denote by the same letter S. After all, in the dice game, the caster is directly concerned only with the probabilities associated with X. Thus, for convenience, in many instances the reader can think of the space of X as being the outcome space.

Let X denote a random variable with space S. Suppose that we know how the probability is distributed over the various subsets A of S; that is, we can compute $P(X \in A)$. In this sense, we speak of the distribution of the random variable X, meaning, of course, the distribution of probability associated with the space S of X.

Let X denote a random variable with one-dimensional space S, a subset of the real numbers. Suppose that the space S contains a countable number of points; that is, either S contains a finite number of points, or the points of S can be put into a one-to-one correspondence with the positive integers. Such a set S is called a set of discrete points or simply a discrete outcome space. Furthermore, the random variable X is called a random variable of the **discrete type**, and X is said to have a distribution of the discrete type.

For a random variable X of the discrete type, the probability $P(X = x)$ is frequently denoted by $f(x)$, and this function $f(x)$ is called the **probability mass function**. Note that some authors refer to $f(x)$ as the probability function, the frequency function, or the probability density function. In the discrete case, we shall use "probability mass function," and it is hereafter abbreviated p.m.f.

Let $f(x)$ be the p.m.f. of the random variable X of the discrete type, and let S be the space of X. Since $f(x) = P(X = x)$ for $x \in S$, $f(x)$ must be nonnegative for $x \in S$, and we want all these probabilities to add to 1 because each $P(X = x)$ represents the fraction of times x can be expected to occur. Moreover, to determine the probability associated with the event $A \in S$, we would sum the probabilities of the x values in A. This idea leads us to the following definition.

DEFINITION 2.1-2   The **probability mass function** (p.m.f.) $f(x)$ of a discrete random variable X is a function that satisfies the following properties:

(a) $f(x) > 0$, $x \in S$;

(b) $\displaystyle\sum_{x \in S} f(x) = 1$;

(c) $P(X \in A) = \displaystyle\sum_{x \in A} f(x)$,   where $A \subset S$.

Of course, we usually let $f(x) = 0$ when $x \notin S$; thus, the domain of $f(x)$ is the set of real numbers. When we define the p.m.f. $f(x)$ and do not say "zero elsewhere," we tacitly mean that $f(x)$ has been defined at all x's in the space S and it is assumed that $f(x) = 0$ elsewhere; that is, $f(x) = 0$ when $x \notin S$. Since the probability $P(X = x) = f(x) > 0$ when $x \in S$, and since S contains all the probabilities associated with X, we sometimes refer to S as the **support** of X as well as the space of X.

When a p.m.f. is constant on the space or support, we say that the distribution is **uniform** over that space. As an illustration, in Example 2.1-2 X has a discrete uniform distribution on $S = \{1, 2, 3, 4, 5, 6\}$ and its p.m.f. is

$$f(x) = \frac{1}{6}, \qquad x = 1, 2, 3, 4, 5, 6.$$

We can generalize this result by letting X have a discrete uniform distribution over the first m positive integers, so that its p.m.f. is

$$f(x) = \frac{1}{m}, \qquad x = 1, 2, 3, \cdots, m.$$

We now give an example in which X does not have a uniform distribution.

EXAMPLE 2.1-3 Roll a four-sided die twice, and let X equal the larger of the two outcomes if they are different and the common value if they are the same. The outcome space for this experiment is $S_0 = \{(d_1, d_2): d_1 = 1, 2, 3, 4; d_2 = 1, 2, 3, 4\}$, where we assume that each of these 16 points has probability 1/16. Then $P(X = 1) = P[(1,1)] = 1/16$, $P(X = 2) = P[\{(1,2), (2,1), (2,2)\}] = 3/16$, and similarly $P(X = 3) = 5/16$ and $P(X = 4) = 7/16$. That is, the p.m.f. of X can be written simply as

$$f(x) = P(X = x) = \frac{2x - 1}{16}, \qquad x = 1, 2, 3, 4. \tag{2.1-1}$$

We could add that $f(x) = 0$ elsewhere; but if we do not, the reader should take $f(x)$ to equal zero when $x \notin S = \{1, 2, 3, 4\}$. ■

A better understanding of a particular probability distribution can often be obtained with a graph that depicts the p.m.f. of X. Note that the graph of the p.m.f. when $f(x) > 0$ would be simply the set of points $\{[x, f(x)]: x \in S\}$, where S is the space of X. Two types of graphs can be used to give a better visual appreciation of the p.m.f.: a bar graph and a probability histogram. A **bar graph** of the p.m.f. $f(x)$ of the random variable X is a graph having a vertical line segment drawn from $(x, 0)$ to $[x, f(x)]$ at each x in S, the space of X. If X can assume only integer values, a **probability histogram** of the p.m.f. $f(x)$ is a graphical representation that has a rectangle of height $f(x)$ and a base of length 1, centered at x for each $x \in S$, the space of X. Thus, the area of each rectangle is equal to the respective probability $f(x)$, and the total area of a probability histogram is 1.

Figure 2.1-1 displays a bar graph and a probability histogram for the p.m.f. $f(x)$ defined in Equation 2.1-1.

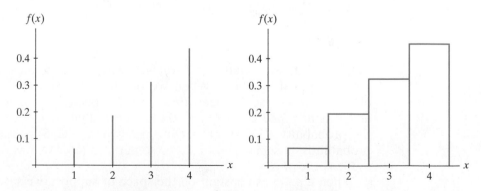

FIGURE 2.1-1: Bar graph and probability histogram

Our next probability model uses the material in Section 1.3 on methods of enumeration. Consider a collection of $N = N_1 + N_2$ similar objects, N_1 of them belonging to one of two dichotomous classes (red chips, say) and N_2 of them belonging to the second class (blue chips, say). A collection of n objects is selected from these

N objects at random and without replacement. Find the probability that exactly x (where the nonnegative integer x satisfies $x \leq n, x \leq N_1$, and $n - x \leq N_2$) of these n objects are red (i.e., x belong to the first class and $n - x$ belong to the second). Of course, we can select x red chips in any one of $\binom{N_1}{x}$ ways and $n - x$ blue chips in any one of $\binom{N_2}{n-x}$ ways. By the multiplication principle, the product $\binom{N_1}{x}\binom{N_2}{n-x}$ equals the number of ways the joint operation can be performed. If we assume that each of the $\binom{N}{n}$ ways of selecting n objects from $N = N_1 + N_2$ objects has the same probability, it follows that the probability of selecting exactly x red chips is

$$f(x) = P(X = x) = \frac{\binom{N_1}{x}\binom{N_2}{n-x}}{\binom{N}{n}},$$

where the space S is the collection of nonnegative integers x that satisfies the inequalities $x \leq n, x \leq N_1$, and $n - x \leq N_2$. We say that the random variable X has a **hypergeometric distribution**.

EXAMPLE 2.1-4 Some examples of hypergeometric probability histograms are given in Figure 2.1-2. The values of N_1, N_2, and n are given with each figure. ■

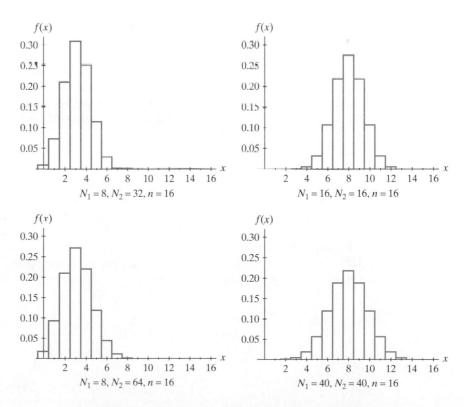

FIGURE 2.1-2: Hypergeometric probability histograms

EXAMPLE 2.1-5 In a small pond there are 50 fish, 10 of which have been tagged. If a fisherman's catch consists of 7 fish selected at random and without replacement, and X denotes the number of tagged fish, the probability that exactly 2 tagged fish are caught is

$$P(X = 2) = \frac{\binom{10}{2}\binom{40}{5}}{\binom{50}{7}} = \frac{(45)(658{,}008)}{99{,}884{,}400} = \frac{246{,}753}{832{,}370} \approx 0.2964.$$ ∎

EXAMPLE 2.1-6 A lot (collection) consisting of 100 fuses is inspected by the following procedure: Five fuses are chosen at random and tested; if all 5 blow at the correct amperage, the lot is accepted. Suppose that the lot contains 20 defective fuses. If X is a random variable equal to the number of defective fuses in the sample of 5, the probability of accepting the lot is

$$P(X = 0) = \frac{\binom{20}{0}\binom{80}{5}}{\binom{100}{5}} = \frac{19{,}513}{61{,}110} \approx 0.3193.$$

More generally, the p.m.f. of X is

$$f(x) = P(X = x) = \frac{\binom{20}{x}\binom{80}{5 - x}}{\binom{100}{5}}, \qquad x = 0, 1, 2, 3, 4, 5.$$ ∎

EXAMPLE 2.1-7 A supplier ships parts to another company in lots (collections) of 25, some of which could be defective. The receiving company has an *acceptance sampling plan* which states that $n = 5$ parts are to be taken at random and without replacement from each lot. If there are zero defectives among those 5 parts, the entire lot is accepted; otherwise, the lot is rejected. That is, if one or more defectives are found among the 5 in the sample, the lot is rejected. If X is the number of defectives among the 5 sampled parts, X must have a hypergeometric distribution. But here we do not know the values of N_1, the number of defectives in the lot, and N_2, the number of good parts in the lot; we know only that $N_1 + N_2 = 25$. Clearly, we want the probability of accepting the lot—namely, $P(X = 0)$—to be large if N_1 is very small. This probability is usually called the *operating characteristic curve* when treated as a function of N_1, or equivalently, the fraction defective, $p = N_1/25$, in the lot. That is, the operating characteristic curve is

$$\text{OC}(p) = P(X = 0) = \frac{\binom{N_1}{0}\binom{25 - N_1}{5}}{\binom{25}{5}},$$

where $p = N_1/25$. Letting $N_1 = 0, 1, 2, \ldots$ so that $p = 0.00, 0.04, 0.08, \ldots$, we use Table I in the appendix to find that

$$OC(0.00) = \frac{\binom{0}{0}\binom{25}{5}}{\binom{25}{5}} = 1.00,$$

$$OC(0.04) = \frac{\binom{1}{0}\binom{24}{5}}{\binom{25}{5}} = 0.800,$$

$$OC(0.08) = \frac{\binom{2}{0}\binom{23}{5}}{\binom{25}{5}} = 0.633,$$

$$OC(0.12) = \frac{\binom{3}{0}\binom{22}{5}}{\binom{25}{5}} = 0.496,$$

$$OC(0.16) = \frac{\binom{4}{0}\binom{21}{5}}{\binom{25}{5}} = 0.383,$$

and so on. Upon observing these probabilities, the company might find this acceptance sampling plan unsatisfactory because, possibly, with $N_1 = (25)(0.04) = 1$, $OC(0.04) = P(X = 0) = 0.8$ is too low and, with $N_1 = (25)(0.16) = 4$, $OC(0.16) = P(X = 0) = 0.383$ is too high. If this is so, the plan must be changed. In industrial practice, usually lot sizes are much larger than 25, sometimes running into the thousands, and the sample sizes might be in the hundreds rather than $n = 5$. We kept the numbers small here so that we could use Table I, but we illustrate what is actually done in practice with those larger values in Section 2.6 on the Poisson distribution. (See Example 2.6-8.)   ∎

In Section 1.2, we discussed the relationship between the probability $P(A)$ of an event A and the relative frequency $\mathcal{N}(A)/n$ of occurrences of event A in n repetitions of an experiment. We shall now extend those ideas.

Suppose that a random experiment is repeated n independent times. Let $A = \{X = x\}$, the event that x is the outcome of the experiment. Then we would expect the relative frequency $\mathcal{N}(A)/n$ to be close to $f(x)$. The next example illustrates this property.

EXAMPLE 2.1-8   A tetrahedron (four-sided die with outcomes 1, 2, 3, and 4) is rolled twice. Let X equal the sum of the two outcomes. Then the possible values of X are 2, 3, 4, 5, 6, 7, and 8. The following argument suggests that the p.m.f. of X is given by

	TABLE 2.1-1: Sum of Two Tetrahedral Dice		
x	Number of Observations of x	Relative Frequency of x	Probability of $\{X = x\}$, $f(x)$
2	71	0.071	0.0625
3	124	0.124	0.1250
4	194	0.194	0.1875
5	258	0.258	0.2500
6	177	0.177	0.1875
7	122	0.122	0.1250
8	54	0.054	0.0625

$f(x) = (4 - |x - 5|)/16$, for $x = 2, 3, 4, 5, 6, 7, 8$ (i.e., $f(2) = 1/16$, $f(3) = 2/16$, $f(4) = 3/16$, $f(5) = 4/16$, $f(6) = 3/16$, $f(7) = 2/16$, and $f(8) = 1/16$): Intuitively, these probabilities seem correct if we think of the 16 points (result on first roll, result on second roll) and assume that each has probability 1/16. Then note that $X = 2$ only for the point $(1, 1)$, $X = 3$ for the two points $(2, 1)$ and $(1, 2)$, and so on. This experiment was simulated 1000 times on a computer. Table 2.1-1 lists the results and compares the relative frequencies with the corresponding probabilities.

A graph can be used to display the results shown in Table 2.1-1. The probability histogram of the p.m.f. $f(x)$ of X is given in Figure 2.1-3. It is superimposed over the shaded histogram that represents the observed relative frequencies of the corresponding x values. The shaded histogram is the relative frequency histogram. For random experiments of the discrete type, this relative frequency histogram of a set of data gives an estimate of the probability histogram of the associated random variable when the latter is unknown. (Estimation is considered in detail later in the book.) ∎

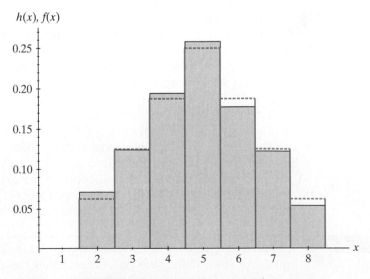

FIGURE 2.1-3: Sum of two tetrahedral dice

EXERCISES

2.1-1. Let the p.m.f. of X be defined by $f(x) = x/9$, $x = 2, 3, 4$.

(a) Draw a bar graph for this p.m.f.

(b) Draw a probability histogram for this p.m.f.

2.1-2. Let a chip be taken at random from a bowl that contains six white chips, three red chips, and one blue chip. Let the random variable $X = 1$ if the outcome is a white chip, let $X = 5$ if the outcome is a red chip, and let $X = 10$ if the outcome is a blue chip.

(a) Find the p.m.f. of X.

(b) Graph the p.m.f. as a bar graph.

2.1-3. For each of the following, determine the constant c so that $f(x)$ satisfies the conditions of being a p.m.f. for a random variable X, and then depict each p.m.f. as a bar graph:

(a) $f(x) = x/c$, $x = 1, 2, 3, 4$.

(b) $f(x) = cx$, $x = 1, 2, 3, \ldots, 10$.

(c) $f(x) = c(1/4)^x$, $x = 1, 2, 3, \ldots$.

(d) $f(x) = c(x + 1)^2$, $x = 0, 1, 2, 3$.

(e) $f(x) = x/c$, $x = 1, 2, 3, \ldots, n$.

(f) $f(x) = \dfrac{c}{(x + 1)(x + 2)}$, $x = 0, 1, 2, 3, \ldots$.

2.1-4. The state of Michigan generates a three-digit number at random six days a week for its daily lottery. The numbers are generated one digit at a time. Consider the following set of 50 three-digit numbers as 150 one-digit integers that were generated at random:

169	938	506	757	594	656	444	809	321	545
732	146	713	448	861	612	881	782	209	752
571	701	852	924	766	633	696	023	601	789
137	098	534	826	642	750	827	689	979	000
933	451	945	464	876	866	236	617	418	988

Let X denote the outcome when a single digit is generated.

(a) With true random numbers, what is the p.m.f. of X? Draw the probability histogram.

(b) For the 150 observations, determine the relative frequencies of 0, 1, 2, 3, 4, 5, 6, 7, 8, and 9, respectively.

(c) Draw the relative frequency histogram of the observations on the same graph paper as that of the probability histogram. Use a colored or dashed line for the relative frequency histogram.

2.1-5. The p.m.f. of X is $f(x) = (5 - x)/10$, $x = 1, 2, 3, 4$.

(a) Graph the p.m.f. as a bar graph.

(b) Use the following independent observations of X, simulated on a computer, to construct a table like Table 2.1-1:

3	1	2	2	3	2	2	2	1	3	3	2	3	2	4	4	2	1	1	3
3	1	2	2	1	1	4	2	3	1	1	1	2	1	3	1	1	3	3	1
1	1	1	1	1	4	1	3	1	2	4	1	1	2	3	4	3	1	4	2
2	1	3	2	1	4	1	1	1	2	1	3	4	3	2	1	4	4	1	3
2	2	2	1	2	3	1	1	4	2	1	4	2	1	2	3	1	4	2	3

(c) Construct a probability histogram and a relative frequency histogram like Figure 2.1-3.

2.1-6. Let a random experiment be the casting of a pair of unbiased dice, each having six faces, and let the random variable X denote the sum of the dice.

(a) With reasonable assumptions, determine the p.m.f. $f(x)$ of X. HINT: Picture the sample space consisting of the 36 points (result on first die, result on second die), and assume that each has probability 1/36. Find the probability of each possible outcome of X, namely, $x = 2, 3, 4, \ldots, 12$.

(b) Draw a bar graph for $f(x)$.

2.1-7. Let a random experiment be the casting of a pair of unbiased six-sided dice, and let X equal the smaller of the outcomes if they are different and the common value if they are equal.

(a) With reasonable assumptions, find the p.m.f. of X.

(b) Draw a probability histogram of the p.m.f. of X.

(c) Let Y equal the range of the two outcomes (i.e., the absolute value of the difference of the largest and the smallest outcomes). Determine the p.m.f. $g(y)$ of Y for $y = 0, 1, 2, 3, 4, 5$.

(d) Draw a probability histogram for $g(y)$.

2.1-8. An unbiased four-sided die X has two faces numbered 0 and two faces numbered 2. An unbiased four-sided die Y has its faces numbered 0, 1, 4, and 5. The two dice are rolled. Let $W = X + Y$.

(a) Determine the p.m.f. of W.

(b) Draw a probability histogram of the p.m.f. of W.

2.1-9. Let the p.m.f. of X be defined by $f(x) = (1 + |x - 3|)/11$, $x = 1, 2, 3, 4, 5$. Graph the p.m.f. of X as a bar graph.

2.1-10. Suppose there are 3 defective items in a lot (collection) of 50 items. A sample of size 10 is taken at random and without replacement. Let X denote the number of defective items in the sample. Find the probability that the sample contains

(a) Exactly one defective item.

(b) At most one defective item.

2.1-11. In a lot (collection) of 100 light bulbs, there are 5 bad bulbs. An inspector inspects 10 bulbs selected at random. Find the probability of finding at least one defective bulb. HINT: First compute the probability of finding no defectives in the sample.

2.1-12. A lot (collection) of $N_1 + N_2 = 25$ items is accepted if the number of defectives, X, among $n = 5$ items taken at random and without replacement from the lot is less than or equal to 1. The operating characteristic curve is defined as the probability of accepting the lot, namely,

$$OC(p) = P(X \leq 1) = P(X = 0) + P(X = 1),$$

where $p = N_1/25$. Determine the probabilities $OC(0.04)$, $OC(0.08)$, $OC(0.12)$, and $OC(0.16)$ (and compare them with those in Example 2.1-7).

2.1-13. Let X be the number of accidents per week in a factory. Let the p.m.f. of X be

$$f(x) = \frac{1}{(x + 1)(x + 2)}, \qquad x = 0, 1, 2, \cdots.$$

Find the conditional probability of $X \geq 4$, given that $X \geq 1$.

HINT: Write $f(x) = 1/(x + 1) - 1/(x + 2)$.

2.1-14. Often in buying a product at a supermarket, there is a concern about the item being underweight. Suppose there are 20 "one-pound" packages of frozen ground turkey on display and 3 of them are underweight. A consumer group buys 5 of the 20 packages at random. What is the probability of at least 1 of the 5 being underweight?

2.1-15. A professor gave her students six essay questions, from which she will select three for a test. A student has time to study for only three of these questions. What is the probability that, of the questions studied,

(a) at least one is selected for the test?

(b) all three are selected?

(c) exactly two are selected?

2.1-16. An urn contains 20 chips, 10 marked H and 10 marked C. Randomly select chips, one at a time, without replacement. After 10 chips have been selected, let X equal the difference between the numbers of H and C chips that have been selected.

(a) Define the p.m.f. of X.

(b) What value of X is most likely? That is, what is the mode of X?

REMARK Hope College (H chips) and Calvin College (C chips) began playing basketball in 1920–21. After 166 games, each team had won 83. The 20 chips correspond to the first 20 games only, and assume that Hope and Calvin were tied after 20 games in order to keep the numbers reasonable. ■

2.1-17. Five cards are selected at random without replacement from a standard, thoroughly shuffled 52-card deck of playing cards. Let X equal the number of face cards (kings, queens, jacks) in the hand. Forty observations of X yielded the following data:

$$
\begin{array}{cccccccccccccccccccc}
2 & 1 & 2 & 1 & 0 & 0 & 1 & 0 & 1 & 1 & 0 & 2 & 0 & 2 & 3 & 0 & 1 & 1 & 0 & 3 \\
1 & 2 & 0 & 2 & 0 & 2 & 0 & 1 & 0 & 1 & 1 & 2 & 1 & 0 & 1 & 1 & 2 & 1 & 1 & 0
\end{array}
$$

(a) Argue that the p.m.f. of X is

$$f(x) = \frac{\binom{12}{x}\binom{40}{5 - x}}{\binom{52}{5}}, \qquad x = 0, 1, 2, 3, 4, 5$$

and thus, that $f(0) = 2109/8330, f(1) = 703/1666, f(2) = 209/833, f(3) = 55/833, f(4) = 165/21,658,$ and $f(5) = 33/108,290.$

(b) Draw a probability histogram for this distribution.

(c) Determine the relative frequencies of $0, 1, 2, 3,$ and superimpose the relative frequency histogram on your probability histogram.

2.1-18. (Michigan Mathematics Prize Competition, 1992, Part II) From the set $\{1, 2, 3, \ldots, n\}$, k distinct integers are selected at random and arranged in numerical order (from lowest to highest). Let $P(i, r, k, n)$ denote the probability that integer i is in position r. For example, observe that $P(1, 2, k, n) = 0$, as it is impossible for the number 1 to be in the second position after ordering.

(a) Compute $P(2, 1, 6, 10)$.

(b) Find a general formula for $P(i, r, k, n)$.

2.1-19. A bag contains 144 ping-pong balls. More than half of the balls are painted orange and the rest are painted blue. Two balls are simultaneously drawn at random. The probability of drawing two balls of the same color is the same as the probability of drawing two balls of different colors. How many orange balls are in the bag?

2.2 MATHEMATICAL EXPECTATION

An extremely important concept in summarizing important characteristics of distributions of probability is that of mathematical expectation, which we introduce with an example.

EXAMPLE 2.2-1

An enterprising young man who needs a little extra money devises a game of chance in which some of his friends might wish to participate. The game that he proposes is to let the participant cast an unbiased die and then receive a payment according to the following schedule: If the event $A = \{1, 2, 3\}$ occurs, he receives 1¢; if $B = \{4, 5\}$ occurs, he receives 5¢; and if $C = \{6\}$ occurs, he receives 35¢. The probabilities of the respective events are assumed to be 3/6, 2/6, and 1/6, since the die is unbiased. The problem that now faces the young man is the determination of the amount that should be charged for the opportunity of playing the game. He reasons, correctly, that if the game is played a large number of times, about 3/6 of the trials will require a payment of 1¢, about 2/6 of them will require one of 5¢, and about 1/6 of them will require one of 35¢. Thus, the approximate average payment is

$$(1)\left(\frac{3}{6}\right) + (5)\left(\frac{2}{6}\right) + (35)\left(\frac{1}{6}\right) = 8.$$

That is, he expects to pay 8¢ "on the average." Note that he never pays exactly 8¢; the payment is either 1¢, 5¢, or 35¢. However, the "weighted average" of 1, 5, and 35, in which the weights are the respective probabilities 3/6, 2/6, and 1/6, equals 8. Such a weighted average is called the mathematical expectation of payment. Thus, if the young man decides to charge 10¢ per play, he would make 2¢ per play "on the average." Since the most that a player would lose at the charge of 10¢ per play is 9¢, the young man might find that several players are attracted by the possible gain of 25¢. ∎

A more mathematical way of formulating the preceding example would be to let X be the random variable defined by the outcome of the cast of the die. Thus, the p.m.f. of X is the uniform one given by

$$f(x) = \frac{1}{6}, \qquad x = 1, 2, 3, 4, 5, 6.$$

In terms of the observed value x, the payment is given by the function

$$u(x) = \begin{cases} 1, & x = 1, 2, 3, \\ 5, & x = 4, 5, \\ 35, & x = 6. \end{cases}$$

The mathematical expectation of payment is then equal to

$$\sum_{x=1}^{6} u(x)f(x) = (1)\left(\frac{1}{6}\right) + (1)\left(\frac{1}{6}\right) + (1)\left(\frac{1}{6}\right) + (5)\left(\frac{1}{6}\right) + (5)\left(\frac{1}{6}\right) + (35)\left(\frac{1}{6}\right)$$

$$= (1)\left(\frac{3}{6}\right) + (5)\left(\frac{2}{6}\right) + (35)\left(\frac{1}{6}\right)$$

$$= 8.$$

This discussion suggests the more general definition of mathematical expectation of a function of X.

DEFINITION 2.2-1   If $f(x)$ is the p.m.f. of the random variable X of the discrete type with space S, and if the summation

$$\sum_{x \in S} u(x)f(x), \qquad \text{which is sometimes written} \qquad \sum_{S} u(x)f(x),$$

exists, then the sum is called the **mathematical expectation** or the **expected value** of the function $u(X)$, and it is denoted by $E[u(X)]$. That is,

$$E[u(X)] = \sum_{x \in S} u(x)f(x).$$

We can think of the expected value $E[u(X)]$ as a weighted mean of $u(x)$, $x \in S$, where the weights are the probabilities $f(x) = P(X = x)$, $x \in S$.

REMARK   The usual definition of mathematical expectation of $u(X)$ requires that the sum converge absolutely—that is, that

$$\sum_{x \in S} |u(x)| f(x)$$

converge and be finite. However, in this book, each $u(x)$ is such that the convergence is absolute, and we do not burden the student with this additional requirement. Moreover, sometimes $E[u(X)]$ is called, more simply, the expectation of $u(X)$. ∎

There is another important observation that must be made about the consistency of Definition 2.1-1. Certainly, this function $u(X)$ of the random variable X is itself a random variable—say, Y. Suppose that we find the p.m.f. of Y to be $g(y)$ on the support S_1. Then $E(Y)$ is given by the summation

$$\sum_{y \in S_1} y g(y).$$

In general, it is true that

$$\sum_{x \in S} u(x)f(x) = \sum_{y \in S_1} y\,g(y).$$

That is, the same expectation is obtained by either formula. We do not prove this general result, but only illustrate it in the next example.

EXAMPLE 2.2-2 Let the random variable X have the p.m.f.

$$f(x) = \frac{1}{3}, \qquad x \in S,$$

where $S = \{-1, 0, 1\}$. Let $u(X) = X^2$. Then

$$\sum_{x \in S} x^2 f(x) = (-1)^2 \left(\frac{1}{3}\right) + (0)^2 \left(\frac{1}{3}\right) + (1)^2 \left(\frac{1}{3}\right) = \frac{2}{3}.$$

However, the support of the random variable $Y = X^2$ is $S_1 = \{0, 1\}$ and

$$P(Y = 0) = P(X = 0) = \frac{1}{3},$$

$$P(Y = 1) = P(X = -1) + P(X = 1) = \frac{1}{3} + \frac{1}{3} = \frac{2}{3}.$$

That is,

$$g(y) = \begin{cases} \dfrac{1}{3}, & y = 0, \\ \dfrac{2}{3}, & y = 1; \end{cases}$$

and $S_1 = \{0, 1\}$. Hence,

$$\sum_{y \in S_1} y\,g(y) = (0)\left(\frac{1}{3}\right) + (1)\left(\frac{2}{3}\right) = \frac{2}{3},$$

which illustrates the preceding observation. ∎

Before presenting additional examples, we list some useful facts about mathematical expectation in the following theorem.

THEOREM 2.2-1 When it exists, the mathematical expectation E satisfies the following properties:

(a) If c is a constant, then $E(c) = c$.

(b) If c is a constant and u is a function, then

$$E[c\,u(X)] = cE[u(X)].$$

(c) If c_1 and c_2 are constants and u_1 and u_2 are functions, then

$$E[c_1 u_1(X) + c_2 u_2(X)] = c_1 E[u_1(X)] + c_2 E[u_2(X)].$$

Proof. First, for the proof of (a), we have

$$E(c) = \sum_{x \in S} cf(x) = c \sum_{x \in S} f(x) = c$$

because

$$\sum_{x \in S} f(x) = 1.$$

Next, to prove (b), we see that

$$E[cu(X)] = \sum_{x \in S} cu(x)f(x)$$
$$= c \sum_{x \in S} u(x)f(x)$$
$$= cE[u(X)].$$

Finally, the proof of (c) is given by

$$E[c_1 u_1(X) + c_2 u_2(X)] = \sum_{x \in S} [c_1 u_1(x) + c_2 u_2(x)]f(x)$$
$$= \sum_{x \in S} c_1 u_1(x)f(x) + \sum_{x \in S} c_2 u_2(x)f(x).$$

By applying (b), we obtain

$$E[c_1 u_1(X) + c_2 u_2(X)] = c_1 E[u_1(X)] + c_2 E[u_2(X)]. \qquad \square$$

Property (c) can be extended to more than two terms by mathematical induction; that is, we have

$$(c)' \quad E\left[\sum_{i=1}^{k} c_i u_i(X))\right] = \sum_{i=1}^{k} c_i E[u_i(X)].$$

Because of property $(c)'$, the mathematical expectation E is often called a **linear** or **distributive** operator.

EXAMPLE 2.2-3 Let X have the p.m.f.

$$f(x) = \frac{x}{10}, \qquad x = 1, 2, 3, 4.$$

Then

$$E(X) = \sum_{x=1}^{4} x\left(\frac{x}{10}\right)$$
$$= (1)\left(\frac{1}{10}\right) + (2)\left(\frac{2}{10}\right) + (3)\left(\frac{3}{10}\right) + (4)\left(\frac{4}{10}\right) = 3,$$

$$E(X^2) = \sum_{x=1}^{4} x^2\left(\frac{x}{10}\right)$$
$$= (1)^2\left(\frac{1}{10}\right) + (2)^2\left(\frac{2}{10}\right) + (3)^2\left(\frac{3}{10}\right) + (4)^2\left(\frac{4}{10}\right) = 10,$$

and

$$E[X(5 - X)] = 5E(X) - E(X^2) = (5)(3) - 10 = 5. \qquad \blacksquare$$

EXAMPLE 2.2-4 Let $u(x) = (x - b)^2$, where b is not a function of X, and suppose $E[(X - b)^2]$ exists. To find that value of b for which $E[(X - b)^2]$ is a minimum, we write

$$g(b) = E[(X - b)^2] = E[X^2 - 2bX + b^2]$$
$$= E(X^2) - 2bE(X) + b^2$$

because $E(b^2) = b^2$. To find the minimum, we differentiate $g(b)$ with respect to b, set $g'(b) = 0$, and solve for b as follows:

$$g'(b) = -2E(X) + 2b = 0,$$
$$b = E(X).$$

Since $g''(b) = 2 > 0$, $E(X)$ is the value of b that minimizes $E[(X - b)^2]$.   ∎

EXAMPLE 2.2-5 Let X have a hypergeometric distribution in which n objects are selected from $N = N_1 + N_2$ objects as described in Section 2.1. Then

$$E(X) = \sum_{x \in S} x \frac{\binom{N_1}{x}\binom{N_2}{n - x}}{\binom{N}{n}}.$$

Since the first term of this summation equals zero when $x = 0$, and since

$$\binom{N}{n} = \left(\frac{N}{n}\right)\binom{N - 1}{n - 1},$$

we can write

$$E(X) = \sum_{0 < x \in S} x \frac{N_1!}{x!(N_1 - x)!} \frac{\binom{N_2}{n - x}}{\left(\dfrac{N}{n}\right)\binom{N - 1}{n - 1}}.$$

Of course, $x/x! = 1/(x - 1)!$ when $x \neq 0$; thus,

$$E(X) = \left(\frac{n}{N}\right) \sum_{0 < x \in S} \frac{(N_1)(N_1 - 1)!}{(x - 1)!(N_1 - x)!} \frac{\binom{N_2}{n - x}}{\binom{N - 1}{n - 1}}$$

$$= n\left(\frac{N_1}{N}\right) \sum_{0 < x \in S} \frac{\binom{N_1 - 1}{x - 1}\binom{N_2}{n - 1 - (x - 1)}}{\binom{N - 1}{n - 1}}.$$

However, when $x > 0$, the summand of this last expression represents the probability of obtaining, say, $x - 1$ red chips if $n - 1$ chips are selected from $N_1 - 1$ red chips and N_2 blue chips. Since the summation is over all possible values of $x - 1$, it must sum to 1, as it is the sum of all possible probabilities of $x - 1$. Thus,

$$E(X) = n\left(\frac{N_1}{N}\right),$$

which is a result that agrees with our intuition: We expect the number X of red chips to equal the product of the number n of selections and the fraction N_1/N of red chips in the original collection. ■

EXERCISES

2.2-1. Find $E(X)$ for each of the distributions given in Exercise 2.1-3.
   HINT: Part (c). Note that the difference $E(X) - (1/4)E(X)$ is equal to the sum of a geometric series.

2.2-2. Let the random variable X have the p.m.f.

$$f(x) = \frac{(|x| + 1)^2}{9}, \qquad x = -1, 0, 1.$$

Compute $E(X), E(X^2)$, and $E(3X^2 - 2X + 4)$.

2.2-3. In a particular lottery, 3 million tickets are sold each week for 50¢ apiece. Out of the 3 million tickets, 12,006 are drawn at random and without replacement and awarded prizes: twelve thousand $25 prizes, four $10,000 prizes, one $50,000 prize, and one $200,000 prize. If you purchased a single ticket each week, what is the expected value of this game to you?

2.2-4. In a state lottery, a three-digit integer is selected at random. A player bets $1 on a particular number, and if that number is selected, the payoff is $500 minus the $1 paid for the ticket. Let X equal the payoff to the bettor, namely, $-$1$ or $499, and find $E(X)$.

2.2-5. Let the random variable X be the number of days that a certain patient needs to be in the hospital. Suppose X has the p.m.f.

$$f(x) = \frac{5 - x}{10}, \qquad x = 1, 2, 3, 4.$$

If the patient is to receive $200 from an insurance company for each of the first two days in the hospital and $100 for each day after the first two days, what is the expected payment for the hospitalization?

2.2-6. An insurance company sells an automobile policy with a deductible of one unit. Let X be the amount of the loss having p.m.f.

$$f(x) = \begin{cases} 0.9, & x = 0, \\ \dfrac{c}{x}, & x = 1, 2, 3, 4, 5, 6, \end{cases}$$

where c is a constant. Determine c and the expected value of the amount the insurance company must pay.

2.2-7. In the gambling game chuck-a-luck, for a $1 bet it is possible to win $1, $2, or $3 with respective probabilities 75/216, 15/216, and 1/216. One dollar is lost with probability 125/216. Let X equal the payoff for this game and find $E(X)$. Note that when a bet is won, the $1 that was bet, in addition to the $1, $2, or $3 that is won, is returned to the bettor.

2.2-8. Let the p.m.f. of X be defined by $f(x) = 6/(\pi^2 x^2), x = 1, 2, 3, \ldots$. Show that $E(X)$ does not exist in this case. (Can you show that $f(x)$ is a p.m.f.?)

2.2-9. Let us select at random a number from the first n positive integers. If the payment is equal to the reciprocal of the number, find an expression for the expected payment. Evaluate this number when $n = 5$. Approximate this number when $n = 100$, or find the exact value using the computer. HINT: For the approximation, use a modification of the integral test for testing the convergence of a series.

2.2-10. Let X be a random variable with support $\{1, 2, 3, 5, 15, 25, 50\}$, each point of which has the same probability 1/7. Argue that $c = 5$ is the value that minimizes $h(c) = E(|X - c|)$. Compare c with the value of b that minimizes $g(b) = E[(X - b)^2]$.

2.2-11. A roulette wheel used in a U. S. casino has 38 slots, of which 18 are red, 18 are black, and 2 are green. A roulette wheel used in a French casino has 37 slots, of which 18 are red, 18 are black, and 1 is green. A ball is rolled around the wheel and ends up in one of the slots with equal probability. Suppose that a player bets on red. If a $1 bet is placed, the player wins $1 if the ball ends up in a red slot. (The player's $1 bet is returned.) If the ball ends up in a black or green slot, the player loses $1. Find the expected value of this game to the player in

(a) The United States.
(b) France.

2.2-12. In the casino game called **high−low**, there are three possible bets. Assume that $1 is the size of the bet. A pair of fair six-sided dice is rolled and their sum is calculated. If you bet **low**, you win $1 if the sum of the dice is $\{2, 3, 4, 5, 6\}$. If you bet

high, you win $1 if the sum of the dice is $\{8, 9, 10, 11, 12\}$. If you bet on $\{7\}$, you win $4 if a sum of 7 is rolled. Otherwise you lose on each of the three bets. In all three cases, your original dollar is returned if you win. Find the expected value of the game to the bettor for each of these three bets.

2.2-13. In the gambling game craps (see Exercise 1.4-13), the player wins $1 with probability 0.49293 and loses $1 with probability 0.50707 for each $1 bet. What is the expected value of the game to the player?

2.2-14. Suppose that a school has 20 classes: 16 with 25 students in each, three with 100 students in each, and one with 300 students, for a total of 1000 students.

(a) What is the average class size?

(b) Select a student randomly out of the 1000 students. Let the random variable X equal the size of the class to which this student belongs, and define the p.m.f. of X.

(c) Find $E(X)$, the expected value of X. Does this answer surprise you?

2.3   THE MEAN, VARIANCE, AND STANDARD DEVIATION

Let us consider an example in which $x \in \{1, 2, 3\}$ and the p.m.f. is given by $f(1) = 3/6$, $f(2) = 2/6$, $f(3) = 1/6$. That is, the probability that the random variable X equals 1, denoted by $P(X = 1)$, is $f(1) = 3/6$. Likewise, $P(X = 2) = f(2) = 2/6$ and $P(X = 3) = f(3) = 1/6$. Of course, $f(x) > 0$ when $x \in S$, and it is always the case that

$$\sum_{x \in S} f(x) = f(1) + f(2) + f(3) = 1.$$

We can think of the points 1, 2, 3 having respective weights (probabilities) 3/6, 2/6, 1/6, and their weighted mean (weighted average) is

$$1 \cdot \frac{3}{6} + 2 \cdot \frac{2}{6} + 3 \cdot \frac{1}{6} = \frac{10}{6},$$

which, in this illustration, does not equal one of the x values in S. As a matter of fact, it is two-thirds of the way between $x = 1$ and $x = 2$. We denote this weighted average or mean by the Greek letter μ (mu). In our example,

$$\mu = \sum_{x \in S} x f(x) = \frac{10}{6}.$$

In general, suppose the random variable X has the space $S = \{u_1, u_2, \ldots, u_k\}$ and these points have respective probabilities $P(X = u_i) = f(u_i) > 0$, where $f(x)$ is the p.m.f. Of course,

$$\sum_{x \in S} f(x) = 1$$

and the **mean** of the random variable X (or of its distribution) is

$$\mu = \sum_{x \in S} x f(x) = u_1 f(u_1) + u_2 f(u_2) + \cdots + u_k f(u_k).$$

That is, in the notation of Section 2.2, $\mu = E(X)$.

Now, u_i is the distance of that ith point from the origin. In mechanics, the product of a distance and its weight is called a moment, so $u_i f(u_i)$ is a moment having a moment arm of length u_i. The sum of such products would be the moment of the system of distances and weights. Actually, it is called the first moment about the origin, since the distances are simply to the first power and the lengths of the arms (distances) are measured from the origin. However, if we compute the first moment about the mean μ, then, since here a moment arm equals $(x - \mu)$, we have

$$\sum_{x \in S} (x - \mu)f(x) = E[(X - \mu)] = E(X) - \mu$$
$$= \mu - \mu = 0.$$

That is, that first moment about μ is equal to zero. In mechanics μ is called the centroid. The last equation implies that if a fulcrum is placed at the centroid μ, then the system of weights would balance, as the sum of the positive moments (when $x > \mu$) about μ equals the sum of the negative moments (when $x < \mu$). In our first illustration, $\mu = 10/6$ is the centroid, so the negative moment

$$\left(1 - \frac{10}{6}\right) \cdot \frac{3}{6} = -\frac{12}{36}$$

equals the sum of the two positive moments

$$\left(2 - \frac{10}{6}\right) \cdot \frac{2}{6} + \left(3 - \frac{10}{6}\right) \cdot \frac{1}{6} = \frac{12}{36}.$$

Since $\mu = E(X)$, it follows from Example 2.2-4 that $b = \mu$ minimizes $E[(X - b)^2]$. Also, Example 2.2-5 shows that

$$\mu = n\left(\frac{N_1}{N}\right)$$

is the mean of the hypergeometric distribution.

Statisticians often find it valuable to compute the second moment about the mean μ. It is called the second moment because the distances are raised to the second power and is equal to

$$\sum_{x \in S} (x - \mu)^2 f(x) = (u_1 - \mu)^2 f(u_1) + (u_2 - \mu)^2 f(u_2) + \cdots + (u_k - \mu)^2 f(u_k).$$

This weighted mean of the squares of those distances is called the **variance** of the random variable X (or of its distribution). The positive square root of the variance is called the **standard deviation** of X and is denoted by the Greek letter σ (sigma). Thus, the variance is σ^2, sometimes denoted by $\text{Var}(X)$. That is, $\sigma^2 = E[(X - \mu)]^2 = \text{Var}(X)$. In our first illustration, since $\mu = 10/6$, the variance equals

$$\sigma^2 = \text{Var}(X) = \left(1 - \frac{10}{6}\right)^2 \cdot \frac{3}{6} + \left(2 - \frac{10}{6}\right)^2 \cdot \frac{2}{6} + \left(3 - \frac{10}{6}\right)^2 \cdot \frac{1}{6} = \frac{120}{216}.$$

Hence, the standard deviation is

$$\sigma = \sqrt{\sigma^2} = \sqrt{\frac{120}{216}} = 0.745.$$

It is worth noting that the variance can be computed in another way, because

$$\sigma^2 = E[(X - \mu)^2] = E[X^2 - 2\mu X + \mu^2]$$
$$= E(X^2) - 2\mu E(X) + \mu^2$$
$$= E(X^2) - \mu^2.$$

That is, the variance σ^2 equals the difference of the second moment about the origin and the square of the mean. For our first illustration,

$$\sigma^2 = \sum_{x=1}^{3} x^2 f(x) - \mu^2$$

$$= 1^2\left(\frac{3}{6}\right) + 2^2\left(\frac{2}{6}\right) + 3^2\left(\frac{1}{6}\right) - \left(\frac{10}{6}\right)^2 = \frac{20}{6} - \frac{100}{36} = \frac{120}{216},$$

which agrees with our previous computation.

EXAMPLE 2.3-1 Let X equal the number of spots on the side facing upwards after a six-sided die (one of a pair of dice) is cast at random. If everything is fair about this experiment, a reasonable probability model is given by the p.m.f.

$$f(x) = P(X = x) = \frac{1}{6}, \qquad x = 1, 2, 3, 4, 5, 6.$$

The mean of X is

$$\mu = E(X) = \sum_{x=1}^{6} x\left(\frac{1}{6}\right) = \frac{1 + 2 + 3 + 4 + 5 + 6}{6} = \frac{7}{2}.$$

The second moment about the origin is

$$E(X^2) = \sum_{x=1}^{6} x^2\left(\frac{1}{6}\right) = \frac{1^2 + 2^2 + 3^2 + 4^2 + 5^2 + 6^2}{6} = \frac{91}{6}.$$

Thus, the variance equals

$$\sigma^2 = \frac{91}{6} - \left(\frac{7}{2}\right)^2 = \frac{182 - 147}{12} = \frac{35}{12}.$$

The standard deviation is $\sigma = \sqrt{35/12} = 1.708$. ∎

EXAMPLE 2.3-2 Let the p.m.f. of X be defined by $f(x) = x/6, x = 1, 2, 3$. The mean of X is

$$\mu = E(X) = 1\left(\frac{1}{6}\right) + 2\left(\frac{2}{6}\right) + 3\left(\frac{3}{6}\right) = \frac{7}{3}.$$

To find the variance and standard deviation of X, we first find

$$E(X^2) = 1^2\left(\frac{1}{6}\right) + 2^2\left(\frac{2}{6}\right) + 3^2\left(\frac{3}{6}\right) = \frac{36}{6} = 6.$$

Thus, the variance of X is

$$\sigma^2 = E(X^2) - \mu^2 = 6 - \left(\frac{7}{3}\right)^2 = \frac{5}{9}$$

and the standard deviation of X is

$$\sigma = \sqrt{5/9} = 0.745.$$ ∎

Although most students understand that $\mu = E(X)$ is, in some sense, a measure of the middle of the distribution of X, it is difficult to get much of a feeling for the variance and the standard deviation. The next example illustrates that the standard deviation is a measure of the dispersion, or spread, of the points belonging to the space S.

EXAMPLE 2.3-3   Let X have the p.m.f. $f(x) = 1/3$, $x = -1, 0, 1$. Here the mean is

$$\mu = \sum_{x=-1}^{1} xf(x) = (-1)\left(\frac{1}{3}\right) + (0)\left(\frac{1}{3}\right) + (1)\left(\frac{1}{3}\right) = 0.$$

Accordingly, the variance, denoted by σ_X^2, is

$$\sigma_X^2 = E[(X - 0)^2]$$
$$= \sum_{x=-1}^{1} x^2 f(x)$$
$$= (-1)^2\left(\frac{1}{3}\right) + (0)^2\left(\frac{1}{3}\right) + (1)^2\left(\frac{1}{3}\right)$$
$$= \frac{2}{3},$$

so the standard deviation is $\sigma_X = \sqrt{2/3}$. Next, let another random variable Y have the p.m.f. $g(y) = 1/3$, $y = -2, 0, 2$. Its mean is also zero, and it is easy to show that $\text{Var}(Y) = 8/3$, so the standard deviation of Y is $\sigma_Y = 2\sqrt{2/3}$. Here the standard deviation of Y is twice that of the standard deviation of X, reflecting the fact that the probability of Y is spread out twice as much as that of X. ∎

EXAMPLE 2.3-4   The mean of X, which has a uniform distribution on the first m positive integers, is given by

$$\mu = E(X) = \sum_{x=1}^{m} x\left(\frac{1}{m}\right) = \frac{1}{m}\sum_{x=1}^{m} x$$
$$= \left(\frac{1}{m}\right)\frac{m(m + 1)}{2} = \frac{m + 1}{2}.$$

To find the variance of X, we first find

$$E(X^2) = \sum_{x=1}^{m} x^2\left(\frac{1}{m}\right) = \frac{1}{m}\sum_{x=1}^{m} x^2$$
$$= \left(\frac{1}{m}\right)\frac{m(m + 1)(2m + 1)}{6} = \frac{(m + 1)(2m + 1)}{6}.$$

Thus, the variance of X is

$$\sigma^2 = \text{Var}(X) = E[(X - \mu)^2]$$
$$= E(X^2) - \mu^2 = \frac{(m + 1)(2m + 1)}{6} - \left(\frac{m + 1}{2}\right)^2$$
$$= \frac{m^2 - 1}{12}.$$

For example, we find that if X equals the outcome when rolling a fair six-sided die, the p.m.f. of X is

$$f(x) = \frac{1}{6}, \qquad x = 1, 2, 3, 4, 5, 6;$$

the respective mean and variance of X are

$$\mu = \frac{6+1}{2} = 3.5 \qquad \text{and} \qquad \sigma^2 = \frac{6^2 - 1}{12} = \frac{35}{12},$$

which agrees with calculations of Example 2.3-1.   ∎

Now let X be a random variable with mean μ_X and variance σ_X^2. Of course, $Y = aX + b$, where a and b are constants, is a random variable, too. The mean of Y is

$$\mu_Y = E(Y) = E(aX + b) = aE(X) + b = a\mu_X + b.$$

Moreover, the variance of Y is

$$\sigma_Y^2 = E[(Y - \mu_Y)^2] = E[(aX + b - a\mu_X - b)^2] = E[a^2(X - \mu_X)^2] = a^2\sigma_X^2.$$

Thus, $\sigma_Y = |a|\sigma_X$. To illustrate, note in Example 2.3-3 that the relationship between the two distributions could be explained by $Y = 2X$, so that $\sigma_Y^2 = 4\sigma_X^2$ and consequently $\sigma_Y = 2\sigma_X$, which we had observed there. In addition, we see that adding or subtracting a constant from X does not change the variance. In our example, $\text{Var}(X - 1) = \text{Var}(X)$, because $a = 1$ and $b = -1$.

Let r be a positive integer. If

$$E(X^r) = \sum_{x \in S} x^r f(x)$$

is finite, it is called the rth **moment** of the distribution about the origin. In addition, the expectation

$$E[(X - b)^r] = \sum_{x \in S} (x - b)^r f(x)$$

is called the rth moment of the distribution about b.

For a given positive integer r,

$$E[(X)_r] = E[X(X - 1)(X - 2) \cdots (X - r + 1)]$$

is called the rth **factorial moment**. We note that the second factorial moment is equal to the difference of the second and first moments:

$$E[X(X - 1)] = E(X^2) - E(X).$$

There is another formula that can be used to compute the variance. This formula uses the second factorial moment and sometimes simplifies the calculations. First find the values of $E(X)$ and $E[X(X - 1)]$. Then

$$\sigma^2 = E[X(X - 1)] + E(X) - [E(X)]^2,$$

since, by the distributive property of E, this becomes

$$\sigma^2 = E(X^2) - E(X) + E(X) - [E(X)]^2 = E(X^2) - \mu^2.$$

EXAMPLE 2.3-5 In Example 2.2-5 concerning the hypergeometric distribution, we found that the mean of that distribution is

$$\mu = E(X) = n\left(\frac{N_1}{N}\right) = np,$$

where $p = N_1/N$, the fraction of red chips in the N chips. In Exercise 2.3-17, it is determined that

$$E[X(X - 1)] = \frac{(n)(n - 1)(N_1)(N_1 - 1)}{N(N - 1)}.$$

Thus, the variance of X is $E[X(X - 1)] + E(X) - [E(X)]^2$, namely,

$$\sigma^2 = \frac{n(n - 1)(N_1)(N_1 - 1)}{N(N - 1)} + \frac{nN_1}{N} - \left(\frac{nN_1}{N}\right)^2.$$

After some straightforward algebra, we find that

$$\sigma^2 = n\left(\frac{N_1}{N}\right)\left(\frac{N_2}{N}\right)\left(\frac{N - n}{N - 1}\right) = np(1 - p)\left(\frac{N - n}{N - 1}\right). \qquad \blacksquare$$

Suppose that we now consider the situation in which we actually perform a certain random experiment n times, obtaining n observed values of the random variable—say, $x_1, x_2, \cdots, x_n$. Often the collection is referred to as a **sample**. It is possible that some of these values might be the same, but we do not worry about this at this time. We artificially create a probability distribution by placing the weight $1/n$ on each of these x values. Note that these weights are positive and sum to 1, so we have a distribution we call the **empirical distribution**, since it is determined by the data $x_1, x_2, \cdots, x_n$. The mean of the empirical distribution is

$$\sum_{i=1}^{n} x_i\left(\frac{1}{n}\right) = \frac{1}{n}\sum_{i=1}^{n} x_i,$$

which is the arithmetic mean of the observations $x_1, x_2, \cdots, x_n$. We denote this mean by $\bar{x}$ and call it the **sample mean** (or mean of the sample $x_1, x_2, \cdots, x_n$). That is, the sample mean is

$$\bar{x} = \frac{1}{n}\sum_{i=1}^{n} x_i,$$

which is, in some sense, an estimate of μ if the latter is unknown.

Likewise, the **variance of the empirical distribution** is

$$v = \sum_{i=1}^{n} (x_i - \bar{x})^2\left(\frac{1}{n}\right) = \frac{1}{n}\sum_{i=1}^{n} (x_i - \bar{x})^2,$$

which can be written as

$$v = \sum_{i=1}^{n} x_i^2\left(\frac{1}{n}\right) - \bar{x}^2 = \frac{1}{n}\sum_{i=1}^{n} x_i^2 - \bar{x}^2,$$

that is, the second moment about the origin, minus the square of the mean. However, v is not called the sample variance, but

$$s^2 = \left[\frac{n}{n-1} \right] v = \frac{1}{n-1} \sum_{i=1}^{n} (x_i - \bar{x})^2$$

is, because we will see later that, in some sense, s^2 is a better estimate of an unknown σ^2 than is v. Thus, the **sample variance** is

$$s^2 = \frac{1}{n-1} \sum_{i=1}^{n} (x_i - \bar{x})^2.$$

REMARK   It is easy to expand the sum of squares; we have

$$\sum_{i=1}^{n} (x_i - \bar{x})^2 = \sum_{i=1}^{n} x_i^2 - \frac{(\sum_{i=1}^{n} x_i)^2}{n}.$$

Many find that the right-hand expression makes the computation easier than first taking the n differences, $x_i - \bar{x}, i = 1, 2, \ldots, n$; squaring them; and then summing. There is another advantage when $\bar{x}$ has many digits to the right of the decimal point. If that is the case, then $x_i - \bar{x}$ must be rounded off, and that creates an error in the sum of squares. In the easier form, that rounding off is not necessary until the computation is completed. Of course, if you are using a statistical calculator or statistics package on the computer, all of these computations are done for you.   ∎

The **sample standard deviation**, $s = \sqrt{s^2} \geq 0$, is a measure of how dispersed the data are from the sample mean. At this stage of your study of statistics, it is difficult to get a good understanding or meaning of the standard deviation s, but you can roughly think of it as the average distance of the values $x_1, x_2, \ldots, x_n$ from the mean $\bar{x}$. This is not true exactly, for, in general,

$$s > \frac{1}{n} \sum_{i=1}^{n} |x_i - \bar{x}|,$$

but it is fair to say that s is somewhat larger, yet of the same magnitude, as the average of the distances of $x_1, x_2, \cdots, x_n$ from $\bar{x}$.

EXAMPLE 2.3-6   Rolling a fair six-sided die five times could result in the following sample of $n = 5$ observations:

$$x_1 = 3, \quad x_2 = 1, \quad x_3 = 2, \quad x_4 = 6, \quad x_5 = 3.$$

In this case,

$$\bar{x} = \frac{3 + 1 + 2 + 6 + 3}{5} = 3$$

and

$$s^2 = \frac{(3-3)^2 + (1-3)^2 + (2-3)^2 + (6-3)^2 + (3-3)^2}{4} = \frac{14}{4} = 3.5.$$

It follows that $s = \sqrt{14/4} = 1.87$. We had noted that s can roughly be thought of as the average distance that the x-values are away from the sample mean $\bar{x}$. In this example, the distances from $\bar{x} = 3$ are 0, 2, 1, 3, 0, with an average of 1.2, which is less than $s = 1.87$. In general, s will be somewhat larger than this average distance. ∎

There is an alternative way of computing s^2, because $s^2 = [n/(n-1)]v$ and

$$v = \frac{1}{n}\sum_{i=1}^{n}(x_i - \bar{x})^2 = \frac{1}{n}\sum_{i=1}^{n}x_i^2 - \bar{x}^2.$$

It follows that

$$s^2 = \frac{\sum_{i=1}^{n}x_i^2 - n\bar{x}^2}{n-1} = \frac{\sum_{i=1}^{n}x_i^2 - \frac{1}{n}\left(\sum_{i=1}^{n}x_i\right)^2}{n-1}.$$

COMPUTATIONAL COMMENTS (**CAS**) When it comes to computing probabilities, means, and variances in the discrete case, we need to evaluate summations and could use a computer algebra system (CAS) such as *Maple* or *Mathematica*. The same statement could be made about the continuous case in Chapter 3, with integrals replacing summations.

For example, let us consider the hypergeometric distribution:

```
> f:=binomial(N[1],x)*binomial(N[2],n-x)/binomial(N[1]+N[2],n);
```

$$f := \frac{\text{binomial}(N_1, x)\,\text{binomial}(N_2, n - x)}{\text{binomial}(N_1 + N_2, n)}$$

```
> simplify(sum(f, x = 0 .. N[1]));
```

$$1$$

```
> mu := simplify(sum(x*f, x = 0 .. N[1]));
```

$$\mu := \frac{N_1 n}{N_1 + N_2}$$

```
> var := simplify(sum(x^2*f, x = 0 .. N[1]) - mu^2);
```

$$var := -\frac{N_2 N_1 (n - N_1 - N_2) n}{(N_1 + N_2)^2 (N_1 - 1 + N_2)}$$

```
> N[1] := 20: N[2] := 30: n := 10:
> mu := sum(x*f, x = 0 .. N[1]);
```

$$\mu := 4$$

```
> var := sum((x - mu)^2*f, x = 0 .. N[1]);
```

$$var := \frac{96}{49}$$

```
> sum(f, x = 3 .. 6);
```

$$\frac{8469282699}{10272278170}$$

```
> evalf(%);
```

$$0.8244794931$$

Maple does not always write the output in the standard way, so we note that usually we write

$$\mu = n\left(\frac{N_1}{N}\right) = np$$

and

$$\sigma^2 = n\left(\frac{N_1}{N}\right)\left(\frac{N_2}{N}\right)\left(\frac{N-n}{N-1}\right) = np(1-p)\left(\frac{N-n}{N-1}\right).$$

The last two commands in the *Maple* output give the values of $P(3 \le X \le 6)$, as a fraction and as a decimal. It is also possible to find these probabilities with the use of a program like Minitab—or some calculators can even do this. You are encouraged to use whatever computing facilities are available to you.

REMARK Note that we have introduced two types of means and standard deviations. The first pair is associated with the probability distribution and is denoted by the Greek symbols (μ, σ). The second pair is associated with data and is denoted by the Roman letters $(\bar{x}, s)$. ■

EXERCISES

2.3-1. Find the mean and variance for the following discrete distributions:

(a) $f(x) = \dfrac{1}{5}, \qquad x = 5, 10, 15, 20, 25.$

(b) $f(x) = 1, \qquad x = 5.$

(c) $f(x) = \dfrac{4-x}{6}, \qquad x = 1, 2, 3.$

2.3-2. For each of the following distributions, find $\mu = E(X), E[X(X-1)],$ and $\sigma^2 = E[X(X-1)] + E(X) - \mu^2$:

(a) $f(x) = \dfrac{3!}{x!(3-x)!}\left(\dfrac{1}{4}\right)^x\left(\dfrac{3}{4}\right)^{3-x},$

$\qquad x = 0, 1, 2, 3.$

(b) $f(x) = \dfrac{4!}{x!(4-x)!}\left(\dfrac{1}{2}\right)^4,$

$\qquad x = 0, 1, 2, 3, 4.$

2.3-3. Given $E(X+4) = 10$ and $E[(X+4)^2] = 116$, determine (a) $\mathrm{Var}(X+4)$, (b) μ, and (c) σ^2.

2.3-4. Let μ and σ^2 denote the mean and variance of the random variable X. Determine $E[(X-\mu)/\sigma]$ and $E\{[(X-\mu)/\sigma]^2\}$.

2.3-5. Consider an experiment that consists of selecting a card at random from an ordinary deck of cards. Let the random variable X equal the value of the selected card, where Ace $= 1$, Jack $= 11$, Queen $= 12$, and King $= 13$. Thus, the space of X is $S = \{1, 2, 3, \cdots, 13\}$. If the experiment is performed in an unbiased manner, assign probabilities to these 13 outcomes and compute the mean μ of this probability distribution.

2.3-6. Place eight chips in a bowl: Three have the number 1 on them, two have the number 2, and three have the number 3. Say each chip has a probability of 1/8 of being drawn at random. Let the random variable X equal the number on the chip that is selected, so that the space of X is $S = \{1, 2, 3\}$. Make reasonable probability assignments to each of these three outcomes, and compute the mean μ and the variance σ^2 of this probability distribution.

2.3-7. A fair coin is flipped successively at random until the first head is observed. Let the random variable X denote the number of flips of the coin that are required. Then the space of X is $S = \{x : x = 1, 2, 3, 4, \ldots\}$. Later we learn that, under certain conditions, we can assign probabilities to these outcomes in S with the function $f(x) = (1/2)^x, x = 1, 2, 3, 4, \cdots$. Compute the mean μ. HINT: Write out the series for μ, and then construct the series for $(1/2)\mu$ and take the difference. An alternative method would be to compare the series for μ with that of the negative binomial $(1-z)^{-2}$, with $z = 1/2$.

2.3-8. Let X equal the number of calls per hour received by 911 between midnight and noon and reported in the *Holland Sentinel*. On October 29 and October 30, the following numbers of calls were reported:

Oct. 29: 0 1 1 1 0 1 2 1 4 1 2 3
Oct. 30: 0 3 0 1 0 1 1 2 3 0 2 2

(a) Find the sample mean.

(b) Find the sample variance.

2.3-9. Let X equal an integer selected at random from the first m positive integers, $\{1, 2, \ldots, m\}$.

(a) Give the values of $E(X)$ and $\text{Var}(X)$.

(b) Find the value of m for which $E(X) = \text{Var}(X)$. (See Zerger in the references.)

2.3-10. Students in a statistics class were asked to report the number of pets in their family. The following data were collected:

3 3 2 4 4 4 3 2 3 2 2 3 4 2 2 3
2 3 4 4 2 3 2 3 3 2 4 3 3 3 2 2
2 2 1 4 4 2 3 3 1 2 2 2 2 3 1

(a) Find the frequencies of 1, 2, 3, 4.

(b) Calculate the sample mean and the sample standard deviation.

(c) Construct a histogram and locate the mean on the histogram.

2.3-11. Calculate the mean $\bar{x}$ and the variance s^2 of the samples given in

(a) Exercise 1.1-3,

(b) Exercise 1.1-4,

(c) Exercise 1.1-5,

(d) Exercise 1.1-7.

2.3-12. Referring to Exercise 1.1-6, compute the sample mean $\bar{x}$. Could a statistician tell that family about how many boxes of cereal they could expect to purchase to obtain all four prizes?

2.3-13. A measure of **skewness** is defined by

$$\frac{E[(X - \mu)^3]}{\{E[(X - \mu)^2]\}^{3/2}} = \frac{E[(X - \mu)^3]}{(\sigma^2)^{3/2}}$$
$$= \frac{E[(X - \mu)^3]}{\sigma^3}.$$

When a distribution is symmetrical about the mean, the skewness is equal to zero. If the probability histogram has a longer "tail" to the right than to the left, the measure of skewness is positive, and we say that the distribution is skewed positively, or to the right. If the probability histogram has a longer tail to the left than to the right, the measure of skewness is negative, and we say that the distribution is skewed negatively, or to the left. If the p.m.f. of X is given by $f(x)$, (i) depict the p.m.f. as a probability histogram and find the values of (ii) the mean, (iii) the standard deviation, and (iv) the skewness.

(a)
$$f(x) = \begin{cases} \dfrac{2^{6-x}}{64}, & x = 1, 2, 3, 4, 5, 6, \\ \dfrac{1}{64}, & x = 7. \end{cases}$$

(b)
$$f(x) = \begin{cases} \dfrac{1}{64}, & x = 1, \\ \dfrac{2^{x-2}}{64}, & x = 2, 3, 4, 5, 6, 7. \end{cases}$$

2.3-14. In LOTTO 49, Michigan's lottery game, a player selects 6 integers out of the first 49 positive integers. The state then randomly selects 6 out of the first 49 positive integers. Cash prizes are given to a player who matches 4, 5, or 6 integers. Let X equal the number of integers selected by a player that match integers selected by the state.

(a) State the p.m.f. of X.

(b) Calculate the mean, variance, and standard deviation of X.

(c) What value of X is most likely to occur?

(d) On February 25, 1995, the jackpot was worth $45 million. When the prize is this large, many bets are placed. Out of the 25 million bets that were placed, 3 people matched all six numbers, with each winning $15 million (most of which was paid by losers during preceding games); 390 matched five numbers, to win $2500 each and; 22,187 matched four numbers, to win $100 each. Are these numbers of winners consistent with the probability model?

(e) A mathematics professor convinced some colleagues to pool their LOTTO bets, so that they were able to purchase 138 tickets together. They let the state computer randomly select the numbers on which they placed their bets. Among their 138 bets, 65 matched none of the winning LOTTO numbers, $\{3, 20, 33, 34, 43, 46\}$; 55 matched one, 16 matched two, and 2 matched three numbers. How do these results compare with what they could have expected?

2.3-15. Let X equal the larger outcome when a pair of four-sided dice is rolled. The p.m.f. of X is

$$f(x) = \frac{2x - 1}{16}, \qquad x = 1, 2, 3, 4.$$

(a) Find the mean, variance, and standard deviation of X.

(b) Calculate the sample mean, sample variance, and sample standard deviation of the following 100 simulated observations of X, and compare these answers with those found in part (a):

4	4	4	4	2	2	2	3	1	4	3	3	2	3	2	4	4	2	3	4
4	3	4	3	4	3	3	2	4	4	3	2	3	3	3	2	4	4	3	4
1	4	3	4	4	4	3	2	4	4	4	3	1	3	2	4	4	4	4	1
3	4	3	2	4	4	3	3	1	3	3	3	3	2	2	2	3	4	3	3
2	4	2	3	3	2	4	4	3	4	4	4	4	3	4	4	4	4	4	4

(c) Draw the graphs of the probability histogram and the relative frequency histogram on the same figure.

(d) Generate your own data, either rolling (four-sided or six-sided) dice or simulating this experiment on a computer.

2.3-16. A Bingo card has 25 squares with numbers on 24 of them, the center being a free square. The integers that are placed on the Bingo card are selected randomly and without replacement from 1 to 75, inclusive. When a game called "cover-up" is played, balls numbered from 1 to 75, inclusive, are selected randomly and without replacement until a player covers each of the numbers on a card. Let X equal the number of balls that must be drawn to cover all the numbers on a single card.

(a) Argue that the p.m.f. of X, for $x = 24, 25, \ldots, 75$, is

$$f(x) = \frac{\binom{24}{23}\binom{51}{x-24}}{\binom{75}{x-1}} \cdot \frac{1}{75-(x-1)} = \frac{\binom{51}{x-24}}{\binom{75}{x}} \cdot \frac{24}{x} = \frac{\binom{x}{24}}{\binom{75}{24}} \cdot \frac{24}{x}.$$

(b) What value of X is most likely to occur? In other words, what is the mode of this distribution?

(c) To show that the mean of X is $(24)(76)/25 = 72.96$, use the combinatorial identity

$$\binom{k+n+1}{k+1} = \sum_{x=k}^{n+k}\binom{x}{k}.$$

(d) Show that $E[X(X+1)] = \dfrac{24 \cdot 77 \cdot 76}{26} = 5401.8462$.

(e) Calculate $\text{Var}(X) = E[X(X+1)] - E[X] - [E(X)]^2 = \dfrac{46,512}{8125} = 5.7246$ and $\sigma = 2.39$.

(f) The following 100 observations of X were simulated on the computer:

75	73	74	74	73	75	75	71	71	74	75	74	74	73	75	
75	75	71	68	75	74	72	73	75	74	67	73	71	74	74	
73	75	71	73	71	68	74	75	66	75	75	74	75	71	75	
72	75	74	75	75	72	75	75	74	73	75	62	64	75	72	
74	74	75	73	75	75	73	74	75	68	75	69	74	75	61	
73	73	73	72	74	68	74	71	73	75	73	74	72	74	73	
75	73	71	70	62	74	74	72	75	74						

Compare these data with the probability model. In particular, make the following comparisons: (i) $\bar{x}$ with μ, (ii) s^2 with σ^2, (iii) s with σ, (iv) $(1/100)\sum_{i=1}^{100} x_i(x_i+1)$ with $E[X(X+1)]$.

(g) Compare the probability histogram with the relative frequency histogram.

2.3-17. To find the variance of a hypergeometric random variable in Example 2.3-5, we used the fact that

$$E[X(X-1)] = \frac{N_1(N_1-1)(n)(n-1)}{N(N-1)}.$$

Prove this result by making the change of variables $k = x - 2$ and noting that

$$\binom{N}{n} = \frac{N(N-1)}{n(n-1)}\binom{N-2}{n-2}.$$

2.3-18. A deck of $n = 10$ cards is numbered from 1 to 10. The cards are shuffled and laid down from left to right, face up. Order each of the five successive pairs of cards. Each of these five pairs determines a random interval. Let X equal the number of these five random intervals that intersect each of the other four intervals. It can be shown that

$$P(X \geq k) = \frac{2^k}{\dbinom{2k+1}{k}}, \qquad k = 0, 1, 2, 3, 4, 5.$$

(a) Find the probability that at least one interval intersects all the other intervals. (You could also simulate this.)

(b) Prove that $\mu = E(X) = \sum_{k=1}^{5} P(X \geq k)$.

(c) Find the value of $\mu = E(X)$ with $n = 10$ cards.

(d) Let the number of cards increase without bound. Find the value of

$$\mu = \sum_{k=0}^{\infty} \frac{2^k}{\dbinom{2k+1}{k}}.$$

2.3-19. A warranty is written on a product worth \$10,000 so that the buyer is given \$8000 if it fails in the first year, \$6000 if it fails in the second, \$4000 if it fails in the third, \$2000 if it fails in the fourth, and zero after that. The probability of the product's failing in a year is 0.1; failures are independent of those of other years. What is the expected value of the warranty?

2.4 BERNOULLI TRIALS AND THE BINOMIAL DISTRIBUTION

The probability models for random experiments that will be described in this section occur frequently in applications.

A **Bernoulli experiment** is a random experiment, the outcome of which can be classified in one of two mutually exclusive and exhaustive ways—say, success or failure (e.g., female or male, life or death, nondefective or defective). A sequence of **Bernoulli trials** occurs when a Bernoulli experiment is performed several *independent* times so that the probability of success—say, p—remains the *same* from trial to trial. That is, in such a sequence we let p denote the probability of success on each trial. In addition, we shall frequently let $q = 1 - p$ denote the probability of failure; that is, we shall use q and $1 - p$ interchangeably.

EXAMPLE 2.4-1 Suppose that the probability of germination of a beet seed is 0.8 and the germination of a seed is called a success. If we plant 10 seeds and can assume that the germination of one seed is independent of the germination of another seed, this would correspond to 10 Bernoulli trials with $p = 0.8$. ∎

EXAMPLE 2.4-2 In the Michigan daily lottery, the probability of winning when placing a six-way boxed bet is 0.006. A bet placed on each of 12 successive days would correspond to 12 Bernoulli trials with $p = 0.006$. ∎

Let X be a random variable associated with a Bernoulli trial by defining it as follows:

$$X(\text{success}) = 1 \qquad \text{and} \qquad X(\text{failure}) = 0.$$

That is, the two outcomes, success and failure, are denoted by one and zero, respectively. The p.m.f. of X can be written as

$$f(x) = p^x (1 - p)^{1-x}, \qquad x = 0, 1,$$

and we say that X has a **Bernoulli distribution**. The expected value of X is

$$\mu = E(X) = \sum_{x=0}^{1} x p^x (1 - p)^{1-x} = (0)(1 - p) + (1)(p) = p,$$

and the variance of X is

$$\sigma^2 = \text{Var}(X) = \sum_{x=0}^{1} (x - p)^2 p^x (1 - p)^{1-x}$$
$$= (0 - p)^2 (1 - p) + (1 - p)^2 p = p(1 - p) = pq.$$

It follows that the standard deviation of X is

$$\sigma = \sqrt{p(1 - p)} = \sqrt{pq}.$$

In a sequence of n Bernoulli trials, we shall let X_i denote the Bernoulli random variable associated with the ith trial. An observed sequence of n Bernoulli trials will then be an n-tuple of zeros and ones, and we often call this collection a **random sample** of size n from a Bernoulli distribution.

EXAMPLE 2.4-3   Out of millions of instant lottery tickets, suppose that 20% are winners. If five such tickets are purchased, then $(0, 0, 0, 1, 0)$ is a possible observed sequence in which the fourth ticket is a winner and the other four are losers. Assuming independence among winning and losing tickets, we observe that the probability of this outcome is

$$(0.8)(0.8)(0.8)(0.2)(0.8) = (0.2)(0.8)^4.$$ ■

EXAMPLE 2.4-4   If five beet seeds are planted in a row, a possible observed sequence would be $(1, 0, 1, 0, 1)$, in which the first, third, and fifth seeds germinated and the other two did not. If the probability of germination is $p = 0.8$, the probability of this outcome is, assuming independence,

$$(0.8)(0.2)(0.8)(0.2)(0.8) = (0.8)^3 (0.2)^2.$$ ■

In a sequence of Bernoulli trials, we are often interested in the total number of successes and not in the order of their occurrence. If we let the random variable X equal the number of observed successes in n Bernoulli trials, then the possible values of X are $0, 1, 2, \dots, n$. If x successes occur, where $x = 0, 1, 2, \dots, n$, then $n - x$ failures occur. The number of ways of selecting x positions for the x successes in the n trials is

$$\binom{n}{x} = \frac{n!}{x!(n - x)!}.$$

Since the trials are independent and since the probabilities of success and failure on each trial are, respectively, p and $q = 1 - p$, the probability of each of these ways is $p^x (1 - p)^{n-x}$. Thus, $f(x)$, the p.m.f. of X, is the sum of the probabilities of the $\binom{n}{x}$ mutually exclusive events; that is,

$$f(x) = \binom{n}{x} p^x (1 - p)^{n-x}, \qquad x = 0, 1, 2, \dots, n.$$

These probabilities are called binomial probabilities, and the random variable X is said to have a **binomial distribution**.

Summarizing, a binomial experiment satisfies the following properties:

1. A Bernoulli (success–failure) experiment is performed n times.
2. The trials are independent.
3. The probability of success on each trial is a constant p; the probability of failure is $q = 1 - p$.
4. The random variable X equals the number of successes in the n trials.

A binomial distribution will be denoted by the symbol $b(n,p)$, and we say that the distribution of X is $b(n,p)$. The constants n and p are called the **parameters** of the binomial distribution; they correspond to the number n of independent trials and the probability p of success on each trial. Thus, if we say that the distribution of X is $b(12, 1/4)$, we mean that X is the number of successes in a random sample of size $n = 12$ from a Bernoulli distribution with $p = 1/4$.

EXAMPLE 2.4-5 In the instant lottery with 20% winning tickets, if X is equal to the number of winning tickets among $n = 8$ that are purchased, then the probability of purchasing two winning tickets is

$$f(2) = P(X = 2) = \binom{8}{2}(0.2)^2(0.8)^6 = 0.2936.$$

The distribution of the random variable X is $b(8, 0.2)$. ∎

EXAMPLE 2.4-6 In order to obtain a better feeling for the effect of the parameters n and p on the distribution of probabilities, four probability histograms are displayed in Figure 2.4-1. ∎

EXAMPLE 2.4-7 In Example 2.4-1, the number X of seeds that germinate in $n = 10$ independent trials is $b(10, 0.8)$; that is,

$$f(x) = \binom{10}{x}(0.8)^x(0.2)^{10-x}, \qquad x = 0, 1, 2, \ldots, 10.$$

In particular,

$$P(X \leq 8) = 1 - P(X = 9) - P(X = 10)$$
$$= 1 - 10(0.8)^9(0.2) - (0.8)^{10} = 0.6242.$$

Also, with a little more work, we could compute

$$P(X \leq 6) = \sum_{x=0}^{6} \binom{10}{x}(0.8)^x(0.2)^{10-x}. \qquad ∎$$

Cumulative probabilities like those in the previous example are often of interest. We call the function defined by

$$F(x) = P(X \leq x), \qquad -\infty < x < \infty,$$

the **cumulative distribution function** or, more simply, the **distribution function** of the random variable X. Values of the distribution function of a random variable X that is $b(n,p)$ are given in Table II in the appendix for selected values of n and p.

FIGURE 2.4-1: Binomial probability histograms

For the binomial distribution given in Example 2.4-7, namely, the $b(10, 0.8)$ distribution, the distribution function is defined by

$$F(x) = P(X \leq x) = \sum_{y=0}^{\lfloor x \rfloor} \binom{10}{y}(0.8)^y(0.2)^{10-y},$$

where $\lfloor x \rfloor$ is the floor or greatest integer in x. A graph of this distribution function is shown in Figure 2.4-2 on the next page. Note that the jumps at the integers in this step function are equal to the probabilities associated with those respective integers.

EXAMPLE 2.4-8 Leghorn chickens are raised for laying eggs. Let $p = 0.5$ be the probability of a female chick hatching. Assuming independence, let X equal the number of female chicks out of 10 newly hatched chicks selected at random. Then the distribution of X is $b(10, 0.5)$. From Table II in the appendix, the probability of 5 or fewer female chicks is

$$P(X \leq 5) = 0.6230.$$

The probability of exactly 6 female chicks is

$$P(X = 6) = \binom{10}{6}\left(\frac{1}{2}\right)^6\left(\frac{1}{2}\right)^4$$
$$= P(X \leq 6) - P(X \leq 5)$$
$$= 0.8281 - 0.6230 = 0.2051,$$

FIGURE 2.4-2: Distribution function for the $b(10, 0.8)$ distribution

since $P(X \leq 6) = 0.8281$. The probability of at least 6 female chicks is

$$P(X \geq 6) = 1 - P(X \leq 5) = 1 - 0.6230 = 0.3770.$$ ■

COMPUTATIONAL COMMENTS   Here is the Minitab solution for Example 2.4-8: To find the preceding probabilities, you use

Calc $\rightarrow$ Probability Distributions $\rightarrow$ Binomial

and input the necessary information, namely,

```
Cumulative Distribution Function
Binomial with n = 10 and p = 0.5
x P( X <= x )
5 0.623047

Probability Density Function
Binomial with n = 10 and p = 0.5
x P(X = x )
6 0.205078
```

Although probabilities for the binomial distribution $b(n, p)$ are given in Table II in the appendix for selected values of p that are less than or equal to 0.5, the next example demonstrates that this table can also be used for values of p that are greater than 0.5. In later sections we learn how to approximate certain binomial probabilities with those of other distributions. In addition, you may use your calculator and/or statistical package such as Minitab to find binomial probabilities.

EXAMPLE 2.4-9   Suppose that we are in one of those rare times when 65% of the American public approve of the way the President of the United States is handling his job. Take a random sample of $n = 8$ Americans and let Y equal the number who give approval. Then the distribution of Y is $b(8, 0.65)$. To find $P(Y \geq 6)$, note that

$$P(Y \geq 6) = P(8 - Y \leq 8 - 6) = P(X \leq 2),$$

where $X = 8 - Y$ counts the number who disapprove. Since $q = 1 - p = 0.35$ equals the probability of disapproval by each person selected, the distribution of X

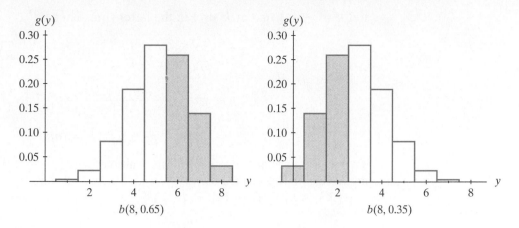

FIGURE 2.4-3: Presidential approval histogram

is $b(8, 0.35)$. (See Figure 2.4-3.) From Table II in the appendix, since $P(X \leq 2) = 0.4278$, it follows that $P(Y \geq 6) = 0.4278$.

Similarly,

$$P(Y \leq 5) = P(8 - Y \geq 8 - 5)$$
$$= P(X \geq 3) = 1 - P(X \leq 2)$$
$$= 1 - 0.4278 = 0.5722$$

and

$$P(Y = 5) = P(8 - Y = 8 - 5)$$
$$= P(X = 3) = P(X \leq 3) - P(X \leq 2)$$
$$= 0.7064 - 0.4278 = 0.2786.$$ ■

Recall that if n is a positive integer, then

$$(a + b)^n = \sum_{x=0}^{n} \binom{n}{x} b^x a^{n-x}.$$

Thus, if we use this binomial expansion with $b = p$ and $a = 1 - p$, then the sum of the binomial probabilities is

$$\sum_{x=0}^{n} \binom{n}{x} p^x (1 - p)^{n-x} = [(1 - p) + p]^n = 1,$$

a result that had to follow from the fact that $f(x)$ is a p.m.f.

We use the binomial expansion to find the mean and the variance of a binomial random variable X that is $b(n, p)$. The mean is given by

$$\mu = E(X) = \sum_{x=0}^{n} x \frac{n!}{x!(n - x)!} p^x (1 - p)^{n-x}$$

$$= \sum_{x=1}^{n} \frac{n!}{(x - 1)!(n - x)!} p^x (1 - p)^{n-x}.$$

Let $k = x - 1$ or $x = k + 1$ in the latter sum. Then

$$\mu = E(X) = \sum_{k=0}^{n-1} \frac{n!}{k!(n-k-1)!} p^{k+1}(1-p)^{n-k-1}$$

$$= np \sum_{k=0}^{n-1} \frac{(n-1)!}{k!(n-1-k)!} p^k (1-p)^{n-1-k}$$

$$= np(1 - p + p)^{n-1} = np.$$

To find the variance, we first find the value of $E[X(X-1)]$, the second factorial moment. Using the second factorial moment, we find that the variance of X is given by $\mathrm{Var}(X) = E[X(X-1)] + E(X) - \mu^2$. Now,

$$E[X(X-1)] = \sum_{x=0}^{n} x(x-1) \frac{n!}{x!(n-x)!} p^x (1-p)^{n-x}$$

$$= \sum_{x=2}^{n} \frac{n!}{(x-2)!(n-x)!} p^x (1-p)^{n-x}.$$

Letting $k = x - 2$ or $x = k + 2$, we obtain

$$E[X(X-1)] = \sum_{k=0}^{n-2} \frac{n!}{k!(n-k-2)!} p^{k+2}(1-p)^{n-k-2}$$

$$= n(n-1)p^2 \sum_{k=0}^{n-2} \frac{(n-2)!}{k!(n-2-k)!} p^k (1-p)^{n-2-k}$$

$$= n(n-1)p^2 (1 - p + p)^{n-2} = n(n-1)p^2.$$

Thus,

$$\sigma^2 = \mathrm{Var}(X) = E(X^2) - [E(X)]^2 = E[X(X-1)] + E(X) - [E(X)]^2$$

$$= n(n-1)p^2 + np - (np)^2 = np(1-p).$$

Summarizing, if X is $b(n,p)$, then

$$\mu = np \qquad \text{and} \qquad \sigma^2 = np(1-p) = npq.$$

We find the mean and variance with the use of the moment-generating function in Example 2.5-3. (See page 91.)

Note that when p is the probability of success on each trial, the expected number of successes in n trials is np, a result that agrees with our intuition. Of course, the standard deviation is

$$\sigma = \sqrt{np(1-p)}.$$

EXAMPLE 2.4-10 Suppose that observation over a long period of time has disclosed that, on the average, 1 out of 10 items produced by a process is defective. Select 5 items independently from the production line and test them. Let X denote the number of defective items among the $n = 5$ items. Then X is $b(5, 0.1)$. Furthermore,

$$E(X) = 5(0.1) = 0.5, \ \mathrm{Var}(X) = 5(0.1)(0.9) = 0.45.$$

For example, the probability of observing at most one defective item is

$$P(X \le 1) = \binom{5}{0}(0.1)^0(0.9)^5 + \binom{5}{1}(0.1)^1(0.9)^4 = 0.9185. \qquad \blacksquare$$

The next example shows the relationship between the binomial probability model and a set of observed data. Replications of this example would of course yield different results, in which the fits could be better or worse than the given one.

EXAMPLE 2.4-11   Consider the simple experiment of flipping a fair coin five independent times. If X equals the number of heads observed, then X is $b(5, 0.5)$, $\mu = 2.5$, $\sigma^2 = 1.25$, and $\sigma = 1.118$. This experiment was simulated 100 times, yielding the following data:

```
2  3  2  4  1  2  1  1  4  2  4  2  0  4  4  2  4  4  3  4
2  2  4  4  1  1  3  3  1  4  2  3  1  2  4  1  2  5  3  2
4  3  2  2  2  3  5  2  0  3  2  1  3  4  2  2  4  0  2  1
3  3  2  3  2  1  3  2  2  2  1  1  3  3  1  1  4  2  1  5
3  2  3  0  3  5  3  2  4  3  3  5  2  3  3  1  3  2  1  1
```

For these data, $\bar{x} = 2.47$, $s^2 = 1.5243$, and $s = 1.235$. Figure 2.4-4 gives the probability histogram and the relative frequency histogram (shaded). $\blacksquare$

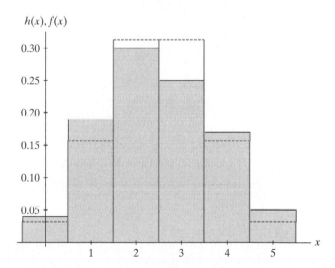

FIGURE 2.4-4: Number of heads when five coins are flipped

Suppose that an urn contains N_1 success balls and N_2 failure balls. Now, let $p = N_1/(N_1 + N_2)$, and let X equal the number of success balls in a random sample of size n that is taken from this urn. If the sampling is done one at a time with replacement, then the distribution of X is $b(n, p)$; if the sampling is done without replacement, then X has a hypergeometric distribution with p.m.f.

$$f(x) = \frac{\binom{N_1}{x}\binom{N_2}{n-x}}{\binom{N_1 + N_2}{n}},$$

where x is a nonnegative integer such that $x \leq n$, $x \leq N_1$, and $n - x \leq N_2$. When $N_1 + N_2$ is large and n is relatively small, it makes little difference if the sampling is done with or without replacement. In Figure 2.4-5, the probability histograms are compared for different combinations of n, N_1, and N_2. You are asked to compute some of these probabilities in Exercise 2.4-17.

FIGURE 2.4-5: Binomial and hypergeometric (shaded) probability histograms

EXERCISES

2.4-1. An urn contains 7 red and 11 white balls. Draw one ball at random from the urn. Let $X = 1$ if a red ball is drawn, and let $X = 0$ if a white ball is drawn. Give the p.m.f., mean, and variance of X.

2.4-2. Suppose that in Exercise 2.4-1, $X = 1$ if a red ball is drawn and $X = -1$ if a white ball is drawn. Give the p.m.f., mean, and variance of X.

2.4-3. On a six-question multiple-choice test there are five possible answers for each question, of which one is correct (C) and four are incorrect (I). If a student guesses randomly and independently, find the probability of

(a) Being correct only on questions 1 and 4 (i.e., scoring C, I, I, C, I, I).

(b) Being correct on two questions.

2.4-4. Suppose that the percentage of college students who engaged in binge drinking, which is defined as having five drinks for male students and four drinks for female students at one "drinking occasion" during the previous two weeks, is approximately 40%. Let X equal the number of students in a random sample of size $n = 12$ who binge drink.

(a) Find the probability that X is at most 5.

(b) Find the probability that X is at least 6.

(c) Find the probability that X is equal to 7.

(d) Give the mean, variance, and standard deviation of X.

2.4-5. In a lab experiment involving inorganic syntheses of molecular precursors to organometallic ceramics, the final step of a five-step reaction involves the formation of a metal–metal bond. The probability of such a bond forming is $p = 0.20$. Let X equal the number of successful reactions out of $n = 25$ such experiments.

(a) Find the probability that X is at most 4.
(b) Find the probability that X is at least 5.
(c) Find the probability that X is equal to 6.
(d) Give the mean, variance, and standard deviation of X.

2.4-6. It is claimed that 15% of the ducks in a particular region have patent schistosome infection. Suppose that seven ducks are selected at random. Let X equal the number of ducks that are infected.

(a) Assuming independence, how is X distributed?
(b) Find (i) $P(X \geq 2)$, (ii) $P(X = 1)$, and (iii) $P(X \leq 3)$.

2.4-7. Suppose that 2000 points are selected independently and at random from the unit square $\{(x, y) : 0 \leq x < 1, 0 \leq y < 1\}$. Let W equal the number of points that fall into $A = \{(x, y) : x^2 + y^2 < 1\}$.

(a) How is W distributed?
(b) Give the mean, variance, and standard deviation of W.
(c) What is the expected value of $W/500$?
(d) Use the computer to select 2000 pairs of random numbers. Determine the value of W and use that value to find an estimate for π. (Of course, we know the real value of π, and more will be said about estimation later on in this text.)
(e) How could you extend part (d) to estimate the volume $V = (4/3)\pi$ of a ball of radius 1 in 3-space?
(f) How could you extend these techniques to estimate the "volume" of a ball of radius 1 in n-space?

2.4-8. It is believed that approximately 65% of Americans under the age of 65 have private health

insurance. Suppose this is true, and let X equal the number of Americans under age 65 in a random sample of $n = 15$ with private health insurance.

(a) Find the probability that X is at least 10.
(b) Find the probability that X is at most 10.
(c) Find the probability that X is equal to 10.
(d) How is X distributed?
(e) Give the mean, variance, and standard deviation of X.

2.4-9. Suppose that the percentage of American drivers who are multitaskers (e.g., talk on cell phones, eat a snack, or text message at the same time they are driving) is approximately 80%. In a random sample of $n = 20$ drivers, let X equal the number of multitaskers.

(a) How is X distributed?
(b) Give the values of the mean, variance, and standard deviation of X.
(c) Find (i) $P(X = 15)$, (ii) $P(X > 15)$, and (iii) $P(X \leq 15)$.

2.4-10. A boiler has four relief valves. The probability that each opens properly is 0.99.

(a) Find the probability that at least one opens properly.
(b) Find the probability that all four open properly.

2.4-11. A random variable X has a binomial distribution with mean 6 and variance 3.6. Find $P(X = 4)$.

2.4-12. A certain type of mint has a label weight of 20.4 grams. Suppose that the probability is 0.90 that a mint weighs more than 20.7 grams. Let X equal the number of mints that weigh more than 20.7 grams in a sample of eight mints selected at random.

(a) How is X distributed if we assume independence?
(b) Find (i) $P(X = 8)$, (ii) $P(X \leq 6)$, and (iii) $P(X \geq 6)$.

2.4-13. It is given that the probability of germination of each beet seed is 0.8. Thus, if X denotes the number of seeds that germinate in a set of nine seeds, then X is $b(9, 0.8)$, provided that the assumption of independence is valid.

(a) Give the values of μ, σ^2, and σ.
(b) Draw the probability histogram for the p.m.f. of X.
(c) If 100 sets of nine beet seeds were planted, calculate $\bar{x}$, s^2, and s for the following 100 observations:

7	9	7	7	9	7	9	8	5	8	8	6	8	7	8	8	9	7	8	8
8	7	8	7	7	7	6	7	9	7	6	7	6	7	6	5	7	8	7	4
8	8	7	6	9	7	8	7	7	6	7	9	6	6	8	8	8	7	9	5
7	4	7	8	8	9	6	7	6	8	7	5	7	7	9	7	7	8	7	9
7	5	9	8	7	7	8	9	8	5	9	9	8	6	9	8	8	8	9	7

(d) Superimpose the relative frequency histogram of these observations on the graph of the probability histogram.

2.4-14. In the casino game chuck-a-luck, three unbiased six-sided dice are rolled. One possible bet is $1 on fives, and the payoff is equal to $1 for each five on that roll. In addition, the dollar bet is returned if at least one five is rolled. The dollar that was bet is lost only if no fives are rolled. Let X denote the payoff for this game. Then X can equal $-1, 1, 2,$ or 3.

(a) Determine the p.m.f. $f(x)$.
(b) Calculate $\mu, \sigma^2,$ and σ.
(c) Depict the p.m.f. as a probability histogram.
(d) Calculate $\bar{x}, s^2,$ and s, using the 100 following simulated observations of X:

-1	1	-1	-1	1	-1	-1	-1	-1	1
-1	1	-1	2	1	-1	-1	-1	1	-1
1	-1	-1	1	1	-1	-1	-1	-1	1
-1	-1	2	1	1	-1	-1	1	-1	1
-1	-1	1	2	-1	-1	1	-1	-1	-1
-1	1	-1	-1	-1	1	-1	1	-1	2
-1	1	-1	1	-1	2	1	-1	-1	1
-1	1	1	-1	1	1	-1	3	1	-1
-1	-1	1	1	1	1	-1	-1	2	-1
1	-1	1	-1	1	2	-1	1	1	-1

(e) Superimpose the relative frequency histogram of these observations on the probability histogram.

2.4-15. It is claimed that for a particular lottery, 1/10 of the 50 million tickets will win a prize. What is the probability of winning at least one prize if you purchase (a) 10 tickets or (b) 15 tickets?

2.4-16. For the lottery described in Exercise 2.4-15, find the smallest number of tickets that must be purchased so that the probability of winning at least one prize is greater than (a) 0.50; (b) 0.95.

2.4-17. Construct a table that gives the probabilities for the $b(8, 0.2)$ distribution, the hypergeometric distribution with $N_1 = 8, N_2 = 32,$ and $n = 8$, and the hypergeometric distribution with $N_1 = 16$, $N_2 = 64,$ and $n = 8$. (See Figure 2.1-2.)

2.4-18. A hospital obtains 40% of its flu vaccine from Company A, 50% from Company B, and 10% from Company C. From past experience, it is known that 3% of the vials from A are ineffective, 2% from B are ineffective, and 5% from C are ineffective. The hospital tests five vials from each shipment. If at least one of the five is ineffective, find the conditional probability of that shipment's having come from C.

2.4-19. Many products can operate if, out of n parts, k of them are working. Say $n = 20$ and $k = 17$, $p = 0.05$ is the probability that a part fails, and assume independence. What is the probability that the product operates?

2.4-20. A company starts a fund of M dollars from which it pays $1000 to each employee who achieves high performance during the year. The probability of each employee achieving this goal is 0.10 and is independent of the probabilities of the other employees doing so. If there are $n = 10$ employees, how much should M equal so that the fund has a probability of at least 99% of covering those payments?

2.4-21. A businessman has an important meeting to attend, but he is running a little late. He can take one route to work that has six stoplights or another, longer route that has two stoplights. He figures that if he stops at more than half of the lights on either route, he will be late for the meeting. Assume independence, and assume that the chance of stopping at each light is $p = 0.5$. Which route should he take?

2.4-22. In group testing for a certain disease, a blood sample was taken from each of n individuals and part of each sample was placed in a common pool. The latter was then tested. If the result was negative, there was no more testing and all n individuals were declared negative with one test. If, however, the combined result was found positive, all individuals were tested, requiring $n + 1$ tests. If $p = 0.05$ is the probability of a person's having the disease and $n = 5$, compute the expected number of tests needed, assuming independence.

2.4-23. Your stockbroker is free to take your calls about 60% of the time; otherwise he is talking to another client or is out of the office. You call him at five random times during a given month. (Assume independence.)

(a) What is the probability that he will take every one of the five calls?
(b) What is the probability that he will accept exactly three of your five calls?
(c) What is the probability that he will accept at least one of the calls?

2.5   THE MOMENT-GENERATING FUNCTION

The mean, variance, and standard deviation are important characteristics of a distribution. For some distributions—the binomial, for instance—it is fairly difficult to directly compute $E(X)$ and $E(X^2)$ to find the mean and the variance. In this section, we define a function of t that will help us generate the moments of a distribution. Thus, this function is called the moment-generating function. Although this generating characteristic is extremely important, there is a uniqueness property that is even more important. We first define the new function of t and then explain this uniqueness property before showing how it can be used to compute the moments of X.

DEFINITION 2.5-1

Let X be a random variable of the discrete type with p.m.f. $f(x)$ and space S. If there is a positive number h such that

$$E(e^{tX}) = \sum_{x \in S} e^{tx} f(x)$$

exists and is finite for $-h < t < h$, then the function of t defined by

$$M(t) = E(e^{tX})$$

is called the **moment-generating function of X** (or of the distribution of X). This function is often abbreviated as m.g.f.

First, it is evident that if we set $t = 0$, we have $M(0) = 1$. Moreover, if the space of S is $\{b_1, b_2, b_3, \ldots\}$, then the moment-generating function is given by the expansion

$$M(t) = e^{tb_1} f(b_1) + e^{tb_2} f(b_2) + e^{tb_3} f(b_3) + \cdots.$$

Thus, the coefficient of e^{tb_i} is the probability

$$f(b_i) = P(X = b_i).$$

Accordingly, if two random variables (or two distributions of probability) have the same moment-generating function, they must have the same distribution of probability. That is, if the two random variables had the two probability mass functions $f(x)$ and $g(y)$, as well as the same space $S = \{b_1, b_2, b_3, \ldots\}$, and if

$$e^{tb_1} f(b_1) + e^{tb_2} f(b_2) + \cdots = e^{tb_1} g(b_1) + e^{tb_2} g(b_2) + \cdots \qquad (2.5\text{-}1)$$

for all t, $-h < t < h$, then mathematical transform theory requires that

$$f(b_i) = g(b_i), \qquad i = 1, 2, 3, \cdots.$$

So we see that the moment-generating function uniquely determines the distribution of a random variable. In other words, if the m.g.f. exists, there is one and only one distribution of probability associated with that m.g.f.

REMARK   From elementary algebra, we can get some understanding of why Equation 2.5-1 requires that $f(b_i) = g(b_i)$. In that equation, let $e^t = w$ and say the points in the support, namely, $b_1, b_2, \ldots, b_k$, are positive integers, the largest of which is m. Then Equation 2.5-1 provides the equality of two mth-degree polynomials in w for an uncountable number of values of w. A fundamental theorem of algebra requires that the corresponding coefficients of the two polynomials be equal; that is, $f(b_i) = g(b_i)$, $i = 1, 2, \ldots, k$.

EXAMPLE 2.5-1 If X has the m.g.f.

$$M(t) = e^t \left(\frac{3}{6}\right) + e^{2t}\left(\frac{2}{6}\right) + e^{3t}\left(\frac{1}{6}\right),$$

then the probabilities are

$$P(X = 1) = \frac{3}{6}, \qquad P(X = 2) = \frac{2}{6}, \qquad P(X = 3) = \frac{1}{6}.$$

We could write this, if we choose to do so, by saying that X has the p.m.f.

$$f(x) = \frac{4 - x}{6}, \qquad x = 1, 2, 3.$$

EXAMPLE 2.5-2 Suppose the m.g.f. of X is

$$M(t) = \frac{e^t/2}{1 - e^t/2}, \qquad t < \ln 2.$$

Until we expand $M(t)$, we cannot detect the coefficients of $e^{b_i t}$. Recalling that

$$(1 - z)^{-1} = 1 + z + z^2 + z^3 + \cdots, \qquad -1 < z < 1,$$

we have

$$\frac{e^t}{2}\left(1 - \frac{e^t}{2}\right)^{-1} = \frac{e^t}{2}\left(1 + \frac{e^t}{2} + \frac{e^{2t}}{2^2} + \frac{e^{3t}}{2^3} + \cdots\right)$$

$$= \left(e^t\right)\left(\frac{1}{2}\right)^1 + \left(e^{2t}\right)\left(\frac{1}{2}\right)^2 + \left(e^{3t}\right)\left(\frac{1}{2}\right)^3 + \cdots$$

when $e^t/2 < 1$ and thus $t < \ln 2$. That is,

$$P(X = x) = \left(\frac{1}{2}\right)^x$$

when x is a positive integer, or, equivalently, the p.m.f. of X is

$$f(x) = \left(\frac{1}{2}\right)^x, \qquad x = 1, 2, 3, \cdots.$$

From the theory of Laplace transforms, it can be shown that the existence of $M(t)$, for $-h < t < h$, implies that derivatives of $M(t)$ of all orders exist at $t = 0$; moreover, it is permissible to interchange differentiation and summation as the series converges uniformly. Thus,

$$M'(t) = \sum_{x \in S} x e^{tx} f(x),$$

$$M''(t) = \sum_{x \in S} x^2 e^{tx} f(x),$$

and for each positive integer r,

$$M^{(r)}(t) = \sum_{x \in S} x^r e^{tx} f(x).$$

Setting $t = 0$, we see that

$$M'(0) = \sum_{x \in S} x f(x) = E(X),$$

$$M''(0) = \sum_{x \in S} x^2 f(x) = E(X^2),$$

and, in general,

$$M^{(r)}(0) = \sum_{x \in S} x^r f(x) = E(X^r).$$

In particular, if the moment-generating function exists, then

$$M'(0) = E(X) = \mu \quad \text{and} \quad M''(0) - [M'(0)]^2 = E(X^2) - [E(X)]^2 = \sigma^2.$$

The preceding argument shows that we can find the moments of X by differentiating $M(t)$. In using this technique, it must be emphasized that first we evaluate the summation representing $M(t)$ to obtain a closed-form solution and then we differentiate that solution to obtain the moments of X. The next example illustrates the use of the moment-generating function for finding the first and second moments and then the mean and variance of the important binomial distribution.

EXAMPLE 2.5-3 The p.m.f. of the binomial distribution is

$$f(x) = \binom{n}{x} p^x (1 - p)^{n-x}, \qquad x = 0, 1, 2, ..., n.$$

Thus, the m.g.f. is

$$M(t) = E(e^{tX}) = \sum_{x=0}^{n} e^{tx} \binom{n}{x} p^x (1 - p)^{n-x}$$

$$= \sum_{x=0}^{n} \binom{n}{x} (pe^t)^x (1 - p)^{n-x}$$

$$= \left[(1 - p) + pe^t \right]^n, \qquad -\infty < t < \infty,$$

from the expansion of $(a + b)^n$ with $a = 1 - p$ and $b = pe^t$. It is interesting to note that here and elsewhere the m.g.f. is usually rather easy to compute if the p.m.f. has a factor involving an exponential, like p^x in the binomial p.m.f.

The first two derivatives of $M(t)$ are

$$M'(t) = n[(1 - p) + pe^t]^{n-1}(pe^t)$$

and

$$M''(t) = n(n - 1)[(1 - p) + pe^t]^{n-2}(pe^t)^2 + n[(1 - p) + pe^t]^{n-1}(pe^t).$$

Thus,

$$\mu = E(X) = M'(0) = np$$

and

$$\sigma^2 = E(X^2) - [E(X)]^2 = M''(0) - [M'(0)]^2$$
$$= n(n - 1)p^2 + np - (np)^2 = np(1 - p),$$

as we showed in Section 2.4 using a longer derivation.

In the special case when $n = 1$, X has a Bernoulli distribution and

$$M(t) = (1 - p) + pe^t$$

for all real values of t, $\mu = p$, and $\sigma^2 = p(1 - p)$. ∎

We turn now to the situation in which we observe a sequence of Bernoulli trials until exactly r successes occur, where r is a fixed positive integer. Let the random variable X denote the number of trials needed to observe the rth success. That is, X is the trial number on which the rth success is observed. By the multiplication rule of probabilities, the p.m.f. of X—say, $g(x)$—equals the product of the probability

$$\binom{x - 1}{r - 1}p^{r-1}(1 - p)^{x-r} = \binom{x - 1}{r - 1}p^{r-1}q^{x-r}$$

of obtaining exactly $r - 1$ successes in the first $x - 1$ trials and the probability p of a success on the rth trial. Thus, the p.m.f. of X is

$$g(x) = \binom{x - 1}{r - 1}p^r(1 - p)^{x-r} = \binom{x - 1}{r - 1}p^r q^{x-r}, \qquad x = r, r + 1, \ldots.$$

We say that X has a **negative binomial distribution**.

REMARK   The reason for calling this distribution the negative binomial distribution is as follows: Consider $h(w) = (1 - w)^{-r}$, the binomial $(1 - w)$ with the negative exponent $-r$. Using Maclaurin's series expansion (see Appendix D.4 on the CD-ROM), we have

$$(1 - w)^{-r} = \sum_{k=0}^{\infty} \frac{h^{(k)}(0)}{k!}w^k = \sum_{k=0}^{\infty} \binom{r + k - 1}{r - 1}w^k, \qquad -1 < w < 1.$$

If we let $x = k + r$ in the summation, then $k = x - r$ and

$$(1 - w)^{-r} = \sum_{x=r}^{\infty} \binom{r + x - r - 1}{r - 1}w^{x-r} = \sum_{x=r}^{\infty} \binom{x - 1}{r - 1}w^{x-r},$$

the summand of which is, except for the factor p^r, the negative binomial probability when $w = q$. In particular, the sum of the probabilities for the negative binomial distribution is 1 because

$$\sum_{x=r}^{\infty} g(x) = \sum_{x=r}^{\infty} \binom{x - 1}{r - 1}p^r q^{x-r} = p^r(1 - q)^{-r} = 1.$$ ∎

If $r = 1$ in the negative binomial distribution, we say that X has a **geometric distribution**, since the p.m.f. consists of terms of a geometric series, namely,

$$g(x) = p(1 - p)^{x-1}, \qquad x = 1, 2, 3, \cdots.$$

Recall that for a geometric series (see Appendix D.4 on the CD-ROM for a review), the sum is given by

$$\sum_{k=0}^{\infty} ar^k = \sum_{k=1}^{\infty} ar^{k-1} = \frac{a}{1-r}$$

when $|r| < 1$. Thus, for the geometric distribution,

$$\sum_{x=1}^{\infty} g(x) = \sum_{x=1}^{\infty} (1-p)^{x-1}p = \frac{p}{1-(1-p)} = 1,$$

so that $g(x)$ does satisfy the properties of a p.m.f.

From the sum of a geometric series, we also note that when k is an integer,

$$P(X > k) = \sum_{x=k+1}^{\infty} (1-p)^{x-1}p = \frac{(1-p)^k p}{1-(1-p)} = (1-p)^k = q^k.$$

Thus, the value of the distribution function at a positive integer k is

$$P(X \le k) = \sum_{x=1}^{k} (1-p)^{x-1}p = 1 - P(X > k) = 1 - (1-p)^k = 1 - q^k.$$

EXAMPLE 2.5-4 Some biology students were checking eye color in a large number of fruit flies. For the individual fly, suppose that the probability of white eyes is 1/4 and the probability of red eyes is 3/4, and that we may treat these observations as having independent Bernoulli trials. The probability that at least four flies have to be checked for eye color to observe a white-eyed fly is given by

$$P(X \ge 4) = P(X > 3) = q^3 = \left(\frac{3}{4}\right)^3 = \frac{27}{64} = 0.4219.$$

The probability that at most four flies have to be checked for eye color to observe a white-eyed fly is given by

$$P(X \le 4) = 1 - q^4 = 1 - \left(\frac{3}{4}\right)^4 = \frac{175}{256} = 0.6836.$$

The probability that the first fly with white eyes is the fourth fly considered is

$$P(X = 4) = q^{4-1}p = \left(\frac{3}{4}\right)^3 \left(\frac{1}{4}\right) = \frac{27}{256} = 0.1055.$$

It is also true that

$$P(X = 4) = P(X \le 4) - P(X \le 3)$$
$$= [1 - (3/4)^4] - [1 - (3/4)^3]$$
$$= \left(\frac{3}{4}\right)^3 \left(\frac{1}{4}\right). \qquad \blacksquare$$

We now show that the mean and the variance of a negative binomial random variable X are, respectively,

$$\mu = E(X) = \frac{r}{p} \qquad \text{and} \qquad \sigma^2 = \frac{rq}{p^2} = \frac{r(1-p)}{p^2}.$$

In particular, if $r = 1$, so that X has a geometric distribution, then

$$\mu = \frac{1}{p} \quad \text{and} \quad \sigma^2 = \frac{q}{p^2} = \frac{1-p}{p^2}.$$

The mean $\mu = 1/p$ agrees with our intuition. Let's check: If $p = 1/6$, then we would expect, on the average, $1/(1/6) = 6$ trials before the first success.

To find these moments, we determine the m.g.f. of the negative binomial distribution. It is

$$M(t) = \sum_{x=r}^{\infty} e^{tx} \binom{x-1}{r-1} p^r (1-p)^{x-r}$$

$$= (pe^t)^r \sum_{x=r}^{\infty} \binom{x-1}{r-1} \left[(1-p)e^t\right]^{x-r}$$

$$= \frac{(pe^t)^r}{[1-(1-p)e^t]^r}, \quad \text{where } (1-p)e^t < 1$$

(or, equivalently, when $t < -\ln(1-p)$). Thus,

$$M'(t) = (pe^t)^r(-r)[1-(1-p)e^t]^{-r-1}[-(1-p)e^t] +$$
$$r(pe^t)^{r-1}(pe^t)[1-(1-p)e^t]^{-r}$$
$$= r(pe^t)^r[1-(1-p)e^t]^{-r-1}$$

and

$$M''(t) = r(pe^t)^r(-r-1)[1-(1-p)e^t]^{-r-2}[-(1-p)e^t] +$$
$$r^2(pe^t)^{r-1}(pe^t)[1-(1-p)e^t]^{-r-1}.$$

Accordingly,

$$M'(0) = rp^r p^{-r-1} = rp^{-1}$$

and

$$M''(0) = r(r+1)p^r p^{-r-2}(1-p) + r^2 p^r p^{-r-1}$$
$$= rp^{-2}[(1-p)(r+1) + rp] = rp^{-2}(r+1-p).$$

Hence, we have

$$\mu = \frac{r}{p} \quad \text{and} \quad \sigma^2 = \frac{r(r+1-p)}{p^2} - \frac{r^2}{p^2} = \frac{r(1-p)}{p^2}.$$

Even these calculations are a little messy, so a somewhat easier way is given in Exercise 2.5-20.

EXAMPLE 2.5-5 Suppose that during practice a basketball player can make a free throw 80% of the time. Furthermore, assume that a sequence of free-throw shooting can be thought of as independent Bernoulli trials. Let X equal the minimum number of free throws that this player must attempt to make a total of 10 shots. The p.m.f. of X is

$$g(x) = \binom{x-1}{10-1}(0.80)^{10}(0.20)^{x-10}, \quad x = 10, 11, 12, \ldots.$$

The mean, variance, and standard deviation of X are, respectively,

$$\mu = 10\left(\frac{1}{0.80}\right) = 12.5, \qquad \sigma^2 = \frac{10(0.20)}{0.80^2} = 3.125, \qquad \text{and} \qquad \sigma = 1.768.$$

And we have, for example,

$$P(X = 12) = g(12) = \binom{11}{9}(0.80)^{10}(0.20)^2 = 0.2362.$$ ∎

EXAMPLE 2.5-6   To consider the effect of p and r on the negative binomial distribution, Figure 2.5-1 gives the probability histograms for four combinations of p and r. Note that since $r = 1$ in the first of these, it represents a geometric p.m.f. ∎

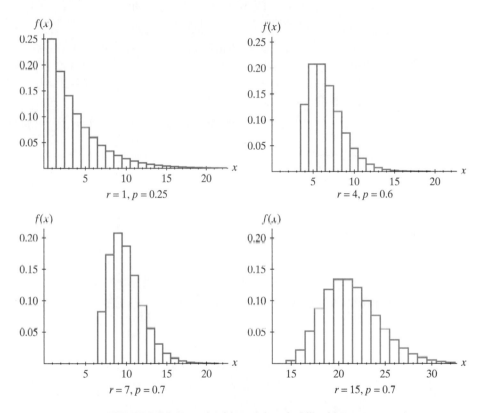

FIGURE 2.5-1: Negative binomial probability histograms

When the moment-generating function exists, derivatives of all orders exist at $t = 0$. Thus, it is possible to represent $M(t)$ as a Maclaurin series, namely,

$$M(t) = M(0) + M'(0)\left(\frac{t}{1!}\right) + M''(0)\left(\frac{t^2}{2!}\right) + M'''(0)\left(\frac{t^3}{3!}\right) + \cdots.$$

If the Maclaurin series expansion of $M(t)$ exists and the moments are given, we can frequently sum the Maclaurin series to obtain the closed form of $M(t)$. This approach is illustrated in the next example.

EXAMPLE 2.5-7 Let the moments of X be defined by

$$E(X^r) = 0.8, \qquad r = 1, 2, 3, \ldots.$$

The moment-generating function of X is then

$$M(t) = M(0) + \sum_{r=1}^{\infty} 0.8 \left(\frac{t^r}{r!} \right) = 1 + 0.8 \sum_{r=1}^{\infty} \frac{t^r}{r!}$$

$$= 0.2 + 0.8 \sum_{r=0}^{\infty} \frac{t^r}{r!} = 0.2 e^{0t} + 0.8 e^{1t}.$$

Thus,

$$P(X = 0) = 0.2 \qquad \text{and} \qquad P(X = 1) = 0.8.$$ ■

COMPUTATIONAL COMMENTS   (**CAS**) We now illustrate how *Maple* can be used to find the moment-generating function for the negative binomial distribution. It is then used to find the mean and the variance. Here is the *Maple* code:

```
>  f := binomial(x - 1, r - 1)*p^r*(1 - p)^(x - r);
```
$$f := \text{binomial}(x - 1, r - 1) \, p^r \, (1 - p)^{(x-r)}$$

```
>  mu := sum(x*f, x = r .. infinity);
```
$$\mu := \frac{r}{p}$$

```
>  var := sum((x - mu)^2*f, x = r .. infinity);
```
$$var := -\frac{r(-1 + p)}{p^2}$$

```
>  M := sum(exp(t*x)*f, x = r .. infinity);
```
$$M := \frac{e^{(tr)} p^r}{(1 - e^t + e^t p)^r}$$

```
>  Mprime := diff(M, t):
>  Mdoubleprime := diff(Mprime, t):
>  mu := subs(t = 0, Mprime):
>  mu := factor(%);
```
$$\mu := \frac{r}{p}$$

```
>  var := subs(t = 0, Mdoubleprime) - mu^2:
>  var := factor(%);
```
$$var := -\frac{r(-1 + p)}{p^2}$$

Note that although the result for the variance is correct, we would probably write it as

$$\sigma^2 = \frac{r(1 - p)}{p^2}.$$

EXERCISES

2.5-1. Define the p.m.f. and give the values of μ, σ^2, and σ when the moment-generating function of X is defined by

(a) $M(t) = 1/3 + (2/3)e^t$.

(b) $M(t) = (0.25 + 0.75e^t)^{12}$.

2.5-2. (i) Give the name of the distribution of X (if it has a name), (ii) find the values of μ and σ^2, and (iii) calculate $P(1 \le X \le 2)$ when the moment-generating function of X is given by

(a) $M(t) = (0.3 + 0.7e^t)^5$.

(b) $M(t) = \dfrac{0.3e^t}{1 - 0.7e^t}$, $t < -\ln(0.7)$.

(c) $M(t) = 0.45 + 0.55e^t$.

(d) $M(t) = 0.3e^t + 0.4e^{2t} + 0.2e^{3t} + 0.1e^{4t}$.

(e) $M(t) = (0.6e^t)^2(1 - 0.4e^t)^{-2}$,

$$t < -\ln(0.4).$$

(f) $M(t) = \sum_{x=1}^{10} (0.1)e^{tx}$.

2.5-3. If the moment-generating function of X is

$$M(t) = \frac{2}{5}e^t + \frac{1}{5}e^{2t} + \frac{2}{5}e^{3t},$$

find the mean, variance, and p.m.f. of X.

2.5-4. Let X equal the number of people selected at random that you must ask in order to find someone with the same birthday as yours. Assume that each day of the year is equally likely, and ignore February 29.

(a) What is the p.m.f. of X?

(b) Give the values of the mean, variance, and standard deviation of X.

(c) Find $P(X > 400)$ and $P(X < 300)$.

2.5-5. For each question on a multiple-choice test, there are five possible answers, of which exactly one is correct. If a student selects answers at random, give the probability that the first question answered correctly is question 4.

2.5-6. The probability that a machine produces a defective item is 0.01. Each item is checked as it is produced. Assume that these are independent trials, and compute the probability that at least 100 items must be checked to find one that is defective.

2.5-7. Apples are packaged automatically in 3-pound bags. Suppose that 4% of the time the bag of apples weighs less than 3 pounds. If you select bags randomly and weigh them in order to discover one underweight bag of apples, find the probability that the number of bags that must be selected is

(a) At least 20.

(b) At most 20.

(c) Exactly 20.

2.5-8. Show that 63/512 is the probability that the fifth head is observed on the tenth independent flip of an unbiased coin.

2.5-9. An excellent free-throw shooter attempts several free throws until she misses.

(a) If $p = 0.9$ is her probability of making a free throw, what is the probability of having the first miss on the 13th attempt or later?

(b) If she continues shooting until she misses three, what is the probability that the third miss occurs on the 30th attempt?

2.5-10. Suppose that a basketball player different from the one in Example 2.5-5 can make a free throw 60% of the time. Let X equal the minimum number of free throws that this player must attempt to make a total of 10 shots.

(a) Give the mean, variance, and standard deviation of X.

(b) Find $P(X = 16)$.

2.5-11. Let X equal the number of flips of a fair coin that are required to observe the same face on consecutive flips.

(a) Find the p.m.f. of X. HINT: Draw a tree diagram.

(b) Find the moment generating function of X.

(c) Use the m.g.f. to find the values of (i) the mean and (ii) the variance of X.

(d) Find the values of (i) $P(X \le 3)$, (ii) $P(X \ge 5)$, and (iii) $P(X = 3)$.

(e) Simulate this experiment to confirm you answers to parts (c) and (d).

2.5-12. Let X have a geometric distribution. Show that

$$P(X > k + j \mid X > k) = P(X > j),$$

where k and j are nonnegative integers. NOTE: We sometimes say that in this situation there has been loss of memory.

2.5-13. Let X equal the number of flips of a fair coin that are required to observe heads–tails on consecutive flips.

(a) Find the p.m.f. of X. HINT: Draw a tree diagram.

(b) Show that the m.g.f. of X is
$$M(t) = e^{2t}/(e^t - 2)^2.$$

(c) Use the m.g.f. to find the values of (i) the mean and (ii) the variance of X.

(d) Find the values of (i) $P(X \leq 3)$, (ii) $P(X \geq 5)$, and (iii) $P(X = 3)$.

(e) Simulate this experiment to confirm you answers to parts (c) and (d).

2.5-14. Let X equal the number of flips of a fair coin that are required to observe heads on consecutive flips. Let f_n equal the nth Fibonacci number, where $f_1 = 1$, $f_2 = 1$, and $f_n = f_{n-1} + f_{n-2}$, $n = 3, 4, 5, \ldots$.

(a) Show that the p.m.f. of X is

$$f(x) = \frac{f_{x-1}}{2^x}, \qquad x = 2, 3, 4, \ldots.$$

HINT: Draw a tree diagram.

(b) Use the fact that

$$f_x = \frac{1}{\sqrt{5}} \left[\left(\frac{1 + \sqrt{5}}{2} \right)^x - \left(\frac{1 - \sqrt{5}}{2} \right)^x \right]$$

to show that $\sum_{x=2}^{\infty} f(x) = 1$.

(c) Show that $\mu = E(X) = 6$.

(d) Show that $E[X(X - 1)] = 52$, so that the variance of X is $\sigma^2 = 22$.

(e) Find the values of (i) $P(X \leq 3)$, (ii) $P(X \geq 5)$, and (iii) $P(X = 3)$.

(f) Simulate this experiment to confirm your answers to parts (c), (d), and (e). Either write a computer program or physically toss a coin to support the answers.

2.5-15. One of four different prizes was randomly put into each box of a cereal. If a family decided to buy this cereal until it obtained at least one of each of the four different prizes, what is the expected number of boxes of cereal that must be purchased?

2.5-16. Suppose an airport metal detector catches a person with metal 99% of the time. That is, it misses detecting a person with metal 1% of the time. Assume independence of people carrying metal. What is the probability that the first person missed (not detected) is among the first 50 persons scanned?

2.5-17. In 2006, Red Rose tea randomly began placing 1 of 10 English porcelain miniature animals in a 100-bag box of the tea, selecting from 10 "Pet Shop Friends."

(a) On the average, how many boxes of tea must be purchased by a customer to obtain a complete collection consisting of 10 different animals?

(b) If the customer uses one tea bag per day, how long can a customer expect to take, on the average, to obtain a complete collection?

(c) From 2002 to 2006, the figurines were part of the Noah's Ark Animal Series, which contains 15 pieces. Assume again that these figurines were selected randomly. How many boxes of tea would have had to be purchased, on the average, to obtain the complete collection?

2.5-18. If $E(X^r) = 5^r, r = 1, 2, 3, \ldots$, find the moment-generating function $M(t)$ of X and the p.m.f. of X.

2.5-19. Let the moment-generating function $M(t)$ of X exist for $-h < t < h$. Consider the function $R(t) = \ln M(t)$. The first two derivatives of $R(t)$ are, respectively,

$$R'(t) = \frac{M'(t)}{M(t)} \qquad \text{and}$$

$$R''(t) = \frac{M(t)M''(t) - [M'(t)]^2}{[M(t)]^2}.$$

Setting $t = 0$, show that

(a) $\mu = R'(0)$.

(b) $\sigma^2 = R''(0)$.

2.5-20. Use the result of Exercise 2.5-19 to find the mean and variance of the

(a) Bernoulli distribution.

(b) Binomial distribution.

(c) Geometric distribution.

(d) Negative binomial distribution.

2.5-21. Given a random permutation of the integers in the set $\{1, 2, 3, 4, 5\}$, let X equal the number of integers that are in their natural position. The moment-generating function of X is

$$M(t) = \frac{44}{120} + \frac{45}{120} e^t + \frac{20}{120} e^{2t} + \frac{10}{120} e^{3t} + \frac{1}{120} e^{5t}.$$

(a) Find the mean and variance of X.

(b) Find the probability that at least one integer is in its natural position.

(c) Draw a graph of the probability histogram of the p.m.f. of X.

2.5-22. The probability that a company's workforce has no accidents in a given month is 0.7. The numbers of accidents from month to month are independent. What is the probability that the third month in a year is the first month that at least one accident occurs?

2.6 THE POISSON DISTRIBUTION

Some experiments result in counting the number of times particular events occur at given times or with given physical objects. For example, we could count the number of phone calls arriving at a switchboard between 9 and 10 A.M., the number of flaws in 100 feet of wire, the number of customers that arrive at a ticket window between 12 noon and 2 P.M., or the number of defects in a 100-foot roll of aluminum screen that is 2 feet wide. Each count can be looked upon as a random variable associated with an approximate Poisson process, provided that the conditions in the following definition are satisfied.

DEFINITION 2.6-1	Let the number of changes that occur in a given continuous interval be counted. Then we have an **approximate Poisson process** with parameter $\lambda > 0$ if the following conditions are satisfied:

(a) The numbers of changes occurring in nonoverlapping intervals are independent.

(b) The probability of exactly one change occurring in a sufficiently short interval of length h is approximately λh.

(c) The probability of two or more changes occurring in a sufficiently short interval is essentially zero.

Suppose that an experiment satisfies the preceding three conditions of an approximate Poisson process. Let X denote the number of changes in an interval of length 1 (where "length 1" represents one unit of the quantity under consideration). We would like to find an approximation for $P(X = x)$, where x is a nonnegative integer. To achieve this, we partition the unit interval into n subintervals of equal length $1/n$. If n is sufficiently large (i.e., much larger than x), we shall approximate the probability that x changes occur in this unit interval by finding the probability that one change occurs in each of exactly x of these n subintervals. The probability of one change occurring in any one subinterval of length $1/n$ is approximately $\lambda(1/n)$, by condition (b). The probability of two or more changes in any one subinterval is essentially zero, by condition (c). So, for each subinterval, exactly one change occurs with a probability of approximately $\lambda(1/n)$. Consider the occurrence or nonoccurrence of a change in each subinterval as a Bernoulli trial. By condition (a), we have a sequence of n Bernoulli trials with probability p approximately equal to $\lambda(1/n)$. Thus, an approximation for $P(X = x)$ is given by the binomial probability

$$\frac{n!}{x!\,(n-x)!}\left(\frac{\lambda}{n}\right)^x\left(1-\frac{\lambda}{n}\right)^{n-x}.$$

If n increases without bound, then

$$\lim_{n\to\infty}\frac{n!}{x!\,(n-x)!}\left(\frac{\lambda}{n}\right)^x\left(1-\frac{\lambda}{n}\right)^{n-x}$$

$$=\lim_{n\to\infty}\frac{n(n-1)\cdots(n-x+1)}{n^x}\frac{\lambda^x}{x!}\left(1-\frac{\lambda}{n}\right)^n\left(1-\frac{\lambda}{n}\right)^{-x}.$$

Now, for fixed x, we have (see Appendix D.3 on the CD-ROM)

$$\lim_{n\to\infty} \frac{n(n-1)\cdots(n-x+1)}{n^x} = \lim_{n\to\infty}\left[(1)\left(1-\frac{1}{n}\right)\cdots\left(1-\frac{x-1}{n}\right)\right] = 1,$$

$$\lim_{n\to\infty}\left(1-\frac{\lambda}{n}\right)^n = e^{-\lambda},$$

$$\lim_{n\to\infty}\left(1-\frac{\lambda}{n}\right)^{-x} = 1.$$

Thus,

$$\lim_{n\to\infty} \frac{n!}{x!\,(n-x)!}\left(\frac{\lambda}{n}\right)^x\left(1-\frac{\lambda}{n}\right)^{n-x} = \frac{\lambda^x e^{-\lambda}}{x!} = P(X=x).$$

The distribution of probability associated with this process has a special name. We say that the random variable X has a **Poisson distribution** if its p.m.f. is of the form

$$f(x) = \frac{\lambda^x e^{-\lambda}}{x!}, \qquad x = 0,1,2,\ldots,$$

where $\lambda > 0$.

It is easy to see that $f(x)$ has the properties of a p.m.f. because, clearly, $f(x) \geq 0$ and, from the Maclaurin series expansion of e^λ (see Appendix D.4 on the CD-ROM), we have

$$\sum_{x=0}^{\infty} \frac{\lambda^x e^{-\lambda}}{x!} = e^{-\lambda}\sum_{x=0}^{\infty} \frac{\lambda^x}{x!} = e^{-\lambda}e^\lambda = 1.$$

To discover the exact role of the parameter $\lambda > 0$, let us find some of the characteristics of the Poisson distribution.

The moment-generating function of X is

$$M(t) = E(e^{tX}) = \sum_{x=0}^{\infty} e^{tx}\frac{\lambda^x e^{-\lambda}}{x!} = e^{-\lambda}\sum_{x=0}^{\infty} \frac{(\lambda e^t)^x}{x!}.$$

From the series representation of the exponential function, we have

$$M(t) = e^{-\lambda}e^{\lambda e^t} = e^{\lambda(e^t-1)}$$

for all real values of t. Now,

$$M'(t) = \lambda e^t e^{\lambda(e^t-1)}$$

and

$$M''(t) = (\lambda e^t)^2 e^{\lambda(e^t-1)} + \lambda e^t e^{\lambda(e^t-1)}.$$

The values of the mean and variance of X are, respectively,

$$\mu = M'(0) = \lambda$$

and

$$\sigma^2 = M''(0) - [M'(0)]^2 = (\lambda^2 + \lambda) - \lambda^2 = \lambda.$$

That is, for the Poisson distribution, $\mu = \sigma^2 = \lambda$.

REMARKS   It is also possible to find the mean and the variance for the Poisson distribution directly, without using the moment-generating function. The mean for the Poisson distribution is given by

$$E(X) = \sum_{x=0}^{\infty} x \frac{\lambda^x e^{-\lambda}}{x!} = e^{-\lambda} \sum_{x=1}^{\infty} \frac{\lambda^x}{(x-1)!}$$

because $(0)f(0) = 0$ and $x/x! = 1/(x-1)!$ when $x > 0$. If we let $k = x - 1$, then

$$E(X) = e^{-\lambda} \sum_{k=0}^{\infty} \frac{\lambda^{k+1}}{k!} = \lambda e^{-\lambda} \sum_{k=0}^{\infty} \frac{\lambda^k}{k!}$$
$$= \lambda e^{-\lambda} e^{\lambda} = \lambda.$$

To find the variance, we first determine the second factorial moment $E[X(X-1)]$. We have

$$E[X(X-1)] = \sum_{x=0}^{\infty} x(x-1) \frac{\lambda^x e^{-\lambda}}{x!} = e^{-\lambda} \sum_{x=2}^{\infty} \frac{\lambda^x}{(x-2)!}$$

because $(0)(0-1)f(0) = 0, (1)(1-1)f(1) = 0$, and $x(x-1)/x! = 1/(x-2)!$ when $x > 1$. If we let $k = x - 2$, then

$$E[X(X-1)] = e^{-\lambda} \sum_{k=0}^{\infty} \frac{\lambda^{k+2}}{k!} = \lambda^2 e^{-\lambda} \sum_{k=0}^{\infty} \frac{\lambda^k}{k!}$$
$$= \lambda^2 e^{-\lambda} e^{\lambda} = \lambda^2.$$

Thus,

$$\mathrm{Var}(X) = E(X^2) - [E(X)]^2 = E[X(X-1)] + E(X) - [E(X)]^2$$
$$= \lambda^2 + \lambda - \lambda^2 = \lambda.$$

We again see that, for the Poisson distribution, $\mu = \sigma^2 = \lambda$. ∎

Table III in the appendix gives values of the distribution function of a Poisson random variable for selected values of λ. This table is illustrated in the next example.

EXAMPLE 2.6-1   Let X have a Poisson distribution with a mean of $\lambda = 5$. Then, using Table III in the appendix, we obtain

$$P(X \le 6) = \sum_{x=0}^{6} \frac{5^x e^{-5}}{x!} = 0.762,$$

$$P(X > 5) = 1 - P(X \le 5) = 1 - 0.616 = 0.384,$$

and

$$P(X = 6) = P(X \le 6) - P(X \le 5) = 0.762 - 0.616 = 0.146.$$ ∎

EXAMPLE 2.6-2   To see the effect of λ on the p.m.f. $f(x)$ of X, Figure 2.6-1 shows the probability histograms of $f(x)$ for four different values of λ. ∎

FIGURE 2.6-1: Poisson probability histograms

EXAMPLE 2.6-3   Let X denote the number of alpha particles emitted by barium-133 in one-tenth of a second and counted by a Geiger counter. One hundred observations of X produced the data in Table 2.6-1. For these data, $\bar{x} = 559/100 = 5.59$ and $s^2 = 3{,}619/99 - (559)^2/9900 = 4.992$. Thus, clearly, the sample mean and sample variance are fairly close to each other. In Figure 2.6-2, the probability histogram for a

TABLE 2.6-1: Barium 133			
Outcome (x)	Frequency (f)	fx	fx^2
1	1	1	1
2	4	8	16
3	13	39	117
4	19	76	304
5	16	80	400
6	15	90	540
7	9	63	441
8	12	96	768
9	7	63	567
10	2	20	200
11	1	11	121
12	1	12	144
	100	559	3619

FIGURE 2.6-2: Barium histograms

Poisson distribution with $\lambda = 5.59$ is drawn. The relative frequency histogram of the data is shaded. Accordingly, it looks as if the Poisson distribution provides a fairly reasonable probability model for this random variable X. (Later in the text, formal tests of the goodness of fit of various models are given.)   ∎

If events in a Poisson process occur at a mean rate of λ per unit, then the expected number of occurrences in an interval of length t is λt. For example, if phone calls arrive at a switchboard following a Poisson process at a mean rate of three per minute, then the expected number of phone calls in a 5-minute period is $(3)(5) = 15$. Or if calls arrive at a mean rate of 22 in a 5-minute period, then the expected number of calls per minute is $\lambda = 22(1/5) = 4.4$. Moreover, the number of occurrences—say, X—in the interval of length t has the Poisson p.m.f.

$$f(x) = \frac{(\lambda t)^x e^{-\lambda t}}{x!}, \qquad x = 0, 1, 2, \ldots.$$

This equation follows if we treat the interval of length t as if it were the "unit interval" with mean λt instead of λ.

EXAMPLE 2.6-4   A USB flash drive is sometimes used to back up computer files. However, in the past, a less reliable backup system that was used was a computer tape, and flaws occurred on these tapes. In a particular situation, flaws (bad records) on a used computer tape occurred on the average of one flaw per 1200 feet. If one assumes a Poisson distribution, what is the distribution of X, the number of flaws in a 4800-foot roll? The expected number of flaws in $4800 = 4(1200)$ feet is 4; that is, $E(X) = 4$. Thus, the p.m.f. of X is

$$f(x) = \frac{4^x e^{-4}}{x!}, \qquad x = 0, 1, 2, \ldots,$$

and, in particular,

$$P(X = 0) = \frac{4^0 e^{-4}}{0!} = e^{-4} = 0.018,$$
$$P(X \leq 4) = 0.629,$$

by Table III in the appendix.   ∎

EXAMPLE 2.6-5 Telephone calls enter a college switchboard on the average of two every 3 minutes. If one assumes an approximate Poisson process, what is the probability of five or more calls arriving in a 9-minute period? Let X denote the number of calls in a 9-minute period. We see that $E(X) = 6$; that is, on the average, six calls will arrive during a 9-minute period. Thus,

$$P(X \geq 5) = 1 - P(X \leq 4) = 1 - \sum_{x=0}^{4} \frac{6^x e^{-6}}{x!}$$
$$= 1 - 0.285 = 0.715,$$

by Table III in the appendix. ∎

Not only is the Poisson distribution important in its own right, but it can also be used to approximate probabilities for a binomial distribution. Earlier we saw that if X has a Poisson distribution with parameter λ, then, with n large,

$$P(X = x) \approx \binom{n}{x} \left(\frac{\lambda}{n}\right)^x \left(1 - \frac{\lambda}{n}\right)^{n-x},$$

where $p = \lambda/n$, so that $\lambda = np$ in the above binomial probability. That is, if X has the binomial distribution $b(n,p)$ with large n and small p, then

$$\frac{(np)^x e^{-np}}{x!} \approx \binom{n}{x} p^x (1 - p)^{n-x}.$$

This approximation is reasonably good if n is large. But since λ was a fixed constant in that earlier argument, p should be small, because $np = \lambda$. In particular, the approximation is quite accurate if $n \geq 20$ and $p \leq 0.05$ or if $n \geq 100$ and $p \leq 0.10$, but it is not bad in other situations violating these bounds somewhat, such as $n = 50$ and $p = 0.12$.

EXAMPLE 2.6-6 A manufacturer of Christmas tree light bulbs knows that 2% of its bulbs are defective. Assuming independence, we have a binomial distribution with parameters $p = 0.02$ and $n = 100$. To approximate the probability that a box of 100 of these bulbs contains at most three defective bulbs, we use the Poisson distribution with $\lambda = 100(0.02) = 2$, which gives

$$\sum_{x=0}^{3} \frac{2^x e^{-2}}{x!} = 0.857,$$

from Table III in the appendix. Using the binomial distribution, we obtain, after some tedious calculations,

$$\sum_{x=0}^{3} \binom{100}{x} (0.02)^x (0.98)^{100-x} = 0.859.$$

Hence, in this case, the Poisson approximation is extremely close to the true value, but much easier to find. ∎

REMARK With the availability of statistical computer packages and statistical calculators, it is often very easy to find binomial probabilities. So do not use the Poisson approximation if you are able to find the probability exactly. ∎

FIGURE 2.6-3: Binomial (shaded) and Poisson probability histograms

EXAMPLE 2.6-7   In Figure 2.6-3, Poisson probability histograms have been superimposed on shaded binomial probability histograms so that we can see whether or not these are close to each other. If the distribution of X is $b(n, p)$, the approximating Poisson distribution has a mean of $\lambda = np$. Note that the approximation is not good when p is large (e.g., $p = 0.30$). ∎

EXAMPLE 2.6-8   A lot (collection) of 1000 parts is shipped to a company. A sampling plan dictates that $n = 100$ parts are to be taken at random and without replacement and the lot accepted if no more than 2 of these 100 parts are defective. Here Ac = 2 is usually called the **acceptance number**. The operating characteristic curve

$$\mathrm{OC}(p) = P(X \le 2),$$

where p is the fraction defective in the lot, is really the sum of the three hypergeometric probabilities

$$P(X = x) = \frac{\dbinom{N_1}{x}\dbinom{1000 - N_1}{100 - x}}{\dbinom{1000}{100}}, \qquad x = 0, 1, 2,$$

where $N_1 = (1000)(p)$. However, we have seen that the hypergeometric distribution can be approximated by the binomial distribution, which in turn can be approximated by the Poisson distribution, when n is large and p is small. This is exactly our situation, since $n = 100$ and since we are interested in values of p in the range from 0.00 to 0.10. Thus,

$$OC(p) = P(X \le 2) \approx \sum_{x=0}^{2} \frac{(100p)^x e^{-100p}}{x!}.$$

Using Table III in the appendix, we find that some approximate values of the operating characteristic curve are $OC(0.01) = 0.920$, $OC(0.02) = 0.677$, $OC(0.03) = 0.423$, $OC(0.05) = 0.125$, and $OC(0.10) = 0.003$, and this curve is plotted in Figure 2.6-4. If this is not a satisfactory operating characteristic curve for the type of parts involved, the sample size n and the acceptance number Ac must be changed. ■

FIGURE 2.6-4: Operating characteristic curve

EXERCISES

2.6-1. Let X have a Poisson distribution with a mean of 4. Find

(a) $P(2 \le X \le 5)$.

(b) $P(X \ge 3)$.

(c) $P(X \le 3)$.

2.6-2. Let X have a Poisson distribution with a variance of 3. Find $P(X = 2)$.

2.6-3. Customers arrive at a travel agency at a mean rate of 11 per hour. Assuming that the number of arrivals per hour has a Poisson distribution, give the probability that more than 10 customers arrive in a given hour.

2.6-4. If X has a Poisson distribution such that

$$3P(X = 1) = P(X = 2),$$ find $P(X = 4)$.

2.6-5. Flaws in a certain type of drapery material appear on the average of one in 150 square feet. If we assume a Poisson distribution, find the probability of at most one flaw appearing in 225 square feet.

2.6-6. A certain type of aluminum screen that is 2 feet wide has, on the average, one flaw in a 100-foot roll. Find the probability that a 50-foot roll has no flaws.

2.6-7. With probability 0.001, a prize of $499 is won in the Michigan Daily Lottery when a $1 straight bet is placed. Let Y equal the number of $499 prizes won by a gambler after placing n straight bets. Note that Y is $b(n, 0.001)$. After placing $n = 2000$ $1 bets, the gambler is behind if $\{Y \le 4\}$. Use the Poisson distribution to approximate $P(Y \le 4)$ when $n = 2000$.

2.6-8. Suppose that the probability of suffering a side effect from a certain flu vaccine is 0.005. If 1000 persons are inoculated, find the approximate probability that

(a) At most 1 person suffers.

(b) 4, 5, or 6 persons suffer.

2.6-9. A store selling newspapers orders only $n = 4$ of a certain newspaper because the manager does not get many calls for that publication. If the number of requests per day follows a Poisson distribution with mean 3,

  (a) What is the expected value of the number sold?

  (b) How many should the manager order so that the chance of running out is less than 0.05?

2.6-10. The mean of a Poisson random variable X is $\mu = 9$. Compute

$$P(\mu - 2\sigma < X < \mu + 2\sigma).$$

2.6-11. A Geiger counter was set up in the physics laboratory to record the number of alpha particle emissions of carbon-14 in 0.5 second. Following are 50 observations:

4	6	6	12	11	11	10	5	10	7
9	6	9	11	9	6	4	9	11	8
10	11	7	4	5	8	6	5	8	7
6	3	4	6	4	12	7	14	5	9
7	10	9	6	6	12	10	7	12	13

  (a) Calculate the sample mean $\bar{x}$ and sample variance s^2 for these data. Are these two measurements close to each other?

  (b) On the same graph, depict the probability histogram for a Poisson distribution with $\lambda = 8$ (we use 8 only because it is the closest value of λ to $\bar{x}$ in Table III in the appendix), with the relative frequency histogram of the data superimposed.

2.6-12. For determining the half-lives of radioactive isotopes, it is important to know what the background radiation is for a given detector over a certain period. A γ-ray detection experiment over 300 one-second intervals yielded the following data:

0	2	4	6	6	1	7	4	6	1	1	2	3	6	4	2	7	4	4	2
2	5	4	4	4	1	2	4	3	2	2	5	0	3	1	1	0	0	5	2
7	1	3	3	3	2	3	1	4	1	3	5	3	5	1	3	3	0	3	2
6	1	1	4	6	3	6	4	4	2	2	4	3	3	6	1	6	2	5	0
6	3	4	3	1	1	4	6	1	5	1	1	4	1	4	1	1	1	3	3
4	3	3	2	5	2	1	3	5	3	2	7	0	4	2	3	3	5	6	1
4	2	6	4	2	0	4	4	7	3	5	2	2	3	1	3	1	3	6	5
4	8	2	2	4	2	2	1	4	7	5	2	1	1	4	1	4	3	6	2
1	1	2	2	2	2	3	5	4	3	2	2	3	3	2	4	4	3	2	2
3	6	1	1	3	3	2	1	4	5	5	1	2	3	3	1	3	7	2	5
4	2	0	6	2	3	2	3	0	4	4	5	2	5	3	0	4	6	2	2
2	2	2	5	2	2	3	4	2	3	7	1	1	7	1	3	6	0	5	3
0	0	3	3	0	2	4	3	1	2	3	3	3	4	3	2	2	7	5	3
5	1	1	2	2	6	1	3	1	4	4	2	3	4	5	1	3	4	3	1
0	3	7	4	0	5	2	5	4	4	2	2	3	2	4	6	5	5	3	4

Do these look like observations of a Poisson random variable with mean $\lambda = 3$? To help answer this question, do the following:

  (a) Find the frequencies of $0, 1, 2, \ldots, 8$.

  (b) Calculate the sample mean and sample variance. Are they approximately equal to each other?

  (c) Construct a probability histogram with $\lambda = 3$ and a relative frequency histogram on the same graph.

  (d) What is your conclusion? (Later in this book we consider tests of hypotheses, which help give a statistical answer.)

2.6-13. Let X equal the number of telephone calls per hour that are received by 911 between midnight and noon and that are reported in the *Holland Sentinel*. On October 29 and October 30, the following numbers of calls were reported:

Oct. 29: 0 1 1 1 0 1 2 1 4 1 2 3
Oct. 30: 0 3 0 1 0 1 1 2 3 0 2 2

  (a) Calculate the sample mean and sample variance for these data. Are they approximately equal to each other? (See Exercise 2.3-8.)

  (b) Assume that $\lambda = 1.3$, so that Table III in the appendix can be used. Draw a probability histogram for the Poisson distribution and a relative frequency histogram of the data on the same graph.

  (c) On the basis of these limited data, does it look like the Poisson distribution with $\lambda = 1.3$ could be a reasonable probability model?

2.6-14. Let X equal the number of green peanut M&M's in packages of size 22. Forty-five observations of X yielded the following frequencies for the possible outcomes of X:

Outcome (x):	0	1	2	3	4	5	6	7	8	9
Frequency:	0	2	4	5	7	9	8	5	3	2

(a) Use these data to construct a relative frequency histogram and superimpose over it a Poisson probability histogram with a mean of $\lambda = 5$.

(b) Do these data appear to be observations of a Poisson random variable?

2.6-15. A student trainer worked in the training room in the physical education building. The number of students coming in to be taped or for treatment of other injuries was recorded each half hour, yielding the following data:

```
3 3 2 0 4 5 6 4 4 3 2 1 2 3 0
5 5 3 2 3 5 4 1 2 0 3 2 4 2 6
```

(a) Find the sample mean, $\bar{x}$.

(b) Find the sample variance, s^2.

(c) Find the sample standard deviation, s.

(d) Draw a relative frequency histogram of these data with an appropriate Poisson probability histogram superimposed.

2.6-16. A lot (collection) of 10,000 items is accepted if a sample of $n = 400$ has no more than Ac = 3 defective items. Determine the approximate values of OC(0.002), OC(0.004), OC(0.006), OC(0.01), and OC(0.02), and plot the operating characteristic curve.

2.6-17. Consider a lot (collection) of 5000 items. Design a sampling plan (i.e., find n and Ac) such that OC(0.02) = 0.95 and OC(0.08) = 0.10, approximately.

HINT: Start with $n = 50$ and Ac = 2 and modify as necessary, noting that increasing n creates a steeper OC curve.

2.6-18. A baseball team loses $10,000 for each consecutive day it rains. Say X, the number of consecutive days it rains at the beginning of the season, has a Poisson distribution with mean 0.2. What is the expected loss before the opening game?

2.6-19. An airline always overbooks if possible. A particular plane has 95 seats on a flight in which a ticket sells for $300. The airline sells 100 such tickets for this flight.

(a) If the probability of an individual not showing up is 0.05, assuming independence, what is the probability that the airline can accommodate all the passengers who do show up?

(b) If the airline must return the $300 price plus a penalty of $400 to each passenger that cannot get on the flight, what is the expected payout (penalty plus ticket refund) that the airline will pay?

2.6-20. Assume that a policyholder is four times more likely to file exactly two claims as to file exactly three claims. Assume also that the number X of claims of this policyholder is Poisson. Determine the expectation $E(X^2)$.

2.6-21. In the decade of the 1980s, the number X of cases of diphtheria reported each year in the United States followed a Poisson distribution with mean 2.5. What is the probability that

(a) No cases were reported in a given year?

(b) Exactly three cases were reported?

(c) At least three cases were reported?

(d) At most three cases were reported?

2.6-22. For determining half-lives of radioactive isotopes, it is important to know what the background radiation is for a given detector over a certain period. A γ-ray detection experiment over 98 ten-second intervals (observations of a random variable X) yielded the following data:

```
58 50 57 58 64 63 54 64 59 41 43 56 60 50
46 59 54 60 59 60 67 52 65 63 55 61 68 58
63 36 42 54 58 54 40 60 64 56 61 51 48 50
60 42 62 67 58 49 66 58 57 59 52 54 53 53
57 43 73 65 45 43 57 55 73 62 68 55 51 55
53 68 58 53 51 73 44 50 53 62 58 47 63 59
59 56 60 59 50 52 62 51 66 51 56 53 59 57
```

(a) Find the sample mean.

(b) Find the sample variance.

HISTORICAL COMMENTS   The next major items advanced in probability theory were by the Bernoullis, a remarkable Swiss family of mathematicians of the late 1600s to the late 1700s. There were eight mathematicians among them, but we shall mention just three of them: Jacob, Nicolaus II, and Daniel. While writing *Ars Conjectandi*

(*The Art of Conjecture*), Jacob died in 1705, and a nephew, Nicolaus II, edited the work for publication. However, it was Jacob who discovered the important law of large numbers, which is included in our Section 10.5.

Another nephew of Jacob, Daniel, noted in his St. Petersburg paper that "expected values are computed by multiplying each possible gain by the number of ways in which it can occur and then dividing the sum of these products by the total number of cases." His cousin, Nicolaus II, then proposed the so-called St. Petersburg paradox: Peter continues to toss a coin until a head first appears—say, on the xth trial—and he then pays Paul 2^{x-1} units (originally ducats, but for convenience we use dollars). With each additional throw, the number of dollars has doubled. How much should another person pay Paul to take his place in this game? Clearly,

$$E(2^{X-1}) = \sum_{x=1}^{\infty} (2^{x-1})\left(\frac{1}{2^x}\right) = \sum_{x=1}^{\infty} \frac{1}{2} = \infty.$$

However, if we consider this as a practical problem, would someone be willing to give Paul \$1000 to take his place even though there is this unlimited expected value? We doubt it and Daniel doubted it, and it made him think about the utility of money. For example, to most of us, \$3 million is not worth three times \$1 million. To convince you of that, suppose you had exactly \$1 million and a very rich man offers to bet you \$2 million against your \$1 million on the flip of a coin. You will have zero or \$3 million after the flip, so your expected value is

$$(\$0)\left(\frac{1}{2}\right) + (\$3,000,000)\left(\frac{1}{2}\right) = \$1,500,000,$$

much more than your \$1 million. Seemingly, then, this is a great bet and one that Bill Gates might take. However, remember you have \$1 million for certain and you could have zero with probability 1/2. None of us with limited resources should consider taking that bet, because the utility of that extra money to us is not worth the utility of the first \$1 million. Now, each of us has our own utility function. Two dollars is worth twice as much as one dollar for practically all of us. But is \$200,000 worth twice as much as \$100,000? It depends upon your situation; so while the utility function is a straight line for the first several dollars, it still increases, but begins to bend downward someplace as the amount of money increases. This occurs at different spots for all of us. Bob Hogg, one of the authors of this text, would bet \$1000 against \$2000 on a flip of the coin anytime, but probably not \$100,000 against \$200,000, so Hogg's utility function has started to bend downward someplace between \$1000 and \$100,000. Daniel Bernoulli made this observation, and it is extremely useful in all kinds of businesses.

As an illustration, in insurance, most of us know that the premium we pay for all types of insurance is greater than what the company expects to pay us; that is how they make money. Seemingly, insurance is a bad bet, but it really isn't always. It is true that we should self-insure less expensive items—those whose value is on that straight part of the utility function. We have even heard the "rule" that you not insure anything worth less than two month's salary; this is a fairly good guide, but each of us has our own utility function and must make that decision. Hogg can afford losses in the \$5000 to \$10,000 range (not that he likes them, of course), but he does not want to pay losses of \$100,000 or more. So his utility function for negative values of the argument follows that straight line for relatively small negative amounts, but again bends down for large negative amounts. If you insure expensive items, you will discover that the expected utility in absolute value will now exceed the premium. This is why most people insure their life, their home, and their car (particularly on the liability side). They should not, however, insure their golf clubs, eyeglasses, furs, or jewelry (unless the latter two items are extremely valuable).

Continuous Distributions

3.1   CONTINUOUS-TYPE DATA

In Chapter 2 we considered probability distributions of random variables whose space S contains a countable number of outcomes: either a finite number of outcomes or outcomes that can be put into a one-to-one correspondence with the positive integers. Such a random variable is said to be of the **discrete type**, and its distribution of probabilities is of the discrete type.

Many experiments or observations of random phenomena do not have integers as outcomes, but instead are measurements selected from an interval of numbers. For example, you could find the length of time that it takes when waiting in line to buy frozen yogurt. Or the weight of a "1-pound" package of hot dogs could be any number between 0.94 pound and 1.25 pounds. The weight of a miniature Baby Ruth candy bar could be any number between 20 and 27 grams. Even though such times and weights could be selected from an interval of values, times and weights are generally rounded off so that the data often look like discrete data. If, conceptually, the measurements could come from an interval of possible outcomes, we call them data from a continuous-type population or, more simply, **continuous-type data**.

Given a set of continuous-type data, we shall group the data into classes and then construct a histogram of the grouped data. This will help us better visualize the data. The following guidelines and terminology will be used to group continuous-type data into classes of equal length (these guidelines can also be used for sets of discrete data that have a large range):

1. Determine the largest (maximum) and smallest (minimum) observations. The **range** is the difference R = maximum − minimum.

2. In general, select from $k = 5$ to $k = 20$ classes, which are usually nonoverlapping intervals of equal length. These classes should cover the interval from the minimum to the maximum.

3. Each interval begins and ends halfway between two possible values of the measurements, which have been rounded off to a given number of decimal places.

4. The first interval should begin about as much below the smallest value as the last interval ends above the largest.

5. The intervals are called **class intervals** and the boundaries are called **class boundaries** or **cutpoints**. We shall denote these k class intervals by

$$(c_0, c_1), (c_1, c_2), \ldots, (c_{k-1}, c_k).$$

6. The **class limits** are the smallest and the largest possible observed (recorded) values in a class.

7. The **class mark** is the midpoint of a class.

A frequency table is constructed that lists the class intervals, the class limits, a tabulation of the measurements in the various classes, the frequency f_i of each class, and the class marks. A column is sometimes used to construct a relative frequency (density) histogram. With class intervals of equal length, a frequency histogram is constructed by drawing, for each class, a rectangle having as its base the class interval and a height equal to the frequency of the class. For the relative frequency histogram, each rectangle has an **area** equal to the relative frequency f_i/n of the observations for the class. That is, the function defined by

$$h(x) = \frac{f_i}{(n)(c_i - c_{i-1})}, \qquad \text{for } c_{i-1} < x \le c_i, \quad i = 1, 2, \ldots, k,$$

is called a **relative frequency histogram** or **density histogram**, where f_i is the frequency of the ith class and n is the total number of observations. Clearly, if the class intervals are of equal length, the relative frequency histogram, $h(x)$, is proportional to the **frequency histogram** f_i, for $c_{i-1} < x \le c_i, i = 1, 2, \ldots, k$. The frequency histogram should be used only in those situations in which the class intervals are of equal length.

EXAMPLE 3.1-1   The weights in grams of 40 miniature Baby Ruth candy bars, with the weights ordered, are given in Table 3.1-1.

TABLE 3.1-1: Candy Bar Weights

20.5	20.7	20.8	21.0	21.0	21.4	21.5	22.0	22.1	22.5
22.6	22.6	22.7	22.7	22.9	22.9	23.1	23.3	23.4	23.5
23.6	23.6	23.6	23.9	24.1	24.3	24.5	24.5	24.8	24.8
24.9	24.9	25.1	25.1	25.2	25.6	25.8	25.9	26.1	26.7

We shall group these data and then construct a histogram to visualize the distribution of weights. The range of the data is $R = 26.7 - 20.5 = 6.2$. The interval $(20.5, 26.7)$ could be covered with $k = 8$ classes of width 0.8 or with $k = 9$ classes of width 0.7. (There are other possibilities.) We shall use $k = 7$ classes of width 0.9. The first class interval will be $(20.45, 21.35)$ and the last class interval will be $(25.85, 26.75)$. The data are grouped in Table 3.1-2.

		TABLE 3.1-2: Frequency Table of Candy Bar Weights				
Class Interval	Class Limits	Tabulation	Frequency (f_i)	$h(x)$	Class Marks
$(20.45, 21.35)$	20.5–21.3	⊮	5	5/36	20.9
$(21.35, 22.25)$	21.4–22.2	||||	4	4/36	21.8
$(22.25, 23.15)$	22.3–23.1	⊮ |||	8	8/36	22.7
$(23.15, 24.05)$	23.2–24.0	⊮ ||	7	7/36	23.6
$(24.05, 24.95)$	24.1–24.9	⊮ |||	8	8/36	24.5
$(24.95, 25.85)$	25.0–25.8	⊮	5	5/36	25.4
$(25.85, 26.75)$	25.9–26.7	|||	3	3/36	26.3

FIGURE 3.1-1: Relative frequency histogram of weights of candy bars

A relative frequency histogram of these data is given in Figure 3.1-1. Note that the total area of this histogram is equal to 1. We could also construct a frequency histogram in which the heights of the rectangles would be equal to the frequencies of the classes. The shape of the two histograms is the same. Later we will see the reason for preferring the relative frequency histogram. In particular, we will be superimposing on the relative frequency histogram the graph of a continuous function, called a probability density function, that corresponds to a probability model. ■

Given a set of measurements, the sample mean is the center of the data such that the deviations from that center sum to zero; that is, $\sum_{i=1}^{n}(x_i - \overline{x}) = 0$. The sample standard deviation gives a measure of how spread out the data are from the sample mean. If the histogram is "mound shaped" or "bell shaped," the following empirical rule gives rough approximations to the percentages of the data that fall between certain points.

Empirical Rule: Let $x_1, x_2, \ldots, x_n$ have a sample mean $\overline{x}$ and sample standard deviation s. If the histogram of these data is "bell shaped," then

- approximately 68% of the data are in the interval $(\overline{x} - s, \overline{x} + s)$.
- approximately 95% of the data are in the interval $(\overline{x} - 2s, \overline{x} + 2s)$.
- approximately 99.7% of the data are in the interval $(\overline{x} - 3s, \overline{x} + 3s)$.

For the data in Example 3.1-1, the sample mean is $\bar{x} = 23.505$ and the standard deviation is $s = 1.641$. The number of weights that fall within one standard deviation of the mean, $(23.505 - 1.641, 23.505 + 1.641)$, is 27, or 67.5%. For these particular weights, 100% fall within two standard deviations of $\bar{x}$. Thus, the histogram is missing part of the "bell" in the tails in order for the empirical rule to hold. The theory needed to explain why this approximation does or does not work is considered in Section 3.6.

When you draw a histogram, it is useful to indicate the location of $\bar{x}$, as well as that of the points $\bar{x} \pm s$ and $\bar{x} \pm 2s$.

There is a refinement of the relative frequency histogram that can be made when the class intervals are of equal length. The **relative frequency polygon** smooths out the corresponding histogram somewhat. To form such a polygon, mark the midpoints at the top of each "bar" of the histogram. Connect adjacent midpoints with straight-line segments. On each of the two end bars, draw a line segment from the top middle mark through the middle point of the outer vertical line of the bar. Of course, if the area underneath the tops of the relative frequency histogram is equal to 1, which it should be, then the area underneath the relative frequency polygon is also equal to 1, because the areas lost and gained cancel out by a consideration of congruent triangles. This idea is made clear in the next example.

EXAMPLE 3.1-2 A manufacturer of fluoride toothpaste regularly measures the concentration of fluoride in the toothpaste to make sure that it is within the specification of 0.85 to 1.10 mg/g. Table 3.1-3 lists 100 such measurements.

The minimum of these measurements is 0.85 and the maximum is 1.06. The range is $1.06 - 0.85 = 0.21$. We shall use $k = 8$ classes of length 0.03. Note that $8(0.03) = 0.24 > 0.21$. We start at 0.835 and end at 1.075. These boundaries are the same distance below the minimum and above the maximum. In Table 3.1-4, we also give the values of the heights of each rectangle in the relative frequency histogram, so that the total area of the histogram is 1. These heights are given by the formula

$$h(x) = \frac{f_i}{(0.03)(100)} = \frac{f_i}{3}.$$

The plots of the relative frequency histogram and polygon are given in Figure 3.1-2.

If you are using a computer program to analyze a set of data, it is very easy to find the sample mean, the sample variance, and the sample standard deviation. However, if you have only grouped data or if you are not using a computer, you can obtain close approximations of these values by computing the mean $\bar{u}$ and variance s_u^2 of the

TABLE 3.1-3: Concentrations of Fluoride in mg/g in Toothpaste

0.98	0.92	0.89	0.90	0.94	0.99	0.86	0.85	1.06	1.01
1.03	0.85	0.95	0.90	1.03	0.87	1.02	0.88	0.92	0.88
0.88	0.90	0.98	0.96	0.98	0.93	0.98	0.92	1.00	0.95
0.88	0.90	1.01	0.98	0.85	0.91	0.95	1.01	0.88	0.89
0.99	0.95	0.90	0.88	0.92	0.89	0.90	0.95	0.93	0.96
0.93	0.91	0.92	0.86	0.87	0.91	0.89	0.93	0.93	0.95
0.92	0.88	0.87	0.98	0.98	0.91	0.93	1.00	0.90	0.93
0.89	0.97	0.98	0.91	0.88	0.89	1.00	0.93	0.92	0.97
0.97	0.91	0.85	0.92	0.87	0.86	0.91	0.92	0.95	0.97
0.88	1.05	0.91	0.89	0.92	0.94	0.90	1.00	0.90	0.93

TABLE 3.1-4: Frequency Table of Fluoride Concentrations				
Class Interval	Class Mark (u_i)	Tabulation	Frequency (f_i)	$h(x) = f_i/3$
$(0.835, 0.865)$	0.85	卌 ||	7	7/3
$(0.865, 0.895)$	0.88	卌 卌 卌 卌	20	20/3
$(0.895, 0.925)$	0.91	卌 卌 卌 卌 卌 ||	27	27/3
$(0.925, 0.955)$	0.94	卌 卌 卌 |||	18	18/3
$(0.955, 0.985)$	0.97	卌 卌 ||||	14	14/3
$(0.985, 1.015)$	1.00	卌 ||||	9	9/3
$(1.015, 1.045)$	1.03	|||	3	3/3
$(1.045, 1.075)$	1.06	||	2	2/3

grouped data, using the class marks weighted with their respective frequencies. We have

$$\bar{u} = \frac{1}{n} \sum_{i=1}^{k} f_i u_i$$

$$= \frac{1}{100} \sum_{i=1}^{8} f_i u_i = \frac{92.83}{100} - 0.9283,$$

$$s_u^2 = \frac{1}{n-1} \sum_{i=1}^{k} f_i(u_i - \bar{u})^2 = \frac{\sum_{i=1}^{k} f_i u_i^2 - \frac{1}{n}\left(\sum_{i=1}^{k} f_i u_i\right)^2}{n-1}$$

$$- \frac{0.237411}{99} - 0.002398.$$

Thus,

$$s_u = \sqrt{0.002398} = 0.04897.$$

These results compare rather favorably with $\bar{x} = 0.9293$ and $s_x = 0.04895$ of the original data. ∎

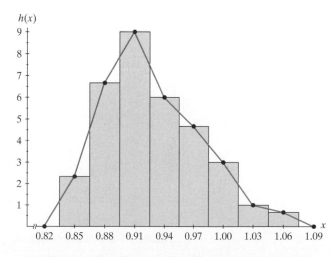

FIGURE 3.1-2: Concentrations of fluoride in toothpaste

In some situations, it is not necessarily desirable to use class intervals of equal widths in the construction of the frequency distribution and histogram. This is particularly true if the data are skewed with a very long tail. We now present an illustration in which it seems desirable to use class intervals of unequal lengths; thus, we cannot use the relative frequency polygon.

EXAMPLE 3.1-3 The following 40 losses, due to wind-related catastrophes, were recorded to the nearest $1 million (these data include only losses of $2 million or more; for convenience, they have been ordered and recorded in millions):

$$
\begin{array}{cccccccccc}
2 & 2 & 2 & 2 & 2 & 2 & 2 & 2 & 2 & 2 \\
2 & 2 & 3 & 3 & 3 & 3 & 4 & 4 & 4 & 5 \\
5 & 5 & 5 & 6 & 6 & 6 & 6 & 8 & 8 & 9 \\
15 & 17 & 22 & 23 & 24 & 24 & 25 & 27 & 32 & 43
\end{array}
$$

The selection of class boundaries is more subjective in this case. It makes sense to let $c_0 = 1.5$ and $c_1 = 2.5$ because only values of $2 million or more are recorded and there are 12 observations equal to 2. We could then let $c_2 = 6.5$, $c_3 = 29.5$, and $c_4 = 49.5$, yielding the following relative frequency histogram:

$$
h(x) = \begin{cases}
\dfrac{12}{40}, & 1.5 < x \le 2.5, \\[2mm]
\dfrac{15}{(40)(4)}, & 2.5 < x \le 6.5, \\[2mm]
\dfrac{11}{(40)(23)}, & 6.5 < x \le 29.5, \\[2mm]
\dfrac{2}{(40)(20)}, & 29.5 < x \le 49.5.
\end{cases}
$$

This histogram is displayed in Figure 3.1-3. It takes some experience before a person can display a relative frequency histogram that is most meaningful.

FIGURE 3.1-3: Relative frequency histogram of losses

The areas of the four rectangles—0.300, 0.375, 0.275, and 0.050—are the respective relative frequencies. It is important to note in the case of unequal lengths among class intervals that the *areas*, not the heights, of the rectangles are proportional to the frequencies. In particular, the first and second classes have frequencies $f_1 = 12$ and $f_2 = 15$, yet the height of the first is greater than the height of the second, while here $f_1 < f_2$. If we have equal lengths among the class intervals, then the heights are proportional to the frequencies. ∎

For continuous-type data, the interval with the largest class height is called the **modal class** and the respective class mark is called the **mode**. Hence, in the last example, $x = 2$ is the mode and $(1.5, 2.5)$ the modal class.

EXERCISES

3.1-1. One characteristic of a car's storage console that is checked by the manufacturer is the time in seconds that it takes for the lower storage compartment door to open completely. A random sample of size $n = 5$ yielded the following times:

$$1.1 \quad 0.9 \quad 1.4 \quad 1.1 \quad 1.0$$

(a) Find the sample mean, $\bar{x}$.
(b) Find the sample variance, s^2.
(c) Find the sample standard deviation, s.

3.1-2. A leakage test was conducted to determine the effectiveness of a seal designed to keep the inside of a plug airtight. An air needle was inserted into the plug, which was then placed underwater. Next, the pressure was increased until leakage was observed. The magnitude of this pressure in psi was recorded for 10 trials:

$$3.1 \quad 3.5 \quad 3.3 \quad 3.7 \quad 4.5 \quad 4.2 \quad 2.8 \quad 3.9 \quad 3.5 \quad 3.3$$

Find the sample mean and sample standard deviation for these 10 measurements.

3.1-3. Before Halloween two bags of miniature Clark bars were purchased and each bar was weighed (in grams). There were 19 bars in the first bag and 23 in the second bag, yielding the following weights:

19.8	20.3	20.2	19.6	20.0	19.4	15.4	19.9	20.5	20.5	19.8
15.5	21.8	20.0	15.4	20.6	15.7	20.4	21.3	16.7	15.6	14.2
15.9	16.8	15.2	15.0	16.0	16.4	15.4	16.0	16.1	17.0	16.0
15.1	14.1	15.8	15.7	16.6	15.2	15.5	14.9	15.3		

(a) Using 13.95 14.95, 14.95 15.95, and so on as class boundaries, group these weights and construct a histogram.
(b) Calculate the values of the sample mean and sample standard deviation.
(c) Locate $\bar{x}$, $\bar{x} \pm s_x$, $\bar{x} \pm 2s_x$ on your histogram.
(d) Give an interpretation of your output.

3.1-4. When a customer purchases a laptop computer, one characteristic of interest is the weight of the laptop. Here are the weights of 100 laptop computers:

9.38	5.20	5.38	6.80	7.97	2.20	6.80	7.10	4.60	6.51
6.50	5.33	8.10	4.90	7.07	6.17	7.70	6.28	6.60	3.90
5.11	5.40	4.90	5.20	2.69	7.97	6.20	5.33	7.97	5.58
5.20	5.21	5.33	5.20	5.90	6.50	5.73	4.60	5.60	5.54
5.38	6.50	5.20	3.13	5.12	4.90	6.27	6.50	7.85	5.20
4.77	4.85	5.20	7.90	9.00	5.20	7.90	6.06	5.73	6.50
5.50	7.85	5.60	3.90	3.90	6.20	4.90	3.57	3.90	4.20
8.20	7.90	5.33	4.77	3.90	4.60	3.50	5.20	9.40	4.30
7.60	8.60	9.38	6.00	3.72	2.69	6.60	6.10	9.50	4.90
3.50	10.10	5.33	6.40	6.50	5.96	4.00	7.90	2.69	4.40

(a) Calculate the sample mean and sample standard deviation.
(b) Use the numbers 1.995, 2.995, ..., 10.995 as cutpoints to construct a histogram of the weights.

3.1-5. During the course of an internship at a company that manufactures diesel engine fuel injector pumps, a student had to measure the category of the plungers that force the fuel out of the pumps. This category is based on a relative scale, measuring the difference in diameter (in microns or micrometers) of a plunger from that of an absolute minimum acceptable diameter. For 96 plungers randomly taken from the production line, the data are as follows:

17.1	19.3	18.0	19.4	16.5	14.4	15.8	16.6	18.5	14.9
14.8	16.3	20.8	17.8	14.8	15.6	16.7	16.1	17.1	16.5
18.8	19.3	18.1	16.1	18.0	17.2	16.8	17.3	14.4	14.1
16.9	17.6	15.5	17.8	17.2	17.4	18.1	18.4	17.8	16.7
17.2	13.7	18.0	15.6	17.8	17.0	17.7	11.9	15.9	17.8
15.5	14.6	15.6	15.1	15.4	16.1	16.6	17.1	19.1	15.0
17.6	19.7	17.1	13.6	15.6	16.3	14.8	17.4	14.8	14.9
14.1	17.8	19.8	18.9	15.6	16.1	15.9	15.7	22.1	16.1
18.9	21.5	17.4	12.3	20.2	14.9	17.1	15.0	14.4	14.7
15.9	19.0	16.6	15.3	17.7	15.8				

(a) Calculate the sample mean and the sample standard deviation of these measurements.

(b) Use the cutpoints $10.95, 11.95, \ldots, 22.95$ to construct a histogram of the data.

3.1-6. Decks have become very popular items for American homeowners. Lumberyards sell many styles of decks during the summer months, but they all use the same type of nail. The question that always arises is how many nails to buy. If 800 nails are needed, how many grams of nails should be purchased? The following weights (in grams) of 50 nails should help to answer the question:

9.42	8.69	8.93	8.27	8.82	8.66	8.90	8.31	9.15	9.63
9.41	8.56	8.82	8.58	8.43	8.05	8.56	8.55	8.88	8.73
8.29	8.79	8.51	8.85	9.34	9.21	8.38	8.51	8.41	8.98
8.58	9.21	8.27	8.76	9.26	8.59	8.36	8.71	8.51	8.88
9.20	8.24	8.57	8.85	8.69	8.85	9.08	9.40	9.25	8.79

(a) Using $(7.995, 8.245)$, $(8.245, 8.495)$, and so on as class intervals, group these data into seven classes.

(b) Construct a histogram of the data.

(c) Calculate the sample mean and sample standard deviation.

(d) How many grams of nails would you recommend?

3.1-7. In the casino game roulette, if a player bets $1 on red, the probability of winning $1 is 18/38 and the probability of losing $1 is 20/38. Let X equal the number of successive $1 bets that a player makes before losing $5. One hundred observations of X were simulated on a computer, yielding the following data:

23	127	877	65	101	45	61	95	21	43
53	49	89	9	75	93	71	39	25	91
15	131	63	63	41	7	37	13	19	413
65	43	35	23	135	703	83	7	17	65
49	177	61	21	9	27	507	7	5	87
13	213	85	83	75	95	247	1815	7	13
71	67	19	615	11	15	7	131	47	25
25	5	471	11	5	13	75	19	307	33
57	65	9	57	35	19	9	33	11	51
27	9	19	63	109	515	443	11	63	9

(a) Find the sample mean and sample standard deviation of these data.

(b) Construct a relative frequency histogram of the data, using about 10 classes. The classes do not need to be of the same length.

(c) Locate $\bar{x}$, $\bar{x} \pm s$, $\bar{x} \pm 2s$, $\bar{x} \pm 3s$ on your histogram .

(d) In your opinion, does the median or sample mean give a better measure of the center of these data?

3.1-8. Ledolter and Hogg (see references) report that a manufacturer of metal alloys is concerned about customer complaints regarding the lack of uniformity in the melting points of one of the firm's alloy filaments. Fifty filaments are selected and their melting points determined. The following results were obtained:

320	326	325	318	322	320	329	317	316	331
320	320	317	329	316	308	321	319	322	335
318	313	327	314	329	323	327	323	324	314
308	305	328	330	322	310	324	314	312	318
313	320	324	311	317	325	328	319	310	324

(a) Construct a frequency distribution, and display the histogram, of the data.

(b) Calculate the sample mean and sample standard deviation.

(c) Locate $\bar{x}$, $\bar{x} \pm s$ on your histogram. How many observations lie within one standard deviation of the mean? How many lie within two standard deviations of the mean?

3.1-9. Ledolter and Hogg (see references) report 64 observations that are a sample of daily weekday afternoon (3 to 7 P.M.) lead concentrations (in micrograms per cubic meter, $\mu g/m^3$). The following data were recorded at an air-monitoring station near the San Diego Freeway in Los Angeles during the fall of 1976:

6.7	5.4	5.2	6.0	8.7	6.0	6.4	8.3	5.3	5.9	7.6
5.0	6.9	6.8	4.9	6.3	5.0	6.0	7.2	8.0	8.1	7.2
10.9	9.2	8.6	6.2	6.1	6.5	7.8	6.2	8.5	6.4	8.1
2.1	6.1	6.5	7.9	14.1	9.5	10.6	8.4	8.3	5.9	6.0
6.4	3.9	9.9	7.6	6.8	8.6	8.5	11.2	7.0	7.1	6.0
9.0	10.1	8.0	6.8	7.3	9.7	9.3	3.2	6.4		

(a) Construct a frequency distribution of the data and display the results in the form of a histogram. Is this distribution symmetric?

(b) Calculate the sample mean and sample standard deviation.

(c) Locate $\bar{x}$, $\bar{x} \pm s$ on your histogram. How many observations lie within one standard deviation of the mean? How many lie within two standard deviations of the mean?

3.1-10. An insurance company experienced the following mobile home losses in 10,000's of dollars for 50 catastrophic events:

1	2	2	3	3	4	4	5	5	5
5	6	7	7	9	9	9	10	11	12
22	24	28	29	31	33	36	38	38	38
39	41	48	49	53	55	74	82	117	134
192	207	224	225	236	280	301	308	351	527

(a) Using class boundaries 0.5, 5.5, 17.5, 38.5, 163.5, and 549.5, group these data into five classes.

(b) Construct a relative frequency histogram of the data.

(c) Describe the distribution of losses.

3.1-11. Old Faithful is a geyser in Yellowstone National Park. Tourists always want to know when the next eruption will occur, so data have been collected to help make those predictions. In the data set on the next page, observations were made on several consecutive days, and the data recorded give the starting time of the eruption (STE); the duration of the eruption, in seconds (DIS); the predicted time until the next eruption, in minutes (PTM); the actual time until the next eruption, in minutes (ATM); and the duration of the eruption, in minutes (DIM).

(a) Construct a histogram of the durations of the eruptions, in seconds. Use 10 to 12 classes.

(b) Calculate the sample mean and locate it on your histogram. Does it give a good measure of the average length of an eruption? Why or why not?

(c) Construct a histogram of the lengths of the times between eruptions. Use 10 to 12 classes.

(d) Calculate the sample mean and locate it on your histogram. Does it give a good measure of the average length of the times between eruptions?

STE	DIS	PTM	ATM	DIM	STE	DIS	PTM	ATM	DIM
706	150	65	72	2.500	1411	110	55	65	1.833
818	268	89	88	4.467	616	289	89	97	4.817
946	140	65	62	2.333	753	114	58	52	1.900
1048	300	95	87	5.000	845	271	89	94	4.517
1215	101	55	57	1.683	1019	120	58	60	2.000
1312	270	89	94	4.500	1119	279	89	84	4.650
651	270	89	91	4.500	1253	109	55	63	1.817
822	125	59	51	2.083	1356	295	95	91	4.917
913	262	89	98	4.367	608	240	85	83	4.000
1051	95	55	59	1.583	731	259	86	84	4.317
1150	270	89	93	4.500	855	128	60	71	2.133
637	273	89	86	4.550	1006	287	92	83	4.783
803	104	55	70	1.733	1129	253	65	70	4.217
913	129	62	63	2.150	1239	284	89	81	4.733
1016	264	89	91	4.400	608	120	58	60	2.000
1147	239	82	82	3.983	708	283	92	91	4.717
1309	106	55	58	1.767	839	115	58	51	1.917
716	259	85	97	4.317	930	254	85	85	4.233
853	115	55	59	1.917	1055	94	55	55	1.567
952	275	89	90	4.583	1150	274	89	98	4.567
1122	110	55	58	1.833	1328	128	64	49	2.133
1220	286	92	98	4.767	557	270	93	85	4.500
735	115	55	55	1.917	722	103	58	65	1.717
830	266	89	107	4.433	827	287	89	102	4.783
1017	105	55	61	1.750	1009	111	55	56	1.850
1118	275	89	82	4.583	1105	275	89	86	4.583
1240	226	79	91	3.767	1231	104	55	62	1.733

3.1-12. A small part for an automobile rearview mirror was produced on two different punch presses. In order to describe the distribution of the weights of those parts, a random sample was selected, and each piece was weighed in grams, resulting in the following data set:

3.968	3.534	4.032	3.912	3.572	4.014	3.682	3.608
3.669	3.705	4.023	3.588	3.945	3.871	3.744	3.711
3.645	3.977	3.888	3.948	3.551	3.796	3.657	3.667
3.799	4.010	3.704	3.642	3.681	3.554	4.025	4.079
3.621	3.575	3.714	4.017	4.082	3.660	3.692	3.905
3.977	3.961	3.948	3.994	3.958	3.860	3.965	3.592
3.681	3.861	3.662	3.995	4.010	3.999	3.993	4.004
3.700	4.008	3.627	3.970	3.647	3.847	3.628	3.646
3.674	3.601	4.029	3.603	3.619	4.009	4.015	3.615
3.672	3.898	3.959	3.607	3.707	3.978	3.656	4.027
3.645	3.643	3.898	3.635	3.865	3.631	3.929	3.635
3.511	3.539	3.830	3.925	3.971	3.646	3.669	3.931
4.028	3.665	3.681	3.984	3.664	3.893	3.606	3.699
3.997	3.936	3.976	3.627	3.536	3.695	3.981	3.587
3.680	3.888	3.921	3.953	3.847	3.645	4.042	3.692
3.910	3.672	3.957	3.961	3.950	3.904	3.928	3.984
3.721	3.927	3.621	4.038	4.047	3.627	3.774	3.983
3.658	4.034	3.778					

(a) Using about 10 (say, 8 to 12) classes, construct a frequency distribution of the data.

(b) Draw a histogram of the data.

(c) Describe the shape of the distribution represented by the histogram.

3.2   EXPLORATORY DATA ANALYSIS

To explore the other characteristics of an unknown distribution, we need to take a sample of n observations, $x_1, x_2, \ldots, x_n$, from that distribution and often need to order them from the smallest to the largest. One convenient way of doing this is to use a stem-and-leaf display, a method that was started by John W. Tukey. (For more details, see the books by Tukey (1977) and Velleman and Hoaglin (1981).)

Possibly the easiest way to begin is with an example to which all of us can relate. Say we have the following 50 test scores on a statistics examination:

93	77	67	72	52	83	66	84	59	63
75	97	84	73	81	42	61	51	91	87
34	54	71	47	79	70	65	57	90	83
58	69	82	76	71	60	38	81	74	69
68	76	85	58	45	73	75	42	93	65

We can do much the same thing as a frequency table and histogram can, but keep the original values, through a **stem-and-leaf display**. For this particular data set, we could use the following procedure: The first number in the set, 93, is recorded by treating the 9 (in the tens place) as the stem and the 3 (in the units place) as the corresponding leaf. Note that this leaf of 3 is the first digit after the stem of 9 in Table 3.2-1. The second number, 77, is then given by the leaf of 7 after the stem of 7; the third number, 67, by the leaf of 7 after the stem of 6; the fourth number, 72, as the leaf of 2 after the stem of 7 (note that this is the second leaf on the 7 stem); and so on. Table 3.2-1 is an example of a stem-and-leaf display. If the leaves are carefully aligned vertically, this table has the same effect as a histogram, but the original numbers are not lost.

TABLE 3.2-1: Stem-and-Leaf Display of Scores from 50 Statistics Examinations

Stems	Leaves	Frequency	Depths
3	4 8	2	2
4	2 7 5 2	4	6
5	2 9 1 4 7 8 8	7	13
6	7 6 3 1 5 9 0 9 8 5	10	23
7	7 2 5 3 1 9 0 6 1 4 6 3 5	13	(13)
8	3 4 4 1 7 3 2 1 5	9	14
9	3 7 1 0 3	5	5

An additional new term in these stem-and-leaf displays is **depths**. In the "depths" column, the frequencies are added together (accumulated) from the low end and the high end until the row is reached that contains the middle value in the ordered display. Later we call that middle value the median of the sample, and the frequency of that row is simply placed in parentheses. That is, the frequency of the row containing this median is not included in the sum from either the low end or the high end.

It is useful to modify the stem-and-leaf display by ordering the leaves in each row from smallest to largest. The resulting stem-and-leaf diagram is called an **ordered stem-and-leaf display**. Table 3.2-2 uses the data from Table 3.2-1 to produce an ordered stem-and-leaf display.

There is another modification that can also be helpful. Suppose that we want two rows of leaves with each original stem. We can do this by recording leaves 0, 1, 2, 3, and 4 with a stem adjoined with an asterisk (∗) and leaves 5, 6, 7, 8, and 9 with a stem

TABLE 3.2-2: Ordered Stem-and-Leaf Display of Statistics Examinations

Stems	Leaves	Frequency	Depths
3	4 8	2	2
4	2 2 5 7	4	6
5	1 2 4 7 8 8 9	7	13
6	0 1 3 5 5 6 7 8 9 9	10	23
7	0 1 1 2 3 3 4 5 5 6 6 7 9	13	(13)
8	1 1 2 3 3 4 4 5 7	9	14
9	0 1 3 3 7	5	5

adjoined with a dot (•). Of course, in our example, by going from 7 original classes to 14 classes, we lose a certain amount of smoothness with this particular data set, as illustrated in Table 3.2-3, which is also ordered.

TABLE 3.2-3: Ordered Stem-and-Leaf Display of Statistics Examinations

Stems	Leaves	Frequency	Depths
3∗	4	1	1
3•	8	1	2
4∗	2 2	2	4
4•	5 7	2	6
5∗	1 2 4	3	9
5•	7 8 8 9	4	13
6∗	0 1 3	3	16
6•	5 5 6 7 8 9 9	7	23
7∗	0 1 1 2 3 3 4	7	(7)
7•	5 5 6 6 7 9	6	20
8∗	1 1 2 3 3 4 4	7	14
8•	5 7	2	7
9∗	0 1 3 3	4	5
9•	7	1	1

Tukey suggested another modification, which is used in the next example.

EXAMPLE 3.2-1   The following numbers represent ACT composite scores for 60 entering freshmen at a certain college:

26	19	22	28	31	29	25	23	20	33	23	26
30	27	26	29	20	23	18	24	29	27	32	24
25	26	22	29	21	24	20	28	23	26	30	19
27	21	32	28	29	23	25	21	28	22	25	24
19	24	35	26	25	20	31	27	23	26	30	29

An ordered stem-and-leaf display of these scores is given in Table 3.2-4, where leaves are recorded as zeros and ones with a stem adjoined with an asterisk (∗), twos and threes with a stem adjoined with t, fours and fives with a stem adjoined with f, sixes and sevens with a stem adjoined with s, and eights and nines with a stem adjoined with a dot (•). ■

TABLE 3.2-4: Ordered Stem-and-Leaf Display of 60 ACT Scores

Stems	Leaves	Frequency	Depths
1•	8 9 9 9	4	4
2*	0 0 0 0 1 1 1	7	11
2t	2 2 2 3 3 3 3 3 3	9	20
2f	4 4 4 4 4 5 5 5 5 5	10	30
2s	6 6 6 6 6 6 7 7 7 7	11	30
2•	8 8 8 8 9 9 9 9 9 9	10	19
3*	0 0 0 1 1	5	9
3t	2 2 3	3	4
3f	5	1	1

There is a reason for constructing ordered stem-and-leaf diagrams. For a sample of n observations, $x_1, x_2, \ldots, x_n$, when the observations are ordered from small to large, the resulting ordered data are called the **order statistics** of the sample. Statisticians have found that order statistics and certain of their functions are extremely valuable. It is very easy to determine the values of the sample, in order, from an ordered stem-and-leaf display. As an illustration, consider the values in Table 3.2-2 or Table 3.2-3. The order statistics of the 50 test scores are given in Table 3.2-5.

TABLE 3.2-5: Order Statistics of 50 Exam Scores

34	38	42	42	45	47	51	52	54	57
58	58	59	60	61	63	65	65	66	67
68	69	69	70	71	71	72	73	73	74
75	75	76	76	77	79	81	81	82	83
83	84	84	85	87	90	91	93	93	97

Sometimes we give ranks to these order statistics and use the rank as the subscript on y. The first order statistic $y_1 = 34$ has rank 1; the second order statistic $y_2 = 38$ has rank 2; the third order statistic $y_3 = 42$ has rank 3; the fourth order statistic $y_4 = 42$ has rank 4, ... ; and the 50th order statistic $y_{50} = 97$ has rank 50. It is also about as easy to determine these values from the ordered stem-and-leaf display. We see that $y_1 \leq y_2 \leq \cdots \leq y_{50}$.

From either these order statistics or the corresponding ordered stem-and-leaf display, it is rather easy to find the **sample percentiles**. If $0 < p < 1$, then the $(100p)$th sample percentile has *approximately np* sample observations less than it and also $n(1 - p)$ sample observations greater than it. One way of achieving this is to take the $(100p)$th sample percentile as the $(n + 1)p$th order statistic, provided that $(n + 1)p$ is an integer. If $(n + 1)p$ is not an integer but is equal to r plus some proper fraction—say, a/b—use a weighted average of the rth and the $(r + 1)$st order statistics. That is, define the $(100p)$th sample percentile as

$$\tilde{\pi}_p = y_r + (a/b)(y_{r+1} - y_r) = (1 - a/b)y_r + (a/b)y_{r+1}.$$

This formula is simply a linear interpolation between y_r and y_{r+1}. (If $p < 1/(n + 1)$ or $p > n/(n + 1)$, that sample percentile is not defined.)

As an illustration, consider the 50 ordered test scores. With $p = 1/2$, we find the 50th percentile by averaging the 25th and 26th order statistics, since

$(n + 1)p = (51)(1/2) = 25.5$. Thus, the 50th percentile is

$$\tilde{\pi}_{0.50} = (1/2)y_{25} + (1/2)y_{26} = (71 + 71)/2 = 71.$$

With $p = 1/4$, we have $(n + 1)p = (51)(1/4) = 12.75$, and the 25th sample percentile is then

$$\tilde{\pi}_{0.25} = (1 - 0.75)y_{12} + (0.75)y_{13} = (0.25)(58) + (0.75)(59) = 58.75.$$

With $p = 3/4$, so that $(n + 1)p = (51)(3/4) = 38.25$, the 75th sample percentile is

$$\tilde{\pi}_{0.75} = (1 - 0.25)y_{38} + (0.25)y_{39} = (0.75)(81) + (0.25)(82) = 81.25.$$

Note that *approximately* 50%, 25%, and 75% of the sample observations are less than 71, 58.75, and 81.25, respectively.

Special names are given to certain percentiles. The 50th percentile is the **median** of the sample. The 25th, 50th, and 75th percentiles are, respectively, the **first, second**, and **third quartiles** of the sample. For notation, we let $\tilde{q}_1 = \tilde{\pi}_{0.25}$, $\tilde{q}_2 = \tilde{m} = \tilde{\pi}_{0.50}$, and $\tilde{q}_3 = \tilde{\pi}_{0.75}$. The 10th, 20th, ..., and 90th percentiles are the **deciles** of the sample, so note that the 50th percentile is also the median, the second quartile, and the fifth decile. With the set of 50 test scores, since $(51)(2/10) = 10.2$ and $(51)(9/10) = 45.9$, the second and ninth deciles are, respectively,

$$\tilde{\pi}_{0.20} = (0.8)y_{10} + (0.2)y_{11} = (0.8)(57) + (0.2)(58) = 57.2$$

and

$$\tilde{\pi}_{0.90} = (0.1)y_{45} + (0.9)y_{46} = (0.1)(87) + (0.9)(90) = 89.7.$$

The second decile is commonly called the 20th percentile, and the ninth decile is the 90th percentile.

EXAMPLE 3.2-2 We illustrate the preceding ideas with the data given in Table 3.1-3. For convenience, we use 0.02 as the length of a class interval. The ordered stem-and-leaf display is given in Table 3.2-6.

TABLE 3.2-6: Ordered Stem-and-Leaf Diagram of Fluoride Concentrations

Stems	Leaves	Frequency	Depths
0.8f	5 5 5 5	4	4
0.8s	6 6 6 7 7 7 7	7	11
0.8●	8 8 8 8 8 8 8 8 9 9 9 9 9 9 9	16	27
0.9*	0 0 0 0 0 0 0 0 0 1 1 1 1 1 1 1 1	17	44
0.9t	2 2 2 2 2 2 2 2 2 2 3 3 3 3 3 3 3 3 3	19	(19)
0.9f	4 4 5 5 5 5 5 5 5	9	37
0.9s	6 6 7 7 7 7	6	28
0.9●	8 8 8 8 8 8 8 8 9 9	10	22
1.0*	0 0 0 0 1 1 1	7	12
1.0t	2 3 3	3	5
1.0f	5	1	2
1.0s	6	1	1

This ordered stem-and-leaf diagram is useful for finding sample percentiles of the data. ■

We now find some of the sample percentiles associated with the fluoride data. Since $n = 100$, $(n + 1)(0.25) = 25.25$, $(n + 1)(0.50) = 50.5$, and $(n + 1)(0.75) = 75.75$, so that the 25th, 50th, and 75th percentiles are, respectively,

$$\tilde{\pi}_{0.25} = (0.75)y_{25} + (0.25)y_{26} = (0.75)(0.89) + (0.25)(0.89) = 0.89,$$
$$\tilde{\pi}_{0.50} = (0.50)y_{50} + (0.50)y_{51} = (0.50)(0.92) + (0.50)(0.92) = 0.92,$$
$$\tilde{\pi}_{0.75} = (0.25)y_{75} + (0.75)y_{76} = (0.25)(0.97) + (0.75)(0.97) = 0.97.$$

These three percentiles are often called the **first quartile**, the **median** or **second quartile**, and the **third quartile**, respectively. Along with the smallest (the **minimum**) and largest (the **maximum**) values, they give the **five-number summary** of a set of data. Furthermore, the difference between the third and first quartiles is called the **interquartile range**, **IQR**. Here, it is equal to

$$\tilde{q}_3 - \tilde{q}_1 = \tilde{\pi}_{0.75} - \tilde{\pi}_{0.25} = 0.97 - 0.89 - 0.08.$$

One graphical means for displaying the five-number summary of a set of data is called a **box-and-whisker diagram**. To construct a horizontal box and whisker diagram, or, more simply, a **box plot**, draw a horizontal axis that is scaled to the data. Above the axis, draw a rectangular box with the left and right sides drawn at $\tilde{q}_1$ and $\tilde{q}_3$ and with a vertical line segment drawn at the median, $\tilde{q}_2 = \tilde{m}$. A left whisker is drawn as a horizontal line segment from the minimum to the midpoint of the left side of the box, and a right whisker is drawn as a horizontal line segment from the midpoint of the right side of the box to the maximum. Note that the length of the box is equal to the IQR. The left and right whiskers represent the first and fourth quarters of the data, while the two middle quarters of the data are represented, respectively, by the two sections of the box, one to the left and one to the right of the median line.

EXAMPLE 3.2-3   Using the fluoride data shown in Table 3.2-6, we found that the five-number summary is given by

$$y_1 = 0.85, \quad \tilde{q}_1 = 0.89, \quad \tilde{q}_2 = \tilde{m} = 0.92, \quad \tilde{q}_3 = 0.97, \quad y_{100} = 1.06.$$

The box plot of these data is given in Figure 3.2-1. The fact that the long whisker is to the right and the right half of the box is larger than the left half of the box leads us to say that these data are slightly *skewed to the right*. Note that this skewness can also be seen in the histogram and in the stem-and-leaf diagram. ■

FIGURE 3.2-1: Box plot of fluoride concentrations

The next example illustrates how the box plot depicts data that are *skewed to the left*.

EXAMPLE 3.2-4 The following data give the ordered weights (in grams) of 39 gold coins that were produced during the reign of Verica, a pre-Roman British king:

4.90	5.06	5.07	5.08	5.15	5.17	5.18	5.19	5.24	5.25
5.25	5.25	5.25	5.27	5.27	5.27	5.27	5.28	5.28	5.28
5.29	5.30	5.30	5.30	5.30	5.31	5.31	5.31	5.31	5.31
5.32	5.32	5.33	5.34	5.35	5.35	5.35	5.36	5.37	

For these data, the minimum is 4.90 and the maximum 5.37. Since

$$(39 + 1)(1/4) = 10, \qquad (39 + 1)(2/4) = 20, \qquad (39 + 1)(3/4) = 30,$$

we have

$$\tilde{q}_1 = y_{10} = 5.25,$$
$$\tilde{m} = y_{20} = 5.28,$$
$$\tilde{q}_3 = y_{30} = 5.31.$$

Thus the five-number summary is given by

$$y_1 = 4.90, \quad \tilde{q}_1 = 5.25, \quad \tilde{q}_2 = \tilde{m} = 5.28, \quad \tilde{q}_3 = 5.31, \quad y_{39} = 5.37.$$

The box plot associated with the given data is shown in Figure 3.2-2. Note that the box plot indicates that the data are skewed to the left. ∎

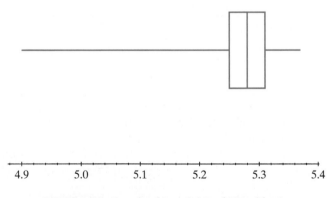

FIGURE 3.2-2: Box plot for weights of 39 gold coins

Sometimes we are interested in picking out observations that seem to be much larger or much smaller than most of the other observations. That is, we are looking for outliers. Tukey suggested a method for defining outliers that is resistant to the effect of one or two extreme values and makes use of the IQR. In a box-and-whisker diagram, construct **inner fences** to the left and right of the box at a distance of 1.5 times the IQR. **Outer fences** are constructed in the same way at a distance of 3 times the IQR. Observations that lie between the inner and outer fences are called **suspected outliers**. Observations that lie beyond the outer fences are called **outliers**. The observations beyond the inner fences are denoted with a circle (•), and the whiskers are drawn only to the extreme values within or on the inner fences. When you are analyzing a set of data, suspected outliers deserve a closer look and outliers should be looked at very carefully. It does not follow that suspected outliers should be removed from the data, unless some error (such as a recording error) has been made. Moreover, it is sometimes important to determine the cause of extreme values, because outliers can often provide useful insights into the situation under consideration (such as a better way of doing things).

EXAMPLE 3.2-5   Continuing with Example 3.2-4, we find that the interquartile range is IQR $=$ $5.31 - 5.25 = 0.06$. Thus, the inner fences would be constructed at a distance of $1.5(0.06) = 0.09$ to the left and right of the box, and the outer fences would be constructed at a distance of $3(0.06) = 0.18$ to the left and right of the box. Figure 3.2-3 shows a box plot with the fences. Of course, since the maximum is 0.06 greater than $\tilde{q}_3$, there are no fences to the right. From this box plot, we see that there are three suspected outliers and two outliers. (You may speculate as to why there are outliers with these data and why they fall to the left—that is, they are lighter than expected.) Note that many computer programs use an asterisk to plot outliers and suspected outliers, and do not print fences.   ∎

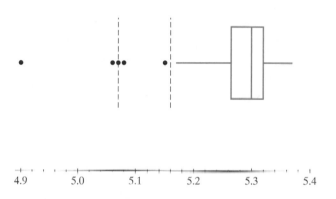

FIGURE 3.2-3: Box plot for weights of 39 gold coins with fences and outliers

Some functions of two or more order statistics are quite important in modern statistics. We mention and illustrate two more, along with the range and the IQR, using the 100 fluoride concentrations shown in Table 3.2-6.

(a) **Midrange** $=$ average of the extremes
$$= \frac{y_1 + y_n}{2} = \frac{0.85 + 1.06}{2} = 0.955,$$

(b) **Trimean** $= \dfrac{(\text{1st quartile}) + 2(\text{2nd quartile}) + (\text{3rd quartile})}{4}$
$$= \frac{\tilde{q}_1 + 2\tilde{q}_2 + \tilde{q}_3}{4} = \frac{0.89 + 2(0.92) + 0.97}{4} = 0.925,$$

(c) **Range** $=$ difference of the extremes
$$= y_n - y_1 = 1.06 - 0.85 = 0.21,$$

(d) **Interquartile range** $=$ difference of third and first quartiles
$$= \tilde{q}_3 - \tilde{q}_1 = 0.97 - 0.89 = 0.08.$$

Thus, we see that the mean, the median, the midrange, and the trimean are measures of the middle of the sample. In some sense, the standard deviation, the range, and the interquartile range provide measures of spread of the sample.

EXERCISES

3.2-1. In Exercise 3.1-5, measurements for 96 plungers are given. Use those measurements to

(a) Construct a stem-and-leaf diagram using integer stems.

(b) Find the five-number summary of the data.

(c) Construct a box-and-whisker diagram. Are there any outliers?

3.2-2. Exercise 3.1-4 gives the weights of 100 laptop computers. Use those weights to

(a) Construct a stem-and-leaf diagram using integer stems.

(b) Find the five-number summary of the weights and construct a box-and-whisker diagram.

3.2-3. When you purchase "1-pound bags" of carrots, you can either buy "baby" carrots or regular carrots. We shall compare the weights of 75 bags of each of these types of carrots. The following table gives the weights of the bags of baby carrots:

1.03	1.03	1.06	1.02	1.03	1.03	1.03	1.02	1.03	1.03
1.06	1.04	1.05	1.03	1.04	1.03	1.05	1.06	1.04	1.04
1.03	1.04	1.04	1.06	1.03	1.04	1.05	1.04	1.04	1.02
1.03	1.05	1.05	1.03	1.04	1.03	1.04	1.04	1.03	1.04
1.03	1.04	1.04	1.04	1.05	1.04	1.04	1.03	1.03	1.05
1.04	1.04	1.05	1.04	1.03	1.03	1.05	1.03	1.04	1.05
1.04	1.04	1.04	1.05	1.03	1.04	1.04	1.04	1.04	1.03
1.05	1.05	1.05	1.03	1.04					

This table gives the weights of the regular-sized carrots:

1.29	1.10	1.28	1.29	1.23	1.20	1.31	1.25	1.13	1.26
1.19	1.33	1.24	1.20	1.26	1.24	1.11	1.14	1.15	1.15
1.19	1.26	1.14	1.20	1.20	1.20	1.24	1.25	1.28	1.24
1.26	1.20	1.30	1.23	1.26	1.16	1.34	1.10	1.22	1.27
1.21	1.09	1.23	1.03	1.32	1.21	1.23	1.34	1.19	1.18
1.20	1.20	1.13	1.43	1.19	1.05	1.16	1.19	1.07	1.21
1.36	1.21	1.00	1.23	1.22	1.13	1.24	1.10	1.18	1.26
1.12	1.10	1.19	1.10	1.24					

(a) Calculate the five-number summary of each set of weights.

(b) On the same graph, construct box plots for each set of weights.

(c) If the carrots are the same price per package, which is the better buy? Which type of carrots would you select?

3.2-4. In the Fifth Third River Bank Run (a 25K race), the times are listed separately for males and females and also by age group. The following table gives the race times, in order in minutes, for 125 50–54 year-old male runners:

96.35	98.40	100.43	101.38	101.48	101.63	103.18	103.97	104.47
105.03	105.23	106.00	106.23	108.82	109.47	109.53	109.70	110.82
110.90	111.80	111.90	112.28	112.55	112.80	112.90	112.92	113.15
113.17	113.25	113.72	114.50	114.60	115.47	115.65	115.73	116.18
116.57	116.93	117.75	118.17	118.18	118.20	118.47	118.50	118.73
119.17	119.45	119.88	120.28	120.32	121.20	122.25	122.28	122.77
122.95	122.98	123.77	124.07	124.10	124.82	124.92	125.23	125.25
125.27	126.15	126.20	126.48	126.72	127.35	127.42	127.95	128.03
128.38	129.22	129.35	129.65	130.02	130.12	130.48	130.57	131.12
131.37	131.97	132.45	132.85	132.97	133.85	134.07	134.67	134.88
135.08	135.68	136.45	136.85	136.87	136.88	137.15	137.22	137.52
137.80	138.02	138.22	138.73	139.57	139.85	140.05	140.08	141.33
142.47	143.13	143.77	144.17	144.58	145.67	147.08	148.20	151.30
151.65	152.00	156.22	160.72	161.20	162.48	166.70	169.90	

The following table gives the race times, in order in minutes, for 76 50–54 year-old female runners:

118.05	180.45	123.80	125.23	125.27	125.90	126.90	128.37	129.30
130.08	131.43	131.72	132.17	133.00	133.25	133.42	133.43	133.50
134.25	136.98	137.10	137.28	138.52	139.80	140.87	142.85	143.75
145.05	145.07	145.48	146.55	146.67	148.07	148.75	148.93	149.77
150.10	150.22	150.62	150.82	150.85	151.10	151.43	152.25	152.60
152.90	155.55	157.02	159.58	159.70	159.80	159.92	159.97	161.33
162.17	163.60	165.45	166.23	168.10	168.60	169.17	169.93	170.88
171.20	173.23	174.22	178.57	178.83	183.08	186.07	186.48	189.77
192.52	197.88	199.80	203.92					

(a) Group each set of data using class boundaries 94.95, 99.95, 104.95, ..., 169.95 for the men and 109.95, 119.95, ..., 209.95 for the women.

(b) Draw histograms of each set of grouped data.

(c) Construct a back-to-back stem-and-leaf display. Place the stems 9•, 10∗, 10•,... down the center, place leaves for the male times to the left, and place leaves for the female times to the right. (See Table 8.5-3 on page 449 for an example.)

(d) Find the five-number summaries of each set of data and draw box-and-whisker diagrams on the same figure.

(e) Use your graphical results to interpret what you see.

3.2-5. In Example 1.1-7, the sample space for underwater weights was given. Here are underwater weights for 82 male students:

3.7	3.6	4.0	4.3	3.8	3.4	4.1	4.0	3.7	3.4	3.5	3.8	3.7	4.9
3.5	3.8	3.3	4.8	3.4	4.6	3.5	5.3	4.4	4.2	2.5	3.1	5.2	3.8
3.3	3.4	4.1	4.6	4.0	1.4	4.3	3.8	4.7	4.4	5.0	3.2	3.1	4.2
4.9	4.5	3.8	4.2	2.7	3.8	3.8	2.0	3.4	4.9	3.3	4.3	5.6	3.2
4.7	4.5	5.2	5.0	5.0	4.0	3.8	5.3	4.5	3.8	3.8	3.4	3.6	3.3
4.2	5.1	4.0	4.7	6.5	4.4	3.6	4.7	4.5	2.3	4.0	3.7		

Here are underwater weights for 100 female students:

2.0	2.0	2.1	1.6	1.9	2.0	2.0	1.3	1.3	1.2	2.3	1.9
2.1	1.2	2.0	1.6	1.1	2.2	2.2	1.4	1.7	2.4	1.8	1.7
2.0	2.1	1.6	1.7	1.8	0.7	1.9	1.7	1.7	1.1	2.0	2.3
0.5	1.3	2.7	1.8	2.0	1.7	1.2	0.7	1.1	1.1	1.7	1.7
1.2	1.2	0.7	2.3	1.7	2.4	1.0	2.4	1.4	1.9	2.5	2.2
2.1	1.4	2.4	1.8	2.5	1.3	0.5	1.7	1.9	1.8	1.3	2.0
2.2	1.7	2.0	2.5	1.2	1.4	1.4	1.2	2.2	2.0	1.8	1.4
1.9	1.4	1.3	2.5	1.2	1.5	0.8	2.0	2.2	1.8	2.0	1.6
1.5	1.6	1.5	2.6								

(a) Group each set of data into classes with a class width of 0.5 kilograms and in which the class marks are 0.5, 1.0, 1.5, ...

(b) Draw histograms of the grouped data.

(c) Construct box-and-whisker diagrams of the data and draw them on the same graph. Describe what this graph shows.

3.2-6. An insurance company experienced the following mobile home losses in 10,000's of dollars for 50 catastrophic events:

1	2	2	3	3	4	4	5	5	5
5	6	7	7	9	9	9	10	11	12
22	24	28	29	31	33	36	38	38	38
39	41	48	49	53	55	74	82	117	134
192	207	224	225	236	280	301	308	351	527

(a) Find the five-number summary of the data and draw a box-and-whisker diagram.

(b) Calculate the IQR and the locations of the inner and outer fences.

(c) Draw a box plot that shows the fences, suspected outliers, and outliers.

(d) Describe the distribution of losses. (See Exercise 3.1-10.)

3.2-7. In Exercise 3.1-7, data are given for the number of $1 bets a player can make in roulette before losing $5. Use those data to respond to the following:

(a) Determine the order statistics.

(b) Find the five-number summary of the data.

(c) Draw a box-and-whisker diagram.

(d) Find the locations of the inner and outer fences, and draw a box plot that shows the fences, the suspected outliers, and the outliers.

(e) In your opinion, does the median or sample mean give a better measure of the center of the data?

3.2-8. In the casino game roulette, if a player bets $1 on red (or on black or on odd or on even), the probability of winning $1 is 18/38 and the probability of losing $1 is 20/38. Suppose that a player begins with $5 and makes successive $1 bets. Let Y equal the player's maximum capital before losing the $5. One hundred observations of Y were simulated on a computer, yielding the following data:

25	9	5	5	5	9	6	5	15	45
55	6	5	6	24	21	16	5	8	7
7	5	5	35	13	9	5	18	6	10
19	16	21	8	13	5	9	10	10	6
23	8	5	10	15	7	5	5	24	9
11	34	12	11	17	11	16	5	15	5
12	6	5	5	7	6	17	20	7	8
8	6	10	11	6	7	5	12	11	18
6	21	6	5	24	7	16	21	23	15
11	8	6	8	14	11	6	9	6	10

(a) Construct an ordered stem-and-leaf display.

(b) Find the five-number summary of the data and draw a box-and-whisker diagram.

(c) Calculate the IQR and the locations of the inner and outer fences.

(d) Draw a box plot that shows the fences, suspected outliers, and outliers.

(e) Find the 90th percentile.

3.2-9. Let X denote the concentration of calcium carbonate ($CaCO_3$) in milligrams per liter. Following are 20 observations of X:

130.8	129.9	131.5	131.2	129.5
132.7	131.5	127.8	133.7	132.2
134.8	131.7	133.9	129.8	131.4
128.8	132.7	132.8	131.4	131.3

(a) Construct an ordered stem-and-leaf display, using stems of 127, 128, . . . , 134.

(b) Find the midrange, range, interquartile range, median, sample mean, and sample variance.

(c) Draw a box-and-whisker diagram.

3.2-10. The weights (in grams) of 25 indicator housings used on gauges are as follows:

102.0	106.3	106.6	108.8	107.7
106.1	105.9	106.7	106.8	110.2
101.7	106.6	106.3	110.2	109.9
102.0	105.8	109.1	106.7	107.3
102.0	106.8	110.0	107.9	109.3

(a) Construct an ordered stem-and-leaf display, using integers as the stems and tenths as the leaves.

(b) Find the five-number summary of the data and draw a box plot.

(c) Are there any suspected outliers? Are there any outliers?

3.2-11. In Exercise 3.1-8, the melting points of a firms's alloy filaments are given for a sample of 50 filaments.

(a) Construct a stem-and-leaf diagram of those melting points, using $30f, 30s, \ldots, 33f$ as stems.

(b) Find the five-number summary for these melting points.

(c) Construct a box-and-whisker diagram.

(d) Describe the symmetry of the data.

3.2-12. In Exercise 3.1-9, lead concentrations near the San Diego Freeway in 1976 are given. During the fall of 1977, the weekday afternoon lead concentrations (in $\mu g/m^3$) at the measurement station near the San Diego Freeway in Los Angeles were as follows:

9.5	10.7	8.3	9.8	9.1	9.4	9.6	11.9	9.5	12.6	10.5
8.9	11.4	12.0	12.4	9.9	10.9	12.3	11.0	9.2	9.3	9.3
10.5	9.4	9.4	8.2	10.4	9.3	8.7	9.8	9.1	2.9	9.8
5.7	8.2	8.1	8.8	9.7	8.1	8.8	10.3	8.6	10.2	9.4
14.8	9.9	9.3	8.2	9.9	11.6	8.7	5.0	9.9	6.3	6.5
10.2	8.8	8.0	8.7	8.9	6.8	6.6	7.3	16.7		

(a) Construct a frequency distribution and display the results in the form of a histogram. Is this distribution symmetric?

(b) Calculate the sample mean and sample standard deviation.

(c) Locate $\bar{x}$, $\bar{x} \pm s$ on your histogram. How many observations lie within one standard deviation of the mean? How many lie within two standard deviations of the mean?

(d) Using the data from Exercise 3.1-9 and the data from this exercise, construct a back-to-back stem-and-leaf diagram with integer stems in the center and the leaves for 1976 going to the left and those for 1977 going to the right.

(e) Construct box-and-whisker displays of both sets of data on the same graph.

(f) Use your numerical and graphical results to interpret what you see.

REMARK   In the spring of 1977, a new traffic lane was added to the freeway. This lane reduced traffic congestion but increased traffic speed.   ■

3.3 RANDOM VARIABLES OF THE CONTINUOUS TYPE

The relative frequency (or density) histogram $h(x)$ associated with n observations of a random variable of the continuous type is a nonnegative function defined so that the total area between its graph and the x-axis equals 1. This statement is also true of the function created by the relative frequency polygon. In addition, $h(x)$ is constructed so that the integral

$$\int_a^b h(x)\,dx$$

is approximately equal to the relative frequency of the interval $\{x: a < x < b\}$ and hence can be thought of as an estimate of the probability that the random variable X falls into the interval (a, b). This probability is often written $P(a < X < b)$.

Let us now consider what happens to the function $h(x)$ in the limit, or what happens to the relative frequency polygon as n increases without bound and as the lengths of the class intervals decrease to zero. It is to be hoped that $h(x)$ and the relative frequency polygon will become closer and closer to some function—say, $f(x)$—that gives the true probabilities, such as $P(a < X < b)$, through the integral

$$P(a < X < b) = \int_a^b f(x)\,dx.$$

That is, $f(x)$ should be a nonnegative function such that the total area between its graph and the x-axis equals 1. Moreover, the probability $P(a < X < b)$ is the area bounded by the graph of $f(x)$, the x-axis, and the lines $x = a$ and $x = b$. Thus, we say that the **probability density function (p.d.f.)** of a random variable X of the **continuous type**, with space S that is an interval or union of intervals, is an integrable function $f(x)$ satisfying the following conditions:

(a) $f(x) > 0, \qquad x \in S$.

(b) $\int_S f(x)\,dx = 1$.

(c) If $(a, b) \subseteq S$, then the probability of the event $\{a < X < b\}$ is

$$P(a < X < b) = \int_a^b f(x)\,dx.$$

The corresponding distribution of probability is said to be of the continuous type.

EXAMPLE 3.3-1 Let the random variable X be the lengths of time **in minutes** between calls to 911 in a small city that were reported in the newspaper on February 26 and 27. Suppose that a reasonable probability model for X is given by the p.d.f.

$$f(x) = \frac{1}{20} e^{-x/20}, \qquad 0 \le x < \infty.$$

Note that $S = \{x : 0 \le x < \infty\}$ and $f(x) > 0$ for $x \in S$. Also,

$$
\begin{aligned}
\int_S f(x)\, dx &= \int_0^\infty \frac{1}{20} e^{-x/20}\, dx \\
&= \lim_{b \to \infty} \left[-e^{-x/20} \right]_0^b \\
&= 1 - \lim_{b \to \infty} e^{-b/20} = 1.
\end{aligned}
$$

The probability that the time between calls is greater than 20 minutes is

$$P(X > 20) = \int_{20}^\infty \frac{1}{20} e^{-x/20}\, dx = e^{-1} = 0.368.$$

The p.d.f. and the probability of interest are depicted in Figure 3.3-1. ■

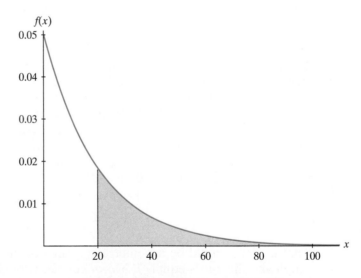

FIGURE 3.3-1: Times between calls to 911

So that we can avoid repeated references to the space (or support) S of the random variable X, we shall adopt the same convention when describing probability density functions of the continuous type as we did in the discrete case. We extend the definition of the p.d.f. $f(x)$ to the entire set of real numbers by letting it equal zero when $x \notin S$. For example,

$$
f(x) =
\begin{cases}
\dfrac{1}{20} e^{-x/20}, & 0 \le x < \infty, \\[2mm]
0, & \text{elsewhere,}
\end{cases}
$$

has the properties of a p.d.f. of a continuous-type random variable X having support $\{x: 0 \leq x < \infty\}$. It will always be understood that $f(x) = 0$ when $x \notin S$, even when this is not explicitly written out.

The **distribution function** of a random variable X of the continuous type, defined in terms of the p.d.f. of X, is given by

$$F(x) = P(X \leq x) = \int_{-\infty}^{x} f(t)\,dt, \qquad -\infty < x < \infty.$$

Here, again, $F(x)$ accumulates (or, more simply, cumulates) all of the probability less than or equal to x. (This function is sometimes called a **cumulative distribution function** or c.d.f.) From the fundamental theorem of calculus, we have, for x values for which the derivative $F'(x)$ exists, $F'(x) = f(x)$.

| EXAMPLE 3.3-2 | Continuing with Example 3.3-1, we note that if the p.d.f. of X is |

$$f(x) = \begin{cases} 0, & -\infty < x < 0, \\ \dfrac{1}{20}e^{-x/20}, & 0 \leq x < \infty, \end{cases}$$

then, for $x < 0$, the distribution function of X is $F(x) = 0$ and for $x \geq 0$,

$$\begin{aligned} F(x) &= \int_{-\infty}^{x} f(t)\,dt = \int_{0}^{x} \frac{1}{20}e^{-t/20}\,dt \\ &= \left[-e^{-t/20}\right]_{0}^{x} = 1 - e^{-x/20}. \end{aligned}$$

Note also that

$$F'(x) = \begin{cases} 0, & -\infty < x < 0, \\ \dfrac{1}{20}e^{-x/20}, & 0 < x < \infty. \end{cases}$$

Furthermore, $F'(0)$ does not exist. (Sketch a graph of $y = F(x)$ to see why this is true.) ∎

Since there are no steps or jumps in $F(x)$, a distribution function of the continuous type, it must be true that

$$P(X = b) = 0$$

for all real values of b. This agrees with the fact that the integral $\int_{b}^{b} f(x)\,dx$ is taken to be zero in calculus. In general,

$$P(a \leq X \leq b) = P(a < X < b) = P(a \leq X < b) = P(a < X \leq b) = F(b) - F(a),$$

provided that X is a random variable of the continuous type. Moreover, we can change the definition of a p.d.f. of a random variable of the continuous type at a finite (actually, countable) number of points without altering the distribution of probability. For example,

$$f(x) = \begin{cases} 0, & -\infty < x < 0, \\ \dfrac{1}{20}e^{-x/20}, & 0 \leq x < \infty, \end{cases}$$

and

$$f(x) = \begin{cases} 0, & -\infty < x \le 0, \\ \dfrac{1}{20} e^{-x/20}, & 0 < x < \infty, \end{cases}$$

are equivalent in the computation of probabilities involving this random variable.

EXAMPLE 3.3-3 The following table lists 105 observations of X, the times between calls to 911:

30	17	65	8	38	35	4	19	7	14	12	4	5	4	2
7	5	12	50	33	10	15	2	10	1	5	30	41	21	31
1	18	12	5	24	7	6	31	1	3	2	22	1	30	2
1	3	12	12	9	28	6	50	63	5	17	11	23	2	46
90	13	21	55	43	5	19	47	24	4	6	27	4	6	37
16	41	68	9	5	28	42	3	42	8	52	2	11	41	4
35	21	3	17	10	16	1	68	105	45	23	5	10	12	17

To help determine visually whether the exponential model in Example 3.3-1 is perhaps appropriate for this situation, we shall look at two graphs. First, we have constructed a relative frequency histogram, $h(x)$, of these data in Figure 3.3-2(a), with $f(x) = (1/20)e^{-x/20}$ superimposed. Second, we have also constructed the empirical distribution function of these data in Figure 3.3-2(b), with the theoretical distribution function superimposed. Note that $F_n(x)$, the **empirical distribution function**, is a step function with a step of size $1/n$ at each observation of X. If k observations are equal, the step at that value is k/n. Later in the text we give statistical tests to help us decide whether this probability model provides a good fit for these data. ∎

(a) Histogram and p.d.f. of X

(b) Theoretical and empirical c.d.f.'s

FIGURE 3.3-2: Times between calls to 911

EXAMPLE 3.3-4 Let Y be a continuous random variable with p.d.f. $g(y) = 2y$, $0 < y < 1$. The distribution function of Y is defined by

$$G(y) = \begin{cases} 0, & y < 0, \\ \displaystyle\int_0^y 2t\, dt = y^2, & 0 \le y < 1, \\ 1, & 1 \le y. \end{cases}$$

FIGURE 3.3-3: Continuous distribution p.d.f. and c.d.f.

Figure 3.3-3 gives the graph of the p.d.f. $g(y)$ and the graph of the distribution function $G(y)$. For examples of computations of probabilities, consider

$$P\left(\frac{1}{2} < Y \le \frac{3}{4}\right) = G\left(\frac{3}{4}\right) - G\left(\frac{1}{2}\right) = \left(\frac{3}{4}\right)^2 - \left(\frac{1}{2}\right)^2 = \frac{5}{16}$$

and

$$P\left(\frac{1}{4} \le Y < 2\right) = G(2) - G\left(\frac{1}{4}\right) = 1 - \left(\frac{1}{4}\right)^2 = \frac{15}{16}. \qquad \blacksquare$$

Recall that the p.m.f. $f(x)$ of a random variable of the discrete type is bounded by 1 because $f(x)$ gives a probability, namely,

$$f(x) = P(X = x).$$

For random variables of the continuous type, the p.d.f. does not have to be bounded. [See Exercises 3.3-2(c) and 3.3-3(c).] The restriction is that *the area between the p.d.f. and the x-axis must equal 1.* Furthermore, the p.d.f. of a random variable X of the continuous type does not need to be a continuous function. For example, the function

$$f(x) = \begin{cases} \dfrac{1}{2}, & 0 < x < 1 \quad \text{or} \quad 2 < x < 3, \\ 0, & \text{elsewhere,} \end{cases}$$

enjoys the properties of a p.d.f. of a distribution of the continuous type and yet has discontinuities at $x = 0, 1, 2,$ and 3. However, the distribution function associated with a distribution of the continuous type is always a continuous function.

For continuous-type random variables, the definitions associated with mathematical expectation are the same as those in the discrete case, except that integrals replace summations. As an illustration, let X be a continuous random variable with a p.d.f. $f(x)$. Then the **expected value of** X, or the **mean of** X, is

$$\mu = E(X) = \int_{-\infty}^{\infty} xf(x)\,dx,$$

the **variance of** X is

$$\sigma^2 = \text{Var}(X) = E[(X - \mu)^2] = \int_{-\infty}^{\infty} (x - \mu)^2 f(x)\,dx,$$

the **standard deviation of** X is

$$\sigma = \sqrt{\text{Var}(X)},$$

and the **moment-generating function**, if it exists, is

$$M(t) = \int_{-\infty}^{\infty} e^{tx} f(x)\, dx, \qquad -h < t < h.$$

Moreover, important results such as $\sigma^2 = E(X^2) - \mu^2$, $\mu = M'(0)$, and $\sigma^2 = M''(0) - [M'(0)]^2$ are still valid. Again, it is important to note that the moment-generating function, if it is finite for $-h < t < h$ for some $h > 0$, completely determines the distribution.

REMARK   In both the discrete and continuous cases, note that if the rth moment, $E(X^r)$, exists and is finite, then the same is true of all lower order moments, $E(X^k)$, $k = 1, 2, \ldots, r - 1$. However, the converse is not true; for example, the first moment can exist and be finite, but the second moment is not necessarily finite. (See Exercise 3.3-9.) Moreover, if $E(e^{tX})$ exists and is finite for $-h < t < h$, then all moments exist and are finite, but the converse is not necessarily true. ∎

EXAMPLE 3.3-5   For the random variable Y in Example 3.3-4,

$$\mu = E(Y) = \int_0^1 y\,(2y)\, dy = \left[\left(\frac{2}{3}\right) y^3\right]_0^1 = \frac{2}{3}$$

and

$$\sigma^2 = \text{Var}(Y) = E(Y^2) - \mu^2$$

$$= \int_0^1 y^2 (2y)\, dy - \left(\frac{2}{3}\right)^2 = \left[\left(\frac{1}{2}\right) y^4\right]_0^1 - \frac{4}{9} = \frac{1}{18},$$

are the mean and variance, respectively, of Y. ∎

EXAMPLE 3.3-6   Let X have the p.d.f.

$$f(x) = \begin{cases} xe^{-x}, & 0 \le x < \infty, \\ 0, & \text{elsewhere.} \end{cases}$$

Then

$$M(t) = \int_0^{\infty} e^{tx} xe^{-x}\, dx = \lim_{b \to \infty} \int_0^b xe^{-(1-t)x}\, dx$$

$$= \lim_{b \to \infty} \left[-\frac{xe^{-(1-t)x}}{1-t} - \frac{e^{-(1-t)x}}{(1-t)^2} \right]_0^b$$

$$= \lim_{b \to \infty} \left[-\frac{be^{-(1-t)b}}{1-t} - \frac{e^{-(1-t)b}}{(1-t)^2} \right] + \frac{1}{(1-t)^2}$$

$$= \frac{1}{(1-t)^2},$$

provided that $t < 1$. Note that $M(0) = 1$, which is true for every moment-generating function. Now,

$$M'(t) = \frac{2}{(1 - t)^3} \qquad \text{and} \qquad M''(t) = \frac{6}{(1 - t)^4}.$$

Thus,

$$\mu = M'(0) = 2$$

and

$$\sigma^2 = M''(0) - [M'(0)]^2 = 6 - 2^2 = 2. \qquad \blacksquare$$

The **(100p)th percentile** is a number π_p such that the area under $f(x)$ to the left of π_p is p. That is,

$$p = \int_{-\infty}^{\pi_p} f(x)\, dx = F(\pi_p).$$

The 50th percentile is called the **median**. We let $m = \pi_{0.50}$. The 25th and 75th percentiles are called the **first** and **third quartiles**, respectively, and are denoted by $q_1 = \pi_{0.25}$ and $q_3 = \pi_{0.75}$. Of course, the median $m = \pi_{0.50} = q_2$ is also called the second quartile.

The $(100p)$th percentile of a distribution is often called the quantile of order p. So if $y_1 \leq y_2 \leq \cdots \leq y_n$ are the order statistics associated with the sample $x_1, x_2, \ldots, x_n$, then y_r is called the **quantile of order r/(n + 1)** as well as the **100r/(n + 1)th percentile**. Also, the percentile π_p of a theoretical distribution is the quantile of order p. Now, suppose the theoretical distribution is a good model for the observations. Then if we plot (y_r, π_p), where $p = r/(n + 1)$, for several values of r (possibly even for all r values, $r = 1, 2, \ldots, n$), we would expect these points (y_r, π_p) to lie close to a line through the origin with slope equal to 1 because $y_r \approx \pi_p$. If they are not close to that line, then we would doubt that the theoretical distribution is a good model for the observations. The plot of (y_r, π_p) for several values of r is called the **quantile–quantile plot** or, more simply, the **q–q plot**. (See Figure 3.6-3 on page 165 for an example.)

EXAMPLE 3.3-7   The time X in months until failure of a certain product has the p.d.f. (of the Weibull type)

$$f(x) = \frac{3x^2}{4^3}\, e^{-(x/4)^3}, \qquad 0 < x < \infty.$$

Its distribution function is

$$F(x) = 1 - e^{-(x/4)^3}, \qquad 0 \leq x < \infty.$$

For example, the 30th percentile, $\pi_{0.3}$, is given by

$$F(\pi_{0.3}) = 0.3$$

or, equivalently,

$$1 - e^{-(\pi_{0.3}/4)^3} = 0.3$$
$$\ln(0.7) = -(\pi_{0.3}/4)^3$$
$$\pi_{0.3} = -4\sqrt[3]{\ln(0.7)} = 2.84.$$

Likewise, $\pi_{0.9}$ is found by

$$F(\pi_{0.9}) = 0.9;$$

so

$$\pi_{0.9} = -4\sqrt[3]{\ln(0.1)} = 5.28.$$

Thus,

$$P(2.84 < X < 5.28) = 0.6.$$

The 30th and 90th percentiles are shown in Figure 3.3-4.

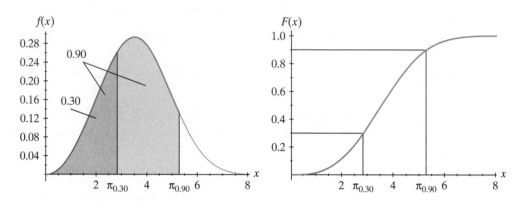

FIGURE 3.3-4: Illustration of percentiles $\pi_{0.30}$ and $\pi_{0.90}$

We conclude this section with an example that reviews its important ideas.

EXAMPLE 3.3-8 Let X have the p.d.f.

$$f(x) = e^{-x-1}, \qquad -1 < x < \infty.$$

Then

$$P(1 \le X) = \int_1^\infty e^{-x-1}\,dx = e^{-1}\left[-e^{-x}\right]_1^\infty = e^{-2} = 0.135.$$

Also,

$$M(t) = E(e^{tX}) = \int_{-1}^\infty e^{tx}e^{-x-1}\,dx$$

$$= e^{-1}\left[\frac{-e^{-(1-t)x}}{1-t}\right]_{-1}^\infty$$

$$= e^{-t}(1-t)^{-1}, \qquad t < 1.$$

Thus, since

$$M'(t) = (e^{-t})(1-t)^{-2} - e^{-t}(1-t)^{-1}$$

and

$$M''(t) = (e^{-t})(2)(1-t)^{-3} - 2e^{-t}(1-t)^{-2} + e^{-t}(1-t)^{-1},$$

we have

$$\mu = M'(0) = 0 \qquad \text{and} \qquad \sigma^2 = M''(0) - [M'(0)]^2 = 1.$$

The distribution function is

$$F(x) = \int_{-1}^{x} e^{-w-1}\,dw = e^{-1}\left[-e^{-w}\right]_{-1}^{x}$$
$$= e^{-1}\left[e^{1} - e^{-x}\right] = 1 - e^{-x-1}, \qquad -1 < x < \infty,$$

and zero for $x \le -1$.

As an illustration, the median, $\pi_{0.5}$, is found by solving

$$F(\pi_{0.5}) = 0.5,$$

which is equivalent to

$$-\pi_{0.5} - 1 = \ln(0.5);$$

so

$$\pi_{0.5} = \ln 2 - 1 = -0.307. \qquad \blacksquare$$

EXERCISES

3.3-1. Let the random variable X have the p.d.f. $f(x) = 2(1 - x), 0 \le x \le 1$, zero elsewhere.

(a) Sketch the graph of this p.d.f.

(b) Determine and sketch the graph of the distribution function of X.

(c) Find (i) $P(0 \le X \le 1/2)$, (ii) $P(1/4 \le X \le 3/4)$, (iii) $P(X = 3/4)$, and (iv) $P(X \ge 3/4)$.

3.3-2. For each of the following functions, (i) find the constant c so that $f(x)$ is a p.d.f. of a random variable X, (ii) find the distribution function, $F(x) = P(X \le x)$, and (iii) sketch graphs of the p.d.f. $f(x)$ and the distribution function $F(x)$:

(a) $f(x) = x^3/4, \qquad 0 < x < c;$

(b) $f(x) = (3/16)x^2, \qquad -c < x < c;$

(c) $f(x) = c/\sqrt{x}, \qquad 0 < x < 1.$ Is this p.d.f. bounded?

3.3-3. For each of the following functions, (i) find the constant c so that $f(x)$ is a p.d.f. of a random variable X, (ii) find the distribution function, $F(x) = P(X \le x)$, and (iii) sketch graphs of the p.d.f. $f(x)$ and the distribution function $F(x)$:

(a) $f(x) = 4x^c, \qquad 0 \le x \le 1,$

(b) $f(x) = c\sqrt{x}, \qquad 0 \le x \le 4,$

(c) $f(x) = c/x^{3/4}, \qquad 0 < x < 1.$

3.3-4. For each of the distributions in Exercise 3.3-2, find μ, σ^2, and σ.

3.3-5. For each of the distributions in Exercise 3.3-3, find μ, σ^2, and σ.

3.3-6. Let $f(x) = (1/2)x^2 e^{-x}, 0 < x < \infty$, be the p.d.f. of X.

(a) Find the m.g.f. $M(t)$.

(b) Find the values of μ and σ^2.

3.3-7. Let $f(x) = (1/2)\sin x, 0 \le x \le \pi$, be the p.d.f. of X.

(a) Find μ and σ^2.

(b) Sketch the graph of the p.d.f. of X.

(c) Determine and sketch the graph of the distribution function of X.

3.3-8. The p.d.f. of X is $f(x) = c/x^2, 1 < x < \infty$.

(a) Calculate the value of c so that $f(x)$ is a p.d.f.

(b) Show that $E(X)$ is not finite.

3.3-9. The p.d.f. of Y is $g(y) = d/y^3, 1 < y < \infty$.

(a) Calculate the value of d so that $g(y)$ is a p.d.f.

(b) Find $E(Y)$.

(c) Show that $\text{Var}(Y)$ is not finite.

3.3-10. Sketch the graphs of the following probability density functions, and find and sketch the graphs of the distribution functions associated with these distributions (note carefully the relationship between the shape of the graph of the p.d.f. and the concavity of the graph of the distribution function):

(a) $f(x) = \left(\dfrac{3}{2}\right)x^2, \qquad -1 < x < 1.$

(b) $f(x) = \dfrac{1}{2}, \qquad -1 < x < 1.$

(c) $f(x) = \begin{cases} x + 1, & -1 < x < 0, \\ 1 - x, & 0 \le x < 1. \end{cases}$

3.3-11. Find the mean and variance for each of the distributions in Exercise 3.3-10.

3.3-12. Let $R(t) = \ln M(t)$, where $M(t)$ is the moment-generating function of a random variable. Show that

(a) $\mu = R'(0)$.

(b) $\sigma^2 = R''(0)$.

3.3-13. If $M(t) = e^{-t}(1 - t)^{-1}$, $t < 1$, use $R(t) = \ln M(t)$ and the result in Exercise 3.3-12 to find (a) μ and (b) σ^2.

3.3-14. Find the moment-generating function $M(t)$ of the distribution with p.d.f. $f(x) = (1/10)e^{-x/10}$, $0 < x < \infty$. Use $M(t)$ or $R(t) = \ln M(t)$ to determine the mean μ and the variance σ^2.

3.3-15. The logistic distribution is associated with the distribution function $F(x) = (1 + e^{-x})^{-1}$, $-\infty < x < \infty$. Find the p.d.f. of the logistic distribution and show that its graph is symmetric about $x = 0$.

3.3-16. Let $f(x) = 1/2$, $0 < x < 1$ or $2 < x < 3$, zero elsewhere, be the p.d.f. of X.

(a) Sketch the graph of this p.d.f.

(b) Define the distribution function of X and sketch its graph.

(c) Find $q_1 = \pi_{0.25}$.

(d) Find $m = \pi_{0.50}$. Is it unique?

(e) Find $q_3 = \pi_{0.75}$.

3.3-17. The life X (in years) of a voltage regulator of a car has the p.d.f.

$$f(x) = \frac{3x^2}{7^3} e^{-(x/7)^3}, \qquad 0 < x < \infty.$$

(a) What is the probability that this regulator will last at least 7 years?

(b) Given that it has lasted at least 7 years, what is the conditional probability that it will last at least another 3.5 years?

3.3-18. Let $f(x) = (x + 1)/2$, $-1 < x < 1$. Find (a) $\pi_{0.64}$, (b) $q_1 = \pi_{0.25}$, and (c) $\pi_{0.81}$.

3.3-19. Let the random variable X_n have the p.d.f. $f(x_n) = n$, $0 < x_n < 1/n$.

(a) Define the distribution function of X_n—say, $F_n(x_n)$.

(b) Graph the p.d.f. and the distribution function for $n = 1, 5$, and 10.

3.3-20. It is interesting to note that back in 1940 if a person in the United States was selected at random,

then his or her age X had a distribution that was almost given by the p.d.f. of the form

$$f(x) = \begin{cases} c, & 0 < x < 35, \\ c\left[1 + \dfrac{3}{140}(35 - x)\right], & 35 \le x \le 81\frac{2}{3}. \end{cases}$$

(a) Find c so that $f(x)$ is a p.d.f.

(b) In 1940, about what percentage of the U.S. population was older than 65?

(c) What was the median age then?

3.3-21. The lifetime X (in years) of a machine has a p.d.f.

$$f(x) = \frac{2x}{\theta^2} e^{-(x/\theta)^2}, \qquad 0 < x < \infty.$$

If $P(X > 5) = 0.01$, determine θ.

3.3-22. The weekly demand X for propane gas (in thousands of gallons) has the p.d.f.

$$f(x) = 4x^3 e^{-x^4}, \qquad 0 < x < \infty.$$

If the stockpile consists of two thousand gallons at the beginning of each week (and nothing extra is received during the week), what is the probability of not being able to meet the demand during a given week?

3.3-23. An insurance agent receives a bonus if the loss ratio L on his business is less than 0.5, where L is the total losses (say, X) divided by the total premiums (say, T). The bonus equals $(0.5 - L)(T/30)$ if $L < 0.5$ and equals zero otherwise. If X (in \$100,000) has the p.d.f.

$$f(x) = \frac{3}{x^4}, \qquad x > 1,$$

and if T (in \$100,000) equals 3, determine the expected value of the bonus.

3.3-24. The p.d.f. of time X to failure of an electronic component is

$$f(x) = \frac{2x}{1000^2} e^{-(x/1000)^2}, \qquad 0 < x < \infty.$$

(a) Compute $P(X > 2000)$.

(b) Determine the 75th percentile, $\pi_{0.75}$, of the distribution.

(c) Find the 10th and 60th percentiles, $\pi_{0.10}$ and $\pi_{0.60}$

3.3-25. The total amount of medical claims (in \$100,000) of the employees of a company has the p.d.f. $f(x) = 30x(1 - x)^4$, $0 < x < 1$. Find

(a) the mean and the standard deviation of the total in dollars,

(b) the probability that the total exceeds \$20,000.

3.3-26. Nicol (see references) lets the p.d.f. of X be defined by

$$f(x) = \begin{cases} x, & 0 \le x \le 1, \\ c/x^3, & 1 \le x < \infty, \\ 0, & \text{elsewhere.} \end{cases}$$

Find

(a) the value of c so that $f(x)$ is a p.d.f.
(b) the mean of X (if it exists).
(c) the variance of X (if it exists).
(d) $P(1/2 \le X \le 2)$.

3.4 THE UNIFORM AND EXPONENTIAL DISTRIBUTIONS

Let the random variable X denote the outcome when a point is selected at random from an interval $[a, b]$, $-\infty < a < b < \infty$. If the experiment is performed in a fair manner, it is reasonable to assume that the probability that the point is selected from the interval $[a, x]$, $a \le x < b$, is $(x - a)/(b - a)$. That is, the probability is proportional to the length of the interval, so the distribution function of X is

$$F(x) = \begin{cases} 0, & x < a, \\ \dfrac{x - a}{b - a}, & a \le x < b, \\ 1, & b \le x. \end{cases}$$

Because X is a continuous-type random variable, $F'(x)$ is equal to the p.d.f. of X whenever $F'(x)$ exists; thus, when $a < x < b$, we have $f(x) = F'(x) = 1/(b - a)$.

The random variable X has a **uniform distribution** if its p.d.f. is equal to a constant on its support. In particular, if the support is the interval $[a, b]$, then

$$f(x) = \frac{1}{b - a}, \qquad a \le x \le b.$$

Moreover, we shall say that X is $U(a, b)$. This distribution is also referred to as **rectangular**, because the graph of $f(x)$ suggests that name. Figure 3.4-1 shows the graph of $f(x)$ and the distribution function $F(x)$ when $a = 0.30$ and $b = 1.55$. Note that we could have taken $f(a) = 0$ or $f(b) = 0$ without altering the probabilities, since this is a continuous-type distribution. We shall do so in some cases.

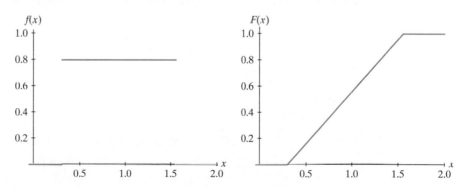

FIGURE 3.4-1: Uniform p.d.f. and distribution function

The mean, variance, and moment-generating function of X are not difficult to calculate. (See Exercise 3.4.) They are, respectively,

$$\mu = \frac{a + b}{2}, \qquad \sigma^2 = \frac{(b - a)^2}{12},$$

$$M(t) = \begin{cases} \dfrac{e^{tb} - e^{ta}}{t(b - a)}, & t \neq 0, \\ 1, & t = 0. \end{cases}$$

An important uniform distribution is that for which $a = 0$ and $b = 1$, namely, $U(0, 1)$. If X is $U(0, 1)$, approximate values of X can be simulated on most computers with the use of a random-number generator. In fact, it should be called a **pseudo-random number generator** because the programs that produce the random numbers are usually such that if the starting number (the seed number) is known, all subsequent numbers in the sequence may be determined by simple arithmetical operations. Yet, despite their deterministic origin, these computer-produced numbers do behave as if they were truly randomly generated, and we shall not encumber our terminology by adding *pseudo*. (Examples of computer-produced random numbers are given in the appendix in Table IX. Place a decimal point in front of each of the four-digit entries so that each is a number between 0 and 1.)

EXAMPLE 3.4-1 Let X have the p.d.f.

$$f(x) = \frac{1}{100}, \qquad 0 < x < 100,$$

so that X is $U(0, 100)$. The mean and the variance are, respectively,

$$\mu = \frac{0 + 100}{2} = 50 \qquad \text{and} \qquad \sigma^2 = \frac{(100 - 0)^2}{12} = \frac{10{,}000}{12}.$$

The standard deviation is $\sigma = 100/\sqrt{12}$, which is 100 times that of the $U(0, 1)$ distribution. This agrees with our intuition, since the standard deviation is a measure of spread and $U(0, 100)$ is clearly spread out 100 times more than $U(0, 1)$. ■

We turn now to a continuous distribution that is related to the Poisson distribution. When previously observing a process of the (approximate) Poisson type, we counted the number of changes occurring in a given interval. This number was a discrete-type random variable with a Poisson distribution. But not only is the number of changes a random variable; the waiting times between successive changes are also random variables. However, the latter are of the continuous type, since each of them can assume any positive value. In particular, let W denote the waiting time until the first change occurs during the observation of a Poisson process in which the mean number of changes in the unit interval is λ. Then W is a continuous-type random variable, and we proceed to find its distribution function.

Because this waiting time is nonnegative, the distribution function $F(w) = 0$, $w < 0$. For $w \geq 0$,

$$\begin{aligned} F(w) = P(W \leq w) &= 1 - P(W > w) \\ &= 1 - P(\text{no changes in } [0, w]) \\ &= 1 - e^{-\lambda w}, \end{aligned}$$

since we previously discovered that $e^{-\lambda w}$ equals the probability of no changes in an interval of length w. That is, if the mean number of changes per unit interval is λ, then the mean number of changes in an interval of length w is proportional to w, and hence is given by λw. Thus, when $w > 0$, the p.d.f. of W is

$$F'(w) = f(w) = \lambda e^{-\lambda w}.$$

We often let $\lambda = 1/\theta$ and say that the random variable X has an **exponential distribution** if its p.d.f. is defined by

$$f(x) = \frac{1}{\theta} e^{-x/\theta}, \qquad 0 \le x < \infty,$$

where the parameter $\theta > 0$. Accordingly, the waiting time W until the first change in a Poisson process has an exponential distribution with $\theta = 1/\lambda$. To determine the exact meaning of the parameter θ, we first find the moment-generating function of X. It is

$$M(t) = \int_0^\infty e^{tx} \left(\frac{1}{\theta} \right) e^{-x/\theta} \, dx = \lim_{b \to \infty} \int_0^b \left(\frac{1}{\theta} \right) e^{-(1-\theta t)x/\theta} \, dx$$

$$= \lim_{b \to \infty} \left[-\frac{e^{-(1-\theta t)x/\theta}}{1 - \theta t} \right]_0^b = \frac{1}{1 - \theta t}, \qquad t < \frac{1}{\theta}.$$

Thus,

$$M'(t) = \frac{\theta}{(1 - \theta t)^2}$$

and

$$M''(t) = \frac{2\theta^2}{(1 - \theta t)^3}.$$

Hence, for an exponential distribution, we have

$$\mu = M'(0) = \theta \qquad \text{and} \qquad \sigma^2 = M''(0) - [M'(0)]^2 = \theta^2.$$

So if λ is the mean number of changes in the unit interval, then $\theta = 1/\lambda$ is the mean waiting time for the first change. In particular, suppose that $\lambda = 7$ is the mean number of changes per minute; then the mean waiting time for the first change is 1/7 of a minute, a result that agrees with our intuition.

EXAMPLE 3.4-2 Let X have an exponential distribution with a mean of $\theta = 20$. Then the p.d.f. of X is that of Example 3.3-1, namely,

$$f(x) = \frac{1}{20} e^{-x/20}, \qquad 0 \le x < \infty.$$

(See Figure 3.3-1 for the graph of this p.d.f.) The probability that X is less than 18 is

$$P(X < 18) = \int_0^{18} \frac{1}{20} e^{-x/20} \, dx = 1 - e^{-18/20} = 0.593. \qquad \blacksquare$$

Let X have an exponential distribution with mean $\mu = \theta$. Then the distribution function of X is

$$F(x) = \begin{cases} 0, & -\infty < x < 0, \\ 1 - e^{-x/\theta}, & 0 \le x < \infty. \end{cases}$$

The p.d.f. and distribution function are graphed in Figure 3.4-2 for $\theta = 5$. The median, m, is found by solving $F(m) = 0.5$. That is,

$$1 - e^{-m/\theta} = 0.5.$$

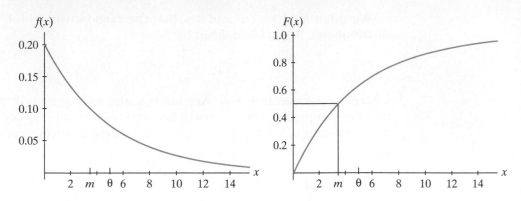

FIGURE 3.4-2: Exponential p.d.f. and distribution function

Thus,

$$m = -\theta \ln(0.5) = \theta \ln(2).$$

So with $\theta = 5$, the median is $m = -5 \ln(0.5) = 3.466$. Both the median and the mean $\theta = 5$ are indicated on the graphs.

It is useful to note that for an exponential random variable, X, we have

$$P(X > x) = 1 - F(x) = 1 - (1 - e^{-x/\theta})$$
$$= e^{-x/\theta} \quad \text{when } x > 0.$$

EXAMPLE 3.4-3 Customers arrive in a certain shop according to an approximate Poisson process at a mean rate of 20 per hour. What is the probability that the shopkeeper will have to wait more than 5 minutes for the arrival of the first customer? Let X denote the waiting time in minutes until the first customer arrives, and note that $\lambda = 1/3$ is the expected number of arrivals per minute. Thus,

$$\theta = \frac{1}{\lambda} = 3$$

and

$$f(x) = \frac{1}{3} e^{-(1/3)x}, \qquad 0 \le x < \infty.$$

Hence,

$$P(X > 5) = \int_5^\infty \frac{1}{3} e^{-(1/3)x} \, dx = e^{-5/3} = 0.1889.$$

The median time until the first arrival is

$$m = -3 \ln(0.5) = 2.0794.$$ ∎

EXAMPLE 3.4-4 Suppose that a certain type of electronic component has an exponential distribution with a mean life of 500 hours. If X denotes the life of this component (or the time to failure of this component), then

$$P(X > x) = \int_x^\infty \frac{1}{500} e^{-t/500} \, dt = e^{-x/500}.$$

If the component has been in operation for 300 hours, the conditional probability that it will last for another 600 hours is

$$P(X > 900 \mid X > 300) = \frac{P(X > 900)}{P(X > 300)} = \frac{e^{-900/500}}{e^{-300/500}} = e^{-6/5}.$$

It is important to note that this conditional probability is exactly equal to $P(X > 600) = e^{-6/5}$. That is, the probability that the component will last an additional 600 hours, given that it has operated for 300 hours, is the same as the probability that it will last 600 hours when first put into operation. Thus, for such components, an old component is as good as a new one, and we say that the failure rate is constant. Certainly, with a constant failure rate, there is no advantage in replacing components that are operating satisfactorily. Obviously, however, this is not true in practice, because most components would have an increasing failure rate with time; hence, the exponential distribution is probably not the best model for the probability distribution of such a life. ∎

REMARK   In Exercise 3.4-11, the result of Example 3.4-4 is generalized; that is, if the component has an exponential distribution, then the probability that it will last a time of at least $x + y$ units, given that it has lasted at least x units, is exactly the same as the probability that it will last at least y units when first put into operation. In effect, this statement says that the exponential distribution has a "forgetfulness" (or "no memory") property. It is also interesting to observe that, for continuous random variables whose support is $(0, \infty)$, the exponential distribution is the only distribution with this forgetfulness property. Recall, however, that when we considered distributions of the discrete type, we noted that the geometric distribution has the property as well. (See Exercise 2.5-12.) ∎

We often want to check whether the order statistics $y_1 \leq y_2 \leq \cdots \leq y_n$ are those of a random sample from some exponential distribution with mean θ. If we knew θ (say, $\theta = \theta_0$), we could construct the q–q plot (y_r, π_p), where $p = r/(n + 1)$ and

$$F(\pi_p) = 1 - e^{-\pi_p/\theta_0} = p$$

or, equivalently,

$$\pi_p = -\theta_0 \ln(1 - p).$$

If θ_0 is the correct value of θ and if the exponential distribution is a suitable model, then (y_r, π_p), $r = 1, 2, \ldots, n$, would lie close to a straight line through the origin with slope 1. However, we do not know θ. Fortunately, there is a way out of our dilemma: Simply plot (y_r, π_p), with $\theta = 1$; that is, $\pi_p = -\ln(1 - p)$. If these points (y_r, π_p) fall close to a straight line through the origin, then the exponential model is appropriate. Moreover, the slope of the line—say, s_0—provides an estimate of θ through $s_0 = 1/\theta$. That is, with $\theta = 1/s_0$, the points $[y_r, \pi_p = -\theta \ln(1 - p)]$ lie close to a straight line through the origin with slope 1.

COMPUTATIONAL COMMENTS   Minitab can be used to find the values of π_p. For example, if $\theta = 1$ and $n = 9$, label two columns p and π_p. Then use

Calc → Make Patterned Data → Simple Set of Numbers

to fill the p column with $0.1, 0.2, \ldots, 0.9$. Next, use the commands

Calc → Probability Distributions → Exponential

to find the values of π_p. Use Inverse cumulative probability with input column p and to fill the column π_p with $0.10536, 0.22314, \ldots, 2.30249$.

EXERCISES

3.4-1. Show that the mean, variance, and moment-generating function of the uniform distribution are as given in this section.

3.4-2. Let $f(x) = 1/2$, $-1 \le x \le 1$, be the p.d.f. of X. Graph the p.d.f. and distribution function, and record the mean and variance of X.

3.4-3. Customers arrive randomly at a bank teller's window. Given that one customer arrived during a particular 10-minute period, let X equal the time within the 10 minutes that the customer arrived. If X is $U(0,10)$, find

(a) the p.d.f. of X,

(b) $P(X \ge 8)$,

(c) $P(2 \le X < 8)$,

(d) $E(X)$, and

(e) $\text{Var}(X)$.

3.4-4. If the moment-generating function of X is

$$M(t) = \frac{e^{5t} - e^{4t}}{t}, \quad t \ne 0, \qquad \text{and} \qquad M(0) = 1,$$

find (a) $E(X)$, (b) $\text{Var}(X)$, and (c) $P(4.2 < X \le 4.7)$.

3.4-5. Let Y have a uniform distribution $U(0,1)$, and let

$$W = a + (b - a)Y, \qquad a < b.$$

(a) Find the distribution function of W.
    HINT: Find $P[a + (b - a)Y \le w]$.

(b) How is W distributed?

3.4-6. Let X have an exponential distribution with a mean of $\theta = 20$.

(a) Compute $P(10 < X < 30)$.

(b) Compute $P(X > 30)$.

(c) Compute $P(X > 40 \mid X > 10)$.

(d) What are the variance and the moment-generating function of X?

(e) The following ordered data were simulated from an exponential distribution with $\theta = 20$:

0.45	0.65	0.66	0.70	0.94	1.05	1.17	1.28	1.32	1.35
1.45	1.52	1.59	1.76	1.92	2.05	2.75	2.88	3.48	3.63
4.01	4.37	4.54	4.84	5.39	5.53	5.95	6.26	6.44	6.79
6.85	7.45	7.51	7.75	8.23	8.41	9.42	9.45	9.57	9.72
9.85	9.97	10.26	10.27	10.34	11.05	11.37	11.42	11.45	11.78
12.26	12.90	13.09	13.19	13.45	16.29	16.55	17.07	17.18	17.32
19.40	19.49	20.04	20.95	21.11	21.13	21.37	22.49	23.34	23.88
25.06	25.41	27.86	27.91	28.32	28.60	29.41	30.84	31.74	33.64
35.42	35.60	36.26	37.06	37.98	39.49	40.13	41.34	44.13	45.11
46.10	47.67	47.68	54.36	55.98	62.81	72.92	76.08	92.16	120.54

Compare the relative frequencies of the appropriate events in the data with the respective probabilities in parts (a), (b), and (c).

3.4-7. The *Holland Sentinel* reported the following numbers of calls per hour received by 911 between noon, February 26, and all day February 27:

3	0	3	4	9	1	6	2	2	5	7	6	4	2	2	4	1	0
3	1	3	3	4	2	1	3	3	2	3	2	5	2	1	0	2	4

The *Sentinel* also reported the following lengths of time per minute between calls:

30	17	65	8	38	35	4	19	7	14	12	4	5	4	2
7	5	12	50	33	10	15	2	10	1	5	30	41	21	31
1	18	12	5	24	7	6	31	1	3	2	22	1	30	2
1	3	12	12	9	28	6	50	63	5	17	11	23	2	46
90	13	21	55	43	5	19	47	24	4	6	27	4	6	37
16	41	68	9	5	28	42	3	42	8	52	2	11	41	4
35	21	3	17	10	16	1	68	105	45	23	5	10	12	17

(a) Show graphically that the numbers of calls per hour have an approximate Poisson distribution with a mean of $\lambda = 3$.

(b) Show that the sample mean and the sample standard deviation of the times between calls are both approximately equal to 20.

(c) If X is an exponential random variable with mean $\theta = 20$, compare the probability $P(X > 15)$ with the proportion of times that are greater than 15.

(d) Compare $P(X > 45.5 \,|\, X > 30.5)$ with the proportion of observations that satisfy this condition.

3.4-8. Telephone calls enter a college switchboard according to a Poisson process on the average of two every 3 minutes. Let X denote the waiting time until the first call that arrives after 10 A.M.

(a) What is the p.d.f. of X?

(b) Find $P(X > 2)$.

3.4-9. What are the p.d.f., the mean, and the variance of X if the moment-generating function of X is given by the following?

(a) $M(t) = \dfrac{1}{1 - 3t}, \qquad t < 1/3.$

(b) $M(t) = \dfrac{3}{3 - t}, \qquad t < 3.$

3.4-10. A biologist is studying the life cycle of the avian schistosome that causes swimmers itch. His study uses Menganser ducks for the adult parasites and aquatic snails as intermediate hosts for the larval stages. The life history is cyclic. (For more information, see http://swimmersitch.org/.) As a part of this study, the biologist and his students used snails from a natural population to measure the distances that snails travel. The conjecture is that snails that had a patent infection would not travel as far as those without such an infection.

Here are the measurements in cm that snails traveled per day. There are 39 in the control group and 31 in the infected group.

Distances for Infected Snail Group (ordered):

263	238	226	220	170	155	139	123	119	107	107	97	90
90	90	79	75	74	71	66	60	55	47	47	47	45
43	41	40	39	38	38	35	32	32	28	19	10	10

Distances for Control Snail Group (ordered):

314	300	274	246	190	186	185	182	180	141	132
129	110	100	95	95	93	83	55	52	50	48
48	44	40	32	30	25	24	18	7		

(a) Find the sample means and sample standard deviations for the two groups of snails.

(b) Make box plots of the two groups of snails on the same graph.

(c) Construct two q–q plots (y_r, π_p), where $p = r/(n + 1)$ and $\pi_p = -\ln(1 - p)$. That is, because we do not know the values of θ, use $\theta = 1$ in constructing the q–q plot.

(d) Do the data appear to be exponentially distributed?

(e) What are your conclusions about the snails?

3.4-11. Let X have an exponential distribution with mean $\theta > 0$. Show that

$$P(X > x + y \,|\, X > x) = P(X > y).$$

3.4-12. Let $F(x)$ be the distribution function of the continuous-type random variable X, and assume that $F(x) = 0$ for $x \leq 0$ and $0 < F(x) < 1$ for $0 < x$. Prove that if

$$P(X > x + y \,|\, X > x) = P(X > y),$$

then

$$F(x) = 1 - e^{-\lambda x}, \qquad 0 < x.$$

HINT: Show that $g(x) = 1 - F(x)$ satisfies the functional equation

$$g(x + y) = g(x)g(y),$$

which implies that $g(x) = a^{cx}$.

3.4-13. Let X equal the number of bad records in each 100 feet of a used computer tape. Assume that X has a Poisson distribution with mean 2.5. Let W equal the number of feet before the first bad record is found.

(a) Give the mean number of flaws per foot.

(b) How is W distributed?

(c) Give the mean and variance of W.

(d) Find (i) $P(W \le 20)$, (ii) $P(W > 40)$, and (iii) $P(W > 60 \mid W > 20)$.

3.4-14. The initial value of an appliance is $700 and its value in the future is given by

$$v(t) = 100 \, (2^{3-t} - 1), \qquad 0 \le t \le 3,$$

where t is time in years. Thus, after the first 3 years the appliance is worth nothing as far as the warranty is concerned. If it fails in the first three years, the warranty pays $v(t)$. Compute the expected value of the payment on the warranty if T has an exponential distribution with mean five.

3.4-15. Let X equal the time (in minutes) between calls that are made over the public safety radio. On four different days (February 14, 21, and 28, and March 6) and during a period of one hour on each day, the following observations of X were made:

5	7	8	20	17	2	24	8	8	6	4
3	42	10	18	5	7	8	4	5	10	

If calls arrive randomly in accordance with an approximate Poisson process, then the distribution of X should be approximately exponential.

(a) Calculate the values of the sample mean and sample standard deviation. Are they close to each other in value?

(b) Construct a q–q plot of the ordered observations versus the respective quartiles of the exponential distribution with a mean of $\theta = 1$. If this plot is approximately linear, the exponential model is supported. Since the mean of these data is not close to 1, the line plotted will not have slope 1, but a linear fit will still indicate an exponential model. What is your conclusion?

3.4-16. A grocery store has n watermelons to sell and makes $1.00 on each sale. Say the number of consumers of these watermelons is a random variable with a distribution that can be approximated by

$$f(x) = \frac{1}{200}, \qquad 0 < x < 200,$$

a p.d.f. of the continuous type. If the grocer does not have enough watermelons to sell to all consumers, she figures that she loses $5.00 in goodwill from each unhappy customer. But if she has surplus watermelons, she loses 50 cents on each extra watermelon. What should n be to maximize profit? HINT: If $X \le n$, then her profit is $(1.00)X + (-0.50)(n - X)$; but if $X > n$, her profit is $(1.00)n + (-5.00)(X - n)$. Find the expected value of profit as a function of n, and then select n to maximize that function.

3.4-17. There are times when a shifted exponential model is appropriate. That is, let the p.d.f. of X be

$$f(x) = \frac{1}{\theta} e^{-(x-\delta)/\theta}, \qquad \delta < x < \infty.$$

(a) Define the distribution function of X.

(b) Calculate the mean and variance of X.

3.4-18. A certain type of aluminum screen 2 feet in width has, on the average, three flaws in a 100-foot roll.

(a) What is the probability that the first 40 feet in a roll contain no flaws?

(b) What assumption did you make to solve part (a)?

3.4-19. Let X have an exponential distribution with mean θ.

(a) Find the first quartile, q_1.

(b) How far is the first quartile below the mean?

(c) Find the third quartile, q_3.

(d) How far is the third quartile above the mean?

3.4-20. Let X have a logistic distribution with p.d.f.

$$f(x) = \frac{e^{-x}}{(1 + e^{-x})^2}, \qquad -\infty < x < \infty.$$

Show that

$$Y = \frac{1}{1 + e^{-X}}$$

has a $U(0,1)$ distribution.

HINT: Find $G(y) = P(Y \le y) = P\left(\dfrac{1}{1 + e^{-X}} \le y\right)$ when $0 < y < 1$.

3.4-21. Suppose that the length of life of a human female, X, is modeled by the exponential p.d.f.

$$f(x) = \frac{1}{80} e^{-x/80}, \qquad 0 < x < \infty.$$

(a) Compute the probability $P(X > 10)$ and also compute the conditional probability $P(X > 90 \mid X > 80)$. Note that the answers are the same, indicating that if length of life has an exponential distribution, then those following it would have a "mathematical fountain of youth." Unfortunately, this is **not** a good model of length of life.

(b) Possibly a more realistic model for length of life is the distribution function

$$F(x) = 1 - \exp[-(a/b)(e^{bx} - 1)],$$
$$0 < x < \infty, \text{ where } a > 0, b > 0.$$

Find and graph $f(x) = F'(x)$

(c) With $a = 0.000025$ and $b = 0.1$, determine the percentile of $x = 70$.

3.4-22. Let the random variable X be equal to the number of days that it takes a high-risk driver to have an accident. Assume that X has an exponential distribution. If $P(X < 50) = 0.25$, compute $P(X > 100 \mid X > 50)$.

3.4-23. A loss (in \$100,000) due to fire in a building has a p.d.f. $f(x) = (1/6)e^{-x/6}, 0 < x < \infty$. Given that the loss is greater than 5, find the probability that it is greater than 8.

3.5 THE GAMMA AND CHI-SQUARE DISTRIBUTIONS

In the (approximate) Poisson process with mean λ, we have seen that the waiting time until the first change has an exponential distribution. We now let W denote the waiting time until the αth change occurs and find the distribution of W.

The distribution function of W when $w \geq 0$ is given by

$$F(w) = P(W \leq w) = 1 - P(W > w)$$
$$= 1 - P(\text{fewer than } \alpha \text{ changes occur in } [0, w])$$
$$= 1 - \sum_{k=0}^{\alpha-1} \frac{(\lambda w)^k e^{-\lambda w}}{k!}, \qquad (3.5\text{-}1)$$

since the number of changes in the interval $[0, w]$ has a Poisson distribution with mean λw. Because W is a continuous-type random variable, $F'(w)$, if it exists, is equal to the p.d.f. of W. Also, provided that $w > 0$, we have

$$F'(w) = \lambda e^{-\lambda w} - e^{-\lambda w} \sum_{k=1}^{\alpha-1} \left[\frac{k(\lambda w)^{k-1} \lambda}{k!} - \frac{(\lambda w)^k \lambda}{k!} \right]$$
$$= \lambda e^{-\lambda w} - e^{-\lambda w} \left[\lambda - \frac{\lambda(\lambda w)^{\alpha-1}}{(\alpha - 1)!} \right]$$
$$= \frac{\lambda(\lambda w)^{\alpha-1}}{(\alpha - 1)!} e^{-\lambda w}.$$

If $w < 0$, then $F(w) = 0$ and $F'(w) = 0$. A p.d.f. of this form is said to be one of the gamma type, and the random variable W is said to have a **gamma distribution**.

Before determining the characteristics of the gamma distribution, let us consider the gamma function for which the distribution is named. The **gamma function** is defined by

$$\Gamma(t) = \int_0^\infty y^{t-1} e^{-y} \, dy, \qquad 0 < t.$$

This integral is positive for $0 < t$ because the integrand is positive. Values of it are often given in a table of integrals. If $t > 1$, integration of the gamma function of t by parts yields

$$\Gamma(t) = \left[-y^{t-1} e^{-y} \right]_0^\infty + \int_0^\infty (t - 1) y^{t-2} e^{-y} \, dy$$
$$= (t - 1) \int_0^\infty y^{t-2} e^{-y} \, dy = (t - 1)\Gamma(t - 1).$$

For example, $\Gamma(6) = 5\Gamma(5)$ and $\Gamma(3) = 2\Gamma(2) = (2)(1)\Gamma(1)$. Whenever $t = n$, a positive integer, we have, by repeated application of $\Gamma(t) = (t - 1)\Gamma(t - 1)$,

$$\Gamma(n) = (n - 1)\Gamma(n - 1) = (n - 1)(n - 2) \cdots (2)(1)\Gamma(1).$$

However,

$$\Gamma(1) = \int_0^\infty e^{-y} \, dy = 1.$$

Thus, when n is a positive integer, we have

$$\Gamma(n) = (n-1)!.$$

For this reason, the gamma function is called the generalized factorial. (Incidentally, $\Gamma(1)$ corresponds to $0!$, and we have noted that $\Gamma(1) = 1$, which is consistent with earlier discussions.)

Let us now formally define the p.d.f. of the gamma distribution and find its characteristics. The random variable X has a **gamma distribution** if its p.d.f. is defined by

$$f(x) = \frac{1}{\Gamma(\alpha)\theta^\alpha} x^{\alpha-1} e^{-x/\theta}, \qquad 0 \le x < \infty.$$

Hence, W, the waiting time until the αth change in a Poisson process, has a gamma distribution with parameters α and $\theta = 1/\lambda$. To see that $f(x)$ actually has the properties of a p.d.f., note that $f(x) \ge 0$ and

$$\int_{-\infty}^\infty f(x) \, dx = \int_0^\infty \frac{x^{\alpha-1} e^{-x/\theta}}{\Gamma(\alpha)\theta^\alpha} \, dx,$$

which, by the change of variables $y = x/\theta$, equals

$$\int_0^\infty \frac{(\theta y)^{\alpha-1} e^{-y}}{\Gamma(\alpha)\theta^\alpha} \theta \, dy = \frac{1}{\Gamma(\alpha)} \int_0^\infty y^{\alpha-1} e^{-y} \, dy = \frac{\Gamma(\alpha)}{\Gamma(\alpha)} = 1.$$

The moment-generating function of X is (Exercise 3.5-3)

$$M(t) = \frac{1}{(1 - \theta t)^\alpha}, \qquad t < 1/\theta.$$

The mean and variance are (Exercise 3.5-4)

$$\mu = \alpha\theta \qquad \text{and} \qquad \sigma^2 = \alpha\theta^2.$$

EXAMPLE 3.5-1 Suppose the number of customers per hour arriving at a shop follows a Poisson process with mean 30. That is, if a minute is our unit, then $\lambda = 1/2$. What is the probability that the shopkeeper will wait more than 5 minutes before both of the first two customers arrive? If X denotes the waiting time in minutes until the second customer arrives, then X has a gamma distribution with $\alpha = 2$, $\theta = 1/\lambda = 2$. Hence,

$$P(X > 5) = \int_5^\infty \frac{x^{2-1} e^{-x/2}}{\Gamma(2)2^2} \, dx = \int_5^\infty \frac{x e^{-x/2}}{4} \, dx$$

$$= \frac{1}{4}\left[(-2)x e^{-x/2} - 4 e^{-x/2}\right]_5^\infty$$

$$= \frac{7}{2} e^{-5/2} = 0.287.$$

We could also have used Equation 3.5-1 with $\lambda = 1/\theta$ because α is an integer. From that equation, we have

$$P(X > x) = \sum_{k=0}^{\alpha-1} \frac{(x/\theta)^k e^{-x/\theta}}{k!}.$$

Thus, with $x = 5$, $\alpha = 2$, and $\theta = 2$, this is equal to

$$P(X > 5) = \sum_{k=0}^{2-1} \frac{(5/2)^k e^{-5/2}}{k!}$$

$$= e^{-5/2}\left(1 + \frac{5}{2}\right) = \left(\frac{7}{2}\right)e^{-5/2}.\qquad\blacksquare$$

EXAMPLE 3.5-2 Telephone calls arrive at a switchboard at a mean rate of $\lambda = 2$ per minute according to a Poisson process. Let X denote the waiting time in minutes until the fifth call arrives. The p.d.f. of X, with $\alpha = 5$ and $\theta = 1/\lambda = 1/2$, is

$$f(x) = \frac{2^5 x^4}{4!} e^{-2x}, \qquad 0 \le x < \infty.$$

The mean and the variance of X are, respectively, $\mu = 5/2$ and $\sigma^2 = 5/4$. $\blacksquare$

In order to see the effect of the parameters on the shape of the p.d.f., several combinations of α and θ have been used for graphs that are displayed in Figure 3.5-1. Note that for a fixed θ, as α increases, the probability moves to the right. The same is true for increasing θ with fixed α. Since $\theta = 1/\lambda$, as θ increases, λ decreases. That is, if $\theta_2 > \theta_1$, then $\lambda_2 = 1/\theta_2 < \lambda_1 = 1/\theta_1$. So if the mean number of changes per unit decreases, the waiting time to observe α changes can be expected to increase.

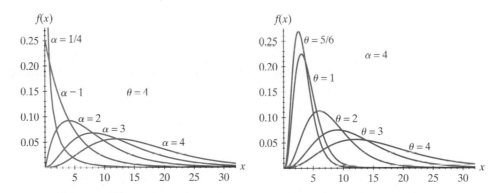

FIGURE 3.5-1: Gamma p.d.f.'s: $\theta = 4$, $\alpha = 1/4, 1, 2, 3, 4$; $\alpha = 4$, $\theta = 5/6, 1, 2, 3, 4$

We now consider a special case of the gamma distribution that plays an important role in statistics. Let X have a gamma distribution with $\theta = 2$ and $\alpha = r/2$, where r is a positive integer. The p.d.f. of X is

$$f(x) = \frac{1}{\Gamma(r/2)2^{r/2}} x^{r/2-1} e^{-x/2}, \qquad 0 \le x < \infty.$$

We say that X has a **chi-square distribution with r degrees of freedom**, which we abbreviate by saying that X is $\chi^2(r)$. The mean and the variance of this chi-square distribution are, respectively,

$$\mu = \alpha\theta = \left(\frac{r}{2}\right)2 = r \qquad \text{and} \qquad \sigma^2 = \alpha\theta^2 = \left(\frac{r}{2}\right)2^2 = 2r.$$

That is, the mean equals the number of degrees of freedom, and the variance equals twice the number of degrees of freedom. An explanation of "number of degrees of freedom" is given later. From the results concerning the more general gamma distribution, we see that its moment-generating function is

$$M(t) = (1 - 2t)^{-r/2}, \qquad t < \frac{1}{2}.$$

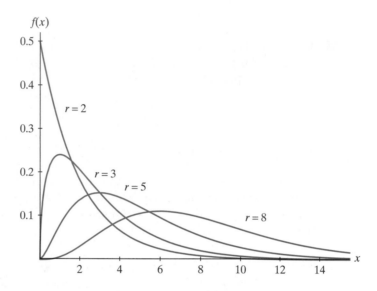

FIGURE 3.5-2: Chi-square p.d.f.'s with $r = 2, 3, 5, 8$

In Figure 3.5-2, the graphs of chi-square p.d.f.'s for $r = 2, 3, 5$, and 8 are given. Note the relationship between the mean $\mu \ (= r)$ and the point at which the p.d.f. attains its maximum. (See Exercise 3.5-13.)

Because the chi-square distribution is so important in applications, tables have been prepared, giving the values of the distribution function

$$F(x) = \int_0^x \frac{1}{\Gamma(r/2)2^{r/2}} w^{r/2-1}e^{-w/2}\,dw$$

for selected values of r and x. (For an example, see Table IV in the appendix.)

EXAMPLE 3.5-3 Let X have a chi-square distribution with $r = 5$ degrees of freedom. Then, using Table IV in the appendix, we obtain

$$P(1.145 \le X \le 12.83) = F(12.83) - F(1.145) = 0.975 - 0.050 = 0.925$$

and

$$P(X > 15.09) = 1 - F(15.09) = 1 - 0.99 = 0.01. \qquad \blacksquare$$

EXAMPLE 3.5-4 If X is $\chi^2(7)$, then two constants, a and b, such that

$$P(a < X < b) = 0.95$$

are $a = 1.690$ and $b = 16.01$. Other constants a and b can be found, and we are restricted in our choices only by the limited table. ∎

Probabilities like that of Example 3.5-4 are so important in statistical applications that we use special symbols for a and b. Let α be a positive probability (that is usually less than 0.5), and let X have a chi-square distribution with r degrees of freedom. Then $\chi^2_\alpha(r)$ is a number such that

$$P[X \geq \chi^2_\alpha(r)] = \alpha.$$

That is, $\chi^2_\alpha(r)$ is the $100(1 - \alpha)$th percentile (or upper 100αth percent point) of the chi-square distribution with r degrees of freedom. Then the 100αth percentile is the number $\chi^2_{1-\alpha}(r)$ such that

$$P[X \leq \chi^2_{1-\alpha}(r)] = \alpha.$$

That is, the probability to the right of $\chi^2_{1-\alpha}(r)$ is $1 - \alpha$. (See Figure 3.5-3.)

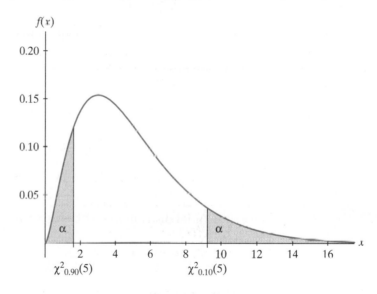

FIGURE 3.5-3: Chi-square tails, $r = 5, \alpha = 0.10$

EXAMPLE 3.5-5 Let X have a chi-square distribution with five degrees of freedom. Then, using Table IV in the appendix, we find that $\chi^2_{0.10}(5) = 9.236$ and $\chi^2_{0.90}(5) = 1.610$. These are the points that are indicated in Figure 3.5-3. ∎

COMPUTATIONAL COMMENTS   Some calculators and most statistical packages give values of the distribution functions and the inverse distribution functions for the gamma and chi-square distributions. For example, here is the Minitab solution for Example 3.5-1: To find the requested probability, select

$$\text{Calc} \rightarrow \text{Probability Distributions} \rightarrow \text{Gamma}$$

and select cumulative probability. Then input the values of the shape parameter $\alpha = 2$, the scale parameter $\theta = 2$, and the constant 5 to obtain

```
Cumulative Distribution Function
Gamma with shape = 2 and scale = 2
x    P( X <= x )
5      0.712703
```

Thus, $P(X > 5) = 1 - P(X \le 5) = 1 - 0.712703 = 0.287297$.

Using Minitab to find the values of $\chi^2_{0.10}(5)$ and $\chi^2_{0.90}(5)$ (see Example 3.5-5), select

$$\text{Calc} \rightarrow \text{Probability Distributions} \rightarrow \text{Chi-Square}$$

and then select inverse cumulative probability with 5 degrees of freedom. First input the value 0.10 and then repeat and input the value 0.90, to obtain

```
Inverse Cumulative Distribution Function
Chi-Square with 5 DF
P( X <= x )          x
      0.1  1.61031
      0.9  9.23636
```

Now see if you can find other values of a and b for Example 3.5-4.

EXAMPLE 3.5-6   If customers arrive at a shop on the average of 30 per hour in accordance with a Poisson process, what is the probability that the shopkeeper will have to wait longer than 9.390 minutes for the first nine customers to arrive? Note that the mean rate of arrivals per minute is $\lambda = 1/2$. Thus, $\theta = 2$ and $\alpha = r/2 = 9$. If X denotes the waiting time until the ninth arrival, then X is $\chi^2(18)$. Hence,

$$P(X > 9.390) = 1 - 0.05 = 0.95.$$ ■

EXAMPLE 3.5-7   If X has an exponential distribution with a mean of 2, then the p.d.f. of X is

$$f(x) = \frac{1}{2} e^{-x/2} = \frac{x^{2/2-1} e^{-x/2}}{\Gamma(2/2)2^{2/2}}, \qquad 0 \le x < \infty.$$

That is, X is $\chi^2(2)$. As an illustration,

$$P(0.051 < X < 7.378) = 0.975 - 0.025 = 0.95.$$ ■

COMPUTATIONAL COMMENTS   A CAS can often be helpful in finding the mean, variance, and moment-generating function. We use *Maple* to illustrate this with the gamma distribution. In the *Maple* code that follows, we first must let *Maple* know the possible values of the parameters. Usually *Maple* denotes assumed variables with a tilde ($\sim$). The first line suppresses this symbol. The mean and variance are calculated by integration. Then the moment-generating function is calculated and is used to find the mean and the variance. Here is the *Maple* code:

```
> interface(showassumed=0);

> assume(alpha > 0): additionally(theta > 0):
```

```
>  f := x^(alpha - 1)*exp(-x/theta)/GAMMA(alpha)/theta^alpha;
```

$$f := \frac{x^{(\alpha-1)} e^{(-\frac{x}{\theta})}}{\Gamma(\alpha)\,\theta^\alpha}$$

```
>  mu := int(x*f, x = 0 .. infinity);
```

$$\mu := \theta\,\alpha$$

```
>  var := factor(int((x - mu)^2*f, x = 0 .. infinity));
```

$$var := \theta^2\,\alpha$$
$$\sigma := \theta\,\sqrt{\alpha}$$

```
>  M := int(exp(t*x)*f, x = 0 .. infinity);
```

$$M := (1 - t\theta)^{(-\alpha)}$$

```
>  Mprime := diff(M, t);
```

$$Mprime := \frac{(1 - t\theta)^{(-\alpha)}\,\alpha\theta}{1 - t\theta}$$

```
>  Mdoubleprime := diff(Mprime, t);
```

$$Mdoubleprime := \frac{(1 - t\theta)^{(-\alpha)}\,\alpha^2\theta^2}{(1 - t\theta)^2} + \frac{(1 - t\theta)^{(-\alpha)}\,\alpha\theta^2}{(1 - t\theta)^2}$$

```
>  mu := subs(t = 0, Mprime);
```

$$\mu := \theta\,\alpha$$

```
>  var := subs(t = 0, Mdoubleprime) - mu^2;
```

$$var := \theta^2\,\alpha$$

It is very easy to find probabilities for the gamma distribution for given values of the parameters.

EXERCISES

3.5-1. Telephone calls enter a college switchboard at a mean rate of two-thirds of a call per minute according to a Poisson process. Let X denote the waiting time until the tenth call arrives.

(a) What is the p.d.f. of X?
(b) What are the moment-generating function, mean, and variance of X?

3.5-2. If X has a gamma distribution with $\theta = 4$ and $\alpha = 2$, find $P(X < 5)$.

3.5-3. Find the moment-generating function for the gamma distribution with parameters α and θ.
HINT: In the integral representing $E(e^{tX})$, change variables by letting $y = (1 - \theta t)x/\theta$, where $1 - \theta t > 0$.

3.5-4. Use the moment-generating function of a gamma distribution to show that $E(X) = \alpha\theta$ and $Var(X) = \alpha\theta^2$.

3.5-5. If the moment-generating function of a random variable W is

$$M(t) = (1 - 7t)^{-20},$$

find the p.d.f., mean, and variance of W.

3.5-6. Let X denote the number of alpha particles emitted by barium-133 and observed by a Geiger counter in a fixed position. Assume that X has a Poisson distribution and $\lambda = 14.7$ is the mean number of counts per second. Let W denote the waiting time to observe 100 counts. Twenty-five independent observations of W are

6.9	7.3	6.7	6.4	6.3
5.9	7.0	7.1	6.5	7.6
7.2	7.1	6.1	7.3	7.6
7.6	6.7	6.3	5.7	6.7
7.5	5.3	5.4	7.4	6.9

(a) Give the p.d.f., mean, and variance of W.
(b) Calculate the sample mean and sample variance of the 25 observations of W.
(c) Use the relative frequency of event $\{W \le 6.6\}$ to approximate $P(W \le 6.6)$.

3.5-7. The waiting times in minutes until two calls to 911, as reported by the *Holland Sentinel* on November 13 between noon and midnight, were

20	28	81	4	9	41	9	11	10	24	20
44	18	30	16	53	15	38	50	84	44	69

Could these times represent a random sample from a gamma distribution with $\alpha = 2$ and $\theta = 120/7$?

(a) Compare the distribution and sample means.
(b) Compare the distribution and sample variances.
(c) Compare $P(X < 35)$ with the proportion of times that are less than 35.
(d) If possible, make some graphical comparisons. For example, if you have access to Minitab, construct a q–q plot for these data.
(e) What is your conclusion?

3.5-8. Let X equal the number of alpha particle emissions of carbon-14 that are counted by a Geiger counter each second. Assume that the distribution of X is Poisson with mean 16. Let W equal the time in seconds before the seventh count is made.

(a) Give the distribution of W.
(b) Find $P(W \le 0.5)$. HINT: Use Equation 3.5-1 with $\lambda w = 8$.

3.5-9. If X is $\chi^2(17)$, find

(a) $P(X < 7.564)$.
(b) $P(X > 27.59)$.
(c) $P(6.408 < X < 27.59)$.
(d) $\chi^2_{0.95}(17)$.
(e) $\chi^2_{0.025}(17)$.

3.5-10. If X is $\chi^2(12)$, find constants a and b such that

$$P(a < X < b) = 0.90 \text{ and } P(X < a) = 0.05.$$

3.5-11. If X is $\chi^2(23)$, find the following:

(a) $P(14.85 < X < 32.01)$.
(b) Constants a and b such that $P(a < X < b) = 0.95$ and $P(X < a) = 0.025$.
(c) The mean and variance of X.
(d) $\chi^2_{0.05}(23)$ and $\chi^2_{0.95}(23)$.

3.5-12. If the moment-generating function of X is $M(t) = (1 - 2t)^{-12}, t < 1/2$, find

(a) $E(X)$.
(b) $\text{Var}(X)$.
(c) $P(15.66 < X < 42.98)$.

3.5-13. Let the distribution of X be $\chi^2(r)$.

(a) Find the point at which the p.d.f. of X attains its maximum when $r \ge 2$. This is the mode of a $\chi^2(r)$ distribution.

(b) Find the points of inflection for the p.d.f. of X.
(c) Use the results of parts (a) and (b) to sketch the p.d.f. of X when $r = 4$ and when $r = 10$.

3.5-14. Cars arrive at a toll booth at a mean rate of five cars every 10 minutes according to a Poisson process. Find the probability that the toll collector will have to wait longer than 26.30 minutes before collecting the eighth toll.

3.5-15. If 15 observations are taken independently from a chi-square distribution with four degrees of freedom, find the probability that at most 3 of the 15 observations exceed 7.779.

3.5-16. If 10 observations are taken independently from a chi-square distribution with 19 degrees of freedom, find the probability that exactly 2 of the 10 sample items exceed 30.14.

3.5-17. In a medical experiment, a rat has been exposed to some radiation. The experimenters believe that the rat's survival time X (in weeks) has the p.d.f.

$$f(x) = \frac{3x^2}{120^3} e^{-(x/120)^3}, \qquad 0 < x < \infty.$$

(a) What is the probability that the rat survives at least 100 weeks?
(b) Find the expected value of the survival time. HINT: In the integral representing $E(X)$, let $y = (x/120)^3$ and get the answer in terms of a gamma function.

3.5-18. Say the serum cholesterol level (X) of U.S. males ages 25–34 follows a translated gamma distribution with p.d.f.

$$f(x) = \frac{x - 80}{50^2} e^{-(x - 80)/50}, \qquad 80 < x < \infty.$$

(a) What are the mean and the variance of this distribution?
(b) What is the mode?
(c) What percentage have the model cholesterol level less than 200? HINT: Integrate by parts.

3.5-19. A bakery sells rolls in units of a dozen. The demand X (in 1000 units) for rolls has a gamma distribution with parameters $\alpha = 3, \theta = 0.5$. It costs 40 cents to make a unit that sells for \$1 on the first day when the rolls are fresh. Any leftover units are sold on the second day at 20 cents. How many units should be made to maximize the expected value of the profit?

3.6 THE NORMAL DISTRIBUTION

Sometimes certain measurements have (approximately) normal distributions. This assumption is sometimes applied too frequently, however; for example, it often bothers us when teachers "grade on the curve" because they believe that their students' grades are normally distributed and frequently they are not. Each student should be able to earn an A grade (or B or C or D) if he or she does the appropriate work; thus, grades should not depend upon what others do. Nevertheless, the normal distribution is extremely important and we must study its major characteristics.

In this section, we give the definition of the p.d.f. for the normal distribution, verify that it is a p.d.f., and then justify the use of μ and σ^2 in its formula. That is, we will show that μ and σ^2 are actually the mean and the variance of this distribution. Toward that end, the random variable X has a **normal distribution** if its p.d.f. is defined by

$$f(x) = \frac{1}{\sigma\sqrt{2\pi}} \exp\left[-\frac{(x-\mu)^2}{2\sigma^2}\right], \qquad -\infty < x < \infty,$$

where μ and σ are parameters satisfying $-\infty < \mu < \infty$ and $0 < \sigma < \infty$, and also where $\exp[v]$ means e^v. Briefly, we say that X is $N(\mu, \sigma^2)$.

Clearly, $f(x) > 0$. We now evaluate the integral

$$I = \int_{-\infty}^{\infty} \frac{1}{\sigma\sqrt{2\pi}} \exp\left[-\frac{(x-\mu)^2}{2\sigma^2}\right] dx$$

and show that it is equal to 1. In I, change the variables of integration by letting $z = (x-\mu)/\sigma$. Then

$$I = \int_{-\infty}^{\infty} \frac{1}{\sqrt{2\pi}} e^{-z^2/2} dz.$$

Since $I > 0$, it follows that if $I^2 = 1$, then $I = 1$. Now,

$$I^2 = \frac{1}{2\pi} \left[\int_{-\infty}^{\infty} e^{-x^2/2} dx\right]\left[\int_{-\infty}^{\infty} e^{-y^2/2} dy\right],$$

or, equivalently,

$$I^2 = \frac{1}{2\pi} \int_{-\infty}^{\infty} \int_{-\infty}^{\infty} \exp\left(-\frac{x^2 + y^2}{2}\right) dx\, dy.$$

Letting $x = r\cos\theta$, $y = r\sin\theta$ (i.e., using polar coordinates), we have

$$I^2 = \frac{1}{2\pi} \int_0^{2\pi} \int_0^{\infty} e^{-r^2/2} r\, dr\, d\theta$$

$$= \frac{1}{2\pi} \int_0^{2\pi} d\theta = \frac{1}{2\pi} 2\pi = 1.$$

Thus, $I = 1$, and we have shown that $f(x)$ has the properties of a p.d.f. The moment-generating function of X is

$$M(t) = \int_{-\infty}^{\infty} \frac{e^{tx}}{\sigma\sqrt{2\pi}} \exp\left[-\frac{(x-\mu)^2}{2\sigma^2}\right] dx$$

$$= \int_{-\infty}^{\infty} \frac{1}{\sigma\sqrt{2\pi}} \exp\left\{-\frac{1}{2\sigma^2}[x^2 - 2(\mu + \sigma^2 t)x + \mu^2]\right\} dx.$$

To evaluate this integral, we complete the square in the exponent:

$$x^2 - 2(\mu + \sigma^2 t)x + \mu^2 = [x - (\mu + \sigma^2 t)]^2 - 2\mu\sigma^2 t - \sigma^4 t^2.$$

Hence,

$$M(t) = \exp\left(\frac{2\mu\sigma^2 t + \sigma^4 t^2}{2\sigma^2}\right) \int_{-\infty}^{\infty} \frac{1}{\sigma\sqrt{2\pi}} \exp\left\{-\frac{1}{2\sigma^2}[x - (\mu + \sigma^2 t)]^2\right\} dx.$$

Note that the integrand in the last integral is like the p.d.f. of a normal distribution with μ replaced by $\mu + \sigma^2 t$. However, the normal p.d.f. integrates to 1 for all real μ—in particular, when μ is replaced by $\mu + \sigma^2 t$. Thus,

$$M(t) = \exp\left(\frac{2\mu\sigma^2 t + \sigma^4 t^2}{2\sigma^2}\right) = \exp\left(\mu t + \frac{\sigma^2 t^2}{2}\right).$$

Now,

$$M'(t) = (\mu + \sigma^2 t) \exp\left(\mu t + \frac{\sigma^2 t^2}{2}\right)$$

and

$$M''(t) = [(\mu + \sigma^2 t)^2 + \sigma^2] \exp\left(\mu t + \frac{\sigma^2 t^2}{2}\right).$$

Consequently,

$$E(X) = M'(0) = \mu,$$
$$\text{Var}(X) = M''(0) - [M'(0)]^2 = \mu^2 + \sigma^2 - \mu^2 = \sigma^2.$$

That is, the parameters μ and σ^2 in the p.d.f. of X are the mean and the variance of X.

EXAMPLE 3.6-1 If the p.d.f. of X is

$$f(x) = \frac{1}{\sqrt{32\pi}} \exp\left[-\frac{(x+7)^2}{32}\right], \qquad -\infty < x < \infty,$$

then X is $N(-7, 16)$. That is, X has a normal distribution with a mean $\mu = -7$, a variance $\sigma^2 = 16$, and the moment-generating function

$$M(t) = \exp(-7t + 8t^2). \qquad \blacksquare$$

EXAMPLE 3.6-2   If the moment-generating function of X is

$$M(t) = \exp(5t + 12t^2),$$

then X is $N(5, 24)$, and its p.d.f. is

$$f(x) = \frac{1}{\sqrt{48\pi}} \exp\left[-\frac{(x-5)^2}{48}\right], \qquad -\infty < x < \infty. \qquad \blacksquare$$

If Z is $N(0, 1)$, we shall say that Z has a **standard normal distribution**. Moreover, the distribution function of Z is

$$\Phi(z) = P(Z \le z) = \int_{-\infty}^{z} \frac{1}{\sqrt{2\pi}} e^{-w^2/2} \, dw.$$

It is not possible to evaluate this integral by finding an antiderivative that can be expressed as an elementary function. However, numerical approximations for integrals of this type have been tabulated and are given in Tables Va and Vb in the appendix. The bell-shaped curved in Figure 3.6-1 represents the graph of the p.d.f. of Z, and the shaded area equals $\Phi(z_0)$.

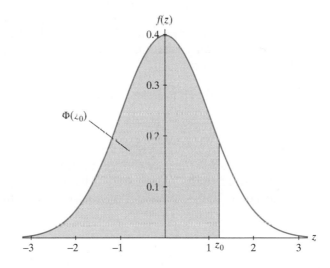

FIGURE 3.6-1: Standard normal p.d.f.

Values of $\Phi(z)$ for $z \ge 0$ are given in Table Va in the appendix. Because of the symmetry of the standard normal p.d.f., it is true that $\Phi(-z) = 1 - \Phi(z)$ for all real z. Thus, Table Va is enough. However, it is sometimes convenient to be able to read $\Phi(-z)$, for $z > 0$, directly from a table. This can be done by using values in Table Vb in the appendix, which lists right-tail probabilities. Again, because of the symmetry of the standard normal p.d.f., when $z > 0$, $\Phi(-z) = P(Z \le -z) = P(Z > z)$ can be read directly from Table Vb.

EXAMPLE 3.6-3   If Z is $N(0, 1)$, then, using Table Va in the appendix, we obtain

$$P(Z \le 1.24) = \Phi(1.24) = 0.8925,$$

$$P(1.24 \leq Z \leq 2.37) = \Phi(2.37) - \Phi(1.24) = 0.9911 - 0.8925 = 0.0986,$$
$$P(-2.37 \leq Z \leq -1.24) = P(1.24 \leq Z \leq 2.37) = 0.0986.$$

Now using Table Vb, we find that

$$P(Z > 1.24) = 0.1075,$$
$$P(Z \leq -2.14) = P(Z \geq 2.14) = 0.0162,$$

and using both tables, we obtain

$$P(-2.14 \leq Z \leq 0.77) = P(Z \leq 0.77) - P(Z \leq -2.14)$$
$$= 0.7794 - 0.0162 = 0.7632. \quad \blacksquare$$

There are times when we want to read the normal probability table in the opposite way, essentially finding the inverse of the standard normal distribution function. That is, given a probability p, we find a constant a so that $P(Z \leq a) = p$. This situation is illustrated in the next example.

EXAMPLE 3.6-4 If the distribution of Z is $N(0,1)$, then, to find constants a and b such that

$$P(Z \leq a) = 0.9147 \qquad \text{and} \qquad P(Z \geq b) = 0.0526,$$

we find the respective probabilities in Tables Va and Vb in the appendix and read off the corresponding values of z. From Table Va, we see that $a = 1.37$, and from Table Vb, we see that $b = 1.62$. $\quad \blacksquare$

In statistical applications, we are often interested in finding a number z_α such that

$$P(Z \geq z_\alpha) = \alpha,$$

where Z is $N(0,1)$ and α is usually less than 0.5. That is, z_α is the $100(1 - \alpha)$th percentile (sometimes called the upper 100α percent point) for the standard normal distribution. (See Figure 3.6-2.) The value of z_α is given in Table Va for selected values of α. For other values of α, z_α can be found in Table Vb.

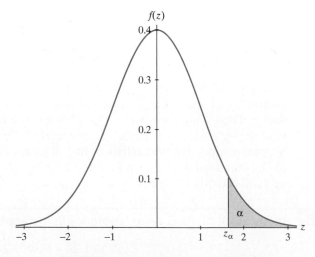

FIGURE 3.6-2: Upper 100α percent point, z_α

Because of the symmetry of the normal p.d.f.,

$$P(Z \le -z_\alpha) = P(Z \ge z_\alpha) = \alpha.$$

Also, since the subscript of z_α is the right-tail probability,

$$z_{1-\alpha} = -z_\alpha.$$

For example,

$$z_{0.95} = z_{1-0.05} = -z_{0.05}.$$

EXAMPLE 3.6-5   To find $z_{0.0125}$, note that

$$P(Z \ge z_{0.0125}) = 0.0125.$$

Thus,

$$z_{0.0125} = 2.24,$$

from Table Vb in the appendix. Also,

$$z_{0.05} = 1.645 \qquad \text{and} \qquad z_{0.025} = 1.960,$$

from the last rows of Table Va.   ∎

REMARK   Recall that the $(100p)$th percentile, π_p, for a random variable X is a number such that $P(X \le \pi_p) = p$. If Z is $N(0,1)$, then since

$$P(Z \ge z_\alpha) = \alpha,$$

it follows that

$$P(Z < z_\alpha) = 1 - \alpha.$$

Thus, z_α is the $100(1-\alpha)$th percentile for the standard normal distribution, $N(0,1)$. For example, $z_{0.05} = 1.645$ is the $100(1-0.05) - 95$th percentile and $z_{0.95} = -1.645$ is the $100(1-0.95) = 5$th percentile.   ∎

The next theorem shows that if X is $N(\mu, \sigma^2)$, then the random variable $(X - \mu)/\sigma$ is $N(0,1)$. Hence, Tables Va and Vb in the appendix can be used to find probabilities relating to X.

THEOREM 3.6-1   If X is $N(\mu, \sigma^2)$, then $Z = (X - \mu)/\sigma$ is $N(0,1)$.

Proof.   The distribution function of Z is

$$P(Z \le z) = P\left(\frac{X - \mu}{\sigma} \le z\right) = P(X \le z\sigma + \mu)$$

$$= \int_{-\infty}^{z\sigma+\mu} \frac{1}{\sigma\sqrt{2\pi}} \exp\left[-\frac{(x-\mu)^2}{2\sigma^2}\right] dx.$$

Now, in the integral representing $P(Z \le z)$, we use the change of variable of integration given by $w = (x - \mu)/\sigma$ (i.e., $x = w\sigma + \mu$) to obtain

$$P(Z \le z) = \int_{-\infty}^{z} \frac{1}{\sqrt{2\pi}} e^{-w^2/2} dw.$$

But this is the expression for $\Phi(z)$, the distribution function of a standardized normal random variable. Hence, Z is $N(0,1)$. □

REMARK   Note that if $E(X) = \mu$ and $E[(X - \mu)^2] = \sigma^2$ exist and $Z = (X - \mu)/\sigma$, then

$$\mu_Z = E(Z) = E\left[\frac{X - \mu}{\sigma}\right] = \frac{E(X) - \mu}{\sigma} = 0$$

and

$$\sigma_Z^2 = E\left[\left(\frac{X - \mu}{\sigma}\right)^2\right] = \frac{E[(X - \mu)^2]}{\sigma^2} = \frac{\sigma^2}{\sigma^2} = 1.$$

That is, the mean and the variance of Z are 0 and 1, respectively, no matter what the distribution of X. The important aspect of the theorem is that if X is normally distributed, then Z is normally distributed—of course with zero mean and unit variance. ■

Theorem 3.6-1 can be used to find probabilities relating to X, which is $N(\mu, \sigma^2)$, as follows:

$$P(a \leq X \leq b) = P\left(\frac{a - \mu}{\sigma} \leq \frac{X - \mu}{\sigma} \leq \frac{b - \mu}{\sigma}\right) = \Phi\left(\frac{b - \mu}{\sigma}\right) - \Phi\left(\frac{a - \mu}{\sigma}\right),$$

since $(X - \mu)/\sigma$ is $N(0,1)$.

EXAMPLE 3.6-6   If X is $N(3,16)$, then

$$P(4 \leq X \leq 8) = P\left(\frac{4 - 3}{4} \leq \frac{X - 3}{4} \leq \frac{8 - 3}{4}\right)$$
$$= \Phi(1.25) - \Phi(0.25) = 0.8944 - 0.5987 = 0.2957,$$

$$P(0 \leq X \leq 5) = P\left(\frac{0 - 3}{4} \leq Z \leq \frac{5 - 3}{4}\right)$$
$$= \Phi(0.5) - \Phi(-0.75) = 0.6915 - 0.2266 = 0.4649,$$

and

$$P(-2 \leq X \leq 1) = P\left(\frac{-2 - 3}{4} \leq Z \leq \frac{1 - 3}{4}\right)$$
$$= \Phi(-0.5) - \Phi(-1.25) = 0.3085 - 0.1056 = 0.2029. ■$$

EXAMPLE 3.6-7   If X is $N(25,36)$, we find a constant c such that

$$P(|X - 25| \leq c) = 0.9544.$$

We want

$$P\left(\frac{-c}{6} \leq \frac{X - 25}{6} \leq \frac{c}{6}\right) = 0.9544.$$

Thus,

$$\Phi\left(\frac{c}{6}\right) - \left[1 - \Phi\left(\frac{c}{6}\right)\right] = 0.9544$$

and

$$\Phi\left(\frac{c}{6}\right) = 0.9772.$$

Hence, $c/6 = 2$ and $c = 12$. That is, the probability that X falls within two standard deviations of its mean is the same as the probability that the standard normal variable Z falls within two units (standard deviations) of zero.   ■

In the next theorem, we give a relationship between the chi-square and normal distributions.

THEOREM 3.6-2	If the random variable X is $N(\mu, \sigma^2)$, $\sigma^2 > 0$, then the random variable $V = (X - \mu)^2/\sigma^2 = Z^2$ is $\chi^2(1)$.

Proof. Because $V = Z^2$, where $Z = (X - \mu)/\sigma$ is $N(0,1)$, the distribution function $G(v)$ of V is, for $v \geq 0$,

$$G(v) = P(Z^2 \leq v) = P(-\sqrt{v} \leq Z \leq \sqrt{v}).$$

That is, with $v \geq 0$,

$$G(v) = \int_{-\sqrt{v}}^{\sqrt{v}} \frac{1}{\sqrt{2\pi}} e^{-z^2/2} \, dz = 2 \int_0^{\sqrt{v}} \frac{1}{\sqrt{2\pi}} e^{-z^2/2} \, dz.$$

If we change the variable of integration by writing $z = \sqrt{y}$, then, since $D_y(z) = 1/(2\sqrt{y})$, we have

$$G(v) = \int_0^{v} \frac{1}{\sqrt{2\pi y}} e^{-y/2} \, dy, \qquad 0 \leq v.$$

Of course, $G(v) = 0$ when $v < 0$. Hence, the p.d.f. $g(v) = G'(v)$ of the continuous-type random variable V is, by one form of the fundamental theorem of calculus,

$$g(v) = \frac{1}{\sqrt{\pi}\sqrt{2}} v^{1/2-1} e^{-v/2}, \qquad 0 < v < \infty.$$

Since $g(v)$ is a p.d.f., it must be true that

$$\int_0^{\infty} \frac{1}{\sqrt{\pi}\sqrt{2}} v^{1/2-1} e^{-v/2} \, dv = 1.$$

The change of variables $x = v/2$ yields

$$1 = \frac{1}{\sqrt{\pi}} \int_0^{\infty} x^{1/2-1} e^{-x} \, dx = \frac{1}{\sqrt{\pi}} \Gamma\left(\frac{1}{2}\right).$$

Hence, $\Gamma(1/2) = \sqrt{\pi}$, and it follows that V is $\chi^2(1)$.   □

EXAMPLE 3.6-8

If Z is $N(0,1)$, then

$$P(|Z| < 1.96 = \sqrt{3.841}\,) = 0.95$$

and, of course,

$$P(Z^2 < 3.841) = 0.95$$

from the chi-square table with $r = 1$.   ∎

Given a set of observations of a random variable X, it is a challenge to determine the distribution of X. In particular, how can we decide whether or not X has an approximate normal distribution? If we have a large number of observations of X, a stem-and-leaf diagram or a histogram of the observations can often be helpful. (See Exercise 3.6-12 and 3.6-19, respectively.) For small samples, a q–q plot can be used to check on whether the sample arises from a normal distribution. For example, suppose the quantiles of a sample were plotted against the corresponding quantiles of a certain normal distribution and the pairs of points generated were on a straight line with slope 1 and intercept 0. Of course, we would then believe that we have an ideal sample from that normal distribution with that certain mean and standard deviation. Such a plot, however, requires that we know the mean and the standard deviation of this normal distribution, and we usually do not. However, since the quantile, q_p, of $N(\mu, \sigma^2)$ is related to the corresponding one, z_{1-p}, of $N(0,1)$ by $q_p = \mu + \sigma z_{1-p}$, we can always plot the quantiles of the sample against the corresponding ones of $N(0,1)$ and get the needed information. That is, if the sample quantiles are plotted as the x-coordinates of the pairs and the $N(0,1)$ quantiles as the y-coordinates, and if the graph is almost a straight line, then it is reasonable to assume that the sample arises from a normal distribution. Moreover, the reciprocal of the slope of that straight line is a good estimate of the standard deviation σ because $z_{1-p} = (q_p - \mu)/\sigma$.

EXAMPLE 3.6-9   In researching groundwater, it is often important to know the characteristics of the soil at a certain site. Many of these characteristics, such as porosity, are at least partially dependent upon the grain size. The diameter of individual grains of soil can be measured. Here are the diameters (in mm) of 30 randomly selected grains:

1.24	1.36	1.28	1.31	1.35	1.20	1.39	1.35	1.41	1.31
1.28	1.26	1.37	1.49	1.32	1.40	1.33	1.28	1.25	1.39
1.38	1.34	1.40	1.27	1.33	1.36	1.43	1.33	1.29	1.34

For these data, $\bar{x} = 1.33$ and $s^2 = 0.0040$. May we assume that these are observations of a random variable X that is $N(1.33, 0.0040)$? To help answer this question, we shall construct a q–q plot of the standard normal quantiles that correspond to $p = 1/31, 2/31, \ldots, 30/31$ versus the ordered observations. To find these quantiles, it is helpful to use the computer. Although we could utilize Minitab, we shall find the quantiles with *Maple* as follows:

k	Diameters in mm (x)	$p = k/31$	z_{1-p}	k	Diameters in mm (x)	$p = k/31$	z_{1-p}
1	1.20	0.0323	−1.85	16	1.34	0.5161	0.04
2	1.24	0.0645	−1.52	17	1.34	0.5484	0.12
3	1.25	0.0968	−1.30	18	1.35	0.5806	0.20
4	1.26	0.1290	−1.13	19	1.35	0.6129	0.29
5	1.27	0.1613	−0.99	20	1.36	0.6452	0.37
6	1.28	0.1935	−0.86	21	1.36	0.6774	0.46
7	1.28	0.2258	−0.75	22	1.37	0.7097	0.55

k	Diameters in mm (x)	$p = k/31$	z_{1-p}	k	Diameters in mm (x)	$p = k/31$	z_{1-p}
8	1.28	0.2581	−0.65	23	1.38	0.7419	0.65
9	1.29	0.2903	−0.55	24	1.39	0.7742	0.75
10	1.31	0.3226	−0.46	25	1.39	0.8065	0.86
11	1.31	0.3548	−0.37	26	1.40	0.8387	0.99
12	1.32	0.3871	−0.29	27	1.40	0.8710	1.13
13	1.33	0.4194	−0.20	28	1.41	0.9032	1.30
14	1.33	0.4516	−0.12	29	1.43	0.9355	1.52
15	1.33	0.4839	−0.04	30	1.49	0.9677	1.85

A q–q plot of these data is shown in Figure 3.6-3. Note that the points do fall close to a straight line, so the normal probability model seems to be appropriate on the basis of these few data.

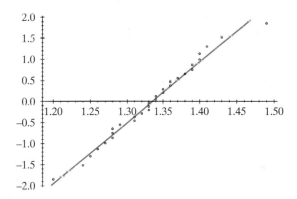

FIGURE 3.6-3: q–q, $N(0, 1)$ quantiles versus grain diameters

EXERCISES

3.6-1. If Z is $N(0,1)$, find

(a) $P(0.53 < Z \le 2.06)$.
(b) $P(-0.79 \le Z < 1.52)$.
(c) $P(Z > -1.77)$.   (d) $P(Z > 2.89)$.
(e) $P(|Z| < 1.96)$.   (f) $P(|Z| < 1)$.
(g) $P(|Z| < 2)$.   (h) $P(|Z| < 3)$.

3.6-2. If Z is $N(0,1)$, find

(a) $P(0 \le Z \le 0.87)$.
(b) $P(-2.64 \le Z \le 0)$.
(c) $P(-2.13 \le Z \le -0.56)$.   (d) $P(|Z| > 1.39)$.
(e) $P(Z < -1.62)$.   (f) $P(|Z| > 1)$.
(g) $P(|Z| > 2)$.   (h) $P(|Z| > 3)$.

3.6-3. Find the values of (a) $z_{0.01}$, (b) $-z_{0.005}$, (c) $z_{0.0475}$, and (d) $z_{0.985}$.

3.6-4. Find the values of (a) $z_{0.10}$, (b) $-z_{0.05}$, (c) $-z_{0.0485}$, and (d) $z_{0.9656}$.

3.6-5. If Z is $N(0,1)$, find values of c such that

(a) $P(Z \ge c) = 0.025$.   (b) $P(|Z| \le c) = 0.95$.
(c) $P(Z > c) = 0.05$.   (d) $P(|Z| \le c) = 0.90$.

3.6-6. If the moment-generating function of X is $M(t) = \exp(166t + 200t^2)$, find

(a) The mean of X.   (b) The variance of X.
(c) $P(170 < X < 200)$.   (d) $P(148 \le X \le 172)$.

3.6-7. If X is normally distributed with a mean of 6 and a variance of 25, find

(a) $P(6 \le X \le 12)$.   (b) $P(0 \le X \le 8)$.
(c) $P(-2 < X \le 0)$.   (d) $P(X > 21)$.
(e) $P(|X - 6| < 5)$.   (f) $P(|X - 6| < 10)$.
(g) $P(|X - 6| < 15)$.   (h) $P(|X - 6| < 12.41)$.

3.6-8. Let the distribution of X be $N(\mu, \sigma^2)$ Show that the points of inflection of the graph of the p.d.f. of X occur at $x = \mu \pm \sigma$.

3.6-9. If X is $N(650, 625)$, find

(a) $P(600 \leq X < 660)$.

(b) A constant $c > 0$ such that
$P(|X - 650| \leq c) = 0.9544$.

3.6-10. If X is $N(\mu, \sigma^2)$, show that $Y = aX + b$ is $N(a\mu + b, a^2\sigma^2)$, $a \neq 0$. HINT: Find the distribution function $P(Y \leq y)$ of Y, and in the resulting integral, let $w = ax + b$ or, equivalently, $x = (w - b)/a$.

3.6-11. Find the distribution of $W = X^2$ when

(a) X is $N(0, 4)$,

(b) X is $N(0, \sigma^2)$.

3.6-12. A company manufactures windows that are inserted into an automobile. Each window has five studs for attaching it. A pullout test is used to determine the force required to pull a stud out of a window. (Note that this is an example of destructive testing.) Let X equal the force required for pulling studs out of position 4. Sixty observations of X were as follows:

159 150 147 160 155 142 143 151 154 133
151 146 140 146 137 148 154 157 142 153
135 144 135 165 118 158 126 147 123 140
125 151 153 158 144 163 150 150 137 164
137 156 139 134 171 144 160 147 155 175
162 160 149 149 158 152 165 131 150 120

(a) Construct an ordered stem-and-leaf diagram, using $11\bullet, 12*, 12\bullet$, and so on as stems.

(b) Construct a q–q plot, using the ordered array and the corresponding quantiles of $N(0, 1)$.

(You will probably want to use a statistical package on the computer to help you out.)

(c) Does it look like X has a normal distribution?

3.6-13. Some measurements (in mm) were made on specimens of the spider Sosippus floridanus, which is native to Florida. Here are the lengths of 9 female spiders and 9 male spiders.

Female spiders	11.06	13.87	12.93	15.08	17.82
	14.14	12.26	17.82	20.17	

Male spiders	12.26	11.66	12.53	13.00	11.79
	12.46	10.65	10.39	12.26	

(a) Construct a q–q plot, $N(0, 1)$ quantiles versus the ordered female spider lengths. Do they appear to be normally distributed?

(b) Construce a q–q plot, $N(0, 1)$ quantiles versus the ordered male spider lengths. Do they appear to be normally distributed?

3.6-14. A candy maker produces mints that have a label weight of 20.4 grams. Assume that the distribution of the weights of these mints is $N(21.37, 0.16)$.

(a) Let X denote the weight of a single mint selected at random from the production line. Find $P(X > 22.07)$.

(b) Suppose that 15 mints are selected independently and weighed. Let Y equal the number of these mints that weigh less than 20.857 grams. Find $P(Y \leq 2)$.

3.6-15. A chemistry major weighed 19 plain M&M's (in grams) on a ± 0.0001 scale. The ordered weights are as follows:

0.7938 0.8032 0.8089 0.8222 0.8268 0.8383 0.8442
0.8490 0.8528 0.8572 0.8674 0.8734 0.8786
0.8850 0.8873 0.8920 0.9069 0.9150 0.9243

(a) Construct a stem-and-leaf diagram using three-digit leaves, with stems $0.7\bullet, 0.8*, 0.8t$, and so on.

(b) Construct a q-q plot of these data and the corresponding quantiles for the standard normal distribution, $N(0, 1)$.

(c) Do the data look like observations from a normal distribution? Why?

3.6-16. If the moment-generating function of X is $M(t) = e^{500t + 5000t^2}$, find $P[27,060 \leq (X - 500)^2 \leq 50,240]$.

3.6-17. If X is $N(7, 4)$, find $P[15.364 \leq (X - 7)^2 \leq 20.096]$.

3.6-18. The strength X of a certain material is such that its distribution is found by $X = e^Y$, where Y is $N(10, 1)$. Find the distribution function and p.d.f.

of X, and compute $P(10{,}000 < X < 20{,}000)$. NOTE: The random variable X is said to have a **lognormal distribution**.

3.6-19. A packaged product has a label weight of 450 grams. The company goal is to fill each package with at least 450 grams, but at most 458 grams. To check these goals, a random sample of 100 packages was selected and weighed, yielding the following weights, rounded to the nearest gram:

457 457 455 457 454 454 457 455 456 459
457 458 456 456 461 457 458 452 457 460
453 458 452 454 454 456 455 456 451 454
456 457 457 453 455 459 458 457 458 457
461 457 455 458 458 455 457 458 456 463
455 455 455 456 456 456 455 456 460 456
456 457 458 454 455 456 459 457 457 451
450 453 453 459 450 453 452 458 456 457
451 458 456 460 455 455 456 460 457 456
457 456 460 459 457 455 461 455 457 457

(a) Construct a frequency table and a corresponding relative frequency histogram.

(b) Do these data seem to come from a normal distribution with mean μ about equal to $\bar{x} = 456.2$ and variance about equal to $s^2 = 5.96$? Sketch the p.d.f. for the normal distribution $N(456.2, 5.96)$ on your histogram.

3.6-20. Nine measurements are taken on the strength of a certain metal. In order, they are 7.2, 8.9, 9.7, 10.5, 10.9, 11.7, 12.9, 13.9, 15.3, and these values correspond to the 10th, 20th, $\cdots$, 90th percentiles of this sample. Plot the measurements against the same percentiles of $N(0, 1)$. Does it seem reasonable that the underlying distribution of strengths could be normal?

3.6-21. The graphs of the moment-generating functions of three normal distributions—$N(0,1)$, $N(-1,1)$, and $N(2,1)$—are given in Figure 3.6-4(a). Identify them.

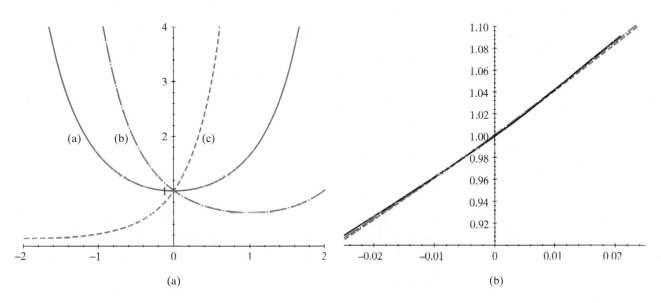

FIGURE 3.6-4: Moment-generating functions

3.6-22. Figure 3.6-4(b) shows the graphs of the following three moment-generating functions:

$$g_1(t) = \frac{1}{1 - 4t}, \qquad t < 1/4,$$

$$g_2(t) = \frac{1}{(1 - 2t)^2}, \qquad t < 1/2,$$

$$g_3(t) = e^{4t + t^2/2}.$$

Why do these three graphs look so similar around $t = 0$?

3.6-23. The serum zinc level X in micrograms per deciliter for males between ages 15 and 17 has a distribution that is approximately normal with $\mu = 90$ and $\sigma = 15$. Compute the conditional probability $P(X > 120 \,|\, X > 105)$.

3.6-24. An interior automotive supplier places several electrical wires in a harness. A pull test measures the force required to pull spliced wires apart. A customer requires that each wire that is spliced into the harness withstand a pull force of 20 pounds. Let X equal the force required to pull the spliced wire apart. The following data give the values of a random sample of $n = 20$ observations of X:

$$
\begin{array}{cccccccccc}
28.8 & 24.4 & 30.1 & 25.6 & 26.4 & 23.9 & 22.1 & 22.5 & 27.6 & 28.1 \\
20.8 & 27.7 & 24.4 & 25.1 & 24.6 & 26.3 & 28.2 & 22.2 & 26.3 & 24.4
\end{array}
$$

(a) Construct a q–q plot, using the ordered array and the corresponding quantiles of $N(0,1)$.

(b) Does it look like X has a normal distribution?

3.6-25. The "fill" problem is important in many industries, such as those making cereal, toothpaste, beer, and so on. If an industry claims that it is selling 12 ounces of its product in a container, it must have a mean greater than 12 ounces, or else the FDA will crack down, although the FDA will allow a very small percentage of the containers to have less than 12 ounces.

(a) If the content X of a container has a $N(12.1, \sigma^2)$ distribution, find σ so that $P(X < 12) = 0.01$.

(b) If $\sigma = 0.05$, find μ so that $P(X < 12) = 0.01$.

3.6-26. The ordered weights (in grams) of 24 bags of peanut M & M's, with a label weight of 49.2 g, are as follows:

$$
\begin{array}{cccccccc}
52.3 & 53.6 & 54.0 & 54.1 & 54.4 & 54.5 & 54.7 & 55.2 \\
55.4 & 55.4 & 55.5 & 55.7 & 55.9 & 56.1 & 56.5 & 56.5 \\
57.0 & 57.0 & 57.3 & 57.3 & 57.3 & 57.8 & 59.6 & 59.7
\end{array}
$$

(a) Calculate the sample mean and sample standard deviation of these weights.

(b) Construct a q–q plot, using the ordered weights and the percentiles of a standard normal distribution.

(c) Do these data look like they come from a normal distribution?

(d) Are the fill weights appropriate for this label weight?

3.6-27. Assume that the fill X of a filling machine for a beverage has a normal distribution with $\mu = 12.2$ and $\sigma = 0.1$, measured in fluid ounces.

(a) Compute $P(X < 12)$.

(b) In 50 independent such measurements, compute the probability that at least 1 is under 12 ounces.

3.7 ADDITIONAL MODELS

The binomial, Poisson, gamma, chi-square, and normal models are frequently used in statistics. However, many other interesting and very useful models can be found. We begin with a modification of one of the postulates of an approximate Poisson process as given in Section 2.6. In that definition, the numbers of changes occurring in nonoverlapping intervals are independent, and the probability of at least two changes in a sufficiently small interval is essentially zero. We continue to use these postulates, but now we say that the probability of exactly one change in a sufficiently short interval of length h is approximately λh, *where λ is a nonnegative function of the position of this interval.* To be explicit, say $p(x, w)$ is the probability of x changes in the interval $(0, w)$, $0 \le w$. Then the last postulate, in more formal terms, becomes

$$
p(x + 1, w + h) - p(x, w) \approx \lambda(w)h,
$$

where $\lambda(w)$ is a nonnegative function of w. This means that if we want the approximate probability of zero changes in the interval $(0, w + h)$, we could take, from the independence of the changes, the probability of zero changes in the interval $(0, w)$ times that of zero changes in the interval $(w, w + h)$. That is,

$$
p(0, w + h) \approx p(0, w)[1 - \lambda(w)h],
$$

because the probability of one or more changes in $(w, w + h)$ is about equal to $\lambda(w)h$. Equivalently,

$$\frac{p(0, w + h) - p(0, w)}{h} \approx -\lambda(w)p(0, w).$$

Taking limits as $h \to 0$, we have

$$D_w[p(0, w)] = -\lambda(w)p(0, w).$$

That is, the resulting differential equation is

$$\frac{D_w[p(0, w)]}{p(0, w)} = -\lambda(w);$$

thus,

$$\ln p(0, w) = -\int \lambda(w)\, dw + c_1.$$

Therefore,

$$p(0, w) = \exp\left[-\int \lambda(w)\, dw + c_1\right] = c_2 \exp\left[-\int \lambda(w)\, dw\right],$$

where $c_2 = e^{c_1}$. However, the boundary condition of the probability of zero changes in an interval of length zero must be 1; that is, $p(0,0) = 1$. So if we select

$$H(w) = \int \lambda(w)\, dw$$

to be such that $H(0) = 0$, then $c_2 = 1$. That is,

$$p(0, w) = e^{-H(w)},$$

where $H'(w) = \lambda(w)$ and $H(0) = 0$. Hence,

$$H(w) = \int_0^w \lambda(t)\, dt.$$

Suppose that we now let the continuous-type random variable W be the interval necessary to produce the first change. Then the distribution function of W is

$$G(w) = P(W \le w) = 1 - P(W > w), \qquad 0 \le w.$$

Because zero changes in the interval $(0, w)$ are the same as $W > w$, then

$$G(w) = 1 - p(0, w) = 1 - e^{-H(w)}, \qquad 0 \le w.$$

The p.d.f. of W is

$$g(w) = G'(w) = H'(w)e^{-H(w)} = \lambda(w)\exp\left[-\int_0^w \lambda(t)\, dt\right], \qquad 0 \le w.$$

From this formula, we see immediately that, in terms of $g(w)$ and $G(w)$,

$$\lambda(w) = \frac{g(w)}{1 - G(w)}.$$

In many applications of this result, W can be thought of as a random time interval. For example, if one change means "death" or "failure" of the item under consideration, then W is actually the length of life of the item. Usually $\lambda(w)$, which is commonly called the **failure rate** or **force of mortality**, is an increasing function of w. That is, the larger w (the older the item), the better is the chance of failure within a short interval of length h, namely, $\lambda(w)h$. As we review the exponential distribution of Section 3.4, we note that there $\lambda(w)$ is a constant; that is, the failure rate or force of mortality does not increase as the item gets older. If this were true in human populations, it would mean that a person 80 years old would have as much chance of living another year as would a person 20 years old (sort of a mathematical "fountain of youth"). However, a constant failure rate (force of mortality) is not the case in most human populations or in most populations of manufactured items. That is, the failure rate $\lambda(w)$ is usually an increasing function of w. We give two important examples of useful probabilistic models.

EXAMPLE 3.7-1 Let

$$H(w) = \left(\frac{w}{\beta}\right)^{\alpha}, \qquad 0 \le w,$$

so that the failure rate is

$$\lambda(w) = H'(w) = \frac{\alpha w^{\alpha-1}}{\beta^{\alpha}},$$

where $\alpha > 0, \beta > 0$. Then the p.d.f. of W is

$$g(w) = \frac{\alpha w^{\alpha-1}}{\beta^{\alpha}} \exp\left[-\left(\frac{w}{\beta}\right)^{\alpha}\right], \qquad 0 \le w.$$

Frequently, in engineering, this distribution, with appropriate values of α and β, is excellent for describing the life of a manufactured item. Often α is greater than 1 but less than 5. This p.d.f. is frequently called that of the **Weibull distribution** and, in model fitting, is a strong competitor of the gamma p.d.f.

The mean and variance of the Weibull distribution are

$$\mu = \beta \Gamma\left(1 + \frac{1}{\alpha}\right),$$

$$\sigma^2 = \beta^2 \left[\Gamma\left(1 + \frac{2}{\alpha}\right) - \left\{\Gamma\left(1 + \frac{1}{\alpha}\right)\right\}^2\right].$$

Some graphs of Weibull p.d.f.s are shown in Figure 3.7-1. ∎

EXAMPLE 3.7-2 People are often shocked to learn that human mortality increases almost exponentially once a person reaches 25 years of age. Depending on which mortality table is used, one finds that the increase is about 10% each year, which means that the rate of

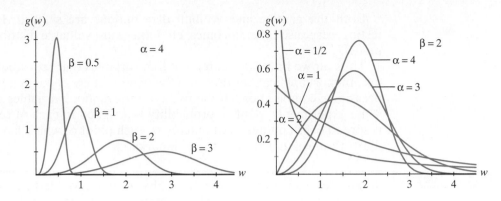

FIGURE 3.7-1: Weibull probability density functions

mortality will double about every 7 years. Although this fact can be shocking, we can be thankful that the force of mortality starts very low. The probability that a person in reasonably good health at age 63 dies within the next year is only about 1%. Now, assuming an exponential force of mortality, we have

$$\lambda(w) = H'(w) = ae^{bw}, \qquad a > 0, \quad b > 0.$$

Thus,

$$H(w) = \int_0^w ae^{bt}\, dt = \frac{a}{b}e^{bw} - \frac{a}{b}.$$

Hence,

$$G(w) = 1 - \exp\left[-\frac{a}{b}e^{bw} + \frac{a}{b}\right], \qquad 0 \le w,$$

and

$$g(w) = ae^{bw}\exp\left[-\frac{a}{b}e^{bw} + \frac{a}{b}\right], \qquad 0 \le w,$$

are, respectively, the distribution function and the p.d.f. associated with the famous **Gompertz law** found in actuarial science. Some graphs of p.d.f.'s associated with the Gompertz law are shown in Figure 3.7-2. Note that the mode of the Gompertz distribution is $\ln(b/a)/b$. ∎

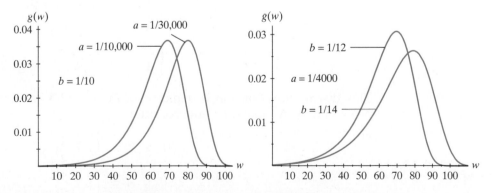

FIGURE 3.7-2: Gompertz law probability density functions

Both the gamma and Weibull distributions are skewed. In many studies (life testing, response times, incomes, etc.), these are valuable distributions for selecting a model.

Thus far we have considered random variables that are either discrete or continuous. In most applications these are the types that are encountered. However, on some occasions, combinations of the two types of random variables are found. That is, in some experiments, positive probability is assigned to each of certain points and also is spread over an interval of outcomes, each point of which has zero probability. An illustration will help clarify these remarks.

EXAMPLE 3.7-3 A bulb for a slide projector is tested by turning it on, letting it burn for 1 hour, and then turning it off. Let X equal the length of time that the bulb performs satisfactorily during this test. There is a positive probability that the bulb will burn out when it is turned on; hence,

$$0 < P(X = 0) < 1,$$

It could also burn out during the 1-hour period during which it is lit; thus,

$$P(0 < X < 1) > 0,$$

with $P(X = x) = 0$ when $x \in (0, 1)$. In addition, $P(X = 1) > 0$. The act of turning the bulb off after 1 hour so that the actual failure time beyond 1 hour is not observed is called censoring, a phenomenon that is considered later in this section. ∎

The distribution function for a distribution of the **mixed type** will be a combination of those for the discrete and continuous types. That is, at each point of positive probability the distribution function will be discontinuous, so that the height of the step there equals the corresponding probability; at all other points the distribution function will be continuous.

EXAMPLE 3.7-4 Let X have a distribution function defined by

$$F(x) = \begin{cases} 0, & x < 0, \\ \dfrac{x^2}{4}, & 0 \le x < 1, \\ \dfrac{1}{2}, & 1 \le x < 2, \\ \dfrac{x}{3}, & 2 \le x < 3, \\ 1, & 3 \le x. \end{cases}$$

This distribution function is depicted in Figure 3.7-3 and can be used to compute probabilities. As an illustration, consider

$$P(0 < X < 1) = \frac{1}{4},$$

$$P(0 < X \le 1) = \frac{1}{2},$$

$$P(X = 1) = \frac{1}{4}.$$
∎

FIGURE 3.7-3: Mixed distribution functions

EXAMPLE 3.7-5 | Consider the following game: An unbiased coin is tossed. If the outcome is heads, the player receives \$2. If the outcome is tails, the player spins a balanced spinner that has a scale from 0 to 1. The player then receives that fraction of a dollar associated with the point selected by the spinner. If X denotes the amount received, the space of X is $S = [0,1) \cup \{2\}$. The distribution function of X is defined by

$$F(x) = \begin{cases} 0, & x < 0, \\ \dfrac{x}{2}, & 0 \leq x < 1, \\ \dfrac{1}{2}, & 1 \leq x < 2, \\ 1, & 2 \leq x. \end{cases}$$

The graph of the distribution function $F(x)$ is given in Figure 3.7-3. ∎

Suppose that the random variable X has a distribution of the mixed type. To find the expectation of the function $u(X)$ of X, a combination of a sum and a Riemann integral is used, as shown in Example 3.7-6.

EXAMPLE 3.7-6 | We shall find the mean and variance of the random variable given in Example 3.7-4. Note that there, $F'(x) = x/2$ when $0 < x < 1$ and $F'(x) = 1/3$ when $2 < x < 3$; also, $P(X = 1) = 1/4$ and $P(X = 2) = 1/6$. Accordingly, we have

$$\mu = E(X) = \int_0^1 x\left(\frac{x}{2}\right) dx + 1\left(\frac{1}{4}\right) + 2\left(\frac{1}{6}\right) + \int_2^3 x\left(\frac{1}{3}\right) dx$$

$$= \left[\frac{x^3}{6}\right]_0^1 + \frac{1}{4} + \frac{1}{3} + \left[\frac{x^2}{6}\right]_2^3$$

$$= \frac{19}{12}$$

and

$$\sigma^2 = E(X^2) - [E(X)]^2$$

$$= \int_0^1 x^2\left(\frac{x}{2}\right) dx + 1^2\left(\frac{1}{4}\right) + 2^2\left(\frac{1}{6}\right) + \int_2^3 x^2\left(\frac{1}{3}\right) dx - \left(\frac{19}{12}\right)^2$$

$$= \frac{31}{48}.$$

∎

Frequently, in life testing, we know that the length of life—say, X—exceeds the number b, but the exact value of X is unknown. This phenomenon is called **censoring**. It can happen, for instance, when a subject in a cancer study simply disappears; the investigator knows that the subject has lived a certain number of months, but the exact length of the subject's life is unknown. Or it might happen when an investigator does not have enough time to observe the moments of deaths of all the animals—say, rats—in some study. Censoring can also occur in the insurance industry, in the case of a loss with a limited-pay policy in which the top amount is exceeded but it is not known by how much.

EXAMPLE 3.7-7 Reinsurance companies are concerned with large losses because they might agree, for example, to cover losses due to wind damages that are between \$2 million and \$10 million. Say that X equals the size of a wind loss in millions of dollars, and suppose that X has the distribution function

$$F(x) = 1 - \left(\frac{10}{10 + x}\right)^3, \qquad 0 \le x < \infty.$$

If losses beyond \$10 million are reported only as 10, then $Y = X$, $X \le 10$, and $Y = 10$, $X > 10$, and the distribution function of this censored distribution is

$$G(y) = \begin{cases} 1 - \left(\dfrac{10}{10 + y}\right)^3, & 0 \le y < 10, \\ 1, & 10 \le y < \infty, \end{cases}$$

which has a jump of $[10/(10 + 10)]^3 = 1/8$ at $y = 10$. ∎

EXAMPLE 3.7-8 A car worth 24 units (1 unit = \$1,000) is insured for a year with a one-unit deductible policy. The probability of no damage in a year is 0.95 and the probability of being totaled is 0.01. If the damage is partial, with probability 0.04, then this damage follows the p.d.f.

$$f(x) = \frac{25}{24} \frac{1}{(x + 1)^2}, \qquad 0 < x < 24.$$

In computing the expected payment, the insurance company recognizes that it will make zero payment if $X \le 1$, $24 - 1 = 23$ if the car is totaled, and $X - 1$ if $1 < X < 24$. Thus, the expected payment is

$$(0)(0.95) + (0)(0.04)\int_0^1 \frac{25}{24} \frac{1}{(x + 1)^2}\, dx + (23)(0.01)$$

$$+ (0.04)\int_1^{24} (x - 1)\frac{25}{24} \frac{1}{(x + 1)^2}\, dx.$$

That is, the answer is

$$0.23 + (0.04)(1.67) = 0.297,$$

because the last integral is equal to

$$\int_1^{24} (x + 1 - 2)\frac{25}{24}\frac{1}{(x + 1)^2}\, dx = (-2)\int_1^{24}\frac{25}{24}\frac{1}{(x + 1)^2}\, dx$$

$$+ \int_1^{24}\frac{25}{24}\frac{1}{(x + 1)}\, dx$$

$$= (-2)\left[\frac{25}{24}\frac{-1}{(x + 1)}\right]_1^{24} + \left[\frac{25}{24}\ln(x + 1)\right]_1^{24}$$

$$= 1.67. \qquad \blacksquare$$

EXERCISES

3.7-1. Let the life W (in years) of the usual family car have a Weibull distribution with $\alpha = 2$.

(a) Show that β must equal 10 for $P(W > 5) = e^{1/4} \approx 0.7788$.
 HINT: $P(W > 5) = e^{-H(5)}$.

(b) Take a sample of size 8 from the uniform distribution $U(0,1)$, and simulate eight values from the Weibull ($\alpha = 2, \beta = 10$) distribution. HINT: You may use the random-number table. Also, recall that if U has that uniform distribution, then $F^{-1}(U)$ has a distribution with distribution function F.

3.7-2. Suppose that the length W of a human life does follow the Gompertz distribution with $\lambda(w) = a(1.1)^w = ae^{(\ln 1.1)w}$, $P(63 < W < 64) = 0.01$.

Determine the constant a and then find $P(W \leq 71 \,|\, 70 < W)$.

3.7-3. Let Y_1 be the smallest observation of three independent random variables W_1, W_2, W_3, each with a Weibull distribution with parameters α and β. Show that Y_1 has a Weibull distribution. What are the parameters of this latter distribution?
HINT:

$$G(y_1) = P(Y_1 \leq y_1) = 1 - P(y_1 < W_i,$$
$$i = 1, 2, 3) = 1 - [P(y_1 < W_1)]^3.$$

3.7-4. A frequent force of mortality used in actuarial science is $\lambda(w) = ae^{bw} + c$. Find the distribution function and p.d.f. associated with this **Makeham's law**.

3.7-5. From the graph of the first distribution function of X in Figure 3.7-4, determine the indicated probabilities:

(a) $P(X < 0)$. (b) $P(X < -1)$.   (c) $P(X \leq -1)$.

(d) $P(X < 1)$. (e) $P\left(-1 \leq X < \frac{1}{2}\right)$. (f) $P(-1 < X \leq 1)$.

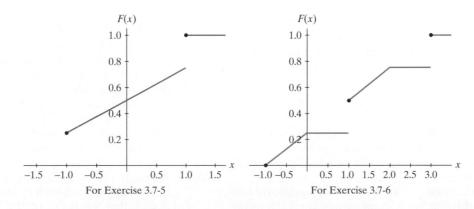

FIGURE 3.7-4: Mixed distribution functions

3.7-6. Determine the indicated probabilities from the graph of the second distribution function of X in Figure 3.7-4:

(a) $P\left(-\frac{1}{2} \leq X \leq \frac{1}{2}\right)$. (b) $P\left(\frac{1}{2} < X < 1\right)$. (c) $P\left(\frac{3}{4} < X < 2\right)$.

(d) $P(X > 1)$. (e) $P(2 < X < 3)$. (f) $P(2 < X \leq 3)$.

3.7-7. Let X be a random variable of the mixed type having the distribution function

$$F(x) = \begin{cases} 0, & x < 0, \\ \dfrac{x^2}{4}, & 0 \leq x < 1, \\ \dfrac{x+1}{4}, & 1 \leq x < 2, \\ 1, & 2 \leq x. \end{cases}$$

(a) Carefully sketch the graph of $F(x)$.

(b) Find the mean and the variance of X.

(c) Find $P(1/4 < X < 1)$, $P(X = 1)$, $P(X = 1/2)$, and $P(1/2 \leq X < 2)$.

3.7-8. Find the mean and variance of X if the distribution function of X is

$$F(x) = \begin{cases} 0, & x < 0, \\ 1 - \left(\dfrac{2}{3}\right)e^{-x}, & 0 \leq x, \end{cases}$$

3.7-9. Consider the following game: A fair die is rolled. If the outcome is even, the player receives a number of dollars equal to the outcome on the die. If the outcome is odd, a number is selected at random from the interval $(0,1)$ with a balanced spinner, and the player receives that fraction of a dollar associated with the point selected.

(a) Define and sketch the distribution function of X, the amount received.

(b) Find the expected value of X.

(c) Simulate 10 plays of this game.

(d) Is the average of your 10 plays close to the expected value calculated in part (b)?

3.7-10. The weekly gravel demand X (in tons) follows the p.d.f.

$$f(x) = \left(\frac{1}{5}\right)e^{-x/5}, \qquad 0 < x < \infty.$$

However, the owner of the gravel pit can produce at most only 4 tons of gravel per week. Compute the expected value of the tons sold per week by the owner.

3.7-11. The lifetime X of a certain device has an exponential distribution with mean five years. However, the device is not observed on a continuous basis

until after three years. Hence, we actually observe $Y = \max(X, 3)$. Compute $E(Y)$.

3.7-12. Let X have an exponential distribution with $\theta = 1$; that is, the p.d.f. of X is $f(x) = e^{-x}$, $0 < x < \infty$. Let T be defined by $T = \ln X$, so that the distribution function of T is

$$G(t) = P(\ln X \leq t) = P(X \leq e^t).$$

(a) Show that the p.d.f. of T is

$$g(t) = e^t e^{-e^t}, \qquad -\infty < x < \infty,$$

which is the p.d.f. of an extreme-value distribution.

(b) Let W be defined by $T = \alpha + \beta \ln W$, where $-\infty < \alpha < \infty$ and $\beta > 0$. Show that W has a Weibull distribution.

3.7-13. A loss X on a car has a mixed distribution with $p = 0.95$ on zero and $p = 0.05$ on an exponential distribution with mean of \$5000. If the loss X on a car is greater than the deductible of \$500, the difference $X - 500$ is paid to the owner of the car. Considering zero (if $X \leq 500$) as a possible payment, determine the mean and the standard deviation of the payment.

3.7-14. A customer buys a \$1000 deductible policy on her \$31,000 car. The probability of having an accident in which the loss is greater than \$1000 is 0.03, and then that loss, as a fraction of the value of the car minus the deductible, has the p.d.f. $f(x) = 6(1 - x)^5$, $0 < x < 1$.

(a) What is the probability that the insurance company must pay the customer more than \$2000?

(b) What does the company expect to pay?

3.7-15. A certain machine has a life X that has an exponential distribution with mean 10. The warranty is such that 100% of the price is returned if the machine fails in the first year, 50% of the price is returned for a failure during the second year, and nothing is returned after that. If the machine cost \$2500, what are the expected value and the standard deviation of the return on the warranty?

3.7-16. A certain machine has a life X that has an exponential distribution with mean 10. The warranty is such that \$m$ is returned if the machine fails in the first year, $(0.5)m$ of the price is returned for a failure during the second year, and nothing is returned after that. If the machine cost $2500, find m so that the expected payment is $200.

3.7-17. Some dental insurance policies cover the insurer only up to a certain amount, say, M. (This seems to us to be a dumb type of insurance policy because most people should want to protect themselves against large losses.) Say the dental expense X is a random variable with p.d.f. $f(x) = (0.001)e^{-x/1000}$, $0 < x < \infty$. Find M so that $P(X < M) = 0.08$.

3.7-18. The time X to failure of a machine has p.d.f. $f(x) = (x/4)^3 e^{-(x/4)^4}$, $0 < x < \infty$. Compute $P(X > 5 \mid X > 4)$.

3.7-19. Some banks now compound daily, but report only on a quarterly basis. It seems to us that it would be easier to compound every instant, for then a dollar invested at an annual rate of i for t years would be worth e^{ti}. (You might find it interesting to prove this statement by taking the limit of $(1 + i/n)^{nt}$ as $n \to \infty$.) If X is a random rate with p.d.f. $f(x) = ce^{-x}$, $0.04 < x < 0.08$, find the p.d.f. of the value of one dollar after three years invested at the rate of X.

3.7-20. Let X be the failure time (in months) of a certain insulating material. The distribution of X is modeled by the p.d.f.

$$f(x) = \frac{2x}{50^2} e^{-(x/50)^2}, \qquad 0 < x < \infty.$$

Find

(a) $P(40 < X < 60)$.
(b) $P(X > 80)$.

3.7-21. Sometimes the failure rate of an item is exponential, like $\lambda(x) = ae^{bx}$, where $a > 0$ and $b > 0$. The corresponding distribution function is

$$F(x) = 1 - \exp\left[-(a/b)(e^{bx} - 1)\right], \; 0<x<\infty,$$

and the p.d.f. is

$$f(x) = ae^{bx} \exp\left[-(a/b)(e^{bx} - 1)\right], \; 0<x<\infty.$$

Actuaries frequently use $f(x)$ as a good model for the length X of a life of a human. Find a and b so that $P(X < 70) = 0.4$ and $P(X < 78) = 0.6$. What is the mode of the resulting distribution? (See Figure 3.7-2.)

3.7-22. Let Y, with p.d.f. $f(y)$ and c.d.f. $F(y)$, be a random variable that represents the length of life (of a product, of a human, etc.). Given that the life exceeds some number x, the expected length of the remaining life is given by

$$e(x) = \frac{\int_x^\infty (y - x)f(y)dy}{1 - F(x)},$$

since $f(y)/[1 - F(x)]$ is the conditional p.d.f. of Y, when $Y > x$. The function $e(x)$ is called the **mean residual life**. For each of the following probability density functions, compute the mean residual life.

(a) $f(y) = \dfrac{1}{100}$, $\quad 0 < y < 100$,

(b) $f(y) = \dfrac{50}{y^2}$, $\quad 50 < y < \infty$,

(c) $f(y) = \dfrac{1}{75} e^{-y/75}$, $\quad 0 < y < \infty$.

3.7-23. Suppose the birth weight (X) in grams of U.S. infants has an approximate Weibull model with p.d.f.

$$f(x) = \frac{3x^2}{3500^3} e^{-(x/3500)^3}, \qquad 0 < x < \infty.$$

Given that a birth weight is greater than 3000, what is the conditional probability that it exceeds 4000?

HISTORICAL COMMENTS In this chapter we studied several continuous distributions, including the very important normal distribution. Actually, the true importance of the normal distribution is given in Chapter 5, where we consider the central limit theorem and its generalizations. Together, that theorem and its generalizations imply that the sum of several random influences on some measurement suggests that the measurement has an approximate normal distribution. For example, in a study of the length of chicken eggs, different hens produce different eggs, the person measuring the eggs makes a difference, the way the egg is placed in a "holder" is a factor, the caliper used is important, and so on. Thus, the length of an egg might have an approximate normal distribution.

Sometimes instructors force grades to be normally distributed because they "grade on a (normal) curve." This is done too often, and it means that a certain percentage of the students should get A's, a certain percentage B's, etc. We believe that all students should be able to earn A's if they satisfy certain appropriate criteria. Thus, we think that it is wrong to restrict grades to a normal curve.

The normal distribution is symmetric, but many important distributions, like the gamma and Weibull, are skewed. We learned that the Weibull distribution has a failure rate equal to $\lambda(x) = \alpha x^{\alpha-1}/\beta^\alpha$, for $\alpha \geq 1$, and this distribution is appropriate for the length of the life of many manufactured products. It is interesting to note that if $\alpha = 1$, the failure rate is a constant, meaning that an old part is as good as a new one. If this were true for the lives of humans, an old man would have the same chance of living 50 more years as would a young man. That is, we would have a "mathematical fountain of youth." Unfortunately, as we learned in this chapter, the failure rate of humans is increasing with age and is close to being exponential (say, $\lambda(x) = ae^{bx}$, $a > 0$, $b > 0$), leading to the Gompertz distribution. As a matter of fact, most would find that force of mortality is such that it increases about 10 percent each year; so by the rule of 72, it would double about every $72/10 = 7.2$ years. (See Equation D.4-1 in Appendix D on the CD-ROM.) Fortunately, it is very small for persons in their twenties.

The reason we try to model any random phenomenon with a probability distribution is that if our model is reasonably good, we know the approximate percentages of being above or below certain marks. Having such information helps us make certain decisions—sometimes very important ones.

With these models, we learned how to simulate random variables having certain distributions. In many situations in practice, we cannot calculate exact solutions of equations that have numerous random variables. Thus, we simulate the random variables in question many times, leading to an approximate distribution of the random solution. "Monte Carlo" is a term often attached to such a simulation, and we believe that it was first used in a computer simulation of nuclear fission associated with the atom bomb in World War II. Of course, the name "Monte Carlo" was taken from that city, which is famous for gambling in its casinos.

Also in this chapter, we studied certain averages, including the mean of a distribution and a random sample. There is a story that statisticians tell about Ralph Sampson, who was an excellent basketball player at the University of Virginia in the early 1980s and later was drafted by the Houston Rockets. He supposedly majored in communication studies at Virginia, and it is reported that the department there said that the average starting salary of their majors was much higher than those in the sciences; that happened because of Sampson's high starting salary with the Rockets. If this story is true, it would have been much more appropriate to report the median starting salary of the communication majors and this median salary would have been much lower than the median starting salaries in the sciences.

The famous statistician John Tukey died a few years ago. Not only did he originate stem-and-leaf displays, but he also coined the terms "software" (as opposed to hardware) and "bit" (for binary digit). He rediscovered and made popular the fast Fourier transform. However, one of his notable quotes was "Far better an approximate answer to the right question, which is often vague, than the exact answer to the wrong question, which can always be made precise." We think that he was taking a crack at pure mathematicians, although Tukey was excellent in math, too.

We close these remarks with an observation about Sir Francis Galton, who everyone agrees was a great scientist. However, he contributed to statistics as well. One of his rules was "Whenever you can, count." Some might say that, on the basis of the data he collected, he was the first statistician, although that would be difficult to prove, as many before him collected lots of important data.

Bivariate Distributions

4.1 DISTRIBUTIONS OF TWO RANDOM VARIABLES
4.2 THE CORRELATION COEFFICIENT
4.3 CONDITIONAL DISTRIBUTIONS
4.4 THE BIVARIATE NORMAL DISTRIBUTION

4.1 DISTRIBUTIONS OF TWO RANDOM VARIABLES

So far, we have taken only one measurement on a single item under observation. However, it is clear in many practical cases that it is possible, and often very desirable, to take more than one measurement of a random observation. Suppose, for example, that we are observing female college students to obtain information about some of their physical characteristics, such as height, x, and weight, y, because we are trying to determine a relationship between those two characteristics. For instance, there may be some pattern between height and weight that can be described by an appropriate curve $y = u(x)$. Certainly, not all the points observed will be on this curve, but we want to attempt to find the "best" curve to describe the relationship and then say something about the variation of the points around the curve.

Another example might concern high school rank—say, x—and the ACT (or SAT) score, say, y—of incoming college students. What is the relationship between these two characteristics? More importantly, how can we use those measurements to predict a third one, such as first-year college GPA—say, z—with a function $z = v(x, y)$? This is a very important problem for college admission offices, particularly when it comes to awarding an athletic scholarship, because the incoming student–athlete must satisfy certain conditions before receiving such an award.

REMARK   We begin with bivariate distributions for the discrete case. The continuous case is essentially the same, but with integrals replacing summations.   ∎

DEFINITION
4.1-1

Let X and Y be two random variables defined on a discrete probability space. Let S denote the corresponding two-dimensional space of X and Y, the two random variables of the discrete type. The probability that $X = x$ and $Y = y$ is denoted by $f(x, y) = P(X = x, Y = y)$. The function $f(x, y)$ is called the **joint probability mass function** (joint p.m.f.) of X and Y and has the following properties:

(a) $0 \leq f(x, y) \leq 1$.

(b) $\displaystyle\sum\sum_{(x,y)\in S} f(x, y) = 1$.

(c) $P[(X, Y) \in A] = \displaystyle\sum\sum_{(x,y)\in A} f(x, y)$, where A is a subset of the space S.

The following example will make this definition more meaningful.

EXAMPLE 4.1-1 Roll a pair of unbiased dice. For each of the 36 sample points with probability 1/36, let X denote the smaller and Y the larger outcome on the dice. For example, if the outcome is $(3, 2)$, then the observed values are $X = 2$, $Y = 3$. The event $\{X = 2, Y = 3\}$ could occur in one of two ways—$(3, 2)$ or $(2, 3)$—so its probability is

$$\frac{1}{36} + \frac{1}{36} = \frac{2}{36}.$$

If the outcome is $(2, 2)$, then the observed values are $X = 2$, $Y = 2$. Since the event $\{X = 2, Y = 2\}$ can occur in only one way, $P(X = 2, Y = 2) = 1/36$. The joint p.m.f. of X and Y is given by the probabilities

$$f(x,y) = \begin{cases} \dfrac{1}{36}, & 1 \leq x = y \leq 6, \\[2mm] \dfrac{2}{36}, & 1 \leq x < y \leq 6, \end{cases}$$

when x and y are integers. Figure 4.1-1 depicts the probabilities of the various points of the space S. ∎

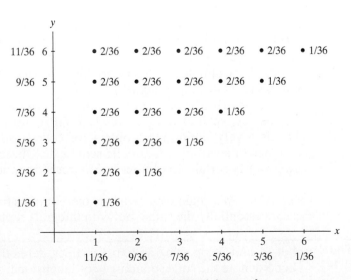

FIGURE 4.1-1: Discrete joint p.m.f.

Notice that certain numbers have been recorded in the bottom and left-hand margins of Figure 4.1-1. These numbers are the respective column and row totals of the probabilities. The column totals are the respective probabilities that X will assume the values in the x space $S_1 = \{1,2,3,4,5,6\}$, and the row totals are the respective probabilities that Y will assume the values in the y space $S_2 = \{1,2,3,4,5,6\}$. That is, the totals describe probability mass functions of X and Y, respectively. Since each collection of these probabilities is frequently recorded in the margins and satisfies the properties of a p.m.f. of one random variable, each is called a **marginal p.m.f.**

DEFINITION 4.1-2	Let X and Y have the joint probability mass function $f(x, y)$ with space S. The probability mass function of X alone, which is called the **marginal probability mass function of X**, is defined by

$$f_1(x) = \sum_y f(x, y) = P(X = x), \qquad x \in S_1,$$

where the summation is taken over all possible y values for each given x in the x space S_1. That is, the summation is over all (x, y) in S with a given x value. Similarly, the **marginal probability mass function of Y** is defined by

$$f_2(y) = \sum_x f(x, y) = P(Y = y), \qquad y \in S_2,$$

where the summation is taken over all possible x values for each given y in the y space S_2. The random variables X and Y are **independent** if and only if

$$P(X = x, Y = y) \equiv P(X = x)P(Y = y)$$

or, equivalently,

$$f(x, y) = f_1(x)f_2(y), \qquad x \in S_1, \qquad y \in S_2;$$

otherwise, X and Y are said to be **dependent**.

We note in Example 4.1-1 that X and Y are dependent because there are many x and y values for which $f(x, y) \neq f_1(x)f_2(y)$. For instance,

$$f_1(1)f_2(1) = \left(\frac{11}{36}\right)\left(\frac{1}{36}\right) \neq \frac{1}{36} = f(1, 1).$$

EXAMPLE 4.1-2   Let the joint p.m.f. of X and Y be defined by

$$f(x, y) = \frac{x + y}{21}, \qquad x = 1, 2, 3, \qquad y = 1, 2.$$

Then

$$f_1(x) = \sum_y f(x, y) = \sum_{y=1}^{2} \frac{x + y}{21}$$

$$= \frac{x + 1}{21} + \frac{x + 2}{21} = \frac{2x + 3}{21}, \qquad x = 1, 2, 3,$$

and

$$f_2(y) = \sum_x f(x, y) = \sum_{x=1}^{3} \frac{x+y}{21} = \frac{6+3y}{21}, \qquad y = 1, 2.$$

Note that both $f_1(x)$ and $f_2(y)$ satisfy the properties of a probability mass function. Since $f(x, y) \not\equiv f_1(x)f_2(y)$, X and Y are dependent. ∎

EXAMPLE 4.1-3 Let the joint p.m.f. of X and Y be

$$f(x, y) = \frac{xy^2}{30}, \qquad x = 1, 2, 3, \qquad y = 1, 2.$$

The marginal probability mass functions are

$$f_1(x) = \sum_{y=1}^{2} \frac{xy^2}{30} = \frac{x}{6}, \qquad x = 1, 2, 3,$$

and

$$f_2(y) = \sum_{x=1}^{3} \frac{xy^2}{30} = \frac{y^2}{5}, \qquad y = 1, 2.$$

Then $f(x, y) \equiv f_1(x)f_2(y)$ for $x = 1, 2, 3$ and $y = 1, 2$; thus, X and Y are independent. (See Figure 4.1-2.) ∎

FIGURE 4.1-2: Joint p.m.f. $f(x, y) = \dfrac{xy^2}{30}$, $x = 1, 2, 3$; $y = 1, 2$

EXAMPLE 4.1-4 Let the joint p.m.f. of X and Y be

$$f(x, y) = \frac{xy^2}{13}, \qquad (x, y) = (1, 1), (1, 2), (2, 2).$$

Then the p.m.f. of X is

$$f_1(x) = \begin{cases} \dfrac{5}{13}, & x = 1, \\[2mm] \dfrac{8}{13}, & x = 2, \end{cases}$$

and that of Y is

$$f_2(y) = \begin{cases} \dfrac{1}{13}, & y = 1, \\[2mm] \dfrac{12}{13}, & y = 2. \end{cases}$$

X and Y are dependent because $f(x, y) \neq f_1(x)f_2(y)$ for $x = 1, 2$ and $y = 1, 2$.   ■

Note that in Example 4.1-4 the support S of X and Y is "triangular." Whenever the support S is not "rectangular," the random variables must be dependent, because S cannot then equal the product set $\{(x, y): x \in S_1, y \in S_2\}$. That is, if we observe that the support S of X and Y is not a product set, then X and Y must be dependent. For example, in Example 4.1-4, X and Y are dependent because $S = \{(1, 1), (1, 2), (2, 2)\}$ is not a product set. On the other hand, if S equals the product set $\{(x, y): x \in S_1, y \in S_2\}$ and if the formula for $f(x, y)$ is the product of an expression in x alone and an expression in y alone, then X and Y are independent, as shown in Example 4.1-3. Example 4.1-2 illustrates the fact that the support can be rectangular but the formula for $f(x, y)$ is not such a product and thus X and Y are dependent.

It is possible to define a probability histogram for a joint p.m.f. just as we did for a p.m.f. for a single random variable. Suppose that X and Y have a joint p.m.f. $f(x, y)$ with space S, where S is a set of pairs of integers. At a point (x, y) in S, construct a "rectangular column" that is centered at (x, y) and has a one-unit-by-one-unit base and a height equal to $f(x, y)$. Note that $f(x, y)$ is equal to the "volume" of this rectangular column. Furthermore, the sum of the volumes of the rectangular columns in this probability histogram is equal to 1.

EXAMPLE 4.1-5   Let the joint p.m.f. of X and Y be

$$f(x, y) = \frac{xy^2}{30}, \qquad x = 1, 2, 3, \qquad y = 1, 2.$$

The probability histogram is shown in Figure 4.1-3 on the next page.   ■

Sometimes it is convenient to replace the symbols X and Y representing random variables by X_1 and X_2. This is particularly true in situations in which we have more than two random variables; so we use X and Y sometimes and then X_1 and X_2 at other times. The reader will see the advantage of the use of subscripts as we go further in the text.

Let X_1 and X_2 be random variables of the discrete type with the joint p.m.f. $f(x_1, x_2)$ on the space S. If $u(X_1, X_2)$ is a function of these two random variables, then

$$E[u(X_1, X_2)] = \sum_{(x_1, x_2) \in S} \sum u(x_1, x_2)f(x_1, x_2),$$

if it exists, is called the **mathematical expectation** (or **expected value**) of $u(X_1, X_2)$.

FIGURE 4.1-3: Joint p.m.f. $f(x, y) = \dfrac{xy^2}{30}$, $x = 1, 2, 3$; $y = 1, 2$

REMARK The same remarks can be made here that were made in the univariate case, namely, that

$$\sum \sum_{(x_1, x_2) \in S} |u(x_1, x_2)| f(x_1, x_2)$$

must converge and be finite. Also, $Y = u(X_1, X_2)$ is a random variable—say, with p.m.f. $g(y)$ on space S_1—and it is true that

$$\sum \sum_{(x_1, x_2) \in S} u(x_1, x_2) f(x_1, x_2) = \sum_{y \in S_1} y g(y).$$ ∎

EXAMPLE 4.1-6 There are eight similar chips in a bowl: three marked $(0, 0)$, two marked $(1, 0)$, two marked $(0, 1)$, and one marked $(1, 1)$. A player selects a chip at random and is given the sum of the two coordinates in dollars. If X_1 and X_2 represent those two coordinates, respectively, their joint p.m.f. is

$$f(x_1, x_2) = \frac{3 - x_1 - x_2}{8}, \qquad x_1 = 0, 1 \qquad \text{and} \qquad x_2 = 0, 1.$$

Thus,

$$E(X_1 + X_2) = \sum_{x_2=0}^{1} \sum_{x_1=0}^{1} (x_1 + x_2) \frac{3 - x_1 - x_2}{8}$$

$$= (0)\left(\frac{3}{8}\right) + (1)\left(\frac{2}{8}\right) + (1)\left(\frac{2}{8}\right) + (2)\left(\frac{1}{8}\right) = \frac{3}{4}.$$

That is, the expected payoff is 75¢. ∎

The following mathematical expectations, if they exist, have special names:

(a) If $u_1(X_1, X_2) = X_i$, then

$$E[u_1(X_1, X_2)] = E(X_i) = \mu_i$$

is called the **mean** of X_i, $i = 1, 2$.

(b) If $u_2(X_1, X_2) = (X_i - \mu_i)^2$, then

$$E[u_2(X_1, X_2)] = E[(X_i - \mu_i)^2] = \sigma_i^2 = \text{Var}(X_i)$$

is called the **variance** of X_i, $i = 1, 2$.

The mean μ_i and the variance σ_i^2 can be computed from the joint p.m.f. $f(x_1, x_2)$ or the marginal p.m.f. $f_i(x_i)$, $i = 1, 2$.

The idea of joint distributions of two random variables of the discrete type can be extended to that of two random variables of the continuous type. The definitions are really the same, except that integrals replace summations. The **joint probability density function** (joint p.d.f.) of two continuous-type random variables is an integrable function $f(x, y)$ with the following properties:

(a) $f(x, y) \geq 0$, where $f(x, y) = 0$ when (x, y) is not in the support (space) S of X and Y.

(b) $\int_{-\infty}^{\infty} \int_{-\infty}^{\infty} f(x, y)\, dx\, dy = 1$.

(c) $P[(X, Y) \in A] = \int \int_A f(x, y)\, dx\, dy$, where $\{(X, Y) \in A\}$ is an event defined in the plane.

Property (c) implies that $P[(X, Y) \in A]$ is the volume of the solid over the region A in the xy-plane and bounded by the surface $z = f(x, y)$.

EXAMPLE 4.1-7 Let X and Y have the joint p.d.f.

$$f(x, y) = \frac{3}{2}x^2(1 - |y|), \qquad -1 < x < 1, \qquad 1 < y < 1.$$

The graph of $z = f(x, y)$ is given in Figure 4.1-4. Let $A = \{(x, y): 0 < x < 1, 0 < y < x\}$. Then the probability that (X, Y) falls into A is given by

$$P[(X, Y) \in A] = \int_0^1 \int_0^x \frac{3}{2}x^2(1 - y)\, dy\, dx = \int_0^1 \frac{3}{2}x^2 \left[y - \frac{y^2}{2} \right]_0^x dx$$

$$= \int_0^1 \frac{3}{2}\left(x^3 - \frac{x^4}{2} \right) dx = \frac{3}{2}\left[\frac{x^4}{4} - \frac{x^5}{10} \right]_0^1$$

$$= \frac{9}{40}.$$ ■

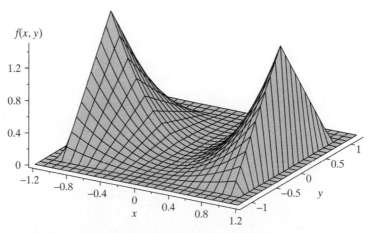

FIGURE 4.1-4: Joint p.d.f. $f(x, y) = \frac{3}{2}x^2(1 - |y|)$

The respective **marginal p.d.f.'s** of continuous-type random variables X and Y are given by

$$f_1(x) = \int_{-\infty}^{\infty} f(x, y)\, dy, \qquad x \in S_1,$$

and

$$f_2(y) = \int_{-\infty}^{\infty} f(x, y)\, dx, \qquad y \in S_2,$$

where S_1 and S_2 are the respective spaces of X and Y. The definitions associated with mathematical expectations in the continuous case are the same as those associated with the discrete case after replacing the summations with integrations.

For the random variables X and Y in Example 4.1-7, the marginal p.d.f.'s are

$$f_1(x) = \int_{-1}^{1} \frac{3}{2} x^2 (1 - |y|)\, dy = \frac{3}{2} x^2, \qquad -1 < x < 1,$$

and

$$f_2(y) = \int_{-1}^{1} \frac{3}{2} x^2 (1 - |y|)\, dx = 1 - |y|, \qquad -1 < y < 1.$$

EXAMPLE 4.1-8 Let X and Y have the joint p.d.f.

$$f(x, y) = 2, \qquad 0 \le x \le y \le 1.$$

Then $S = \{(x, y) : 0 \le x \le y \le 1\}$ is the support; for example,

$$P\left(0 \le X \le \frac{1}{2}, 0 \le Y \le \frac{1}{2}\right) = P\left(0 \le X \le Y, 0 \le Y \le \frac{1}{2}\right)$$

$$= \int_0^{1/2} \int_0^y 2\, dx\, dy = \int_0^{1/2} 2y\, dy = \frac{1}{4}.$$

You should draw a figure to illustrate the set of points for which $f(x, y) > 0$ and then shade the region over which the integral is taken. The given probability is the volume above this shaded region under the surface $z = 2$. The marginal p.d.f.'s are given by

$$f_1(x) = \int_x^1 2\, dy = 2(1 - x), \qquad 0 \le x \le 1,$$

and

$$f_2(y) = \int_0^y 2\, dx = 2y, \qquad 0 \le y \le 1.$$

Three illustrations of expected values are

$$E(X) = \int_0^1 \int_x^1 2x\, dy\, dx = \int_0^1 2x(1 - x)\, dx = \frac{1}{3},$$

$$E(Y) = \int_0^1 \int_0^y 2y\, dx\, dy = \int_0^1 2y^2\, dy = \frac{2}{3},$$

and

$$E(Y^2) = \int_0^1 \int_0^y 2y^2 \, dx \, dy = \int_0^1 2y^3 \, dy = \frac{1}{2}.$$

From these calculations we see that $E(X)$, $E(Y)$, and $E(Y^2)$ could be calculated from the marginal p.d.f.'s instead of the joint one. ∎

The definition of independent random variables of the continuous type carries over naturally from the discrete case. That is, X and Y are **independent** if and only if the joint p.d.f. factors into the product of their marginal p.d.f.'s; namely,

$$f(x,y) \equiv f_1(x)f_2(y), \qquad x \in S_1, \, y \in S_2.$$

Thus, the random variables X and Y in Example 4.1-7 are independent. In addition, the rules that allow us to easily determine dependent and independent random variables are valid here. So X and Y in Example 4.1-8 are dependent because the support S is not a product space, since it is bounded by the diagonal line $y = x$.

We give extensions of two important univariate distributions—the hypergeometric distribution and the binomial distribution—through examples.

EXAMPLE 4.1-9 Consider a population of 200 students who have just finished a first course in calculus. Of these 200, 40 have earned A's, 60 B's, and 100 C's, D's, or F's. A sample of size 25 is taken at random and without replacement from this population in a way that each possible sample has probability

$$\frac{1}{\binom{200}{25}}$$

of being selected. Within the sample of 25, let X_1 be the number of A students, X_2 the number of B students, and $25 - X_1 - X_2$ the number of other students. The space S of (X_1, X_2) is defined by the collection of nonnegative integers (x_1, x_2) such that $x_1 + x_2 \le 25$. The joint p.m.f. of X_1, X_2 is

$$f(x_1, x_2) = \frac{\binom{40}{x_1}\binom{60}{x_2}\binom{100}{25 - x_1 - x_2}}{\binom{200}{25}},$$

for $(x_1, x_2) \in S$, where it is understood that $\binom{k}{j} = 0$ if $j > k$. Without actually summing, we know that the marginal p.m.f. of X_1 is

$$f_1(x_1) = \frac{\binom{40}{x_1}\binom{160}{25 - x_1}}{\binom{200}{25}}, \qquad x_1 = 0, 1, 2, \ldots, 25,$$

since X_1 alone has a hypergeometric distribution. Of course, the function $f_2(x_2)$ is a hypergeometric p.m.f. and

$$f(x_1, x_2) \neq f_1(x_1)f_2(x_2),$$

so X_1 and X_2 are dependent. Note that the space S is not "rectangular," which implies that the random variables are dependent. ∎

We now extend the binomial distribution to a trinomial distribution. Here we have three mutually exclusive and exhaustive ways for an experiment to terminate: perfect, "seconds," and defective. We repeat the experiment n independent times, and the probabilities $p_1, p_2, p_3 = 1 - p_1 - p_2$ of perfect, seconds, and defective, respectively, remain the same from trial to trial. In the n trials, let $X_1 =$ number of perfect items, $X_2 =$ number of seconds, and $X_3 = n - X_1 - X_2 =$ number of defectives. If x_1 and x_2 are nonnegative integers such that $x_1 + x_2 \leq n$, then the probability of having x_1 perfects, x_2 seconds, and $n - x_1 - x_2$ defectives, in that order, is

$$p_1^{x_1} p_2^{x_2} (1 - p_1 - p_2)^{n - x_1 - x_2}.$$

However, if we want $P(X_1 = x_1, X_2 = x_2)$, then we must recognize that $X_1 = x_1, X_2 = x_2$ can be achieved in

$$\binom{n}{x_1, x_2, n - x_1 - x_2} = \frac{n!}{x_1! x_2! (n - x_1 - x_2)!}$$

different ways. Hence, the **trinomial** p.m.f. is

$$
\begin{aligned}
f(x_1, x_2) &= P(X_1 = x_1, X_2 = x_2) \\
&= \frac{n!}{x_1! x_2! (n - x_1 - x_2)!} p_1^{x_1} p_2^{x_2} (1 - p_1 - p_2)^{n - x_1 - x_2},
\end{aligned}
$$

where x_1 and x_2 are nonnegative integers such that $x_1 + x_2 \leq n$. Without summing, we know that X_1 is $b(n, p_1)$ and X_2 is $b(n, p_2)$; thus, X_1 and X_2 are dependent, as the product of these marginal probability mass functions is not equal to $f(x_1, x_2)$.

EXAMPLE 4.1-10 In manufacturing a certain item, it is found that in normal production about 95% of the items are good ones, 4% are "seconds," and 1% are defective. A company has a program of quality control by statistical methods, and each hour an online inspector observes 20 items selected at random, counting the number X of seconds and the number Y of defectives. Suppose that the production is normal. Let us find the probability that, in this sample of size $n = 20$, at least two seconds or at least two defective items are discovered. If we let $A = \{(x, y) : x \geq 2 \text{ or } y \geq 2\}$, then

$$
\begin{aligned}
P(A) &= 1 - P(A') \\
&= 1 - P(X = 0 \text{ or } 1 \text{ and } Y = 0 \text{ or } 1) \\
&= 1 - \frac{20!}{0!0!20!}(0.04)^0(0.01)^0(0.95)^{20} - \frac{20!}{1!0!19!}(0.04)^1(0.01)^0(0.95)^{19} \\
&\quad - \frac{20!}{0!1!19!}(0.04)^0(0.01)^1(0.95)^{19} - \frac{20!}{1!1!18!}(0.04)^1(0.01)^1(0.95)^{18} \\
&= 0.204.
\end{aligned}
$$
∎

EXAMPLE 4.1-11 Let X and Y have a trinomial distribution with parameters $p_1 = 1/5$, $p_2 = 2/5$, and $n = 5$. The probability histogram for the joint p.m.f. of X and Y is shown in Figure 4.1-5. ∎

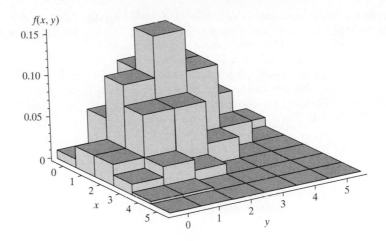

FIGURE 4.1-5: Trinomial distribution, $p_1 = 1/5$, $p_2 = 2/5$, and $n = 5$

EXERCISES

4.1-1. Let the joint p.m.f. of X and Y be defined by

$$f(x,y) = \frac{x + y}{32}, \qquad x = 1, 2, \ y = 1, 2, 3, 4.$$

  (a) Find $f_1(x)$, the marginal p.m.f. of X.
  (b) Find $f_2(y)$, the marginal p.m.f. of Y.
  (c) Find $P(X > Y)$.
  (d) Find $P(Y = 2X)$.
  (e) Find $P(X + Y = 3)$.
  (f) Find $P(X \le 3 - Y)$.
  (g) Are X and Y independent or dependent? Why or why not?

4.1-2. Roll a pair of four-sided dice, one red and one black, each of which has possible outcomes 1, 2, 3, 4 that have equal probabilities. Let X equal the outcome on the red die, and let Y equal the outcome on the black die.

  (a) On graph paper, show the space of X and Y.
  (b) Define the joint p.m.f. on the space (similar to Figure 4.1-1).
  (c) Give the marginal p.m.f. of X in the margin.
  (d) Give the marginal p.m.f. of Y in the margin.
  (e) Are X and Y dependent or independent? Why or why not?

4.1-3. Roll a pair of four-sided dice, one red and one black. Let X equal the outcome on the red die and let Y equal the sum of the two dice.

  (a) On graph paper, describe the space of X and Y.
  (b) Define the joint p.m.f. on the space (similar to Figure 4.1-1).

  (c) Give the marginal p.m.f. of X in the margin.
  (d) Give the marginal p.m.f. of Y in the margin.
  (e) Are X and Y dependent or independent? Why or why not?

4.1-4. Select an (even) integer randomly from the set $\{0, 2, 4, 6, 8\}$. Then select an integer randomly from the set $\{0, 1, 2, 3, 4\}$. Let X equal the integer that is selected from the first set and let Y equal the sum of the two integers.

  (a) Show the joint p.m.f. of X and Y on the space of X and Y.
  (b) Compute the marginal p.m.f.'s.
  (c) Are X and Y independent? Why or why not?

4.1-5. A particle starts at (0,0) and moves in one-unit independent steps with equal probabilities of 1/4 in each of the four directions: north, south, east, and west. Let S equal the east–west position and T the north–south position after n steps.

  (a) Define the joint p.m.f. of S and T with $n = 3$. On a two–dimensional graph, give the probabilities of the joint p.m.f. and the marginal p.m.f.'s (similar to Figure 4.1-1).
  (b) Let $X = S + 3$ and let $Y = T + 3$. How are X and Y distributed?

4.1-6. The torque required to remove bolts in a steel plate is rated as very high, high, average, and low, and these occur about 30%, 40%, 20%, and 10% of the time, respectively. Suppose $n = 25$ bolts are rated; what is the probability of rating 7 very high, 8 high, 6 average, and 4 low? Assume independence of the 25 trials.

4.1-7. Two construction companies make bids of X and Y (in \$100,000's) on a remodeling project. The joint p.d.f. of X and Y is uniform on the space $2 < x < 2.5, 2 < y < 2.3$. If X and Y are within 0.1 of each other, the companies will be asked to rebid; otherwise the low bidder will be awarded the contract. What is the probability that they will be asked to rebid?

4.1-8. In a smoking survey among boys between the ages of 12 and 17, 78% prefer to date nonsmokers, 1% prefer to date smokers, and 21% don't care. Suppose seven such boys are selected randomly. Let X equal the number who prefer to date nonsmokers and Y equal the number who prefer to date smokers.

(a) Determine the joint p.m.f. of X and Y. Be sure to include the support of the p.m.f.

(b) Find the marginal p.m.f. of X. Again include the support.

4.1-9. A manufactured item is classified as good, a "second," or defective with probabilities $6/10, 3/10$, and $1/10$, respectively. Fifteen such items are selected at random from the production line. Let X denote the number of good items, Y the number of seconds, and $15 - X - Y$ the number of defective items.

(a) Give the joint p.m.f. of X and $Y, f(x, y)$.

(b) Sketch the set of points for which $f(x, y) > 0$. From the shape of this region, can X and Y be independent? Why or why not?

(c) Find $P(X = 10, Y = 4)$.

(d) Give the marginal p.m.f. of X.

(e) Find $P(X \le 11)$.

4.1-10. Let $f(x, y) = 3/2, x^2 \le y \le 1, 0 \le x \le 1$, be the joint p.d.f. of X and Y.

(a) Find $P(0 \le X \le 1/2)$.

(b) Find $P(1/2 \le Y \le 1)$.

(c) Find $P(1/2 \le X \le 1, 1/2 \le Y \le 1)$.

(d) Find $P(X \ge 1/2, Y \ge 1/2)$.

(e) Are X and Y independent? Why or why not?

4.1-11. Let $f(x, y) = 2e^{-x-y}, 0 \le x \le y < \infty$, be the joint p.d.f. of X and Y. Find $f_1(x)$ and $f_2(y)$, the marginal p.d.f.'s of X and Y, respectively. Are X and Y independent?

4.1-12. Let X and Y have the joint p.d.f. $f(x, y) = x + y$, $0 \le x \le 1, 0 \le y \le 1$.

(a) Find the marginal p.d.f.'s $f_1(x)$ and $f_2(y)$ and show that $f(x, y) \ne f_1(x)f_2(y)$. Thus, X and Y are dependent.

(b) Compute (i) μ_X; (ii) μ_Y; (iii) σ_X^2; and (iv) σ_Y^2.

4.1-13. Let $f(x, y) = (3/16)xy^2, 0 \le x \le 2, 0 \le y \le 2$, be the joint p.d.f. of X and Y. Find $f_1(x)$ and $f_2(y)$, the marginal probability density functions. Are the two random variables independent? Why or why not?

4.1-14. Let T_1 and T_2 be random times for a company to complete two steps in a certain process. Say T_1 and T_2 are measured in days and they have the joint p.d.f. that is uniform over the space $1 < t_1 < 10, 2 < t_2 < 6, t_1 + 2t_2 < 14$. What is $P(T_1 + T_2 > 10)$?

4.1-15. Let $f(x, y) = 4/3, 0 < x < 1, x^3 < y < 1$, zero elsewhere.

(a) Sketch the region where $f(x, y) > 0$.

(b) Find $P(X > Y)$.

4.2 THE CORRELATION COEFFICIENT

In Section 4.1, we introduced the mathematical expectation of a function of two random variables—say, X_1, X_2. We gave the respective special names of mean and variance of X_i to

$$\mu_i = E(X_i) \qquad \text{and} \qquad \sigma_i^2 = E[(X_i - \mu_i)^2], \qquad i = 1, 2.$$

We introduce two more special names:

(a) If $u_3(X_1, X_2) = (X_1 - \mu_1)(X_2 - \mu_2)$, then

$$E[u_3(X_1, X_2)] = E[(X_1 - \mu_1)(X_2 - \mu_2)] = \sigma_{12} = \text{Cov}(X_1, X_2)$$

is called the **covariance** of X_1 and X_2.

(b) If the standard deviations σ_1 and σ_2 are positive, then

$$\rho = \frac{\text{Cov}(X_1, X_2)}{\sigma_1 \sigma_2} = \frac{\sigma_{12}}{\sigma_1 \sigma_2}$$

is called the **correlation coefficient** of X_1 and X_2.

It is convenient that the mean and the variance of X_1 can be computed from either the joint p.m.f. or the marginal p.m.f. of X_1. For example, in the discrete case,

$$\mu_1 = E(X_1) = \sum_{x_1} \sum_{x_2} x_1 f(x_1, x_2)$$

$$= \sum_{x_1} x_1 \left[\sum_{x_2} f(x_1, x_2) \right] = \sum_{x_1} x_1 f_1(x_1).$$

However, to compute the covariance, we need the joint p.m.f.

Before considering the significance of the covariance and the correlation coefficient, let us note a few simple facts. First,

$$E[(X_1 - \mu_1)(X_2 - \mu_2)] = E(X_1 X_2 - \mu_1 X_2 - \mu_2 X_1 + \mu_1 \mu_2)$$

$$= E(X_1 X_2) - \mu_1 E(X_2) - \mu_2 E(X_1) + \mu_1 \mu_2,$$

because, even in the bivariate situation, E is still a linear or distributive operator. (See Exercise 4.2-4.) Thus,

$$\text{Cov}(X_1, X_2) = E(X_1 X_2) - \mu_1 \mu_2 - \mu_2 \mu_1 + \mu_1 \mu_2 = E(X_1 X_2) - \mu_1 \mu_2.$$

Since $\rho = \text{Cov}(X_1, X_2)/\sigma_1 \sigma_2$, we also have

$$E(X_1 X_2) = \mu_1 \mu_2 + \rho \sigma_1 \sigma_2.$$

That is, the expected value of the product of two random variables is equal to the product $\mu_1 \mu_2$ of their expectations, plus their covariance $\rho \sigma_1 \sigma_2$.

A simple example at this point would be helpful.

EXAMPLE 4.2-1 Let X_1 and X_2 have the joint p.m.f.

$$f(x_1, x_2) = \frac{x_1 + 2x_2}{18}, \qquad x_1 = 1, 2, \qquad x_2 = 1, 2.$$

The marginal probability mass functions are, respectively,

$$f_1(x_1) = \sum_{x_2 = 1}^{2} \frac{x_1 + 2x_2}{18} = \frac{2x_1 + 6}{18}, \qquad x_1 = 1, 2,$$

and

$$f_2(x_2) = \sum_{x_1 = 1}^{2} \frac{x_1 + 2x_2}{18} = \frac{3 + 4x_2}{18}, \qquad x_2 = 1, 2.$$

Since $f(x_1, x_2) \neq f_1(x_1) f_2(x_2)$, X_1 and X_2 are dependent. The mean and the variance of X_1 are, respectively,

$$\mu_1 = \sum_{x_1 = 1}^{2} x_1 \frac{2x_1 + 6}{18} = (1) \left(\frac{8}{18} \right) + (2) \left(\frac{10}{18} \right) = \frac{14}{9}$$

and

$$\sigma_1^2 = \sum_{x_1 = 1}^{2} x_1^2 \frac{2x_1 + 6}{18} - \left(\frac{14}{9} \right)^2 = \frac{24}{9} - \frac{196}{81} = \frac{20}{81}.$$

The mean and the variance of X_2 are, respectively,

$$\mu_2 = \sum_{x_2=1}^{2} x_2 \frac{3 + 4x_2}{18} = (1)\left(\frac{7}{18}\right) + (2)\left(\frac{11}{18}\right) = \frac{29}{18}$$

and

$$\sigma_2^2 = \sum_{x_2=1}^{2} x_2^2 \frac{3 + 4x_2}{18} - \left(\frac{29}{18}\right)^2 = \frac{51}{18} - \frac{841}{324} = \frac{77}{324}.$$

The covariance of X_1 and X_2 is

$$\text{Cov}(X_1, X_2) = \sum_{x_2=1}^{2} \sum_{x_1=1}^{2} x_1 x_2 \frac{x_1 + 2x_2}{18} - \left(\frac{14}{9}\right)\left(\frac{29}{18}\right)$$

$$= (1)(1)\left(\frac{3}{18}\right) + (2)(1)\left(\frac{4}{18}\right) + (1)(2)\left(\frac{5}{18}\right)$$

$$+ (2)(2)\left(\frac{6}{18}\right) - \left(\frac{14}{19}\right)\left(\frac{29}{18}\right)$$

$$= \frac{45}{18} - \frac{406}{162} = -\frac{1}{162}.$$

Hence, the correlation coefficient is

$$\rho = \frac{-1/162}{\sqrt{(20/81)(77/324)}} = \frac{-1}{\sqrt{1540}} = -0.025. \qquad \blacksquare$$

Insight into the correlation coefficient ρ of two discrete random variables X and Y may be gained by thoughtfully examining the definition of ρ, namely,

$$\rho = \frac{\sum_x \sum_y (x - \mu_X)(y - \mu_Y) f(x, y)}{\sigma_X \sigma_Y},$$

where μ_X, μ_Y, σ_X, and σ_Y denote the respective means and standard deviations. If positive probabilities are assigned to pairs (x, y) in which both x and y are either simultaneously above or simultaneously below their respective means, then the corresponding terms in the summation that defines ρ are positive because both factors $(x - \mu_X)$ and $(y - \mu_Y)$ will be positive or both will be negative. If, on the one hand, pairs (x, y), which yield large positive products $(x - \mu_X)(y - \mu_Y)$, contain most of the probability of the distribution, then the correlation coefficient will tend to be positive. If, on the other hand, the points (x, y), in which one component is below its mean and the other above its mean, have most of the probability, then the coefficient of correlation will tend to be negative because the products $(x - \mu_X)(y - \mu_Y)$ having higher probabilities are negative. (See Exercise 4.2-6.) This interpretation of the sign of the correlation coefficient will play an important role in subsequent work.

To gain additional insight into the meaning of the correlation coefficient ρ, consider the following problem: Think of the points (x, y) in the space S and their corresponding probabilities. Let us consider all possible lines in two-dimensional space, each with finite slope, that pass through the point associated with the means, namely, (μ_X, μ_Y). These lines are of the form $y - \mu_Y = b(x - \mu_X)$ or, equivalently,

$y = \mu_Y + b(x - \mu_X)$. For each point in S—say, (x_0, y_0), so that $f(x_0, y_0) > 0$—consider the vertical distance from that point to one of the aforesaid lines. Since y_0 is the height of the point above the x-axis and $\mu_Y + b(x_0 - \mu_X)$ is the height of the point on the line that is directly above or below the point (x_0, y_0), the absolute value of the difference of these two heights is the vertical distance from the point (x_0, y_0) to the line $y = \mu_Y + b(x - \mu_X)$. That is, the required distance is $|y_0 - \mu_Y - b(x_0 - \mu_X)|$. Let us now square this distance and take the weighted average of all such squares; in other words, let us consider the mathematical expectation

$$E\{[(Y - \mu_Y) - b(X - \mu_X)]^2\} = K(b).$$

The problem is to find that line (or that b) which minimizes this expectation of the square $\{Y - \mu_Y - b(X - \mu_X)\}^2$. This is an application of the principle of least squares, and the line is sometimes called the least squares regression line.

The solution of the problem is very easy, since

$$K(b) = E\{(Y - \mu_Y)^2 - 2b(X - \mu_X)(Y - \mu_Y) + b^2(X - \mu_X)^2\}$$
$$= \sigma_Y^2 - 2b\rho\sigma_X\sigma_Y + b^2\sigma_X^2,$$

because E is a linear operator and $E[(X - \mu_X)(Y - \mu_Y)] = \rho\sigma_X\sigma_Y$. Accordingly, the derivative

$$K'(b) = -2\rho\sigma_X\sigma_Y + 2b\sigma_X^2$$

equals zero at $b = \rho\sigma_Y/\sigma_X$, and we see that $K(b)$ obtains its minimum for that b, since $K''(b) = 2\sigma_X^2 > 0$. Consequently, the **least squares regression line** (the line of the given form that is the best fit in the foregoing sense) is

$$y = \mu_Y + \rho\frac{\sigma_Y}{\sigma_X}(x - \mu_X).$$

Of course, if $\rho > 0$, the slope of the line is positive; but if $\rho < 0$, the slope is negative.

It is also instructive to note the value of the minimum of

$$K(b) = E\{[(Y - \mu_Y) - b(X - \mu_X)]^2\} = \sigma_Y^2 - 2b\rho\sigma_X\sigma_Y + b^2\sigma_X^2.$$

It is

$$K\left(\rho\frac{\sigma_Y}{\sigma_X}\right) = \sigma_Y^2 - 2\rho\frac{\sigma_Y}{\sigma_X}\rho\sigma_X\sigma_Y + \left(\rho\frac{\sigma_Y}{\sigma_X}\right)^2\sigma_X^2$$
$$= \sigma_Y^2 - 2\rho^2\sigma_Y^2 + \rho^2\sigma_Y^2 = \sigma_Y^2(1 - \rho^2).$$

Since $K(b)$ is the expected value of a square, it must be nonnegative for all b, and we see that $\sigma_Y^2(1 - \rho^2) \geq 0$; that is, $\rho^2 \leq 1$, and hence $-1 \leq \rho \leq 1$, which is an important property of the correlation coefficient ρ. On the one hand, if $\rho = 0$, then $K(\rho\sigma_Y/\sigma_X) = \sigma_Y^2$; on the other hand, if ρ is close to 1 or -1, then $K(\rho\sigma_Y/\sigma_X)$ is relatively small. That is, the vertical deviations of the points with positive probability from the line $y = \mu_Y + \rho(\sigma_Y/\sigma_X)(x - \mu_X)$ are small if ρ is close to 1 or -1 because $K(\rho\sigma_Y/\sigma_X)$ is the expectation of the square of those deviations. Thus, ρ measures, in this sense, the amount of *linearity* in the probability distribution. As a matter of fact, in the discrete case, all the points of positive probability lie on this straight line if and only if ρ is equal to 1 or -1.

REMARK   More generally, we could have fitted the line $y = a + bx$ by the same application of the principle of least squares. We would then have proved that the "best" line actually passes through the point (μ_X, μ_Y). Recall that, in the preceding discussion, we assumed our line to be of that form. Students will find this derivation to be an interesting exercise using partial derivatives. (See Exercise 4.2-5.)   ∎

The next example illustrates a joint discrete distribution for which ρ is negative. In Figure 4.2-1, the line of best fit, or the least squares regression line, is also drawn.

EXAMPLE 4.2-2 Let X equal the number of ones and Y the number of twos and threes when a pair of fair four-sided dice is rolled. Then X and Y have a trinomial distribution with joint p.m.f.

$$f(x,y) = \frac{2!}{x!\,y!\,(2 - x - y)!}\left(\frac{1}{4}\right)^x\left(\frac{2}{4}\right)^y\left(\frac{1}{4}\right)^{2-x-y}, \qquad 0 \le x + y \le 2,$$

where x and y are nonnegative integers. Since the marginal p.m.f. of X is $b(2, 1/4)$ and the marginal p.m.f. of Y is $b(2, 1/2)$, it follows that $\mu_X = 1/2$, $\mathrm{Var}(X) = 6/16$, $\mu_Y = 1$, and $\mathrm{Var}(Y) = 1/2$. Also, since $E(XY) = (1)(1)(4/16) = 4/16$, we have $\mathrm{Cov}(X, Y) = 4/16 - (1/2)(1) = -4/16$; therefore, the correlation coefficient is $\rho = -1/\sqrt{3}$. Using these values for the parameters, we obtain the line of best fit, namely,

$$y = 1 + \left(-\frac{1}{\sqrt{3}}\right)\sqrt{\frac{1/2}{3/8}}\left(x - \frac{1}{2}\right) = -\frac{2}{3}x + \frac{4}{3}.$$

The joint p.m.f. is displayed in Figure 4.2-1 along with the line of best fit.   ∎

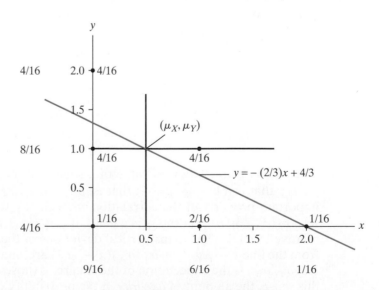

FIGURE 4.2-1: Trinomial distribution

Suppose that X and Y are independent, so that $f(x, y) \equiv f_1(x)f_2(y)$. Suppose also that we want to find the expected value of the product $u(X)v(Y)$. Subject to the existence of the expectations, we know that

$$E[u(X)v(Y)] = \sum_{S_1} \sum_{S_2} u(x)v(y)f(x,y)$$

$$= \sum_{S_1} \sum_{S_2} u(x)v(y)f_1(x)f_2(y)$$

$$= \sum_{S_1} u(x)f_1(x) \sum_{S_2} v(y)f_2(y)$$

$$= E[u(X)]E[v(Y)].$$

This formula can be used to show that the correlation coefficient of two independent variables is zero. For, in standard notation, we have

$$\text{Cov}(X, Y) = E[(X - \mu_X)(Y - \mu_Y)]$$
$$= E(X - \mu_X)E(Y - \mu_Y) = 0.$$

The converse of this equation is not necessarily true, however: Zero correlation does not, in general, imply independence. It is most important to keep the relationship straight: Independence implies zero correlation, but zero correlation does not necessarily imply independence. We now illustrate the latter proposition.

EXAMPLE 4.2-3 Let X and Y have the joint p.m.f.

$$f(x,y) = \frac{1}{3}, \qquad (x,y) = (0,1),(1,0),(2,1).$$

Since the support is not "rectangular," X and Y must be dependent. The means of X and Y are $\mu_X = 1$ and $\mu_Y = 2/3$, respectively. Hence,

$$\text{Cov}(X, Y) = E(XY) - \mu_X\mu_Y$$

$$= (0)(1)\left(\frac{1}{3}\right) + (1)(0)\left(\frac{1}{3}\right) + (2)(1)\left(\frac{1}{3}\right) - (1)\left(\frac{2}{3}\right) = 0.$$

That is, $\rho = 0$, but X and Y are dependent. ∎

Although we have used discrete random variables to define the correlation coefficient and related concepts, these ideas carry over to the continuous case with the usual modifications—in particular, with integrals replacing summations. We illustrate the continuous relationships in Exercises 4.2-10, 4.2-11, and 4.2-12.

EXERCISES

4.2-1. Let the random variables X and Y have the joint p.m.f

$$f(x,y) = \frac{x + y}{32}, \qquad x = 1,2, \ y = 1,2,3,4.$$

Find the means μ_X and μ_Y, the variances σ_X^2 and σ_Y^2, and the correlation coefficient ρ. Are X and Y independent or dependent?

4.2-2. Let X and Y have the joint p.m.f. defined by $f(0,0) = f(1,2) = 0.2, f(0,1) = f(1,1) = 0.3$.

(a) Depict the points and corresponding probabilities on a graph.

(b) Give the marginal p.m.f.'s in the "margins."

(c) Compute $\mu_X, \mu_Y, \sigma_X^2, \sigma_Y^2, \text{Cov}(X, Y)$, and ρ.

(d) Find the equation of the least squares regression line and draw it on your graph. Does the line make sense to you intuitively?

4.2-3. Roll a fair four-sided die twice. Let X equal the outcome on the first roll and let Y equal the sum of the two rolls.

(a) Display the joint p.m.f. on a graph along with the marginal probabilities.

(b) Determine μ_X, μ_Y, σ_X^2, σ_Y^2, Cov(X, Y), and ρ.

(c) Find the equation of the least squares regression line and draw it on your graph. Does the line make sense to you intuitively?

4.2-4. Show that, in the bivariate situation, E is a linear or distributive operator. That is, show that

$$E[a_1 u_1(X_1, X_2) + a_2 u_2(X_1, X_2)]$$
$$= a_1 E[u_1(X_1, X_2)] + a_2 E[u_2(X_1, X_2)].$$

4.2-5. Let X and Y be random variables with respective means μ_X and μ_Y, respective variances σ_X^2 and σ_Y^2, and correlation coefficient ρ. Fit the line $y = a + bx$ by the method of least squares to the probability distribution by minimizing the expectation

$$K(a,b) = E[(Y - a - bX)^2]$$

with respect to a and b. HINT: Set $\partial K/\partial a = 0$ and $\partial K/\partial b = 0$, and solve simultaneously.

4.2-6. Let X and Y have a trinomial distribution with parameters $n = 3$, $p_1 = 1/6$, and $p_2 = 1/2$. Find

(a) $E(X)$.

(b) $E(Y)$.

(c) Var(X).

(d) Var(Y).

(e) Cov(X, Y).

(f) ρ.

Note that $\rho = -\sqrt{p_1 p_2/(1 - p_1)(1 - p_2)}$ in this case. (Indeed, the formula holds in general for the trinomial distribution; see Example 4.3-3.)

4.2-7. Let the joint p.m.f. of X and Y be

$$f(x,y) = 1/4,$$
$$(x,y) \in S = \{(0,0),(1,1),(1,-1),(2,0)\}.$$

(a) Are X and Y independent?

(b) Calculate Cov(X, Y) and ρ.

This exercise also illustrates the fact that dependent random variables can have a correlation coefficient of zero.

4.2-8. The joint p.m.f. of X and Y is $f(x,y) = 1/6$, $0 \leq x + y \leq 2$, where x and y are nonnegative integers.

(a) Sketch the support of X and Y.

(b) Record the marginal p.m.f.'s $f_1(x)$ and $f_2(y)$ in the "margins."

(c) Compute Cov(X, Y).

(d) Determine ρ, the correlation coefficient.

(e) Find the best-fitting line and draw it on your figure.

4.2-9. A certain raw material is classified as to moisture content X (in percent) and impurity Y (in percent). Let X and Y have the joint p.m.f. given by

			x	
y	1	2	3	4
2	0.10	0.20	0.30	0.05
1	0.05	0.05	0.15	0.10

(a) Find the marginal p.m.f.'s, the means, and the variances.

(b) Find the covariance and the correlation coefficient of X and Y.

(c) If additional heating is needed with high moisture content and additional filtering with high impurity such that the additional cost is given by the function $C = 2X + 10Y^2$ in dollars, find $E(C)$.

4.2-10. Let X and Y be random variables of the continuous type having the joint p.d.f.

$$f(x,y) = 2, \qquad 0 \leq y \leq x \leq 1.$$

Draw a graph that illustrates the domain of this p.d.f.

(a) Find the marginal p.d.f.'s of X and Y.

(b) Compute μ_X, μ_Y, σ_X^2, σ_Y^2, Cov(X, Y), and ρ.

(c) Determine the equation of the least squares regression line and draw it on your graph. Does the line make sense to you intuitively?

4.2-11. Let X and Y be random variables of the continuous type having the joint p.d.f.

$$f(x,y) = x + y, \qquad 0 < x < 1, \ 0 < y < 1.$$

Draw a graph that illustrates the domain of this p.d.f.

(a) Find the marginal p.d.f.'s of X and Y.

(b) Compute μ_X, μ_Y, σ_X^2, σ_Y^2, Cov(X, Y), and ρ.

(c) Determine the equation of the least squares regression line and draw it on your graph. Does the line make sense to you intuitively?

4.2-12. Let X and Y be random variables of the continuous type having the joint p.d.f.

$$f(x,y) = 8xy, \qquad 0 \leq x \leq y \leq 1.$$

Draw a graph that illustrates the domain of this p.d.f.

(a) Find the marginal p.d.f.'s of X and Y.

(b) Compute $\mu_X, \mu_Y, \sigma_X^2, \sigma_Y^2, \text{Cov}(X, Y)$, and ρ.

(c) Determine the equation of the least squares regression line and draw it on your graph. Does the line make sense to you intuitively?

4.2-13. A car dealer sells X cars each day and always tries to sell an extended warranty on each of these cars. (In our opinion, most of these warranties are not good deals.) Let Y be the number of extended warranties sold; then $Y \leq X$. The joint p.m.f. of X and Y is given by

$$f(x, y) = c(x + 1)(4 - x)(y + 1)(3 - y),$$

$x = 0, 1, 2, 3; \; y = 0, 1, 2$ with $y \leq x$.

(a) Find the value of c.

(b) Sketch the support of X and Y.

(c) Record the marginal p.m.f.'s $f_1(x)$ and $f_2(y)$ in the "margins."

(d) Are X and Y independent?

(e) Compute μ_X and σ_X^2.

(f) Compute μ_Y and σ_Y^2.

(g) Compute $\text{Cov}(X, Y)$.

(h) Determine ρ, the correlation coefficient.

(i) Find the best-fitting line and draw it on your figure.

4.3 CONDITIONAL DISTRIBUTIONS

Let X and Y have a joint discrete distribution with p.m.f. $f(x, y)$ on space S. Say the marginal probability mass functions are $f_1(x)$ and $f_2(y)$ with spaces S_1 and S_2, respectively. Let event $A = \{X = x\}$ and event $B = \{Y = y\}$, $(x, y) \in S$. Thus, $A \cap B = \{X = x, Y = y\}$. Because

$$P(A \cap B) = P(X = x, Y = y) = f(x, y)$$

and

$$P(B) = P(Y = y) = f_2(y) > 0 \qquad \text{(since } y \in S_2 \text{)},$$

the conditional probability of event A given event B is

$$P(A \mid B) = \frac{P(A \cap B)}{P(B)} = \frac{f(x, y)}{f_2(y)}.$$

This formula leads to the following definition.

DEFINITION 4.3-1

The **conditional probability mass function of** X, given that $Y = y$, is defined by

$$g(x \mid y) = \frac{f(x, y)}{f_2(y)}, \qquad \text{provided that } f_2(y) > 0.$$

Similarly, the **conditional probability mass function of** Y, given that $X = x$, is defined by

$$h(y \mid x) = \frac{f(x, y)}{f_1(x)}, \qquad \text{provided that } f_1(x) > 0.$$

EXAMPLE 4.3-1   Let X and Y have the joint p.m.f.

$$f(x, y) = \frac{x + y}{21}, \qquad x = 1, 2, 3, \qquad y = 1, 2.$$

In Example 4.1-2, we showed that

$$f_1(x) = \frac{2x + 3}{21}, \qquad x = 1, 2, 3,$$

and

$$f_2(y) = \frac{3y + 6}{21}, \qquad y = 1, 2.$$

Thus, the conditional p.m.f. of X, given that $Y = y$, is equal to

$$g(x \mid y) = \frac{(x + y)/21}{(3y + 6)/21} = \frac{x + y}{3y + 6}, \qquad x = 1, 2, 3, \text{ when } y = 1 \text{ or } 2.$$

For example,

$$P(X = 2 \mid Y = 2) = g(2 \mid 2) = \frac{4}{12} = \frac{1}{3}.$$

Similarly, the conditional p.m.f. of Y, given that $X = x$, is equal to

$$h(y \mid x) = \frac{x + y}{2x + 3}, \qquad y = 1, 2, \text{ when } x = 1, 2, \text{ or } 3.$$

The joint p.m.f. $f(x, y)$ is depicted in Figure 4.3-1(a) along with the marginal p.m.f.'s. Now, if $y = 2$, we would expect the outcomes of x—namely, 1, 2, and 3—to occur in the ratio 3:4:5. This is precisely what $g(x \mid y)$ does:

$$g(1 \mid 2) = \frac{1 + 2}{12}, \qquad g(2 \mid 2) = \frac{2 + 2}{12}, \qquad g(3 \mid 2) = \frac{3 + 2}{12}.$$

Figure 4.3-1(b) displays $g(x \mid 1)$ and $g(x \mid 2)$, while Figure 4.3-1(c) gives $h(y \mid 1)$, $h(y \mid 2)$, and $h(y \mid 3)$. Compare the probabilities in Figure 4.3-1(c) with those in Figure 4.3-1(a). They should agree with your intuition as well as with the formula for $h(y \mid x)$. ■

Note that $0 \leq h(y \mid x)$. If we sum $h(y \mid x)$ over y for that fixed x, we obtain

$$\sum_y h(y \mid x) = \sum_y \frac{f(x, y)}{f_1(x)} = \frac{f_1(x)}{f_1(x)} = 1.$$

Thus, $h(y \mid x)$ satisfies the conditions of a probability mass function, and we can compute conditional probabilities such as

$$P(a < Y < b \mid X = x) = \sum_{\{y : a < y < b\}} h(y \mid x)$$

and conditional expectations such as

$$E[u(Y) \mid X = x] = \sum_y u(y) h(y \mid x)$$

in a manner similar to those associated with unconditional probabilities and expectations.

Two special conditional expectations are the **conditional mean** of Y, given that $X = x$, defined by

$$\mu_{Y \mid x} = E(Y \mid x) = \sum_y y h(y \mid x),$$

and the **conditional variance** of Y, given that $X = x$, defined by

$$\sigma^2_{Y \mid x} = E\{[Y - E(Y \mid x)]^2 \mid x\} = \sum_y [y - E(Y \mid x)]^2 h(y \mid x),$$

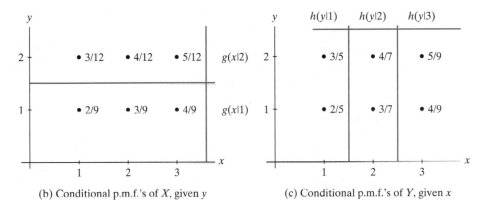

FIGURE 4.3-1: Joint, marginal, and conditional p.m.f.'s

which can be computed with

$$\sigma^2_{Y|x} = E(Y^2 \,|\, x) \;-\; [E(Y\,|\,x)]^2.$$

The conditional mean $\mu_{X|y}$ and the conditional variance $\sigma^2_{X|y}$ are given by similar expressions.

EXAMPLE 4.3-2   We use the background of Example 4.3-1 and compute $\mu_{Y|x}$ and $\sigma^2_{Y|x}$ when $x = 3$:

$$\mu_{Y|3} = E(Y \,|\, X = 3) = \sum_{y=1}^{2} y\, h(y\,|\,3)$$

$$= \sum_{y=1}^{2} y\left(\frac{3+y}{9}\right) = 1\left(\frac{4}{9}\right) + 2\left(\frac{5}{9}\right) = \frac{14}{9},$$

and

$$\sigma^2_{Y|3} = E\left[\left(Y - \frac{14}{9}\right)^2 \middle| X = 3\right] = \sum_{y=1}^{2} \left(y - \frac{14}{9}\right)^2\left(\frac{3+y}{9}\right)$$

$$= \frac{25}{81}\left(\frac{4}{9}\right) + \frac{16}{81}\left(\frac{5}{9}\right) = \frac{20}{81}.$$

∎

The conditional mean of X, given that $Y = y$, is a function of y alone; the conditional mean of Y, given that $X = x$, is a function of x alone. Suppose that the latter conditional mean is a linear function of x; that is, $E(Y|x) = a + bx$. Let us find the constants a and b in terms of characteristics μ_X, μ_Y, σ_X^2, σ_Y^2, and ρ. This development will shed additional light on the correlation coefficient ρ; accordingly, we assume that the respective standard deviations σ_X and σ_Y are both positive, so that the correlation coefficient will exist.

It is given that

$$\sum_y y\, h(y\,|\,x) = \sum_y y\, \frac{f(x,y)}{f_1(x)} = a + bx, \qquad \text{for } x \in S_1,$$

where S_1 is the space of X and S_2 is the space of Y. Hence,

$$\sum_y y f(x,y) = (a + bx)f_1(x), \qquad \text{for } x \in S_1, \tag{4.3-1}$$

and

$$\sum_{x \in S_1} \sum_y y f(x,y) = \sum_{x \in S_1} (a + bx)f_1(x).$$

That is, with μ_X and μ_Y representing the respective means, we have

$$\mu_Y = a + b\mu_X. \tag{4.3-2}$$

In addition, if we multiply both members of Equation 4.3-1 by x and sum the resulting product, we obtain

$$\sum_{x \in S_1} \sum_y xy f(x,y) = \sum_{x \in S_1} (ax + bx^2)f_1(x).$$

That is,

$$E(XY) = aE(X) + bE(X^2)$$

or, equivalently,

$$\mu_X \mu_Y + \rho \sigma_X \sigma_Y = a\mu_X + b(\mu_X^2 + \sigma_X^2). \tag{4.3-3}$$

The solution of Equations 4.3-2 and 4.3-3 is

$$a = \mu_Y - \rho\,\frac{\sigma_Y}{\sigma_X}\,\mu_X \qquad \text{and} \qquad b = \rho\,\frac{\sigma_Y}{\sigma_X},$$

which implies that if $E(Y|x)$ is linear, it is given by

$$E(Y|x) = \mu_Y + \rho\,\frac{\sigma_Y}{\sigma_X}\,(x - \mu_X).$$

So if the conditional mean of Y, given that $X = x$, is linear, it is exactly the same as the best-fitting line (least squares regression line) considered in Section 4.2.

By symmetry, if the conditional mean of X, given that $Y = y$, is linear, then

$$E(X|y) = \mu_X + \rho\,\frac{\sigma_X}{\sigma_Y}\,(y - \mu_Y).$$

We see that the point $[x = \mu_X, E(Y|x) = \mu_Y]$ satisfies the expression for $E(Y|x)$ and $[E(X|y) = \mu_X, y = \mu_Y]$ satisfies the expression for $E(X|y)$. That is, the point (μ_X, μ_Y) is on each of the two lines. In addition, we note that the product of the coefficient of x in $E(Y \mid x)$ and the coefficient of y in $E(X \mid y)$ equals ρ^2 and the ratio of these two coefficients equals σ_Y^2 / σ_X^2. These observations sometimes prove useful in particular problems.

EXAMPLE 4.3-3   Let X and Y have the trinomial p.m.f. with parameters n, p_1, p_2, and $1 - p_1 - p_2 = p_3$. That is,

$$f(x, y) = \frac{n!}{x! \, y! \, (n - x - y)!} \, p_1^x p_2^y p_3^{n-x-y},$$

where x and y are nonnegative integers such that $x + y \le n$. From the development of the trinomial distribution, we note that X and Y have marginal binomial distributions $b(n, p_1)$ and $b(n, p_2)$, respectively. Thus,

$$h(y \mid x) = \frac{f(x, y)}{f_1(x)} = \frac{(n - x)!}{y! \, (n - x - y)!} \left(\frac{p_2}{1 - p_1} \right)^y \left(\frac{p_3}{1 - p_1} \right)^{n-x-y},$$

$$y = 0, 1, 2, \ldots, n - x.$$

That is, the conditional p.m.f. of Y, given that $X = x$, is binomial, or

$$b\left[n - x, \frac{p_2}{1 - p_1} \right],$$

and thus has conditional mean

$$E(Y \mid x) = (n - x) \frac{p_2}{1 - p_1}.$$

In a similar manner, we obtain

$$E(X \mid y) = (n - y) \frac{p_1}{1 - p_2}.$$

Since each of the conditional means is linear, the product of the respective coefficients of x and y is

$$\rho^2 = \left(\frac{-p_2}{1 - p_1} \right) \left(\frac{-p_1}{1 - p_2} \right) = \frac{p_1 p_2}{(1 - p_1)(1 - p_2)}.$$

However, ρ must be negative, because the coefficients of x and y are negative; thus,

$$\rho = -\sqrt{\frac{p_1 p_2}{(1 - p_1)(1 - p_2)}}. \qquad \blacksquare$$

In the next example we again look at conditional p.m.f.'s when X and Y have a trinomial distribution.

EXAMPLE 4.3-4   Let X and Y have a trinomial distribution with $p_1 = 1/3$, $p_2 = 1/3$, and $n = 5$. Using a result from the last example, we find that the conditional distribution of Y, given that $X = x$, is $b(5 - x, (1/3)/(1 - 1/3))$, or $b(5 - x, 1/2)$. The p.m.f.'s, $h(y \mid x)$ for $x = 0, 1, \ldots, 5$ are plotted in Figure 4.3-2(a) on the next page. Note that the orientation of the axes was selected so that the shapes of these p.m.f.'s can be

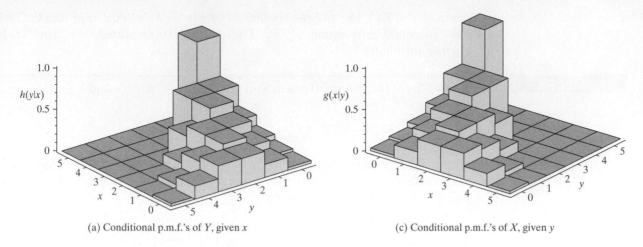

(a) Conditional p.m.f.'s of Y, given x (c) Conditional p.m.f.'s of X, given y

FIGURE 4.3-2: Conditional p.m.f.'s for the trinomial distribution

seen. Similarly, the conditional distribution of X, given that $Y = y$, is $b(n - y, 1/2)$. The p.m.f.'s, $g(x \mid y)$ for $y = 0, 1, \ldots, 5$ are shown in Figure 4.3-2(b). ∎

Although we have used random variables of the discrete type to introduce the new definitions, they also hold for random variables of the continuous type. Let X and Y have a distribution of the continuous type with joint p.d.f. $f(x, y)$ and marginal p.d.f.'s $f_1(x)$ and $f_2(y)$, respectively. Then the conditional p.d.f., mean, and variance of Y, given that $X = x$, are, respectively,

$$h(y \mid x) = \frac{f(x, y)}{f_1(x)}, \qquad \text{provided that } f_1(x) > 0;$$

$$E(Y \mid x) = \int_{-\infty}^{\infty} y \, h(y \mid x) \, dy;$$

and

$$\begin{aligned}
\text{Var}(Y \mid x) &= E\{[Y - E(Y \mid x)]^2 \mid x\} \\
&= \int_{-\infty}^{\infty} [y - E(Y \mid x)]^2 \, h(y \mid x) \, dy \\
&= E[Y^2 \mid x] - [E(Y \mid x)]^2.
\end{aligned}$$

Similar expressions are associated with the conditional distribution of X, given that $Y = y$.

EXAMPLE 4.3-5 Let X and Y be the random variables of Example 4.1-7. Thus,

$$\begin{aligned}
f(x, y) &= 2, & 0 \le x \le y \le 1, \\
f_1(x) &= 2(1 - x), & 0 \le x \le 1, \\
f_2(y) &= 2y, & 0 \le y \le 1.
\end{aligned}$$

Before we actually find the conditional p.d.f. of Y, given that $X = x$, we shall give an intuitive argument. The joint p.d.f. is constant over the triangular region bounded by $y = x$, $y = 1$, and $x = 0$. If the value of X is known (say, $X = x$), then the possible values of Y are between x and 1. Furthermore, we would expect Y to be uniformly distributed on the interval $[x, 1]$. That is, we would anticipate that $h(y \mid x) = 1/(1 - x), x \le y \le 1$.

More formally now, by definition, we have

$$h(y \mid x) = \frac{f(x, y)}{f_1(x)} = \frac{2}{2(1 - x)} = \frac{1}{1 - x}, \qquad x \le y \le 1, \qquad 0 \le x \le 1.$$

The conditional mean of Y, given that $X = x$, is

$$E(Y \mid x) = \int_x^1 y \frac{1}{1 - x} \, dy = \left[\frac{y^2}{2(1 - x)} \right]_x^1 = \frac{1 + x}{2}, \qquad 0 \le x \le 1.$$

Similarly, it can be shown that

$$E(X \mid y) = \frac{y}{2}, \qquad 0 \le y \le 1.$$

The conditional variance of Y, given that $X = x$, is

$$E\{[Y - E(Y \mid x)]^2 \mid x\} = \int_x^1 \left(y - \frac{1 + x}{2} \right)^2 \frac{1}{1 - x} \, dy$$

$$= \left[\frac{1}{3(1 - x)} \left(y - \frac{1 + x}{2} \right)^3 \right]_x^1$$

$$= \frac{(1 - x)^2}{12}.$$

Recall that if a random variable W is $U(a, b)$, then $E(W) = (a + b)/2$ and $\mathrm{Var}(W) = (b - a)^2/12$. Since the conditional distribution of Y, given that $X = x$, is $U(x, 1)$, we could have inferred immediately that $E(Y \mid x) = (x + 1)/2$ and $\mathrm{Var}(Y \mid x) = (1 - x)^2/12$.

An illustration of a computation of a conditional probability is

$$P\left(\frac{3}{4} \le Y \le \frac{7}{8} \,\middle|\, X = \frac{1}{4} \right) = \int_{3/4}^{7/8} h\left(y \,\middle|\, \frac{1}{4} \right) dy$$

$$= \int_{3/4}^{7/8} \frac{1}{3/4} \, dy = \frac{1}{6}. \qquad \blacksquare$$

In general, if $E(Y \mid x)$ is linear, then

$$E(Y \mid x) = \mu_Y + \rho\left(\frac{\sigma_Y}{\sigma_X} \right)(x - \mu_X).$$

If $E(X \mid y)$ is linear, then

$$E(X \mid y) = \mu_X + \rho\left(\frac{\sigma_X}{\sigma_Y} \right)(y - \mu_Y).$$

Thus, in Example 4.3-5, the product of the coefficients of x in $E(Y \mid x)$ and y in $E(X \mid y)$ is $\rho^2 = 1/4$. It then follows that $\rho = 1/2$, since each coefficient is positive. Because the ratio of those coefficients is equal to $\sigma_Y^2/\sigma_X^2 = 1$, we have $\sigma_X^2 = \sigma_Y^2$.

STATISTICAL COMMENTS (**More on Simpson's Paradox**) Let us say that a small business employs 100 men and 100 women and the salary for each person is either $20,000, $30,000, or $50,000. For the men, 20 make $20,000, 20 make $30,000, and 60 make $50,000; the men's average is then

$$\frac{(20)(20{,}000) + (20)(30{,}000) + (60)(50{,}000)}{100} = \$40{,}000.$$

Among the women, 60 make $20,000, 20 make $30,000, and 20 make $50,000; the women's average is then

$$\frac{(60)(20{,}000) + (20)(30{,}000) + (20)(50{,}000)}{100} = \$28{,}000.$$

This seems to be a fairly large gap in these two average salaries.

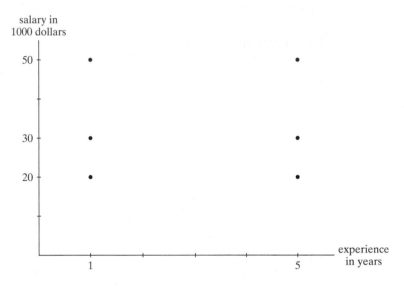

FIGURE 4.3-3: Salary versus experience

It turns out that there is another variable to consider, however, namely, experience. For simplicity, say each employee has either one year or five years of experience. Thus, we have, for the men (women), a bivariate distribution with space given by Figure 4.3-3. Now, on the one hand, we do not know the joint probabilities and hence cannot compute the conditional means, given one year and then five years of experience. On the other hand, we do have the data associated with these 100 men and 100 women and can therefore estimate the conditional means, namely, there are 20 men with one year of experience: 15 earn $20,000 and 5 earn $30,000, giving a mean of $22,500. The other 80 men have five years of experience: 5 earn $20,000, 15 earn $30,000, and 60 earn $50,000, for an average of $44,375. As for the women, there are 80 with one year of experience: 60 earn $20,000, 15 earn $30,000, and 5 earn $50,000, averaging $23,750. Also, 20 women have five years of experience: 5 earn $30,000 and 15 earn $50,000, for an average of $45,000. So while there is a large gap between the $28,000 earned, on the average, by women and the $40,000 earned by men, the women seem to have a slight advantage by a comparison of the conditional averages: $23,750 to $22,500, given one year of experience; and $45,000 to $44,375, given five years of experience. The preceding information is summarized in Table 4.3-1.

		Years of Experience		
	Salary	One year	Five years	Total
Men	$20,000	15	5	20
	$30,000	5	15	20
	$50,000	0	60	60
Average Salary		$22,500	$44,375	$40,000
Women	$20,000	60	0	60
	$30,000	15	5	20
	$50,000	5	15	20
Average Salary		$23,750	$45,000	$28,000

TABLE 4.3-1: Summary of Salaries

This reversal is another illustration of Simpson's paradox. We are not arguing that there has not been discrimination in the pay for different genders, for there has certainly been a great deal of that in the past. Although we know in our business (academia) that most of this discrimination has been eliminated, we still insist that experience must be considered: not only years of teaching, but publications, professional field, activity in professional societies, service on committees, and so on. Clearly, we support the position that a woman and a man with the same credentials should receive the same rank and pay.

EXERCISES

4.3-1. Let X and Y have the joint p.m.f.

$$f(x, y) = \frac{x + y}{32}, \qquad x = 1, 2, \qquad y = 1, 2, 3, 4.$$

(a) Display the joint p.m.f. and the marginal p.m.f.'s on a graph like Figure 4.3-1(a).

(b) Find $g(x \mid y)$ and draw a figure like Figure 4.3-1(b), depicting the conditional p.m.f.'s for $y = 1, 2, 3$, and 4.

(c) Find $h(y \mid x)$ and draw a figure like Figure 4.3-1(c), depicting the conditional p.m.f.'s for $x = 1$ and 2.

(d) Find $P(1 \leq Y \leq 3 \mid X = 1), P(Y \leq 2 \mid X = 2)$, and $P(X = 2 \mid Y = 3)$.

(e) Find $E(Y \mid X = 1)$ and $\mathrm{Var}(Y \mid X = 1)$.

4.3-2. Let the joint p.m.f. $f(x, y)$ of X and Y be given by the following:

(x, y)	$f(x, y)$
$(1, 1)$	3/8
$(2, 1)$	1/8
$(1, 2)$	1/8
$(2, 2)$	3/8

Find the two conditional probability mass functions and the corresponding means and variances.

4.3-3. Let W equal the weight of laundry soap in a 1-kilogram box that is distributed in Southeast Asia. Suppose that $P(W < 1) = 0.02$ and $P(W > 1.072) = 0.08$. Call a box of soap light, good, or heavy, depending on whether $\{W < 1\}$, $\{1 \leq W \leq 1.072\}$, or $\{W > 1.072\}$, respectively. In $n = 50$ independent observations of these boxes, let X equal the number of light boxes and Y the number of good boxes.

(a) What is the joint p.m.f. of X and Y?

(b) Give the name of the distribution of Y along with the values of the parameters of this distribution.

(c) Given that $X = 3$, how is Y distributed conditionally?

(d) Determine $E(Y \mid X = 3)$.

(e) Find ρ, the correlation coefficient of X and Y.

4.3-4. The genes for eye color in a certain male fruit fly are (R, W). The genes for eye color in the mating female fruit fly are (R, W). Their offspring receive one gene for eye color from each parent. If an offspring ends up with either (R, R), (R, W), or

(W, R), its eyes will look red. Let X equal the number of offspring having red-eyes. Let Y equal the number of red eyed offspring having (R, W) or (W, R) genes.

(a) If the total number of offspring is $n = 400$, how is X distributed?

(b) Give the values of $E(X)$ and Var(X).

(c) Given that $X = 300$, how is Y distributed?

(d) Give the values of $E(Y|X = 300)$ and Var$(Y|X = 300)$.

4.3-5. Let X and Y have a trinomial distribution with $n = 2$, $p_1 = 1/4$, and $p_2 = 1/2$.

(a) Give $E(Y|x)$.

(b) Compare your answer in part (a) with the equation of the line of best fit in Example 4.2-2. Are they the same? Why or why not?

4.3-6. An insurance company sells both homeowners insurance and automobile deductible insurance. Let X be the deductible on the homeowners insurance and Y the deductible on automobile insurance. Among those who take both types of insurance with this company, we find the following probabilities:

y	x		
	100	500	1000
1000	0.05	0.10	0.15
500	0.10	0.20	0.05
100	0.20	0.10	0.05

(a) Compute the following probabilities: $P(X = 500)$, $P(Y = 500)$, $P(Y = 500|X = 500)$, $P(Y = 100|X = 500)$.

(b) Compute the means μ_X, μ_Y, and the variances σ_X^2, σ_Y^2.

(c) Compute the conditional means $E(X|Y = 100)$, $E(Y|X = 500)$.

(d) Compute Cov(X, Y).

(e) Compute the correlation coefficient $\rho = $ Cov$(X, Y)/\sigma_X\sigma_Y$.

4.3-7. Using the joint p.m.f. from Exercise 4.2-3, find the value of $E(Y|x)$ for $x = 1, 2, 3, 4$. Do the points $[x, E(Y|x)]$ lie on the best-fitting line?

4.3-8. An unbiased six-sided die is cast 30 independent times. Let X be the number of ones and Y the number of twos.

(a) What is the joint p.m.f. of X and Y?

(b) Find the conditional p.m.f. of X, given $Y = y$.

(c) Compute $E(X^2 - 4XY + 3Y^2)$.

4.3-9. Let X and Y have a uniform distribution on the set of points with integer coordinates in $S = \{(x, y): 0 \le x \le 7, x \le y \le x + 2\}$. That is, $f(x, y) = 1/24$, $(x, y) \in S$, and both x and y are integers. Find

(a) $f_1(x)$.

(b) $h(y|x)$.

(c) $E(Y|x)$.

(d) $\sigma_{Y|x}^2$.

(e) $f_2(y)$.

4.3-10. Let $f_1(x) = 1/10$, $x = 0, 1, 2, \ldots, 9$, and $h(y|x) = 1/(10 - x)$, $y = x, x + 1, \ldots, 9$. Find

(a) $f(x, y)$.

(b) $f_2(y)$.

(c) $E(Y|x)$.

4.3-11. An automobile repair shop makes an initial estimate X (in thousands of dollars) of the amount of money needed to fix a car after an accident. Say X has the p.d.f.

$$f(x) = 2e^{-2(x-0.2)}, \qquad 0.2 < x < \infty.$$

Given that $X = x$, the final payment Y has a uniform distribution between $x - 0.1$ and $x + 0.1$. What is the expected value of Y?

4.3-12. For the random variables defined in Example 4.3-5, calculate the correlation coefficient directly from the definition

$$\rho = \frac{\text{Cov}(X, Y)}{\sigma_X\sigma_Y}.$$

4.3-13. Let $f(x, y) = 1/40$, $0 \le x \le 10$, $10 - x \le y \le 14 - x$, be the joint p.d.f. of X and Y.

(a) Sketch the region for which $f(x, y) > 0$.

(b) Find $f_1(x)$, the marginal p.d.f. of X.

(c) Determine $h(y|x)$, the conditional p.d.f. of Y, given that $X = x$.

(d) Calculate $E(Y|x)$, the conditional mean of Y, given that $X = x$.

4.3-14. Let $f(x, y) = 1/8$, $0 \le y \le 4$, $y \le x \le y + 2$, be the joint p.d.f. of X and Y.

(a) Sketch the region for which $f(x, y) > 0$.

(b) Find $f_1(x)$, the marginal p.d.f. of X.

(c) Find $f_2(y)$, the marginal p.d.f. of Y.

(d) Determine $h(y|x)$, the conditional p.d.f. of Y, given that $X = x$.

(e) Determine $g(x|y)$, the conditional p.d.f. of X, given that $Y = y$.

(f) Compute $y = E(Y|x)$, the conditional mean of Y, given that $X = x$.

(g) Compute $x = E(X|y)$, the conditional mean of X, given that $Y = y$.

(h) Graph $y = E(Y|x)$ on your sketch in part (a). Is $y = E(Y|x)$ linear?

(i) Graph $x = E(X|y)$ on your sketch in part (a). Is $x = E(X|y)$ linear?

4.3-15. Let X have a uniform distribution $U(0,2)$, and let the conditional distribution of Y, given that $X = x$, be $U(0, x^2)$.

(a) Determine $f(x,y)$, the joint p.d.f. of X and Y.

(b) Calculate $f_2(y)$, the marginal p.d.f. of Y.

(c) Compute $E(X|y)$, the conditional mean of X, given that $Y = y$.

(d) Find $E(Y|x)$, the conditional mean of Y, given that $X = x$.

4.3-16. Let X have a uniform distribution on the interval $(0, 1)$. Given that $X = x$, let Y have a uniform distribution on the interval $(0, x)$.

(a) Define the conditional p.d.f. of Y, given that $X = x$. Be sure to include the domain.

(b) Find $E(Y|x)$.

(c) Determine the joint p.d.f. of X and Y.

(d) Find the marginal p.d.f. of Y.

4.3-17. The marginal distribution of X is $U(0,1)$. The conditional distribution of Y, given that $X = x$, is $U(0, e^x)$.

(a) Determine $h(y|x)$, the conditional p.d.f. of Y, given that $X = x$.

(b) Find $E(Y|x)$.

(c) Display the joint p.d.f. of X and Y. Sketch the region where $f(x, y) > 0$.

(d) Find $f_2(y)$, the marginal p.d.f. of Y.

4.3-18. Let X have a uniform distribution on the interval $(0,1)$. Given that $X = x$, let Y have a uniform distribution on the interval $(0, x + 1)$.

(a) Find the joint p.d.f. of X and Y. Sketch the region where $f(x,y) > 0$.

(b) Find $E(Y|x)$, the conditional mean of Y, given that $X = x$. Draw this line on the region sketched in part (a).

(c) Find $f_2(y)$, the marginal p.d.f. of Y. Be sure to include the domain.

4.3-19. Let X and Y have the joint p.d.f. $f(x, y) = cx(1 - y)$, $0 < y < 1$ and $0 < x < 1 - y$.

(a) Determine c.

(b) Compute $P(Y < X|X \le 1/4)$.

4.3-20. Select x and y to create a triangle of perimeter 1 that has sides of lengths x, y, and $1 - x - y$. By Heron's formula, the area of such a triangle is

$$T = \frac{1}{4}\sqrt{(2x + 2y - 1)(1 - 2x)(1 - 2y)}.$$

If x and y are values of the jointly distributed random variables X and Y, then T is a random variable that can be thought of as the area of a "random triangle."

(a) Determine the possible values of (X, Y) and graph this region in the xy-plane. Call the region R.

(b) Select the random point (X, Y) uniformly from R. That is, the joint p.d.f. of (X, Y) is $f(x, y) = 1/A(R)$, where $A(R)$ is the area of R. Show that $E(T) = \pi/105$ and the variance of T is

$$E(T^2) - [E(T)]^2 = 1/960 - (\pi/105)^2.$$

HINT: Use Maple or some other computer algebra system.

(c) Find the marginal p.d.f. of X and the conditional p.d.f. of Y, given that $X = x$.

(d) Simulate 5000 pairs of observations of X and Y and then calculate the areas of the simulated triangles. Show that the sample mean and the sample variance of the areas of these 5000 triangles are close to the theoretical values.

(e) Now select (X, Y) as follows: The random variable X is selected randomly. That is, $f_1(x) = c$, where c is selected appropriately. Be sure to define the domain for this p.d.f. Given that $X = x$, Y is selected randomly (uniformly) from the appropriate interval. Define the joint p.d.f. of X and Y [the domain is R from part (a)], and again find $E(T)$ and the variance of T. Compare the theoretical and the sample values.

Remark   This exercise is based on research at Hope College by students Andrea Douglass, Courtney Fitzgerald, and Scott Mihalik. (See references.)   ∎

4.4   THE BIVARIATE NORMAL DISTRIBUTION

Let X and Y be random variables with joint p.d.f. $f(x,y)$ of the continuous type. Many applications are concerned with the conditional distribution of one of the random variables—say, Y, given that $X = x$. For example, X and Y might be a student's grade point averages from high school and from the first year in college, respectively. Persons in the field of educational testing and measurement are extremely interested in the conditional distribution of Y, given that $X = x$, in such situations.

Suppose that we have an application in which we can make the following three assumptions about the conditional distribution of Y, given that $X = x$:

(a) It is normal for each real x.

(b) Its mean, $E(Y|x)$, is a linear function of x.

(c) Its variance is constant; that is, it does not depend upon the given value of x.

Of course, assumption (b), along with a result given in Section 4.3, implies that

$$E(Y|x) = \mu_Y + \rho \frac{\sigma_Y}{\sigma_X}(x - \mu_X).$$

Let us consider the implication of assumption (c). The conditional variance is given by

$$\sigma_{Y|x}^2 = \int_{-\infty}^{\infty} \left[y - \mu_Y - \rho \frac{\sigma_Y}{\sigma_X}(x - \mu_X) \right]^2 h(y|x)\, dy,$$

where $h(y|x)$ is the conditional p.d.f. of Y given that $X = x$. Multiply each member of this equation by $f_1(x)$ and integrate on x. Since $\sigma_{Y|x}^2$ is a constant, the left-hand member is equal to $\sigma_{Y|x}^2$. Thus, we have

$$\sigma_{Y|x}^2 = \int_{-\infty}^{\infty} \int_{-\infty}^{\infty} \left[y - \mu_Y - \rho \frac{\sigma_Y}{\sigma_X}(x - \mu_X) \right]^2 h(y|x) f_1(x)\, dy\, dx. \quad (4.4\text{-}1)$$

However, $h(y|x)f_1(x) = f(x,y)$; hence, the right-hand member is just an expectation and Equation 4.4-1 can be written as

$$\sigma_{Y|x}^2 = E\left\{ (Y - \mu_Y)^2 - 2\rho \frac{\sigma_Y}{\sigma_X}(X - \mu_X)(Y - \mu_Y) + \rho^2 \frac{\sigma_Y^2}{\sigma_X^2}(X - \mu_X)^2 \right\}.$$

But using the fact that the expectation E is a linear operator and recalling that $E[(X - \mu_X)(Y - \mu_Y)] = \rho\sigma_X\sigma_Y$, we have

$$\sigma_{Y|x}^2 = \sigma_Y^2 - 2\rho \frac{\sigma_Y}{\sigma_X} \rho\sigma_X\sigma_Y + \rho^2 \frac{\sigma_Y^2}{\sigma_X^2} \sigma_X^2$$

$$= \sigma_Y^2 - 2\rho^2\sigma_Y^2 + \rho^2\sigma_Y^2 = \sigma_Y^2(1 - \rho^2).$$

That is, the conditional variance of Y, for each given x, is $\sigma_Y^2(1 - \rho^2)$. These facts about the conditional mean and variance, along with assumption (a), require that the conditional p.d.f. of Y, given that $X = x$, be

$$h(y|x) = \frac{1}{\sigma_Y \sqrt{2\pi}\sqrt{1 - \rho^2}} \exp\left[-\frac{[y - \mu_Y - \rho(\sigma_Y/\sigma_X)(x - \mu_X)]^2}{2\sigma_Y^2(1 - \rho^2)} \right],$$

$-\infty < y < \infty$, for every real x.

Before we make any assumptions about the distribution of X, we give an example and a figure to illustrate the implications of our current assumptions.

EXAMPLE 4.4-1 Let $\mu_X = 10, \sigma_X^2 = 9, \mu_Y = 12, \sigma_Y^2 = 16$, and $\rho = 0.6$. We have seen that assumptions (a), (b), and (c) imply that the conditional distribution of Y, given that $X = x$, is

$$N\left[12 + (0.6)\left(\frac{4}{3}\right)(x - 10), 16(1 - 0.6^2)\right].$$

In Figure 4.4-1, the conditional mean line

$$E(Y \mid x) = 12 + (0.6)\left(\frac{4}{3}\right)(x - 10) = 0.8x + 4$$

has been graphed. For each of $x = 5, 10$, and 15, the conditional p.d.f. of Y, given that $X = x$, is displayed. ■

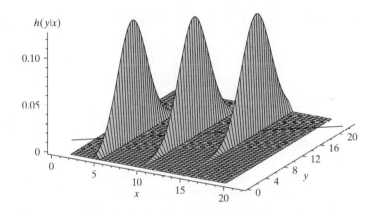

FIGURE 4.4-1: Conditional p.d.f. of Y, given that $x = 5, 10, 15$

Up to this point, nothing has been said about the distribution of X other than that it has mean μ_X and positive variance σ_X^2. Suppose, in addition, we assume that this distribution is also normal; that is, the marginal p.d.f. of X is

$$f_1(x) = \frac{1}{\sigma_X \sqrt{2\pi}} \exp\left[-\frac{(x - \mu_X)^2}{2\sigma_X^2}\right], \qquad -\infty < x < \infty.$$

Hence, the joint p.d.f. of X and Y is given by the product

$$f(x,y) = h(y \mid x)f_1(x) = \frac{1}{2\pi\sigma_X\sigma_Y\sqrt{1 - \rho^2}} \exp\left[-\frac{q(x,y)}{2}\right], \qquad (4.4\text{-}2)$$

where it can be shown (see Exercise 4.4-2) that

$$q(x,y) = \frac{1}{1 - \rho^2}\left[\left(\frac{x - \mu_X}{\sigma_X}\right)^2 - 2\rho\left(\frac{x - \mu_X}{\sigma_X}\right)\left(\frac{y - \mu_Y}{\sigma_Y}\right) + \left(\frac{y - \mu_Y}{\sigma_Y}\right)^2\right].$$

A joint p.d.f. of this form is called a **bivariate normal p.d.f.**

EXAMPLE 4.4-2 Let us assume that in a certain population of college students, the respective grade point averages—say, X and Y—in high school and the first year in college have an approximate bivariate normal distribution with parameters $\mu_X = 2.9$, $\mu_Y = 2.4$, $\sigma_X = 0.4$, $\sigma_Y = 0.5$, and $\rho = 0.8$.

Then, for example,

$$P(2.1 < Y < 3.3) = P\left(\frac{2.1 - 2.4}{0.5} < \frac{Y - 2.4}{0.5} < \frac{3.3 - 2.4}{0.5}\right)$$

$$= \Phi(1.8) - \Phi(-0.6) = 0.6898.$$

Since the conditional p.d.f. of Y, given that $X = 3.2$, is normal with mean

$$2.4 + (0.8)\left(\frac{0.5}{0.4}\right)(3.2 - 2.9) = 2.7$$

and standard deviation $(0.5)\sqrt{1 - 0.64} = 0.3$, we have

$$P(2.1 < Y < 3.3 \,|\, X = 3.2)$$

$$= P\left(\frac{2.1 - 2.7}{0.3} < \frac{Y - 2.7}{0.3} < \frac{3.3 - 2.7}{0.3} \,\middle|\, X = 3.2\right)$$

$$= \Phi(2) - \Phi(-2) = 0.9544.$$

From a practical point of view, however, the reader should be warned that the correlation coefficient of these grade point averages is, in many instances, much smaller than 0.8. ∎

Since x and y enter the bivariate normal p.d.f. in a similar manner, the roles of X and Y could have been interchanged. That is, Y could have been assigned the marginal normal p.d.f. $N(\mu_Y, \sigma_Y^2)$, and the conditional p.d.f. of X, given that $Y = y$, would have then been normal, with mean $\mu_X + \rho(\sigma_X/\sigma_Y)(y - \mu_Y)$ and variance $\sigma_X^2(1 - \rho^2)$. Although this property is fairly obvious, we do want to make special note of it.

In order to have a better understanding of the geometry of the bivariate normal distribution, consider the graph of $z = f(x, y)$, where $f(x, y)$ is given by Equation 4.4-2. If we intersect this surface with planes parallel to the yz-plane (i.e., with $x = x_0$), we have

$$f(x_0, y) = f_1(x_0)h(y \,|\, x_0).$$

In this equation, $f_1(x_0)$ is a constant and $h(y \,|\, x_0)$ is a normal p.d.f. Thus, $z = f(x_0, y)$ is bell shaped; that is, has the shape of a normal p.d.f. However, note that it is not necessarily a p.d.f., because of the factor $f_1(x_0)$. Similarly, intersections of the surface $z = f(x, y)$ with planes $y = y_0$ parallel to the xz-plane will be bell shaped.

If

$$0 < z_0 < \frac{1}{2\pi\sigma_X\sigma_Y\sqrt{1 - \rho^2}},$$

then

$$0 < z_0 2\pi\sigma_X\sigma_Y\sqrt{1 - \rho^2} < 1.$$

If we intersect $z = f(x, y)$ with the plane $z = z_0$, which is parallel to the xy-plane, then we have

$$z_0 2\pi\sigma_X\sigma_Y\sqrt{1 - \rho^2} = \exp\left[\frac{-q(x, y)}{2}\right].$$

We show that these intersections are ellipses by taking the natural logarithm of each side to obtain

$$\left(\frac{x - \mu_X}{\sigma_X}\right)^2 - 2\rho\left(\frac{x - \mu_X}{\sigma_X}\right)\left(\frac{y - \mu_Y}{\sigma_Y}\right) + \left(\frac{y - \mu_Y}{\sigma_Y}\right)^2$$

$$= -2(1 - \rho^2)\ln(z_0 2\pi\sigma_X\sigma_Y\sqrt{1 - \rho^2}).$$

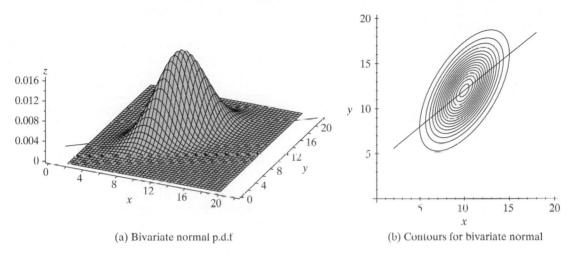

(a) Bivariate normal p.d.f

(b) Contours for bivariate normal

FIGURE 4.4-2: Bivariate normal, $\mu_X = 10$, $\sigma_X^2 = 9$, $\mu_Y = 12$, $\sigma_Y^2 = 16$, $\rho = 0.6$

EXAMPLE 4.4-3 With $\mu_X = 10$, $\sigma_X^2 = 9$, $\mu_Y = 12$, $\sigma_Y^2 = 16$, and $\rho = 0.6$, the bivariate normal p.d.f. has been graphed in Figure 4.4-2(a). For $\rho = 0.6$, level curves, or contours, are given in Figure 4.4-2(b). The conditional mean line,

$$E(Y \mid x) = 12 + (0.6)\left(\frac{4}{3}\right)(x - 10) = 0.8x + 4,$$

is also drawn on Figure 4.4-2(b). Note that this line intersects the level curves at points through which vertical tangents can be drawn to the ellipses. ∎

We close this section by observing another important property of the correlation coefficient ρ if X and Y have a bivariate normal distribution. In Equation 4.4-2 of the product $h(y \mid x)f_1(x)$, let us consider the factor $h(y \mid x)$ if $\rho = 0$. We see that this product, which is the joint p.d.f. of X and Y, equals $f_1(x)f_2(y)$ because when $\rho = 0$, $h(y \mid x)$ is a normal p.d.f. with mean μ_Y and variance σ_Y^2. That is, if $\rho = 0$, then the joint p.d.f. factors into the product of the two marginal probability density functions and hence X and Y are independent random variables. Of course, if X and Y are *any* independent random variables (not necessarily normal), then ρ, if it exists, is always equal to zero. Hence, we have proved the following theorem.

THEOREM 4.4-1 If X and Y have a bivariate normal distribution with correlation coefficient ρ, then X and Y are independent if and only if $\rho = 0$.

Thus, in the bivariate normal case, $\rho = 0$ does imply independence of X and Y. Note that these characteristics of the bivariate normal distribution can be extended to the trivariate normal distribution or, more generally, the multivariate normal distribution. This is done in more advanced texts that assume some knowledge of matrices [e.g., Hogg, McKean, and Craig (2005)].

EXERCISES

4.4-1. Let X and Y have a bivariate normal distribution with parameters $\mu_X = -3$, $\mu_Y = 10$, $\sigma_X^2 = 25$, $\sigma_Y^2 = 9$, and $\rho = 3/5$. Compute

(a) $P(-5 < X < 5)$.
(b) $P(-5 < X < 5 \,|\, Y = 13)$.
(c) $P(7 < Y < 16)$.
(d) $P(7 < Y < 16 \,|\, X = 2)$.

4.4-2. Show that the expression in the exponent of Equation 4.4-2 is equal to the function $q(x, y)$ given in the text.

4.4-3. Let X and Y have a bivariate normal distribution with parameters $\mu_X = 2.8$, $\mu_Y = 110$, $\sigma_X^2 = 0.16$, $\sigma_Y^2 = 100$, and $\rho = 0.6$. Compute

(a) $P(106 < Y < 124)$.
(b) $P(106 < Y < 124 \,|\, X = 3.2)$.

4.4-4. Let X and Y have a bivariate normal distribution with $\mu_X = 70$, $\sigma_X^2 = 100$, $\mu_Y = 80$, $\sigma_Y^2 = 169$, and $\rho = 5/13$. Find

(a) $E(Y \,|\, X = 72)$.
(b) $\mathrm{Var}(Y \,|\, X = 72)$.
(c) $P(Y \le 84 \,|\, X = 72)$.

4.4-5. Let X denote the height in centimeters and Y the weight in kilograms of male college students. Assume that X and Y have a bivariate normal distribution with parameters $\mu_X = 185$, $\sigma_X^2 = 100$, $\mu_Y = 84$, $\sigma_Y^2 = 64$, and $\rho = 3/5$.

(a) Determine the conditional distribution of Y, given that $X = 190$.
(b) Find $P(86.4 < Y < 95.36 \,|\, X = 190)$.

4.4-6. For a freshman taking introductory statistics and majoring in psychology, let X equal the student's ACT mathematics score and Y the student's ACT verbal score. Assume that X and Y have a bivariate normal distribution with $\mu_X = 22.7$, $\sigma_X^2 = 17.64$, $\mu_Y = 22.7$, $\sigma_Y^2 = 12.25$, and $\rho = 0.78$.

(a) Find $P(18.5 < Y < 25.5)$.
(b) Find $E(Y \,|\, x)$.
(c) Find $\mathrm{Var}(Y \,|\, x)$.
(d) Find $P(18.5 < Y < 25.5 \,|\, X = 23)$.
(e) Find $P(18.5 < Y < 25.5 \,|\, X = 25)$.

(f) For $x = 21$, 23, and 25, draw a graph of $z = h(y \,|\, x)$ similar to Figure 4.4-1.

4.4-7. For a pair of gallinules, let X equal the weight in grams of the male and Y the weight in grams of the female. Assume that X and Y have a bivariate normal distribution with $\mu_X = 415$, $\sigma_X^2 = 611$, $\mu_Y = 347$, $\sigma_Y^2 = 689$, and $\rho = -0.25$. Find

(a) $P(309.2 < Y < 380.6)$.
(b) $E(Y \,|\, x)$.
(c) $\mathrm{Var}(Y \,|\, x)$.
(d) $P(309.2 < Y < 380.6 \,|\, X = 385.1)$.

4.4-8. Let X and Y have a bivariate normal distribution with parameters $\mu_X = 10$, $\sigma_X^2 = 9$, $\mu_Y = 15$, $\sigma_Y^2 = 16$, and $\rho = 0$. Find

(a) $P(13.6 < Y < 17.2)$.
(b) $E(Y \,|\, x)$.
(c) $\mathrm{Var}(Y \,|\, x)$.
(d) $P(13.6 < Y < 17.2 \,|\, X = 9.1)$.

4.4-9. Let X and Y have a bivariate normal distribution. Find two different lines, $a(x)$ and $b(x)$, parallel to and equidistant from $E(Y \,|\, x)$, such that

$$P[a(x) < Y < b(x) \,|\, X = x] = 0.9544$$

for all real x. Plot $a(x)$, $b(x)$, and $E(Y \,|\, x)$ when $\mu_X = 2$, $\mu_Y = -1$, $\sigma_X = 3$, $\sigma_Y = 5$, and $\rho = 3/5$.

4.4-10. In a college health fitness program, let X denote the weight in kilograms of a male freshman at the beginning of the program and Y denote his weight change during a semester. Assume that X and Y have a bivariate normal distribution with $\mu_X = 72.30$, $\sigma_X^2 = 110.25$, $\mu_Y = 2.80$, $\sigma_Y^2 = 2.89$, and $\rho = -0.57$. (The lighter students tend to gain weight, while the heavier students tend to lose weight.) Find

(a) $P(2.80 \le Y \le 5.35)$.
(b) $P(2.76 \le y \le 5.34 \,|\, X = 82.3)$.

4.4-11. For a female freshman in a health fitness program, let X equal her percentage of body fat at the beginning of the program and Y equal the change in her percentage of body fat measured at the end of the program. Assume that X and Y have a bivariate normal distribution with $\mu_X = 24.5$, $\sigma_X^2 = 4.8^2 = 23.04$, $\mu_Y = -0.2$, $\sigma_Y^2 = 3.0^2 = 9.0$, and $\rho = -0.32$. Find

(a) $P(1.3 \le Y \le 5.8)$.

(b) $\mu_{Y|x}$, the conditional mean of Y, given that $X = x$.

(c) $\sigma_{Y|x}^2$, the conditional variance of Y, given that $X = x$.

(d) $P(1.3 \le Y \le 5.8 \,|\, X = 18)$.

4.4-12. For a male freshman in a health fitness program, let X equal his percentage of body fat at the beginning of the program and Y equal the change in his percentage of body fat measured at the end of the program. Assume that X and Y have a bivariate normal distribution with $\mu_X = 15.00$, $\sigma_X^2 = 4.5^2$, $\mu_Y = -1.55$, $\sigma_Y^2 = 1.5^2$, and $\rho = -0.60$. Find

(a) $P(0.205 \le Y \le 0.805)$.

(b) $P(0.21 \le Y \le 0.81 \,|\, X = 20)$.

4.4-13. The concentration (X) and the viscosity (Y) of a chemical product have a bivariate normal distribution with parameters $\mu_X = 3$, $\mu_Y = 2$, $\sigma_X = 2$, $\sigma_Y = 1$, and $\rho = 0.6$.

(a) What is $P(X + Y \ge 4)$? HINT: Find the moment-generating function of $Z = X + Y$ and show that it has a normal distribution.

(b) Compute the conditional probability $P(X \ge 3.5 \,|\, Y = 2.5)$.

(c) Compute

$$P\left\{\left(\frac{1}{0.64}\right)\left[\left(\frac{X-3}{2}\right)^2 - (2)(0.6)\left(\frac{X-3}{2}\right)(Y-2) + (Y-2)^2\right] \ge 5.99\right\}.$$

HINT: Show that the moment-generating function of the second degree function $q(X, Y)$ in the probability statement is equal to $(1 - 2t)^{-1}$.

4.4-14. Suppose that in a certain population of cigarettes the tar (X) per cigarette in milligrams and the nicotine (Y) have a bivariate normal distribution with parameters $\mu_X = 14.1$, $\sigma_X = 2.5$, $\mu_Y = 1.3$, $\sigma_Y = 0.1$, and $\rho - 0.8$. Compute

(a) $P(Y > 1.4 \,|\, X = 15)$,

(b) $P(X > 15 \,|\, Y = 1.4)$.

4.4-15. An obstetrician does ultrasound examinations on her patients between their 16th and 25th weeks of pregnancy to check the growth of each fetus. Let X equal the widest diameter of the fetal head, and let Y equal the length of the femur, both measurements in mm. Assume that X and Y have a bivariate normal distribution with $\mu_X = 60.6$, $\sigma_X - 11.2$, $\mu_Y - 16.8$, $\sigma_Y - 8.4$, and $\rho = 0.94$.

(a) Find $P(40.5 < Y < 48.9)$.

(b) Find $P(40.5 < Y < 48.9 \,|\, X = 68.6)$.

HISTORICAL COMMENTS   Now that we have studied conditional distributions, it might be appropriate to point out that there is a group of statisticians called Bayesians who believe in the following approach (it is considered again in Chapter 9): They treat the parameter θ (like μ, σ^2, α, and β in the various distributions) as a random variable with a p.m.f. (or p.d.f.), say, $g(\theta)$. Suppose another random variable X, given θ, has the p.m.f. (or p.d.f.) $f(x \,|\, \theta)$. Say the prior probabilities can be described by $g(\theta)$ so that the joint p.m.f. (or p.d.f.) is given by the sum (or integral)

$$h(x) = \sum_{\theta} g(\theta) f(x \,|\, \theta).$$

Thus, the conditional p.m.f. (or p.d.f.) of θ, given that $X = x$, is

$$k(\theta \,|\, x) = \frac{g(\theta) f(x \,|\, \theta)}{h(x)} = \frac{g(\theta) f(x \,|\, \theta)}{\sum_{\theta} g(\theta) f(x \,|\, \theta)}.$$

With a little thought you can recognize this formula as Bayes's theorem. Here the posterior probabilities, $k(\theta \,|\, x)$, of θ change from the prior probabilities given by $g(\theta)$, after X is observed to be x. Repeating the experiment n independent times (see Chapter 5), we obtain n values of x—say, $x_1, x_2, \ldots, x_n$. The Bayesians use the posterior distribution $k(\theta \mid x_1, x_2, \ldots, x_n)$ to make their inferences about the parameter θ because they then know the conditional probabilities of θ, given $x_1, x_2, \ldots, x_n$.

It is interesting to note that the Reverend Thomas Bayes, a minister who started this method of thinking, never published a mathematical article in his lifetime, and his famous paper was published about two years after his death. Clearly, he was not working for tenure at some university! More will be noted about Bayesian methods in Chapter 9.

Distributions of Functions of Random Variables

5.1   FUNCTIONS OF ONE RANDOM VARIABLE

Let X be a random variable of the continuous type. If we consider a function of X—say, $Y = u(X)$—then Y must also be a random variable that has its own distribution. If we can find its distribution function, say,

$$G(y) = P(Y \leq y) = P[u(X) \leq y],$$

then its p.d.f. is given by $g(y) = G'(y)$. We now illustrate the **distribution function technique** by two examples.

EXAMPLE 5.1-1 Let X have a gamma distribution with p.d.f.

$$f(x) = \frac{1}{\Gamma(\alpha)\theta^\alpha} x^{\alpha-1} e^{-x/\theta}, \qquad 0 < x < \infty,$$

where $\alpha > 0$, $\theta > 0$. Let $Y = e^X$, so that the support of Y is $1 < y < \infty$. For each y in the support, the distribution function of Y is

$$G(y) = P(Y \leq y) = P(e^X \leq y) = P(X \leq \ln y).$$

That is,

$$G(y) = \int_0^{\ln y} \frac{1}{\Gamma(\alpha)\theta^\alpha} x^{\alpha-1} e^{-x/\theta} dx$$

FIGURE 5.1-1: Loggamma probability density functions

and thus the p.d.f. $g(y) = G'(y)$ of Y is

$$g(y) = \frac{1}{\Gamma(\alpha)\theta^\alpha} (\ln y)^{\alpha-1} e^{-(\ln y)/\theta} \left(\frac{1}{y}\right), \qquad 1 < y < \infty.$$

Equivalently, we have

$$g(y) = \frac{1}{\Gamma(\alpha)\theta^\alpha} \frac{(\ln y)^{\alpha-1}}{y^{1+1/\theta}}, \qquad 1 < y < \infty,$$

which is called a **loggamma** p.d.f. (See Figure 5.1-1 for some graphs.) Note that $\alpha\theta$ and $\alpha\theta^2$ are the mean and the variance, not of Y, but of the original random variable $X = \ln Y$. For the loggamma distribution,

$$\mu = \frac{1}{(1-\theta)^\alpha}, \qquad \theta < 1,$$

$$\sigma^2 = \frac{1}{(1-2\theta)^\alpha} - \frac{1}{(1-\theta)^{2\alpha}}, \qquad \theta < \frac{1}{2}. \qquad \blacksquare$$

There is another interesting distribution, this one involving a transformation of a uniform random variable.

EXAMPLE 5.1-2 A spinner is mounted at the point $(0, 1)$. Let w be the smallest angle between the y-axis and the spinner. (See Figure 5.1-2.) Assume that w is the value of a random variable W that has a uniform distribution on the interval $(-\pi/2, \pi/2)$. That is, W is $U(-\pi/2, \pi/2)$, and the distribution function of W is

$$P(W \le w) = F(w) = \begin{cases} 0, & -\infty < w < -\dfrac{\pi}{2}, \\[2mm] \left(w + \dfrac{\pi}{2}\right)\left(\dfrac{1}{\pi}\right), & -\dfrac{\pi}{2} \le w < \dfrac{\pi}{2}, \\[2mm] 1, & \dfrac{\pi}{2} \le w < \infty. \end{cases}$$

The relationship between x and w is given by $x = \tan w$; that is, x is the point on the x-axis which is the intersection of that axis and the linear extension of the spinner. To find the distribution of the random variable $X = \tan W$, we note that the distribution

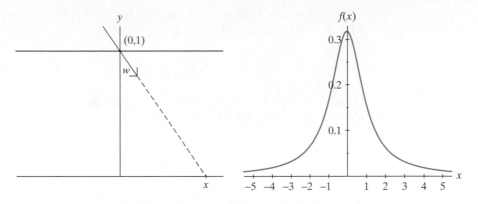

FIGURE 5.1-2: Spinner and Cauchy p.d.f.

function of X is given by

$$G(x) = P(X \le x) = P(\tan W \le x) = P(W \le \arctan x)$$

$$= F(\arctan x) = \left(\arctan x + \frac{\pi}{2}\right)\left(\frac{1}{\pi}\right), \qquad -\infty < x < \infty.$$

The last equality follows because $-\pi/2 < w = \arctan x < \pi/2$. The p.d.f. of X is given by

$$g(x) = G'(x) = \frac{1}{\pi(1 + x^2)}, \qquad -\infty < x < \infty.$$

Figure 5.1-2 shows the graph of this **Cauchy p.d.f.** In Exercise 5.1-12, you will be asked to show that $E(X)$ does not exist because the tails of the Cauchy p.d.f. contain too much probability for this p.d.f. to "balance" at $x = 0$. ■

Thus far, the examples have illustrated the use of the distribution function technique. By making a simple observation, we can sometimes shortcut the distribution function technique by using what is frequently called the **change-of-variable technique.** Let X be a continuous-type random variable with p.d.f. $f(x)$ with support $c_1 < x < c_2$. We begin this discussion by taking $Y = u(X)$ as a continuous increasing function of X with inverse function $X = v(Y)$. Say the support of X, namely, $c_1 < x < c_2$, maps onto $d_1 = u(c_1) < y < d_2 = u(c_2)$, the support of Y. Then the distribution function of Y is

$$G(y) = P(Y \le y) = P[u(X) \le y] = P[X \le v(y)], \qquad d_1 < y < d_2,$$

since u and v are continuous increasing functions. Of course, $G(y) = 0$, $y \le d_1$, and $G(y) = 1$, $y \ge d_2$. Thus,

$$G(y) = \int_{c_1}^{v(y)} f(x)\, dx, \qquad d_1 < y < d_2.$$

Recall from calculus that the derivative, $G'(y) = g(y)$, of such an expression is given by

$$G'(y) = g(y) = f[v(y)][v'(y)], \qquad d_1 < y < d_2.$$

Of course, $G'(y) = g(y) = 0$ if $y < d_1$ or $y > d_2$. We may let $g(d_1) = g(d_2) = 0$.

To illustrate this change-of-variable technique, let us consider again Example 5.1-1 with $Y = e^X$, where X has p.d.f.

$$f(x) = \frac{1}{\Gamma(\alpha)\theta^\alpha} x^{\alpha-1} e^{-x/\theta}, \qquad 0 < x < \infty.$$

Here $c_1 = 0$ and $c_2 = \infty$; thus, $d_1 = 1$ and $d_2 = \infty$. Also, $X = \ln Y = v(Y)$. Since $v'(y) = 1/y$, the p.d.f. of Y is

$$g(y) = \frac{1}{\Gamma(\alpha)\theta^\alpha} (\ln y)^{\alpha-1} e^{-(\ln y)/\theta} \left(\frac{1}{y}\right), \qquad 1 < y < \infty,$$

which is the same result as obtained in that Example 5.1-1.

Suppose now the function $Y = u(X)$ and its inverse $X = v(Y)$ are continuous *decreasing* functions. Then the mapping of $c_1 < x < c_2$ is $d_1 = u(c_1) > y > d_2 = u(c_2)$.

Since u and v are decreasing functions, we have

$$G(y) = P(Y \le y) = P[u(X) \le y] = P[X \ge v(y)] = \int_{v(y)}^{c_2} f(x)\, dx, \qquad d_2 < y < d_1.$$

Accordingly, from calculus, we obtain

$$G'(y) = g(y) = f[v(y)][-v'(y)], \qquad d_2 < y < d_1,$$

and $G'(y) = g(y) = 0$ elsewhere. Note that in both the increasing and decreasing cases we could write

$$g(y) = f[v(y)] |v'(y)|, \qquad y \in S_y,$$

where S_y is the support of Y found by mapping the support of X (say, S_x) onto S_y. The absolute value $|v'(y)|$ assures that $g(y)$ is nonnegative.

EXAMPLE 5.1-3 Let X have the p.d.f.

$$f(x) = 3(1 - x)^2, \qquad 0 < x < 1.$$

Say $Y = (1 - X)^3 = u(X)$, a decreasing function of X. Thus, $X = 1 - Y^{1/3} = v(Y)$ and $0 < x < 1$ is mapped onto $0 < y < 1$. Since

$$v'(y) = -\frac{1}{3y^{2/3}},$$

we have

$$g(y) = 3[1 - (1 - y^{1/3})]^2 \left| \frac{-1}{3y^{2/3}} \right|, \qquad 0 < y < 1.$$

That is,

$$g(y) = 3y^{2/3} \left(\frac{1}{3y^{2/3}} \right) = 1, \qquad 0 < y < 1;$$

so $Y = (1 - X)^3$ has the uniform distribution $U(0,1)$. ∎

As we have seen, it is sometimes easier to use the change-of-variable technique than the distribution function technique. However, there are many occasions on which the latter is more convenient to use. As a matter of fact, we had to use the distribution function in finding the gamma distribution from the Poisson process. (See Section 3.5.) We again use the distribution function technique to prove two theorems involving the uniform distribution.

THEOREM 5.1-1   Let Y have a distribution that is $U(0,1)$. Let $F(x)$ have the properties of a distribution function of the continuous type with $F(a) = 0, F(b) = 1$, and suppose that $F(x)$ is strictly increasing on the support $a < x < b$, where a and b could be $-\infty$ and ∞, respectively. Then the random variable X defined by $X = F^{-1}(Y)$ is a continuous-type random variable with distribution function $F(x)$.

Proof. The distribution function of X is

$$P(X \leq x) = P[F^{-1}(Y) \leq x], \qquad a < x < b.$$

Since $F(x)$ is strictly increasing, $\{F^{-1}(Y) \leq x\}$ is equivalent to $\{Y \leq F(x)\}$. It follows that

$$P(X \leq x) = P[Y \leq F(x)], \qquad a < x < b.$$

But Y is $U(0,1)$; so $P(Y \leq y) = y$ for $0 < y < 1$, and accordingly,

$$P(X \leq x) = P[Y \leq F(x)] = F(x), \qquad 0 < F(x) < 1.$$

That is, the distribution function of X is $F(x)$.   □

The next two examples illustrate how Theorem 5.1-1 can be used to simulate observations from a given distribution.

EXAMPLE 5.1-4   To see how we can simulate observations from an exponential distribution with a mean of $\theta = 10$, note that the distribution function of X is $F(x) = 1 - e^{-x/10}$ when $0 \leq x < \infty$. Solving $y = F(x)$ for x yields $x = F^{-1}(y) = -10\ln(1 - y)$. So, given random numbers $y_1, y_2, \ldots, y_n$ from the $U(0,1)$ distribution, we would expect $x_i = -10\ln(1 - y_i), i = 1, 2, \ldots, n$, to represent n observations of an exponential random variable X with mean $\theta = 10$. Table 5.1-1 gives the values of 15 random numbers, y_i, along with the values of $x_i = -10\ln(1 - y_i)$.

As a rough check on whether these 15 observations seem to come from an exponential distribution with mean 10, let us construct a q–q plot. We must order the 15 observations and plot them against the corresponding quantiles q_i of the exponential distribution with mean $\theta = 10$. The resulting quantiles are the solution of

$$F(q_i) = 1 - e^{-q_i/10} = \frac{i}{16}$$

or, equivalently,

$$q_i = -10\ln\left(1 - \frac{i}{16}\right).$$

The ordered x_i's and the respective quantiles, q_i, are given in Table 5.1-1.

Figure 5.1-3 shows the q–q plot for these data. The linearity of this plot confirms the possibility of the exponential distribution being the correct model. The sample

TABLE 5.1-1: Random Exponential Observations and Exponential Quantiles			
y	$x = -10\ln(1 - y)$	Ordered x's	Quantiles
0.1514	1.6417	0.0431	0.6454
0.6697	11.0775	0.5414	1.3353
0.0527	0.5414	1.0569	2.0764
0.4749	6.4417	1.6417	2.8768
0.2900	3.4249	2.6840	3.7469
0.2354	2.6840	3.4249	4.7000
0.9662	33.8729	6.4417	5.7536
0.0043	0.0431	6.8736	6.9315
0.1003	1.0569	8.9526	8.2668
0.9192	25.1578	11.0775	9.8083
0.4971	6.8736	13.0674	11.6315
0.7293	13.0674	17.2878	13.8629
0.9118	24.2815	24.2815	16.7398
0.8225	17.2878	25.1578	20.7944
0.5915	8.9526	33.8729	27.7259

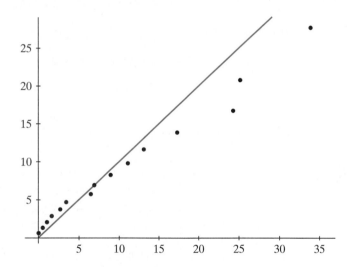

FIGURE 5.1-3: Exponential q–q plot

mean is $\bar{x} = 10.4270$ and the sample standard deviation is $s = 10.4311$, both of which are close to $\theta = 10$. We caution the reader that in other simulations the fit is sometimes better and sometimes not as good. ∎

EXAMPLE 5.1-5 To help appreciate the large probability in the tails of the Cauchy distribution, it is useful to simulate some observations of a Cauchy random variable. We can begin with a random number, Y, that is an observation from the $U(0, 1)$ distribution. From the distribution function of X, namely, $G(x)$, which is that of the Cauchy distribution given in Example 5.1-2, we have

$$y = G(x) = \left(\arctan x + \frac{\pi}{2}\right)\left(\frac{1}{\pi}\right), \qquad -\infty < x < \infty,$$

or, equivalently,

$$x = \tan\left(\pi y - \frac{\pi}{2}\right). \tag{5.1-1}$$

The latter expression provides observations of X.

TABLE 5.1-2: Cauchy Observations			
y	x	y	x
0.1514	−1.9415	0.2354	−1.0962
0.6697	0.5901	0.9662	9.3820
0.0527	−5.9847	0.0043	−74.0211
0.4749	−0.0790	0.1003	−3.0678
0.2900	−0.7757	0.9192	3.8545

In Table 5.1-2, the values of y are the first 10 random numbers in the last column of Table IX in the appendix. The corresponding values of x are given by Equation 5.1-1. Although most of these observations from the Cauchy distribution are relatively small in magnitude, we see that a very large value (in magnitude) occurs occasionally. Another way of looking at this situation is by considering sightings (or firing of a gun) from an observation tower, here with coordinates $(0, 1)$, at independent random angles, each with the uniform distribution $U(-\pi/2, \pi/2)$; the target points would then be at Cauchy observations. ∎

The following probability integral transformation theorem is the converse of Theorem 5.1-1.

THEOREM 5.1-2   Let X have the distribution function $F(x)$ of the continuous type that is strictly increasing on the support $a < x < b$. Then the random variable Y, defined by $Y = F(X)$, has a distribution that is $U(0, 1)$.

Proof. Since $F(a) = 0$ and $F(b) = 1$, the distribution function of Y is

$$P(Y \leq y) = P[F(X) \leq y], \qquad 0 < y < 1.$$

However, $\{F(X) \leq y\}$ is equivalent to $\{X \leq F^{-1}(y)\}$; thus,

$$P(Y \leq y) = P[X \leq F^{-1}(y)], \qquad 0 < y < 1.$$

Since $P(X \leq x) = F(x)$, we have

$$P(Y \leq y) = P[X \leq F^{-1}(y)] = F[F^{-1}(y)] = y, \qquad 0 < y < 1,$$

which is the distribution function of a $U(0, 1)$ random variable.   □

REMARK   Although, in our statements and proofs of Theorems 5.1-1 and 5.1-2, we required $F(x)$ to be strictly increasing, this restriction can be dropped and both theorems are still true. In our exposition, we did not want to bother students with certain difficulties that are experienced if $F(x)$ is not strictly increasing. ∎

Another observation concerns the situation in which the transformation $Y = u(X)$ is not one-to-one, as it has been up to this point in this section. For example, let $Y = X^2$, where X is Cauchy. Here $-\infty < x < \infty$ maps onto $0 \leq y < \infty$, so

$$G(y) = P(X^2 \leq y) = P(-\sqrt{y} \leq X \leq \sqrt{y})$$

$$= \int_{-\sqrt{y}}^{\sqrt{y}} f(x)\, dx, \qquad 0 \leq y < \infty,$$

where

$$f(x) = \frac{1}{\pi(1 + x^2)}, \qquad -\infty < x < \infty.$$

Thus,

$$G'(y) = g(y) = f(\sqrt{y})\left|\frac{1}{2\sqrt{y}}\right| + f(-\sqrt{y})\left|\frac{-1}{2\sqrt{y}}\right|$$

$$= \frac{1}{\pi(1 + y)\sqrt{y}}, \qquad 0 \leq y < \infty.$$

That is, in this case of a two-to-one transformation, there is a need to sum two terms, each of which is similar to a counterpart term in the one-to-one case; but here $x_1 = \sqrt{y}$ and $x_2 = -\sqrt{y}$, $0 < y < \infty$, give the two inverse functions, respectively.

With careful thought, we can handle many situations that generalize this particular example.

EXAMPLE 5.1-6   Let X have the p.d.f.

$$f(x) = \frac{x^2}{3}, \qquad -1 < x < 2.$$

Then the random variable $Y = X^2$ will have the support $0 \leq y < 4$. Now, on the one hand, for $0 < y < 1$, we obtain the two-to-one transformation represented by $x_1 = -\sqrt{y}$ for $-1 < x_1 < 0$ and by $x_2 = \sqrt{y}$ for $0 < x_2 < 1$. On the other hand, if $1 < y < 4$, the one-to-one transformation is represented by $x_2 = \sqrt{y}, 1 < x_2 < 2$. Since

$$\frac{dx_1}{dy} = \frac{-1}{2\sqrt{y}} \qquad \text{and} \qquad \frac{dx_2}{dy} = \frac{1}{2\sqrt{y}},$$

it follows that the p.d.f. of $Y = X^2$ is

$$g(y) = \begin{cases} \dfrac{(-\sqrt{y})^2}{3}\left|\dfrac{-1}{2\sqrt{y}}\right| + \dfrac{(\sqrt{y})^2}{3}\left|\dfrac{1}{2\sqrt{y}}\right| = \dfrac{\sqrt{y}}{3}, & 0 < y < 1, \\[2em] \dfrac{(\sqrt{y})^2}{3}\left|\dfrac{1}{2\sqrt{y}}\right| = \dfrac{\sqrt{y}}{6}, & 1 < y < 4. \end{cases}$$

Note that if $0 < y < 1$, the p.d.f. is the sum of two terms, but if $1 < y < 4$, then there is only one term. These different expressions in $g(y)$ correspond to the two types of transformations, namely, the two-to-one and the one-to-one transformations, respectively. ∎

The change-of-variable technique can be used for a variable X of the discrete type, but there is one major difference: The p.m.f. $f(x) = P(X = x)$, $x \in S_x$, represents probability. Note that the support S_x consists of a countable number of points, say, $c_1, c_2, c_3, \ldots$. Let $Y = u(X)$ be a one-to-one transformation with inverse $X = v(Y)$. The function $y = u(x)$ maps S_x onto $d_1 = u(c_1)$, $d_2 = u(c_2)$, $d_3 = u(c_3), \ldots$, which we denote by S_y. Hence, the p.m.f. of Y is

$$g(y) = P(Y = y) = P[u(X) = y] = P[X = v(y)], \qquad y \in S_y.$$

Since $P(X = x) = f(x)$, we have $g(y) = f[v(y)]$, $y \in S_y$. Note that, in this discrete case, the value of the derivative, namely, $|v'(y)|$, is not needed.

EXAMPLE 5.1-7 Let X have a Poisson distribution with $\lambda = 4$; thus, the p.m.f. is

$$f(x) = \frac{4^x e^{-4}}{x!}, \qquad x = 0, 1, 2, \ldots.$$

If $Y = \sqrt{X}$, then, since $X = Y^2$, we have

$$g(y) = \frac{4^{y^2} e^{-4}}{(y^2)!}, \qquad y = 0, 1, \sqrt{2}, \sqrt{3}, \ldots. \qquad \blacksquare$$

EXAMPLE 5.1-8 Let the distribution of X be binomial with parameters n and p. Since X has a discrete distribution, $Y = u(X)$ will also have a discrete distribution, with the same probabilities as those in the support of X. For example, with $n = 3$, $p = 1/4$, and $Y = X^2$, we have

$$g(y) = \binom{3}{\sqrt{y}} \left(\frac{1}{4}\right)^{\sqrt{y}} \left(\frac{3}{4}\right)^{3-\sqrt{y}}, \qquad y = 0, 1, 4, 9. \qquad \blacksquare$$

EXERCISES

5.1-1. Let X have the p.d.f. $f(x) = 4x^3, 0 < x < 1$. Find the p.d.f. of $Y = X^2$.

5.1-2. Let X have the p.d.f. $f(x) = xe^{-x^2/2}, 0 < x < \infty$. Find the p.d.f. of $Y = X^2$.

5.1-3. Let X have a gamma distribution with $\alpha = 3$ and $\theta = 2$. Determine the p.d.f. of $Y = \sqrt{X}$.

5.1-4. The p.d.f. of X is $f(x) = 2x, 0 < x < 1$.

(a) Find the distribution function of X.
(b) Describe how an observation of X can be simulated.
(c) Simulate 10 observations of X.
(d) Use the ordered observations and the corresponding quantiles for the given distribution to construct a q–q plot.

5.1-5. The p.d.f. of X is $f(x) = \theta x^{\theta-1}, 0 < x < 1, 0 < \theta < \infty$. Let $Y = -2\theta \ln X$. How is Y distributed?

5.1-6. Let X have a **logistic distribution** with p.d.f.

$$f(x) = \frac{e^{-x}}{(1 + e^{-x})^2}, \qquad -\infty < x < \infty.$$

Show that

$$Y = \frac{1}{1 + e^{-X}}$$

has a $U(0, 1)$ distribution.

5.1-7. A sum of \$50,000 is invested at a rate R, selected from a uniform distribution on the interval (0.03, 0.07). Once R is selected, the sum is compounded instantaneously for a year, so that $X = 50000\, e^R$ dollars is the amount at the end of that year.

(a) Find the distribution function and p.d.f. of X.
(b) Verify that $X = 50000\, e^R$ is defined correctly if the compounding is done instantaneously. HINT: Divide the year into n equal parts,

calculate the value of the amount at the end of each part, and then take the limit as $n \to \infty$.

5.1-8. The lifetime (in years) of a manufactured product is $Y = 5X^{0.7}$, where X has an exponential distribution with mean 1. Find the distribution function and p.d.f. of Y.

5.1-9. Statisticians frequently use the **extreme value distribution** given by the distribution function

$$F(x) = 1 - \exp\left[-e^{(x - \theta_1)/\theta_2}\right],$$
$$-\infty < x < \infty.$$

A simple case is when $\theta_1 = 0$ and $\theta_2 = 1$, giving

$$F(x) = 1 - \exp\left[-e^x\right], \qquad -\infty < x < \infty.$$

Let $Y = e^X$ or $X = \ln Y$; then the support of Y is $0 < y < \infty$.

(a) Show that the distribution of Y is exponential when $\theta_1 = 0$ and $\theta_2 = 1$.

(b) Find the distribution function and the p.d.f. of Y when $\theta_1 \neq 0$ and $\theta_2 \neq 1$. What is this distribution?

(c) As suggested by its name, the extreme value distribution can be used to model the longest home run, the deepest mine, the greatest flood, and so on. Suppose the length X (in feet) of the maximum of someone's home runs was modeled by an extreme value distribution with $\theta_1 = 550$ and $\theta_2 = 25$. What is the probability that X exceeds 500 feet?

5.1-10. Let X have the uniform distribution $U(-1, 3)$. Find the p.d.f. of $Y = X^2$.

5.1-11. Let X have a Cauchy distribution. Find

(a) $P(X > 1)$.

(b) $P(X > 5)$.

(c) $P(X > 10)$.

5.1-12. Let $f(x) = 1/[\pi(1 + x^2)]$, $-\infty < x < \infty$, be the p.d.f. of the Cauchy random variable X. Show that $E(X)$ does not exist.

5.1-13. If X is $N(\mu, \sigma^2)$, then $M(t) = E(e^{tX}) = \exp(\mu t + \sigma^2 t^2/2)$. We then say that $Y = e^X$ has a **lognormal distribution**.

(a) Show that the p.d.f. of Y is

$$g(y) = \frac{1}{y\sqrt{2\pi\sigma^2}} \exp[-(\ln y - \mu)^2/2\sigma^2],$$
$$0 < y < \infty.$$

(b) Using $M(t)$, find (i) $E(Y) = E(e^X) = M(1)$, (ii) $E(Y^2) = E(e^{2X}) = M(2)$, and (iii) $\text{Var}(Y)$.

5.1-14. Let X be $N(0, 1)$. Find the p.d.f. of $Y = |X|$, a distribution that is often called the **half normal**. HINT: Here $y \in S_y = \{y: 0 < y < \infty\}$. Consider the two transformations $x_1 = -y$, $-\infty < x_1 < 0$, and $x_2 = y$, $0 < y < \infty$.

5.1-15. Let $Y = X^2$.

(a) Find the p.d.f. of Y when the distribution of X is $N(0, 1)$.

(b) Find the p.d.f. of Y when the p.d.f. of X is $f(x) = (3/2)x^2$, $-1 < x < 1$.

5.2 TRANSFORMATIONS OF TWO RANDOM VARIABLES

In Section 5.1, we considered the transformation of one random variable X with p.d.f. $f(x)$. In particular, in the continuous case, if $Y = u(X)$ was an increasing or decreasing function of X, with inverse $X = v(Y)$, then the p.d.f. of Y was

$$g(y) = |v'(y)|f[v(y)], \qquad c < y < d,$$

where the support $c < y < d$ corresponds to the support of X, say, $a < x < b$, through the transformation $x = v(y)$.

There is one note of warning here: If the function $Y = u(X)$ does not have a single-valued inverse, the determination of the distribution of Y will not be as simple. As a matter of fact, we did consider two examples in Section 5.1 in which there were two inverse functions, and we exercised special care in those examples. Here, we will not consider problems with many inverses; however, such a warning is nonetheless appropriate.

When two random variables are involved, many interesting problems can result. In the case of a single-valued inverse, the rule is about the same as that in the one-variable case, with the derivative being replaced by the Jacobian. That is, if X_1 and X_2 are two continuous-type random variables with joint p.d.f. $f(x_1, x_2)$, and if $Y_1 = u_1(X_1, X_2)$, $Y_2 = u_2(X_1, X_2)$ has the single-valued inverse $X_1 = v_1(Y_1, Y_2)$,

$X_2 = v_2(Y_1, Y_2)$, then the joint p.d.f. of Y_1 and Y_2 is

$$g(y_1, y_2) = |J| f[v_1(y_1, y_2), v_2(y_1, y_2)], \qquad (y_1, y_2) \in S_1,$$

where the Jacobian J is the determinant

$$J = \begin{vmatrix} \dfrac{\partial x_1}{\partial y_1} & \dfrac{\partial x_1}{\partial y_2} \\[2mm] \dfrac{\partial x_2}{\partial y_1} & \dfrac{\partial x_2}{\partial y_2} \end{vmatrix}.$$

Of course, we find the support S_1 of Y_1, Y_2 by considering the mapping of the support S of X_1, X_2 under the transformation $y_1 = u_1(x_1, x_2)$, $y_2 = u_2(x_1, x_2)$. This method of finding the distribution of Y_1 and Y_2 is called the **change-of-variables technique**.

It is often the mapping of the support S of X_1, X_2 into that (say, S_1) of Y_1, Y_2 which causes the biggest challenge. That is, in most cases, it is easy to solve for x_1 and x_2 in terms of y_1 and y_2, say,

$$x_1 = v_1(y_1, y_2), \qquad x_2 = v_2(y_1, y_2),$$

and then to compute the Jacobian

$$J = \begin{vmatrix} \dfrac{\partial v_1(y_1, y_2)}{\partial y_1} & \dfrac{\partial v_1(y_1, y_2)}{\partial y_2} \\[2mm] \dfrac{\partial v_2(y_1, y_2)}{\partial y_1} & \dfrac{\partial v_2(y_1, y_2)}{\partial y_2} \end{vmatrix}.$$

However, the mapping of $(x_1, x_2) \in S$ into $(y_1, y_2) \in S_1$ can be more difficult. Let us consider two simple examples.

EXAMPLE 5.2-1   Let X_1, X_2 have the joint p.d.f.

$$f(x_1, x_2) = 2, \qquad 0 < x_1 < x_2 < 1.$$

Consider the transformation

$$Y_1 = \frac{X_1}{X_2}, \qquad Y_2 = X_2.$$

It is certainly easy enough to solve for x_1 and x_2, namely,

$$x_1 = y_1 y_2, \qquad x_2 = y_2$$

and compute

$$J = \begin{vmatrix} y_2 & y_1 \\ 0 & 1 \end{vmatrix} = y_2.$$

Let us now consider S, which is depicted in Figure 5.2-1(a). The boundaries of S are not part of the support, but let us see how they map. The points for which $x_1 = 0$, $0 < x_2 < 1$, map into $y_1 = 0$, $0 < y_2 < 1$; the points for which $x_2 = 1$, $0 \le x_1 < 1$ map into $y_2 = 1$, $0 \le y_1 < 1$; and $0 < x_1 = x_2 \le 1$ maps into $y_1 = 1$, $0 < y_2 \le 1$.

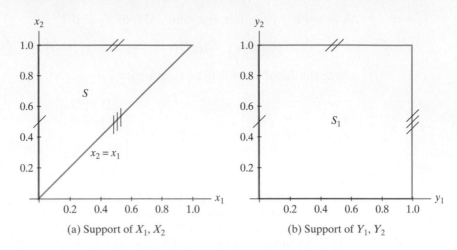

FIGURE 5.2-1: Mapping from x_1, x_2 to y_1, y_2

We depict these line segments in Figure 5.2-1(b) and mark them with the symbols corresponding to the line segments in Figure 5.2-1(a).

We note that the points for which $y_2 = 0$, $0 \leq y_1 < 1$, all map into the single point $x_1 = 0$, $x_2 = 0$. That is, this is a many-to-one mapping, and yet we are restricting ourselves to one-to-one mappings. However, the boundaries are not part of our support! Thus, S_1 is as depicted in Figure 5.2-1(b), and, according to the rule, the joint p.d.f. of Y_1, Y_2 is

$$g(y_1, y_2) = |y_2| \cdot 2 = 2y_2, \qquad 0 < y_1 < 1, \ 0 < y_2 < 1.$$

It is interesting to note that the marginal probability density functions are

$$g_1(y_1) = \int_0^1 2y_2 \, dy_2 = 1, \qquad 0 < y_1 < 1,$$

and

$$g_2(y_2) = \int_0^1 2y_2 \, dy_1 = 2y_2, \qquad 0 < y_2 < 1.$$

Hence, $Y_1 = X_1/X_2$ and $Y_2 = X_2$ are independent. Even though the computation of Y_1 depends very much on the value of Y_2, still Y_1 and Y_2 are independent in the probability sense. ∎

EXAMPLE 5.2-2 Let X_1 and X_2 be independent random variables, each with p.d.f.

$$f(x) = e^{-x}, \qquad 0 < x < \infty.$$

Hence, their joint p.d.f. is

$$f(x_1)f(x_2) = e^{-x_1 - x_2}, \qquad 0 < x_1 < \infty, \ 0 < x_2 < \infty.$$

Let us consider

$$Y_1 = X_1 - X_2, \qquad Y_2 = X_1 + X_2.$$

Thus,

$$x_1 = \frac{y_1 + y_2}{2}, \qquad x_2 = \frac{y_2 - y_1}{2},$$

with

$$J = \begin{vmatrix} \dfrac{1}{2} & \dfrac{1}{2} \\[2mm] -\dfrac{1}{2} & \dfrac{1}{2} \end{vmatrix} = \dfrac{1}{2}.$$

The region S is depicted in Figure 5.2-2(a). The line segments on the boundary, namely, $x_1 = 0$, $0 < x_2 < \infty$, and $x_2 = 0$, $0 < x_1 < \infty$, map into the line segments $y_1 + y_2 = 0$, $y_2 > y_1$ and $y_1 = y_2$, $y_2 > -y_1$, respectively. These are shown in Figure 5.2-2(b) and the support of S_1 is depicted there. Since the region S_1 is not bounded by horizontal and vertical line segments, Y_1 and Y_2 are dependent.

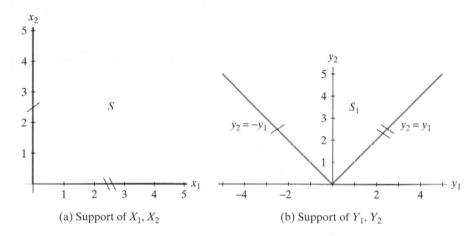

(a) Support of X_1, X_2　　　　　　(b) Support of Y_1, Y_2

FIGURE 5.2-2: Mapping from x_1, x_2 to y_1, y_2

The joint p.d.f. of Y_1 and Y_2 is

$$g(y_1, y_2) = \frac{1}{2} e^{-y_2}, \qquad -y_2 < y_1 < y_2, \ 0 < y_2 < \infty.$$

The marginal p.d.f. of Y_2 is

$$g_2(y_2) = \int_{-y_2}^{y_2} \frac{1}{2} e^{-y_2} \, dy_1 = y_2 e^{-y_2}, \qquad 0 < y_2 < \infty.$$

That of Y_1 is

$$g_1(y_1) = \begin{cases} \displaystyle\int_{-y_1}^{\infty} \frac{1}{2} e^{-y_2} \, dy_2 = \frac{1}{2} e^{y_1}, & -\infty < y_1 < 0, \\[4mm] \displaystyle\int_{y_1}^{\infty} \frac{1}{2} e^{-y_2} \, dy_2 = \frac{1}{2} e^{-y_1}, & 0 < y_1 < \infty. \end{cases}$$

That is, the expression for $g_1(y_1)$ depends on the location of y_1, although this could be written as

$$g_1(y_1) = \frac{1}{2} e^{-|y_1|}, \qquad -\infty < y_1 < \infty,$$

which is called a **double exponential** p.d.f. ∎

We now consider two examples that yield two important distributions. The second of these uses the distribution function technique rather than the change-of-variable method.

EXAMPLE 5.2-3 Let X_1 and X_2 have independent gamma distributions with parameters α, θ and β, θ, respectively. That is, the joint p.d.f. of X_1 and X_2 is

$$f(x_1, x_2) = \frac{1}{\Gamma(\alpha)\Gamma(\beta)\theta^{\alpha+\beta}} x_1^{\alpha-1} x_2^{\beta-1} \exp\left(-\frac{x_1 + x_2}{\theta}\right), \quad 0 < x_1 < \infty, \; 0 < x_2 < \infty.$$

Consider

$$Y_1 = \frac{X_1}{X_1 + X_2}, \qquad Y_2 = X_1 + X_2,$$

or, equivalently,

$$X_1 = Y_1 Y_2, \qquad X_2 = Y_2 - Y_1 Y_2.$$

The Jacobian is

$$J = \begin{vmatrix} y_2 & y_1 \\ -y_2 & 1 - y_1 \end{vmatrix} = y_2(1 - y_1) + y_1 y_2 = y_2.$$

Thus, the joint p.d.f. $g(y_1, y_2)$ of Y_1 and Y_2 is

$$g(y_1, y_2) = |y_2| \frac{1}{\Gamma(\alpha)\Gamma(\beta)\theta^{\alpha+\beta}} (y_1 y_2)^{\alpha-1} (y_2 - y_1 y_2)^{\beta-1} e^{-y_2/\theta},$$

where the support is $0 < y_1 < 1$, $0 < y_2 < \infty$, which is the mapping of $0 < x_i < \infty$, $i = 1, 2$. To see the shape of this joint p.d.f., $z = g(y_1, y_2)$ is graphed in Figure 5.2-3(a) with $\alpha = 4$, $\beta = 7$, and $\theta = 1$ and in Figure 5.2-3(b) with $\alpha = 8$, $\beta = 3$, and $\theta = 1$. To find the marginal p.d.f. of Y_1, we integrate this joint p.d.f. on y_2. We see that the marginal p.d.f. of Y_1 is

$$g_1(y_1) = \frac{y_1^{\alpha-1}(1 - y_1)^{\beta-1}}{\Gamma(\alpha)\Gamma(\beta)} \int_0^\infty \frac{y_2^{\alpha+\beta-1}}{\theta^{\alpha+\beta}} e^{-y_2/\theta} \, dy_2.$$

But the integral in this expression is that of a gamma p.d.f. with parameters $\alpha + \beta$ and θ, except for $\Gamma(\alpha + \beta)$ in the denominator; hence, the integral equals $\Gamma(\alpha + \beta)$, and we have

$$g_1(y_1) = \frac{\Gamma(\alpha + \beta)}{\Gamma(\alpha)\Gamma(\beta)} y_1^{\alpha-1}(1 - y_1)^{\beta-1}, \qquad 0 < y_1 < 1.$$

We say that Y_1 has a **beta p.d.f.** with parameters α and β. (See Figure 5.2-4.) Note the relationship between Figure 5.2-3 and Figure 5.2-4. ∎

The next example illustrates the distribution function technique. You calculate the same results in Exercise 5.2-2, but using the change-of-variable technique.

EXAMPLE 5.2-4 We let

$$F = \frac{U/r_1}{V/r_2},$$

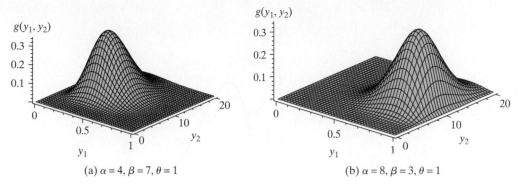

(a) $\alpha = 4, \beta = 7, \theta = 1$ (b) $\alpha = 8, \beta = 3, \theta = 1$

FIGURE 5.2-3: Joint p.d.f. of $z = g(y_1, y_2)$

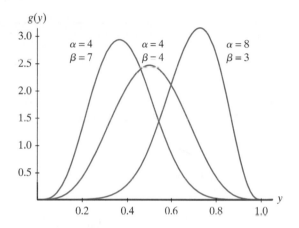

FIGURE 5.2-4: Beta distribution p.d.f.'s

where U and V are independent chi-square variables with r_1 and r_2 degrees of freedom, respectively. Thus, the joint p.d.f. of U and V is

$$g(u, v) = \frac{u^{r_1/2-1}e^{-u/2}}{\Gamma(r_1/2)2^{r_1/2}} \frac{v^{r_2/2-1}e^{-v/2}}{\Gamma(r_2/2)2^{r_2/2}}, \qquad 0 < u < \infty, \ 0 < v < \infty.$$

In this derivation, we let $W = F$ to avoid using f as a symbol for a variable. The distribution function $F(w) = P(W \leq w)$ of W is

$$F(w) = P\left(\frac{U/r_1}{V/r_2} \leq w\right) = P\left(U \leq \frac{r_1}{r_2} w V\right)$$

$$= \int_0^\infty \int_0^{(r_1/r_2)wv} g(u, v) \, du \, dv.$$

That is,

$$F(w) = \frac{1}{\Gamma(r_1/2)\Gamma(r_2/2)} \int_0^\infty \left[\int_0^{(r_1/r_2)wv} \frac{u^{r_1/2-1}e^{-u/2}}{2^{(r_1+r_2)/2}} \, du\right] v^{r_2/2-1}e^{-v/2} \, dv.$$

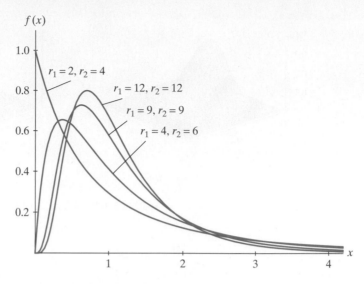

FIGURE 5.2-5: Graphs of F p.d.f.'s

The p.d.f. of W is the derivative of the distribution function; so, applying the fundamental theorem of calculus to the inner integral, we have

$$f(w) = F'(w)$$

$$= \frac{1}{\Gamma(r_1/2)\Gamma(r_2/2)} \int_0^\infty \frac{[(r_1/r_2)vw]^{r_1/2-1}}{2^{(r_1+r_2)/2}} e^{-(r_1/2r_2)(vw)} \left(\frac{r_1}{r_2} v\right) v^{r_2/2-1} e^{-v/2} \, dv$$

$$= \frac{(r_1/r_2)^{r_1/2} w^{r_1/2-1}}{\Gamma(r_1/2)\Gamma(r_2/2)} \int_0^\infty \frac{v^{(r_1+r_2)/2-1}}{2^{(r_1+r_2)/2}} e^{-(v/2)[1+(r_1/r_2)w]} \, dv.$$

In the integral, we make the change of variable

$$y = \left(1 + \frac{r_1}{r_2} w\right) v, \qquad \text{so that} \qquad \frac{dv}{dy} = \frac{1}{1 + (r_1/r_2)w} .$$

Thus, we have

$$f(w) = \frac{(r_1/r_2)^{r_1/2}\Gamma[(r_1 + r_2)/2]w^{r_1/2-1}}{\Gamma(r_1/2)\Gamma(r_2/2)[1 + (r_1 w/r_2)]^{(r_1+r_2)/2}} \int_0^\infty \frac{y^{(r_1+r_2)/2-1}e^{-y/2}}{\Gamma[(r_1 + r_2)/2]2^{(r_1+r_2)/2}} \, dy$$

$$= \frac{(r_1/r_2)^{r_1/2}\Gamma[(r_1 + r_2)/2]w^{r_1/2-1}}{\Gamma(r_1/2)\Gamma(r_2/2)[1 + (r_1 w/r_2)]^{(r_1+r_2)/2}} ,$$

the p.d.f. of the $W = F$ **distribution** with r_1 and r_2 degrees of freedom. Note that the integral in this last expression for $f(w)$ is equal to 1 because the integrand is like a p.d.f. of a chi-square distribution with $r_1 + r_2$ degrees of freedom. Graphs of p.d.f.'s for the F distribution are given in Figure 5.2-5. ∎

To find probabilities for an F random variable with r_1 (numerator) and r_2 (denominator) degrees of freedom, use either your calculator, a computer program, or Table VII in Appendix C. Table VII is limited, but is adequate for most of the applications in this text. For notation, if W has an F distribution with r_1 and r_2 degrees of freedom, we say that the distribution of W is $F(r_1, r_2)$. For a right-tail probability of α, we write

$$P[W \geq F_\alpha(r_1, r_2)] = \alpha.$$

For a left-tail probability of α, where α is generally small, we note that if the distribution of W is $F(r_1, r_2)$, then the distribution of $1/W$ is $F(r_2, r_1)$. Since

$$\alpha = P[W \leq F_{1-\alpha}(r_1, r_2)] = P[1/W \geq 1/F_{1-\alpha}(r_1, r_2)]$$

and

$$P[1/W \geq F_\alpha(r_2, r_1)] = \alpha,$$

it follows that

$$\frac{1}{F_{1-\alpha}(r_1, r_2)} = F_\alpha(r_2, r_1) \quad \text{or} \quad F_{1-\alpha}(r_1, r_2) = \frac{1}{F_\alpha(r_2, r_1)}.$$

EXAMPLE 5.2-5   Let the distribution of W be $F(4, 6)$. From Table VII, we see that

$$F_{0.05}(4, 6) = 4.53;$$

$$P(W \leq 9.15) = 0.99.$$

That is, $F_{0.01}(4, 6) = 9.15$. We also note that

$$F_{0.95}(4, 6) = \frac{1}{F_{0.05}(6, 4)} = \frac{1}{6.16} = 0.1623;$$

$$F_{0.99}(4, 6) = \frac{1}{F_{0.01}(6, 4)} = \frac{1}{15.21}.$$

It follows that

$$P(1/15.21 \leq W \leq 9.15) = 0.98;$$

$$P(1/6.16 \leq W \leq 4.53) = 0.90. \qquad \blacksquare$$

COMPUTATIONAL COMMENTS   We illustrate the use of Minitab to find some of the solutions for Example 5.2-5. To find those probabilities, we use

$$\text{Calc} \rightarrow \text{Probability Distributions} \rightarrow \text{F}$$

and input the necessary information:

```
Inverse Cumulative Distribution Function
F distribution with 4 DF in numerator and 6 DF in denominator

P(X<=x)          x
   0.95    4.53368

P(X<=x)          x
   0.05    0.162255


Cumulative Distribution Function
F distribution with 4 DF in numerator and 6 DF in denominator
```

$$
\begin{array}{cc}
\texttt{x} & \texttt{P(X<=x)} \\
\texttt{9.15} & \texttt{0.990005}
\end{array}
$$

EXAMPLE 5.2-6 (**Box–Muller Transformation**) Consider the following transformation, where X_1 and X_2 are independent, each with the uniform distribution $U(0,1)$: Let

$$
Z_1 = \sqrt{-2 \ln X_1}\, \cos(2\pi X_2), \qquad Z_2 = \sqrt{-2 \ln X_1}\, \sin(2\pi X_2)
$$

or, equivalently,

$$
X_1 = \exp\left(-\frac{Z_1^2 + Z_2^2}{2}\right) = e^{-q/2}, \qquad X_2 = \frac{1}{2\pi}\arctan\left(\frac{Z_2}{Z_1}\right),
$$

which has Jacobian

$$
J = \begin{vmatrix} -z_1 e^{-q/2} & -z_2 e^{-q/2} \\[2mm] \dfrac{-z_2}{2\pi(z_1^2 + z_2^2)} & \dfrac{z_1}{2\pi(z_1^2 + z_2^2)} \end{vmatrix} = \frac{-1}{2\pi} e^{-q/2}.
$$

Since the joint p.d.f. of X_1 and X_2 is

$$
f(x_1, x_2) = 1, \qquad 0 < x_1 < 1, \ 0 < x_2 < 1,
$$

it follows that the joint p.d.f. of Z_1 and Z_2 is

$$
g(z_1, z_2) = \left| \frac{-1}{2\pi} e^{-q/2} \right| (1)
$$

$$
= \frac{1}{2\pi} \exp\left(-\frac{z_1^2 + z_2^2}{2}\right), \qquad -\infty < z_1 < \infty, \ -\infty < z_2 < \infty.
$$

Note that there is some difficulty with the definition of this transformation, particularly when $z_1 = 0$. However, these difficulties occur at events with probability zero and hence cause no problems. (See Exercise 5.2-17.) Summarizing, from two independent $U(0,1)$ random variables we have generated two independent $N(0,1)$ random variables through the **Box–Muller transformation**. ∎

EXERCISES

5.2-1. Let X_1, X_2 denote two independent random variables, each with a $\chi^2(2)$ distribution. Find the joint p.d.f. of $Y_1 = X_1$ and $Y_2 = X_2 + X_1$. Note that the support of Y_1, Y_2 is $0 < y_1 < y_2 < \infty$. Also, find the marginal p.d.f. of each of Y_1 and Y_2. Are Y_1 and Y_2 independent?

5.2-2. Let X_1 and X_2 be independent chi-square random variables with r_1 and r_2 degrees of freedom, respectively. Let $Y_1 = (X_1/r_1)/(X_2/r_2)$ and $Y_2 = X_2$.

(a) Find the joint p.d.f. of Y_1 and Y_2.

(b) Determine the marginal p.d.f. of Y_1 and show that Y_1 has an F distribution. (This is another,

but equivalent, way of finding the p.d.f. of F.)

5.2-3. Find the mean and the variance of an F random variable with r_1 and r_2 degrees of freedom by first finding $E(U)$, $E(1/V)$, $E(U^2)$, and $E(1/V^2)$.

5.2-4. Let the distribution of W be $F(9,24)$. Find the following:

(a) $F_{0.05}(9,24)$.

(b) $F_{0.95}(9,24)$.

(c) $P(0.277 \le W \le 2.70)$.

5.2-5. Let the distribution of W be $F(8,4)$. Find the following:

(a) $F_{0.01}(8,4)$.

(b) $F_{0.99}(8,4)$.

(c) $P(0.198 \le W \le 8.98)$.

5.2-6. Let X_1 and X_2 have independent gamma distributions with parameters α, θ and β, θ, respectively. Let $W = X_1/(X_1 + X_2)$. Use a method similar to that given in the derivation of the F distribution (Examples 5.2-4) to show that the p.d.f. of W is

$$g(w) = \frac{\Gamma(\alpha + \beta)}{\Gamma(\alpha)\Gamma(\beta)} w^{\alpha-1}(1 - w)^{\beta-1}, 0 < w < 1.$$

We say that W has a **beta distribution** with parameters α and β. (See Example 5.2-3.)

5.2-7. Let X_1 and X_2 be independent chi-square random variables with r_1 and r_2 degrees of freedom, respectively. Show that

(a) $U = X_1/(X_1 + X_2)$ has a beta distribution with $\alpha = r_1/2$ and $\beta = r_2/2$.

(b) $V = X_2/(X_1 + X_2)$ has a beta distribution with $\alpha = r_2/2$ and $\beta = r_1/2$.

5.2-8. Let X have a beta distribution with parameters α and β. (See Example 5.2-3.)

(a) Show that the mean and variance of X are, respectively,

$$\mu = \frac{\alpha}{\alpha + \beta}$$

and

$$\sigma^2 = \frac{\alpha\beta}{(\alpha + \beta + 1)(\alpha + \beta)^2}.$$

(b) Show that when $\alpha > 1$ and $\beta > 1$, the mode is at $x = (\alpha - 1)/(\alpha + \beta - 2)$.

5.2-9. Determine the constant c such that $f(x) = cx^3(1 - x)^6, 0 < x < 1$, is a p.d.f.

5.2-10. When α and β are integers and $0 < p < 1$, we have

$$\int_0^p \frac{\Gamma(\alpha + \beta)}{\Gamma(\alpha)\Gamma(\beta)} y^{\alpha-1}(1 - y)^{\beta-1} \, dy$$

$$= \sum_{y=\alpha}^{n} \binom{n}{y} p^y (1 - p)^{n-y},$$

where $n = \alpha + \beta - 1$. Verify this formula when $\alpha = 4$ and $\beta = 3$. HINT: Integrate the left member by parts several times.

5.2-11. Evaluate

$$\int_0^{0.4} \frac{\Gamma(7)}{\Gamma(4)\Gamma(3)} y^3 (1 - y)^2 \, dy$$

(a) Using integration.

(b) Using the result of Exercise 5.2-10.

5.2-12. Let W_1, W_2 be independent, each with a Cauchy distribution. In this exercise, we find the p.d.f. of the sample mean, $(W_1 + W_2)/2$.

(a) Show that the p.d.f. of $X_1 = (1/2)W_1$ is

$$f(x) = \frac{2}{\pi(1 + 4x^2)}, \qquad -\infty < x < \infty.$$

(b) Let $Y_1 = X_1 + X_2 = \overline{W}$ and $Y_2 = X_1$, where $X_2 = (1/2)W_2$. Show that the joint p.d.f. of Y_1 and Y_2 is

$$g(y_1, y_2) = f(y_1 - y_2)f(y_2),$$
$$-\infty < y_1 < \infty, -\infty < y_2 < \infty.$$

(c) Show that the p.d.f. of $Y_1 = \overline{W}$ is given by the **convolution formula**,

$$g_1(y_1) = \int_{-\infty}^{\infty} f(y_1 - y_2)f(y_2) \, dy_2.$$

(d) Show that

$$g_1(y_1) = \frac{1}{\pi(1 + y_1^2)}, \qquad -\infty < y_1 < \infty.$$

That is, the p.d.f. of $\overline{W}$ is the same as that of an individual W.

5.2-13. Let X_1, X_2 be independent random variables representing lifetimes (in hours) of two key components of a device that fails when and only when both components fail. Say each X_i has an exponential distribution with mean 1000. Let $Y_1 = \min(X_1, X_2)$ and $Y_2 = \max(X_1, X_2)$, so that the space of Y_1, Y_2 is $0 < y_1 < y_2 < \infty$.

(a) Find $G(y_1, y_2) = P(Y_1 \le y_1, Y_2 \le y_2)$.

(b) Compute the probability that the device fails after 1200 hours; that is, compute $P(Y_2 > 1200)$.

5.2-14. A company provides earthquake insurance. The premium X is modeled by the p.d.f.

$$f(x) = \frac{x}{5^2} e^{-x/5}, \qquad 0 < x < \infty,$$

while the claims Y have the p.d.f.

$$g(y) = \frac{1}{5} e^{-y/5}, \qquad 0 < y < \infty.$$

If X and Y are independent, find the p.d.f. of $Z = X/Y$.

5.2-15. Let X be the fraction of a 24-hour period that a person is sleeping. We are trying to model the distribution of X with a beta p.d.f. Which of the following do you believe to be the best model for a person who sleeps for 8 hours, on the average: (1) $\alpha = 8, \beta = 16$, (2) $\alpha = 16, \beta = 32$, (3) $\alpha = 24, \beta = 48$? Why? Can you construct a better model?

5.2-16. Let X and Y be the fractions of a 24-hour period that a person spends sleeping and eating, respectively. We wish to model the distribution of X and Y by the joint p.d.f.

$$f(x, y) =$$
$$\frac{\Gamma(\alpha + \beta + \gamma)}{\Gamma(\alpha)\Gamma(\beta)\Gamma(\gamma)} \, x^{\alpha-1} y^{\beta-1} (1 - x - y)^{\gamma-1},$$
$$0 < x, 0 < y, x + y < 1.$$

Select α, β, and γ so that this joint p.d.f. is a reasonable model for someone who sleeps for 8 hours and eats for 2 hours on the average. HINT: X has a beta distribution with parameters $\alpha, \beta + \gamma$, and Y has a beta distribution with parameters $\beta, \alpha + \gamma$; thus, we want $\alpha/(\alpha + \beta + \gamma) = 1/3$ and $\beta/(\alpha + \beta + \gamma) = 1/12$. Now consider the standard deviations.

5.2-17. In Example 5.2-6, verify that the given transformation maps $\{(x_1, x_2) : 0 < x_1 < 1, 0 < x_2 < 1\}$ onto $\{(z_1, z_2) : -\infty < z_1 < \infty, -\infty < z_2 < \infty\}$, except for a set of points that has probability 0. HINT: What is the image of vertical line segments? What is the image of horizontal line segments?

5.3 SEVERAL INDEPENDENT RANDOM VARIABLES

In Section 4.1, we introduced several distributions concerning two random variables. Each of these random variables could be thought of as a measurement in some random experiment. In this section, we consider the possibility of performing several random experiments or one random experiment several times in which each trial results in one measurement that can be considered a random variable. That is, we obtain one random variable from each experiment, and thus we obtain a collection of several random variables from the several experiments. Further, suppose that these experiments are performed in such a way that the events associated with any one of them are independent of the events associated with others, and hence the corresponding random variables are, in the probabilistic sense, independent.

Recall from Section 4.1 that if X_1 and X_2 are random variables of the discrete type with probability mass functions $f_1(x_1)$ and $f_2(x_2)$, respectively, and if

$$P(X_1 = x_1 \text{ and } X_2 = x_2) = P(X_1 = x_1) P(X_2 = x_2)$$
$$= f_1(x_1) f_2(x_2), \qquad x_1 \in S_1, \ x_2 \in S_2,$$

then X_1 and X_2 are said to be independent and the joint p.m.f. is $f_1(x_1) f_2(x_2)$.

Sometimes the two random experiments are exactly the same. For example, we could cast the die twice, resulting first in X_1 and then in X_2. It is reasonable to say that the p.m.f. of X_1 is $f(x_1) = 1/6$, $x_1 = 1, 2, \ldots, 6$, and the p.m.f. of X_2 is $f(x_2) = 1/6$, $x_2 = 1, 2, \ldots, 6$. Assuming independence, which would be a fair way to perform the experiments, the joint p.m.f. is then

$$f(x_1) f(x_2) = \left(\frac{1}{6}\right)\left(\frac{1}{6}\right) = \frac{1}{36}, \qquad x_1 = 1, 2, \ldots, 6; \ x_2 = 1, 2, \ldots, 6.$$

In general, if the p.m.f. $f(x)$ of the independent random variables X_1 and X_2 is the same, then the joint p.m.f. is $f(x_1) f(x_2)$. Moreover, in this case, the collection of the two random variables X_1, X_2 is called a **random sample** of size $n = 2$ from a distribution with p.m.f. $f(x)$. Hence, in the two casts of the die, we say that we have a random sample of size $n = 2$ from the uniform distribution on the space $\{1, 2, 3, 4, 5, 6\}$.

EXAMPLE 5.3-1 Let X_1 and X_2 be two independent random variables resulting from two casts of an unbiased die. That is, X_1, X_2 is a random sample of size $n = 2$ from a distribution with p.m.f. $f(x) = 1/6$, $x = 1, 2, \ldots, 6$. We have

$$E(X_1) = E(X_2) = \sum_{x=1}^{6} xf(x) = 3.5,$$

where $f(x) = 1/6$, $x = 1, 2, \ldots, 6$. Moreover,

$$\mathrm{Var}(X_1) = \mathrm{Var}(X_2) = \sum_{x=1}^{6} (x - 3.5)^2 f(x) = \frac{35}{12}.$$

In addition, from independence,

$$E(X_1 X_2) = E(X_1)E(X_2) = (3.5)(3.5) = 12.25$$

and

$$E[(X_1 - 3.5)(X_2 - 3.5)] = E(X_1 - 3.5)E(X_2 - 3.5) = 0.$$

If $Y = X_1 + X_2$, then

$$E(Y) = E(X_1) + E(X_2) = 3.5 + 3.5 = 7$$

and

$$\begin{aligned}
\mathrm{Var}(Y) &= E[(X_1 + X_2 - 7)^2] = E\{[(X_1 - 3.5) + (X_2 - 3.5)]^2\} \\
&= E[(X_1 - 3.5)^2] + E[2(X_1 - 3.5)(X_2 - 3.5)] + E[(X_2 - 3.5)^2] \\
&= \mathrm{Var}(X_1) + (2)(0) + \mathrm{Var}(X_2) \\
&= (2)\left(\frac{35}{12}\right) = \frac{35}{6}.
\end{aligned}$$

■

In Example 5.3-1, we can find the p.m.f. $g(y)$ of $Y = X_1 + X_2$. Since the space of Y is $\{2, 3, 4, \ldots, 12\}$, we have, by a rather straightforward calculation,

$$g(2) = P(X_1 = 1, X_2 = 1) = f(1)f(1) = \left(\frac{1}{6}\right)\left(\frac{1}{6}\right) = \frac{1}{36},$$

$$g(3) = P(X_1 = 1, X_2 = 2 \text{ or } X_1 = 2, X_2 = 1) = \left(\frac{1}{6}\right)\left(\frac{1}{6}\right) + \left(\frac{1}{6}\right)\left(\frac{1}{6}\right) = \frac{2}{36},$$

$$g(4) = P(X_1 = 1, X_2 = 3 \text{ or } X_1 = 2, X_2 = 2 \text{ or } X_1 = 3, X_2 = 1) = \frac{3}{36},$$

and so on. This results in the p.m.f. given by

y	2	3	4	5	6	7	8	9	10	11	12
$g(y)$	$\frac{1}{36}$	$\frac{2}{36}$	$\frac{3}{36}$	$\frac{4}{36}$	$\frac{5}{36}$	$\frac{6}{36}$	$\frac{5}{36}$	$\frac{4}{36}$	$\frac{3}{36}$	$\frac{2}{36}$	$\frac{1}{36}$

With this p.m.f., it is simple to calculate

$$E(Y) = \sum_{y=2}^{12} y\, g(y) = 7$$

and

$$\text{Var}(Y) = \sum_{y=2}^{12} (y - 7)^2 g(y) = \frac{35}{6},$$

which agrees with the results of Example 5.3-1.

All of the definitions and results concerning two random variables of the discrete type can be carried over to two random variables of the continuous type. Moreover, the notions about two independent random variables can be extended to n independent random variables, which can be thought of as measurements on the outcomes of n random experiments. That is, if $X_1, X_2, \ldots, X_n$ are **independent**, then the joint p.d.f. is the product of the respective p.d.f.'s, namely, $f_1(x_1)f_2(x_2) \cdots f_n(x_n)$.

If all n of the distributions are the same, then the collection of n independent and identically distributed random variables, $X_1, X_2, \ldots, X_n$, is said to be a **random sample of size n from that common distribution**. If $f(x)$ is the common p.m.f. or p.d.f. of these n random variables, then the joint p.m.f. or p.d.f. is $f(x_1)f(x_2) \cdots f(x_n)$.

EXAMPLE 5.3-2 Let X_1, X_2, X_3 be a random sample from a distribution with p.d.f.

$$f(x) = e^{-x}, \qquad 0 < x < \infty.$$

The joint p.d.f. of these three random variables is

$$(e^{-x_1})(e^{-x_2})(e^{-x_3}) = e^{-x_1-x_2-x_3}, \qquad 0 < x_i < \infty, \ i = 1, 2, 3.$$

The probability

$$P(0 < X_1 < 1, 2 < X_2 < 4, 3 < X_3 < 7)$$

$$= \left(\int_0^1 e^{-x_1} dx_1 \right) \left(\int_2^4 e^{-x_2} dx_2 \right) \left(\int_3^7 e^{-x_3} dx_3 \right)$$

$$= (1 - e^{-1})(e^{-2} - e^{-4})(e^{-3} - e^{-7}),$$

because of the independence of X_1, X_2, X_3. ∎

EXAMPLE 5.3-3 An electronic device runs until one of its three components fails. The lifetimes (in weeks), X_1, X_2, X_3, of these components are independent, and each has the Weibull p.d.f.

$$f(x) = \frac{2x}{25} e^{-(x/5)^2}, \qquad 0 < x < \infty.$$

The probability that the device stops running in the first three weeks is equal to

$$1 - P(X_1 > 3, X_2 > 3, X_3 > 3) = 1 - P(X_1 > 3)P(X_2 > 3)P(X_3 > 3)$$

$$= 1 - \left(\int_3^\infty f(x)dx \right)^3$$

$$= 1 - \left(\left[-e^{-(x/5)^2} \right]_3^\infty \right)^3$$

$$= 1 - \left[e^{-(3/5)^2} \right]^3 = 0.660. \qquad ∎$$

The special result following Example 5.3-1 (see also Exercise 5.3-5) with a linear function of two random variables extends to more general functions of several random variables. We accept the following theorem without proof. (See Hogg, McKean, and Craig, 2005.)

THEOREM 5.3-1   Let $X_1, X_2, \ldots, X_n$ be n independent random variables that have the joint p.m.f. $f_1(x_1)f_2(x_2)\cdots f_n(x_n)$. Let the random variable $Y = u(X_1, X_2, \ldots, X_n)$ have the p.m.f. $g(y)$. Then, in the discrete case,

$$E(Y) = \sum_y y\, g(y) = \sum_{x_1}\sum_{x_2}\cdots\sum_{x_n} u(x_1, x_2, \ldots, x_n)f_1(x_1)f_2(x_2)\cdots f_n(x_n),$$

provided that these summations exist. For random variables of the continuous type, obvious changes are made; in particular, integrals replace the summations.

The next theorem proves that the expected value of the product of functions of n independent random variables is the product of their expected values.

THEOREM 5.3-2   Say $X_1, X_2, \ldots, X_n$ are independent random variables and the random variable $Y = u_1(X_1)u_2(X_2)\cdots u_n(X_n)$. If $E[u_i(X_i)]$, $i = 1, 2, \ldots, n$, exist, then

$$E(Y) = E[u_1(X_1)u_2(X_2)\cdots u_n(X_n)] = E[u_1(X_1)]E[u_2(X_2)]\cdots E[u_n(X_n)].$$

Proof. In the discrete case, we have

$$E[u_1(X_1)u_2(X_2)\cdots u_n(X_n)]$$
$$= \sum_{x_1}\sum_{x_2}\cdots\sum_{x_n} u_1(x_1)\, u_2(x_2)\cdots u_n(x_n)f_1(x_1)f_2(x_2)\cdots f_n(x_n)$$
$$= \sum_{x_1} u_1(x_1)f_1(x_1) \sum_{x_2} u_2(x_2)f_2(x_2)\cdots \sum_{x_n} u_n(x_n)f_n(x_n)$$
$$= E[u_1(X_1)]E[u_2(x_2)]\cdots E[u_n(X_n)].$$

In the proof of the continuous case, obvious changes are made; in particular, integrals replace summations.   □

REMARK Sometimes students recognize that $X^2 = (X)(X)$ and thus believe that $E(X^2)$ is equal to $[E(X)][E(X)] = [E(X)]^2$ because Theorem 5.3-2 states that the expected value of the product is the product of the expected values. However, note the hypothesis of independence in the theorem, and certainly X is not independent of itself. Incidentally, if $E(X^2)$ did equal $[E(X)]^2$, then the variance of X, or

$$\sigma^2 = E(X^2) - [E(X)]^2,$$

would always equal zero. This happens only in the case of degenerate (one-point) distributions.   ■

We now prove an important theorem about the mean and the variance of a linear combination of random variables.

THEOREM 5.3-3

If $X_1, X_2, \ldots, X_n$ are n independent random variables with respective means $\mu_1, \mu_2, \ldots, \mu_n$ and variances $\sigma_1^2, \sigma_2^2, \ldots, \sigma_n^2$, then the mean and the variance of $Y = \sum_{i=1}^{n} a_i X_i$, where $a_1, a_2, \ldots, a_n$ are real constants, are, respectively,

$$\mu_Y = \sum_{i=1}^{n} a_i \mu_i \qquad \text{and} \qquad \sigma_Y^2 = \sum_{i=1}^{n} a_i^2 \sigma_i^2.$$

Proof. We have

$$\mu_Y = E(Y) = E\left(\sum_{i=1}^{n} a_i X_i\right) = \sum_{i=1}^{n} a_i E(X_i) = \sum_{i=1}^{n} a_i \mu_i,$$

because the expected value of the sum is the sum of the expected values (i.e., E is a linear operator). Also,

$$\sigma_Y^2 = E[(Y - \mu_Y)^2] = E\left[\left(\sum_{i=1}^{n} a_i X_i - \sum_{i=1}^{n} a_i \mu_i\right)^2\right]$$

$$= E\left\{\left[\sum_{i=1}^{n} a_i(X_i - \mu_i)\right]^2\right\} = E\left[\sum_{i=1}^{n}\sum_{j=1}^{n} a_i a_j (X_i - \mu_i)(X_j - \mu_j)\right].$$

Again using the fact that E is a linear operator, we obtain

$$\sigma_Y^2 = \sum_{i=1}^{n}\sum_{j=1}^{n} a_i a_j E[(X_i - \mu_i)(X_j - \mu_j)].$$

However, if $i \neq j$, then from the independence of X_i and X_j, we have

$$E[(X_i - \mu_i)(X_j - \mu_j)] = E(X_i - \mu_i)E(X_j - \mu_j) = (\mu_i - \mu_i)(\mu_j - \mu_j) = 0.$$

Thus, the variance can be written as

$$\sigma_Y^2 = \sum_{i=1}^{n} a_i^2 E[(X_i - \mu_i)^2] = \sum_{i=1}^{n} a_i^2 \sigma_i^2. \qquad \square$$

REMARK   Although Theorem 5.3-3 gives the mean and the variance of a linear function of independent random variables, the proof can easily be modified to the case in which X_i and X_j are correlated. Then

$$E[(X_i - \mu_i)(X_j - \mu_j)] = \rho_{ij}\sigma_i\sigma_j,$$

instead of zero, where ρ_{ij} is the correlation coefficient of X_i and X_j. Thus,

$$\sigma_Y^2 = \sum_{i=1}^{n} a_i^2 \sigma_i^2 + 2\sum\sum_{i<j} a_i a_j \rho_{ij}\sigma_i\sigma_j,$$

where the factor 2 appears because the sum is over $i < j$ and

$$a_i a_j \rho_{ij} \sigma_i \sigma_j = a_j a_i \rho_{ji} \sigma_j \sigma_i.$$

The mean of Y is still the same in both cases, namely,

$$\mu_Y = \sum_{i=1}^{n} a_i \mu_i.$$

■

We give two illustrations of the theorem.

EXAMPLE 5.3-4 Let the independent random variables X_1 and X_2 have respective means $\mu_1 = -4$ and $\mu_2 = 3$ and variances $\sigma_1^2 = 4$ and $\sigma_2^2 = 9$. Then the mean and the variance of $Y = 3X_1 - 2X_2$ are, respectively,

$$\mu_Y = (3)(-4) + (-2)(3) = -18$$

and

$$\sigma_Y^2 = (3)^2(4) + (-2)^2(9) = 72.$$

■

EXAMPLE 5.3-5 Let X_1, X_2 be a random sample from a distribution with mean μ and variance σ^2. Let $Y = X_1 - X_2$; then

$$\mu_Y = \mu - \mu = 0$$

and

$$\sigma_Y^2 = (1)^2\sigma^2 + (-1)^2\sigma^2 = 2\sigma^2.$$

■

Now consider the **mean of a random sample**, $X_1, X_2, \ldots, X_n$, from a distribution with mean μ and variance σ^2, namely,

$$\overline{X} = \frac{X_1 + X_2 + \cdots + X_n}{n},$$

which is a linear function with each $a_i = 1/n$. Then

$$\mu_{\overline{X}} = \sum_{i=1}^{n} \left(\frac{1}{n}\right)\mu = \mu \qquad \text{and} \qquad \sigma_{\overline{X}}^2 = \sum_{i=1}^{n} \left(\frac{1}{n}\right)^2 \sigma^2 = \frac{\sigma^2}{n}.$$

That is, the mean of $\overline{X}$ is that of the distribution from which the sample arose, but the variance of $\overline{X}$ is that of the underlying distribution divided by n. Any function of the sample observations, $X_1, X_2, \ldots, X_n$, is called a **statistic,** so here $\overline{X}$ is a statistic and also an **estimator** of the distribution mean μ.

EXERCISES

5.3-1. Let X_1 and X_2 be independent Poisson random variables with respective means $\lambda_1 = 2$ and $\lambda_2 = 3$. Find

(a) $P(X_1 = 3, X_2 = 5)$.

(b) $P(X_1 + X_2 = 1)$.

HINT: Note that this event can occur if and only if $\{X_1 = 1, X_2 = 0\}$ or $\{X_1 = 0, X_2 = 1\}$.

5.3-2. Let X_1 and X_2 be independent random variables with respective binomial distributions $b(3, 1/2)$ and $b(5, 1/2)$. Determine

(a) $P(X_1 = 2, X_2 = 4)$.

(b) $P(X_1 + X_2 = 7)$.

5.3-3. Let X_1 and X_2 be independent random variables with probability density functions

$f_1(x_1) = 2x_1$, $0 < x_1 < 1$, and $f_2(x_2) = 4x_2^3$, $0 < x_2 < 1$, respectively. Compute

(a) $P(0.5 < X_1 < 1 \text{ and } 0.4 < X_2 < 0.8)$.

(b) $E(X_1^2 X_2^3)$.

5.3-4. Let X_1 and X_2 be a random sample of size $n = 2$ from the exponential distribution with p.d.f. $f(x) = 2e^{-2x}$, $0 < x < \infty$. Find

(a) $P(0.5 < X_1 < 1.0, 0.7 < X_2 < 1.2)$.

(b) $E[X_1(X_2 - 0.5)^2]$.

5.3-5. Let X_1 and X_2 be observations of a random sample of size $n = 2$ from a distribution with p.m.f. $f(x) = x/6$, $x = 1, 2, 3$. Then find the p.m.f. of $Y = X_1 + X_2$. Determine the mean and the variance of the sum in two ways.

5.3-6. Let X_1 and X_2 be a random sample of size $n = 2$ from a distribution with p.d.f. $f(x) = 6x(1 - x)$, $0 < x < 1$. Find the mean and the variance of $Y = X_1 + X_2$.

5.3-7. The distributions of incomes in two cities follow the two Pareto-type p.d.f.'s

$$f(x) = \frac{2}{x^3}, \quad 1 < x < \infty,$$

and

$$g(y) = \frac{3}{y^4}, \quad 1 < y < \infty,$$

respectively. Here one unit represents \$20,000. One person with income is selected at random from each city. Let X and Y be their respective incomes. Compute $P(X < Y)$.

5.3-8. Suppose two independent claims are made on two insured homes, where each claim has p.d.f.

$$f(x) = \frac{4}{x^5}, \quad 1 < x < \infty,$$

in which the unit is \$1000. Find the expected value of the larger claim.

HINT: If X_1 and X_2 are the two independent claims and $Y = \max(X_1, X_2)$, then

$$G(y) = P(Y \le y) = P(X_1 \le y)P(X_2 \le y)$$
$$= [P(X \le y)]^2.$$

Find $g(y) = G'(y)$ and $E(Y)$.

5.3-9. Let X_1, X_2 be a random sample of size $n = 2$ from a distribution with p.d.f. $f(x) = 3x^2$, $0 < x < 1$. Determine

(a) $P(\max X_i < 3/4) =$
$P(X_1 < 3/4, X_2 < 3/4)$.

(b) The mean and the variance of $Y = X_1 + X_2$.

5.3-10. Let X_1, X_2, X_3 denote a random sample of size $n = 3$ from a distribution with the geometric

p.m.f.

$$f(x) = \left(\frac{3}{4}\right)\left(\frac{1}{4}\right)^{x-1}, \quad x = 1, 2, 3, \ldots.$$

(a) Compute $P(X_1 = 1, X_2 = 3, X_3 = 1)$.

(b) Determine $P(X_1 + X_2 + X_3 = 5)$.

(c) If Y equals the maximum of X_1, X_2, X_3, find

$$P(Y \le 2) = P(X_1 \le 2)P(X_2 \le 2)P(X_3 \le 2).$$

5.3-11. Let X_1, X_2, X_3 be three independent random variables with binomial distributions $b(4, 1/2)$, $b(6, 1/3)$, and $b(12, 1/6)$, respectively. Find

(a) $P(X_1 = 2, X_2 = 2, X_3 = 5)$.

(b) $E(X_1 X_2 X_3)$.

(c) The mean and the variance of $Y = X_1 + X_2 + X_3$.

5.3-12. Let X_1, X_2, X_3 be a random sample of size $n = 3$ from the exponential distribution with p.d.f. $f(x) = e^{-x}$, $0 < x < \infty$. Find

$$P(1 < \min X_i) = P(1 < X_1, 1 < X_2, 1 < X_3).$$

5.3-13. Flip $n = 8$ fair coins and remove all that came up heads. Flip the remaining coins (that came up tails) and remove the heads again. Continue flipping the remaining coins until each has come up heads. We shall find the p.m.f. of Y, the number of trials needed. Let X_i equal the number of flips required to observe heads on coin i, $i = 1, 2, \ldots, 8$. Then $Y = \max(X_1, X_2, \ldots, X_8)$.

(a) Show that $P(Y \le y) = [1 - (1/2)^y]^8$.

(b) Show that $P(Y = y) = [1 - (1/2)^y]^8 - [1 - (1/2)^{y-1}]^8$, $y = 1, 2, \ldots$.

(c) Use a computer algebra system such as *Maple* or *Mathematica* to show that $E(Y) = 13{,}315{,}424/3{,}011{,}805 = 4.421$.

(d) What happens to the expected value of Y as the number of coins is doubled?

5.3-14. Construct a sequence of squares with one vertex at the origin and having lengths $1 - 1/2^x$, $x = 1, 2, \ldots$. Select a point randomly from the unit square with one vertex at the origin and sides of length 1. Let the random variable X equal x if the point is in the region between the squares with sides of lengths $1 - 1/2^x$ and $1 - 1/2^{x-1}$, $x = 1, 2, 3, \ldots$.

(a) Draw a figure illustrating this exercise.

(b) Show that

$$f(x) = P(X = x)$$

$$= \left(1 - \frac{1}{2^x}\right)^2 - \left(1 - \frac{1}{2^{x-1}}\right)^2$$

$$= \frac{2^{x+1} - 3}{2^{2x}}, x = 1, 2, 3, \dots.$$

(c) Show that $f(x)$ is a p.m.f. (See Equation D.4-1 on the CD-ROM.)

(d) Find the mean of X. (See Equation D.4-2 on the CD-ROM.)

(e) Find the variance of X. (See Equation D.4-3 on the CD-ROM.)

(f) Use simulation to confirm your answers.

5.3-15. Construct a sequence of cubes with one vertex at the origin and having lengths $1/2^x$, $x = 1, 2, \dots$. Select a point randomly from the unit cube with one vertex at the origin and sides of length 1. Let the random variable X equal x if the point is in the region between the cubes with sides of lengths $1 - 1/2^x$ and $1 - 1/2^{x-1}$, $x = 1, 2, 3, \dots$.

Another way to state this problem is as follows: A 3-tuple (u_1, u_2, u_3) of random numbers u_1, u_2, and u_3 is selected from the unit cube. The random variable X is defined as follows:

If $u_1 \leq 1 - 1/2$, $u_2 < 1 - 1/2$, and $u_3 \leq 1 - 1/2$, then $X = 1$;

else if $u_1 \leq 1 - 1/2^2$, $u_2 \leq 1 - 1/2^2$, and $u_3 \leq 1 - 1/2^2$, then $X = 2$;

else if $u_1 \leq 1 - 1/2^3$, $u_2 \leq 1 - 1/2^3$, and $u_3 \leq 1 - 1/2^3$, then $X = 3$; and so forth.

(a) Find the p.m.f. of X.

(b) Verify that what you found in part (a) is a p.m.f., and find its mean and variance. HINT: Use a CAS.

(c) Use simulation to compare the theoretical and empirical characteristics.

(d) Extend this problem to higher dimensions.

(e) How is the answer to this exercise related to the answer to Exercise 5.3-13?

5.3-16. Let X_1, X_2, X_3 be independent random variables that represent lifetimes (in hours) of three key components of a device. Say their respective distributions are exponential with means $1000, 1500$, and 2000. Let Y be the minimum of X_1, X_2, X_3 and compute $P(Y > 1000)$.

5.3-17. A device contains three components, each of which has a lifetime in hours with the p.d.f.

$$f(x) = \frac{2x}{10^2} e^{-(x/10)^2}, \qquad 0 < x < \infty.$$

The device fails with the failure of one of the components. Assuming independent lifetimes, what

is the probability that the device fails in the first hour of its operation?

5.3-18. Three drugs are being tested for use as the treatment of a certain disease. Let p_1, p_2, and p_3 represent the probabilities of success for the respective drugs. As three patients come in, each is given one of the drugs in a random order. After $n = 10$ "triples" and assuming independence, compute the probability that the maximum number of successes with one of the drugs exceeds eight if in fact $p_1 = p_2 = p_3 = 0.7$.

5.3-19. Three components are placed in series. The time in hours to failure of each has the p.d.f.

$$f(x) = \frac{x}{500^2} e^{-x/500}, \qquad 0 < x < \infty.$$

Since they are in series, we are concerned about the minimum time Y to failure of the three. Assuming independence, find the c.d.f. and the p.d.f. of Y and compute $P(Y \leq 300)$. HINT: $G(y) = P(Y \leq y) = 1 - P(Y > y) = 1 - P(\text{all three} > y)$.

5.3-20. Each of eight bearings in a bearing assembly has a diameter (in millimeters) that has the p.d.f.

$$f(x) = 10x^9, \qquad 0 < x < 1.$$

Assuming independence, find the distribution function and the p.d.f. of the maximum diameter (say, Y) of the eight bearings and compute $P(0.9999 < Y < 1)$.

5.3-21. In considering medical insurance for a certain operation, let X equal the amount (in dollars) paid for the doctor and let Y equal the amount paid to the hospital. In the past, the variances have been $\text{Var}(X) = 8100$, $\text{Var}(Y) = 10,000$, and $\text{Var}(X + Y) = 20,000$. Due to increased expenses, it was decided to increase the doctor's fee by \$500 and increase the hospital charge Y by 8 percent. Calculate the variance of $X + 500 + (1.08)Y$, the new total claim.

5.3-22. The lifetime in months of a certain part has a gamma distribution with $\alpha = \theta = 2$. A company buys three such parts and uses one until it fails, replacing it with a second part. When the latter fails, it is replaced by the third part. What are the mean and the variance of the total lifetime associated with this situation?

5.3-23. Two components operate in parallel in a device, so the device fails when and only when both components fail. The lifetimes, X_1 and X_2, of the respective components are independent and identically distributed with a gamma distribution, $\alpha = 1, \theta = 2$. The cost of operating the device is $Z = 2Y_1 + Y_2$, where $Y_1 = \min(X_1, X_2)$ and $Y_2 = \max(X_1, X_2)$. Compute $E(Z)$.

5.4 THE MOMENT-GENERATING FUNCTION TECHNIQUE

To illustrate the moment-generating function technique, we begin with a simple example.

EXAMPLE 5.4-1 Let X_1 and X_2 be independent random variables with uniform distributions on $\{1, 2, 3, 4\}$. Let $Y = X_1 + X_2$. For example, Y could equal the sum when two four-sided dice are rolled. The moment-generating function of Y is

$$M_Y(t) = E(e^{tY}) = E[e^{t(X_1+X_2)}] = E(e^{tX_1}e^{tX_2}).$$

The independence of X_1 and X_2 implies that

$$M_Y(t) = E(e^{tX_1})E(e^{tX_2}).$$

In this example, X_1 and X_2 have the same p.m.f., namely,

$$f(x) = \frac{1}{4}, \qquad x = 1, 2, 3, 4,$$

and thus the same moment-generating function,

$$M_X(t) = \frac{1}{4}e^t + \frac{1}{4}e^{2t} + \frac{1}{4}e^{3t} + \frac{1}{4}e^{4t}.$$

It then follows that $M_Y(t) = [M_X(t)]^2$ equals

$$\frac{1}{16}e^{2t} + \frac{2}{16}e^{3t} + \frac{3}{16}e^{4t} + \frac{4}{16}e^{5t} + \frac{3}{16}e^{6t} + \frac{2}{16}e^{7t} + \frac{1}{16}e^{8t}.$$

Note that the coefficient of e^{bt} is equal to the probability $P(Y = b)$; for example, $4/16 = P(Y = 5)$. Thus, we can find the distribution of Y by determining its moment-generating function. ■

In some applications, it is sufficient to know the mean and variance of a linear combination of random variables, say, Y. However, it is often helpful to know exactly how Y is distributed. The next theorem can frequently be used to find the distribution of a linear combination of independent random variables.

THEOREM 5.4-1 If $X_1, X_2, \ldots, X_n$ are independent random variables with respective moment-generating functions $M_{X_i}(t)$, $i = 1, 2, 3, \ldots, n$, then the moment-generating function of $Y = \sum_{i=1}^{n} a_i X_i$ is

$$M_Y(t) = \prod_{i=1}^{n} M_{X_i}(a_i t).$$

Proof. From Theorem 5.3-2, the moment-generating function of Y is given by

$$M_Y(t) = E[e^{tY}] = E[e^{t(a_1 X_1 + a_2 X_2 + \cdots + a_n X_n)}]$$

$$= E[e^{a_1 t X_1} e^{a_2 t X_2} \cdots e^{a_n t X_n}]$$

$$= E[e^{a_1 t X_1}]E[e^{a_2 t X_2}] \cdots E[e^{a_n t X_n}].$$

However, since

$$E(e^{tX_i}) = M_{X_i}(t),$$

it follows that

$$E(e^{a_i t X_i}) = M_{X_i}(a_i t).$$

Thus, we have

$$M_Y(t) = M_{X_1}(a_1 t) M_{X_2}(a_2 t) \cdots M_{X_n}(a_n t) = \prod_{i=1}^{n} M_{X_i}(a_i t). \qquad \square$$

A corollary follows immediately, and it will be used in some important examples.

COROLLARY 5.4-1   If $X_1, X_2, \ldots, X_n$ are observations of a random sample from a distribution with moment-generating function $M(t)$, then

(a) the moment-generating function of $Y = \sum_{i=1}^{n} X_i$ is

$$M_Y(t) = \prod_{i=1}^{n} M(t) = [M(t)]^n;$$

(b) the moment-generating function of $\overline{X} = \sum_{i=1}^{n} (1/n) X_i$ is

$$M_{\overline{X}}(t) = \prod_{i=1}^{n} M\left(\frac{t}{n}\right) = \left[M\left(\frac{t}{n}\right)\right]^n.$$

Proof. For (a), let $a_i = 1$, $i = 1, 2, \ldots, n$, in Theorem 5.4-1. For (b), take $a_i = 1/n$, $i = 1, 2, \ldots, n$. $\qquad \square$

The next two examples and the exercises give some important applications of Theorem 5.4-1 and its corollary.

EXAMPLE 5.4-2   Let $X_1, X_2, \ldots, X_n$ denote the outcomes of n Bernoulli trials. The moment-generating function of X_i, $i = 1, 2, \ldots, n$, is

$$M(t) = q + pe^t.$$

If

$$Y = \sum_{i=1}^{n} X_i,$$

then

$$M_Y(t) = \prod_{i=1}^{n} (q + pe^t) = (q + pe^t)^n.$$

Thus, we again see that Y is $b(n, p)$.   ∎

EXAMPLE 5.4-3   Let X_1, X_2, X_3 be the observations of a random sample of size $n = 3$ from the exponential distribution having mean θ and, of course, moment-generating function

$M(t) = 1/(1 - \theta t), t < 1/\theta$. The moment-generating function of the random variable $Y = X_1 + X_2 + X_3$ is

$$M_Y(t) = [(1 - \theta t)^{-1}]^3 = (1 - \theta t)^{-3}, \qquad t < 1/\theta,$$

which is that of a gamma distribution with parameters $\alpha = 3$ and θ. Thus, Y has this distribution. On the other hand, the moment-generating function of $\overline{X}$ is

$$M_{\overline{X}}(t) = \left[\left(1 - \frac{\theta t}{3}\right)^{-1}\right]^3 = \left(1 - \frac{\theta t}{3}\right)^{-3}, \qquad t < 3/\theta.$$

Hence, the distribution of $\overline{X}$ is gamma with the parameters $\alpha = 3$ and $\theta/3$, respectively. ∎

THEOREM 5.4-2

Let $X_1, X_2, \ldots, X_n$ be independent chi-square random variables with $r_1, r_2, \ldots, r_n$ degrees of freedom, respectively. Then the distribution of the random variable $Y = X_1 + X_2 + \cdots + X_n$ is $\chi^2(r_1 + r_2 + \cdots + r_n)$.

Proof. By Theorem 5.4-1 with each $a = 1$, the m.g.f. of Y is

$$M_Y(t) = \prod_{i=1}^n M_{X_i}(t) = (1 - 2t)^{-r_1/2}(1 - 2t)^{-r_2/2}\cdots(1 - 2t)^{-r_n/2}$$
$$= (1 - 2t)^{-\Sigma r_i/2}, \qquad \text{with } t < 1/2,$$

which is the m.g.f. of a $\chi^2(r_1 + r_2 + \cdots + r_n)$ random variable. Thus, the distribution of Y is $\chi^2(r_1 + r_2 + \cdots + r_n)$. □

The next two corollaries combine and extend the results of Theorems 3.6-2 and 5.4-2 and give one interpretation of degrees of freedom.

COROLLARY 5.4-2

Let $Z_1, Z_2, \ldots, Z_n$ have standard normal distributions, $N(0,1)$. If these random variables are independent, then $W = Z_1^2 + Z_2^2 + \cdots + Z_n^2$ has a distribution that is $\chi^2(n)$.

Proof. By Theorem 3.6-2, Z_i^2 is $\chi^2(1)$ for $i = 1, 2, \ldots, n$. From Theorem 5.4-2, with $k = n$, $Y = W$, and $r_i = 1$, it follows that W is $\chi^2(n)$. □

COROLLARY 5.4-3

If $X_1, X_2, \ldots, X_n$ are independent and have normal distributions $N(\mu_i, \sigma_i^2)$, $i = 1, 2, \ldots, n$, respectively, then the distribution of

$$W = \sum_{i=1}^n \frac{(X_i - \mu_i)^2}{\sigma_i^2}$$

is $\chi^2(n)$.

Proof. This follows from Corollary 5.4-2, since $Z_i = (X_i - \mu_i)/\sigma_i$ is $N(0,1)$, $i = 1, 2, \ldots, n$. □

Note that the number of terms in the summation and the number of degrees of freedom are equal in Theorem 5.4-2 and the corollary.

EXERCISES

5.4-1. Let X_1, X_2, X_3 be a random sample of size 3 from the distribution with p.m.f. $f(x) = 1/4$, $x = 1, 2, 3, 4$. For example, observe three independent rolls of a fair four-sided die.

(a) Find the p.m.f. of $Y = X_1 + X_2 + X_3$.

(b) Sketch a bar graph of the p.m.f. of Y.

5.4-2. Let X_1 and X_2 have independent distributions $b(n_1, p)$ and $b(n_2, p)$. Find the moment-generating function of $Y = X_1 + X_2$. How is Y distributed?

5.4-3. Let X_1, X_2, X_3 be mutually independent random variables with Poisson distributions having means 2, 1, and 4, respectively.

(a) Find the moment-generating function of the sum $Y = X_1 + X_2 + X_3$.

(b) How is Y distributed?

(c) Compute $P(3 < Y < 9)$.

5.4-4. Generalize Exercise 5.4-3 by showing that the sum of n independent Poisson random variables with respective means $\mu_1, \mu_2, \ldots, \mu_n$ is Poisson with mean

$$\mu_1 + \mu_2 + \cdots + \mu_n.$$

5.4-5. Let $Z_1, Z_2, \ldots, Z_7$ be a random sample from the standard normal distribution $N(0, 1)$. Let $W = Z_1^2 + Z_2^2 + \cdots + Z_7^2$. Find $P(1.69 < W < 14.07)$.

5.4-6. Let X_1, X_2, X_3, X_4, X_5 be a random sample of size 5 from a geometric distribution with $p = 1/3$.

(a) Find the moment-generating function of $Y = X_1 + X_2 + X_3 + X_4 + X_5$.

(b) How is Y distributed?

5.4-7. Let X_1, X_2, X_3 denote a random sample of size 3 from a gamma distribution with $\alpha = 7$ and $\theta = 5$.

(a) Find the moment-generating function of $Y = X_1 + X_2 + X_3$.

(b) How is Y distributed?

5.4-8. Let $W = X_1 + X_2 + \cdots + X_h$, a sum of h mutually independent and identically distributed exponential random variables with mean θ. Show that W has a gamma distribution with mean $h\theta$.

5.4-9. Let X equal the outcome when a four-sided die is rolled. Let Y equal the outcome when a six-sided die is rolled. Let $W = X + Y$. Assume that X and Y are independent.

(a) Find the moment-generating function of W.

(b) Give the p.m.f. of W.

5.4-10. Let X equal the outcome when a fair four-sided die that has its faces numbered 0, 1, 2, and 3 is rolled. Let Y equal the outcome when a fair four-sided die that has its faces numbered 0, 4, 8, and 12 is rolled.

(a) Define the moment-generating function of X.

(b) Define the moment-generating function of Y.

(c) Let $W = X + Y$, the sum when the pair of dice is rolled. Find the moment-generating function of W.

(d) Give the p.m.f. of W; that is, determine $P(W = w)$, $w = 0, 1, \ldots, 15$, from the moment-generating function of W.

5.4-11. Let X and Y, with respective p.m.f.'s $f(x)$ and $g(y)$, be independent discrete random variables, each of whose support is a subset of the nonnegative integers $0, 1, 2, \ldots$. Show that the p.m.f. of $W = X + Y$ is given by the **convolution formula**

$$h(w) = \sum_{x=0}^{w} f(x)g(w - x), \qquad w = 0, 1, 2, \ldots.$$

HINT: Argue that $h(w) = P(W = w)$ is the probability of the $w + 1$ mutually exclusive events $(x, y = w - x)$, $x = 0, 1, \ldots, w$.

5.4-12. Let X and Y equal the outcomes when two fair six-sided dice are rolled. Let $W = X + Y$. Assuming independence, find the p.m.f. of W when

(a) The first die has three faces numbered 0 and three faces numbered 2, and the second die has its faces numbered 0, 1, 4, 5, 8, and 9.

(b) The faces on the first die are numbered 0, 1, 2, 3, 4, and 5, and the faces on the second die are numbered 0, 6, 12, 18, 24, and 30.

5.4-13. Let X and Y be the outcomes when a pair of fair eight-sided dice is rolled. Let $W = X + Y$. How should the faces of the dice be numbered so that W has a uniform distribution on $0, 1, \ldots, 15$?

5.4-14. Some states—for example, Texas and New York—have added a new extension to Daily 3 and Daily 4 games. The player selects a three-digit or a four-digit number. The player can bet on the sum of the digits selected in a game called Sum It Up. For the questions that follow, assume that the size of the bet is $1. Recall that the state does not return this dollar. HINT: Use either Exercise 5.4-11 or the moment-generating function.

(a) In the three-digit game, what is the expected value for the player if the prizes for sums of 0, 1, and 2 are $500, $166, and $83, respectively?

(b) To be consistent, what prize should the state offer for a sum of 3?

(c) In the four-digit game, what is the expected value for the player if the prizes for sums of 0, 1, and 2 are $5000, $1250, and $500, respectively?

(d) To be consistent, what prize should the state offer for a sum of 3?

5.4-15. Let $X_1, X_2, \ldots, X_8$ be a random sample from a distribution having p.m.f. $f(x) = (x + 1)/6$, $x = 0, 1, 2$.

(a) Use Exercise 5.4-11 to find the p.m.f. of $W_1 = X_1 + X_2$.

(b) What is the p.m.f. of $W_2 = X_3 + X_4$?

(b) Now find the p.m.f. of $W = W_1 + W_2 = X_1 + X_2 + X_3 + X_4$.

(c) Find the p.m.f. of $Y = X_1 + X_2 + \cdots + X_8$.

(d) Construct probability histograms for X_1, W_1, W, and Y. Are these histograms skewed or symmetric?

5.4-16. The number of accidents in a period of one week follows a Poisson distribution with mean 2. The numbers of accidents from week to week are independent. What is the probability of exactly seven accidents in a given three weeks. HINT: See Exercise 5.4-4.

5.4-17. Given a fair four-sided die, let Y equal the number of rolls needed to observe each face at least once.

(a) Argue that $Y = X_1 + X_2 + X_3 + X_4$, where X_i has a geometric distribution with $p_i = (5 - i)/4$, $i = 1, 2, 3, 4$, and X_1, X_2, X_3, X_4 are independent.

(b) Find the mean and variance of Y.

(c) Find $P(Y = y)$, $y = 4, 5, 6, 7$.

5.4-18. The number X of sick days taken during a year by an employee follows a Poisson distribution with mean 2. Let us observe four such employees. Assuming independence, compute the probability that their total number of sick days exceeds 10.

5.4-19. In a study concerning a new treatment of a certain disease, two groups of 25 participants in each were followed for five years. Those in one group took the old treatment and those in the other took the new treatment. The dropout rate for an individual was 50% in both groups over that 5-year period. Let X be the number that dropped out in the first group and Y the number in the second group. Assuming independence where needed, give the sum that equals the probability that $Y \geq X + 2$. HINT: What is the distribution of $Y - X + 25$?

5.4-20. The number of cracks on a highway averages 0.5 per mile and follows a Poisson distribution. Assuming independence (which may not be a good assumption; why?), what is the probability that, in a 40-mile stretch of that highway, there are fewer than 15 cracks?

5.4-21. Let X be the number of flaws on the outside of a new washing machine. Say $X_1, X_2, \ldots, X_{10}$ is a random sample of 10 such X values that has a Poisson distribution with mean 1.

(a) The manufacturer gives a discount to a buyer of any machine with two or more flaws. Among the 10 machines, what is the probability of giving at least three discounts?

(b) Among the 10 machines, what is the probability of fewer than 13 flaws?

5.4-22. The time X in minutes of a visit to a cardiovascular disease specialist by a patient is modeled by a gamma p.d.f. with $\alpha = 1.5$ and $\theta = 10$. Suppose that you are such a patient and have four patients ahead of you. Assuming independence, what integral gives the probability that you will wait more than 90 minutes?

5.4-23. A doorman at a hotel is trying to get three taxicabs for three different couples. The arrival of empty cabs has an exponential distribution with mean 2 minutes. Assuming independence, what is the probability that the doorman will get all three couples taken care of within 6 minutes?

5.5 RANDOM FUNCTIONS ASSOCIATED WITH NORMAL DISTRIBUTIONS

In statistical applications, it is often assumed that the population from which a sample is taken is normally distributed, $N(\mu, \sigma^2)$. There is then interest in estimating the parameters μ and σ^2 or in testing conjectures about these parameters. The usual statistics that are used in these activities are the sample mean $\overline{X}$ and the sample variance S^2; thus, we need to know something about the distribution of these statistics or functions of these statistics.

We now use the moment-generating function technique of Section 5.4 to prove a theorem that deals with linear functions of independent normally distributed random variables.

THEOREM
5.5-1

If $X_1, X_2, \ldots, X_n$ are n mutually independent normal variables with means $\mu_1, \mu_2, \ldots, \mu_n$ and variances $\sigma_1^2, \sigma_2^2, \ldots, \sigma_n^2$, respectively, then the linear function

$$Y = \sum_{i=1}^{n} c_i X_i$$

has the normal distribution

$$N\left(\sum_{i-1}^{n} c_i \mu_i, \sum_{i-1}^{n} c_i^2 \sigma_i^2 \right).$$

Proof. By Theorem 5.4-1, we have

$$M_Y(t) = \prod_{i=1}^{n} M_{X_i}(c_i t) = \prod_{i=1}^{n} \exp(\mu_i c_i t + \sigma_i^2 c_i^2 t^2 / 2)$$

because $M_{X_i}(t) = \exp(\mu_i t + \sigma_i^2 t^2 / 2)$, $i = 1, 2, \ldots, n$. Thus,

$$M_Y(t) = \exp\left[\left(\sum_{i=1}^{n} c_i \mu_i \right) t + \left(\sum_{i=1}^{n} c_i^2 \sigma_i^2 \right) \left(\frac{t^2}{2} \right) \right].$$

This is the moment-generating function of a distribution that is

$$N\left(\sum_{i=1}^{n} c_i \mu_i, \sum_{i=1}^{n} c_i^2 \sigma_i^2 \right).$$

Thus, Y has this normal distribution.   □

From Theorem 5.5-1, we observe that the difference of two independent normally distributed random variables, say, $Y = X_1 - X_2$, has the normal distribution $N(\mu_1 - \mu_2, \sigma_1^2 + \sigma_2^2)$.

EXAMPLE 5.5-1

Let X_1 and X_2 equal the number of pounds of butterfat produced by two Holstein cows (one selected at random from those on the Koopman farm and one selected at random from those on the Vliestra farm, respectively) during the 305-day lactation period following the births of calves. Assume that the distribution of X_1 is $N(693.2, 22820)$ and the distribution of X_2 is $N(631.7, 19205)$. Moreover, let X_1 and X_2 be independent. We shall find $P(X_1 > X_2)$. That is, we shall find the probability that the butterfat produced by the Koopman farm cow exceeds that produced by the Vliestra farm cow. (Sketch p.d.f.'s on the same graph for these two normal distributions.) If we let $Y = X_1 - X_2$, then the distribution of Y is $N(693.2 - 631.7, 22820 + 19205)$. Thus,

$$P(X_1 > X_2) = P(Y > 0) = P\left(\frac{Y - 61.5}{\sqrt{42025}} > \frac{0 - 61.5}{205} \right)$$
$$= P(Z > -0.30) = 0.6179.$$ ■

COROLLARY
5.5-1

If $X_1, X_2, \ldots, X_n$ are observations of a random sample of size n from the normal distribution $N(\mu, \sigma^2)$, then the distribution of the sample mean $\overline{X} = (1/n) \sum_{i=1}^{n} X_i$ is $N(\mu, \sigma^2/n)$.

Proof. Let $c_i = 1/n$, $\mu_i = \mu$, and $\sigma_i^2 = \sigma^2$ in Theorem 5.5-1. □

Corollary 5.5-1 shows that if $X_1, X_2, \ldots, X_n$ is a random sample from the normal distribution, $N(\mu, \sigma^2)$, then the probability distribution of $\overline{X}$ is also normal with the same mean μ but a variance σ^2/n. This means that $\overline{X}$ has a greater probability of falling into an interval containing μ than does a single observation—say, X_1. For example, if $\mu = 50$, $\sigma^2 = 16$, and $n = 64$, then $P(49 < \overline{X} < 51) = 0.9544$, whereas $P(49 < X_1 < 51) = 0.1974$. This property is illustrated again in the next example.

EXAMPLE 5.5-2

Let $X_1, X_2, \ldots, X_n$ be a random sample from the $N(50, 16)$ distribution. We know that the distribution of $\overline{X}$ is $N(50, 16/n)$. To illustrate the effect of n, the graph of the p.d.f. of $\overline{X}$ is given in Figure 5.5-1 for $n = 1, 4, 16,$ and 64. When $n = 64$, compare the areas that represent $P(49 < \overline{X} < 51)$ and $P(49 < X_1 < 51)$. ■

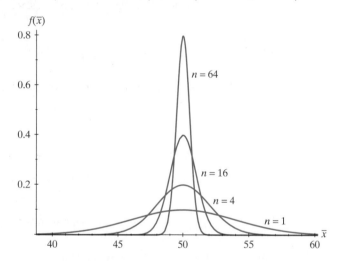

FIGURE 5.5-1: p.d.f.'s of means of samples from $N(50, 16)$

The next theorem gives an important result that will be used in statistical applications. In connection with those applications, we will use the sample variance s^2 to estimate the variance, σ^2, when sampling from the normal distribution, $N(\mu, \sigma^2)$. (More will be said about s^2 at the time of its use.)

THEOREM
5.5-2

Let $X_1, X_2, \ldots, X_n$ be observations of a random sample of size n from the normal distribution $N(\mu, \sigma^2)$. Then the sample mean,

$$\overline{X} = \frac{1}{n} \sum_{i=1}^{n} X_i,$$

and the sample variance,

$$S^2 = \frac{1}{n-1} \sum_{i=1}^{n} (X_i - \overline{X})^2,$$

are independent and

$$\frac{(n-1)S^2}{\sigma^2} = \frac{\sum_{i=1}^{n}(X_i - \overline{X})^2}{\sigma^2} \text{ is } \chi^2(n-1).$$

Proof. We are not prepared to prove the independence of $\overline{X}$ and S^2 at this time (see Section 10.1 for a proof), so we accept it without proof here. To prove the second part, note that

$$W = \sum_{i=1}^{n}\left(\frac{X_i - \mu}{\sigma}\right)^2 = \sum_{i=1}^{n}\left[\frac{(X_i - \overline{X}) + (\overline{X} - \mu)}{\sigma}\right]^2$$

$$= \sum_{i=1}^{n}\left(\frac{X_i - \overline{X}}{\sigma}\right)^2 + \frac{n(\overline{X} - \mu)^2}{\sigma^2} \qquad (5.5\text{-}1)$$

because the cross-product term is equal to

$$2\sum_{i=1}^{n}\frac{(\overline{X} - \mu)(X_i - \overline{X})}{\sigma^2} = \frac{2(\overline{X} - \mu)}{\sigma^2}\sum_{i=1}^{n}(X_i - \overline{X}) = 0.$$

But $Y_i = (X_i - \mu)/\sigma$, $i = 1, 2, \ldots n$, are standardized normal variables that are independent. Hence, $W = \sum_{i=1}^{n}Y_i^2$ is $\chi^2(n)$ by Corollary 5.4-3. Moreover, since $\overline{X}$ is $N(\mu, \sigma^2/n)$, it follows that

$$Z^2 = \left(\frac{\overline{X} - \mu}{\sigma/\sqrt{n}}\right)^2 = \frac{n(\overline{X} - \mu)^2}{\sigma^2}$$

is $\chi^2(1)$ by Theorem 3.6-2. In this notation, Equation 5.5-1 becomes

$$W = \frac{(n-1)S^2}{\sigma^2} + Z^2.$$

However, from the fact that $\overline{X}$ and S^2 are independent, it follows that Z^2 and S^2 are also independent. In the moment-generating function of W, this independence permits us to write

$$E[e^{tW}] = E[e^{t\{(n-1)S^2/\sigma^2 + Z^2\}}] = E[e^{t(n-1)S^2/\sigma^2}e^{tZ^2}]$$

$$= E[e^{t(n-1)S^2/\sigma^2}]E[e^{tZ^2}].$$

Since W and Z^2 have chi-square distributions, we can substitute their moment-generating functions to obtain

$$(1 - 2t)^{-n/2} = E[e^{t(n-1)S^2/\sigma^2}](1 - 2t)^{-1/2}.$$

Equivalently, we have

$$E[e^{t(n-1)S^2/\sigma^2}] = (1 - 2t)^{-(n-1)/2}, \qquad t < \frac{1}{2}.$$

This, of course, is the moment-generating function of a $\chi^2(n-1)$ variable; accordingly, $(n-1)S^2/\sigma^2$ has that distribution. □

Combining the results of Corollary 5.4-3 and Theorem 5.5-2, we see that when sampling is from a normal distribution,

$$U = \sum_{i=1}^{n} \frac{(X_i - \mu)^2}{\sigma^2}$$

is $\chi^2(n)$ and

$$W = \sum_{i=1}^{n} \frac{(X_i - \overline{X})^2}{\sigma^2}$$

is $\chi^2(n-1)$. That is, when the population mean, μ, in $\sum_{i=1}^{n}(X_i - \mu)^2$ is replaced by the sample mean, $\overline{X}$, one degree of freedom is lost. There are more general situations in which a degree of freedom is lost for each parameter estimated in certain chi-square random variables.

EXAMPLE 5.5-3 Let X_1, X_2, X_3, X_4 be a random sample of size 4 from the normal distribution, $N(76.4, 383)$. Then

$$U = \sum_{i=1}^{4} \frac{(X_i - 76.4)^2}{383} \quad \text{is} \quad \chi^2(4),$$

$$W = \sum_{i=1}^{4} \frac{(X_i - \overline{X})^2}{383} \quad \text{is} \quad \chi^2(3),$$

and, for examples,

$$P(0.711 \leq U \leq 7.779) = 0.90 - 0.05 = 0.85,$$
$$P(0.352 \leq W \leq 6.251) = 0.90 - 0.05 = 0.85.$$ ■

In later sections, we shall illustrate the importance of the chi-square distribution in applications.

We now prove a theorem that is the basis for some of the most important inferences in statistics.

THEOREM 5.5-3 **(Student's t distribution)** Let

$$T = \frac{Z}{\sqrt{U/r}},$$

where Z is a random variable that is $N(0,1)$, U is a random variable that is $\chi^2(r)$, and Z and U are independent. Then T has a t distribution with p.d.f.

$$f(t) = \frac{\Gamma((r+1)/2)}{\sqrt{\pi r}\,\Gamma(r/2)} \frac{1}{(1 + t^2/r)^{(r+1)/2}}, \quad -\infty < t < \infty.$$

Proof. The joint p.d.f. of Z and U is

$$g(z,u) = \frac{1}{\sqrt{2\pi}} e^{-z^2/2} \frac{1}{\Gamma(r/2)2^{r/2}} u^{r/2-1} e^{-u/2}, \qquad -\infty < z < \infty, \, 0 < u < \infty.$$

The distribution function $F(t) = P(T \leq t)$ of T is given by

$$\begin{aligned} F(t) &= P(Z/\sqrt{U/r} \leq t) \\ &= P(Z \leq \sqrt{U/r}\, t) \\ &= \int_0^\infty \int_{-\infty}^{\sqrt{(u/r)}\, t} g(z,u)\, dz\, du. \end{aligned}$$

That is,

$$F(t) = \frac{1}{\sqrt{\pi}\, \Gamma(r/2)} \int_0^\infty \left[\int_{-\infty}^{\sqrt{(u/r)}\, t} \frac{e^{-z^2/2}}{2^{(r+1)/2}} dz \right] u^{r/2-1} e^{-u/2}\, du.$$

The p.d.f. of T is the derivative of the distribution function; so, applying the fundamental theorem of calculus to the inner integral, we find that

$$\begin{aligned} f(t) = F'(t) &= \frac{1}{\sqrt{\pi}\, \Gamma(r/2)} \int_0^\infty \frac{e^{-(u/2)(t^2/r)}}{2^{(r+1)/2}} \sqrt{\frac{u}{r}}\, u^{r/2-1} e^{-u/2}\, du \\ &= \frac{1}{\sqrt{\pi r}\, \Gamma(r/2)} \int_0^\infty \frac{u^{(r+1)/2-1}}{2^{(r+1)/2}} e^{-(u/2)(1+t^2/r)}\, du. \end{aligned}$$

In the integral, make the change of variables

$$y = (1 + t^2/r)u, \qquad \text{so that} \qquad \frac{du}{dy} = \frac{1}{1 + t^2/r},$$

Thus,

$$f(t) = \frac{\Gamma((r+1)/2)}{\sqrt{\pi r}\, \Gamma(r/2)} \left[\frac{1}{(1 + t^2/r)^{(r+1)/2}} \right] \int_0^\infty \frac{y^{(r+1)/2-1}}{\Gamma((r+1)/2)\, 2^{(r+1)/2}} e^{-y/2}\, dy.$$

The integral in this last expression for $f(t)$ is equal to 1 because the integrand is like the p.d.f. of a chi-square distribution with $r + 1$ degrees of freedom. Hence, the p.d.f. is

$$f(t) = \frac{\Gamma((r+1)/2)}{\sqrt{\pi r}\, \Gamma(r/2)} \frac{1}{(1 + t^2/r)^{(r+1)/2}}, \qquad -\infty < t < \infty. \qquad \square$$

Graphs of the p.d.f. of T when $r = 1, 3$, and 7, along with the $N(0,1)$ p.d.f., are given in Figure 5.5-2(a) on page 252. In this figure, we see that the tails of the t distribution are heavier than those of a normal one; that is, there is more extreme probability in the t distribution than in the standardized normal one.

To find probabilities for a t random variable with r degrees of freedom, use either your calculator, a computer program, or Table VI in Appendix B. If T has a t distribution with r degrees of freedom, we say that the distribution of T is $t(r)$. Furthermore, right-tail probabilities of size α are denoted by $t_\alpha(r)$. [See Figure 5.5-2(b).]

FIGURE 5.5-2: t distribution p.d.f.'s and right-tail probability

---

EXAMPLE 5.5-4 Let the distribution of T be $t(11)$. Then

$$t_{0.05}(11) = 1.796 \qquad \text{and} \qquad -t_{0.05}(11) = -1.796.$$

Thus,

$$P(-1.796 \le T \le 1.796) = 0.90.$$

We can also find values of the distribution function such as

$$P(T \le 2.201) = 0.975 \qquad \text{and} \qquad P(T \le -1.363) = 0.10. \qquad \blacksquare$$

COMPUTATIONAL COMMENTS We illustrate the use of Minitab to find the solutions for Example 5.5-4. To find those probabilities, we use

$$\text{Calc} \rightarrow \text{Probability Distributions} \rightarrow \text{t}$$

and input the necessary information:

```
Inverse Cumulative Distribution Function
Student's t distribution with 11 DF

P(X<=x)          x
  0.95   1.79588

P(X<=x)          x
  0.05  -1.79588

Cumulative Distribution Function
Student's t distribution with 11 DF

      x    P(X<=x)
  2.201   0.975001

      x    P(X<=x)
 -1.363   0.100066
```

We can use the results of Corollary 5.5-1 and Theorems 5.5-2 and 5.5-3 to construct an important T random variable. Given a random sample $X_1, X_2, \ldots, X_n$ from a normal distribution, $N(\mu, \sigma^2)$, let

$$Z = \frac{\overline{X} - \mu}{\sigma/\sqrt{n}} \quad \text{and} \quad U = \frac{(n-1)S^2}{\sigma^2}.$$

Then the distribution of Z is $N(0,1)$ by Corollary 5.5-1. Theorem 5.5-2 tells us that the distribution of U is $\chi^2(n-1)$ and that Z and U are independent. Thus,

$$T = \frac{\dfrac{\overline{X} - \mu}{\sigma/\sqrt{n}}}{\sqrt{\dfrac{(n-1)S^2}{\sigma^2} \Big/ (n-1)}} = \frac{\overline{X} - \mu}{S/\sqrt{n}} \tag{5.5-2}$$

has a Student's t distribution (see subsequent remark) with $r = n - 1$ degrees of freedom by Theorem 5.5-3. We use this T in Section 6.2 to construct confidence intervals for an unknown mean μ of a normal distribution. (See also Exercise 5.5-20.)

REMARK   The foregoing distribution was first discovered by W. S. Gosset when he was working for an Irish brewery. Because Gosset published under the pseudonym Student, this distribution is sometimes known as Student's t distribution. Actually, Gosset was, in a sense, lucky to find the distribution of the T given in Equation 5.5-2. The distribution of $\overline{X}$ was well known; but to determine the distribution of U, he found the first four moments of U and guessed that the distribution was in the Pearson family and that led to the $\chi^2(n-1)$ distribution. While he proved that $\overline{X}$ and U were uncorrelated, he guessed that they were independent. He then went on to find the distribution of T of Equation 5.5-2. So two educated guesses were needed in this process. Later Fisher gave the real proof, but he also gave credit to Gosset for all of his insight into the problem. This discovery was a giant leap in statistical theory. ∎

EXERCISES

5.5-1. Let $X_1, X_2, \ldots, X_{16}$ be a random sample from a normal distribution $N(77, 25)$. Compute

(a) $P(77 < \overline{X} < 79.5)$.   (b) $P(74.2 < \overline{X} < 78.4)$.

5.5-2. Let X be $N(50, 36)$. Using the same set of axes, sketch the graphs of the probability density functions of

(a) X.

(b) $\overline{X}$, the mean of a random sample of size 9 from this distribution.

(c) $\overline{X}$, the mean of a random sample of size 36 from this distribution.

5.5-3. Let X equal the widest diameter (in millimeters) of the fetal head measured between the 16th and 25th weeks of pregnancy. Assume that the distribution of X is $N(46.58, 40.96)$. Let $\overline{X}$ be the sample mean of a random sample of $n = 16$ observations of X.

(a) Give the values of $E(\overline{X})$ and $\text{Var}(\overline{X})$.

(b) Find $P(44.42 \leq \overline{X} \leq 48.98)$.

5.5-4. Let X equal the weight of the soap in a "6-pound" box. Assume that the distribution of X is $N(6.05, 0.0004)$.

(a) Find $P(X < 6.0171)$.

(b) If nine boxes of soap are selected at random from the production line, find the probability that at most two boxes weigh less than 6.0171 pounds each. HINT: Let Y equal the number of boxes that weigh less than 6.0171 pounds.

(c) Let $\overline{X}$ be the sample mean of the nine boxes. Find $P(\overline{X} \leq 6.035)$.

5.5-5. If $X_1, X_2, \ldots, X_{16}$ is a random sample of size $n = 16$ from the normal distribution $N(50, 100)$, determine

(a) $P\left(796.2 \leq \sum_{i=1}^{16} (X_i - 50)^2 \leq 2630\right)$.

(b) $P\left(726.1 \leq \sum_{i=1}^{16} (X_i - \overline{X})^2 \leq 2500\right)$.

5.5-6. At a heat-treating company, iron castings and steel forgings are heat treated to achieve desired mechanical properties and machinability. One steel forging is annealed to soften the part for each machining. Two lots of this part, made of 1020 steel, are heat treated in two different furnaces. The specification for this part is 36–66 on the Rockwell G scale. Let X_1 and X_2 equal the respective hardness measurements for parts selected randomly from furnaces 1 and 2. Assume that the distributions of X_1 and X_2 are $N(47.88, 2.19)$ and $N(43.04, 14.89)$, respectively.

(a) Sketch the p.d.f.'s of X_1 and X_2 on the same graph.

(b) Compute $P(X_1 > X_2)$, assuming independence of X_1 and X_2.

5.5-7. Let X equal the weight (in grams) of a nail of the type that is used for making decks. Assume that the distribution of X is $N(8.78, 0.16)$. Let $\overline{X}$ be the mean of a random sample of the weights of $n = 9$ nails.

(a) Sketch, on the same set of axes, the graphs of the p.d.f.'s of X and of $\overline{X}$.

(b) Let S^2 be the sample variance of the nine weights. Find constants a and b so that $P(a \leq S^2 \leq b) = 0.90$.

HINT: $P(a \leq S^2 \leq b)$ is equivalent to $P(8a/0.16 \leq 8S^2/0.16 \leq 8b/0.16)$, and $8S^2/0.16$ is $\chi^2(8)$. Find $8a/0.16$ and $8b/0.16$ from Table IV in Appendix B.

5.5-8. Suppose that the distribution of the weight of a prepackaged "1-pound bag" of carrots is $N(1.18, 0.07^2)$ and the distribution of the weight of a prepackaged "3-pound bag" of carrots is $N(3.22, 0.09^2)$. Selecting bags at random, find the probability that the sum of three 1-pound bags exceeds the weight of one 3-pound bag.

HINT: First determine the distribution of Y, the sum of the three, and then compute $P(Y > W)$, where W is the weight of the 3-pound bag.

5.5-9. Let X equal the force required to pull a stud out of a window that is to be inserted into an automobile. Assume that the distribution of X is $N(147.8, 12.3^2)$.

(a) Find $P(X < 163.3)$.

(b) If $\overline{X}$ is the mean and S^2 is the variance of a random sample of size $n = 25$ from this distribution of X, determine $P(\overline{X} \leq 150.9)$.

(c) Find constants a and b so that $P(a \leq S^2 \leq b) = 0.90$. (See the hint in Exercise 5.5-7.)

5.5-10. Let X denote the wing length in millimeters of a male gallinule and Y the wing length in millimeters of a female gallinule. Assume that X is $N(184.09, 39.37)$ and Y is $N(171.93, 50.88)$ and that X and Y are independent. If a male and a female gallinule are captured, what is the probability that X is greater than Y?

5.5-11. Suppose that the length of life in hours (say, X) of a light bulb manufactured by company A is $N(800, 14400)$ and the length of life in hours (say, Y) of a light bulb manufactured by company B is $N(850, 2500)$. One bulb is selected from each company and is burned until "death."

(a) Find the probability that the length of life of the bulb from company A exceeds the length of life of the bulb from company B by at least 15 hours.

(b) Find the probability that at least one of the bulbs "lives" for at least 920 hours.

5.5-12. Let X and Y equal the number of miles per gallon achieved by compact cars and midsized cars, respectively, as reported in fuel economy ratings. Assume that $\mu_X = 24.5$, $\sigma_X = 3.8$, $\mu_Y = 21.3$, and $\sigma_Y = 2.7$. Let $\overline{X}$ and $\overline{Y}$ be the sample means of independent random samples of eight observations of X and Y, respectively.

(a) What are the values of the means and variances of $\overline{X}$ and $\overline{Y}$?

(b) Assuming that $\overline{X}$ and $\overline{Y}$ are each (approximately) normally distributed, how is $\overline{X} - \overline{Y}$ distributed?

(c) Find the (approximate) probability $P(\overline{X} > \overline{Y})$.

5.5-13. Let X equal the weight of a fat-free Fig Newton cookie. Assume that the distribution of X is $N(14.22, 0.0854)$. These cookies are sold in packages that have a label weight of 340 grams. The number of cookies in a package is usually 24, 25, or 26. Assuming that a package is filled with a random sample of cookies, how many cookies should be put into a package to be quite certain (say, with a probability of at least 0.95) that the total weight of the cookies exceeds 340 grams? (Keep in mind that extra cookies in a package decrease profit.)

5.5-14. A consumer buys n light bulbs, each of which has a lifetime that has a mean of 800 hours, a standard deviation of 100 hours, and a normal distribution. A light bulb is replaced by another as soon as it burns out. Assuming independence of the lifetimes, find the smallest n so that the succession of light bulbs produces light for at least 10,000 hours with a probability of 0.90.

5.5-15. A marketing research firm suggests to a company that two possible competing products can generate incomes X and Y (in millions) that are $N(3, 1)$ and $N(3.5, 4)$, respectively. Clearly, $P(X < Y) > 1/2$. However, the company would prefer the one with the smaller variance if in fact $P(X > 2) > P(Y > 2)$. Which product does the company select?

5.5-16. Let the independent random variables X_1 and X_2 be $N(0,1)$ and $\chi^2(r)$, respectively. Let $Y_1 = X_1/\sqrt{X_2/r}$ and $Y_2 = X_2$.

(a) Find the joint p.d.f. of Y_1 and Y_2.

(b) Determine the marginal p.d.f. of Y_1 and show that Y_1 has a t distribution. (This is another, equivalent, way of finding the p.d.f. of T.)

5.5-17. Let T have a t distribution with r degrees of freedom. Show that $E(T) = 0$, $r \geq 2$, and $\mathrm{Var}(T) = r/(r-2)$, provided that $r \geq 3$, by first finding $E(Z)$, $E(1/\sqrt{U})$, $E(Z^2)$, and $E(1/U)$.

5.5-18. Let the distribution of T be $t(23)$. Find

(a) $t_{0.05}(23)$.

(b) $t_{0.90}(23)$.

(c) $P(-2.069 \leq T \leq 2.500)$.

5.5-19. Let the distribution of T be $t(17)$. Find

(a) $t_{0.01}(17)$.

(b) $t_{0.95}(17)$.

(c) $P(-1.740 \leq T \leq 1.740)$.

5.5-20. Let $n = 9$ in the T statistic defined in Equation 5.5-2.

(a) Find $t_{0.025}$ so that $P(-t_{0.025} \leq T \leq t_{0.025}) = 0.95$.

(b) Solve the inequality $[-t_{0.025} \leq T \leq t_{0.025}]$ so that μ is in the middle.

5.5-21. Let Z_1, Z_2, and Z_3 have independent standard normal distributions, $N(0,1)$.

(a) Find the distribution of

$$W = \frac{Z_1}{\sqrt{(Z_2^2 + Z_3^2)/2}}.$$

(b) Show that

$$V = \frac{Z_1}{\sqrt{(Z_1^2 + Z_2^2)/2}}$$

has p.d.f. $f(v) = 1/(\pi\sqrt{2 - v^2})$, $-\sqrt{2} < v < \sqrt{2}$.

(c) Find the mean of V.

(d) Find the standard deviation of V.

(e) Why are the distributions of W and V so different?

5.6 THE CENTRAL LIMIT THEOREM

In Section 5.4, we found that the mean $\overline{X}$ of a random sample of size n from a distribution with mean μ and variance $\sigma^2 > 0$ is a random variable with the properties that

$$E(\overline{X}) = \mu \quad \text{and} \quad \mathrm{Var}(\overline{X}) = \frac{\sigma^2}{n}.$$

As n increases, the variance of $\overline{X}$ decreases. Consequently, the distribution of $\overline{X}$ clearly depends on n, and we see that we are dealing with sequences of distributions. In Theorem 5.5-1, we considered the p.d.f. of $\overline{X}$ when sampling is from the normal distribution $N(\mu, \sigma^2)$. We showed that the distribution of $\overline{X}$ is $N(\mu, \sigma^2/n)$, and in Figure 5.5-1, by graphing the p.d.f.'s for several values of n, we illustrated the property that as n increases, the probability becomes concentrated in a small interval centered at μ. That is, as n increases, $\overline{X}$ tends to converge to μ, or $(\overline{X} - \mu)$ tends to converge to 0 in a probability sense.

In general, if we let

$$W = \frac{\sqrt{n}}{\sigma}(\overline{X} - \mu) = \frac{\overline{X} - \mu}{\sigma/\sqrt{n}} = \frac{Y - n\mu}{\sqrt{n}\,\sigma},$$

where Y is the sum of a random sample of size n from some distribution with mean μ and variance σ^2, then, for each positive integer n,

$$E(W) = E\left[\frac{\overline{X} - \mu}{\sigma/\sqrt{n}}\right] = \frac{E(\overline{X}) - \mu}{\sigma/\sqrt{n}} = \frac{\mu - \mu}{\sigma/\sqrt{n}} = 0$$

and

$$\text{Var}(W) = E(W^2) = E\left[\frac{(\overline{X} - \mu)^2}{\sigma^2/n}\right] = \frac{E[(\overline{X} - \mu)^2]}{\sigma^2/n} = \frac{\sigma^2/n}{\sigma^2/n} = 1.$$

Thus, while $\overline{X} - \mu$ tends to "degenerate" to zero, the factor $\sqrt{n}/\sigma$ in $\sqrt{n}(\overline{X} - \mu)/\sigma$ "spreads out" the probability enough to prevent this degeneration. What, then, is the distribution of W as n increases? One observation that might shed some light on the answer to this question can be made immediately. If the sample arises from a normal distribution, then, from Theorem 5.5-1, we know that $\overline{X}$ is $N(\mu, \sigma^2/n)$, and hence W is $N(0,1)$ for each positive n. Thus, in the limit, the distribution of W must be $N(0,1)$. So if the solution of the question does not depend on the underlying distribution (i.e., it is unique), the answer must be $N(0,1)$. As we will see, that is exactly the case, and this result is so important that it is called the central limit theorem, the proof of which is given in Section 10.6.

THEOREM 5.6-1

(Central Limit Theorem) If $\overline{X}$ is the mean of a random sample $X_1, X_2, \ldots, X_n$ of size n from a distribution with a finite mean μ and a finite positive variance σ^2, then the distribution of

$$W = \frac{\overline{X} - \mu}{\sigma/\sqrt{n}} = \frac{\sum_{i=1}^{n} X_i - n\mu}{\sqrt{n}\,\sigma}$$

is $N(0,1)$ in the limit as $n \to \infty$.

When n is "sufficiently large," a practical use of the central limit theorem is approximating the distribution function of W, namely,

$$P(W \le w) \approx \int_{-\infty}^{w} \frac{1}{\sqrt{2\pi}} e^{-z^2/2}\, dz = \Phi(w).$$

We present some illustrations of this application, discuss the notion of "sufficiently large," and try to give an intuitive feeling for the central limit theorem.

EXAMPLE 5.6-1 Let $\overline{X}$ be the mean of a random sample of $n = 25$ currents (in milliamperes) in a strip of wire in which each measurement has a mean of 15 and a variance of 4. Then $\overline{X}$ has an approximate $N(15, 4/25)$ distribution. As an illustration,

$$P(14.4 < \overline{X} < 15.6) = P\left(\frac{14.5 - 15}{0.4} < \frac{\overline{X} - 15}{0.4} < \frac{15.6 - 15}{0.4}\right)$$

$$\approx \Phi(1.5) - \Phi(-1.5) = 0.9332 - 0.0668 = 0.8664. \quad \blacksquare$$

EXAMPLE 5.6-2 Let $X_1, X_2, \ldots, X_{20}$ denote a random sample of size 20 from the uniform distribution $U(0, 1)$. Here $E(X_i) = 1/2$ and $\text{Var}(X_i) = 1/12$, for $i = 1, 2, \ldots, 20$. If $Y = X_1 + X_2 + \cdots + X_{20}$, then

$$P(Y \le 9.1) = P\left(\frac{Y - 20(1/2)}{\sqrt{20/12}} \le \frac{9.1 - 10}{\sqrt{20/12}} \right) = P(W \le -0.697)$$

$$\approx \Phi(-0.697)$$

$$= 0.2423.$$

Also,

$$P(8.5 \le Y \le 11.7) = P\left(\frac{8.5 - 10}{\sqrt{5/3}} \le \frac{Y - 10}{\sqrt{5/3}} \le \frac{11.7 - 10}{\sqrt{5/3}} \right)$$

$$= P(-1.162 \le W \le 1.317)$$

$$\approx \Phi(1.317) - \Phi(-1.162)$$

$$= 0.9061 - 0.1226 = 0.7835.$$

∎

EXAMPLE 5.6-3 Let $\overline{X}$ denote the mean of a random sample of size 25 from the distribution whose p.d.f. is $f(x) = x^3/4, 0 < x < 2$. It is easy to show that $\mu = 8/5 = 1.6$ and $\sigma^2 = 8/75$. Thus,

$$P(1.5 \le \overline{X} \le 1.65) = P\left(\frac{1.5 - 1.6}{\sqrt{8/75}/\sqrt{25}} \le \frac{\overline{X} - 1.6}{\sqrt{8/75}/\sqrt{25}} \le \frac{1.65 - 1.6}{\sqrt{8/75}/\sqrt{25}} \right)$$

$$= P(-1.531 \le W \le 0.765)$$

$$\approx \Phi(0.765) - \Phi(-1.531)$$

$$= 0.7779 - 0.0629 = 0.7150.$$

∎

These examples show how the central limit theorem can be used for approximating certain probabilities concerning the mean $\overline{X}$ or the sum $Y = \sum_{i=1}^{n} X_i$ of a random sample. That is, $\overline{X}$ is approximately $N(\mu, \sigma^2/n)$, and Y is approximately $N(n\mu, n\sigma^2)$, when n is "sufficiently large," where μ and σ^2 are, respectively, the mean and the variance of the underlying distribution from which the sample arose. Generally, if n is greater than 25 or 30, these approximations will be good. However, if the underlying distribution is symmetric, unimodal, and of the continuous type, a value of n as small as 4 or 5 can yield an adequate approximation. Moreover, if the original distribution is approximately normal, $\overline{X}$ would have a distribution very close to normal when n equals 2 or 3. In fact, we know that if the sample is taken from $N(\mu, \sigma^2)$, $\overline{X}$ is exactly $N(\mu, \sigma^2/n)$ for every $n = 1, 2, 3, \ldots$.

The examples that follow will help to illustrate the previous remarks and will give the reader a better intuitive feeling about the central limit theorem. In particular, we shall see how the size of n affects the distribution of $\overline{X}$ and $Y = \sum_{i=1}^{n} X_i$ for samples from several underlying distributions.

EXAMPLE 5.6-4 Let X_1, X_2, X_3, X_4 be a random sample of size 4 from the uniform distribution $U(0, 1)$ with p.d.f. $f(x) = 1, 0 < x < 1$. Then $\mu = 1/2$ and $\sigma^2 = 1/12$. We shall compare the graph of the p.d.f. of

$$Y = \sum_{i=1}^{n} X_i$$

with the graph of the $N[n(1/2), n(1/12)]$ p.d.f. for $n = 2$ and 4, respectively.

By methods given in Section 5.2, we can determine that the p.d.f. of $Y = X_1 + X_2$ is

$$g(y) = \begin{cases} y, & 0 < y \le 1, \\ 2 - y, & 1 < y < 2. \end{cases}$$

This is the triangular p.d.f. that is graphed in Figure 5.6-1(a). In this figure, the $N[2(1/2), 2(1/12)]$ p.d.f. is also graphed.

Moreover, the p.d.f. of $Y = X_1 + X_2 + X_3 + X_4$ is

$$g(y) = \begin{cases} \dfrac{y^3}{6}, & 0 \le y < 1, \\[2mm] \dfrac{-3y^3 + 12y^2 - 12y + 4}{6}, & 1 \le y < 2, \\[2mm] \dfrac{3y^3 - 24y^2 + 60y - 44}{6}, & 2 \le y < 3, \\[2mm] \dfrac{-y^3 + 12y^2 - 48y + 64}{6}, & 3 \le y \le 4. \end{cases}$$

This p.d.f. is graphed in Figure 5.6-1(b) along with the $N[4(1/2), 4(1/12)]$ p.d.f. If we are interested in finding $P(1.7 \le Y \le 3.2)$, we can do so by evaluating

$$\int_{1.7}^{3.2} g(y)\, dy,$$

which is tedious. (See Exercise 5.6-10.) It is much easier to use a normal approximation, which results in a number close to the exact value. ∎

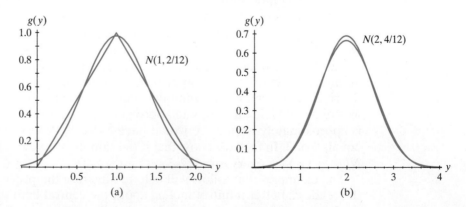

FIGURE 5.6-1: p.d.f.'s of sums of uniform random variables

In Example 5.6-4 and Exercise 5.6-10, we show that even for a small value of n, such as $n = 4$, the sum of the sample items has an approximate normal distribution. The next example illustrates that, for some underlying distributions (particularly skewed ones), n must be quite large in order for us to obtain a satisfactory approximation. To keep the scale on the horizontal axis the same for each value of n, we will use the following result: Let $f(x)$ and $F(x)$ respectively be the p.d.f. and distribution function of a random variable, X, of the continuous type having mean μ and variance σ^2. Let $W = (X - \mu)/\sigma$. The distribution function of W is given by

$$G(w) = P(W \le w) = P\left(\frac{X - \mu}{\sigma} \le w\right)$$
$$= P(X \le \sigma w + \mu) = F(\sigma w + \mu).$$

Thus, the p.d.f. of W is given by

$$g(w) = F'(\sigma w + \mu) = \sigma f(\sigma w + \mu).$$

EXAMPLE 5.6-5   Let $X_1, X_2, \ldots, X_n$ be a random sample of size n from a chi-square distribution with one degree of freedom. If

$$Y = \sum_{i=1}^{n} X_i,$$

then Y is $\chi^2(n)$, and it follows that $E(Y) = n$ and $\mathrm{Var}(Y) = 2n$. Let

$$W = \frac{Y - n}{\sqrt{2n}}.$$

The p.d.f. of W is given by

$$g(w) = \sqrt{2n}\,\frac{(\sqrt{2n}\,w + n)^{n/2-1}}{\Gamma\left(\dfrac{n}{2}\right)2^{n/2}}\,e^{-(\sqrt{2n}\,w+n)/2}, \qquad -n/\sqrt{2n} < w < \infty.$$

Note that $w > n/\sqrt{2n}$ corresponds to $y > 0$. In Figures 5.6-2, the graph of the p.d.f. of W is given along with the $N(0,1)$ p.d.f. for $n = 20$ and $n = 100$.   ∎

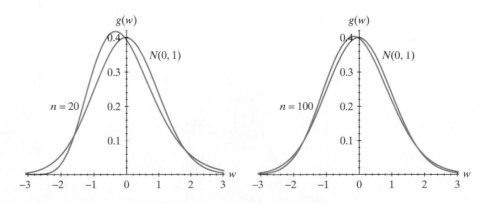

FIGURE 5.6-2: p.d.f.'s of sums of chi-square random variables

In order to gain an intuitive idea about how the sample size n affects the distribution of $W = (\overline{X} - \mu)/(\sigma/\sqrt{n})$, it is helpful to simulate values of W on a computer using different values of n and different underlying distributions. The next example illustrates this simulation.

REMARK   Recall that we simulate observations from a distribution of X having a continuous-type distribution function $F(x)$ as follows: Suppose $F(a) = 0$, $F(b) = 1$, and $F(x)$ is strictly increasing for $a < x < b$. Let $Y = F(x)$ and let the distribution of Y be $U(0,1)$. If y is an observed value of Y, then $x = F^{-1}(y)$ is an observed value of X. (See Section 5.1.) Thus, if y is the value of a computer-generated random number, then $x = F^{-1}(y)$ is the simulated value of X. ∎

EXAMPLE 5.6-6   It is often difficult to find the exact distribution of the random variable $W = (\overline{X} - \mu)/(\sigma/\sqrt{n})$, unless you use a computer algebra system such as *Maple*. In this example, we give some empirical evidence about the distribution of W by simulating random samples on the computer. We also superimpose the theoretical p.d.f. of W, which we found by using *Maple*. Let $X_1, X_2, \ldots, X_n$ denote a random sample of size n from the distribution with p.d.f. $f(x)$, distribution function $F(x)$, mean μ, and variance σ^2. We simulated 1000 random samples of size $n = 2$ and $n = 7$ from each of two distributions. We then computed the value of W for each sample, thus obtaining 1000 observed values of W. Next, we constructed a histogram of these 1000 values by using 21 intervals of equal length. A relative frequency histogram of the observations of W, the p.d.f. for the standard normal distribution, and the theoretical p.d.f. of W are given in Figures 5.6-3 and 5.6-4.

(a) In Figure 5.6-3, $f(x) = (x + 1)/2$ and $F(x) = (x + 1)^2/4$ for $-1 < x < 1$; $\mu = 1/3$, $\sigma^2 = 2/9$; and $n = 2$ and 7. This underlying distribution is skewed to the left.

(b) In Figure 5.6-4, $f(x) = (3/2)x^2$ and $F(x) = (x^3 + 1)/2$ for $-1 < x < 1$; $\mu = 0$, $\sigma^2 = 3/5$; and $n = 2$ and 7. (Sketch the graph of $y = f(x)$. Give an argument as to why the histogram for $n = 2$ looks the way it does.) This underlying distribution is U shaped; thus, W does not follow a normal distribution with small n. ∎

Note that these examples have not *proved* anything. They are presented to give *evidence* of the truth of the central limit theorem, and they do give a nice feeling for what is happening.

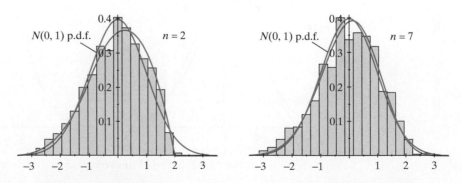

FIGURE 5.6-3: p.d.f.'s of $(\overline{X} - \mu)/(\sigma/\sqrt{n})$, underlying distribution triangular

FIGURE 5.6-4: p.d.f.'s of $(\overline{X} - \mu)/(\sigma/\sqrt{n}\,)$, underlying distribution U shaped

So far, all the illustrations have concerned distributions of the continuous type. However, the hypotheses for the central limit theorem do not require the distribution to be continuous. We shall consider applications of the central limit theorem for discrete-type distributions in the next section.

EXERCISES

5.6-1. Let $\overline{X}$ be the mean of a random sample of size 12 from the uniform distribution on the interval $(0, 1)$. Approximate $P(1/2 \le \overline{X} \le 2/3)$.

5.6-2. Let $Y = X_1 + X_2 + \cdots + X_{15}$ be the sum of a random sample of size 15 from the distribution whose p.d.f. is $f(x) = (3/2)x^2$, $-1 < x < 1$. Using the p.d.f. of Y, we find that $P(-0.3 \le Y \le 1.5) = 0.22788$. Use the central limit theorem to approximate this probability.

5.6-3. Let $\overline{X}$ be the mean of a random sample of size 36 from an exponential distribution with mean 3. Approximate $P(2.5 \le \overline{X} \le 4)$.

5.6-4. Approximate $P(39.75 \le \overline{X} \le 41.25)$, where $\overline{X}$ is the mean of a random sample of size 32 from a distribution with mean $\mu = 40$ and variance $\sigma^2 = 8$.

5.6-5. Let $X_1, X_2, \ldots, X_{18}$ be a random sample of size 18 from a chi-square distribution with $r = 1$. Recall that $\mu = 1$ and $\sigma^2 = 2$.

(a) How is $Y = \sum_{i=1}^{18} X_i$ distributed?

(b) Using the result of part (a), we see from Table IV in the appendix that

$$P(Y \le 9.390) = 0.05$$

and

$$P(Y \le 34.80) = 0.99.$$

Compare these two probabilities with the approximations found with the use of the central limit theorem.

5.6-6. A random sample of size $n = 18$ is taken from the distribution with p.d.f. $f(x) = 1 - x/2$, $0 \le x \le 2$.

(a) Find μ and σ^2. (b) Find, approximately, $P(2/3 \le \overline{X} \le 5/6)$

5.6-7. Let X equal the maximal oxygen intake of a human on a treadmill, where the measurements are in milliliters of oxygen per minute per kilogram of weight. Assume that, for a particular population, the mean of X is $\mu = 54.030$ and the standard deviation is $\sigma = 5.8$. Let $\overline{X}$ be the sample mean of a random sample of size $n = 47$. Find $P(52.761 \le \overline{X} \le 54.453)$, approximately.

5.6-8. Let X equal the weight in grams of a miniature candy bar. Assume that $\mu = E(X) = 24.43$ and $\sigma^2 = \mathrm{Var}(X) = 2.20$. Let $\overline{X}$ be the sample mean of a random sample of $n = 30$ candy bars. Find

(a) $E(\overline{X})$. (b) $\mathrm{Var}(\overline{X})$. (c) $P(24.17 \le \overline{X} \le 24.82)$, approximately.

5.6-9. Let X equal the birth weight in grams of a baby born in the Sudan. Assume that $E(X) = 3320$ and $\mathrm{Var}(X) = 660^2$. Let $\overline{X}$ be the sample mean of a random sample of size $n = 225$. Find $P(3233.76 \le \overline{X} \le 3406.24)$, approximately.

5.6-10. In Example 5.6-4, with $n = 4$, compute $P(1.7 \le Y \le 3.2)$ and compare your answer with the normal approximation of this probability.

5.6-11. Five measurements in "ohms per square unit of area" of the electronic conductive coating (which allows light to pass through it) on a thin, clear piece of glass were made, and the average was calculated. This procedure was repeated 150 times, yielding the following averages:

83.86	75.86	79.65	90.57	95.37	97.97	77.00	80.80	83.53	83.17
80.20	81.42	89.14	88.68	85.15	90.11	89.03	85.00	82.57	79.46
81.20	82.80	81.28	74.64	69.85	75.60	76.60	78.30	88.51	86.32
84.03	95.27	90.05	77.50	75.14	81.33	86.12	78.30	80.30	84.96
79.30	88.96	82.76	83.13	79.60	86.20	85.16	87.86	91.00	91.10
84.04	92.10	79.20	83.76	87.87	86.71	81.89	85.72	75.84	74.27
93.08	81.75	75.66	75.35	76.55	84.86	90.68	91.02	90.97	98.30
91.84	97.41	73.60	90.65	80.20	74.75	90.35	79.66	86.88	83.00
86.24	80.50	74.25	91.20	70.16	78.40	85.60	80.82	75.95	80.75
81.86	82.18	82.98	84.00	76.85	85.00	79.50	86.56	83.30	72.40
79.20	86.20	82.36	84.08	86.11	88.25	88.93	93.12	78.30	77.24
82.52	81.37	83.72	86.90	84.37	92.60	95.01	78.95	81.40	88.40
76.10	85.33	82.95	80.20	88.21	83.49	81.00	82.72	81.12	83.62
91.18	85.90	79.01	77.56	81.13	80.60	81.65	70.70	69.36	79.09
71.35	67.20	67.43	69.95	66.76	76.35	69.45	80.13	84.26	88.13

(a) Using 10 intervals of equal length, and using (66.495, 69.695) for the first class interval, construct a frequency table for these 150 observations. Which is the modal class?

(b) Construct a relative frequency histogram for the grouped data.

(c) Using the sample mean 82.661 and the sample variance 42.2279 of these 150 values as the distribution mean and variance, superimpose a normal p.d.f. on the histogram you constructed in part (b). Do these 150 means of five measurements seem to be normally distributed?

5.6-12. A church has pledges (in dollars) with a mean of 2000 and a standard deviation of 500. A random sample of size $n = 25$ is taken from all of this church's pledges, and the sample mean $\overline{X}$ is considered. Approximate $P(\overline{X} > 2050)$.

5.6-13. A company has a one-year group life policy that divides its employees into two classes as follows:

Class	Probability of Death	Benefit	Number in Class
A	0.01	$20,000	1000
B	0.03	$10,000	500

The insurance company wants to collect a premium that equals the 90th percentile of the distribution of the total claims. What should that premium be?

5.6-14. Let X and Y equal the respective numbers of hours a randomly selected child watches movies or cartoons on TV during a certain month. From experience, it is known that $E(X) = 30$, $E(Y) = 50$, $\text{Var}(X) = 52$, $\text{Var}(Y) = 64$, and $\text{Cov}(X, Y) = 14$. Twenty-five children are selected at random. Let Z equal the total number of hours these 25 children watch TV movies or cartoons in the next month. Approximate $P(1970 < Z < 2090)$. HINT: Use the remark after Theorem 5.3-3.

5.6-15. The tensile strength X of paper, in pounds per square inch, has $\mu = 30$ and $\sigma = 3$. A random sample of size $n = 100$ is taken from the distribution of tensile strengths. Compute the probability that the sample mean $\overline{X}$ is greater than 29.5 pounds per square inch.

5.6-16. At certain times during the year, a bus company runs a special van holding 10 passengers from Iowa City to Chicago. After the opening of sales of the tickets, the time (in minutes) between sales of tickets for the trip has a gamma distribution with $\alpha = 3$ and $\theta = 2$.

(a) Assuming independence, record an integral that gives the probability of being sold out within one hour.

(b) Approximate the answer in part (a).

5.6-17. Let X_1, X_2, X_3, X_4 represent the random times in days needed to complete four steps of a project. These times are independent and have gamma distributions with common $\theta = 2$ and $\alpha_1 = 3, \alpha_2 = 2, \alpha_3 = 5, \alpha_4 = 3$, respectively. One step must be completed before the next can be started. Let Y equal the total time needed to complete the project.

(a) Find an integral that represents $P(Y \le 25)$.

(b) Using a normal distribution, approximate the answer to part (a). Is this approach justified?

5.6-18. Suppose that the sick leave taken by the typical worker per year has $\mu = 10$, $\sigma = 2$, measured in days. A firm has $n = 20$ employees. Assuming independence, how many sick days should the firm budget if the financial officer wants the probability of exceeding the number of days budgeted to be less than 20%?

5.7   APPROXIMATIONS FOR DISCRETE DISTRIBUTIONS

In this section, we illustrate how the normal distribution can be used to approximate probabilities for certain discrete-type distributions. One of the more important discrete distributions is the binomial distribution. To see how the central limit theorem can be applied, recall that a binomial random variable can be described as the sum of Bernoulli random variables. That is, let $X_1, X_2, \ldots, X_n$ be a random sample from a Bernoulli distribution with mean $\mu = p$ and variance $\sigma^2 = p(1 - p)$, where $0 < p < 1$. Then $Y = \sum_{i=1}^{n} X_i$ is $b(n, p)$. The central limit theorem states that the distribution of

$$W = \frac{Y - np}{\sqrt{np(1 - p)}} = \frac{\overline{X} - p}{\sqrt{p(1 - p)/n}}$$

is $N(0, 1)$ in the limit as $n \to \infty$. Thus, if n is "sufficiently large," the distribution of Y is approximately $N[np, np(1 - p)]$, and probabilities for the binomial distribution $b(n, p)$ can be approximated with this normal distribution. A rule often stated is that n is sufficiently large if $np \geq 5$ and $n(1 - p) \geq 5$. This rule can be used as a rough guide, although as p deviates more and more from 0.5, we need larger and larger sample sizes because the underlying Bernoulli distribution becomes more skewed.

Note that we shall be approximating probabilities for a discrete distribution with probabilities for a continuous distribution. Let us discuss a reasonable procedure in this situation. If V is $N(\mu, \sigma^2)$, then $P(a < V < b)$ is equivalent to the area bounded by the p.d.f. of V, the v-axis, $v = a$, and $v = b$. Now recall that, for a Y that is $b(n, p)$, the probability histogram for Y was defined as follows: For each y such that $k - 1/2 < y = k < k + 1/2$, let

$$f(k) = \frac{n!}{k!(n - k)!} p^k (1 - p)^{n-k}, \qquad k = 0, 1, 2, \ldots, n.$$

Then $P(Y = k)$ can be represented by the area of the rectangle with a height of $P(Y = k)$ and a base of length 1 centered at k. Figure 5.7-1 on the next page shows the graph of the probability histogram for the binomial distribution $b(4, 1/4)$. In using the normal distribution to approximate probabilities for the binomial distribution, areas under the p.d.f. for the normal distribution will be used to approximate areas of rectangles in the probability histogram for the binomial distribution.

EXAMPLE 5.7-1   Let the distribution of Y be $b(10, 1/2)$. Then, by the central limit theorem, $P(a < Y < b)$ can be approximated with the use of the normal distribution with mean $10(1/2) = 5$ and variance $10(1/2)(1/2) = 5/2$. Figure 5.7-2(a) on the next page shows the graph of the probability histogram for $b(10, 1/2)$ and the graph of the p.d.f. of the normal distribution $N(5, 5/2)$. Note that the area of the rectangle whose base is

$$\left(k - \frac{1}{2}, k + \frac{1}{2} \right)$$

and the area under the normal curve between $k - 1/2$ and $k + 1/2$ are approximately equal for each integer k.   ∎

FIGURE 5.7-1: Probability histogram for $b(4, 1/4)$

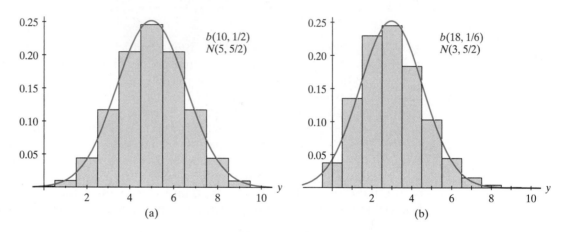

FIGURE 5.7-2: Normal approximation for the binomial distribution

EXAMPLE 5.7-2 Let Y be $b(18, 1/6)$. Because $np = 18(1/6) = 3 < 5$, the normal approximation is not as good here. Figure 5.7-2(b) illustrates this by depicting the skewed probability histogram for $b(18, 1/6)$ and the symmetric p.d.f. of the normal distribution $N(3, 5/2)$. ∎

EXAMPLE 5.7-3 Let Y have the binomial distribution of Example 5.7-1 and Figure 5.7-2(a), namely, $b(10, 1/2)$. Then

$$P(3 \leq Y < 6) = P(2.5 \leq Y \leq 5.5),$$

because $P(Y = 6)$ is not in the desired answer. But the latter equals

$$P\left(\frac{2.5 - 5}{\sqrt{10/4}} \leq \frac{Y - 5}{\sqrt{10/4}} \leq \frac{5.5 - 5}{\sqrt{10/4}} \right) \approx \Phi(0.316) - \Phi(-1.581)$$

$$= 0.6240 - 0.0570 = 0.5670.$$

Using Table II in the appendix, we find that $P(3 \leq Y < 6) = 0.5683$. ∎

EXAMPLE 5.7-4   Let Y be $b(36, 1/2)$. Then, since

$$\mu = (36)(1/2) = 18 \quad \text{and} \quad \sigma^2 = (36)(1/2)(1/2) = 9,$$

it follows that

$$P(12 < Y \leq 18) = P(12.5 \leq Y \leq 18.5)$$

$$= P\left(\frac{12.5 - 18}{\sqrt{9}} \leq \frac{Y - 18}{\sqrt{9}} \leq \frac{18.5 - 18}{\sqrt{9}}\right)$$

$$\approx \Phi(0.167) - \Phi(-1.833)$$

$$= 0.5329.$$

Note that 12 was increased to 12.5 because $P(Y = 12)$ is not included in the desired probability. Using the binomial formula, we find that

$$P(12 < Y \leq 18) = P(13 \leq Y \leq 18) = 0.5334.$$

(You may verify this answer with your calculator or Minitab.) Also,

$$P(Y = 20) = P(19.5 \leq Y \leq 20.5)$$

$$= P\left(\frac{19.5 - 18}{\sqrt{9}} \leq \frac{Y - 18}{\sqrt{9}} \leq \frac{20.5 - 18}{\sqrt{9}}\right)$$

$$\approx \Phi(0.833) - \Phi(0.5)$$

$$= 0.1060.$$

Using the binomial formula, we have $P(Y = 20) = 0.1063$. So, in this situation, the approximations are extremely good. ∎

Note that, in general, if Y is $b(n, p)$, n is large, and $k = 0, 1, \ldots, n$, then

$$P(Y \leq k) \approx \Phi\left(\frac{k + 1/2 - np}{\sqrt{npq}}\right)$$

and

$$P(Y < k) \approx \Phi\left(\frac{k - 1/2 - np}{\sqrt{npq}}\right),$$

because in the first case k is included and in the second it is not.

We now show how the Poisson distribution with large enough mean can be approximated with the use of a normal distribution.

EXAMPLE 5.7-5   A random variable having a Poisson distribution with mean 20 can be thought of as the sum Y of the observations of a random sample of size 20 from a Poisson distribution with mean 1. Thus,

$$W = \frac{Y - 20}{\sqrt{20}}$$

FIGURE 5.7-3: Normal approximation of Poisson, $\lambda = 20$

has a distribution that is approximately $N(0, 1)$, and the distribution of Y is approximately $N(20, 20)$. (See Figure 5.7-3.) For example,

$$P(16 < Y \le 21) = P(16.5 \le Y \le 21.5)$$

$$= P\left(\frac{16.5 - 20}{\sqrt{20}} \le \frac{Y - 20}{\sqrt{20}} \le \frac{21.5 - 20}{\sqrt{20}}\right)$$

$$\approx \Phi(0.335) - \Phi(-0.783)$$

$$= 0.4142.$$

Note that 16 is increased to 16.5 because $Y = 16$ is not included in the event $\{16 < Y \le 21\}$. The answer obtained with the Poisson formula is 0.4226. ∎

In general, if Y has a Poisson distribution with mean λ, then the distribution of

$$W = \frac{Y - \lambda}{\sqrt{\lambda}}$$

is approximately $N(0, 1)$ when λ is sufficiently large.

REMARK   If you have a statistical calculator or a statistics computer package, use it to compute discrete probabilities. However, it is important to learn how to apply the central limit theorem. ∎

EXERCISES

5.7-1. Let the distribution of Y be $b(25, 1/2)$. Find the given probabilities in two ways: exactly, using Table II in the appendix; and approximately, using the central limit theorem. Compare the two results in each of the three cases.

(a) $P(10 < Y \le 12)$. (b) $P(12 \le Y < 15)$.
(c) $P(Y = 12)$.

5.7-2. Suppose that among gifted seventh-graders who score very high on a mathematics exam, approximately 20% are left handed or ambidextrous. Let X equal the number of left-handed or ambidextrous students among a random sample of $n = 25$ gifted seventh-graders. Find $P(2 < X < 9)$

(a) Using Table II in the appendix.

(b) Approximately, using the central limit theorem.

Remark Since X has a skewed distribution, the approximation is not as good as that for the symmetrical distribution where $p = 0.50$, even though $np = 5$. ∎

5.7-3. A public opinion poll in Southern California was conducted to determine whether southern Californians are prepared for the "big earthquake" that experts predict will devastate the region sometime in the next 50 years. It was learned that "60% have not secured objects in their homes that might fall and cause injury and damage during a temblor." In a random sample of $n = 864$ southern Californians, let X equal the number who "have not secured objects in their homes." Find $P(496 \leq X \leq 548)$, approximately.

5.7-4. Let X equal the number out of $n = 48$ mature aster seeds that will germinate when $p = 0.75$ is the probability that a particular seed germinates. Approximate $P(35 \leq X \leq 40)$.

5.7-5. Let p equal the proportion of all college students who would say yes to the question, "Would you drink from the same glass as your friend if you suspected that this friend were an AIDS virus carrier?" Assume that $p = 0.10$. Let X equal the number of students out of a random sample of size $n = 100$ who would say yes to the question. Approximate $P(X \leq 11)$.

Remark The value used for p was based on a poll conducted in a class at San Diego State University. It would be interesting for you to conduct a survey at your college to estimate the value of p. ∎

5.7-6. In adults, the pneumococcus bacterium causes 70% of pneumonia cases. In a random sample of $n = 84$ adults who have pneumonia, let X equal the number whose pneumonia was caused by the pneumococcus bacterium. Use the normal distribution to find $P(X \leq 52)$, approximately.

5.7-7. Let $X_1, X_2, \ldots, X_{48}$ be a random sample of size 48 from the distribution with p.d.f. $f(x) = 1/x^2$, $1 < x < \infty$. Approximate the probability that at most 10 of these random variables have values greater than 4. HINT: Let the ith trial be a success if $X_i > 4$, $i = 1, 2, \ldots, 48$, and let Y equal the number of successes.

5.7-8. A candy maker produces mints that have a label weight of 20.4 grams. Assume that the distribution of the weights of these mints is $N(21.37, 0.16)$.

(a) Let X denote the weight of a single mint selected at random from the production line. Find $P(X < 20.857)$.

(b) During a particular shift, 100 mints are selected at random and weighed. Let Y equal the number of these mints that weigh less than 20.857 grams. Approximate $P(Y \leq 5)$.

(c) Let $\overline{X}$ equal the sample mean of the 100 mints selected and weighed on a particular shift. Find $P(21.31 \leq \overline{X} \leq 21.39)$.

5.7-9. Let X equal the number of alpha particles emitted by barium-133 per second and counted by a Geiger counter. Assume that X has a Poisson distribution with $\lambda = 49$. Approximate $P(45 < X < 60)$.

5.7-10. Let X equal the number of alpha particles counted by a Geiger counter during 30 seconds. Assume that the distribution of X is Poisson with a mean of 4829. Determine (approximately) $P(4776 \leq X \leq 4857)$.

5.7-11. Let $X_1, X_2, \ldots, X_{30}$ be a random sample of size 30 from a Poisson distribution with a mean of 2/3. Approximate

(a) $P\left(15 < \sum_{i=1}^{30} X_i \leq 22\right)$.

(b) $P\left(21 \leq \sum_{i=1}^{30} X_i < 27\right)$.

5.7-12. In the casino game roulette, the probability of winning with a bet on red is $p = 18/38$. Let Y equal the number of winning bets out of 1000 independent bets that are placed. Find $P(Y > 500)$, approximately.

5.7-13. About 60% of all Americans have a sedentary lifestyle. Select $n = 96$ Americans at random. (Assume independence.) What is the probability that between 50 and 60, inclusive, do not exercise regularly?

5.7-14. If X is $b(100, 0.1)$, find the approximate value of $P(12 \leq X \leq 14)$, using

(a) The normal approximation.

(b) The Poisson approximation.

(c) The binomial.

5.7-15. Let $X_1, X_2, \ldots, X_{36}$ be a random sample of size 36 from the geometric distribution with p.m.f. $f(x) = (1/4)^{x-1}(3/4)$, $x = 1, 2, 3, \ldots$. Approximate

(a) $P\left(46 \leq \sum_{i=1}^{36} X_i \leq 49\right)$.

(b) $P(1.25 \leq \overline{X} \leq 1.50)$.

HINT: Observe that the distribution of the sum is of the discrete type.

5.7-16. A die is rolled 24 independent times. Let Y be the sum of the 24 resulting values. Recalling that Y is a random variable of the discrete type, approximate

(a) $P(Y \geq 86)$. (b) $P(Y < 86)$.
(c) $P(70 < Y \leq 86)$.

5.7-17. In the United States, the probability that a child dies in his or her first year of life is about $p = 0.01$. (It is actually slightly less than this.) Consider a group of 5000 such infants. What is the probability that between 45 and 53, inclusive, die in the first year of life?

5.7-18. Let Y equal the sum of $n = 100$ Bernoulli trials. That is, Y is $b(100, p)$. For each of (i) $p = 0.1$, (ii) $p = 0.5$, and (iii) $p = 0.8$,

(a) Draw the approximating normal p.d.f.'s, all on the same graph.
(b) Find $P(|Y/100 - p| \leq 0.015)$, approximately.

5.7-19. The number of trees in one acre has a Poisson distribution with mean 60. Assuming independence, compute $P(5950 \leq X \leq 6100)$, approximately, where X is the number of trees in 100 acres.

5.7-20. Assume that the background noise X of a digital signal has a normal distribution with $\mu = 0$ volts and $\sigma = 0.5$ volt. If we observe $n = 100$ independent measurements of this noise, what is the probability that at least 7 of them exceed 0.98 in absolute value?

(a) Use the Poisson distribution to approximate this probability.
(b) Use the normal distribution to approximate this probability.

(c) Use the binomial distribution to find the exact probability.

5.7-21. The number X of flaws on a certain tape of length 1 yard follows a Poisson distribution with mean 0.3. We examine $n = 100$ such tapes and count the total number Y of flaws.

(a) Assuming independence, what is the distribution of Y?
(b) Approximate $P(Y \leq 25)$.

5.7-22. One of the lottery games that New York State offers is called Quick Draw. In the 4-spot game, the player selects 4 numbers from 1 to 80, inclusive. Every 4 minutes, the Lottery's computer randomly selects 20 numbers from 1 to 80, inclusive. The prize is determined by the number of matches between the player's 4 numbers and the 20 that the computer selected. For a \$1 bet, the prizes are \$1 for two matches, \$5 for three matches, and \$55 for four matches. But the \$1 bet is not returned to the player. During a November promotion in 1997, the State doubled the prizes every Wednesday afternoon.

(a) Calculate the expected value of the regular game and of the double-payoff game.
(b) Calculate the standard deviation for the regular game and for the double-payoff game.
(c) Suppose that you are willing to "invest" \$2000 in 2000 independent bets. Let Y equal the sum of your "winnings." Estimate $P(Y > 0)$ for the regular game and for the promotion game.
(d) If possible, simulate your answers to this question.

HISTORICAL COMMENTS   In this chapter, we have discussed the t and F distributions, among many other important ones. However, we should make a few comments about both of them. We mentioned that W. S. Gosset published his work on the t-distribution under the pseudonym "A Student" because Guinness did not want other breweries to know that they were using statistical methods. We have also heard the story that Gosset did not want Guinness to know that he was spending all his extra time on statistics; so he used "A Student," and it has become Student's t ever since. Whichever account is true, Gosset, in a sense, was lucky to discover the t-distribution, because, as mentioned in an earlier remark, he made two educated guesses, one of which involved the Pearson family of distributions. Incidentally, Karl Pearson, another famous statistician, had proposed this family a few years earlier. A few years later, the great statistician Sir Ronald A. Fisher (possibly the greatest statistician) actually proved that T had the t-distribution that Gosset had discovered. Concerning the F distribution, Fisher had worked with a function of what is now called F. It was George Snedecor of Iowa State University who put it in its present form and called it F (probably to honor Fisher), and Snedecor's was a much more useful form, as we will see later. We should also note that Fisher had been knighted, but so were two other statisticians: Sir Maurice G. Kendall and Sir David R. Cox. (The latter is still alive.) Their knighthood at least proves that the monarch of England appreciated some statistical efforts.

Another important person in the history of probability and statistics is Abraham de Moivre, who was born in France in 1667, but, as a Protestant, he did not fare very well in that Catholic country. As a matter of fact, he was imprisoned for about two years for his beliefs. After his release, he went to England, but led a gloomy life there, as he could not find an academic position. So de Moivre supported himself by tutoring or consulting with gamblers or insurance brokers. After publishing *The Doctrine of Chance*, he turned to a project that Nicolaus Bernoulli had suggested to him. Using the fact that Y is $b(n, p)$, he discovered that the relative frequency of successes, namely, Y/n, which Jacob Bernoulli had proved converged in probability to p, had an interesting approximating distribution itself. De Moivre had discovered the well-known bell-shaped curve called the normal distribution, and a special case of the central limit theorem. Although de Moivre did not have computers in his day, we show that if X is $b(100, 1/2)$, then $X/100$ has the p.m.f. displayed in Figure 5.7-4.

This distribution allowed de Moivre to determine a measure of spread, which we now call a standard deviation. Also, he could determine the approximate probability of Y/n falling into given intervals containing p. De Moivre was truly impressed with the orderliness of these random relative frequencies, which he attributed to the plan of the Almighty. Despite his great works, Abraham de Moivre died as a bitter and antisocial man, blind and in poverty, at the age of 87.

Two additional persons whom we would like to mention in the history of probability are Carl Friedrich Gauss and Marquis Pierre Simon de Laplace. Gauss was 29 years junior to Laplace, and Gauss was so secretive about his work that it is difficult to tell who discovered the central limit theorem first. The theorem was a generalization of de Moivre's result. In de Moivre's case, he was sampling from a Bernoulli distribution where $X_i = 1$ or 0 on the ith trial. Then

$$\frac{Y}{n} = \sum_{i=1}^{n} \frac{X_i}{n} = \overline{X},$$

the relative frequency of success, has that approximate normal distribution of de Moivre's. Laplace and Gauss were sampling from any distribution, provided that the second moment existed, and they found that the sample mean $\overline{X}$ had an approximate normal distribution. Seemingly, the central limit theorem was published in 1809 by

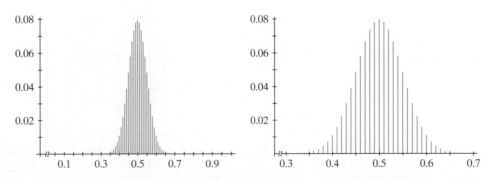

FIGURE 5.7-4: Bar graphs for the distribution of $X/100$

Laplace, just before Gauss's *Theoria Motus* in 1810. For some reason, the normal distribution is often referred to as the Gaussian distribution; people seem to forget about Laplace's contribution and (worse than that) de Moivre's original work 83 years earlier. Since then, there have been many more generalizations of the central limit theorem; in particular, most estimators of parameters in the regular case have approximate normal distributions. (See Section 10.7.)

There are many good histories of probability and statistics. However, the two that we find particularly interesting are Peter L. Bernstein's *Against the Gods: The Remarkable Story of Risk* (New York: John Wiley & Sons, Inc., 1996) and Stephen M. Stigler's *The History of Statistics: The Measurement of Uncertainty Before 1900* (Cambridge, MA: Harvard University Press, 1986).

STATISTICAL COMMENTS **(Simulation: Central Limit Theorem)** As we think about what de Moivre, Laplace, and Gauss discovered in their times, we see that it was truly amazing. Of course, de Moivre could compute the probabilities associated with various binomial distributions and see how they "piled up" in that bell shape, and he came up with the normal formula. Now, Laplace and Gauss had an even tougher task, as they could not easily find the probabilities associated with the sample mean $\overline{X}$, even with simple underlying distributions. As an illustration, suppose the random sample $X_1, X_2, \ldots, X_n$ arises from a uniform distribution on the space $[0,1)$. It is extremely difficult to compute probabilities about $\overline{X}$ unless n is very small, such as $n = 2$ or $n = 3$. Today, of course, we can use a CAS to simulate the distribution of $\overline{X}$ for any sample size and get fairly accurate estimates of the probabilities associated with $\overline{X}$. We did this 10,000 times for $n = 6$, which resulted in Figure 5.7-5. In this chapter, we learned that $\overline{X}$ has an approximate normal distribution with mean 1/2 and variance 1/72. We could then superimpose this normal p.d.f. on the graph of the histogram for comparison. (Rather than superimposing the normal p.d.f., we used the actual p.d.f. $\overline{X}$ that is shown in the display that follows.) From the histogram, we see the bell-shaped curve resulting from this simulation. None of these three outstanding mathematicians had the advantage of the computer. Today, a researcher with a new idea about a probability distribution can check it easily with a simulation to see if the idea is worth devoting more time to it.

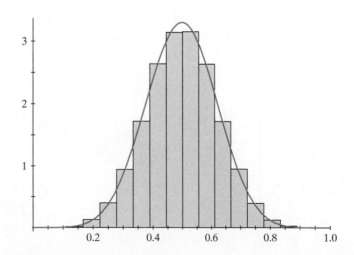

FIGURE 5.7-5: Simulation of 10,000 $\overline{x}$'s for samples of size 6 from $U(0, 1)$

The simulated $\bar{x}$'s were grouped into 18 classes with equal lengths of 1/18. The frequencies of the 18 classes are

0, 1, 9, 72, 223, 524, 957, 1467, 1744, 1751, 1464, 954, 528, 220, 69, 15, 2, 0.

Using these data, we can estimate certain probabilities—for example,

$$P\left(\frac{1}{6} \le \overline{X} \le \frac{2}{6}\right) \approx \frac{72 + 223 + 524}{10,000} = 0.0819$$

and

$$P\left(\frac{11}{18} \le \overline{X} \le 1\right) \approx \frac{954 + 528 + 220 + 69 + 15 + 2 + 0}{10,000} = 0.1788.$$

With a CAS, it is sometimes possible to find p.d.f.'s that involve rather complex calculations. For example, *Maple* was used to find the actual p.d.f. of $\overline{X}$ for this example. It is this p.d.f. that is superimposed on the histogram. Letting $u = \bar{x}$, we find that the p.d.f. is given by

$$g(u) = \begin{cases} 6\left(\dfrac{324u^5}{5}\right), & 0 < u < 1/6, \\[2mm] 6\left(\dfrac{1}{20} - 324u^5 + 324u^4 - 108u^3 + 18u^2 - \dfrac{3u}{2}\right), & 1/6 \le u < 2/6, \\[2mm] 6\left(-\dfrac{79}{20} + \dfrac{117u}{2} + 648u^5 - 1296u^4 + 972u^3 - 342u^2\right), & 2/6 \le u < 3/6, \\[2mm] 6\left(\dfrac{731}{20} - \dfrac{693u}{2} - 648u^5 + 1944u^4 - 2268u^3\right), & 3/6 \le u < 4/6, \\[2mm] 6\left(-\dfrac{1829}{20} + \dfrac{1227u}{2} - 1602u^2 + 2052u^3 \right. \\[1mm] \qquad \left. + 324u^5 - 1296u^4\right), & 4/6 \le u < 5/6, \\[2mm] 6\left(\dfrac{324}{5} - 324u + 648u^2 - 648u^3 + 324u^4 - \dfrac{324u^5}{5}\right), & 5/6 \le u < 1. \end{cases}$$

We can also calculate

$$\int_{1/6}^{2/6} g(u)\, du = \frac{19}{240} = 0.0792$$

and

$$\int_{11/18}^{1} g(u)\, du = \frac{5,818}{32,805} = 0.17735.$$

Although these integrations are not difficult, they are tedious to do by hand.

Estimation

6.1 POINT ESTIMATION

In earlier chapters, we alluded to estimating characteristics of the distribution from the corresponding ones of the sample, hoping that the latter would be reasonably close to the former. For example, the sample mean $\bar{x}$ can be thought of as an estimate of the distribution mean μ, and the sample variance s^2 can be used as an estimate of the distribution variance σ^2. Even the relative frequency histogram associated with a sample can be taken as an estimate of the p.d.f. of the underlying distribution. But how good are these estimates? What makes an estimate good? Can we say anything about the closeness of an estimate to an unknown parameter?

In this section, we consider random variables for which the functional form of the p.m.f. or p.d.f. is known, but the distribution depends on an unknown parameter (say, θ) that may have any value in a set (say, Ω) called the **parameter space**. For example, perhaps it is known that $f(x; \theta) = (1/\theta)e^{-x/\theta}$, $0 < x < \infty$, and that $\theta \in \Omega = \{\theta : 0 < \theta < \infty\}$. In certain instances, it might be necessary for the experimenter to select precisely one member of the family $\{f(x, \theta), \theta \in \Omega\}$ as the most likely p.d.f. of the random variable. That is, the experimenter needs a point estimate of the parameter θ, namely, the value of the parameter that corresponds to the selected p.d.f.

In estimation, we take a random sample from the distribution to elicit some information about the unknown parameter θ. That is, we repeat the experiment n independent times, observe the sample, $X_1, X_2, \ldots, X_n$, and try to estimate the value of θ by using the observations $x_1, x_2, \ldots, x_n$. The function of $X_1, X_2, \ldots, X_n$ used to estimate θ—say, the statistic $u(X_1, X_2, \ldots, X_n)$—is called an **estimator** of θ. We want it to be such that the computed **estimate** $u(x_1, x_2, \ldots, x_n)$ is usually close to θ. Since we are estimating one member of $\theta \in \Omega$, such an estimator is often called a **point estimator**.

The following example should help motivate one principle that is often used in finding point estimates: Suppose that X is $b(1, p)$, so that the p.m.f. of X is

$$f(x; p) = p^x(1 - p)^{1-x}, \qquad x = 0, 1, \qquad 0 \le p \le 1.$$

We note that $p \in \Omega = \{p : 0 \le p \le 1\}$, where Ω represents the parameter space—that is, the space of all possible values of the parameter p. Given a random sample $X_1, X_2, \ldots, X_n$, the problem is to find an estimator $u(X_1, X_2, \ldots, X_n)$ such that $u(x_1, x_2, \ldots, x_n)$ is a good point estimate of p, where $x_1, x_2, \ldots, x_n$ are the observed values of the random sample. Now, the probability that $X_1, X_2, \ldots, X_n$ takes these particular values is (with Σx_i denoting $\Sigma_{i=1}^n x_i$)

$$P(X_1 = x_1, \ldots, X_n = x_n) = \prod_{i=1}^n p^{x_i}(1 - p)^{1-x_i} = p^{\Sigma x_i}(1 - p)^{n - \Sigma x_i},$$

which is the joint p.m.f. of $X_1, X_2, \ldots, X_n$ evaluated at the observed values. One reasonable way to proceed toward finding a good estimate of p is to regard this probability (or joint p.m.f.) as a function of p and find the value of p that maximizes it. That is, we find the p value most likely to have produced these sample values. The joint p.m.f., when regarded as a function of p, is frequently called the **likelihood function**. Thus, here the likelihood function is

$$
\begin{aligned}
L(p) &= L(p; x_1, x_2, \ldots, x_n) \\
&= f(x_1; p)f(x_2; p) \cdots f(x_n; p) \\
&= p^{\Sigma x_i}(1 - p)^{n - \Sigma x_i}, \qquad 0 \le p \le 1.
\end{aligned}
$$

To find the value of p that maximizes $L(p)$, we first take its derivative when $0 < p < 1$:

$$L'(p) = (\Sigma x_i)p^{\Sigma x_i - 1}(1 - p)^{n - \Sigma x_i} - (n - \Sigma x_i)p^{\Sigma x_i}(1 - p)^{n - \Sigma x_i - 1}.$$

Setting this first derivative equal to zero gives us, with the restriction that $0 < p < 1$,

$$p^{\Sigma x_i}(1 - p)^{n - \Sigma x_i}\left[\frac{\Sigma x_i}{p} - \frac{n - \Sigma x_i}{1 - p}\right] = 0.$$

Since $0 < p < 1$, the preceding equation equals zero when

$$\frac{\Sigma x_i}{p} - \frac{n - \Sigma x_i}{1 - p} = 0. \tag{6.1-1}$$

Multiplying each member of Equation 6.1-1 by $p(1 - p)$ and simplifying, we obtain

$$\sum_{i=1}^n x_i - np = 0$$

or, equivalently,

$$p = \frac{\Sigma_{i=1}^n x_i}{n} = \bar{x}.$$

It can be shown that $L''(\bar{x}) < 0$, so that $L(\bar{x})$ is a maximum. The corresponding statistic, namely, $(\Sigma_{i=1}^n X_i)/n = \bar{X}$, is called the **maximum likelihood estimator** and is denoted by $\widehat{p}$; that is,

$$\widehat{p} = \frac{1}{n} \sum_{i=1}^{n} X_i = \overline{X}.$$

When finding a maximum likelihood estimator, it is often easier to find the value of the parameter that maximizes the natural logarithm of the likelihood function rather than the value of the parameter that maximizes the likelihood function itself. Because the natural logarithm function is an increasing function, the solutions will be the same. To see this, note that for $0 < p < 1$, the example we have been considering gives us

$$\ln L(p) = \left(\sum_{i=1}^{n} x_i \right) \ln p + \left(n - \sum_{i=1}^{n} x_i \right) \ln(1 - p).$$

To find the maximum, we set the first derivative equal to zero to obtain

$$\frac{d[\ln L(p)]}{dp} = \left(\sum_{i=1}^{n} x_i \right)\left(\frac{1}{p} \right) + \left(n - \sum_{i=1}^{n} x_i \right)\left(\frac{-1}{1 - p} \right) = 0,$$

which is the same as Equation 6.1-1. Thus, the solution is $p = \overline{x}$ and the maximum likelihood estimator for p is $\widehat{p} = \overline{X}$.

Motivated by the preceding example, we present the formal definition of maximum likelihood estimators (this definition is used in both the discrete and continuous cases):

Let $X_1, X_2, \ldots, X_n$ be a random sample from a distribution that depends on one or more unknown parameters $\theta_1, \theta_2, \ldots, \theta_m$ with p.m.f. or p.d.f. denoted by $f(x; \theta_1, \theta_2, \ldots, \theta_m)$. Suppose that $(\theta_1, \theta_2, \ldots, \theta_m)$ is restricted to a given parameter space Ω. Then the joint p.m.f. or p.d.f. of $X_1, X_2, \ldots, X_n$, namely,

$$L(\theta_1, \theta_2, \ldots, \theta_m) = f(x_1; \theta_1, \ldots, \theta_m) f(x_2; \theta_1, \ldots, \theta_m)$$
$$\cdots f(x_n; \theta_1, \ldots, \theta_m), \qquad (\theta_1, \theta_2, \ldots, \theta_m) \in \Omega,$$

when regarded as a function of $\theta_1, \theta_2, \ldots, \theta_m$, is called the **likelihood function**. Say

$$[u_1(x_1, \ldots, x_n), u_2(x_1, \ldots, x_n), \ldots, u_m(x_1, \ldots, x_n)]$$

is that m-tuple in Ω that maximizes $L(\theta_1, \theta_2, \ldots, \theta_m)$. Then

$$\widehat{\theta_1} = u_1(X_1, \ldots, X_n),$$
$$\widehat{\theta_2} = u_2(X_1, \ldots, X_n),$$
$$\vdots$$
$$\widehat{\theta_m} = u_m(X_1, \ldots, X_n)$$

are **maximum likelihood estimators** of $\theta_1, \theta_2, \ldots, \theta_m$, respectively; and the corresponding observed values of these statistics, namely,

$$u_1(x_1, \ldots, x_n), u_2(x_1, \ldots, x_n), \ldots, u_m(x_1, \ldots, x_n),$$

are called **maximum likelihood estimates**. In many practical cases, these estimators (and estimates) are unique.

For many applications, there is just one unknown parameter. In these cases, the likelihood function is given by

$$L(\theta) = \prod_{i=1}^{n} f(x_i; \theta).$$

Some additional examples will help clarify these definitions.

EXAMPLE 6.1-1 Let $X_1, X_2, \ldots, X_n$ be a random sample from the exponential distribution with p.d.f.

$$f(x; \theta) = \frac{1}{\theta} e^{-x/\theta}, \qquad 0 < x < \infty, \qquad \theta \in \Omega = \{\theta : 0 < \theta < \infty\}.$$

The likelihood function is given by

$$L(\theta) = L(\theta; x_1, x_2, \ldots, x_n)$$

$$= \left(\frac{1}{\theta} e^{-x_1/\theta} \right) \left(\frac{1}{\theta} e^{-x_2/\theta} \right) \cdots \left(\frac{1}{\theta} e^{-x_n/\theta} \right)$$

$$= \frac{1}{\theta^n} \exp\left(\frac{-\sum_{i=1}^{n} x_i}{\theta} \right), \qquad 0 < \theta < \infty.$$

The natural logarithm of $L(\theta)$ is

$$\ln L(\theta) = -(n) \ln(\theta) - \frac{1}{\theta} \sum_{i=1}^{n} x_i, \qquad 0 < \theta < \infty.$$

Thus,

$$\frac{d[\ln L(\theta)]}{d\theta} = \frac{-n}{\theta} + \frac{\sum_{i=1}^{n} x_i}{\theta^2} = 0.$$

The solution of this equation for θ is

$$\theta = \frac{1}{n} \sum_{i=1}^{n} x_i = \bar{x}.$$

Note that

$$\frac{d[\ln L(\theta)]}{d\theta} = \frac{1}{\theta} \left(-n + \frac{n\bar{x}}{\theta} \right) \begin{array}{ll} > 0, & \theta < \bar{x}, \\ = 0, & \theta = \bar{x}, \\ < 0, & \theta > \bar{x}. \end{array}$$

Hence, $\ln L(\theta)$ does have a maximum at $\bar{x}$, and it follows that the maximum likelihood estimator for θ is

$$\widehat{\theta} = \bar{X} = \frac{1}{n} \sum_{i=1}^{n} X_i. \qquad \blacksquare$$

EXAMPLE 6.1-2 Let $X_1, X_2, \ldots, X_n$ be a random sample from the geometric distribution with p.m.f. $f(x; p) = (1 - p)^{x-1} p$, $x = 1, 2, 3, \ldots$. The likelihood function is given by

$$L(p) = (1 - p)^{x_1-1} p (1 - p)^{x_2-1} p \cdots (1 - p)^{x_n-1} p$$

$$= p^n (1 - p)^{\sum x_i - n}, \qquad 0 \le p \le 1.$$

The natural logarithm of $L(p)$ is

$$\ln L(p) = n \ln p + \left(\sum_{i=1}^{n} x_i - n \right) \ln(1 - p), \qquad 0 < p < 1.$$

Thus, restricting p to $0 < p < 1$, so as to be able to take the derivative, we have

$$\frac{d \ln L(p)}{dp} = \frac{n}{p} - \frac{\displaystyle\sum_{i=1}^{n} x_i - n}{1 - p} = 0.$$

Solving for p, we obtain

$$p = \frac{n}{\displaystyle\sum_{i=1}^{n} x_i} = \frac{1}{\bar{x}},$$

and this solution provides a maximum. So the maximum likelihood estimator of p is

$$\hat{p} = \frac{n}{\displaystyle\sum_{i=1}^{n} X_i} = \frac{1}{\bar{X}}.$$

This estimator agrees with our intuition because, in n observations of a geometric random variable, there are n successes in the $\sum_{i=1}^{n} x_i$ trials. Thus, the estimate of p is the number of successes divided by the total number of trials.   ■

In the following important example, we find the maximum likelihood estimators of the parameters associated with the normal distribution.

EXAMPLE 6.1-3   Let $X_1, X_2, \ldots, X_n$ be a random sample from $N(\theta_1, \theta_2)$, where

$$\Omega = \{(\theta_1, \theta_2): -\infty < \theta_1 < \infty, 0 < \theta_2 < \infty\}.$$

That is, here we let $\theta_1 = \mu$ and $\theta_2 = \sigma^2$. Then

$$L(\theta_1, \theta_2) = \prod_{i=1}^{n} \frac{1}{\sqrt{2\pi\theta_2}} \exp\left[-\frac{(x_i - \theta_1)^2}{2\theta_2} \right]$$

or, equivalently,

$$L(\theta_1, \theta_2) = \left(\frac{1}{\sqrt{2\pi\theta_2}} \right)^n \exp\left[\frac{-\sum_{i=1}^{n}(x_i - \theta_1)^2}{2\theta_2} \right], \qquad (\theta_1, \theta_2) \in \Omega.$$

The natural logarithm of the likelihood function is

$$\ln L(\theta_1, \theta_2) = -\frac{n}{2} \ln(2\pi\theta_2) - \frac{\displaystyle\sum_{i=1}^{n}(x_i - \theta_1)^2}{2\theta_2}.$$

The partial derivatives with respect to θ_1 and θ_2 are

$$\frac{\partial(\ln L)}{\partial\theta_1} = \frac{1}{\theta_2} \sum_{i=1}^{n}(x_i - \theta_1)$$

and

$$\frac{\partial (\ln L)}{\partial \theta_2} = \frac{-n}{2\theta_2} + \frac{1}{2\theta_2^2} \sum_{i=1}^{n} (x_i - \theta_1)^2.$$

The equation $\partial (\ln L)/\partial \theta_1 = 0$ has the solution $\theta_1 = \bar{x}$. Setting $\partial (\ln L)/\partial \theta_2 = 0$ and replacing θ_1 by $\bar{x}$ yields

$$\theta_2 = \frac{1}{n} \sum_{i=1}^{n} (x_i - \bar{x})^2.$$

By considering the usual condition on the second partial derivatives, we see that these solutions do provide a maximum. Thus, the maximum likelihood estimators of $\mu = \theta_1$ and $\sigma^2 = \theta_2$ are

$$\widehat{\theta_1} = \overline{X} \qquad \text{and} \qquad \widehat{\theta_2} = \frac{1}{n} \sum_{i=1}^{n} (X_i - \overline{X})^2 = V. \qquad \blacksquare$$

It is interesting to note that in our first illustration, where $\widehat{p} = \overline{X}$, and in Example 6.1-1, where $\widehat{\theta} = \overline{X}$, the expected value of the estimator is equal to the corresponding parameter. This observation leads to the following definition.

**DEFINITION
6.1-1**

If $E[u(X_1, X_2, \ldots, X_n)] = \theta$, then the statistic $u(X_1, X_2, \ldots X_n)$ is called an **unbiased estimator** of θ. Otherwise, it is said to be **biased**.

EXAMPLE 6.1-4   We have shown that when sampling from $N(\theta_1 = \mu, \theta_2 = \sigma^2)$, one finds that the maximum likelihood estimators of μ and σ^2 are

$$\widehat{\theta_1} = \widehat{\mu} = \overline{X} \qquad \text{and} \qquad \widehat{\theta_2} = \widehat{\sigma^2} = \frac{(n-1)S^2}{n}.$$

Recalling that the distribution of $\overline{X}$ is $N(\mu, \sigma^2/n)$, we see that $E(\overline{X}) = \mu$; thus, $\overline{X}$ is an unbiased estimator of μ.

In Theorem 5.5-2, we showed that the distribution of $(n-1)S^2/\sigma^2$ is $\chi^2(n-1)$. Hence,

$$E(S^2) = E\left[\frac{\sigma^2}{n-1} \frac{(n-1)S^2}{\sigma^2} \right] = \frac{\sigma^2}{n-1} (n-1) = \sigma^2.$$

That is, the sample variance

$$S^2 = \frac{1}{n-1} \sum_{i=1}^{n} (X_i - \overline{X})^2$$

is an unbiased estimator of σ^2. Consequently, since

$$E(\widehat{\theta_2}) = \frac{n-1}{n} E(S^2) = \frac{n-1}{n} \sigma^2,$$

$\widehat{\theta_2}$ is a biased estimator of $\theta_2 = \sigma^2$. $\qquad \blacksquare$

Sometimes it is impossible to find maximum likelihood estimators in a convenient closed form, and numerical methods must be used to maximize the likelihood function. For example, suppose that $X_1, X_2, \ldots, X_n$ is a random sample from a gamma distribution with parameters $\alpha = \theta_1$ and $\beta = \theta_2$, where $\theta_1 > 0$, $\theta_2 > 0$. It is difficult to maximize

$$L(\theta_1, \theta_2; x_1, \ldots, x_n) = \left[\frac{1}{\Gamma(\theta_1)\theta_2^{\theta_1}} \right]^n (x_1 x_2 \cdots x_n)^{\theta_1 - 1} \exp\left(-\sum_{i=1}^{n} x_i/\theta_2 \right)$$

with respect to θ_1 and θ_2, owing to the presence of the gamma function $\Gamma(\theta_1)$. Thus, numerical methods must be used to maximize L once $x_1, x_2, \ldots, x_n$ are observed.

There are other ways, however, to easily obtain point estimates of θ_1 and θ_2. One of the early methods was to simply equate the first sample moment to the first theoretical moment. Next, if needed, the two second moments are equated, then the third moments, and so on, until we have enough equations to solve for the parameters. As an illustration, in the gamma distribution situation, let us simply equate the first two moments of the distribution to the corresponding moments of the empirical distribution. This seems like a reasonable way in which to find estimators, since the empirical distribution converges in some sense to the probability distribution, and hence corresponding moments should be about equal. In this situation, we have

$$\theta_1 \theta_2 = \overline{X}, \qquad \theta_1 \theta_2^2 = V,$$

the solutions of which are

$$\widetilde{\theta}_1 = \frac{\overline{X}^2}{V} \qquad \text{and} \qquad \widetilde{\theta}_2 = \frac{V}{\overline{X}}.$$

We say that these latter two statistics, $\widetilde{\theta}_1$ and $\widetilde{\theta}_2$, are respective estimators of θ_1 and θ_2 found by the **method of moments**.

To generalize this discussion, let $X_1, X_2, \ldots, X_n$ be a random sample of size n from a distribution with p.d.f. $f(x; \theta_1, \theta_2, \ldots, \theta_r)$, $(\theta_1, \ldots, \theta_r) \in \Omega$. The expectation $E(X^k)$ is frequently called the kth moment of the distribution, $k = 1, 2, 3, \ldots$. The sum $M_k = \sum_{i=1}^{n} X_i^k/n$ is the kth moment of the sample, $k = 1, 2, 3, \ldots$. The method of moments can be described as follows: Equate $E(X^k)$ to M_k, beginning with $k = 1$ and continuing until there are enough equations to provide unique solutions for $\theta_1, \theta_2, \ldots, \theta_r$—say, $h_i(M_1, M_2, \ldots)$, $i = 1, 2, \ldots, r$, respectively. Note that this could be done in an equivalent manner by equating $\mu = E(X)$ to $\overline{X}$ and $E[(X - \mu)^k]$ to $\sum_{i=1}^{n} (X_i - \overline{X})^k/n$, $k = 2, 3$, and so on, until unique solutions for $\theta_1, \theta_2, \ldots, \theta_r$ are obtained. This alternative procedure was used in the preceding illustration. In most practical cases, the estimator $\widetilde{\theta}_i = h_i(M_1, M_2, \ldots)$ of θ_i, found by the method of moments, is an estimator of θ_i that in some sense gets close to that parameter when n is large, $i = 1, 2, \ldots, r$.

The next two examples—the first for a one-parameter family and the second for a two-parameter family—illustrate the method-of-moments technique for finding estimators.

EXAMPLE 6.1-5   Let $X_1, X_2, \ldots, X_n$, be a random sample of size n from the distribution with p.d.f. $f(x; \theta) = \theta x^{\theta-1}$, $0 < x < 1$, $0 < \theta < \infty$. Sketch the graphs of this p.d.f. for $\theta = 1/4$, 1, and 4. Note that sets of observations for these three values of θ would look very different. How do we estimate the value of θ? The mean of this distribution is given by

$$E(X) = \int_0^1 x\,\theta\,x^{\theta-1}\,dx = \frac{\theta}{\theta + 1}.$$

We shall set the distribution mean equal to the sample mean and solve for θ. We have

$$\bar{x} = \frac{\theta}{\theta + 1}.$$

Solving for θ, we obtain the method-of-moments estimator,

$$\tilde{\theta} = \frac{\overline{X}}{1 - \overline{X}}.$$

Thus, an estimate of θ by the method of moments is $\bar{x}/(1 - \bar{x})$.   ∎

Recall that in the method of moments, if two parameters have to be estimated, the first two sample moments are set equal to the first two distribution moments that are given in terms of the unknown parameters. These two equations are then solved simultaneously for the unknown parameters.

EXAMPLE 6.1-6   Let the distribution of X be $N(\mu, \sigma^2)$. Then

$$E(X) = \mu \qquad \text{and} \qquad E(X^2) = \sigma^2 + \mu^2.$$

For a random sample of size n, the first two moments are given by

$$m_1 = \frac{1}{n}\sum_{i=1}^n x_i \qquad \text{and} \qquad m_2 = \frac{1}{n}\sum_{i=1}^n x_i^2.$$

We set $m_1 = E(X)$ and $m_2 = E(X^2)$ and solve for μ and σ^2. That is,

$$\frac{1}{n}\sum_{i=1}^n x_i = \mu \qquad \text{and} \qquad \frac{1}{n}\sum_{i=1}^n x_i^2 = \sigma^2 + \mu^2.$$

The first equation yields $\bar{x}$ as the estimate of μ. Replacing μ^2 with $\bar{x}^2$ in the second equation and solving for σ^2, we obtain

$$\frac{1}{n}\sum_{i=1}^n x_i^2 - \bar{x}^2 = \sum_{i=1}^n \frac{(x_i - \bar{x})^2}{n} = v$$

as the solution of σ^2. Thus, the method-of-moments estimators for μ and σ^2 are $\tilde{\mu} = \overline{X}$ and $\tilde{\sigma}^2 = V$, which are the same as the maximum likelihood estimators. Of course, $\tilde{\mu} = \overline{X}$ is unbiased whereas $\tilde{\sigma}^2 = V$ is biased.   ∎

In Example 6.1-4, we showed that $\overline{X}$ and S^2 are unbiased estimators of μ and σ^2, respectively, when one is sampling from a normal distribution. This is also true when one is sampling from any distribution with a finite variance σ^2. That is, $E(\overline{X}) = \mu$ and $E(S^2) = \sigma^2$, provided that the sample arises from a distribution with variance $\sigma^2 < \infty$. (See Exercise 6.1-13.) Although S^2 is an unbiased estimator of σ^2, S is a biased estimator of σ. In Exercise 6.1-14, you are asked to show that, when one is sampling from a normal distribution, cS is an unbiased estimator of σ, where

$$c = \frac{\sqrt{n-1}\,\Gamma\!\left(\dfrac{n-1}{2}\right)}{\sqrt{2}\,\Gamma\!\left(\dfrac{n}{2}\right)}.$$

EXERCISES

6.1-1. Let $X_1, X_2, \ldots, X_n$ be a random sample from $N(\mu, \sigma^2)$, where the mean $\theta = \mu$ is such that $-\infty < \theta < \infty$ and σ^2 is a known positive number. Show that the maximum likelihood estimator for θ is $\widehat{\theta} = \overline{X}$.

6.1-2. A random sample $X_1, X_2, \ldots, X_n$ of size n is taken from $N(\mu, \sigma^2)$, where the variance $\theta = \sigma^2$ is such that $0 < \theta < \infty$ and μ is a known real number. Show that the maximum likelihood estimator for θ is $\widehat{\theta} = (1/n)\sum_{i=1}^{n}(X_i - \mu)^2$ and that this estimator is an unbiased estimator of θ.

6.1-3. A random sample $X_1, X_2, \ldots, X_n$ of size n is taken from a Poisson distribution with a mean of λ, $0 < \lambda < \infty$.

(a) Show that the maximum likelihood estimator for λ is $\widehat{\lambda} = \overline{X}$.

(b) Let X equal the number of flaws per 100 feet of a used computer tape. Assume that X has a Poisson distribution with a mean of λ. If 40 observations of X yielded 5 zeros, 7 ones, 12 twos, 9 threes, 5 fours, 1 five, and 1 six, find the maximum likelihood estimate of λ.

6.1-4. For determining half-lives of radioactive isotopes, it is important to know what the background radiation is in a given detector over a specific period. The following data were taken in a γ-ray detection experiment over 98 ten-second intervals:

58	50	57	58	64	63	54	64	59	41	43	56	60	50
46	59	54	60	59	60	67	52	65	63	55	61	68	58
63	36	42	54	58	54	40	60	64	56	61	51	48	50
60	42	62	67	58	49	66	58	57	59	52	54	53	53
57	43	73	65	45	43	57	55	73	62	68	55	31	55
53	68	58	53	51	73	44	50	53	62	58	47	63	59
59	56	60	59	50	52	62	51	66	51	56	53	59	57

Assume that these data are observations of a Poisson random variable with mean λ.

(a) Find the values of $\overline{x}$ and s^2 if you did not do this in Exercise 2.6-22.

(b) What is the value of the maximum likelihood estimator of λ?

(c) Is S^2 an unbiased estimator of λ?

(d) Which of $\overline{x}$ and s^2 would you recommend for estimating λ? Why? You could compare the variance of $\overline{X}$ with the variance of S^2, which is

$$\mathrm{Var}(S^2) = \frac{\lambda(2\lambda n + n - 1)}{n(n-1)}.$$

6.1-5. Let $X_1, X_2, \ldots, X_n$ be a random sample from distributions with the given probability density functions. In each case, find the maximum likelihood estimator $\widehat{\theta}$.

(a) $f(x; \theta) = (1/\theta^2)\,x\,e^{-x/\theta}$, $0 < x < \infty$, $0 < \theta < \infty$.

(b) $f(x; \theta) = (1/2\theta^3)\,x^2\,e^{-x/\theta}$, $0 < x < \infty$, $0 < \theta < \infty$.

(c) $f(x; \theta) = (1/2)\,e^{-|x-\theta|}$, $-\infty < x < \infty$, $-\infty < \theta < \infty$.

HINT: Finding θ involves minimizing $\sum |x_i - \theta|$, which is a difficult problem. When $n = 5$, do it for $x_1 = 6.1$, $x_2 = -1.1$, $x_3 = 3.2$, $x_4 = 0.7$, and $x_5 = 1.7$, and you will see the answer. (See also Exercise 2.2-10.)

6.1-6. Find the maximum likelihood estimates for $\theta_1 = \mu$ and $\theta_2 = \sigma^2$ if a random sample of size 15 from $N(\mu, \sigma^2)$ yielded the following values:

31.5	36.9	33.8	30.1	33.9
35.2	29.6	34.4	30.5	34.2
31.6	36.7	35.8	34.5	32.7

6.1-7. Let $f(x; \theta) = \theta x^{\theta-1}$, $0 < x < 1$, $\theta \in \Omega = \{\theta : 0 < \theta < \infty\}$. Let $X_1, X_2, \ldots, X_n$ denote a random sample of size n from this distribution.

(a) Sketch the p.d.f. of X for (i) $\theta = 1/2$, (ii) $\theta = 1$, and (iii) $\theta = 2$.

(b) Show that $\widehat{\theta} = -n/\ln\left(\prod_{i=1}^{n} X_i\right)$ is the maximum likelihood estimator of θ.

(c) For each of the following three sets of 10 observations from the given distribution, calculate the values of the maximum likelihood estimate and the method-of-moments estimate of θ:

(i)	0.0256	0.3051	0.0278	0.8971	0.0739
	0.3191	0.7379	0.3671	0.9763	0.0102
(ii)	0.9960	0.3125	0.4374	0.7464	0.8278
	0.9518	0.9924	0.7112	0.2228	0.8609
(iii)	0.4698	0.3675	0.5991	0.9513	0.6049
	0.9917	0.1551	0.0710	0.2110	0.2154

6.1-8. Let $f(x; \theta) = (1/\theta)x^{(1-\theta)/\theta}$, $0 < x < 1$, $0 < \theta < \infty$.

(a) Show that the maximum likelihood estimator of θ is $\widehat{\theta} = -(1/n)\sum_{i=1}^{n} \ln X_i$.

(b) Show that $E(\widehat{\theta}) = \theta$ and thus that $\widehat{\theta}$ is an unbiased estimator of θ.

6.1-9. Let $X_1, X_2, \ldots, X_n$ be a random sample of size n from the exponential distribution whose p.d.f. is $f(x; \theta) = (1/\theta)e^{-x/\theta}$, $0 < x < \infty$, $0 < \theta < \infty$.

(a) Show that $\overline{X}$ is an unbiased estimator of θ.

(b) Show that the variance of $\overline{X}$ is θ^2/n.

(c) What is a good estimate of θ if a random sample of size 5 yielded the sample values 3.5, 8.1, 0.9, 4.4, and 0.5?

6.1-10. Let $X_1, X_2, \ldots, X_n$ be a random sample of size n from a geometric distribution for which p is the probability of success.

(a) Use the method of moments to find a point estimate for p.

(b) Explain intuitively why your estimate makes good sense.

(c) Use the following data to give a point estimate of p:

3	34	7	4	19	2	1	19	43	2
22	4	19	11	7	1	2	21	15	16

6.1-11. Out of 50,000,000 instant-winner lottery tickets, the proportion of winning tickets is p. Each day, for 20 consecutive days, a bettor purchased tickets, one at a time, until a winning ticket was

purchased. The numbers of tickets that were purchased each day to obtain the winning ticket were

1	26	19	6	6	1	2	3	1	23
19	3	6	8	4	1	18	34	1	8

By making reasonable assumptions, find the maximum likelihood estimate of p on the basis of these data.

6.1-12. Let $X_1, X_2, \ldots, X_n$ be a random sample from $b(1, p)$ (i.e., n Bernoulli trials). Thus,

$$Y = \sum_{i=1}^{n} X_i \text{ is } b(n, p).$$

(a) Show that $\overline{X} = Y/n$ is an unbiased estimator of p.

(b) Show that $\mathrm{Var}(\overline{X}) = p(1 - p)/n$.

(c) Show that $E[\overline{X}(1 - \overline{X})/n] = (n - 1)[p(1 - p)/n^2]$.

(d) Find the value of c so that $c\overline{X}(1 - \overline{X})$ is an unbiased estimator of $\mathrm{Var}(\overline{X}) = p(1 - p)/n$.

6.1-13. Let $X_1, X_2, \ldots, X_n$ be a random sample from a distribution having finite variance σ^2. Show that

$$S^2 = \sum_{i=1}^{n} \frac{(X_i - \overline{X})^2}{n - 1}$$

is an unbiased estimator of σ^2. HINT: Write

$$S^2 = \frac{1}{n - 1}\left(\sum_{i=1}^{n} X_i^2 - n\overline{X}^2\right)$$

and compute $E(S^2)$.

6.1-14. Let $X_1, X_2, \ldots, X_n$ be a random sample of size n from a normal distribution.

(a) Show that an unbiased estimator of σ is cS, where

$$c = \frac{\sqrt{n - 1}\,\Gamma\left(\dfrac{n - 1}{2}\right)}{\sqrt{2}\,\Gamma\left(\dfrac{n}{2}\right)}.$$

HINT: Recall that the distribution of $(n - 1)S^2/\sigma^2$ is $\chi^2(n - 1)$.

(b) Find the value of c when $n = 5$; when $n = 6$.

(c) Graph c as a function of n. What is the limit of c as n increases without bound?

6.1-15. Let $X_1, X_2, \ldots, X_n$ be a random sample from a uniform distribution on the interval $(\theta - 1, \theta + 1)$.

(a) Find the method-of-moments estimator of θ.

(b) Is your estimator in part (a) an unbiased estimator of θ?

(c) Given the following $n = 5$ observations of X, give a point estimate of θ:

6.61   7.70   6.98   8.36   7.26

(d) The method-of-moments estimator actually has greater variance than the estimator

$[\min(X_i) + \max(X_i)]/2$, which is a maximum likelihood estimator of θ. Compute the value of the latter estimator for the $n = 5$ observations in (c).

6.1-16. Given the following 25 observations from a gamma distribution with mean $\mu = \alpha\theta$ and variance $\sigma^2 = \alpha\theta^2$, use the method-of-moments estimators to find point estimates of α and θ:

6.9	7.3	6.7	6.4	6.3	5.9	7.0	7.1	6.5	7.6	7.2	7.1	6.1
7.3	7.6	7.6	6.7	6.3	5.7	6.7	7.5	5.3	5.4	7.4	6.9	

6.1-17. Let X equal the number of telephone calls per five minutes that are received at Great Lakes Pizza Co. in the evening. Assume that the distribution of X is Poisson with mean λ.

(a) Given n observations of X, find the method-of-moments estimate of λ.

(b) Give a point estimate of λ, using the following 12 observations of X:

1 2 1 1 2 4 0 1 0 1 1 0

(c) Compare the values of $\bar{x}$ and s^2. Does the information yielded by the comparison of x and s^2 support the assumption that X has a Poisson distribution?

6.1-18. An urn contains 64 balls, of which N_1 are orange and N_2 are blue. A random sample of $n = 8$ balls is selected from the urn without replacement, and X is equal to the number of orange balls in the sample. This experiment was repeated 30 times (the 8 balls being returned to the urn before each repetition), yielding the following data:

3 0 0 1 1 1 1 3 1 1 2 0 1 3 1
0 1 0 2 1 1 2 3 2 2 4 3 1 1 2

Using these data, guess the value of N_1 and give a reason for your guess.

6.1-19. A Geiger counter was set up in the physics laboratory to record the number of alpha particle emissions of carbon-14 in half a second. Ten observations yielded the following data:

4 6 9 6 10 11 6 3 7 10

Give the value of an unbiased estimate of λ, the mean number of counts per half second, assuming that these data are observations of a Poisson random variable.

6.1-20. Let the p.d.f. of X be defined by

$$f(x) = \begin{cases} \left(\dfrac{4}{\theta^2}\right)x, & 0 < x \le \dfrac{\theta}{2}, \\[2mm] -\left(\dfrac{4}{\theta^2}\right)x + \dfrac{4}{\theta}, & \dfrac{\theta}{2} < x \le \theta, \\[2mm] 0, & \text{elsewhere}, \end{cases}$$

where $\theta \in \Omega = \{\theta : 0 < \theta \le 2\}$.

(a) Sketch the graph of this p.d.f. when $\theta = 1/2$, $\theta = 1$, and $\theta = 2$.

(b) Find an estimator of θ by the method of moments.

(c) For the following observations of X, give a point estimate of θ:

0.3206	0.2408	0.2577	0.3557	0.4188
0.5601	0.0240	0.5422	0.4532	0.5592

6.2   CONFIDENCE INTERVALS FOR MEANS

Given a random sample $X_1, X_2, \ldots, X_n$ from a normal distribution $N(\mu, \sigma^2)$, we shall now consider the closeness of $\overline{X}$, the unbiased estimator of μ, to the unknown mean μ. To do this, we use the error structure (distribution) of $\overline{X}$, namely, that $\overline{X}$ is $N(\mu, \sigma^2/n)$ (see Corollary 5.5-1), to construct what is called a confidence interval for the unknown parameter μ when the variance σ^2 is known. For the probability $1 - \alpha$, we can find a number $z_{\alpha/2}$ from Table V in the appendix such that

$$P\left(-z_{\alpha/2} \le \frac{\overline{X} - \mu}{\sigma/\sqrt{n}} \le z_{\alpha/2}\right) = 1 - \alpha.$$

For example, if $1 - \alpha = 0.95$, then $z_{\alpha/2} = z_{0.025} = 1.96$, and if $1 - \alpha = 0.90$, then $z_{\alpha/2} = z_{0.05} = 1.645$. Now, recalling that $\sigma > 0$, we see that the following inequalities are equivalent:

$$-z_{\alpha/2} \leq \frac{\overline{X} - \mu}{\sigma/\sqrt{n}} \leq z_{\alpha/2},$$

$$-z_{\alpha/2}\left(\frac{\sigma}{\sqrt{n}}\right) \leq \overline{X} - \mu \leq z_{\alpha/2}\left(\frac{\sigma}{\sqrt{n}}\right),$$

$$-\overline{X} - z_{\alpha/2}\left(\frac{\sigma}{\sqrt{n}}\right) \leq -\mu \leq -\overline{X} + z_{\alpha/2}\left(\frac{\sigma}{\sqrt{n}}\right),$$

$$\overline{X} + z_{\alpha/2}\left(\frac{\sigma}{\sqrt{n}}\right) \geq \mu \geq \overline{X} - z_{\alpha/2}\left(\frac{\sigma}{\sqrt{n}}\right).$$

Thus, since the probability of the first of these is $1 - \alpha$, the probability of the last must also be $1 - \alpha$, because the latter is true if and only if the former is true. That is, we have

$$P\left[\overline{X} - z_{\alpha/2}\left(\frac{\sigma}{\sqrt{n}}\right) \leq \mu \leq \overline{X} + z_{\alpha/2}\left(\frac{\sigma}{\sqrt{n}}\right)\right] = 1 - \alpha.$$

So the probability that the random interval

$$\left[\overline{X} - z_{\alpha/2}\left(\frac{\sigma}{\sqrt{n}}\right), \overline{X} + z_{\alpha/2}\left(\frac{\sigma}{\sqrt{n}}\right)\right]$$

includes the unknown mean μ is $1 - \alpha$.

Once the sample is observed and the sample mean computed to equal $\bar{x}$, the interval $[\bar{x} - z_{\alpha/2}(\sigma/\sqrt{n}), \bar{x} + z_{\alpha/2}(\sigma/\sqrt{n})]$ becomes known. Since the probability that the random interval covers μ before the sample is drawn is equal to $1 - \alpha$, we now call the computed interval, $\bar{x} \pm z_{\alpha/2}(\sigma/\sqrt{n})$ (for brevity), a $100(1 - \alpha)\%$ **confidence interval** for the unknown mean μ. For example, $\bar{x} \pm 1.96(\sigma/\sqrt{n})$ is a 95% confidence interval for μ. The number $100(1 - \alpha)\%$, or equivalently, $1 - \alpha$, is called the **confidence coefficient**.

We see that the confidence interval for μ is centered at the point estimate $\bar{x}$ and is completed by subtracting and adding the quantity $z_{\alpha/2}(\sigma/\sqrt{n})$. Note that as n increases, $z_{\alpha/2}(\sigma/\sqrt{n})$ decreases, resulting in a shorter confidence interval with the same confidence coefficient $1 - \alpha$. A shorter confidence interval indicates that we have more credence in $\bar{x}$ as an estimate of μ. Statisticians who are not restricted by time, money, effort, or the availability of observations can obviously make the confidence interval as short as they like by increasing the sample size n. For a fixed sample size n, the length of the confidence interval can also be shortened by decreasing the confidence coefficient $1 - \alpha$. But if this is done, we achieve a shorter confidence interval at the expense of losing some confidence.

EXAMPLE 6.2-1 Let X equal the length of life of a 60-watt light bulb marketed by a certain manufacturer. Assume that the distribution of X is $N(\mu, 1296)$. If a random sample of $n = 27$ bulbs is tested until they burn out, yielding a sample mean of $\bar{x} = 1478$ hours, then a 95% confidence interval for μ is

$$\left[\bar{x} - z_{0.025}\left(\frac{\sigma}{\sqrt{n}}\right), \bar{x} + z_{0.025}\left(\frac{\sigma}{\sqrt{n}}\right) \right]$$

$$= \left[1478 - 1.96\left(\frac{36}{\sqrt{27}}\right), 1478 + 1.96\left(\frac{36}{\sqrt{27}}\right) \right]$$

$$= [1478 - 13.58, 1478 + 13.58]$$

$$= [1464.42, 1491.58].$$

The next example will help to give a better intuitive feeling for the interpretation of a confidence interval.

EXAMPLE 6.2-2   Let $\bar{x}$ be the observed sample mean of five observations of a random sample from the normal distribution $N(\mu, 16)$. A 90% confidence interval for the unknown mean μ is

$$\left[\bar{x} - 1.645\sqrt{\frac{16}{5}}, \bar{x} + 1.645\sqrt{\frac{16}{5}} \right].$$

For a particular sample, this interval either does or does not contain the mean μ. However, if many such intervals were calculated, about 90% of them should contain the mean μ. Fifty random samples of size 5 from the normal distribution $N(50, 16)$ were simulated on a computer. A 90% confidence interval was calculated for each random sample, as if the mean were unknown. Figure 6.2-1(a) depicts each of these 50 intervals as a line segment. Note that 45 (or 90%) of them contain the mean, $\mu = 50$. In other simulations of 50 confidence intervals, the number of 90% confidence intervals containing the mean could be larger or smaller. [In fact, if W is a random variable that counts the number of 90% confidence intervals containing the mean, then the distribution of W is $b(50, 0.90)$.]

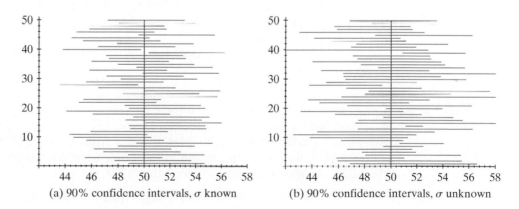

(a) 90% confidence intervals, σ known   (b) 90% confidence intervals, σ unknown

FIGURE 6.2-1: Confidence intervals using z and t

If we cannot assume that the distribution from which the sample arose is normal, we can still obtain an approximate confidence interval for μ. By the central limit theorem, provided that n is large enough, the ratio $(\overline{X} - \mu)/(\sigma/\sqrt{n})$ has the approximate normal distribution $N(0, 1)$ when the underlying distribution is not normal. In this case,

$$P\left(-z_{\alpha/2} \leq \frac{\overline{X} - \mu}{\sigma/\sqrt{n}} \leq z_{\alpha/2}\right) \approx 1 - \alpha,$$

and

$$\left[\overline{x} - z_{\alpha/2}\left(\frac{\sigma}{\sqrt{n}}\right), \overline{x} + z_{\alpha/2}\left(\frac{\sigma}{\sqrt{n}}\right)\right]$$

is an approximate $100(1 - \alpha)\%$ confidence interval for μ.

The closeness of the approximate probability $1 - \alpha$ to the exact probability depends on both the underlying distribution and the sample size. When the underlying distribution is unimodal (has only one mode) and continuous, the approximation is usually quite good even for small n, such as $n = 5$. As the underlying distribution becomes "less normal" (i.e., badly skewed or discrete), a larger sample size might be required to keep a reasonably accurate approximation. But, in almost all cases, an n of at least 30 is usually adequate.

EXAMPLE 6.2-3 Let X equal the amount of orange juice (in grams per day) consumed by an American. Suppose it is known that the standard deviation of X is $\sigma = 96$. To estimate the mean μ of X, an orange growers' association took a random sample of $n = 576$ Americans and found that they consumed, on the average, $\overline{x} = 133$ grams of orange juice per day. Thus, an approximate 90% confidence interval for μ is

$$133 \pm 1.645\left(\frac{96}{\sqrt{576}}\right), \quad \text{or} \quad [133 - 6.58, 133 + 6.58] = [126.42, 139.58]. \quad \blacksquare$$

If σ^2 is unknown and the sample size n is 30 or greater, we shall use the fact that the ratio $(\overline{X} - \mu)/(S/\sqrt{n})$ has an approximate normal distribution $N(0, 1)$. This statement is true whether or not the underlying distribution is normal. However, if the underlying distribution is badly skewed or contaminated with occasional outliers, most statisticians would prefer to have a larger sample size—say, 50 or more—and even that might not produce good results. After this next example, we consider what to do when n is small.

EXAMPLE 6.2-4 Lake Macatawa, an inlet lake on the east side of Lake Michigan, is divided into an east basin and a west basin. To measure the effect on the lake of salting city streets in the winter, students took 32 samples of water from the west basin and measured the amount of sodium in parts per million in order to make a statistical inference about the unknown mean μ. They obtained the following data:

13.0	18.5	16.4	14.8	19.4	17.3	23.2	24.9
20.8	19.3	18.8	23.1	15.2	19.9	19.1	18.1
25.1	16.8	20.4	17.4	25.2	23.1	15.3	19.4
16.0	21.7	15.2	21.3	21.5	16.8	15.6	17.6

For these data, $\bar{x} = 19.07$ and $s^2 = 10.60$. Thus, an approximate 95% confidence interval for μ is

$$\bar{x} \pm 1.96\left(\frac{s}{\sqrt{n}}\right), \qquad \text{or} \qquad 19.07 \pm 1.96\sqrt{\frac{10.60}{32}}, \qquad \text{or} \qquad [17.94, 20.20]. \quad \blacksquare$$

So we have found a confidence interval for the mean μ of a normal distribution, assuming that the value of the standard deviation σ is known or assuming that σ is unknown but the sample size is large. However, in many applications, the sample sizes are small and we do not know the value of the standard deviation, although in some cases we might have a very good idea about its value. For example, a manufacturer of light bulbs probably has a good notion from past experience of the value of the standard deviation of the length of life of different types of light bulbs. But certainly, most of the time, the investigator will not have any more idea about the standard deviation than about the mean—and frequently less. Let us consider how to proceed under these circumstances.

If the random sample arises from a normal distribution, we use the fact that

$$T = \frac{\overline{X} - \mu}{S/\sqrt{n}}$$

has a t distribution with $r = n - 1$ degrees of freedom (see Corollary 5.5-1, Theorem 5.5-2, and Equation 5.5-2), where S^2 is the usual unbiased estimator of σ^2. Select $t_{\alpha/2}(n-1)$ so that $P[T \geq t_{\alpha/2}(n-1)] = \alpha/2$. (See Figure 5.5-2(b) and Table VI in the appendix.) Then

$$1 - \alpha = P\left[-t_{\alpha/2}(n-1) \leq \frac{\overline{X} - \mu}{S/\sqrt{n}} \leq t_{\alpha/2}(n-1)\right]$$

$$= P\left[-t_{\alpha/2}(n-1)\left(\frac{S}{\sqrt{n}}\right) \leq \overline{X} - \mu \leq t_{\alpha/2}(n-1)\left(\frac{S}{\sqrt{n}}\right)\right]$$

$$= P\left[-\overline{X} - t_{\alpha/2}(n-1)\left(\frac{S}{\sqrt{n}}\right) \leq -\mu \leq -\overline{X} + t_{\alpha/2}(n-1)\left(\frac{S}{\sqrt{n}}\right)\right]$$

$$= P\left[\overline{X} - t_{\alpha/2}(n-1)\left(\frac{S}{\sqrt{n}}\right) \leq \mu \leq \overline{X} + t_{\alpha/2}(n-1)\left(\frac{S}{\sqrt{n}}\right)\right].$$

Thus, the observations of a random sample provide $\bar{x}$ and s^2, and

$$\left[\bar{x} - t_{\alpha/2}(n-1)\left(\frac{s}{\sqrt{n}}\right), \bar{x} + t_{\alpha/2}(n-1)\left(\frac{s}{\sqrt{n}}\right)\right]$$

is a $100(1 - \alpha)\%$ confidence interval for μ.

EXAMPLE 6.2-5 Let X equal the amount of butterfat in pounds produced by a typical cow during a 305-day milk production period between her first and second calves. Assume that the distribution of X is $N(\mu, \sigma^2)$. To estimate μ, a farmer measured the butterfat production for $n = 20$ cows and obtained the following data:

$$481 \quad 537 \quad 513 \quad 583 \quad 453 \quad 510 \quad 570 \quad 500 \quad 457 \quad 555$$
$$618 \quad 327 \quad 350 \quad 643 \quad 499 \quad 421 \quad 505 \quad 637 \quad 599 \quad 392$$

For these data, $\bar{x} = 507.50$ and $s = 89.75$. Thus, a point estimate of μ is $\bar{x} = 507.50$. Since $t_{0.05}(19) = 1.729$, a 90% confidence interval for μ is

$$507.50 \pm 1.729 \left(\frac{89.75}{\sqrt{20}} \right)$$

$$507.50 \pm 34.70, \qquad \text{or equivalently,} \qquad [472.80, 542.20]. \qquad \blacksquare$$

Let T have a t distribution with $n - 1$ degrees of freedom. Then $t_{\alpha/2}(n-1) > z_{\alpha/2}$. Consequently, we would expect the interval $\bar{x} \pm z_{\alpha/2}(\sigma/\sqrt{n})$ to be shorter than the interval $\bar{x} \pm t_{\alpha/2}(n-1)(s/\sqrt{n})$. After all, we have more information, namely, the value of σ, in constructing the first interval. However, the length of the second interval is very much dependent on the value of s. If the observed s is smaller than σ, a shorter confidence interval could result by the second procedure. But on the average, $\bar{x} \pm z_{\alpha/2}(\sigma/\sqrt{n})$ is the shorter of the two confidence intervals (Exercise 6.2-16).

EXAMPLE 6.2-6 In Example 6.2-2, 50 confidence intervals were simulated for the mean of a normal distribution, assuming that the variance was known. For those same data, since $t_{0.05}(4) = 2.132$, $\bar{x} \pm 2.132(s/\sqrt{5})$ was used to calculate a 90% confidence interval for μ. For those particular 50 intervals, 46 contained the mean $\mu = 50$. These 50 intervals are depicted in Figure 6.2-1(b). Note the different lengths of the intervals. Some are longer and some are shorter than the corresponding z intervals. The average length of the 50 t intervals is 7.137, which is quite close to the expected length of such an interval: 7.169. (See Exercise 6.2-16.) The length of the intervals that use z and $\sigma = 4$ is 5.885. $\qquad \blacksquare$

If we are not able to assume that the underlying distribution is normal, but μ and σ are both unknown, approximate confidence intervals for μ can still be constructed with the formula

$$T = \frac{\bar{X} - \mu}{S/\sqrt{n}},$$

which now only has an approximate t distribution. Generally, this approximation is quite good (i.e., it is robust) for many nonnormal distributions; in particular, it works well if the underlying distribution is symmetric, unimodal, and of the continuous type. However, if the distribution is highly skewed, there is great danger in using that approximation. In such a situation, it would be safer to use certain nonparametric methods for finding a confidence interval for the median of the distribution, one of which is given in Section 8.4.

There is one other aspect of confidence intervals that should be mentioned. So far, we have created only what are called **two-sided confidence intervals** for the mean μ. Sometimes, however, we might want only a lower (or upper) bound on μ. We proceed as follows.

Say $\overline{X}$ is the mean of a random sample of size n from the normal distribution $N(\mu, \sigma^2)$, where, for the moment, assume that σ^2 is known. Then

$$P\left(\frac{\overline{X} - \mu}{\sigma/\sqrt{n}} \leq z_\alpha\right) = 1 - \alpha,$$

or equivalently,

$$P\left[\overline{X} - z_\alpha\left(\frac{\sigma}{\sqrt{n}}\right) \leq \mu\right] = 1 - \alpha.$$

Once $\overline{X}$ is observed to be equal to $\overline{x}$, it follows that $[\overline{x} - z_\alpha(\sigma/\sqrt{n}), \infty)$ is a $100(1 - \alpha)\%$ **one-sided confidence interval** for μ. That is, with the confidence coefficient $1 - \alpha$, $\overline{x} - z_\alpha(\sigma/\sqrt{n})$ is a lower bound for μ. Similarly, $(-\infty, \overline{x} + z_\alpha(\sigma/\sqrt{n})]$ is a one-sided confidence interval for μ and $\overline{x} + z_\alpha(\sigma/\sqrt{n})$ provides an upper bound for μ with confidence coefficient $1 - \alpha$.

When σ is unknown, we would use $T = (\overline{X} - \mu)/(S/\sqrt{n})$ to find the corresponding lower or upper bounds for μ, namely, $\overline{x} - t_\alpha(n - 1)(s/\sqrt{n})$ and $\overline{x} + t_\alpha(n - 1)(s/\sqrt{n})$.

EXERCISES

6.2-1. A random sample of size 16 from the normal distribution $N(\mu, 25)$ yielded $\overline{x} = 73.8$. Find a 95% confidence interval for μ.

6.2-2. A random sample of size 8 from $N(\mu, 72)$ yielded $\overline{x} = 85$. Find the following confidence intervals for μ:

(a) 99%. (b) 95%. (c) 90%. (d) 80%.

6.2-3. To determine the effect of 100% nitrate on the growth of pea plants, several specimens were planted and then watered with 100% nitrate every day. At the end of two weeks, the plants were measured. Here are data on seven of them:

$$17.5 \quad 14.5 \quad 15.2 \quad 14.0 \quad 17.3 \quad 18.0 \quad 13.8$$

Assume that these data are observations from a normal distribution $N(\mu, \sigma^2)$.

(a) Find the value of a point estimate of μ.
(b) Find the value of a point estimate of σ.
(c) Give the endpoints for a 90% confidence interval for μ.

6.2-4. Let X equal the weight in grams of a "52-gram" snack pack of candies. Assume that the distribution of X is $N(\mu, 4)$. A random sample of $n = 10$ observations of X yielded the following data:

$$55.95 \quad 56.54 \quad 57.58 \quad 55.13 \quad 57.48$$
$$56.06 \quad 59.93 \quad 58.30 \quad 52.57 \quad 58.46$$

(a) Give a point estimate for μ.

(b) Find the endpoints for a 95% confidence interval for μ.
(c) On the basis of these very limited data, what is the probability that an individual snack pack selected at random is filled with less than 52 grams of candy?

6.2-5. As a clue to the amount of organic waste in Lake Macatawa (see Example 6.2-4), a count was made of the number of bacteria colonies in 100 milliliters of water. The number of colonies, in hundreds, for $n = 30$ samples of water from the east basin yielded

$$\begin{array}{cccccccccc} 93 & 140 & 8 & 120 & 3 & 120 & 33 & 70 & 91 & 61 \\ 7 & 100 & 19 & 98 & 110 & 23 & 14 & 94 & 57 & 9 \\ 66 & 53 & 28 & 76 & 58 & 9 & 73 & 49 & 37 & 92 \end{array}$$

Find an approximate 90% confidence interval for the mean number (say, μ_E) of colonies in 100 milliliters of water in the east basin.

6.2-6. To determine whether the bacteria count was lower in the west basin of Lake Macatawa than in the east basin, $n = 37$ samples of water were taken from the west basin and the number of bacteria colonies in 100 milliliters of water was counted. The sample characteristics were $\overline{x} = 11.95$ and $s = 11.80$, measured in hundreds of colonies. Find the approximate 95% confidence interval for the mean number of colonies (say, μ_W) in 100 milliliters of water in the west basin.

6.2-7. Thirteen tons of cheese including "22-pound" wheels (label weight), is stored in some old gypsum mines. A random sample of $n = 9$ of these wheels yielded the following weights in pounds:

$$21.50 \quad 18.95 \quad 18.55 \quad 19.40 \quad 19.15$$
$$22.35 \quad 22.90 \quad 22.20 \quad 23.10$$

Assuming that the distribution of the weights of the wheels of cheese is $N(\mu, \sigma^2)$, find a 95% confidence interval for μ.

6.2-8. Assume that the yield per acre for a particular variety of soybeans is $N(\mu, \sigma^2)$. For a random sample of $n = 5$ plots, the yields in bushels per acre were 37.4, 48.8, 46.9, 55.0, and 44.0.

(a) Give a point estimate for μ.
(b) Find a 90% confidence interval for μ.

6.2-9. The grip strengths of 32 Hope College coeds were measured at the beginning of a health fitness program. The following data were obtained:

$$43 \quad 43 \quad 49 \quad 55 \quad 45 \quad 33 \quad 56 \quad 43 \quad 45 \quad 40 \quad 42 \quad 53 \quad 50 \quad 39 \quad 43 \quad 33$$
$$47 \quad 35 \quad 21 \quad 35 \quad 37 \quad 30 \quad 36 \quad 44 \quad 37 \quad 35 \quad 29 \quad 39 \quad 49 \quad 27 \quad 24 \quad 29$$

(a) Find the sample mean and the sample standard deviation for these grip strengths.
(b) Construct a (an approximate) 95% confidence interval for the mean, μ.

6.2-10. During the Friday night shift, $n = 28$ mints were selected at random from a production line and weighed. They had an average weight of $\bar{x} = 21.45$ grams and $s = 0.31$ grams. Give the lower endpoint of a 90% one-sided confidence interval for μ, the mean weight of all the mints.

6.2-11. In Exercise 3.1-5, 96 fuel injector pumps were selected randomly from the production line. Plungers force the fuel out of the pumps. The differences in diameter of the plungers from that of an absolute minimum were measured in microns. Use those data for this question.

(a) Give the value of a point estimate of μ.
(b) Determine point estimates of σ^2 and σ.
(c) Find a 90% confidence interval for μ.

6.2-12. A leakage test was conducted to determine the effectiveness of a seal designed to keep the inside of a plug airtight. An air needle was inserted into the plug, and the plug and needle were placed under water. The pressure was then increased until leakage was observed. Let X equal the pressure in pounds per square inch. Assume that the distribution of X is $N(\mu, \sigma^2)$. The following $n = 10$ observations of X were obtained:

$$3.1 \quad 3.3 \quad 4.5 \quad 2.8 \quad 3.5 \quad 3.5 \quad 3.7 \quad 4.2 \quad 3.9 \quad 3.3$$

Use the observations to

(a) Find a point estimate of μ.
(b) Find a point estimate of σ.
(c) Find a 95% one-sided confidence interval for μ that provides an upper bound for μ.

6.2-13. In groundwater research, it is often important to know the characteristics of the soil at a certain site. For example, let X equal the diameter of an individual grain of soil, and assume that the distribution of X is $N(\mu, \sigma^2)$. (See Example 3.6-9, where this assumption is discussed.) We again give the diameters of $n = 30$ individual grains:

$$1.24 \quad 1.36 \quad 1.28 \quad 1.31 \quad 1.35 \quad 1.20 \quad 1.39 \quad 1.35 \quad 1.41 \quad 1.31$$
$$1.28 \quad 1.26 \quad 1.37 \quad 1.49 \quad 1.32 \quad 1.40 \quad 1.33 \quad 1.28 \quad 1.25 \quad 1.39$$
$$1.38 \quad 1.34 \quad 1.40 \quad 1.27 \quad 1.33 \quad 1.36 \quad 1.43 \quad 1.33 \quad 1.29 \quad 1.34$$

Find a 90% confidence interval for μ.

6.2-14. In nuclear physics, detectors are often used to measure the energy of a particle. To calibrate a detector, particles of known energy are directed into it. The values of signals from 15 different detectors, for the same energy, are

$$260 \quad 216 \quad 259 \quad 206 \quad 265 \quad 284 \quad 291 \quad 229$$
$$232 \quad 250 \quad 225 \quad 242 \quad 240 \quad 252 \quad 236$$

(a) Find a 95% confidence interval for μ, assuming that these are observations from a $N(\mu, \sigma^2)$ distribution.

(b) Construct a box-and-whisker diagram of the data.
(c) Are these detectors doing a good job or a poor job of putting out the same signal for the same input energy?

6.2-15. Students took $n = 35$ samples of water from the east basin of Lake Macatawa (see Example 6.2-4) and measured the amount of sodium in parts per million. For their data, they calculated $\bar{x} = 24.11$ and $s^2 = 24.44$. Find an approximate

90% confidence interval for μ, the mean of the amount of sodium in parts per million.

6.2-16. Let $X_1, X_2, \ldots, X_n$ be a random sample of size n from the normal distribution $N(\mu, \sigma^2)$. Calculate the expected length of a 95% confidence interval for μ, assuming that $n = 5$ and the variance is

(a) known,

(b) unknown.

HINT: To find $E(S)$, first determine $E[\sqrt{(n-1)S^2/\sigma^2}]$, recalling that $(n-1)S^2/\sigma^2$ is $\chi^2(n-1)$. (See also Exercise 6.1-14.)

6.2-17. Let X equal the weight (in pounds) of a "12-ounce" can of buttermilk biscuits. Assume that

the distribution of X is $N(\mu, \sigma^2)$. Use the following 18 net weights to find a 95% one-sided confidence interval for μ that provides a lower bound for μ:

0.75 0.78 0.77 0.75 0.77 0.76 0.78 0.76 0.78
0.78 0.75 0.74 0.78 0.75 0.74 0.79 0.75 0.77

6.2-18. A manufacturer of soap powder packages the soap in "6-pound" boxes. To check the filling machine, a sample of $n = 1219$ boxes was weighed. Assuming that $\bar{x} = 6.05$ pounds and $s = 0.02$ pounds, give the endpoints for a 99% confidence interval for μ, the mean weight of the boxes of soap filled by this machine.

6.2-19. A study was conducted to measure (1) the amount of cervical spine movement induced by different methods of gaining access to the mouth and nose to begin resuscitation of a football player who is wearing a helmet and (2) the time it takes to complete each method. One method involves using a manual screwdriver to remove the side clips holding the face mask in place and then flipping the mask up. Twelve measured times in seconds for the manual screwdriver are

$$33.8 \quad 31.6 \quad 28.5 \quad 29.9 \quad 29.8 \quad 26.0 \quad 35.7 \quad 27.2 \quad 29.1 \quad 32.1 \quad 26.1 \quad 24.1$$

Assume that these are independent observations of a normally distributed random variable that is $N(\mu, \sigma^2)$.

(a) Find point estimates of μ and σ.

(b) Find a 95% one-sided confidence interval for μ that provides an upper bound for μ.

(c) Does the assumption of normality seem to be justified? Why?

6.2-20. Let X equal the weight of an unbreaded $2.00 fish fry at a local restaurant in East Greenbush, New York. Assume that the distribution of X is $N(\mu, \sigma^2)$. A random sample of $n = 24$ weights (in ounces) was

4.4 3.8 5.1 4.6 4.5 4.5 4.8 4.1
3.9 4.2 4.4 4.9 5.0 4.3 4.4 3.6
5.2 4.8 4.4 4.6 4.6 5.0 4.0 4.5

(a) Find point estimates of μ, σ^2, and σ.

(b) Find a 95% one-sided confidence interval for μ that gives a lower bound for μ.

(c) Does the assumption that these weights are normal seem to be justified? Why?

6.2-21. An automotive supplier of interior parts places several electrical wires in a harness. A pull test measures the force required to pull spliced wires apart. A customer requires that each wire spliced into the harness must withstand a pull force of 20 pounds. Let X equal the force required to pull 20 gauge wires apart. Assume that the distribution of X is $N(\mu, \sigma^2)$. The following data give 20 observations of X:

$$28.8 \quad 24.4 \quad 30.1 \quad 25.6 \quad 26.4 \quad 23.9 \quad 22.1 \quad 22.5 \quad 27.6 \quad 28.1$$
$$20.8 \quad 27.7 \quad 24.4 \quad 25.1 \quad 24.6 \quad 26.3 \quad 28.2 \quad 22.2 \quad 26.3 \quad 24.4$$

(a) Find point estimates for μ and σ.

(b) Find a 99% one-sided confidence interval for μ that provides a lower bound for μ.

6.2-22. The weights of 100 laptop computers are given in Exercise 3.1-4. Use those 100 weights as a random sample of weights from the population of this type of computer.

(a) Find point estimates for the mean and standard deviation of laptop computer weights.

(b) Find an approximate 95% confidence interval for the mean weight of a laptop computer.

6.3 CONFIDENCE INTERVALS FOR THE DIFFERENCE OF TWO MEANS

Suppose that we are interested in comparing the means of two normal distributions. Let $X_1, X_2, \ldots, X_n$ and $Y_1, Y_2, \ldots, Y_m$ be, respectively, two independent random samples of sizes n and m from the two normal distributions $N(\mu_X, \sigma_X^2)$ and $N(\mu_Y, \sigma_Y^2)$.

Suppose, for now, that σ_X^2 and σ_Y^2 are known. The random samples are independent; thus, the respective sample means $\overline{X}$ and $\overline{Y}$ are also independent and have distributions $N(\mu_X, \sigma_X^2/n)$ and $N(\mu_Y, \sigma_Y^2/m)$. Consequently, the distribution of $W = \overline{X} - \overline{Y}$ is $N(\mu_X - \mu_Y, \sigma_X^2/n + \sigma_Y^2/m)$ and

$$P\left(-z_{\alpha/2} \leq \frac{(\overline{X} - \overline{Y}) - (\mu_X - \mu_Y)}{\sqrt{\sigma_X^2/n + \sigma_Y^2/m}} \leq z_{\alpha/2}\right) = 1 - \alpha,$$

which can be rewritten as

$$P[(\overline{X} - \overline{Y}) - z_{\alpha/2}\sigma_W \leq \mu_X - \mu_Y \leq (\overline{X} - \overline{Y}) + z_{\alpha/2}\sigma_W] = 1 - \alpha,$$

where $\sigma_W = \sqrt{\sigma_X^2/n + \sigma_Y^2/m}$ is the standard deviation of $\overline{X} - \overline{Y}$. Once the experiments have been performed and the means $\bar{x}$ and $\bar{y}$ computed, the interval

$$[\bar{x} - \bar{y} - z_{\alpha/2}\sigma_W, \bar{x} - \bar{y} + z_{\alpha/2}\sigma_W]$$

or, equivalently, $\bar{x} - \bar{y} \pm z_{\alpha/2}\sigma_W$ provides a $100(1 - \alpha)\%$ confidence interval for $\mu_X - \mu_Y$. Note that this interval is centered at the point estimate $\bar{x} - \bar{y}$ of $\mu_X - \mu_Y$ and is completed by subtracting and adding the product of $z_{\alpha/2}$ and the standard deviation of the point estimator.

EXAMPLE 6.3-1 In the preceding discussion, let $n = 15, m = 8, \bar{x} = 70.1, \bar{y} = 75.3, \sigma_X^2 = 60, \sigma_Y^2 = 40$, and $1 - \alpha = 0.90$. Thus, $1 - \alpha/2 = 0.95 = \Phi(1.645)$. Hence,

$$1.645\sigma_W = 1.645\sqrt{\frac{60}{15} + \frac{40}{8}} = 4.935,$$

and, since $\bar{x} - \bar{y} = -5.2$, it follows that

$$[-5.2 - 4.935, -5.2 + 4.935] = [-10.135, -0.265]$$

is a 90% confidence interval for $\mu_X - \mu_Y$. Because the confidence interval does not include zero, we suspect that μ_Y is greater than μ_X. ∎

If the sample sizes are large and σ_X and σ_Y are unknown, we can replace σ_X^2 and σ_Y^2 with s_x^2 and s_y^2, where s_x^2 and s_y^2 are the values of the respective unbiased estimates of the variances. This means that

$$\bar{x} - \bar{y} \pm z_{\alpha/2}\sqrt{\frac{s_x^2}{n} + \frac{s_y^2}{m}}$$

serves as an approximate $100(1 - \alpha)\%$ confidence interval for $\mu_X - \mu_Y$.

Now consider the problem of constructing confidence intervals for the difference of the means of two normal distributions when the variances are unknown but the sample sizes are small. Let $X_1, X_2, \ldots, X_n$ and $Y_1, Y_2, \ldots, Y_m$ be two independent random samples from the distributions $N(\mu_X, \sigma_X^2)$ and $N(\mu_Y, \sigma_Y^2)$, respectively. If the sample sizes are not large (say, considerably smaller than 30), this problem can be a difficult one. However, even in these cases, if we can assume common, but unknown, variances (say, $\sigma_X^2 = \sigma_Y^2 = \sigma^2$), there is a way out of our difficulty.

We know that

$$Z = \frac{\overline{X} - \overline{Y} - (\mu_X - \mu_Y)}{\sqrt{\sigma^2/n + \sigma^2/m}}$$

is $N(0, 1)$. Moreover, since the random samples are independent,

$$U = \frac{(n-1)S_X^2}{\sigma^2} + \frac{(m-1)S_Y^2}{\sigma^2}$$

is the sum of two independent chi-square random variables; thus, the distribution of U is $\chi^2(n + m - 2)$. In addition, the independence of the sample means and sample variances implies that Z and U are independent. According to the definition of a T random variable,

$$T = \frac{Z}{\sqrt{U/(n + m - 2)}}$$

has a t distribution with $n + m - 2$ degrees of freedom. That is,

$$T = \frac{\dfrac{\overline{X} - \overline{Y} - (\mu_X - \mu_Y)}{\sqrt{\sigma^2/n + \sigma^2/m}}}{\sqrt{\left[\dfrac{(n-1)S_X^2}{\sigma^2} + \dfrac{(m-1)S_Y^2}{\sigma^2}\right]\Big/ (n + m - 2)}}$$

$$= \frac{\overline{X} - \overline{Y} - (\mu_X - \mu_Y)}{\sqrt{\left[\dfrac{(n-1)S_X^2 + (m-1)S_Y^2}{n + m - 2}\right]\left[\dfrac{1}{n} + \dfrac{1}{m}\right]}}$$

has a t distribution with $r = n + m - 2$ degrees of freedom. Thus, with $t_0 = t_{\alpha/2}(n + m - 2)$, we have

$$P(-t_0 \leq T \leq t_0) = 1 - \alpha.$$

Solving the inequalities for $\mu_X - \mu_Y$ yields

$$P\left(\overline{X} - \overline{Y} - t_0 S_P\sqrt{\frac{1}{n} + \frac{1}{m}} \leq \mu_X - \mu_Y \leq \overline{X} - \overline{Y} + t_0 S_P\sqrt{\frac{1}{n} + \frac{1}{m}}\right),$$

where the pooled estimator of the common standard deviation is

$$S_P = \sqrt{\frac{(n-1)S_X^2 + (m-1)S_Y^2}{n + m - 2}}.$$

If $\overline{x}, \overline{y}$, and s_p are the observed values of $\overline{X}, \overline{Y}$, and S_P, then

$$\left[\overline{x} - \overline{y} - t_0 s_p\sqrt{\frac{1}{n} + \frac{1}{m}}, \overline{x} - \overline{y} + t_0 s_p\sqrt{\frac{1}{n} + \frac{1}{m}}\right]$$

is a $100(1 - \alpha)\%$ confidence interval for $\mu_X - \mu_Y$.

EXAMPLE 6.3-2 Suppose that scores on a standardized test in mathematics taken by students from large and small high schools are $N(\mu_X, \sigma^2)$ and $N(\mu_Y, \sigma^2)$, respectively, where σ^2 is unknown. If a random sample of $n = 9$ students from large high schools yielded $\bar{x} = 81.31$, $s_x^2 = 60.76$, and a random sample of $m = 15$ students from small high schools yielded $\bar{y} = 78.61$, $s_y^2 = 48.24$, then the endpoints for a 95% confidence interval for $\mu_X - \mu_Y$ are given by

$$81.31 - 78.61 \pm 2.074 \sqrt{\frac{8(60.76) + 14(48.24)}{22}} \sqrt{\frac{1}{9} + \frac{1}{15}}$$

because $t_{0.025}(22) = 2.074$. The 95% confidence interval is $[-3.65, 9.05]$. ∎

REMARK   The assumption of equal variances, namely, $\sigma_X^2 = \sigma_Y^2$, can be modified somewhat so that we are still able to find a confidence interval for $\mu_X - \mu_Y$. That is, if we know the ratio σ_X^2/σ_Y^2 of the variances, we can still make this type of statistical inference by using a random variable with a t distribution. (See Exercise 6.3-10.) However, if we do not know the ratio of the variances and yet suspect that the unknown σ_X^2 and σ_Y^2 differ by a great deal, what do we do? It is safest to return to

$$\frac{\overline{X} - \overline{Y} - (\mu_X - \mu_Y)}{\sqrt{\sigma_X^2/n + \sigma_Y^2/m}}$$

for the inference about $\mu_X - \mu_Y$ but replacing σ_X^2 and σ_Y^2 by their respective estimators S_X^2 and S_Y^2. That is, consider

$$W = \frac{\overline{X} - \overline{Y} - (\mu_X - \mu_Y)}{\sqrt{S_X^2/n + S_Y^2/m}}.$$

What is the distribution of W? As before, we note that if n and m are large enough and the underlying distributions are close to normal (or at least not badly skewed), then W has an approximate normal distribution and a confidence interval for $\mu_X - \mu_Y$ can be found by considering

$$P(-z_{\alpha/2} \leq W \leq z_{\alpha/2}) \approx 1 - \alpha.$$

However, for smaller n and m, Welch has proposed a Student's t distribution as the approximating one for W. Welch's proposal was later modified by Aspin. [See A. A. Aspin, "Tables for Use in Comparisons Whose Accuracy Involves Two Variances, Separately Estimated," *Biometrika*, **36** (1949), pp. 290–296, with an appendix by B. L. Welch in which he makes the suggestion used here.] The approximating Student's t distribution has r degrees of freedom, where

$$\frac{1}{r} = \frac{c^2}{n-1} + \frac{(1-c)^2}{m-1} \quad \text{and} \quad c = \frac{s_x^2/n}{s_x^2/n + s_y^2/m}.$$

An equivalent formula for r is

$$r = \frac{\left(\dfrac{s_x^2}{n} + \dfrac{s_y^2}{m}\right)^2}{\dfrac{1}{n-1}\left(\dfrac{s_x^2}{n}\right)^2 + \dfrac{1}{m-1}\left(\dfrac{s_y^2}{m}\right)^2}. \tag{6.3-1}$$

In particular, the assignment of r by this rule provides protection in the case in which the smaller sample size is associated with the larger variance by greatly reducing the number of degrees of freedom from the usual $n + m - 2$. Of course, this reduction increases the value of $t_{\alpha/2}$. If r is not an integer, then use the greatest integer in r; that is, $\lfloor r \rfloor$, the "floor," or greatest integer in r, is the number of degrees of freedom associated with the approximating Student's t distribution. An approximate $100(1 - \alpha)$ percent confidence interval for $\mu_X - \mu_Y$ is given by

$$\bar{x} - \bar{y} \pm t_{\alpha/2}(r) \sqrt{\frac{s_x^2}{n} + \frac{s_y^2}{m}}.$$

It is interesting to consider the two-sample T in more detail. It is

$$T = \frac{\overline{X} - \overline{Y} - (\mu_X - \mu_Y)}{\sqrt{\dfrac{(n-1)S_X^2 + (m-1)S_Y^2}{n + m - 2}\left(\dfrac{1}{n} + \dfrac{1}{m}\right)}} \tag{6.3-2}$$

$$= \frac{\overline{X} - \overline{Y} - (\mu_X - \mu_Y)}{\sqrt{\left[\dfrac{(n-1)S_X^2}{nm} + \dfrac{(m-1)S_Y^2}{nm}\right]\left[\dfrac{n+m}{n+m-2}\right]}}.$$

Now, since $(n-1)/n \approx 1$, $(m-1)/m \approx 1$, and $(n+m)/(n+m-2) \approx 1$, we have

$$T \approx \frac{\overline{X} - \overline{Y} - (\mu_X - \mu_Y)}{\sqrt{\dfrac{S_X^2}{m} + \dfrac{S_Y^2}{n}}}.$$

We note that, in this form, each variance is divided by the wrong sample size! That is, if the sample sizes are large or the variances known, we would like

$$\sqrt{\frac{S_X^2}{n} + \frac{S_Y^2}{m}} \qquad \text{or} \qquad \sqrt{\frac{\sigma_X^2}{n} + \frac{\sigma_Y^2}{m}}$$

in the denominator; so T seems to change the sample sizes. Thus, using this T is particularly bad when the sample sizes and the variances are unequal; hence, caution must be taken in using that T to construct a confidence interval for $\mu_X - \mu_Y$. That is, if $n < m$ and $\sigma_X^2 < \sigma_Y^2$, then T does not have a distribution which is close to that of a Student t-distribution with $n + m - 2$ degrees of freedom: Instead, its spread is much less than the Student t's, as the term s_y^2/n in the denominator is much larger than it should be. By contrast, if $m < n$ and $\sigma_X^2 < \sigma_Y^2$, then $s_x^2/m + s_y^2/n$ is generally smaller than it should be and the distribution of T is spread out more than that of the Student t.

There is a way out of this difficulty, however: When the underlying distributions are close to normal, but the sample sizes and the variances are seemingly much different, we suggest the use of

$$W = \frac{\overline{X} - \overline{Y} - (\mu_X - \mu_Y)}{\sqrt{\dfrac{S_X^2}{n} + \dfrac{S_Y^2}{m}}}, \tag{6.3-3}$$

where Welch proved that W has an approximate t distribution with $\lfloor r \rfloor$ degrees of freedom, with the number of degrees of freedom given by Equation 6.3-1. ∎

EXAMPLE 6.3-3 To help understand the preceding remarks, a simulation was done with *Maple*. In order to obtain a q–q plot of the quantiles of a t distribution, a CAS or some type of computer program is very important because of the challenge in finding these quantiles.

Maple was used to simulate $N = 500$ observations of T (Equation 6.3-2) and $N = 500$ observations of W (Equation 6.3-3). In Figure 6.3-1, $n = 6$, $m = 18$, the X

T Observations, T(22) p.d.f. Superimposed

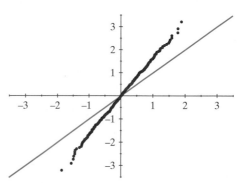

T(22) Quantiles versus T Order Statistics

W Observations, T(19) p.d.f. Superimposed

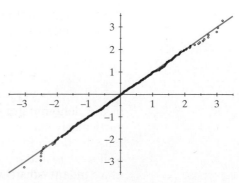

T(19) Quantiles versus W Order Statistics

FIGURE 6.3-1: Observations of T and of W, $n = 6$, $m = 18$, $\sigma_X^2 = 1$, $\sigma_Y^2 = 36$

T Observations, $T(22)$ p.d.f. Superimposed

$T(22)$ Quantiles versus T Order Statistics

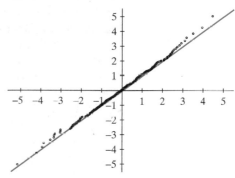

W Observations, $T(5)$ p.d.f. Superimposed

$T(5)$ Quantiles versus W Order Statistics

FIGURE 6.3-2: Observations of T and of W, $n = 18$, $m = 6$, $\sigma_X^2 = 1$, $\sigma_Y^2 = 36$

observations were generated from the $N(0,1)$ distribution, and the Y observations were generated from the $N(0,36)$ distribution. For the value of r for Welch's approximate t distribution, we used the distribution variances rather than the sample variances so that we could use the same r for each of the 500 values of W.

For the simulation results shown in Figure 6.3-2, $n = 18$, $m = 6$, the X observations were generated from the $N(0,1)$ distribution, and the Y observations were generated from the $N(0,36)$ distribution. In both cases, Welch's W with a corrected number of r degrees of freedom is much better than the usual T when the variances and sample sizes are unequal, as they are in these examples.   ■

In some applications, two measurements—say, X and Y—are taken on the same subject. In these cases, X and Y are dependent random variables. Many times these are "before" and "after" measurements, such as weight before and after participating in a diet-and-exercise program. To compare the means of X and Y, it is not permissible to use the t statistics and confidence intervals that we just developed, because in that situation X and Y are independent. Instead, we proceed as follows.

Let $(X_1, Y_1), (X_2, Y_2), \ldots, (X_n, Y_n)$ be n pairs of dependent measurements. Let $D_i = X_i - Y_i$, $i = 1, 2, \ldots, n$. Suppose that $D_1, D_2, \ldots, D_n$ can be thought of as a random sample from $N(\mu_D, \sigma_D^2)$, where μ_D and σ_D are the mean and standard deviation of each difference. To form a confidence interval for $\mu_X - \mu_Y$, use

$$T = \frac{\overline{D} - \mu_D}{S_D/\sqrt{n}},$$

where $\overline{D}$ and S_D are, respectively, the sample mean and sample standard deviation of the n differences. Thus, T is a t statistic with $n - 1$ degrees of freedom. The endpoints for a $100(1 - \alpha)\%$ confidence interval for $\mu_D = \mu_X - \mu_Y$ are then

$$\overline{d} \pm t_{\alpha/2}(n - 1)\frac{s_d}{\sqrt{n}},$$

where $\overline{d}$ and s_d are the observed mean and standard deviation of the sample of the D values. Of course, this is like the confidence interval for a single mean, presented in the last section.

EXAMPLE 6.3-4 An experiment was conducted to compare people's reaction times to a red light versus a green light. When signaled with either the red or the green light, the subject was asked to hit a switch to turn off the light. When the switch was hit, a clock was turned off and the reaction time in seconds was recorded. The following results give the reaction times for eight subjects:

Subject	Red (X)	Green (Y)	$D = X - Y$
1	0.30	0.43	−0.13
2	0.23	0.32	−0.09
3	0.41	0.58	−0.17
4	0.53	0.46	0.07
5	0.24	0.27	−0.03
6	0.36	0.41	−0.05
7	0.38	0.38	0.00
8	0.51	0.61	−0.10

For these data, $\overline{d} = -0.0625$ and $s_d = 0.0765$. To form a 95% confidence interval for $\mu_D = \mu_X - \mu_Y$, we find, from Table VI in the appendix, that $t_{0.025}(7) = 2.365$. Thus, the endpoints for the confidence interval are

$$-0.0625 \pm 2.365\frac{0.0765}{\sqrt{8}}, \qquad \text{or} \qquad [-0.1265, 0.0015].$$

In this very limited data set, zero is included in the confidence interval, but is close to the endpoint 0.0015. We suspect that if more data were taken, zero might not be included in the confidence interval. If that actually were to happen, it would seem that people react faster to a red light. ■

Of course, we can find one-sided confidence intervals for the difference of the means, $\mu_X - \mu_Y$. Suppose we believe that we have changed some characteristic of the X distribution and created a Y distribution such that we think that $\mu_X > \mu_Y$. Let us find a one-sided 95% confidence interval that is a lower bound for $\mu_X - \mu_Y$. Say this lower bound is greater than zero. Then we would feel 95% confident that the mean μ_X is larger than the mean μ_Y. That is, the change that was made seemed to decrease the mean; this would be good in some cases, such as golf or racing. In other cases, in which we hope the change would be such that $\mu_X < \mu_Y$, we would find a one-sided confidence interval which is an upper bound for $\mu_X - \mu_Y$, and we would hope that it would be less than zero. These ideas are illustrated in Exercises 6.3-5, 6.3-12, and 6.3-13.

EXERCISES

6.3-1. The length of life of brand X light bulbs is assumed to be $N(\mu_X, 784)$. The length of life of brand Y light bulbs is assumed to be $N(\mu_Y, 627)$ and independent of X. If a random sample of $n = 56$ brand X light bulbs yielded a mean of $\bar{x} = 937.4$ hours and a random sample of size $m = 57$ brand Y light bulbs yielded a mean of $\bar{y} = 988.9$ hours, find a 90% confidence interval for $\mu_X - \mu_Y$.

6.3-2. Let $X_1, X_2, \ldots, X_5$ be a random sample of SAT mathematics scores, assumed to be $N(\mu_X, \sigma^2)$, and let $Y_1, Y_2, \ldots, Y_8$ be an independent random sample of SAT verbal scores, assumed to be $N(\mu_Y, \sigma^2)$. If the following data are observed, find a 90% confidence interval for $\mu_X - \mu_Y$:

$$x_1 = 644 \quad x_2 = 493 \quad x_3 = 532 \quad x_4 = 462 \quad x_5 = 565$$
$$y_1 = 623 \quad y_2 = 472 \quad y_3 = 492 \quad y_4 = 661 \quad y_5 = 540$$
$$y_6 = 502 \quad y_7 = 549 \quad y_8 = 518$$

6.3-3. Independent random samples of the heights of adult males living in two countries yielded the following results: $n = 12$, $\bar{x} = 65.7$ inches, $s_x = 4$ inches; and $m = 15$, $\bar{y} = 68.2$ inches, $s_y = 3$ inches. Find an approximate 98% confidence interval for the difference $\mu_X - \mu_Y$ of the means of the populations of heights. Assume that $\sigma_X^2 = \sigma_Y^2$.

6.3-4. [*Medicine and Science in Sports and Exercise* (January 1990)] Let X and Y equal, respectively, the blood volumes in milliliters for a male who is a paraplegic and participates in vigorous physical activities and for a male who is able bodied and participates in everyday, ordinary activities. Assume that X is $N(\mu_X, \sigma_X^2)$ and Y is $N(\mu_Y, \sigma_Y^2)$. Following are $n = 7$ observations of X:

1612 1352 1456 1222 1560 1456 1924

Following are $m = 10$ observations of Y:

1082 1300 1092 1040 910
1248 1092 1040 1092 1288

Use the observations of X and Y to

(a) Give a point estimate for $\mu_X - \mu_Y$.
(b) Find a 95% confidence interval for $\mu_X - \mu_Y$. Since the variances σ_X^2 and σ_Y^2 might not be equal, use Welch's T.

6.3-5. A biologist who studies spiders was interested in comparing the lengths of female and male green lynx spiders. Assume that the length X of the male spider is approximately $N(\mu_X, \sigma_X^2)$ and the length Y of the female spider is approximately $N(\mu_Y, \sigma_Y^2)$. Following are $n = 30$ observations of X:

5.20	4.70	5.75	7.50	6.45	6.55
4.70	4.80	5.95	5.20	6.35	6.95
5.70	6.20	5.40	6.20	5.85	6.80
5.65	5.50	5.65	5.85	5.75	6.35
5.75	5.95	5.90	7.00	6.10	5.80

Following are $m = 30$ observations of Y:

8.25	9.95	5.90	7.05	8.45	7.55
9.80	10.80	6.60	7.55	8.10	9.10
6.10	9.30	8.75	7.00	7.80	8.00
9.00	6.30	8.35	8.70	8.00	7.50
9.50	8.30	7.05	8.30	7.95	9.60

The units of measurement for both sets of observations are millimeters. Find an approximate one-sided 95% confidence interval that is an upper bound for $\mu_X - \mu_Y$.

6.3-6. Consider the butterfat production (in pounds) of a cow during a 305-day milk production period following the birth of a calf. Let X and Y equal the butterfat production for such cows on a farm in Wisconsin and a farm in Michigan. Twelve observations of X are

649	657	714	877	975	468
567	849	721	791	874	405

Sixteen observations of Y are

699	891	632	815	589	764	524	727
597	868	652	978	479	733	549	790

(a) Assuming that X is $N(\mu_X, \sigma^2)$ and Y is $N(\mu_Y, \sigma^2)$, find a 95% confidence interval for $\mu_X - \mu_Y$.
(b) Construct box-and-whisker diagrams for these two sets of data on the same graph.
(c) Does there seem to be a significant difference in butterfat production for cows on these two farms?

6.3-7. An automotive supplier is considering changing its electrical wire harness to save money. The idea is to replace a current 20-gauge wire with a 22-gauge wire. Since not all wires in the harness can be changed, the new wire must work with the current wire splice process. To determine whether the new wire is compatible, random samples were selected and measured with with a pull test. A pull test measures the force required to pull the spliced wires apart. The minimum pull force required by the customer is 20 pounds. Twenty observations of the forces needed for the current wire are

28.8 24.4 30.1 25.6 26.4 23.9 22.1 22.5 27.6 28.1
20.8 27.7 24.4 25.1 24.6 26.3 28.2 22.2 26.3 24.4

Twenty observations of the forces needed for the new wire are

14.1 12.2 14.0 14.6 8.5 12.6 13.7 14.8 14.1 13.2
12.1 11.4 10.1 14.2 13.6 13.1 11.9 14.8 11.1 13.5

(a) Does the current wire meet the customer's specifications?

(b) Find a 90% confidence interval for the difference of the means for these two sets of wire.

(c) Construct box-and-whisker diagrams of the two sets of data on the same figure.

(d) What is your recommendation for this company?

6.3-8. A test was conducted to determine whether a wedge on the end of a plug fitting designed to hold a seal onto the plug was doing its job. The data taken were in the form of measurements of the force required to remove a seal from the plug with the wedge in place (say, X) and the force required without the plug (say, Y). Assume that the distributions of X and Y are $N(\mu_X, \sigma^2)$ and $N(\mu_Y, \sigma^2)$, respectively. Ten independent observations of X are

$$3.26 \quad 2.26 \quad 2.62 \quad 2.62 \quad 2.36 \quad 3.00 \quad 2.62 \quad 2.40 \quad 2.30 \quad 2.40$$

Ten independent observations of Y are

$$1.80 \quad 1.46 \quad 1.54 \quad 1.42 \quad 1.32 \quad 1.56 \quad 1.36 \quad 1.64 \quad 2.00 \quad 1.54$$

(a) Find a 95% confidence interval for $\mu_X - \mu_Y$.

(b) Construct box-and-whisker diagrams of these data on the same figure.

(c) Is the wedge necessary?

6.3-9. Students are weighed (in kilograms) at the beginning and the end of a semester-long health-fitness program. Let the random variable D equal the weight change for a student (post-program weight minus pre-program weight). Assume that the distribution of D is $N(\mu_D, \sigma_D^2)$. A random sample of $n = 12$ female students yielded the following observations of D:

2.0	−0.5	1.4	−2.2	0.3	−0.8
3.7	−0.1	0.6	0.2	0.9	−0.1

(a) Give a point estimate of μ_D.

(b) Find a 95% confidence interval for μ_D.

6.3-10. Let $\overline{X}$, $\overline{Y}$, S_X^2, and S_Y^2 be the respective sample means and unbiased estimates of the variances obtained from independent samples of sizes n and m from the normal distributions $N(\mu_X, \sigma_X^2)$ and $N(\mu_Y, \sigma_Y^2)$, where μ_X, μ_Y, σ_X^2, and σ_Y^2 are unknown. If $\sigma_X^2/\sigma_Y^2 = d$, a known constant,

(a) Argue that $\dfrac{(\overline{X} - \overline{Y}) - (\mu_X - \mu_Y)}{\sqrt{d\sigma_Y^2/n + \sigma_Y^2/m}}$ is $N(0,1)$.

(b) Argue that $\dfrac{(n-1)S_X^2}{d\sigma_Y^2} + \dfrac{(m-1)S_Y^2}{\sigma_Y^2}$ is $\chi^2(n + m - 2)$.

(c) Argue that the two random variables in (a) and (b) are independent.

(d) With these results, construct a random variable (not depending upon σ_Y^2) that has a t distribution and that can be used to construct a confidence interval for $\mu_X - \mu_Y$.

6.3-11. Students in a semester-long health-fitness program have their percentage of body fat measured at the beginning of the semester and at the end of the semester. The following measurements give these percentages for 10 men and for 10 women:

Males		Females	
Pre-program %	Post-program %	Pre-program %	Post-program %
11.10	9.97	22.90	22.89
19.50	15.80	31.60	33.47
14.00	13.02	27.70	25.75
8.30	9.28	21.70	19.80
12.40	11.51	19.36	18.00

Males		Females	
Pre-program %	Post-program %	Pre-program %	Post-program %
7.89	7.40	25.03	22.33
12.10	10.70	26.90	25.26
8.30	10.40	25.75	24.90
12.31	11.40	23.63	21.80
10.00	11.95	25.06	24.28

(a) Find a 90% confidence interval for the mean of the difference in the percentages for the males.

(b) Find a 90% confidence interval for the mean of the difference in the percentages for the females.

(c) On the basis of these data, have these percentages decreased?

(d) If possible, check whether each set of differences comes from a normal distribution.

6.3-12. Twenty-four 9th- and 10th-grade high school girls were put on an ultraheavy rope-jumping program. The following data give the time difference for each girl ("before program time" minus "after program time") for the 40-yard dash:

$$\begin{array}{cccccccc}
0.28 & 0.01 & 0.13 & 0.33 & -0.03 & 0.07 & -0.18 & -0.14 \\
-0.33 & 0.01 & 0.22 & 0.29 & -0.08 & 0.23 & 0.08 & 0.04 \\
-0.30 & -0.08 & 0.09 & 0.70 & 0.33 & -0.34 & 0.50 & 0.06
\end{array}$$

(a) Give a point estimate of μ_D, the mean of the difference in race times.

(b) Find a one-sided 95% confidence interval that is a lower bound for μ_D.

(c) Does it look like the rope-jumping program was effective?

6.3-13. The Biomechanics Lab at Hope College tested healthy old women and healthy young women to discover whether or not lower extremity response time to a stimulus is a function of age. Let X and Y respectively equal the independent response times for these two groups when taking steps in the anterior direction. Find a one-sided 95% confidence interval that is a lower bound for $\mu_X - \mu_Y$ if $n = 60$ observations of X yielded $\bar{x} = 671$ and $s_x = 129$ while $m = 60$ observations of Y yielded $\bar{y} = 480$ and $s_y = 93$.

6.3-14. Let X and Y equal the hardness of the hot and cold water, respectively, in a campus building. Hardness is measured in terms of the calcium ion concentration (in ppm). The following data were collected ($n = 12$ observations of X and $m = 10$ observations of Y):

X: 133.5 137.2 136.3 133.3 137.5 135.4
    138.4 137.1 136.5 139.4 137.9 136.8

Y: 134.0 134.7 136.0 132.7 134.6 135.2
    135.9 135.6 135.8 134.2

(a) Calculate the sample means and the sample variances of these data.

(b) Construct a 95% confidence interval for $\mu_X - \mu_Y$, assuming that the distributions of X and Y are $N(\mu_X, \sigma_X^2)$ and $N(\mu_Y, \sigma_Y^2)$, respectively.

(c) Construct box plots of the two sets of data on the same graph.

(d) Do the means seem to be equal or different?

6.3-15. Ledolter and Hogg (see references) report that two rubber compounds were tested for tensile strength. Rectangular materials were prepared and pulled in a longitudinal direction. A sample of 14 specimens, 7 from compound A and 7 from compound B, was prepared, but it was later found that two B specimens were defective and they had to be removed from the test. The tensile strength (in units of 100 pounds per square inch) of the remaining specimens are as follows:

A: 32 30 33 32 29 34 32
B: 33 35 36 37 35

Calculate a 95 percent confidence interval for the difference of the mean tensile strengths of the two rubber compounds. State your assumptions.

6.3-16. Ledolter and Hogg (see references) report that two different fabrics—say, X and Y—are compared on a Martindale wear tester that can compare two materials in a single run. The weights losses (in milligrams) from seven runs are as follows:

X: 36 26 31 38 28 37 22
Y: 39 27 35 42 31 39 21

Analyze these data by constructing a 90% confidence interval for the difference of the means. State your assumptions.

6.4 CONFIDENCE INTERVALS FOR VARIANCES

In this section, we find confidence intervals for the variance of a normal distribution and for the ratio of the variances of two normal distributions. The confidence interval for the variance σ^2 is based on the sample variance

$$S^2 = \frac{1}{n-1} \sum_{i=1}^{n} (X_i - \overline{X})^2.$$

We use the fact that the distribution of $(n-1)S^2/\sigma^2$ is $\chi^2(n-1)$ to find a confidence interval for σ^2. We select constants a and b from Table IV in the appendix with $n-1$ degrees of freedom such that

$$P\left(a \le \frac{(n-1)S^2}{\sigma^2} \le b\right) = 1 - \alpha.$$

One way to do this is by selecting a and b so that $a = \chi^2_{1-\alpha/2}(n-1)$ and $b = \chi^2_{\alpha/2}(n-1)$. That is, we select a and b so that the probabilities in the two tails are equal. Then, solving the inequalities, we have

$$1 - \alpha = P\left(\frac{a}{(n-1)S^2} \le \frac{1}{\sigma^2} \le \frac{b}{(n-1)S^2}\right)$$

$$= P\left(\frac{(n-1)S^2}{b} \le \sigma^2 \le \frac{(n-1)S^2}{a}\right).$$

Thus, the probability that the random interval $[(n-1)S^2/b, (n-1)S^2/a]$ contains the unknown σ^2 is $1 - \alpha$. Once the values of $X_1, X_2, \ldots, X_n$ are observed to be $x_1, x_2, \ldots, x_n$ and s^2 is computed, the interval $[(n-1)s^2/b, (n-1)s^2/a]$ is a $100(1-\alpha)\%$ confidence interval for σ^2. It follows that a $100(1-\alpha)\%$ confidence interval for σ, the standard deviation, is given by

$$\left[\sqrt{\frac{(n-1)s^2}{b}}, \sqrt{\frac{(n-1)s^2}{a}}\right] = \left[\sqrt{\frac{n-1}{b}}\, s, \sqrt{\frac{n-1}{a}}\, s\right].$$

EXAMPLE 6.4-1 Assume that the time in days required for maturation of seeds of a species of Guardiola, a flowering plant found in Mexico, is $N(\mu, \sigma^2)$. A random sample of $n = 13$ seeds, both parents having narrow leaves, yielded $\overline{x} = 18.97$ days and

$$12s^2 = \sum_{i=1}^{13} (x_i - \overline{x})^2 = 128.41.$$

A 90% confidence interval for σ^2 is

$$\left[\frac{128.41}{21.03}, \frac{128.41}{5.226}\right] = [6.11, 24.57]$$

because $5.226 = \chi^2_{0.95}(12)$ and $21.03 = \chi^2_{0.05}(12)$, from Table IV in the appendix. The corresponding 90% confidence interval for σ is

$$[\sqrt{6.11}, \sqrt{24.57}] = [2.47, 4.96].$$

∎

Although a and b are generally selected so that the probabilities in the two tails are equal, the resulting $100(1 - \alpha)\%$ confidence interval is not the shortest that can be formed with the available data. Table X in the appendix gives solutions for a and b that yield confidence intervals of minimum length for the standard deviation. (See Exercise 6.4-16.)

EXAMPLE 6.4-2   Using the data in Example 6.4-1, we find that a 90% confidence interval for σ of minimum length is

$$\left[\sqrt{\frac{128.41}{24.202}}, \sqrt{\frac{128.41}{5.940}} \right] = [2.30, 4.65]$$

(since $a = 5.940$ and $b = 24.202$ from Table X). The length of this interval is 2.35, whereas the length of the interval given in Example 6.4-1 is 2.49. To see why this new interval is shorter, carefully sketch a graph of the $\chi^2(12)$ p.d.f. and compare $\chi^2_{0.95}(12)$ with a and $\chi^2_{0.05}(12)$ with b on the graph.   ■

There are occasions when it is of interest to compare the variances of two normal distributions. We do this by finding a confidence interval for σ^2_X/σ^2_Y, using the ratio of S^2_X/σ^2_X to S^2_Y/σ^2_Y, where S^2_X and S^2_Y are the two sample variances based on two independent samples of sizes n and m from $N(\mu_X, \sigma^2_X)$ and $N(\mu_Y, \sigma^2_Y)$, respectively. However, the reciprocal of that ratio can be rewritten as follows:

$$\frac{\frac{S^2_Y}{\sigma^2_Y}}{\frac{S^2_X}{\sigma^2_X}} = \frac{\left[\frac{(m-1)S^2_Y}{\sigma^2_Y}\right] / (m-1)}{\left[\frac{(n-1)S^2_X}{\sigma^2_X}\right] / (n-1)}.$$

Since $(m-1)S^2_Y/\sigma^2_Y$ and $(n-1)S^2_X/\sigma^2_X$ are independent chi-square variables with $(m-1)$ and $(n-1)$ degrees of freedom, respectively, we know from Example 5.2-4 that the distribution of this ratio is $F(m-1, n-1)$. That is,

$$F = \frac{\frac{(m-1)S^2_Y}{\sigma^2_Y(m-1)}}{\frac{(n-1)S^2_X}{\sigma^2_X(n-1)}} = \frac{\frac{S^2_Y}{\sigma^2_Y}}{\frac{S^2_X}{\sigma^2_X}}$$

has an F distribution with $r_1 = m - 1$ and $r_2 = n - 1$ degrees of freedom. This is the ratio that we want to use to find a confidence interval for σ^2_X/σ^2_Y.

To form the desired confidence interval, we select constants c and d from Table VII in the appendix so that

$$1 - \alpha = P\left(c \leq \frac{S^2_Y/\sigma^2_Y}{S^2_X/\sigma^2_X} \leq d \right)$$

$$= P\left(c \frac{S^2_X}{S^2_Y} \leq \frac{\sigma^2_X}{\sigma^2_Y} \leq d \frac{S^2_X}{S^2_Y} \right).$$

Because of the limitations of Table VII, we generally let $c = F_{1-\alpha/2}(m-1, n-1) = 1/F_{\alpha/2}(n-1, m-1)$ and $d = F_{\alpha/2}(m-1, n-1)$. If s_x^2 and s_y^2 are the observed values of S_X^2 and S_Y^2, respectively, then

$$\left[\frac{1}{F_{\alpha/2}(n-1,\, m-1)} \frac{s_x^2}{s_y^2}, \; F_{\alpha/2}(m-1, n-1) \frac{s_x^2}{s_y^2} \right]$$

is a $100(1-\alpha)\%$ confidence interval for σ_X^2/σ_Y^2. By taking square roots of both endpoints, we would obtain a $100(1-\alpha)\%$ confidence interval for σ_X/σ_Y.

EXAMPLE 6.4-3 In Example 6.4-1, denote σ^2 by σ_X^2. In that example, $(n-1)s_x^2 = 12s^2 = 128.41$. Assume that the time in days required for the maturation of seeds of a species of Guardiola, both parents having broad leaves, is $N(\mu_Y, \sigma_Y^2)$. A random sample of size $m = 9$ seeds yielded $\bar{y} = 23.20$ and

$$8s_y^2 = \sum_{i=1}^{8} (y_i - \bar{y})^2 = 36.72.$$

A 98% confidence interval for σ_X^2/σ_Y^2 is given by

$$\left[\left(\frac{1}{5.67} \right) \frac{(128.41)/12}{(36.72)/8}, \; (4.50) \frac{(128.41)/12}{(36.72)/8} \right] = [0.41, 10.49]$$

because $F_{0.01}(12, 8) = 5.67$ and $F_{0.01}(8, 12) = 4.50$. It follows that a 98% confidence interval for σ_X/σ_Y is

$$[\sqrt{0.41}, \; \sqrt{10.49}] = [0.64, 3.24]. \qquad \blacksquare$$

Although we are able to formally find a confidence interval for the ratio of two distribution variances or standard deviations, we should point out that these intervals are generally not too useful because they are often very wide. Moreover, they are not very robust. That is, the confidence coefficients are not very accurate if we deviate much from underlying normal distributions, because, in those instances, the distribution of $(n-1)S^2/\sigma^2$ could deviate greatly from $\chi^2(n-1)$. This same statement is not true for confidence intervals for μ or $\mu_X - \mu_Y$ based upon the sample means, because the sample means have approximate normal distributions. Hence, the confidence coefficients are reasonably accurate in the situations associated with means. Having said that, we do warn the reader that in cases which are highly skewed or have very heavy tails, there are better ways of finding confidence intervals for the "middles" than using the sample means. We consider a few of these later in the text.

EXERCISES

6.4-1. Let X equal the length (in centimeters) of a certain species of fish caught in the springtime. A random sample of $n = 13$ observations of X is

13.1	5.1	18.0	8.7	16.5	9.8	6.8
12.0	17.8	25.4	19.2	15.8	23.0	

(a) Give a point estimate of the standard deviation σ of this species of fish.

(b) Find a 95% confidence interval for σ.

6.4-2. A random sample of $n = 9$ wheels of cheese yielded the following weights in pounds, assumed to be $N(\mu, \sigma^2)$:

$$21.50 \quad 18.95 \quad 18.55 \quad 19.40 \quad 19.15$$
$$22.35 \quad 22.90 \quad 22.20 \quad 23.10$$

(a) Give a point estimate of σ.
(b) Find a 95% confidence interval for σ.
(c) Find a 90% confidence interval for σ.

6.4-3. A student who works in a blood lab tested 25 men for cholesterol levels and found the following values:

$$164 \quad 272 \quad 261 \quad 248 \quad 235 \quad 192 \quad 203 \quad 278 \quad 268$$
$$230 \quad 242 \quad 305 \quad 286 \quad 310 \quad 345 \quad 289 \quad 326$$
$$335 \quad 297 \quad 328 \quad 400 \quad 228 \quad 194 \quad 338 \quad 252$$

Assume that these values represent observations of a random sample taken from $N(\mu, \sigma^2)$.

(a) Calculate the sample mean and sample variance for these data.
(b) Find a 90% confidence interval for σ^2.
(c) Find a 90% confidence interval for σ.
(d) Find a 90% confidence interval for σ that has minimum length.
(e) Does the assumption of normality seem to be valid?

6.4-4. The data in Exercise 6.2-12 that give results of a leakage test are repeated here:

$$3.1 \quad 3.3 \quad 4.5 \quad 2.8 \quad 3.5 \quad 3.5 \quad 3.7 \quad 4.2 \quad 3.9 \quad 3.3$$

Use these data to find a 95% confidence interval for

(a) σ^2.
(b) σ.
(c) σ having minimum length.

6.4-5. Let $X_1, X_2, X_3, \ldots, X_n$ be a random sample from $N(\mu, \sigma^2)$ with known mean μ. Describe how you would construct a confidence interval for the unknown variance σ^2.
HINT: Use the fact that $\sum_{i=1}^{n}(X_i - \mu)^2/\sigma^2$ is $\chi^2(n)$.

6.4-6. Let $X_1, X_2, \ldots, X_n$ be a random sample of size n from an exponential distribution with unknown mean of $\mu = \theta$.

(a) Show that the distribution of the random variable $W = (2/\theta)\sum_{i=1}^{n} X_i$ is $\chi^2(2n)$.
HINT: Find the moment-generating function of W.
(b) Use W to construct a $100(1 - \alpha)\%$ confidence interval for θ.
(c) If $n = 7$ and $\bar{x} = 93.6$, give the endpoints for a 90% confidence interval for the mean θ.

6.4-7. Let X equal the time in seconds between phone calls at Great Lakes Pizza Company. Sixteen observations of X are

$$82 \quad 42 \quad 185 \quad 66 \quad 384 \quad 27 \quad 334 \quad 545$$
$$650 \quad 127 \quad 35 \quad 45 \quad 285 \quad 133 \quad 120 \quad 471$$

(a) Assuming that X has an exponential distribution with mean θ, use Exercise 6.4-6 to construct a 95% confidence interval for θ.
(b) Calculate the values of $\bar{x}$ and s. Do they support the assumption that X has an exponential distribution?
(c) Construct a q–q plot or a box-and-whiskers display to confirm or deny the assumption that X has an exponential distribution.

6.4-8. In Exercise 3.4-10, data are given for the distances that two sets of snails move. Assume that those distances are observations of an exponential random variable.

(a) Use Exercise 6.4-6 to construct a 95% confidence interval for the mean distance traveled by the infected snails. HINT: Either estimate the chi-square values or use a computer program such as Minitab.
(b) Use Exercise 6.4-6 to construct a 95% confidence interval for the mean distance traveled by the control snails. HINT: Either estimate the chi-square values or use a computer program such as Minitab.

6.4-9. Let $X_1, X_2, \ldots, X_n$ be a random sample of size n from an exponential distribution. If n is sufficiently large, the distribution of

$$Z = \frac{\bar{X} - \theta}{\theta/\sqrt{n}}$$

is approximately $N(0, 1)$. Thus,

$$P(-z_{\alpha/2} \leq Z \leq z_{\alpha/2}) \approx 1 - \alpha.$$

(a) Show that a solution of the inequality yields

$$\bar{x} \pm z_{\alpha/2}(\theta/\sqrt{n})$$

as endpoints for an approximate $1 - \alpha$ confidence interval for θ.
(b) Since θ is unknown, what are two possible statistics that could be used to estimate θ in this confidence interval?
(c) Show that an exact solution of the inequality for θ yields

$$\left[\frac{\bar{x}}{1 + z_{\alpha/2}/\sqrt{n}}, \frac{\bar{x}}{1 - z_{\alpha/2}/\sqrt{n}}\right]$$

as endpoints.

(d) Using the data in Exercise 6.4-7, compare the 95% confidence interval in that exercise with the approximate 95% confidence intervals in this exercise. Which of these intervals would you recommend and why?

6.4-10. Let X and Y equal the weights of a phosphorus-free laundry detergent in a "6-pound" box and a "12-pound" box, respectively. Assume that the distributions of X and Y are $N(\mu_X, \sigma_X^2)$ and $N(\mu_Y, \sigma_Y^2)$, respectively. A random sample of $n = 10$ observations of X yielded a sample mean of $\bar{x} = 6.10$ pounds with a sample variance of $s_x^2 = 0.0040$, while an independent random sample of $m = 9$ observations of Y yielded a sample mean of $\bar{y} = 12.10$ pounds with a sample variance of $s_y^2 = 0.0076$.

(a) Give a point estimate of σ_X^2/σ_Y^2.
(b) Find a 95% confidence interval for σ_X^2/σ_Y^2.

6.4-11. Let X and Y equal the number of milligrams of tar in filtered and unfiltered cigarettes, respectively. Assume that the distributions of X and Y are $N(\mu_X, \sigma_X^2)$ and $N(\mu_Y, \sigma_Y^2)$, respectively. A random sample of $n = 9$ observations of X was

0.9 1.1 0.1 0.7 0.3 0.9 0.8 1.0 0.4

and an independent random sample of $m = 11$ observations of Y was

1.5 0.9 1.6 0.5 1.4 1.9 1.0 1.2 1.3 1.6 2.1

(a) Give a point estimate of σ_X^2/σ_Y^2.
(b) Find a one-sided 95% confidence interval that is an upper bound for σ_X^2/σ_Y^2. Does this confidence interval include the number 1?

6.4-12. A candy maker produces mints that have a label weight of 20.4 grams. For quality assurance, $n = 16$ mints were selected at random from the Wednesday morning shift, resulting in the statistics $\bar{x} = 21.95$ grams and $s_x = 0.197$. On Wednesday afternoon $m = 13$ mints were selected at random, giving $\bar{y} = 21.88$ grams and $s_y = 0.318$. Find a 90% confidence interval for σ_X/σ_Y, the ratio of the standard deviations of the mints produced by the morning and by the afternoon shifts, respectively.

6.4-13. Let X and Y equal the concentration in parts per billion of chromium in the blood for healthy persons and for persons with a suspected disease, respectively. Assume that the distributions of X and Y are $N(\mu_X, \sigma_X^2)$ and $N(\mu_Y, \sigma_Y^2)$, respectively. Following are $n = 8$ observations of X:

15 23 12 18 9 28 11 10

Following are $m = 10$ observations of Y:

25 20 35 15 40 16 10 22 18 32

Use these observations of X and Y to

(a) Give a point estimate of σ_X^2/σ_Y^2.
(b) Find a one-sided 95% confidence interval that is an upper bound for σ_X^2/σ_Y^2.

6.4-14. Some nurses were interested in the effect of pre-natal care on the birth weight of babies. Mothers were divided into two groups, and their babies' weights were compared. The birth weight, in ounces, of babies of mothers who had received five or fewer prenatal visits were

49   108   110   82   93   114   134
114   96   52   101   114   120   116

and the birth weights of babies of mothers who had received six or more prenatal visits were

133   108   93   119   119   98   106
87   153   116   129   97   110   131

Assuming that these are independent observations of X and Y that are $N(\mu_X, \sigma_X^2)$ and $N(\mu_Y, \sigma_Y^2)$, respectively, find a 95% confidence interval for

(a) σ_X^2/σ_Y^2.
(b) σ_X/σ_Y.

6.4-15. Using a random sample of size n from the normal distribution $N(\mu, \sigma^2)$, find the $100(1 - \alpha)\%$ confidence interval for μ of minimum length based on the statistic $T = (\bar{X} - \mu)/(S/\sqrt{n})$, where $\bar{X}$ and S^2 are the unbiased estimators of μ and σ^2, respectively.

HINT: (i) Show that

$$1 - \alpha = P(a \le T \le b)$$
$$= P[\bar{X} - b(S/\sqrt{n}) \le \mu \le \bar{X} - a(S/\sqrt{n})]$$

and (ii) the length of the resulting confidence interval is $L = (s/\sqrt{n})(b - a)$. (iii) Minimize L subject to the condition $\int_a^b g(t)\, dt = 1 - \alpha$, where $g(t)$ is the p.d.f. of a t random variable with $r = n - 1$ degrees of freedom.

6.4-16. Let $X_1, X_2, \ldots, X_n$ be a random sample of size n from a normal distribution $N(\mu, \sigma^2)$. Select a and b so that

$$P\left(a \le \frac{(n - 1)S^2}{\sigma^2} \le b\right) = 1 - \alpha.$$

So a $100(1 - \alpha)\%$ confidence interval for σ is $[\sqrt{(n - 1)/b}\, s, \sqrt{(n - 1)/a}\, s]$. Find values of a

and b that minimize the length of this confidence interval. That is, minimize

$$k = s\sqrt{n-1}\left(\frac{1}{\sqrt{a}} - \frac{1}{\sqrt{b}}\right)$$

under the restriction

$$G(b) - G(a) = \int_a^b g(u)\,du = 1 - \alpha,$$

where $G(u)$ and $g(u)$ are, respectively, the distribution function and p.d.f. of a $\chi^2(n-1)$ distribution. HINT: Due to the restriction, b is a function of a. In particular, by taking derivatives of the restricting equation with respect to a, show that $\dfrac{db}{da} = \dfrac{g(a)}{g(b)}$. Determine $\dfrac{dk}{da}$. By setting $\dfrac{dk}{da} = 0$, show that a and b must satisfy

$$a^{n/2}e^{-a/2} - b^{n/2}e^{-b/2} = 0.$$

This condition, along with the restriction, was used to calculate the values in Table X in the appendix.

6.4-17. In Section 6.3, confidence intervals were found for the difference of the means for two normal distributions. It was sometimes assumed that the variances were equal. Find point estimates and 90% confidence intervals for the ratios of the variances for the data in

  (a) Exercise 6.3-14.
  (b) Exercise 6.3-15.
  (c) Exercise 6.3-16.

6.4-18. Find a 95% confidence interval for the ratio of the variances in

  (a) Exercise 6.3-6,
  (b) Exercise 6.3-7,
  (c) Exercise 6.3-8.

6.4-19. Measurements in millimeters were made on a species of spider named *Sosippus floridanus* that are native to Florida. Nine female spiders and nine male spiders were used. The lengths of the female spiders are

$$11.06 \; 13.87 \; 12.93 \; 15.08 \; 17.82 \; 14.14 \; 12.26 \; 17.82 \; 20.17$$

The lengths of the male spiders are

$$12.26 \; 11.66 \; 12.53 \; 13.00 \; 11.79 \; 12.46 \; 10.65 \; 10.39 \; 12.26$$

Assume that these lengths are $N(\mu_X, \sigma_X^2)$ and $N(\mu_Y, \sigma_Y^2)$, respectively.

  (a) Find point estimates for the sample variances for the female and the male spiders.
  (b) Find a 95% confidence interval for the ratio of the variances.
  (c) Is the assumption of normality valid? Why or why not?

6.5 CONFIDENCE INTERVALS FOR PROPORTIONS

We have suggested that the histogram is a good description of how the observations of a random sample are distributed. We might naturally inquire about the accuracy of those relative frequencies (or percentages) associated with the various classes. To illustrate, in Example 3.1-1 concerning the weights of $n = 40$ candy bars, we found that the relative frequency of the class interval $(22.25, 23.15)$ was $8/40 = 0.20$, or 20%. If we think of this collection of 40 weights as a random sample observed from a larger population of candy bar weights, how close is 20% to the true percentage (or 0.20 to the true proportion) of weights in that class interval for the entire population of weights for this type of candy bar?

In considering this problem, we generalize it somewhat by treating the class interval $(22.25, 23.15)$ as "success." That is, there is some true probability of success, p—namely, the proportion of the population in that interval. Let Y equal the frequency of measurements in the interval out of the n observations, so that (under the assumptions of independence and constant probability p) Y has the binomial distribution $b(n, p)$. Thus, the problem is to determine the accuracy of the relative frequency Y/n as an estimator of p. We solve this problem by finding, for the unknown p, a confidence interval based on Y/n.

In general, when observing n Bernoulli trials with probability p of success on each trial, we shall find a confidence interval for p based on Y/n, where Y is the number of successes and Y/n is an unbiased point estimator for p.

In Section 5.7, we noted that

$$\frac{Y - np}{\sqrt{np(1 - p)}} = \frac{(Y/n) - p}{\sqrt{p(1 - p)/n}}$$

has an approximate normal distribution $N(0, 1)$, provided that n is large enough. This means that, for a given probability $1 - \alpha$, we can find a $z_{\alpha/2}$ in Table V in the appendix such that

$$P\left[-z_{\alpha/2} \leq \frac{(Y/n) - p}{\sqrt{p(1 - p)/n}} \leq z_{\alpha/2}\right] \approx 1 - \alpha. \tag{6.5-1}$$

If we proceed as we did when we found a confidence interval for μ in Section 6.2, we would obtain

$$P\left[\frac{Y}{n} - z_{\alpha/2}\sqrt{\frac{p(1 - p)}{n}} \leq p \leq \frac{Y}{n} + z_{\alpha/2}\sqrt{\frac{p(1 - p)}{n}}\right] \approx 1 - \alpha.$$

Unfortunately, the unknown parameter p appears in the endpoints of this inequality. There are two ways out of this dilemma. First, we could make an additional approximation, namely, replacing p with Y/n in $p(1 - p)/n$ in the endpoints. That is, if n is large enough, it is still true that

$$P\left[\frac{Y}{n} - z_{\alpha/2}\sqrt{\frac{(Y/n)(1 - Y/n)}{n}} \leq p \leq \frac{Y}{n} + z_{\alpha/2}\sqrt{\frac{(Y/n)(1 - Y/n)}{n}}\right] \approx 1 - \alpha.$$

Thus, for large n, if the observed Y equals y, then the interval

$$\left[\frac{y}{n} - z_{\alpha/2}\sqrt{\frac{(y/n)(1 - y/n)}{n}}, \frac{y}{n} + z_{\alpha/2}\sqrt{\frac{(y/n)(1 - y/n)}{n}}\right]$$

serves as an approximate $100(1 - \alpha)\%$ confidence interval for p. Frequently, this interval is written as

$$\frac{y}{n} \pm z_{\alpha/2}\sqrt{\frac{(y/n)(1 - y/n)}{n}} \tag{6.5-2}$$

for brevity. This formulation clearly notes, as does $\bar{x} \pm z_{\alpha/2}(\sigma/\sqrt{n})$ in Section 6.2, the reliability of the estimate y/n, namely, that we are $100(1 - \alpha)\%$ confident that p is within $z_{\alpha/2}\sqrt{(y/n)(1 - y/n)/n}$ of $\hat{p} = y/n$.

A second way to solve for p in the inequality in Equation 6.5-1 is to note that

$$\frac{|Y/n - p|}{\sqrt{p(1 - p)/n}} \leq z_{\alpha/2}$$

is equivalent to

$$H(p) = \left(\frac{Y}{n} - p\right)^2 - \frac{z_{\alpha/2}^2 \, p(1-p)}{n} \le 0. \qquad (6.5\text{-}3)$$

But $H(p)$ is a quadratic expression in p. Thus, we can find those values of p for which $H(p) \le 0$ by finding the two zeros of $H(p)$. Letting $\widehat{p} = Y/n$ and $z_0 = z_{\alpha/2}$ in Equation 6.5-3, we have

$$H(p) = \left(1 + \frac{z_0^2}{n}\right)p^2 - \left(2\widehat{p} + \frac{z_0^2}{n}\right)p + \widehat{p}^{\,2}.$$

By the quadratic formula, the zeros of $H(p)$ are, after simplification,

$$\frac{\widehat{p} + z_0^2/(2n) \pm z_0\sqrt{\widehat{p}\,(1-\widehat{p}\,)/n + z_0^2/(4n^2)}}{1 + z_0^2/n}, \qquad (6.5\text{-}4)$$

and these zeros give the endpoints for an approximate $100(1 - \alpha)\%$ confidence interval for p. If n is large, $z_0^2/(2n)$, $z_0^2/(4n^2)$, and z_0^2/n are small. Thus, the confidence intervals given by Equation 6.5-2 and Equation 6.5-4 are approximately equal when n is large.

EXAMPLE 6.5-1   Let us return to the example of the histogram of the candy bar weights, Example 3.1-1, with $n = 40$ and $y/n = 8/40 = 0.2$. If $1 - \alpha = 0.90$, so that $z_{\alpha/2} = 1.645$, then, using Equation 6.5-2, we find that the endpoints

$$0.2 \pm 1.645\sqrt{\frac{(0.2)(0.8)}{40}}$$

serve as an approximate 90% confidence interval for the true fraction p. That is, $[0.096, 0.304]$, which is the same as $[9.6\%, 30.4\%]$, is an approximate 90% confidence interval for the percentage of weights of the entire population in the interval $(22.25, 23.15)$. If we had used the endpoints given by Equation 6.5-4, the confidence interval would be $[0.117, 0.321]$. Because of the small sample size, there is a difference in the lengths of these intervals. If the sample size had been $n = 400$, the two 90% confidence intervals would have been $[0.167, 0.233]$ and $[0.169, 0.235]$, respectively, which differ very little. ∎

EXAMPLE 6.5-2   In a certain political campaign, one candidate has a poll taken at random among the voting population. The results are that $y = 185$ out of $n = 351$ voters favor this candidate. Even though $y/n = 185/351 = 0.527$, should the candidate feel very confident of winning? From Equation 6.5-2, an approximate 95% confidence interval for the fraction p of the voting population who favor the candidate is

$$0.527 \pm 1.96\sqrt{\frac{(0.527)(0.473)}{351}}$$

or, equivalently, $[0.475, 0.579]$. Thus, there is a good possibility that p is less than 50%, and the candidate should certainly take this possibility into account in campaigning. ∎

One-sided confidence intervals are sometimes appropriate for p. For example, we may be interested in an upper bound on the proportion of defectives in manufacturing some item. Or we may be interested in a lower bound on the proportion of voters who favor a particular candidate. The one-sided confidence interval for p given by

$$\left[0, \frac{y}{n} + z_\alpha \sqrt{\frac{y/n(1 - y/n)}{n}} \right]$$

provides an upper bound for p, while

$$\left[\frac{y}{n} - z_\alpha \sqrt{\frac{y/n(1 - y/n)}{n}}, 1 \right]$$

provides a lower bound for p.

REMARK   Sometimes the confidence intervals suggested here are not very close to having the stated confidence coefficient. This is particularly true if n is small or if one of Y or $n - Y$ is close to zero. It is obvious that something is wrong if $Y = 0$ or $n - Y = 0$, because the radical is then equal to zero.

It has been suggested (see, for example, Agresti and Coull) that we use $\widetilde{p} = (Y + 2)/(n + 4)$ as an estimator for p in those cases because the results are usually much better. It is true that $\widetilde{p}$ is a biased estimator of p, but we will see later that it is a certain kind of Bayes shrinkage estimator. In those cases in which n is small or Y or $n - Y$ is close to zero,

$$\widetilde{p} \pm z_{\alpha/2} \sqrt{\widetilde{p}(1 - \widetilde{p})/n} \tag{6.5-5}$$

provides a much better $100(1 - \alpha)\%$ confidence interval for p. A similar statement can be made about one-sided confidence intervals.

Look again at Equation 6.5-4. If we form a 95% confidence interval using this equation, we find that $z_0 = 1.96 \approx 2$. Thus, a 95% confidence interval is centered approximately at

$$\frac{\widehat{p} + z_0^2/(2n)}{1 + z_0^2/n} = \frac{y + z_0^2/2}{n + z_0^2} \approx \frac{y + 2}{n + 4}.$$

This result is consistent with Equation 6.5-5 for 95% confidence intervals.   ■

EXAMPLE 6.5-3   Returning to the data in Example 6.5-1, and using Equation 6.5-5, we have $\widetilde{p} = (8 + 2)/(40 + 4) = 0.227$. Thus, a 90% confidence interval is

$$0.227 \pm 1.645 \sqrt{\frac{(0.227)(0.773)}{40}},$$

or $[0.118, 0.336]$. If it had been true that $y = 80$ and $n = 400$, the confidence interval given by Equation 6.5-5 would have been $[0.170, 0.236]$.   ■

Frequently, there are two (or more) possible independent ways of performing an experiment; suppose these have probabilities of success p_1 and p_2, respectively. Let

n_1 and n_2 be the number of independent trials associated with these two methods, and let us say that they result in Y_1 and Y_2 successes, respectively. In order to make a statistical inference about the difference $p_1 - p_2$, we proceed as follows.

Since the independent random variables Y_1/n_1 and Y_2/n_2 have respective means p_1 and p_2 and variances $p_1(1-p_1)/n_1$ and $p_2(1-p_2)/n_2$, we know from Section 5.4 that the difference $Y_1/n_1 - Y_2/n_2$ must have mean $p_1 - p_2$ and variance

$$\frac{p_1(1-p_1)}{n_1} + \frac{p_2(1-p_2)}{n_2}.$$

(Recall that the variances are added to get the variance of a difference of two independent random variables.) Moreover, the fact that Y_1/n_1 and Y_2/n_2 have approximate normal distributions would suggest that the difference

$$\frac{Y_1}{n_1} - \frac{Y_2}{n_2}$$

would have an approximate normal distribution with the above mean and variance. (See Theorem 5.5-1.) That is,

$$\frac{(Y_1/n_1) - (Y_2/n_2) - (p_1 - p_2)}{\sqrt{p_1(1-p_1)/n_1 + p_2(1-p_2)/n_2}}$$

has an approximate normal distribution $N(0,1)$. If we now replace p_1 and p_2 in the denominator of this ratio by Y_1/n_1 and Y_2/n_2, respectively, it is still true for large enough n_1 and n_2 that the new ratio will be approximately $N(0,1)$. Thus, for a given $1 - \alpha$, we can find $z_{\alpha/2}$ from Table V in the appendix, so that

$$P\left[-z_{\alpha/2} \le \frac{(Y_1/n_1) - (Y_2/n_2) - (p_1 - p_2)}{\sqrt{(Y_1/n_1)(1 - Y_1/n_1)/n_1 + (Y_2/n_2)(1 - Y_2/n_2)/n_2}} \le z_{\alpha/2}\right] \approx 1 - \alpha.$$

Once Y_1 and Y_2 are observed to be y_1 and y_2, respectively, this approximation can be solved to obtain an approximate $100(1 - \alpha)\%$ confidence interval

$$\frac{y_1}{n_1} - \frac{y_2}{n_2} \pm z_{\alpha/2}\sqrt{\frac{(y_1/n_1)(1 - y_1/n_1)}{n_1} + \frac{(y_2/n_2)(1 - y_2/n_2)}{n_2}}$$

for the unknown difference $p_1 - p_2$. Note again how this form indicates the reliability of the estimate $y_1/n_1 - y_2/n_2$ of the difference $p_1 - p_2$.

EXAMPLE 6.5-4 Two detergents were tested for their ability to remove stains of a certain type. An inspector judged the first one to be successful on 63 out of 91 independent trials and the second one to be successful on 42 out of 79 independent trials. The respective relative frequencies of success are 0.692 and 0.532. An approximate 90% confidence interval for the difference $p_1 - p_2$ of the two detergents is

$$0.692 - 0.532 \pm 1.645\sqrt{\frac{(0.692)(0.308)}{91} + \frac{(0.532)(0.468)}{79}}$$

or, equivalently, $[0.038, 0.282]$. Accordingly, since this interval does not include zero, it seems that the first detergent is definitely better than the second one for removing the type of stains in question.   ∎

EXERCISES

6.5-1. A machine shop manufactures toggle levers. A lever is flawed if a standard nut cannot be screwed onto the threads. Let p equal the proportion of flawed toggle levers that the shop manufactures. If there were 24 flawed levers out of a sample of 642 that were selected randomly from the production line,

   (a) Give a point estimate of p.

   (b) Use Equation 6.5-2 to find an approximate 95% confidence interval for p.

   (c) Use Equation 6.5-4 to find an approximate 95% confidence interval for p.

   (d) Use Equation 6.5-5 to find an approximate 95% confidence interval for p.

   (e) Find a one-sided 95% confidence interval for p that provides an upper bound for p.

6.5-2. Let p equal the proportion of letters mailed in the Netherlands that are delivered the next day. Suppose that $y = 142$ out of a random sample of $n = 200$ letters were delivered the day after they were mailed.

   (a) Give a point estimate of p.

   (b) Use Equation 6.5-2 to find an approximate 90% confidence interval for p.

   (c) Use Equation 6.5-4 to find an approximate 90% confidence interval for p.

   (d) Use Equation 6.5-5 to find an approximate 90% confidence interval for p.

   (e) Find a one-sided 90% confidence interval for p that provides a lower bound for p.

6.5-3. Let p equal the proportion of adult Americans who favor a law requiring a teenager to have her parents' consent before having an abortion. In a survey of 1000 adult Americans (conducted by *Time*/CNN and reported in *Time* on July 9, 1990), 690 said they favored such a law.

   (a) Give a point estimate of p.

   (b) Find an approximate 95% confidence interval for p.

6.5-4. Let p equal the proportion of Americans who favor the death penalty. If a random sample of $n = 1234$ Americans yielded $y = 864$ who favored the death penalty, find an approximate 95% confidence interval for p.

6.5-5. Let p equal the proportion of triathletes who suffered a training-related overuse injury during the past year. Out of 330 triathletes who responded to a survey, 167 indicated that they had suffered such an injury during the past year.

   (a) Use these data to give a point estimate of p.

   (b) Use these data to find an approximate 90% confidence interval for p.

   (c) Do you think that the 330 triathletes who responded to the survey may be considered a random sample from the population of triathletes?

6.5-6. Let p equal the proportion of Americans who select jogging as one of their recreational activities. If 1497 out of a random sample of 5757 selected jogging, find an approximate 98% confidence interval for p.

6.5-7. In order to estimate the proportion, p, of a large class of college freshmen that had high school GPAs from 3.2 to 3.6, inclusive, a sample of $n = 50$ students was taken. It was found that $y = 9$ students fell into this interval.

   (a) Give a point estimate of p.

   (b) Use Equation 6.5-2 to find an approximate 95% confidence interval for p.

   (c) Use Equation 6.5-4 to find an approximate 95% confidence interval for p.

   (d) Use Equation 6.5-5 to find an approximate 95% confidence interval for p.

6.5-8. A proportion, p, that many public opinion polls estimate is the number of Americans who would say yes to the question, "If something were to happen to the President of the United States, do you think that the Vice President would be qualified to take over as President?" In one such random sample of 1022 adults, 388 said yes.

   (a) On the basis of the given data, find a point estimate of p.

   (b) Find an approximate 90% confidence interval for p.

   (c) Give updated answers to this question if new poll results are available.

6.5-9. To obtain an estimate of the proportion, p, of New York City residents who feel that the quality of life in New York City has become worse in the past few years, a telephone poll by *Time*/CNN on August 2–5, 1990, revealed that 686 out of 1009 residents said that life has become worse.

   (a) Give a point estimate of p.

   (b) Find an approximate 98% confidence interval for p.

6.5-10. The January 17, 1994, issue of *Time* magazine reported that 58% of adult Americans said yes to the question "If you or your spouse were pregnant, would you want the unborn child tested for genetic defects?'' These results were based on a telephone poll of 500 adult Americans. Let p equal the proportion of all adult Americans who would say yes to this question.

(a) Give the endpoints for a 90% confidence interval for p.

(b) The poll takers claimed that their sampling error was 4.5%. What is their confidence coefficient?

6.5-11. The January 29, 1996, issue of *Time* magazine reported that 48% of adult Americans "like the principle of a flat tax." It further claimed that this estimate had a sampling error of ±3%. Let p equal the proportion of all adult Americans who like the principle of a flat tax.

(a) Give the endpoints for a 95% confidence interval for p, given that the sample size was $n = 800$.

(b) Given that the reported sampling error was ±3% with $n = 800$, what is *Time*'s confidence coefficient?

6.5-12. The March 29, 1993, issue of *Time* magazine reported the proportions of adult Americans who favor "stricter gun-control laws." A telephone poll of 800 adult Americans, of whom 374 were gun owners and 426 did not own guns, showed that 206 gun owners and 338 non–gun owners favored stricter gun-control laws. Let p_1 and p_2 be the respective proportions of gun owners and non–gun owners who favor stricter gun-control laws.

(a) Give point estimates of p_1 and p_2.

(b) Find a 95% confidence interval for $p_1 - p_2$.

6.5-13. In developing countries in Africa and the Americas, let p_1 and p_2 be the respective proportions of women with nutritional anemia. Find an approximate 90% confidence interval for $p_1 - p_2$, given that a random sample of $n_1 = 2100$ African women yielded $y_1 = 840$ with nutritional anemia and a random sample of $n_2 = 1900$ women from the Americas yielded $y_2 = 323$ women with nutritional anemia.

6.5-14. A candy manufacturer selects mints at random from the production line and weighs them. For one week, the day shift weighed $n_1 = 194$ mints and the night shift weighed $n_2 = 162$ mints. The numbers of these mints that weighed at most 21 grams was $y_1 = 28$ for the day shift and $y_2 = 11$ for the night shift. Let p_1 and p_2 denote the proportions of mints that weigh at most 21 grams for the day and night shifts, respectively.

(a) Give a point estimate of p_1.

(b) Give the endpoints for a 95% confidence interval for p_1.

(c) Give a point estimate of $p_1 - p_2$.

(d) Find a one-sided 95% confidence interval that gives a lower bound for $p_1 - p_2$.

6.5-15. Consider the following two groups of women: Group 1 consists of women who spend less than $500 annually on clothes; group 2 comprises women who spend over $1000 annually on clothes. Let p_1 and p_2 equal the proportions of women in these two groups, respectively, who believe that clothes are too expensive. If 1009 out of a random sample of 1230 women from group 1 and 207 out of a random sample 340 from group 2 believe that clothes are too expensive,

(a) Give a point estimate of $p_1 - p_2$.

(b) Find an approximate 95% confidence interval for $p_1 - p_2$.

6.5-16. For developing countries in Asia (excluding China) and Africa, let p_1 and p_2 be the respective proportions of preschool children with chronic malnutrition (stunting). If respective random samples of $n_1 = 1300$ and $n_2 = 1100$ yielded $y_1 = 520$ and $y_2 = 385$ children with chronic malnutrition, find an approximate 95% confidence interval for $p_1 - p_2$.

6.5-17. The following question was asked in a *Newsweek* poll. "Would you prefer to live in a neighborhood with mostly whites, with mostly blacks, or in a neighborhood mixed half and half?" Let p_1 and p_2 equal the proportion of black and white adult respondents, respectively, who prefer "half and half." If 207 out of 305 black adults and 291 out of 632 white adults prefer "half and half,"

(a) Give a point estimate of $p_1 - p_2$.

(b) Find an approximate one-sided 90% confidence interval that gives a lower bound for $p_1 - p_2$.

6.5-18. An environmental survey contained a question asking what respondents thought was the major cause of air pollution in this country, giving the choices "automobiles," "factories," and "incinerators." Two versions of the test, A and B, were used. Let p_A and p_B be the respective proportions of people using forms A and B who select "factories." If 170 out of 460 people who used version A chose "factories" and 141 out of 440 people who used version B chose "factories,"

(a) Find a 95% confidence interval for $p_A - p_B$.

(b) Do the versions seem to be consistent concerning this answer? Why or why not?

6.5-19. Let p_1 equal the proportion of 30-year-old men in the United States who weigh over 200 pounds, and let p_2 be the proportion of 30-year-old men in Vietnam who weigh over 200 pounds. Suppose that in a random sample of 766 Americans who were asked if they weighed more than 200 pounds, 586 responded yes, and in a random sample of 725 Vietnamese who were asked the same question, 357 responded yes.

(a) Give a point estimate for $p_1 - p_2$.
(b) Give the endpoints for a 95% confidence interval for $p_1 - p_2$.

6.6 SAMPLE SIZE

In statistical consulting, the first question frequently asked is "How large should the sample size be to estimate a mean?" In order to convince the inquirer that the answer will depend on the variation associated with the random variable under observation, the statistician could correctly respond, "Only one observation is needed, provided that the standard deviation of the distribution is zero." That is, if σ equals zero, then the value of that one observation would necessarily equal the unknown mean of the distribution. This, of course, is an extreme case and one that is not met in practice; however, it should help convince people that the smaller the variance, the smaller is the sample size needed to achieve a given degree of accuracy. This assertion will become clearer as we consider several examples. Let us begin with a problem that involves a statistical inference about the unknown mean of a distribution.

EXAMPLE 6.6-1   A mathematics department wishes to evaluate a new method of teaching calculus with a computer. At the end of the course, the evaluation will be made on the basis of scores of the participating students on a standard test. There is particular interest in estimating μ, the mean score for students taking the course. Thus, there is a desire to determine the number of students, n, who are to be selected at random from a larger group of students to take the course. Since new computing equipment must be purchased, the department cannot afford to let all of the school's students take calculus the new way. In addition, some of the staff question the value of this approach and hence do not want to expose every student to this new procedure. So, let us find the sample size n such that we are fairly confident that $\bar{x} \pm 1$ contains the unknown test mean μ. From past experience, it is believed that the standard deviation associated with this type of test is about 15. (The mean is also known when students take the standard calculus course.) Accordingly, using the fact that the sample mean of the test scores, $\overline{X}$, is approximately $N(\mu, \sigma^2/n)$, we see that the interval given by $\bar{x} \pm 1.96(15/\sqrt{n})$ will serve as an approximate 95% confidence interval for μ. That is, we want

$$1.96\left(\frac{15}{\sqrt{n}}\right) = 1$$

or, equivalently,

$$\sqrt{n} = 29.4 \qquad \text{and thus} \qquad n \approx 864.36,$$

or $n = 865$ because n must be an integer.   ∎

It is quite likely that, in the preceding example, it had not been anticipated that as many as 865 students would be needed in this study. If that is the case, the statistician must discuss with those involved in the experiment whether or not the accuracy and the confidence level could be relaxed some. For example, rather than requiring $\bar{x} \pm 1$ to be a 95% confidence interval for μ, possibly $\bar{x} \pm 2$ would be a satisfactory 80%

one. If this modification is acceptable, we now have

$$1.282\left(\frac{15}{\sqrt{n}}\right) = 2$$

or, equivalently,

$$\sqrt{n} = 9.615 \qquad \text{so that} \qquad n \approx 92.4.$$

Since n must be an integer, we would probably use 93 in practice. Most likely, the persons involved in the project would find that a more reasonable sample size. Of course, any sample size greater than 93 could be used. Then either the length of the confidence interval could be decreased from $\bar{x} \pm 2$, or the confidence coefficient could be increased from 80%, or a combination of both approaches could be taken. Also, since there might be some question as to whether the standard deviation σ actually equals 15, the sample standard deviation s would no doubt be used in the construction of the interval. For instance, suppose that the sample characteristics observed are

$$n = 145, \qquad \bar{x} = 77.2, \qquad s = 13.2;$$

then

$$\bar{x} \pm \frac{1.282s}{\sqrt{n}}, \qquad \text{or} \qquad 77.2 \pm 1.41,$$

provides an approximate 80% confidence interval for μ.

In general, if we want the $100(1 - \alpha)\%$ confidence interval for μ, $\bar{x} \pm z_{\alpha/2}(\sigma/\sqrt{n})$, to be no longer than that given by $\bar{x} \pm \varepsilon$, then the sample size n is the solution of

$$\varepsilon = \frac{z_{\alpha/2}\sigma}{\sqrt{n}}, \qquad \text{where} \qquad \Phi(z_{\alpha/2}) = 1 - \frac{\alpha}{2}.$$

That is,

$$n = \frac{z_{\alpha/2}^2\sigma^2}{\varepsilon^2}, \tag{6.6-1}$$

where it is assumed that σ^2 is known. We sometimes call $\varepsilon = z_{\alpha/2}(\sigma/\sqrt{n})$ the **maximum error of the estimate**. If the experimenter has no idea about the value of σ^2, it may be necessary to first take a preliminary sample to estimate σ^2.

The type of statistic we see most often in newspapers and magazines is an estimate of a proportion p. We might, for example, want to know the percentage of the labor force that is unemployed or the percentage of voters favoring a certain candidate. Sometimes, extremely important decisions are made on the basis of these estimates. If this is the case, we would most certainly desire short confidence intervals for p with large confidence coefficients. We recognize that these conditions will require a large sample size. If, to the contrary, the fraction p being estimated is not too important, an estimate associated with a longer confidence interval with a smaller confidence coefficient is satisfactory, and in that case a smaller sample size can be used.

EXAMPLE 6.6-2 Suppose we know that the unemployment rate has been about 8% (0.08). However, we wish to update our estimate in order to make an important decision about the national economic policy. Accordingly, let us say we wish to be 99% confident that the new estimate of p is within 0.001 of the true p. If we assume Bernoulli trials (an assumption that might be questioned), the relative frequency y/n, based upon a large sample size n, provides the approximate 99% confidence interval:

$$\frac{y}{n} \pm 2.576 \sqrt{\frac{(y/n)(1 - y/n)}{n}}.$$

Although we do not know y/n exactly before sampling, since y/n will be near 0.08 we do know that

$$2.576 \sqrt{\frac{(y/n)(1 - y/n)}{n}} \approx 2.576 \sqrt{\frac{(0.08)(0.92)}{n}},$$

and we want this number to equal 0.001. That is,

$$2.576 \sqrt{\frac{(0.08)(0.92)}{n}} = 0.001$$

or, equivalently,

$$\sqrt{n} = 2576 \sqrt{0.0736}, \qquad \text{and then} \qquad n \approx 488,394.$$

That is, under our assumptions, such a sample size is needed in order to achieve the reliability and the accuracy desired. Because n is so large, we would probably be willing to increase the error, say, to 0.01, and perhaps reduce the confidence level to 98%. In that case,

$$\sqrt{n} = (2.326/0.01)\sqrt{0.0736} \qquad \text{and} \qquad n \approx 3,982,$$

which is a more reasonable sample size.   ■

From the preceding example, we hope that the student will recognize how important it is to know the sample size (or the length of the confidence interval and the confidence coefficient) before he or she can place much weight on a statement such as "Fifty-one percent of the voters seem to favor candidate A, 46% favor candidate B, and 3% are undecided." Is this statement based on a sample of 100 or 2000 or 10,000 voters? If we assume Bernoulli trials, the approximate 95% confidence intervals for the fraction of voters favoring candidate A in these cases are, respectively, [0.41, 0.61], [0.49, 0.53], and [0.50, 0.52]. Quite obviously, the first interval, with $n = 100$, does not assure candidate A of the support of at least half the voters, whereas the interval with $n = 10,000$ is more convincing.

In general, to find the required sample size to estimate p, recall that the point estimate of p is $\widehat{p} = y/n$ and an approximate $1 - \alpha$ confidence interval for p is

$$\widehat{p} \pm z_{\alpha/2} \sqrt{\frac{\widehat{p}(1 - \widehat{p})}{n}}.$$

Suppose we want an estimate of p that is within ε of the unknown p with $100(1 - \alpha)\%$ confidence, where $\varepsilon = z_{\alpha/2}\sqrt{\widehat{p}(1 - \widehat{p})/n}$ is the **maximum error of the point estimate** $\widehat{p} = y/n$. Since $\widehat{p}$ is unknown before the experiment is run, we cannot use the value of $\widehat{p}$ in our determination of n. However, if it is known that p is about equal to p^*, the necessary sample size n is the solution of

$$\varepsilon = \frac{z_{\alpha/2}\sqrt{p^*(1 - p^*)}}{\sqrt{n}} .$$

That is,

$$n = \frac{z_{\alpha/2}^2 p^*(1 - p^*)}{\varepsilon^2} . \qquad (6.6\text{-}2)$$

Often, however, we do not have a strong prior idea about p, as we did in Example 6.6-2 about the rate of unemployment. It is interesting to observe that no matter what value p takes between 0 and 1, it is always true that $p^*(1 - p^*) \leq 1/4$. Hence,

$$n = \frac{z_{\alpha/2}^2 p^*(1 - p^*)}{\varepsilon^2} \leq \frac{z_{\alpha/2}^2}{4\varepsilon^2} .$$

Thus, if we want the $100(1 - \alpha)\%$ confidence interval for p to be no longer than $y/n \pm \varepsilon$, a solution for n that provides this protection is

$$n = \frac{z_{\alpha/2}^2}{4\varepsilon^2} . \qquad (6.6\text{-}3)$$

REMARK   Up to this point in the text, we have used the "hat" ($\frown$) notation to indicate an estimator, as in $\widehat{p} = Y/n$ and $\widehat{\mu} = \overline{X}$. Note, however, that in the previous discussion we used $\widehat{p} = y/n$, an estimate of p. Occasionally, statisticians find it convenient to use the "hat" notation for an estimate as well as an estimator. It is usually clear from the context which is being used.   ■

EXAMPLE 6.6-3   A possible gubernatorial candidate wants to assess initial support among the voters before making an announcement about her candidacy. If the fraction p of voters who are favorable, without any advance publicity, is around 0.15, the candidate will enter the race. From a poll of n voters selected at random, the candidate would like the estimate y/n to be within 0.03 of p. That is, the decision will be based on a 95% confidence interval of the form $y/n \pm 0.03$. Since the candidate has no idea about the magnitude of p, a consulting statistician formulates the equation

$$n = \frac{(1.96)^2}{4(0.03)^2} = 1067.11.$$

Thus, the sample size should be around 1068 to achieve the desired reliability and accuracy. Suppose that 1068 voters around the state were selected at random and interviewed and $y = 214$ express support for the candidate. Then $\widehat{p} = 214/1068 = 0.20$ is a point estimate of p, and an approximate 95% confidence interval for p is

$$0.20 \pm 1.96\sqrt{(0.20)(0.80)/n}, \qquad \text{or} \qquad 0.20 \pm 0.024.$$

That is, we are 95% confident that p belongs to the interval $[0.176, 0.224]$. On the basis of this sample, the candidate decided to run for office. Note that, for a confidence coefficient of 95%, we found a sample size so that the maximum error of the estimate would be 0.03. From the data that were collected, the maximum error of the estimate is only 0.024. We ended up with a smaller error because we found the sample size assuming that $p = 0.50$ while, in fact, p is closer to 0.20. ∎

Suppose that you want to estimate the proportion p of a student body that favors a new policy. How large should the sample be? If p is close to $1/2$ and you want to be 95% confident that the maximum error of the estimate is $\varepsilon = 0.02$, then

$$n = \frac{(1.96)^2}{4(0.02)^2} = 2401.$$

Such a sample size makes sense at a large university. However, if you are a student at a small college, the entire enrollment could be less than 2401. Thus, we now give a procedure that can be used to determine the sample size when the population is small relative to the desired sample size.

Let N equal the size of a population, and assume that N_1 individuals in the population have a certain characteristic C (e.g., favor a new policy). Let $p = N_1/N$, the proportion with this characteristic. Then $1 - p = 1 - N_1/N$. If we take a sample of size n without replacement, then X, the number of observations with the characteristic C, has a hypergeometric distribution. The mean and variance of X are, respectively,

$$\mu = n\left(\frac{N_1}{N}\right) = np$$

and

$$\sigma^2 = n\left(\frac{N_1}{N}\right)\left(1 - \frac{N_1}{N}\right)\left(\frac{N - n}{N - 1}\right) = np(1 - p)\left(\frac{N - n}{N - 1}\right).$$

The mean and variance of X/n are, respectively,

$$E\left(\frac{X}{n}\right) = \frac{\mu}{n} = p$$

and

$$\mathrm{Var}\left(\frac{X}{n}\right) = \frac{\sigma^2}{n^2} = \frac{p(1 - p)}{n}\left(\frac{N - n}{N - 1}\right).$$

To find an approximate confidence interval for p, we can use the normal approximation:

$$P\left[-z_{\alpha/2} \le \frac{(X/n) - p}{\sqrt{\dfrac{p(1 - p)}{n}\left(\dfrac{N - n}{N - 1}\right)}} \le z_{\alpha/2}\right] \approx 1 - \alpha.$$

Thus, $1 - \alpha \approx$

$$P\left[\frac{X}{n} - z_{\alpha/2}\sqrt{\frac{p(1 - p)}{n}\left(\frac{N - n}{N - 1}\right)} \le p \le \frac{X}{n} + z_{\alpha/2}\sqrt{\frac{p(1 - p)}{n}\left(\frac{N - n}{N - 1}\right)}\right].$$

Replacing p under the radical with $\widehat{p} = x/n$, we find that an approximate $1 - \alpha$ confidence interval for p is

$$\widehat{p} \pm z_{\alpha/2}\sqrt{\frac{\widehat{p}\,(1 - \widehat{p}\,)}{n}\left(\frac{N - n}{N - 1}\right)}.$$

This is similar to the confidence interval for p when the distribution of X is $b(n, p)$. If N is large relative to n, then

$$\frac{N - n}{N - 1} = \frac{1 - n/N}{1 - 1/N} \approx 1,$$

so in this case the two intervals are essentially equal.

Suppose now that we are interested in determining the sample size n that is required to have $1 - \alpha$ confidence that the maximum error of the estimate of p is ε. We let

$$\varepsilon = z_{\alpha/2}\sqrt{\frac{p(1 - p)}{n}\left(\frac{N - n}{N - 1}\right)}$$

and solve for n. After some simplification, we obtain

$$n = \frac{Nz_{\alpha/2}^2 p(1 - p)}{(N - 1)\varepsilon^2 + z_{\alpha/2}^2 p(1 - p)}$$

$$= \frac{z_{\alpha/2}^2 p(1 - p)/\varepsilon^2}{\dfrac{N - 1}{N} + \dfrac{z_{\alpha/2}^2 p(1 - p)/\varepsilon^2}{N}}.$$

If we let

$$m = \frac{z_{\alpha/2}^2 p^*(1 - p^*)}{\varepsilon^2},$$

which is the n value given by Equation 6.6-2, then we choose

$$n = \frac{m}{1 + \dfrac{m - 1}{N}}$$

for our sample size n.

If we know nothing about p, we set $p^* = 1/2$ to determine m. For example, if the size of the student body is $N = 4000$ and $1 - \alpha = 0.95$, $\varepsilon = 0.02$, and we let $p^* = 1/2$, then $m = 2401$ and

$$n = \frac{2401}{1 + 2400/4000} = 1501,$$

rounded up to the nearest integer. Thus, we would sample approximately 37.5% of the student body.

EXAMPLE 6.6-4   Suppose that a college of $N = 3000$ students is interested in assessing student support for a new form for teacher evaluation. To estimate the proportion p in favor of the new form, how large a sample is required so that the maximum error of the estimate of p is $\varepsilon = 0.03$ with 95% confidence? If we assume that p is completely unknown, we use $p^* = 1/2$ to obtain

$$m = \frac{(1.96)^2}{4(0.03)^2} = 1068,$$

rounded up to the nearest integer. Thus, the desired sample size is

$$n = \frac{1068}{1 + 1067/3000} = 788,$$

rounded up to the nearest integer.   ■

EXERCISES

6.6-1. Let X equal the tarsus length for a male grackle. Assume that the distribution of X is $N(\mu, 4.84)$. Find the sample size n that is needed so that we are 95% confident that the maximum error of the estimate of μ is 0.4.

6.6-2. Let X equal the excess weight of soap in a "1000-gram" bottle. Assume that the distribution of X is $N(\mu, 169)$. What sample size is required so that we have 95% confidence that the maximum error of the estimate of μ is 1.5?

6.6-3. A company packages powdered soap in "6-pound" boxes. The sample mean and standard deviation of the soap in these boxes are currently 6.09 pounds and 0.02 pound, respectively. If the mean fill can be lowered by 0.01 pound, $14,000 would be saved per year. Adjustments were made in the filling equipment, but it can be assumed that the standard deviation remains unchanged.

(a) How large a sample is needed so that the maximum error of the estimate of the new μ is $\varepsilon = 0.001$ with 90% confidence?

(b) A random sample of size $n = 1219$ yielded $\bar{x} = 6.048$ and $s = 0.022$. Calculate a 90% confidence interval for μ.

(c) Estimate the savings per year with these new adjustments.

(d) Estimate the proportion of boxes that will now weigh less than 6 pounds.

6.6-4. Measurements of the length in centimeters of $n = 29$ fish (species nezumia) yielded an average length of $\bar{x} = 16.82$ and $s^2 = 34.9$. Determine the size of a new sample so that $\bar{x} \pm 0.5$ is an approximate 95% confidence interval for μ.

6.6-5. A quality engineer wanted to be 98% confident that the maximum error of the estimate of the mean strength, μ, of the left hinge on a vanity cover molded by a machine is 0.25. A preliminary sample of size $n = 32$ parts yielded a sample mean of $\bar{x} = 35.68$ and a standard deviation of $s = 1.723$.

(a) How large a sample is required?

(b) Does this seem to be a reasonable sample size? (Note that destructive testing is needed to obtain the data.)

6.6-6. A manufacturer sells a light bulb that has a mean life of 1450 hours with a standard deviation of 33.7 hours. A new manufacturing process is being tested, and there is interest in knowing the mean life μ of the new bulbs. How large a sample is required so that $\bar{x} \pm 5$ is a 95% confidence interval for μ? You may assume that the change in the standard deviation is minimal.

6.6-7. For a public opinion poll for a close presidential election, let p denote the proportion of voters who favor candidate A. How large a sample should be taken if we want the maximum error of the estimate of p to be equal to

(a) 0.03 with 95% confidence?

(b) 0.02 with 95% confidence?

(c) 0.03 with 90% confidence?

6.6-8. Let p equal the proportion of all college and university students who would say yes to the question, "Would you drink from the same glass as your friend if you suspected that this friend were an AIDS virus carrier?" Find the sample size required to be 95% confident that the maximum error of the estimate of p is 0.025. (For your information, it was reported in *Sociology and Social Research* **72**, 2 [January 1988] that 30 out of 375 San Diego State University students answered yes to this question.)

6.6-9. A die has been loaded to change the probability of rolling a 6. In order to estimate p, the new probability of rolling a 6, how many times must the die be rolled so that we are 99% confident that the maximum error of the estimate of p is $\varepsilon = 0.02$?

6.6-10. Some college professors and students examined 137 Canadian geese for patent schistosome in the year they hatched. Of these 137 birds, 54 were infected. The professors and students were interested in estimating p, the proportion of infected birds of this type. For future studies, determine the sample size n so that the estimate of p is within $\varepsilon = 0.04$ of the unknown p with 90% confidence.

6.6-11. According to the Department of Health Care of the Elderly at Sherwood Hospital, Nottingham, England, only 25% of the patients who were using canes had canes of the correct length. Suppose that you were interested in estimating the proportion p of Americans who used canes of the correct length. How large a sample is required so that, with 95% confidence, the maximum error of the estimate of p is 0.04?

6.6-12. A seed distributor claims that 80% of its beet seeds will germinate. How many seeds must be tested for germination in order to estimate p, the true proportion that will germinate, so that the maximum error of the estimate is $\varepsilon = 0.03$ with 90% confidence?

6.6-13. Some dentists were interested in studying the fusion of embryonic rat palates by a standard transplantation technique. When no treatment is used, the probability of fusion equals approximately 0.89. The dentists would like to estimate p, the probability of fusion, when vitamin A is lacking.

(a) How large a sample n of rat embryos is needed for $y/n \pm 0.10$ to be a 95% confidence interval for p?

(b) If $y = 44$ out of $n = 60$ palates showed fusion, give a 95% confidence interval for p.

6.6-14. Let p equal the proportion of New York City residents who feel that the quality of life in New York City has become worse in the past few years. To update the estimate given in Exercise 6.5-9, how large a sample is required to be 98% confident that the maximum error of the estimate of p is 0.025?

6.6-15. Let p equal the proportion of triathletes who suffered a training-related overuse injury during the past year. (See Exercise 6.5-5.) How large a sample would be required to estimate p so that the maximum error of the estimate of p is 0.04 with 95% confidence?

6.6-16. Let p equal the proportion of college students who favor a new policy for alcohol consumption on campus. How large a sample is required to estimate p so that the maximum error of the estimate of p is 0.04 with 95% confidence when the size of the student body is

(a) $N = 1500$?

(b) $N = 15,000$?

(c) $N = 25,000$?

6.6-17. Out of 1000 welds that have been made on a tower, it is suspected that 15% are defective. To estimate p, the proportion of defective welds, how many welds must be inspected to have approximately 95% confidence that the maximum error of the estimate of p is 0.04?

6.6-18. If Y_1/n and Y_2/n are the respective independent relative frequencies of success associated with the two binomial distributions $b(n, p_1)$ and $b(n, p_2)$, compute n such that the approximate probability that the random interval $Y_1/n - Y_2/n \pm 0.05$ covers $p_1 - p_2$ is at least 0.80. HINT: Take $p_1^* = p_2^* = 1/2$ to provide an upper bound for n.

6.6-19. If $\overline{X}$ and $\overline{Y}$ are the respective means of two independent random samples of the same size n, find n if we want $\overline{x} - \overline{y} \pm 4$ to be a 90% confidence interval for $\mu_X - \mu_Y$ if the standard deviations are known to be $\sigma_X = 15$ and $\sigma_Y = 25$.

6.7   A SIMPLE REGRESSION PROBLEM

There is often interest in the relation between two variables—for example, a student's scholastic aptitude test score in mathematics and this same student's grade in calculus. Frequently, one of these variables, say, x, is known in advance of the other, so there is interest in predicting a future random variable Y. Since Y is a random variable, we cannot predict its future observed value $Y = y$ with certainty. Let us first concentrate on the problem of estimating the mean of Y—that is, $E(Y)$. Now, $E(Y)$ is usually a function of x. For example, in our illustration with the calculus grade, say, Y, we would expect $E(Y)$ to increase with increasing mathematics aptitude score x. Sometimes $E(Y) = \mu(x)$ is assumed to be of a given form, such as linear, quadratic, or exponential; that is, $\mu(x)$ could be assumed to be equal to $\alpha + \beta x$, $\alpha + \beta x + \gamma x^2$, or $\alpha e^{\beta x}$. To estimate $E(Y) = \mu(x)$, or, equivalently, the parameters α, β, and γ, we observe the random variable Y for each of n different values of x—say, $x_1, x_2, \ldots, x_n$. Once the n independent experiments have been performed, we have n pairs of known numbers $(x_1, y_1), (x_2, y_2), \ldots, (x_n, y_n)$. These pairs are then used to estimate the mean $E(Y)$. Problems like this are often classified under **regression** because $E(Y) = \mu(x)$ is frequently called a regression curve.

REMARK   A model for the mean that is of the form $\alpha + \beta x + \gamma x^2$ is called a linear model because it is linear in the parameters, α, β, and γ. Thus, $\alpha e^{\beta x}$ is not a linear model, because it is not linear in α and β. ■

FIGURE 6.7-1: Scatter plot and least squares regression line

Let us begin with the case in which $E(Y) = \mu(x)$ is a linear function. The data points are $(x_1, y_1), (x_2, y_2), \ldots, (x_n, y_n)$, so the first problem is that of fitting a straight line to the set of data. (See Figure 6.7-1.) In addition to assuming that the mean of Y is a linear function, we assume that, for a particular value of x, the value of Y will differ from its mean by a random amount ε. We further assume that the distribution of ε is $N(0, \sigma^2)$. So we have, for our linear model,

$$Y_i = \alpha_1 + \beta x_i + \varepsilon_i,$$

where ε_i, for $i = 1, 2, \ldots, n$, are independent and $N(0, \sigma^2)$.

We shall now find point estimates for α_1, β, and σ^2. For convenience, we let $\alpha_1 = \alpha - \beta\bar{x}$, so that

$$Y_i = \alpha + \beta(x_i - \bar{x}) + \varepsilon_i, \text{ where } \bar{x} = \frac{1}{n}\sum_{i=1}^{n} x_i.$$

Then Y_i is equal to a constant, $\alpha + \beta(x_i - \bar{x})$, plus a normal random variable ε_i. Hence, $Y_1, Y_2, \ldots, Y_n$ are mutually independent normal variables with respective means $\alpha + \beta(x_i - \bar{x})$, $i = 1, 2, \ldots, n$, and unknown variance σ^2. Their joint p.d.f. is therefore the product of the individual probability density functions; that is, the likelihood function equals

$$L(\alpha, \beta, \sigma^2) = \prod_{i=1}^{n} \frac{1}{\sqrt{2\pi\sigma^2}} \exp\left\{-\frac{[y_i - \alpha - \beta(x_i - \bar{x})]^2}{2\sigma^2}\right\}$$

$$= \left(\frac{1}{2\pi\sigma^2}\right)^{n/2} \exp\left\{-\frac{\sum_{i=1}^{n}[y_i - \alpha - \beta(x_i - \bar{x})]^2}{2\sigma^2}\right\}.$$

To maximize $L(\alpha, \beta, \sigma^2)$ or, equivalently, to minimize

$$- \ln L(\alpha, \beta, \sigma^2) = \frac{n}{2} \ln(2\pi\sigma^2) + \frac{\sum_{i=1}^{n} [y_i - \alpha - \beta(x_i - \bar{x})]^2}{2\sigma^2},$$

we must select α and β to minimize

$$H(\alpha, \beta) = \sum_{i=1}^{n} [y_i - \alpha - \beta(x_i - \bar{x})]^2.$$

Since $|y_i - \alpha - \beta(x_i - \bar{x})| = |y_i - \mu(x_i)|$ is the vertical distance from the point (x_i, y_i) to the line $y = \mu(x)$, we note that $H(\alpha, \beta)$ represents the sum of the squares of those distances. Thus, selecting α and β so that the sum of the squares is minimized means that we are fitting the straight line to the data by the **method of least squares**.

To minimize $H(\alpha, \beta)$, we find the two first partial derivatives

$$\frac{\partial H(\alpha, \beta)}{\partial \alpha} = 2 \sum_{i=1}^{n} [y_i - \alpha - \beta(x_i - \bar{x})](-1)$$

and

$$\frac{\partial H(\alpha, \beta)}{\partial \beta} = 2 \sum_{i=1}^{n} [y_i - \alpha - \beta(x_i - \bar{x})][-(x_i - \bar{x})].$$

Setting $\partial H(\alpha, \beta)/\partial \alpha = 0$, we obtain

$$\sum_{i=1}^{n} y_i - n\alpha - \beta \sum_{i=1}^{n} (x_i - \bar{x}) = 0.$$

Since

$$\sum_{i=1}^{n} (x_i - \bar{x}) = 0,$$

we have

$$\sum_{i=1}^{n} y_i - n\alpha = 0;$$

thus,

$$\hat{\alpha} = \bar{Y}.$$

With α replaced by $\bar{y}$, the equation $\partial H(\alpha, \beta)/\partial \beta = 0$ yields,

$$\sum_{i=1}^{n} (y_i - \bar{y})(x_i - \bar{x}) - \beta \sum_{i=1}^{n} (x_i - \bar{x})^2 = 0$$

or, equivalently,

$$\hat{\beta} = \frac{\sum_{i=1}^{n} (Y_i - \bar{Y})(x_i - \bar{x})}{\sum_{i=1}^{n} (x_i - \bar{x})^2} = \frac{\sum_{i=1}^{n} Y_i(x_i - \bar{x})}{\sum_{i=1}^{n} (x_i - \bar{x})^2}.$$

Hence, to find the mean line of best fit, $\mu(x) = \alpha + \beta(x_i - \bar{x})$, we use

$$\widehat{\alpha} = \bar{y} \qquad (6.7\text{-}1)$$

and

$$\widehat{\beta} = \frac{\sum_{i=1}^{n} y_i(x_i - \bar{x})}{\sum_{i=1}^{n}(x_i - \bar{x})^2} = \frac{\sum_{i=1}^{n} x_i y_i - \left(\frac{1}{n}\right)\left(\sum_{i=1}^{n} x_i\right)\left(\sum_{i=1}^{n} y_i\right)}{\sum_{i=1}^{n} x_i^2 - \left(\frac{1}{n}\right)\left(\sum_{i=1}^{n} x_i\right)^2}. \qquad (6.7\text{-}2)$$

To find the maximum likelihood estimator of σ^2, consider the partial derivative

$$\frac{\partial[-\ln L(\alpha, \beta, \sigma^2)]}{\partial(\sigma^2)} = \frac{n}{2\sigma^2} - \frac{\sum_{i=1}^{n}[y_i - \alpha - \beta(x_i - \bar{x})]^2}{2(\sigma^2)^2}.$$

Setting this equal to zero and replacing α and β by their solutions $\widehat{\alpha}$ and $\widehat{\beta}$, we obtain

$$\widehat{\sigma^2} = \frac{1}{n}\sum_{i=1}^{n}[Y_i - \widehat{\alpha} - \widehat{\beta}(x_i - \bar{x})]^2. \qquad (6.7\text{-}3)$$

A formula that is useful in calculating $n\widehat{\sigma^2}$ is

$$n\widehat{\sigma^2} = \sum_{i=1}^{n} y_i^2 - \frac{1}{n}\left(\sum_{i=1}^{n} y_i\right)^2 - \widehat{\beta}\sum_{i=1}^{n} x_i y_i + \widehat{\beta}\left(\frac{1}{n}\right)\left(\sum_{i=1}^{n} x_i\right)\left(\sum_{i=1}^{n} y_i\right). \qquad (6.7\text{-}4)$$

Note that the summand in Equation 6.7-3 for $\widehat{\sigma^2}$ is the square of the difference between the value of Y_i and the predicted mean of Y_i. Let $\widehat{Y}_i = \widehat{\alpha} + \widehat{\beta}(x_i - \bar{x})$, the predicted mean value of Y_i. The difference

$$Y_i - \widehat{Y}_i = Y_i - \widehat{\alpha} - \widehat{\beta}(x_i - \bar{x})$$

is called the ith **residual**, $i = 1, 2, \ldots, n$. The maximum likelihood estimate of σ^2 is then the sum of the squares of the residuals divided by n. It should always be true that the sum of the residuals is equal to zero. However, in practice, due to rounding off, the sum of the observed residuals, $y_i - \widehat{y}_i$, sometimes differs slightly from zero. A graph of the residuals plotted as a scatter plot of the points $x_i, y_i - \widehat{y}_i, i = 1, 2, \ldots, n$, can show whether or not linear regression provides the best fit.

EXAMPLE 6.7-1 The data plotted in Figure 6.7-1 are 10 pairs of test scores of 10 students in a psychology class, x being the score on a preliminary test and y the score on the final examination. The values of x and y are shown in Table 6.7-1. The sums that are needed to calculate estimates of the parameters are also given. Of course, the estimates of α and β have to be found before the residuals can be calculated.

Thus, $\widehat{\alpha} = 813/10 = 81.3$, and

$$\widehat{\beta} = \frac{56,089 - (683)(813)/10}{47,405 - (683)(683)/10} = \frac{561.1}{756.1} = 0.742.$$

			TABLE 6.7-1: Calculations for Test Score Data				
x	y	x^2	xy	y^2	$\widehat{y}$	$y - \widehat{y}$	$(y - \widehat{y})^2$
70	77	4,900	5,390	5,929	82.561566	−5.561566	30.931016
74	94	5,476	6,956	8,836	85.529956	8.470044	71.741645
72	88	5,184	6,336	7,744	84.045761	3.954239	15.636006
68	80	4,624	5,440	6,400	81.077371	−1.077371	1.160728
58	71	3,364	4,118	5,041	73.656395	−2.656395	7.056434
54	76	2,916	4,104	5,776	70.688004	5.311996	28.217302
82	88	6,724	7,216	7,744	91.466737	−3.466737	12.018265
64	80	4,096	5,120	6,400	78.108980	1.891020	3.575957
80	90	6,400	7,200	8,100	89.982542	0.017458	0.000305
61	69	3,721	4,209	4,761	75.882687	−6.882687	47.371380
683	813	47,405	56,089	66,731		0.000001	217.709038

Since $\bar{x} = 683/10 = 68.3$, the least squares regression line is

$$\widehat{y} = 81.3 + (0.742)(x - 68.3)$$

The maximum likelihood estimate of σ^2 is

$$\widehat{\sigma^2} = \frac{217.709038}{10} = 21.7709.$$

A plot of the residuals for these data is shown in Figure 6.7-2. ■

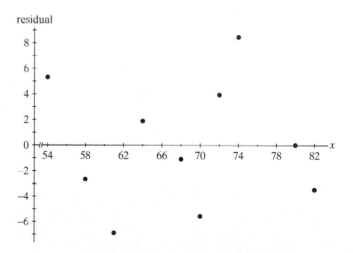

FIGURE 6.7-2: Residuals plot for data in Table 6.7-1

We shall now consider the problem of finding the distributions of $\widehat{\alpha}$, $\widehat{\beta}$, and $\widehat{\sigma^2}$ (or distributions of functions of these estimators). We would like to be able to say something about the error of the estimates to find confidence intervals for the parameters.

The preceding discussion treated $x_1, x_2, \ldots, x_n$ as constants. Of course, many times they can be set by the experimenter; for example, an experimental chemist might produce a compound at many different temperatures. But these numbers might instead be observations on an earlier random variable, such as an SAT score or a preliminary test grade (as in Example 6.7-1), but we consider the problem on the condition that the x values are given in either case. Thus, in finding the distributions of $\widehat{\alpha}$, $\widehat{\beta}$, and $\widehat{\sigma^2}$, the only random variables are $Y_1, Y_2, \ldots, Y_n$.

Since $\widehat{\alpha}$ is a linear function of independent and normally distributed random variables, $\widehat{\alpha}$ has a normal distribution with mean

$$
E(\widehat{\alpha}) = E\left(\frac{1}{n}\sum_{i=1}^{n} Y_i\right) = \frac{1}{n}\sum_{i=1}^{n} E(Y_i)
$$

$$
= \frac{1}{n}\sum_{i=1}^{n} [\alpha + \beta(x_i - \overline{x})] = \alpha
$$

and variance

$$
\operatorname{Var}(\widehat{\alpha}) = \sum_{i=1}^{n}\left(\frac{1}{n}\right)^2 \operatorname{Var}(Y_i) = \frac{\sigma^2}{n}.
$$

The estimator $\widehat{\beta}$ is also a linear function of $Y_1, Y_2, \ldots, Y_n$ and hence has a normal distribution with mean

$$
E(\widehat{\beta}) = \frac{\displaystyle\sum_{i=1}^{n}(x_i - \overline{x})E(Y_i)}{\sum_{i=1}^{n}(x_i - \overline{x})^2}
$$

$$
= \frac{\displaystyle\sum_{i=1}^{n}(x_i - \overline{x})[\alpha + \beta(x_i - \overline{x})]}{\displaystyle\sum_{i=1}^{n}(x_i - \overline{x})^2}
$$

$$
= \frac{\alpha\displaystyle\sum_{i=1}^{n}(x_i - \overline{x}) + \beta\displaystyle\sum_{i=1}^{n}(x_i - \overline{x})^2}{\displaystyle\sum_{i=1}^{n}(x_i - \overline{x})^2} = \beta
$$

and variance

$$
\operatorname{Var}(\widehat{\beta}) = \sum_{i=1}^{n}\left[\frac{x_i - \overline{x}}{\sum_{j=1}^{n}(x_j - \overline{x})^2}\right]^2 \operatorname{Var}(Y_i)
$$

$$
= \frac{\displaystyle\sum_{i=1}^{n}(x_i - \overline{x})^2}{\left[\displaystyle\sum_{i=1}^{n}(x_i - \overline{x})^2\right]^2}\sigma^2 = \frac{\sigma^2}{\displaystyle\sum_{i=1}^{n}(x_i - \overline{x})^2}.
$$

It can be shown (Exercise 6.7-6) that

$$\sum_{i=1}^{n} [Y_i - \alpha - \beta(x_i - \bar{x})]^2 = \sum_{i=1}^{n} \{(\widehat{\alpha} - \alpha) + (\widehat{\beta} - \beta)(x_i - \bar{x})$$

$$+ [Y_i - \widehat{\alpha} - \widehat{\beta}(x_i - \bar{x})]\}^2$$

$$= n(\widehat{\alpha} - \alpha)^2 + (\widehat{\beta} - \beta)^2 \sum_{i=1}^{n} (x_i - \bar{x})^2$$

$$+ \sum_{i=1}^{n} [Y_i - \widehat{\alpha} - \widehat{\beta}(x_i - \bar{x})]^2. \qquad (6.7\text{-}5)$$

From the fact that Y_i, $\widehat{\alpha}$, and $\widehat{\beta}$ have normal distributions, it follows that each of

$$\frac{[Y_i - \alpha - \beta(x_i - \bar{x})]^2}{\sigma^2}, \qquad \frac{(\widehat{\alpha} - \alpha)^2}{\left[\dfrac{\sigma^2}{n}\right]}, \qquad \text{and} \qquad \frac{(\widehat{\beta} - \beta)^2}{\left[\dfrac{\sigma^2}{\sum_{i=1}^{n}(x_i - \bar{x})^2}\right]}$$

has a chi-square distribution with one degree of freedom. Since $Y_1, Y_2, \ldots, Y_n$ are mutually independent,

$$\frac{\sum_{i=1}^{n} [Y_i - \alpha - \beta(x_i - \bar{x})]^2}{\sigma^2}$$

is $\chi^2(n)$. That is, the left-hand member of Equation 6.7-5 divided by σ^2 is $\chi^2(n)$ and is equal to the sum of two $\chi^2(1)$ variables and

$$\frac{\sum_{i=1}^{n} [Y_i - \widehat{\alpha} - \widehat{\beta}(x_i - \bar{x})]^2}{\sigma^2} = \frac{n\widehat{\sigma^2}}{\sigma^2} \geq 0.$$

Thus, we might guess that $n\widehat{\sigma^2}/\sigma^2$ is $\chi^2(n-2)$. This is true, and moreover, $\widehat{\alpha}$, $\widehat{\beta}$, and $\widehat{\sigma^2}$ are mutually independent. [For a proof, see Hogg, McKean, and Craig, *Introduction to Mathematical Statistics*, 6th ed. (Upper Saddle River, NJ: Prentice Hall, 2005).]

Suppose now that we are interested in forming a confidence interval for β, the slope of the line. We can use the fact that

$$T_1 = \frac{\sqrt{\sum_{i=1}^{n} (x_i - \bar{x})^2} \left(\dfrac{\widehat{\beta} - \beta}{\sigma}\right)}{\sqrt{\dfrac{n\widehat{\sigma^2}}{\sigma^2(n-2)}}} = \frac{\widehat{\beta} - \beta}{\sqrt{\dfrac{n\widehat{\sigma^2}}{(n-2)\sum_{i=1}^{n}(x_i - \bar{x})^2}}}$$

has a t distribution with $n - 2$ degrees of freedom. Therefore,

$$P\left[-t_{\gamma/2}(n - 2) \leq \frac{\widehat{\beta} - \beta}{\sqrt{\dfrac{n\widehat{\sigma^2}}{(n - 2)\sum_{i=1}^{n}(x_i - \overline{x})^2}}} \leq t_{\gamma/2}(n - 2)\right] = 1 - \gamma,$$

and it follows that

$$\left[\widehat{\beta} - t_{\gamma/2}(n - 2)\sqrt{\frac{n\widehat{\sigma^2}}{(n - 2)\sum_{i=1}^{n}(x_i - \overline{x})^2}},\right.$$

$$\left.\widehat{\beta} + t_{\gamma/2}(n - 2)\sqrt{\frac{n\widehat{\sigma^2}}{(n - 2)\sum_{i=1}^{n}(x_i - \overline{x})^2}}\right]$$

is a $100(1 - \gamma)\%$ confidence interval for β.

Similarly,

$$T_2 = \frac{\dfrac{\sqrt{n}(\widehat{\alpha} - \alpha)}{\sigma}}{\sqrt{\dfrac{n\widehat{\sigma^2}}{\sigma^2(n - 2)}}} = \frac{\widehat{\alpha} - \alpha}{\sqrt{\dfrac{\widehat{\sigma^2}}{n - 2}}}$$

has a t distribution with $n - 2$ degrees of freedom. Thus, T_2 can be used to make inferences about α. (See Exercise 6.7-7.) The fact that $n\widehat{\sigma^2}/\sigma^2$ has a chi-square distribution with $n - 2$ degrees of freedom can be used to make inferences about the variance σ^2. (See Exercise 6.7-8.)

STATISTICAL COMMENTS We now give an illustration (see Ledolter and Hogg in the references) using data from the *Challenger* explosion on January 28, 1986. The *Challenger* space shuttle was launched from Cape Kennedy in Florida on a very cold January morning. Meteorologists had forecasted temperatures (as of January 27) in the range of 26°–29° Fahrenheit. The night before the launch there was much debate among engineers and NASA officials whether a launch under such low-temperature conditions would be advisable. Several engineers advised against a launch because they thought that O-ring failures were related to temperature. Data on O-ring failures experienced in previous launches were available and were studied the night before the launch. There were seven previous incidents of known distressed O-rings. Figure 6.7-3(a) displays this information; it is a simple scatter plot of the number of distressed rings per launch against temperature at launch.

From this plot alone, there does not seem to be a strong relationship between the number of O-ring failures and temperature. On the basis of this information, along with many other technical and political considerations, it was decided to launch the *Challenger* space shuttle. As you all know, the launch resulted in disaster: the loss of seven lives and billions of dollars, and a serious setback to the space program.

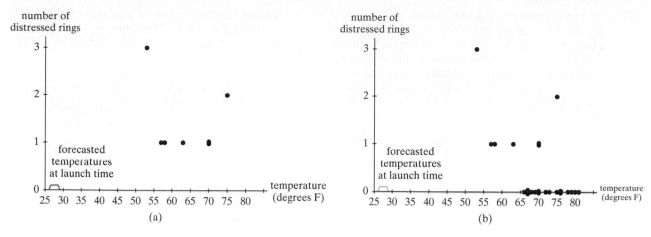

FIGURE 6.7-3: Number of distressed rings per launch versus temperature

One may argue that engineers looked at the scatter plot of the number of failures against temperature, but could not see a relationship. However, this argument misses the fact that engineers did not display *all the data that were relevant to the question.* They looked only at instances in which there were failures; they ignored the cases where there were no failures. In fact, there were 17 previous launches in which no failures occurred. A scatter plot of the number of distressed O-rings per launch against temperature using data from all previous shuttle launches is given in Figure 6.7-3(b).

It is difficult to look at these data and not see a relationship between failures and temperature. Moreover, one recognizes that an extrapolation is required and that an inference about the number of failures outside the observed range of temperature is needed. The actual temperature at launch was 31°F, while the lowest temperature recorded at a previous launch was 53°F. It is always very dangerous to extrapolate inferences to a region for which one does not have data. If NASA officials had looked at this plot, certainly the launch would have been delayed. This example shows why it is important to have statistically minded engineers involved in important decisions.

These comments raise two interesting points: (1) It is important to produce a scatter plot of one variable against another. (2) It is also important to plot *relevant data.* Yes, it is true that some data were used in making the decision to launch the *Challenger.* But not all the relevant data were utilized. To make good decisions, it takes knowledge of statistics as well as subject knowledge, common sense, and an ability to question the relevance of information.

EXERCISES

Remark   In many of these exercises, to find the confidence intervals, we need the distributional assumptions made in this section.   ■

6.7-1. The midterm and final exam scores of 10 students in a statistics course are tabulated as shown.

  (a) Calculate the least squares regression line for these data.

  (b) Plot the points and the least squares regression line on the same graph.

(c) Find the value of $\widehat{\sigma}^2$.

Midterm	Final	Midterm	Final
70	87	67	73
74	79	70	83
80	88	64	79
84	98	74	91
80	96	82	94

6.7-2. The final grade in a calculus course was predicted on the basis of the student's high school grade point average in mathematics, Scholastic Aptitude Test (SAT) score in mathematics, and score on a mathematics entrance examination. The predicted grades x and the earned grades y for 10 students are given (2.0 represents a C, 2.3 a C+, 2.7 a B−, etc.).

(a) Calculate the least squares regression line for these data.

(b) Plot the points and the least squares regression line on the same graph.

(c) Find the value of $\widehat{\sigma^2}$.

x	y	x	y
2.0	1.3	2.7	3.0
3.3	3.3	4.0	4.0
3.7	3.3	3.7	3.0
2.0	2.0	3.0	2.7
2.3	1.7	2.3	3.0

6.7-3. The following table from Ledolter and Hogg (see references) gives the attendance at a racetrack (x) and the amount that was bet (y) on $n = 10$ randomly selected days:

Attendance, x (in hundreds)	Amount Bet, y (in millions of dollars)
117	2.07
128	2.80
122	3.14
119	2.26
131	3.40
135	3.89
125	2.93
120	2.66
130	3.30
127	3.54

(a) Make a scatter plot of y against x.

(b) Fit a simple linear regression model to the data and calculate the least squares regression line.

(c) Make a scatter plot of the residuals.

6.7-4. Chemists often use ion-sensitive electrodes (ISEs) to measure the ion concentration of aqueous solutions. These devices measure the migration of ionic charge and give a reading in millivolts (mV). A standard curve is produced by measuring known concentrations (in ppm) and fitting a line to the millivolt data. The following table gives the concentrations in ppm and the voltage in mV at a calcium ISE:

ppm	mV	ppm	mV
0	1.72	100	2.91
0	1.68	100	3.00
0	1.74	100	2.89
50	2.04	150	4.47
50	2.11	150	4.51
50	2.17	150	4.43
75	2.40	200	6.67
75	2.32	200	6.66
75	2.33	200	6.57

(a) Calculate the least squares regression line for mV versus ppm.

(b) Plot the points and the least squares regression line on the same graph.

(c) Calculate and plot the residuals. Does linear regression seem to be appropriate? (See Exercise 6.8-15.)

6.7-5. A student who considered himself to be a "car guy" was interested in how the horsepower and weight of a car affected the time that it takes the car to go from 0 to 60 mph. The following table gives, for each of 14 cars, the horsepower, the time in seconds to go from 0 to 60 mph, and the weight in pounds:

Horse Power	0–60	Weight	Horse Power	0–60	Weight
230	8.1	3516	282	6.2	3627
225	7.8	3690	300	6.4	3892
375	4.7	2976	220	7.7	3377
322	6.6	4215	250	7.0	3625
190	8.4	3761	315	5.3	3230
150	8.4	2940	200	6.2	2657
178	7.2	2818	300	5.5	3518

(a) Calculate the least squares regression line for "0–60" versus horsepower.

(b) Plot the points and the least squares regression line on the same graph.

(c) Calculate the least squares regression line for "0–60" versus weight.

(d) Plot the points and the least squares regression line on the same graph.

(e) Which of the two variables, horsepower or weight, has the most effect on the "0–60" time?

6.7-6. Show that

$$\sum_{i=1}^{n} [Y_i - \alpha - \beta(x_i - \bar{x})]^2$$

$$= n(\widehat{\alpha} - \alpha)^2 + (\widehat{\beta} - \beta)^2 \sum_{i=1}^{n} (x_i - \bar{x})^2$$

$$+ \sum_{i=1}^{n} [Y_i - \widehat{\alpha} - \widehat{\beta}(x_i - \bar{x})]^2.$$

6.7-7. Show that the endpoints for a $100(1 - \gamma)\%$ confidence interval for α are

$$\widehat{\alpha} \pm t_{\gamma/2}(n - 2) \sqrt{\frac{\widehat{\sigma^2}}{n - 2}}.$$

6.7-8. Show that a $100(1 - \gamma)\%$ confidence interval for σ^2 is

$$\left[\frac{n\widehat{\sigma}^2}{\chi^2_{\gamma/2}(n - 2)}, \frac{n\widehat{\sigma}^2}{\chi^2_{1-\gamma/2}(n - 2)} \right].$$

6.7-9. Find 95% confidence intervals for α, β, and σ^2 for the data in Exercise 6.7-1.

6.7-10. Find 95% confidence intervals for α, β, and σ^2 for the data in Exercise 6.7-2.

6.7-11. In a "48.1-gram" package of candies, let x equal the number of pieces of candy and let y equal the total weight of the candies. For each of five different values of x, four observations of y are given.

x	y	x	y
54	48.8	56	50.8
54	49.4	56	50.9
54	49.2	57	50.7
54	50.4	57	51.6
55	49.5	57	51.3
55	49.0	57	50.8
55	50.2	58	52.1
55	48.9	58	51.3
56	49.9	58	51.4
56	50.1	58	52.0

(a) Calculate the least squares regression line for these data.

(b) Plot the points and the least squares regression line on the same graph.

(c) Find the value of $\widehat{\sigma}^2$.

(d) Find a 95% confidence interval for β under the usual assumptions.

6.7-12. The Federal Trade Commission measured the number of milligrams of tar and carbon monoxide (CO) per cigarette for all domestic cigarettes. Let x and y equal the measurements of tar and CO, respectively, for 100-millimeter filtered and mentholated cigarettes. A sample of 12 brands yielded the following data:

Brand	x	y	Brand	x	y
Capri	9	6	Now	3	4
Carlton	4	6	Salem	17	18
Kent	14	14	Triumph	6	8
Kool Milds	12	12	True	7	8
Marlboro Lights	10	12	Vantage	8	13
Merit Ultras	5	7	Virginia Slims	15	13

(a) Calculate the least squares regression line for these data.

(b) Plot the points and the least squares regression line on the same graph.

(c) Find point estimates for α, β, and σ^2.

(d) Find 95% confidence intervals for α, β, and σ^2 under the usual assumptions.

6.7-13. Let x and y equal the lengths in inches of a foot and a hand, respectively. The following measurements were made on 15 women:

x	y	x	y	x	y
9.00	6.50	10.00	7.00	9.25	7.00
8.50	6.25	9.50	6.50	10.00	7.50
9.25	7.25	9.00	7.00	10.00	7.25
9.75	7.00	9.25	7.00	9.75	7.25
9.00	6.75	9.50	7.00	9.50	7.25

(a) Calculate the least squares regression line for these data.

(b) Plot the points and the least squares regression line on the same graph.

(c) Find point estimates for α, β, and σ^2.

(d) Find 95% confidence intervals for α, β, and σ^2 under the usual assumptions.

6.7-14. Let x and y equal the ACT scores in social science and natural science, respectively, for a student who is applying for admission to a small liberal arts college. A sample of $n = 15$ such students yielded the following data:

x	y	x	y	x	y
32	28	30	27	26	32
23	25	17	23	16	22
23	24	20	30	21	28
23	32	17	18	24	31
26	31	18	18	30	26

(a) Calculate the least squares regression line for these data.

(b) Plot the points and the least squares regression line on the same graph.

(c) Find point estimates for α, β, and σ^2.

(d) Find 95% confidence intervals for α, β, and σ^2 under the usual assumptions.

6.7-15. The data in the following table, part of a set of data collected by Ledolter and Hogg (see references), provide the number of miles per gallon (mpg) for city and highway driving of 2007 midsize-model cars, as well as the curb weight of the cars:

Type	mpg City	mpg Hwy	Curb Weight
Ford Fusion V6 SE	20	28	3230
Chevrolet Sebring Sedan Base	24	32	3287
Toyota Camry Solara SE	24	34	3240
Honda Accord Sedan	20	29	3344
Audi A6 3.2	21	29	3825
BMW 5-series 525i Sedan	20	29	3450
Chrysler PT Cruiser Base	22	29	3076
Mercedes E-Class E350 Sedan	19	26	3740
Volkswagen Passat Sedan 2.0T	23	32	3305
Nissan Altima 2.5	26	35	3055
Kia Optima LX	24	34	3142

(a) Find the least squares regression line for highway mpg (y) and city mpg (x).

(b) Plot the points and the least squares regression line on the same graph.

(c) Find 95% confidence intervals for α, β, and σ^2.

6.7-16. Some measurements in millimeters were made on a species of spider named *Sosippus floridanus* that are native to Florida. Ten female spiders and 10 male spiders were used. The total lengths and the lengths of their front and their back legs are as follows:

Female body lengths	Female front legs	Female back legs	Male body lengths	Male front legs	Male back legs
11.06	15.03	19.29	12.26	20.22	25.54
13.87	17.96	22.74	11.66	18.62	23.94
12.93	17.56	21.28	12.53	18.62	23.94
15.08	21.22	25.54	13.00	19.95	25.80
17.82	22.61	28.86	11.79	19.15	25.40
14.14	20.08	25.14	12.46	19.02	25.27
12.26	16.49	20.22	10.65	17.29	22.21
17.82	18.75	24.61	10.39	17.02	21.81
20.17	23.01	28.46	12.26	18.49	23.41
16.88	22.48	28.60	14.07	22.61	28.86

Calculate the least squares regression line, and plot the points and the least squares regression line on the same graph, for

(a) Female body lengths and the lengths of their front legs.

(b) The lengths of the front and back legs of the female spiders.

(c) Male body lengths and the lengths of their back legs.

(d) The lengths of the front and back legs of the male spiders.

6.7-17. Using an Instron 4204, rectangular strips of Plexiglas® were stretched to failure in a tensile test. The following data give the change in length, in millimeters (mm), before breaking (x) and the cross–sectional area in square millimeters (mm^2) (y):

$(5.28, 52.36)$ $(5.40, 52.58)$ $(4.65, 51.07)$ $(4.76, 52.28)$ $(5.55, 53.02)$
$(5.73, 52.10)$ $(5.84, 52.61)$ $(4.97, 52.21)$ $(5.50, 52.39)$ $(6.24, 53.77)$

(a) Find the equation of the least squares regression line.
(b) Plot the points and the line on the same graph.
(c) Interpret your output.

6.7-18. The "Golden Ratio" is $\phi = (1 + \sqrt{5})/2$. John Putz, a mathematician who is interested in music, analyzed Mozart's sonata movements, which are divided into two distinct sections, both of which are repeated in performance (see references). The length of the "Exposition" in measures is represented by a, and the length of the "Development and Recapitulation" is represented by b. Putz's conjecture was that Mozart divided his movements near to the golden ratio. That is, Putz was interested in studying whether a scatter plot of $a + b$ against b not only would be linear, but also would actually fall along the line $y = \phi x$. Here are the data in tabular form, in which the first column identifies the piece and movement by the Köchel cataloging system:

Köchel	a	b	$a + b$	Köchel	a	b	$a + b$
279, I	38	62	100	279, II	28	46	74
279, III	56	102	158	280, I	56	88	144
280, II	24	36	60	280, III	77	113	190
281, I	40	69	109	281, II	46	60	106
282, I	15	18	33	282, III	39	63	102
283, I	53	67	120	283, II	14	23	37
283, III	102	171	273	284, I	51	76	127
309, I	58	97	155	311, I	39	73	112
310, I	49	84	133	330, I	58	92	150
330, III	68	103	171	332, I	93	136	229
332, III	90	150	245	333, I	63	102	165
333, II	31	50	81	457, I	74	93	167
533, I	102	137	239	533, II	46	76	122
545, I	28	45	73	547a, I	78	118	196
570, I	79	130	209				

(a) Make a scatter plot of the points $a + b$ against the points b. Is this plot linear?
(b) Find the equation of the least squares regression line. Superimpose it on the scatter plot.
(c) On the scatter plot, superimpose the line $y = \phi x$. Compare this line with the least squares regression line (graphically if you wish).
(d) Find the sample mean of the points $(a + b)/b$. Is the mean close to ϕ?

6.8 MORE REGRESSION

In this section, we use the notation and assumptions of Section 6.7. We have noted that $\widehat{Y} = \widehat{\alpha} + \widehat{\beta}(x - \overline{x})$ is a point estimate for the mean of Y for some given x, or we could think of this as a prediction of the value of Y for this given x. But how close is $\widehat{Y}$ to the mean of Y or to Y itself? We shall now find a confidence interval for $\alpha + \beta(x - \overline{x})$ and a prediction interval for Y, given a particular value of x.

To find a confidence interval for

$$E(Y) = \mu(x) = \alpha + \beta(x - \overline{x}),$$

let

$$\widehat{Y} = \widehat{\alpha} + \widehat{\beta}(x - \overline{x}).$$

Recall that $\widehat{Y}$ is a linear combination of normally and independently distributed random variables $\widehat{\alpha}$ and $\widehat{\beta}$, so $\widehat{Y}$ has a normal distribution. Furthermore,

$$E(\widehat{Y}) = E[\widehat{\alpha} + \widehat{\beta}(x - \overline{x})]$$
$$= \alpha + \beta(x - \overline{x})$$

and

$$\text{Var}(\widehat{Y}) = \text{Var}[\widehat{\alpha} + \widehat{\beta}(x - \overline{x})]$$
$$= \frac{\sigma^2}{n} + \frac{\sigma^2}{\sum_{i=1}^{n}(x_i - \overline{x})^2}(x - \overline{x})^2$$
$$= \sigma^2\left[\frac{1}{n} + \frac{(x - \overline{x})^2}{\sum_{i=1}^{n}(x_i - \overline{x})^2}\right].$$

Recall that the distribution of $n\widehat{\sigma}^2/\sigma^2$ is $\chi^2(n-2)$. Since $\widehat{\alpha}$ and $\widehat{\beta}$ are independent of $\widehat{\sigma}^2$, we can form the t statistic

$$T = \frac{\dfrac{\widehat{\alpha} + \widehat{\beta}(x - \overline{x}) - [\alpha + \beta(x - \overline{x})]}{\sigma\sqrt{\dfrac{1}{n} + \dfrac{(x - \overline{x})^2}{\sum_{i=1}^{n}(x_i - \overline{x})^2}}}}{\sqrt{\dfrac{n\widehat{\sigma}^2}{(n-2)\sigma^2}}},$$

which has a t distribution with $r = n - 2$ degrees of freedom. Next, we select $t_{\gamma/2}(n-2)$ from Table VI in the appendix so that

$$P[-t_{\gamma/2}(n-2) \leq T \leq t_{\gamma/2}(n-2)] = 1 - \gamma.$$

This becomes

$$P[\widehat{\alpha} + \widehat{\beta}(x - \overline{x}) - ct_{\gamma/2}(n-2) \leq \alpha + \beta(x - \overline{x}) \leq$$
$$\widehat{\alpha} + \widehat{\beta}(x - \overline{x}) + ct_{\gamma/2}(n-2)] = 1 - \gamma,$$

where

$$c = \sqrt{\frac{n\widehat{\sigma}^2}{n-2}}\sqrt{\frac{1}{n} + \frac{(x - \overline{x})^2}{\sum_{i=1}^{n}(x_i - \overline{x})^2}}.$$

Thus, the endpoints for a $100(1 - \gamma)\%$ confidence interval for $\mu(x) = \alpha + \beta(x - \overline{x})$ are

$$\widehat{\alpha} + \widehat{\beta}(x - \overline{x}) \pm ct_{\gamma/2}(n-2).$$

Note that the width of this interval depends on the particular value of x, because c depends on x. (See Example 6.8-1.)

We have used $(x_1, y_1), (x_2, y_2), \ldots, (x_n, y_n)$ to estimate α and β. Suppose that we are given a value of x, say, x_{n+1}. A point estimate of the corresponding value of Y is

$$\widehat{y}_{n+1} = \widehat{\alpha} + \widehat{\beta}(x_{n+1} - \overline{x}).$$

However, $\widehat{y}_{n+1}$ is just one possible value of the random variable

$$Y_{n+1} = \alpha + \beta(x_{n+1} - \bar{x}) + \varepsilon_{n+1}.$$

What can we say about possible values for Y_{n+1}? We shall now obtain a **prediction interval** for Y_{n+1} when $x = x_{n+1}$ that is similar to the confidence interval for the mean of Y when $x = x_{n+1}$.

We have

$$Y_{n+1} = \alpha + \beta(x_{n+1} - \bar{x}) + \varepsilon_{n+1},$$

where ε_{n+1} is $N(0, \sigma^2)$ and $\bar{x} = (1/n) \sum_{i=1}^{n} x_i$. Now,

$$W = Y_{n+1} - \widehat{\alpha} - \widehat{\beta}(x_{n+1} - \bar{x})$$

is a linear combination of normally and independently distributed random variables, so W has a normal distribution. The mean of W is

$$\begin{aligned} E(W) &= E[Y_{n+1} - \widehat{\alpha} - \widehat{\beta}(x_{n+1} - \bar{x})] \\ &= \alpha + \beta(x_{n+1} - \bar{x}) - \alpha - \beta(x_{n+1} - \bar{x}) = 0. \end{aligned}$$

Since $Y_{n+1}, \widehat{\alpha}$ and $\widehat{\beta}$ are independent, the variance of W is

$$\begin{aligned} \mathrm{Var}(W) &= \sigma^2 + \frac{\sigma^2}{n} + \frac{\sigma^2}{\sum_{i=1}^{n}(x_i - \bar{x})^2}(x_{n+1} - x)^2 \\ &= \sigma^2 \left[1 + \frac{1}{n} + \frac{(x_{n+1} - \bar{x})^2}{\sum_{i=1}^{n}(x_i - \bar{x})^2} \right]. \end{aligned}$$

Recall that $n\widehat{\sigma}^2/[(n-2)\sigma^2]$ is $\chi^2(n-2)$. Since $Y_{n+1}, \widehat{\alpha}$, and $\widehat{\beta}$ are independent of $\widehat{\sigma}^2$, we can form the t statistic

$$T = \frac{\dfrac{Y_{n+1} - \widehat{\alpha} - \widehat{\beta}(x_{n+1} - \bar{x})}{\sigma\sqrt{1 + \dfrac{1}{n} + \dfrac{(x_{n+1} - \bar{x})^2}{\sum_{i=1}^{n}(x_i - \bar{x})^2}}}}{\sqrt{\dfrac{n\widehat{\sigma}^2}{(n-2)\sigma^2}}}$$

which has a t distribution with $r = n - 2$ degrees of freedom. Now we select a constant $t_{\gamma/2}(n-2)$ from Table VI in the appendix so that

$$P[-t_{\gamma/2}(n-2) \leq T \leq t_{\gamma/2}(n-2)] = 1 - \gamma.$$

Solving this inequality for Y_{n+1}, we have

$$P[\widehat{\alpha} + \widehat{\beta}(x_{n+1} - \bar{x}) - d\, t_{\gamma/2}(n-2) \leq Y_{n+1} \leq$$
$$\widehat{\alpha} + \widehat{\beta}(x_{n+1} - \bar{x}) + d\, t_{\gamma/2}(n-2)] = 1 - \gamma,$$

where

$$d = \sqrt{\frac{n\widehat{\sigma^2}}{n-2}} \sqrt{1 + \frac{1}{n} + \frac{(x_{n+1} - \overline{x})^2}{\sum_{i=1}^{n}(x_i - \overline{x})^2}}.$$

Thus, the endpoints for a $100(1 - \gamma)\%$ prediction interval for Y_{n+1} are

$$\widehat{\alpha} + \widehat{\beta}(x_{n+1} - \overline{x}) \pm d t_{\gamma/2}(n - 2).$$

We shall now use the data in Example 6.7-1 to illustrate a 95% confidence interval for $\mu(x)$ and a 95% prediction interval for Y for a given value of x. To find such intervals, we use Equations 6.7-1, 6.7-2, and 6.7-4.

EXAMPLE 6.8-1   To find a 95% confidence interval for $\mu(x)$ using the data in Example 6.7-1, note that we have already found that $\overline{x} = 68.3$, $\widehat{\alpha} = 81.3$, $\widehat{\beta} = 561.1/756.1 = 0.7421$, and $\widehat{\sigma^2} = 21.7709$. We also need

$$\sum_{i=1}^{n}(x_i - \overline{x})^2 = \sum_{i=1}^{n} x_i^2 - \left(\frac{1}{n}\right)\left(\sum_{i=1}^{n} x_i\right)^2$$

$$= 47{,}405 - \frac{683^2}{10} = 756.1.$$

For 95% confidence, $t_{0.025}(8) = 2.306$. When $x = 60$, the endpoints for a 95% confidence interval for $\mu(60)$ are

$$81.3 + 0.7421(60 - 68.3) \pm \left[\sqrt{\frac{10(21.7709)}{8}} \sqrt{\frac{1}{10} + \frac{(60 - 68.3)^2}{756.1}}\right](2.306),$$

or

$$75.1406 \pm 5.2589.$$

Similarly, when $x = 70$, the endpoints for a 95% confidence interval for $\mu(70)$ are

$$82.5616 \pm 3.8761.$$

Note that the lengths of these intervals depend on the particular value of x. A 95% confidence band for $\mu(x)$ is graphed in Figure 6.8-1(a) along with the scatter diagram and $\widehat{y} = \widehat{\alpha} + \widehat{\beta}(x - \overline{x})$.

The endpoints for a 95% prediction interval for Y when $x = 60$ are

$$81.3 + 0.7421(60 - 68.3) \pm \left[\sqrt{\frac{10(21.7709)}{8}} \sqrt{1.1 + \frac{(60 - 68.3)^2}{756.1}}\right](2.306),$$

or

$$75.1406 \pm 13.1289.$$

Note that this interval is much wider than the confidence interval for $\mu(60)$. In Figure 6.8-1(b), the 95% prediction band for Y is graphed along with the scatter diagram and the least squares regression line.   ∎

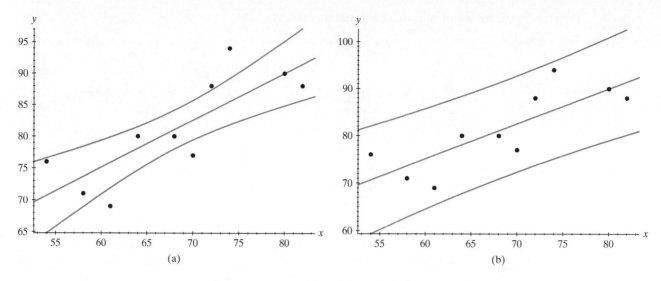

FIGURE 6.8-1: A 95% (a) confidence band for $\mu(x)$ and (b) prediction band for Y

We now generalize the simple regression model to the **multiple regression** case. Suppose we observe several x values—say $x_1, x_2, \ldots, x_k$—along with the y value. For example, suppose that x_1 equals the student's ACT composite score, x_2 equals the student's high school class rank, and y equals the student's first-year GPA in college. We want to estimate a regression function $E(Y) = \mu(x_1, x_2, \ldots, x_k)$ from some observed data. If

$$\mu(x_1, x_2, \ldots, x_k) = \beta_1 x_1 + \beta_2 x_2 + \cdots + \beta_k x_k,$$

then we say that we have a **linear model** because this expression is linear in the coefficients $\beta_1, \beta_2, \ldots, \beta_k$.

To illustrate, note that the model in Section 6.7 is linear in $\alpha = \beta_1$ and $\beta = \beta_2$, with $x_1 = 1$ and $x_2 = x$, giving the mean $\alpha + \beta x$. (For convenience, there the mean of the x values was subtracted from x.) Suppose, however, that we had wished to use the cubic function $\beta_1 + \beta_2 x + \beta_3 x^2 + \beta_4 x^3$ as the mean. This cubic expression still provides a linear model (i.e., linear in the β-values), and we would take $x_1 = 1$, $x_2 = x$, $x_3 = x^2$, and $x_4 = x^3$.

Say our n observation points are

$$(x_{1j}, x_{2j}, \ldots, x_{kj}, y_j), \qquad j = 1, 2, \ldots, n.$$

To fit the linear model $\beta_1 x_1 + \beta_2 x_2 + \cdots + \beta_k x_k$ by the method of least squares, we minimize

$$G = \sum_{j=1}^{n} (y_j - \beta_1 x_{1j} - \beta_2 x_{2j} - \cdots - \beta_k x_{kj})^2.$$

If we equate the k first partial derivatives

$$\frac{\partial G}{\partial \beta_i} = \sum_{j=1}^{n} (-2)(y_j - \beta_1 x_{1j} - \beta_2 x_{2j} - \cdots - \beta_k x_{kj})(x_{ij}), \qquad i = 1, 2, \ldots, k,$$

to zero, we obtain the k **normal equations**

$$\beta_1 \sum_{j=1}^{n} x_{1j}^2 + \beta_2 \sum_{j=1}^{n} x_{1j}x_{2j} + \cdots + \beta_k \sum_{j=1}^{n} x_{1j}x_{kj} = \sum_{j=1}^{n} x_{1j}y_j,$$

$$\beta_1 \sum_{j=1}^{n} x_{2j}x_{1j} + \beta_2 \sum_{j=1}^{n} x_{2j}^2 + \cdots + \beta_k \sum_{j=1}^{n} x_{2j}x_{kj} = \sum_{j=1}^{n} x_{2j}y_j,$$

$$\vdots \qquad\qquad \vdots \qquad\qquad \vdots \qquad\qquad \vdots$$

$$\beta_1 \sum_{j=1}^{n} x_{kj}x_{1j} + \beta_2 \sum_{j=1}^{n} x_{kj}x_{2j} + \cdots + \beta_k \sum_{j=1}^{n} x_{kj}^2 = \sum_{j=1}^{n} x_{kj}y_j.$$

The solution of the preceding k equations provides the least squares estimates of $\beta_1, \beta_2, \ldots, \beta_k$. These estimates are also maximum likelihood estimates of $\beta_1, \beta_2, \ldots, \beta_k$, provided that the random variables $Y_1, Y_2, \ldots, Y_n$ are mutually independent and Y_j is $N(\beta_1 x_{1j} + \beta_2 x_{2j} + \cdots + \beta_k x_{kj}, \sigma^2)$, $j = 1, 2, \ldots, n$.

By the method of least squares, we fit $y = \beta_1 x_1 + \beta_2 x_2 + \beta_3 x_3$ to the five observed points (x_1, x_2, x_3, y):

$$(1,1,0,4), \quad (1,0,1,3), \quad (1,2,3,2), \quad (1,3,0,6), \quad (1,0,0,1).$$

Note that $x_1 = 1$ in each point, so we are really fitting $y = \beta_1 + \beta_2 x_2 + \beta_3 x_3$. Since

$$\sum_{j=1}^{5} x_{1j}^2 = 5, \quad \sum_{j=1}^{5} x_{1j}x_{2j} = 6, \quad \sum_{j=1}^{5} x_{1j}x_{3j} = 4, \quad \sum_{j=1}^{5} x_{1j}y_j = 16,$$

$$\sum_{j=1}^{5} x_{2j}x_{1j} = 6, \quad \sum_{j=1}^{5} x_{2j}^2 = 14, \quad \sum_{j=1}^{5} x_{2j}x_{3j} = 6, \quad \sum_{j=1}^{5} x_{2j}y_j = 26,$$

$$\sum_{j=1}^{5} x_{3j}x_{1j} = 4, \quad \sum_{j=1}^{5} x_{3j}x_{2j} = 6, \quad \sum_{j=1}^{5} x_{3j}^2 = 10, \quad \sum_{j=1}^{5} x_{3j}y_j = 9,$$

the normal equations are

$$5\beta_1 + 6\beta_2 + 4\beta_3 = 16,$$
$$6\beta_1 + 14\beta_2 + 6\beta_3 = 26,$$
$$4\beta_1 + 6\beta_2 + 10\beta_3 = 9.$$

Solving these three linear equations in three unknowns, we obtain

$$\widehat{\beta}_1 = \frac{274}{112}, \qquad \widehat{\beta}_2 = \frac{127}{112}, \qquad \widehat{\beta}_3 = -\frac{85}{112}.$$

Thus, the least squares fit is

$$y = \frac{274x_1 + 127x_2 - 85x_3}{112}.$$

If x_1 always equals 1, then the equation reads

$$y = \frac{274 + 127x_2 - 85x_3}{112}.$$

It is interesting to observe that the usual two-sample problem is actually a linear model. Let $\beta_1 = \mu_1$ and $\beta_2 = \mu_2$, and consider n pairs of (x_1, x_2) that equal $(1, 0)$ and m pairs that equal $(0, 1)$. This would require each of the first n variables $Y_1, Y_2, \ldots, Y_n$ to have the mean

$$\beta_1 \cdot 1 + \beta_2 \cdot 0 = \beta_1 = \mu_1$$

and the next m variables $Y_{n+1}, Y_{n+2}, \ldots, Y_{n+m}$ to have the mean

$$\beta_1 \cdot 0 + \beta_2 \cdot 1 = \beta_2 = \mu_2.$$

This is the background of the two-sample problem, but with the usual $X_1, X_2, \ldots, X_n$ and $Y_1, Y_2, \ldots, Y_m$ replaced by $Y_1, Y_2, \ldots, Y_n$ and $Y_{n+1}, Y_{n+2}, \ldots, Y_{n+m}$, respectively.

EXERCISES

6.8-1. For the data given in Exercise 6.7-1, with the usual assumptions,

(a) Find a 95% confidence interval for $\mu(x)$ when $x = 68, 75$, and 82.

(b) Find a 95% prediction interval for Y when $x = 68, 75$, and 82.

6.8-2. For the data given in Exercise 6.7-2, with the usual assumptions,

(a) Find a 95% confidence interval for $\mu(x)$ when $x = 2, 3$, and 4.

(b) Find a 95% prediction interval for Y when $x = 2, 3$, and 4.

6.8-3. For the data given in Exercise 6.7-3, with the usual assumptions,

(a) Find a 95% confidence interval for $\mu(x)$ when $x = 120, 125$, and 130.

(b) Find a 95% prediction interval for Y when $x = 120, 125$, and 130.

6.8-4. For the candy data given in Exercise 6.7-11, with the usual assumptions,

(a) Find a 90% confidence interval for $\mu(x)$ when $x = 54, 56$, and 58.

(b) Determine a 90% prediction interval for Y when $x = 54, 56$, and 58.

6.8-5. For the cigarette data in Exercise 6.7-12, with the usual assumptions,

(a) Find a 95% confidence interval for $\mu(x)$ when $x = 5, 10$, and 15.

(b) Determine a 95% prediction interval for Y when $x = 5, 10$, and 15.

6.8-6. A computer center recorded the number of programs it maintained during each of 10 consecutive years.

(a) Calculate the least squares regression line for the data shown.

(b) Plot the points and the line on the same graph.

(c) Find a 95% prediction interval for the number of programs in year 11 under the usual assumptions.

Year	Number of Programs
1	430
2	480
3	565
4	790
5	885
6	960
7	1200
8	1380
9	1530
10	1591

6.8-7. For the ACT scores in Exercise 6.7-14, with the usual assumptions,

(a) Find a 95% confidence interval for $\mu(x)$ when $x = 17, 20, 23, 26$, and 29.

(b) Determine a 90% prediction interval for Y when $x = 17, 20, 23, 26$, and 29.

6.8-8. By the method of least squares, fit the regression plane $y = \beta_1 + \beta_2 x_1 + \beta_3 x_2$ to the following 12 observations of (x_1, x_2, y): (1, 1, 6), (0, 2, 3), (3, 0, 10), (−2, 0, −4), (−1, 2, 0), (0, 0, 1), (2, 1, 8), (−1, −1, −2), (0, −3, −3), (2, 1, 5), (1, 1, 1), (−1, 0, −2).

6.8-9. By the method of least squares, fit the cubic equation $y = \beta_1 + \beta_2 x + \beta_3 x^2 + \beta_4 x^3$ to the following 10 observed data points (x, y): (0, 1), (−1, −3), (0, 3), (1, 3), (−1, −1), (2, 10), (0, 0), (−2, −9), (−1, −2), (2, 8).

6.8-10. The data that follow give the times of 22 male swimmers who entered the 50-yard freestyle race in the conference championship meet. For these data, x is the swimmer's best time for the season and y is the swimmer's time in the meet.

x	y	x	y	x	y	x	y
24.97	23.98	24.76	23.63	23.80	23.61	22.84	23.58
24.07	23.57	22.98	23.17	22.93	22.83	23.41	22.75
22.10	22.74	23.08	22.70	23.59	22.62	23.80	22.53
22.38	22.39	22.91	22.34	22.08	21.95	22.46	21.73
21.34	21.68	22.76	21.66	21.82	21.68	21.80	21.58
22.26	21.57	21.36	21.35				

(a) Graph the points and the regression line for the data.
(b) Calculate and plot the residuals. Does linear regression seem to be appropriate?
(c) Find a 90% confidence band for the regression line. Make usual assumptions.
(d) Find a 90% prediction band for these data. Make usual assumptions.
(e) Find 95% confidence intervals for α, β, and σ^2. Make usual assumptions.

6.8-11. Explain why the model $\mu(x) = \beta_1 e^{\beta_2 x}$ is not a linear model. Would the logarithm, $\ln \mu(x)$, be a linear model?

6.8-12. We would like to fit the quadratic curve $y = \beta_1 + \beta_2 x + \beta_3 x^2$ to a set of points $(x_1, y_1), (x_2, y_2), \ldots, (x_n, y_n)$ by the method of least squares. To do this, let

$$h(\beta_1, \beta_2, \beta_3) = \sum_{i=1}^{n} (y_i - \beta_1 - \beta_2 x_i - \beta_3 x_i^2)^2.$$

(a) By setting the three first partial derivatives of h with respect to β_1, β_2, and β_3 equal to 0, show that β_1, β_2, and β_3 satisfy the following set of equations (called normal equations), all of which are sums going from 1 to n:

$$\beta_1 n + \beta_2 \sum x_i + \beta_3 \sum x_i^2 = \sum y_i;$$
$$\beta_1 \sum x_i + \beta_2 \sum x_i^2 + \beta_3 \sum x_i^3 = \sum x_i y_i;$$
$$\beta_1 \sum x_i^2 + \beta_2 \sum x_i^3 + \beta_3 \sum x_i^4 = \sum x_i^2 y_i.$$

(b) For the data

(6.91, 17.52)	(4.32, 22.69)	(2.38, 17.61)	(7.98, 14.29)
(8.26, 10.77)	(2.00, 12.87)	(3.10, 18.63)	(7.69, 16.77)
(2.21, 14.97)	(3.42, 19.16)	(8.18, 11.15)	(5.39, 22.41)
(1.19, 7.50)	(3.21, 19.06)	(5.47, 23.89)	(7.35, 16.63)
(2.32, 15.09)	(7.54, 14.75)	(1.27, 10.75)	(7.33,17.42)
(8.41, 9.40)	(8.72, 9.83)	(6.09, 22.33)	(5.30, 21.37)
(7.30, 17.36)			

$n = 25$, $\sum x_i = 133.34$, $\sum x_i^2 = 867.75$, $\sum x_i^3 = 6197.21$, $\sum x_i^4 = 46,318.88$, $\sum y_i = 404.22$, $\sum x_i y_i = 2138.38$, and $\sum x_i^2 y_i = 13,380.30$. Show that $a = -1.88$, $b = 9.86$, and $c = -0.995$.

(c) Plot the points and the linear regression line for these data.
(d) Calculate and plot the residuals. Does linear regression seem to be appropriate?
(e) Show that the least squares quadratic regression line is $\widehat{y} = -1.88 + 9.86x - 0.995x^2$.
(f) Plot the points and this least squares quadratic regression curve on the same graph.
(g) Plot the residuals for quadratic regression and compare this plot with that in part (d).

6.8-13. (The information presented in this exercise comes from the Westview Blueberry Farm and National Oceanic and Atmospheric Administration Reports [NOAA].) For the following paired data, (x, y), x gives the

Holland, Michigan, rainfall for June and y gives the blueberry production in thousands of pounds from the Westview Blueberry Farm:

$$(4.11, 56.2) \quad (5.49, 45.3) \quad (5.35, 31.0) \quad (6.53, 30.1)$$
$$(5.18, 40.0) \quad (4.89, 38.5) \quad (2.09, 50.0) \quad (1.40, 45.8)$$
$$(4.52, 45.9) \quad (1.11, 32.4) \quad (0.60, 18.2) \quad (3.80, 56.1)$$

The data are from 1971 to 1989 for those years in which the last frost occurred May 10 or earlier.

(a) Find the correlation coefficient for these data.
(b) Find the least squares regression line.
(c) Make a scatter plot of the data with the least squares regression line on the plot.
(d) Calculate and plot the residuals. Does linear regression seem to be appropriate?
(e) Find the least squares quadratic regression curve.
(f) Calculate and plot the residuals. Does quadratic regression seem to be appropriate?
(g) Give a short interpretation of your results.

6.8-14. The March 13, 1994, issue of *Parade* stated, "Heart surgeries, including bypass and angioplasty, totaled 196,000 in 1980 and reached 839,000 in 1991." (*Source*: National Hospital Discharge Survey, 1991.) The data for the years 1980 to 1991 in thousands are

$$(1980, 196) \quad (1981, 217) \quad (1982, 243) \quad (1983, 275) \quad (1984, 314) \quad (1985, 379)$$
$$(1986, 490) \quad (1987, 588) \quad (1988, 674) \quad (1989, 719) \quad (1990, 781) \quad (1991, 839)$$

(a) Make a scatter plot (here a time sequence) of the data with the least squares regression line superimposed.
(b) If possible, find a polynomial that provides a better fit of these data. Construct a scatter plot of the data with this polynomial superimposed.
(c) Calculate and plot the residuals for the polynomial that you select.

6.8-15. If, in Exercise 6.7-4, a linear regression fit does not seem appropriate, fit the quadratic curve $\beta_1 + \beta_2 x + \beta_3 x^2$ to the data. Does this curve seem to provide a better fit?

HISTORICAL COMMENTS   When a statistician thinks of estimation, he or she recalls R. A. Fisher's contributions to many aspects of the subject: maximum likelihood estimation, efficiency (Section 10.7), and sufficiency (Section 10.1). Of course, many more statisticians have contributed to that discipline since the 1920s. It would be an interesting exercise for the reader to go through the tables of contents of the *Journal of the American Statistical Association*, the *Annals of Statistics*, and related journals to observe how many articles are about estimation. Often our friends ask, "What is there left to do in mathematics?" University libraries are full of expanding journals of new mathematics, including statistics.

The other topic, among many important ones, in this chapter is regression, a technique that leads to a mathematical model of the result of some process in terms of some associated (explanatory) variables. We create such models to give us some idea of the value of a response variable if we know the values of certain explanatory variables. If we have an idea of the form of the equation relating these variables, then we can "fit" this model to the data; that is, we can determine approximate values for the unknown constants in the model from the data. Now, no model is exactly correct; but, as the well-known statistician George Box observed, "Some are useful." That is, while models may be wrong and we should check them as best we can, they may be good enough approximations to shed some light on the issues of interest.

Once satisfactory models are found, they may be used

1. to determine the effect of each explanatory variable (some may have very little effect and can be dropped),

2. to estimate the response variable for given values of important explanatory variables,

3. to predict the future, such as upcoming sales (although this sometimes should be done with great care),

4. to often substitute a cheaper explanatory variable for an expensive one that is difficult to obtain [such as chemical oxygen demand (COD) for biological oxygen demand (BOD)].

Finally, we must observe (and we will give a heuristic proof of it in Chapter 10) that most maximum likelihood estimators have approximate normal distributions. (See Section 10.7.) These estimators are of what is called the *regular cases*—in particular, those cases in which the parameters are not in the endpoints of the support of X. Abraham de Moivre proved this theorem for $\widehat{p}$ of the binomial distribution, and Laplace and Gauss did for $\overline{X}$ in a number of other distributions. This is the real reason the normal distribution is so important: Most estimators of parameters have approximate normal distributions, allowing us to construct confidence intervals and perform tests with such estimates.

Tests of Statistical Hypotheses

7.1   TESTS ABOUT PROPORTIONS

A first major area of statistical inference, namely, estimation of parameters, was introduced in Chapter 6. Here we consider a second major area: **tests of statistical hypotheses**. This very important topic is introduced through an illustration.

Suppose a manufacturer of a certain printed circuit observes that about $p = 0.06$ of the circuits fail. An engineer and statistician working together suggest some changes that might improve the design of the product. To test this new procedure, it was agreed that $n = 200$ circuits would be produced by the proposed method and then checked. Let Y equal the number of these 200 circuits that fail. Clearly, if the number of failures, Y, is such that $Y/200$ is about equal to 0.06, then it seems that the new procedure has not resulted in an improvement. Also, on the one hand, if Y is small, so that $Y/200$ is about 0.02 or 0.03, we might believe that the new method is better than the old. On the other hand, if $Y/200$ is 0.09 or 0.10, the proposed method has perhaps caused a greater proportion of failures.

What we need to establish is a formal rule that tells us when to accept the new procedure as an improvement. In addition, we must know the consequences of this rule. As an example of such a rule, we could accept the new procedure as an improvement if $Y \leq 7$ or $Y/n \leq 0.035$. We do note, however, that the probability of failure could still be about $p = 0.06$ even with the new procedure, and yet we could observe 7 or fewer failures in $n = 200$ trials. That is, we would erroneously accept the new method as being an improvement when, in fact, it was not. This decision is a mistake we call a **Type I error**. By contrast, the new procedure might actually improve the product so that p is much smaller, say, $p = 0.03$, and yet we could observe $y = 9$ failures, so that $y/200 = 0.045$. Thus, we would, again erroneously, not accept the new method as resulting in an improvement when in fact it had. This decision is a mistake we call a **Type II error**. We must study the probabilities of these two types of errors to understand fully the consequences of our rule.

Let us begin by modeling the situation. If we believe that these trials, conducted under the new procedure, are independent, and that each trial has about the same probability of failure, then Y is binomial $b(200, p)$. We wish to make a statistical inference about p using the unbiased estimator $\widehat{p} = Y/200$. Of course, we could construct a one-sided confidence interval—say, one that has 95% confidence of providing an upper bound for p—and obtain

$$\left[0, \widehat{p} + 1.645\sqrt{\frac{\widehat{p}(1 - \widehat{p})}{200}} \right].$$

This inference is appropriate and many statisticians simply make it. If the limits of this confidence interval contain 0.06, they would not say that the new procedure is necessarily better, at least until more data are taken. If, however, the upper limit of the confidence interval is less than 0.06, then those same statisticians would feel 95% confident that the true p is now less than 0.06. Hence, they would support the fact that the new procedure has improved the manufacturing of the printed circuits in question.

While this use of confidence intervals is highly appropriate, and later we indicate the relationship of confidence intervals to tests of hypotheses, every student of statistics should also have some understanding of the basic concepts in the latter area. Here, in our illustration, we are testing whether the probability of failure has or has not decreased from 0.06 when the new manufacturing procedure is used. The *no-change* hypothesis, H_0: $p = 0.06$, is called the **null hypothesis**. Since H_0: $p = 0.06$ completely specifies the distribution, it is called a **simple hypothesis**; thus, H_0: $p = 0.06$ is a **simple null hypothesis**. The *research worker's* (here the engineer's or the statistician's) hypothesis H_1: $p < 0.06$ is called the **alternative hypothesis**. Since H_1: $p < 0.06$ does not completely specify the distribution, it is a **composite hypothesis**, because it is composed of many simple hypotheses. Our rule of rejecting H_0 and accepting H_1 if $Y \le 7$, and otherwise accepting H_0, is called a **test of a statistical hypothesis**. We now see that the two types of errors can be recorded as follows:

Type I error: Rejecting H_0 and accepting H_1 when H_0 is true;

Type II error: Failing to reject H_0 when H_1 is true (i.e., when H_0 is false).

Since, in our illustration, we make a Type I error if $Y \le 7$ when in fact $p = 0.06$, we can calculate the probability of this error. We denote that probability by α and call it the **significance level of the test**. Under our assumptions, it is

$$\alpha = P(Y \le 7; p = 0.06) = \sum_{y=0}^{7} \binom{200}{y}(0.06)^y(0.94)^{200-y}.$$

Since n is rather large and p is small, these binomial probabilities can be approximated very well by Poisson probabilities with $\lambda = 200(0.06) = 12$. That is, from the Poisson table, the probability of a Type I error is

$$\alpha \approx \sum_{y=0}^{7} \frac{12^y e^{-12}}{y!} = 0.090.$$

Thus, the approximate significance level of this test is $\alpha = 0.090$. (Using the binomial distribution, we find that the exact value of α is 0.0829, which you can easily verify with Minitab.)

This value of α is reasonably small. However, what about the probability of a Type II error in case p has been improved to, say, 0.03? This error occurs if $Y > 7$ when in fact $p = 0.03$; hence, its probability, denoted by β, is

$$\beta = P(Y > 7; \; p = 0.03) = \sum_{y=8}^{200} \binom{200}{y}(0.03)^y(0.97)^{200-y}.$$

Again we use the Poisson approximation, here with $\lambda = 200(0.03) = 6$, to obtain

$$\beta \approx 1 - \sum_{y=0}^{7} \frac{6^y e^{-6}}{y!} = 1 - 0.744 = 0.256.$$

(The binomial distribution tells us that the exact probability is 0.2539, so the approximation is very good.) The engineer and the statistician who created the new procedure probably are not too pleased with this answer. That is, they might note that if their new procedure of manufacturing circuits has actually decreased the probability of failure to 0.03 from 0.06 (*a big improvement*), there is still a good chance, 0.256, that $H_0: p = 0.06$ is accepted and their improvement rejected. Thus, in their eyes, this test of $H_0: p = 0.06$ against $H_1: p = 0.03$ is unsatisfactory. In Section 10.2, more will be said about modifying tests so that satisfactory values of the probabilities of the two types of errors, namely, α and β, can be obtained; however, to decrease both of them, we need larger sample sizes.

Without worrying more about the probability of the Type II error here, we present a frequently used procedure for testing $H_0: p = p_0$, where p_0 is some specified probability of success. This test is based upon the fact that the number of successes Y in n independent Bernoulli trials is such that Y/n has an approximate normal distribution $N[p_0, p_0(1 - p_0)/n]$, provided that $H_0: p = p_0$ is true and n is large. Suppose the alternative hypothesis is $H_1: p > p_0$; that is, it has been hypothesized by a research worker that something has been done to increase the probability of success. Consider the test of $H_0: p = p_0$ against $H_1: p > p_0$ that rejects H_0 and accepts H_1 if and only if

$$Z = \frac{Y/n - p_0}{\sqrt{p_0(1 - p_0)/n}} \geq z_\alpha.$$

That is, if Y/n exceeds p_0 by z_α standard deviations of Y/n, we reject H_0 and accept the hypothesis $H_1: p > p_0$. Since, under H_0, Z is approximately $N(0,1)$, the approximate probability of this occurring when $H_0: p = p_0$ is true is α. So the significance level of this test is approximately α.

If the alternative is $H_1: p < p_0$ instead of $H_1: p > p_0$, then the appropriate α-level test is given by $Z \leq -z_\alpha$. Hence, if Y/n is smaller than p_0 by z_α standard deviations of Y/n, we accept $H_1: p < p_0$.

EXAMPLE 7.1-1 It was claimed that many commercially manufactured dice are not fair because the "spots" are really indentations, so that, for example, the 6-side is lighter than the 1-side. Let p equal the probability of rolling a 6 with one of these dice. To test H_0: $p = 1/6$ against the alternative hypothesis $H_1: p > 1/6$, several such dice will be rolled to yield a total of $n = 8000$ observations. Let Y equal the number of times that

6 resulted in the 8000 trials. The test statistic is

$$Z = \frac{Y/n - 1/6}{\sqrt{(1/6)(5/6)/n}} = \frac{Y/8000 - 1/6}{\sqrt{(1/6)(5/6)/8000}}.$$

If we use a significance level of $\alpha = 0.05$, the critical region is

$$z \geq z_{0.05} = 1.645.$$

The results of the experiment yielded $y = 1389$, so the calculated value of the test statistic is

$$z = \frac{1389/8000 - 1/6}{\sqrt{(1/6)(5/6)/8000}} = 1.670.$$

Since

$$z = 1.670 > 1.645,$$

the null hypothesis is rejected, and the experimental results indicate that these dice favor a 6 more than a fair die would. (You could perform your own experiment to check out other dice; see also Exercise 7.1-6.) ∎

Tests of $H_0: p = p_0$ against $H_1: p < p_0$ or $H_0: p = p_0$ against $H_1: p > p_0$ are called **one-sided** tests because the alternative hypotheses are **one sided**. There are times when **two-sided** alternatives and tests are appropriate, as in $H_1: p \neq p_0$. For example, suppose that the pass rate in the usual beginning statistics course is p_0. There has been an intervention (say, some new teaching method) and it is not known whether the pass rate will increase, decrease, or stay about the same. Thus, we test the null (no-change) hypothesis $H_0: p = p_0$ against the two-sided alternative $H_1: p \neq p_0$. A test with the approximate significance level α for doing this is to reject $H_0: p = p_0$ if

$$|Z| = \frac{|Y/n - p_0|}{\sqrt{p_0(1 - p_0)/n}} \geq z_{\alpha/2},$$

since, under H_0, $P(|Z| \geq z_{\alpha/2}) \approx \alpha$. These tests of approximate significance level α are summarized in Table 7.1-1. The rejection region for H_0 is often called the **critical region** of the test, and we use that terminology in the table.

TABLE 7.1-1: Tests of Hypotheses for One Proportion		
H_0	H_1	Critical Region
$p = p_0$	$p > p_0$	$z = \dfrac{y/n - p_0}{\sqrt{p_0(1 - p_0)/n}} \geq z_\alpha$
$p = p_0$	$p < p_0$	$z = \dfrac{y/n - p_0}{\sqrt{p_0(1 - p_0)/n}} \leq -z_\alpha$
$p = p_0$	$p \neq p_0$	$|z| = \dfrac{|y/n - p_0|}{\sqrt{p_0(1 - p_0)/n}} \geq z_{\alpha/2}$

For a number of years, there has been another value associated with a statistical test, and most statistical computer programs automatically print this value out; it is called the **probability value** or, for brevity, **p-value**. The p-value associated with a test is the probability, under the null hypothesis H_0, that the test statistic (a random variable) is equal to or exceeds the observed value (a constant) of the test statistic in the direction of the alternative hypothesis. Rather than select the critical region ahead of time, the p-value of a test can be reported and the reader then makes a decision.

In Example 7.1-1, the value of the test statistic was $z = 1.67$. Because the alternative hypothesis was $H_1: p > 1/6$, the p-value is

$$P(Z \geq 1.67) = 0.0475.$$

Note that this p-value is less than $\alpha = 0.05$, which would lead to the rejection of H_0 at an $\alpha = 0.05$ significance level. If the alternative hypothesis were two sided, $H_1: p \neq 1/6$, then the p-value would be $P(|Z| \geq 1.67) = 0.095$ and would not lead to the rejection of H_0 at $\alpha = 0.05$.

Often there is interest in tests about p_1 and p_2, the probabilities of success for two different distributions or the proportions of two different populations having a certain characteristic. For example, if p_1 and p_2 denote the respective proportions of homeowners and renters who vote in favor of a proposal to reduce property taxes, a politician might be interested in testing $H_0: p_1 = p_2$ against the one-sided alternative hypothesis $H_1: p_1 > p_2$.

Let Y_1 and Y_2 represent, respectively, the numbers of observed successes in n_1 and n_2 independent trials with probabilities of success p_1 and p_2. Recall that the distribution of $\widehat{p_1} = Y_1/n_1$ is approximately $N[p_1, p_1(1 - p_1)/n_1]$ and the distribution of $\widehat{p_2} = Y_2/n_2$ is approximately $N[p_2, p_2(1 - p_2)/n_2]$. So the distribution of $\widehat{p_1} - \widehat{p_2} = Y_1/n_1 - Y_2/n_2$ is approximately $N[p_1 - p_2, p_1(1 - p_1)/n_1 + p_2(1 - p_2)/n_2]$. It follows that the distribution of

$$Z = \frac{Y_1/n_1 - Y_2/n_2 - (p_1 - p_2)}{\sqrt{p_1(1 - p_1)/n_1 + p_2(1 - p_2)/n_2}} \tag{7.1-1}$$

is approximately $N(0, 1)$. To test $H_0: p_1 - p_2 = 0$ or, equivalently, $H_0: p_1 = p_2$, let $p = p_1 = p_2$ be the common value under H_0. We shall estimate p with $\widehat{p} = (Y_1 + Y_2)/(n_1 + n_2)$. Replacing p_1 and p_2 in the denominator of Equation 7.1-1 with this estimate, we obtain the test statistic

$$Z = \frac{\widehat{p_1} - \widehat{p_2} - 0}{\sqrt{\widehat{p}(1 - \widehat{p})(1/n_1 + 1/n_2)}},$$

which has an approximate $N(0, 1)$ distribution when the null hypothesis is true.

The three possible alternative hypotheses and their critical regions are summarized in Table 7.1-2.

TABLE 7.1-2: Tests of Hypotheses for Two Proportions

H_0	H_1	Critical Region
$p_1 = p_2$	$p_1 > p_2$	$z = \dfrac{\widehat{p_1} - \widehat{p_2}}{\sqrt{\widehat{p}(1 - \widehat{p})(1/n_1 + 1/n_2)}} \geq z_\alpha$
$p_1 = p_2$	$p_1 < p_2$	$z = \dfrac{\widehat{p_1} - \widehat{p_2}}{\sqrt{\widehat{p}(1 - \widehat{p})(1/n_1 + 1/n_2)}} \leq -z_\alpha$
$p_1 = p_2$	$p_1 \neq p_2$	$|z| = \dfrac{|\widehat{p_1} - \widehat{p_2}|}{\sqrt{\widehat{p}(1 - \widehat{p})(1/n_1 + 1/n_2)}} \geq z_{\alpha/2}$

REMARKS   In testing both $H_0: p = p_0$ and $H_0: p_1 = p_2$, statisticians sometimes use different denominators for z. For tests of single proportions, $\sqrt{p_0(1 - p_0)/n}$ can be replaced by $\sqrt{(y/n)(1 - y/n)/n}$, and for tests of the equality of two proportions, the following denominator can be used:

$$\sqrt{\frac{\widehat{p_1}(1 - \widehat{p_1})}{n_1} + \frac{\widehat{p_2}(1 - \widehat{p_2})}{n_2}}.$$

We do not have a strong preference one way or the other, since the two methods provide about the same numerical result. The substitutions do provide better estimates of the standard deviations of the numerators when the null hypotheses are clearly false. There is some advantage to this result if the null hypothesis is likely to be false. In addition, the substitutions tie together the use of confidence intervals and tests of hypotheses. For example, if the null hypothesis is $H_0: p = p_0$, then the alternative hypothesis $H_1: p < p_0$ is accepted if

$$z = \frac{\widehat{p} - p_0}{\sqrt{\dfrac{\widehat{p}(1 - \widehat{p})}{n}}} \leq -z_\alpha.$$

This formula is equivalent to the statement that

$$p_0 \notin \left[0, \widehat{p} + z_\alpha\sqrt{\frac{\widehat{p}(1 - \widehat{p})}{n}} \right),$$

where the latter is a one-sided confidence interval providing an upper bound for p. Or if the alternative hypothesis is $H_1: p \neq p_0$, then H_0 is rejected if

$$\frac{|\widehat{p} - p_0|}{\sqrt{\dfrac{\widehat{p}(1 - \widehat{p})}{n}}} \geq z_{\alpha/2}.$$

This inequality is equivalent to

$$p_0 \notin \left(\widehat{p} - z_{\alpha/2}\sqrt{\frac{\widehat{p}(1 - \widehat{p})}{n}}, \widehat{p} + z_{\alpha/2}\sqrt{\frac{\widehat{p}(1 - \widehat{p})}{n}} \right),$$

where the latter is a confidence interval for p. However, using the forms given in Tables 7.1-1 and 7.1-2, we do get better approximations to α-level significance tests. Thus, there are trade-offs, and it is difficult to say that one is better than the other. Fortunately, the numerical answers are about the same.

In the second situation in which the estimates of p_1 and p_2 are the observed $\widehat{p_1} = y_1/n_1$ and $\widehat{p_2} = y_2/n_2$, we have, with large values of n_1 and n_2, an approximate 95% confidence interval for $p_1 - p_2$ given by

$$\frac{y_1}{n_1} - \frac{y_2}{n_2} \pm 1.96\sqrt{\frac{(y_1/n_1)(1 - y_1/n_1)}{n_1} + \frac{(y_2/n_2)(1 - y_2/n_2)}{n_2}}.$$

If $p_1 - p_2 = 0$ is not in this interval, we reject $H_0: p_1 - p_2 = 0$ at the $\alpha = 0.05$ significance level. This is equivalent to saying that we reject $H_0: p_1 - p_2 = 0$ if

$$\frac{\left| \dfrac{y_1}{n_1} - \dfrac{y_2}{n_2} \right|}{\sqrt{\dfrac{(y_1/n_1)(1 - y_1/n_1)}{n_1} + \dfrac{(y_2/n_2)(1 - y_2/n_2)}{n_2}}} \geq 1.96.$$

In general, if the estimator $\widehat{\theta}$ (often, the maximum likelihood) of θ has an approximate (sometimes exact) normal distribution $N(\theta, \sigma_{\widehat{\theta}}^2)$, then $H_0: \theta = \theta_0$ is rejected in favor of $H_1: \theta \neq \theta_0$ at the α significance level if

$$\theta_0 \notin \left(\widehat{\theta} - z_{\alpha/2}\sigma_{\widehat{\theta}}, \ \widehat{\theta} + z_{\alpha/2}\sigma_{\widehat{\theta}} \right)$$

or, equivalently,

$$\frac{|\widehat{\theta} - \theta_0|}{\sigma_{\widehat{\theta}}} \geq z_{\alpha/2}.$$

Note that $\sigma_{\widehat{\theta}}$ often depends upon some unknown parameter that must be estimated and substituted in $\sigma_{\widehat{\theta}}$ to obtain $\widehat{\sigma}_{\widehat{\theta}}$. Sometimes $\sigma_{\widehat{\theta}}$ or its estimate is called the **standard error** of $\widehat{\theta}$. This was the case in our last illustration when, with $\theta = p_1 - p_2$ and $\widehat{\theta} = \widehat{p_1} - \widehat{p_2}$, we substituted y_1/n_1 for p_1 and y_2/n_2 for p_2 in

$$\sqrt{\frac{p_1(1 - p_1)}{n_1} + \frac{p_2(1 - p_2)}{n_2}}$$

to obtain the standard error of $\widehat{p_1} - \widehat{p_2} = \widehat{\theta}$.   ■

EXERCISES

7.1-1. Bowl A contains 100 red balls and 200 white balls; bowl B contains 200 red balls and 100 white balls. Let p denote the probability of drawing a red ball from a bowl, but say p is unknown, since it is unknown whether bowl A or bowl B is being used. We shall test the simple null hypothesis $H_0: p = 1/3$ against the simple alternative hypothesis $H_1: p = 2/3$. Draw three balls at random, one at a time and with replacement, from the selected bowl. Let X equal the number of red balls drawn. Then let the critical region be $C = \{x : x = 2, 3\}$. What are the values of α and β, the probabilities of Type I and Type II errors, respectively?

7.1-2. A bowl contains two red balls, two white balls, and a fifth ball that is either red or white. Let p denote the probability of drawing a red ball from the bowl. We shall test the simple null hypothesis H_0: $p = 3/5$ against the simple alternative hypothesis $H_1: p = 2/5$. Draw four balls at random from the bowl, one at a time and with replacement. Let X equal the number of red balls drawn.

(a) Define a critical region C for this test in terms of X.

(b) For the critical region C defined in part (a), find the values of α and β.

7.1-3. Let Y be $b(100, p)$. To test $H_0: p = 0.08$ against $H_1: p < 0.08$, we reject H_0 and accept H_1 if and only if $Y \leq 6$.

(a) Determine the significance level α of the test.

(b) Find the probability of the Type II error if in fact $p = 0.04$.

7.1-4. Let p denote the probability that, for a particular tennis player, the first serve is good. Since $p = 0.40$, this player decided to take lessons in order to increase p. When the lessons are completed, the hypothesis $H_0: p = 0.40$ will be tested against $H_1: p > 0.40$ on the basis of $n = 25$ trials. Let y equal the number of first serves that are good, and let the critical region be defined by $C = \{y: y \geq 13\}$.

(a) Determine $\alpha = P(Y \geq 13; p = 0.40)$. Use Table II in the appendix.

(b) Find $\beta = P(Y < 13)$ when $p = 0.60$; that is, $\beta = P(Y \leq 12; p = 0.60)$. Use Table II.

7.1-5. Let Y be $b(192, p)$. We reject $H_0: p = 0.75$ and accept $H_1: p > 0.75$ if and only if $Y \geq 152$. Use the normal approximation to determine

(a) $\alpha = P(Y \geq 152; p = 0.75)$.

(b) $\beta = P(Y < 152)$ when $p = 0.80$.

7.1-6. To determine whether the 1-side on a commercially manufactured die is heavy and the 6-side is light (see Example 7.1-1), some students kept track of the number of observed "ones" in $n = 8000$ rolls of the dice. Let p equal the probability of rolling a one with such a die. We shall test the null hypothesis $H_0: p = 1/6$ against the alternative hypothesis $H_1: p < 1/6$.

(a) Define the test statistic and an $\alpha = 0.05$ critical region.

(b) If $y = 1265$ ones were observed in 8000 rolls, calculate the value of the test statistic and state your conclusion.

(c) Is 1/6 in the 95% one-sided confidence interval providing an upper bound for p?

7.1-7. If a newborn baby has a birth weight that is less than 2500 grams (5.5 pounds), we say that the baby has a low birth weight. The proportion of babies with a low birth weight is an indicator of lack of nutrition for the mothers. For the United States, approximately 7% of babies have a low birth weight. Let p equal the proportion of babies born in the Sudan who weigh less than 2500 grams. We shall test the null hypothesis $H_0: p = 0.07$ against the alternative hypothesis $H_1: p > 0.07$. In a random sample of $n = 209$ babies, $y = 23$ weighed less than 2500 grams.

(a) What is your conclusion at a significance level of $\alpha = 0.05$?

(b) What is your conclusion at a significance level of $\alpha = 0.01$?

(c) Find the p-value for this test.

7.1-8. It was claimed that 75% of all dentists recommend a certain brand of gum for their gum-chewing patients. A consumer group doubted this claim and decided to test $H_0: p = 0.75$ against the alternative hypothesis $H_1: p < 0.75$, where p is the proportion of dentists who recommend that brand of gum. A survey of 390 dentists found that 273 recommended the given brand of gum.

(a) Which hypothesis would you accept if the significance level is $\alpha = 0.05$?

(b) Which hypothesis would you accept if the significance level is $\alpha = 0.01$?

(c) Find the p-value for this test.

7.1-9. It was claimed that the proportion of Americans who select jogging as one of their recreational activities is $p = 0.25$. A shoe manufacturer thought that p was larger than 0.25 and decided to test the null hypothesis $H_0: p = 0.25$ against the alternative hypothesis $H_1: p > 0.25$. Out of a random sample of $n = 5757$, $y = 1497$ selected jogging.

(a) What is your conclusion at a significance level of $\alpha = 0.05$?

(b) What is your conclusion at a significance level of $\alpha = 0.025$?

(c) Find the p-value for this test.

7.1-10. Let p equal the proportion of drivers who use a seat belt in a state that does not have a mandatory seat belt law. It was claimed that $p = 0.14$. An advertising campaign was conducted to increase this proportion. Two months after the campaign, $y = 104$ out of a random sample of $n = 590$ drivers were wearing their seat belts. Was the campaign successful?

(a) Define the null and alternative hypotheses.

(b) Define a critical region with an $\alpha = 0.01$ significance level.

(c) What is your conclusion?

7.1-11. The management of the Tigers baseball team decided to sell only low-alcohol beer in their ballpark to help combat rowdy fan conduct. They claimed that more than 40% of the fans would approve of this decision. Let p equal the proportion of Tiger fans on opening day who approved of the decision. We shall test the null hypothesis $H_0: p = 0.40$ against the alternative hypothesis $H_1: p > 0.40$.

(a) Define a critical region that has an $\alpha = 0.05$ significance level.

(b) If, out of a random sample of $n = 1278$ fans, $y = 550$ said that they approved of the new policy, what is your conclusion?

7.1-12. Because of tourism in the State, it was proposed that public schools in Michigan begin after

Labor Day. To determine whether support for this change was greater than 65%, a public poll was taken. Let p equal the proportion of Michigan adults who favor a post–Labor Day start. We shall test $H_0: p = 0.65$ against $H_1: p > 0.65$.

(a) Define a test statistic and an $\alpha = 0.025$ critical region.

(b) Given that 414 out of a sample of 600 favor a post–Labor Day start, calculate the value of the test statistic.

(c) Find the p-value and state your conclusion.

(d) Find a 95% one-sided confidence interval that gives a lower bound for p.

7.1-13. Let p equal the proportion of women who agree that "men are basically selfish and self-centered." Suppose that in the past it was believed that $p = 0.40$. It is now claimed that p has increased.

(a) Define the null and alternative hypotheses, a test statistic, and a critical region that has an approximate $\alpha = 0.01$ significance level. Sketch a standard normal p.d.f. and illustrate this critical region.

(b) The *Detroit Free Press* (April 26, 1990) reported that $y = 1260$ out of a random sample of $n = 3000$ women agree with the aforesaid statement about men. What is the conclusion of your test? Locate the calculated value of your test statistic on the p.d.f. in part (a).

7.1-14. A physician at a local hospital wrote, "In medicine we are now asked to perform quality assurance studies by various regulatory governmental or quasi-governmental agencies." He continued, "Dr. X was notified that for a particular procedure, his complication rate of 20% last year ($n = 20$) was unacceptable for continued privileges. So he took a one month remedial course for improving his technique. His complication rate is now 15% ($n = 15$). Has he really improved?" (You may assume that some rounding off was done in the letter and that the 15% is really 2/15.)

7.1-15. According to a population census in 1986, the percentage of males who are 18 or 19 years old and are married was 3.7%. We shall test whether this percentage increased from 1986 to 1988.

(a) Define the null and alternative hypotheses.

(b) Define a critical region that has an approximate significance level of $\alpha = 0.01$. Sketch a standard normal p.d.f. to illustrate this critical region.

(c) If $y = 20$ out of a random sample of $n = 300$ males, each 18 or 19 years old, were

married (*U.S. Bureau of the Census, Statistical Abstract of the United States: 1988*), what is your conclusion? Show the calculated value of the test statistic on your figure in part (b).

7.1-16. Let p equal the proportion of yellow candies in a package of mixed colors. It is claimed that $p = 0.20$.

(a) Define a test statistic and critical region with a significance level of $\alpha = 0.05$ for testing $H_0: p = 0.20$ against a two-sided alternative hypothesis.

(b) To perform the test, each of 20 students counted the number of yellow candies, y, and the total number of candies, n, in a 48.1-gram package, yielding the following ratios, y/n: 8/56, 13/55, 12/58, 13/56, 14/57, 5/54, 14/56, 15/57, 11/54, 13/55, 10/57, 8/59, 10/54, 11/55, 12/56, 11/57, 6/54, 7/58, 12/58, 14/58. If each individual tests $H_0: p = 0.20$, what proportion of the students rejected the null hypothesis?

(c) If we may assume that the null hypothesis is true, what proportion of the students would you have expected to reject the null hypothesis?

(d) For each of the 20 ratios in part (b), a 95% confidence interval for p can be calculated. What proportion of these 95% confidence intervals contain $p = 0.20$?

(e) If the 20 results are pooled so that $\sum_{i=1}^{20} y_i$ equals the number of yellow candies and $\sum_{i=1}^{20} n_i$ equals the total sample size, do we reject $H_0: p = 0.20$?

7.1-17. A machine shop that manufactures toggle levers has both a day and a night shift. A toggle lever is defective if a standard nut cannot be screwed onto the threads. Let p_1 and p_2 be the proportion of defective levers among those manufactured by the day and night shifts, respectively. We shall test the null hypothesis, $H_0: p_1 = p_2$, against a two-sided alternative hypothesis based on two random samples, each of 1000 levers taken from the production of the respective shifts.

(a) Define the test statistic and a critical region that has an $\alpha = 0.05$ significance level. Sketch a standard normal p.d.f. illustrating this critical region.

(b) If $y_1 = 37$ and $y_2 = 53$ defectives were observed for the day and night shifts, respectively, calculate the value of the test statistic. Locate the calculated test statistic on your figure in part (a) and state your conclusion.

7.1-18. The April 18, 1994, issue of *Time* magazine reported the results of a telephone poll of 800 adult Americans, 605 of them nonsmokers, who were asked the following question: "Should the federal tax on cigarettes be raised by $1.25 to pay for health care reform?" Let p_1 and p_2 equal the proportions of nonsmokers and smokers, respectively, who would say yes to this question. Given that $y_1 = 351$ nonsmokers and $y_2 = 41$ smokers said yes,

(a) With $\alpha = 0.05$, test $H_0: p_1 = p_2$ against $H_1: p_1 \neq p_2$.
(b) Find a 95% confidence interval for $p_1 - p_2$. Is your interval in agreement with the conclusion of part (a)?
(c) Find a 95% confidence interval for p, the proportion of adult Americans who would say yes.

7.1-19. Let p_m and p_f be the respective proportions of male and female white-crowned sparrows that return to their hatching site. Give the endpoints for a 95% confidence interval for $p_m - p_f$ if 124 out of 894 males and 70 out of 700 females returned. (*The Condor*, 1992, pp. 117–133.) Does your result agree with the conclusion of a test of $H_0: p_1 = p_2$ against $H_1: p_1 \neq p_2$ with $\alpha = 0.05$?

7.1-20. For developing countries in Africa and the Americas, let p_1 and p_2 be the respective proportions of babies with a low birth weight (below 2500 grams). We shall test $H_0: p_1 = p_2$ against the alternative hypothesis $H_1: p_1 > p_2$.

(a) Define a critical region that has an $\alpha = 0.05$ significance level.
(b) If respective random samples of sizes $n_1 = 900$ and $n_2 = 700$ yielded $y_1 = 135$ and $y_2 = 77$ babies with a low birth weight, what is your conclusion?
(c) What would your decision be with a significance level of $\alpha = 0.01$?
(d) What is the p-value of your test?

7.1-21. The following text was shown to a large class of students for 30 seconds, and they were told to report the number of F's that they found:

> FINISHED FILES ARE THE RE-
> SULT OF YEARS OF SCIENTIFIC
> STUDY COMBINED WITH THE
> EXPERIENCE OF MANY YEARS.

Let p equal the proportion of students who find six F's. We shall test the null hypothesis

$H_0: p = 0.5$ against the alternative hypothesis $H_1: p < 0.5$.

(a) Given a sample size of $n = 200$, define a critical region with an approximate significance level of $\alpha = 0.05$.
(b) If $y = 88$ students reported that they found six F's, what is your conclusion?
(c) What is the p-value of this test?
(d) Test this hypothesis or a modification of it with a class of students.

7.1-22. Each of six students has a deck of cards and selects a card randomly from his or her deck.

(a) Show that the probability of at least one match is equal to 0.259.
(b) Now let each of the students randomly select an integer from 1–52, inclusive. Let p equal the probability of at least one match. Test the null hypothesis $H_0: p = 0.259$ against an appropriate alternative hypothesis. Give a reason for your alternative.
(c) Perform this experiment a large number of times. What is your conclusion?

7.1-23. In a certain industry, about 15 percent of the workers showed some signs of ill effects due to radiation. After the management claimed that improvements had been made, 140 workers were tested and 19 experienced some ill effects due to radiation. Does this result support the management's claim? Use $\alpha = 0.05$.

7.1-24. Let p be the fraction of engineers who do not understand certain basic statistical concepts. Unfortunately, in the past, this number has been high, about $p = 0.73$. A new program to improve the knowledge of statistical methods has been implemented, and it is expected that under this program p would decrease from the aforesaid 0.73 value. To test $H_0: p = 0.73$ against $H_1: p < 0.73$, 300 engineers in the new program were tested and 204 (i.e., 68%) did not comprehend certain basic statistical concepts. Compute the p-value to determine whether this result indicates progress. That is, can we reject H_0 is favor of H_1? Use $\alpha = 0.05$.

7.1-25. A certain washing machine manufacturer claims that the fraction p_1 of its washing machines that need repairs in the first five years of operation is less than the fraction p_2 of another brand's. To test this claim, we observe $n_1 = n_2 = 200$ machines of each brand. We find that $y_1 = 21$ and $y_2 = 37$ machines need repairs. Do these data support the manufacturer's claim? Use $\alpha = 0.05$.

7.2 TESTS ABOUT ONE MEAN

Many applications assume that we are sampling from a normal distribution. We begin this section with such an application, for which we also assume that the variance is known.

EXAMPLE 7.2-1 Let X equal the breaking strength of a steel bar. If the bar is manufactured by process I, X is $N(50, 36)$. It is hoped that if process II (a new process) is used, X will be $N(55, 36)$. Given a large number of steel bars manufactured by process II, how could we test whether the increase in the mean breaking strength was realized?

In answering this question, we review some of the terminology that was introduced in Section 7.1, because we find that is helpful to most readers. So in this problem, we are assuming that X is $N(\mu, 36)$ and μ is equal to 50 or 55. We want to test the **simple null hypothesis** $H_0: \mu = 50$ against the **simple alternative hypothesis** $H_1: \mu = 55$. Note that each of these hypotheses completely specifies the distribution of X. That is, H_0 states that X is $N(50, 36)$ and H_1 states that X is $N(55, 36)$. (If the alternative hypothesis had been $H_1: \mu > 50$, it would be a **composite hypothesis**, because it is composed of all normal distributions with $\sigma^2 = 36$ and means greater than 50.) In order to test which of the two hypotheses, H_0 or H_1, is true, we shall set up a rule based on the breaking strengths $x_1, x_2, \ldots, x_n$ of n bars (the observed values of a random sample of size n from this new normal distribution). The rule leads to a decision to accept or reject H_0; hence, it is necessary to partition the sample space into two parts—say, C and C'—so that if $(x_1, x_2, \ldots, x_n) \in C$, H_0 is rejected, and if $(x_1, x_2, \ldots, x_n) \in C'$, H_0 is accepted (not rejected). The rejection region C for H_0 is called the **critical region** for the test. Often, the partitioning of the sample space is specified in terms of the values of a statistic called the **test statistic**. In this example, we could let $\overline{X}$ be the test statistic and, say, take $C = \{(x_1, x_2, \ldots, x_n): \overline{x} \geq 53\}$. We could then define the critical region as those values of the test statistic for which H_0 is rejected. That is, the given critical region is equivalent to defining $C = \{\overline{x}: \overline{x} \geq 53\}$ in the $\overline{x}$ space. If $(x_1, x_2, \ldots, x_n) \in C$ when H_0 is true, H_0 would be rejected when it is true, a **Type I error**. If $(x_1, x_2, \ldots, x_n) \in C'$ when H_1 is true, H_0 would be accepted (i.e., not rejected) when in fact H_1 is true, a **Type II error**. Recall that the probability of a Type I error is called the **significance level** of the test and is denoted by α. That is, $\alpha = P[(X_1, X_2, \ldots, X_n) \in C; H_0]$ is the probability that $(X_1, X_2, \ldots, X_n)$ falls into C when H_0 is true. The probability of a Type II error is denoted by β; that is, $\beta = P[(X_1, X_2, \ldots, X_n) \in C'; H_1]$ is the probability of accepting (failing to reject) H_0 when it is false. As an illustration, suppose $n = 16$ bars were tested and $C = \{\overline{x}: \overline{x} \geq 53\}$. Then $\overline{X}$ is $N(50, 36/16)$ when H_0 is true and is $N(55, 36/16)$ when H_1 is true. Thus,

$$\alpha = P(\overline{X} \geq 53; H_0) = P\left(\frac{\overline{X} - 50}{6/4} \geq \frac{53 - 50}{6/4}; H_0\right)$$

$$= 1 - \Phi(2) = 0.0228$$

and

$$\beta = P(\overline{X} < 53; H_1) = P\left(\frac{\overline{X} - 55}{6/4} < \frac{53 - 55}{6/4}; H_1\right)$$

$$= \Phi\left(-\frac{4}{3}\right) = 1 - 0.9087 = 0.0913.$$

Figure 7.2-1 shows the graphs of the probability density functions of $\overline{X}$ when H_0 and H_1, respectively, are true. Note that a decrease in the size of α leads to an increase in the size of β, and vice versa. Both α and β can be decreased if the sample size n is increased. ∎

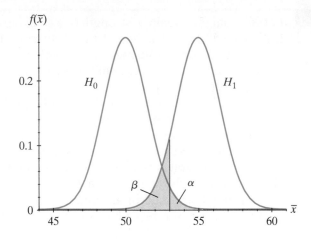

FIGURE 7.2-1: p.d.f. of $\overline{X}$ under H_0 and H_1

The definition of *p-value* is given in Section 7.1. We now give an example of a *p*-value obtained in testing a hypothesis about a mean.

EXAMPLE 7.2-2 Assume that the underlying distribution is normal with unknown mean μ but known variance $\sigma^2 = 100$. Say we are testing $H_0: \mu = 60$ against $H_1: \mu > 60$ with a sample mean $\overline{X}$ based on $n = 52$ observations. Suppose that we obtain the observed sample mean of $\bar{x} = 62.75$. If we compute the probability of obtaining an $\overline{X}$ of that value of 62.75 or greater when $\mu = 60$, then we obtain the *p*-value associated with $\bar{x} = 62.75$. That is,

$$p\text{-value} = P(\overline{X} \geq 62.75; \mu = 60)$$

$$= P\left(\frac{\overline{X} - 60}{10/\sqrt{52}} \geq \frac{62.75 - 60}{10/\sqrt{52}}; \mu = 60\right)$$

$$= 1 - \Phi\left(\frac{62.75 - 60}{10/\sqrt{52}}\right) = 1 - \Phi(1.983) = 0.0237.$$

If this *p*-value is small, we tend to reject the hypothesis $H_0: \mu = 60$. For example, rejecting $H_0: \mu = 60$ if the *p*-value is less than or equal to $\alpha = 0.05$ is exactly the same as rejecting H_0 if

$$\bar{x} \geq 60 + (1.645)\left(\frac{10}{\sqrt{52}}\right) = 62.718.$$

Here

$$p\text{-value} = 0.0237 < \alpha = 0.05 \quad \text{and} \quad \bar{x} = 62.75 > 62.718.$$

To help the reader keep the definition of **p-value** in mind, we note that it can be thought of as that **tail-end probability**, under H_0, of the distribution of the statistic (here $\overline{X}$), beyond the observed value of the statistic. (See Figure 7.2-2 for the p-value associated with $\overline{x} = 62.75$.)

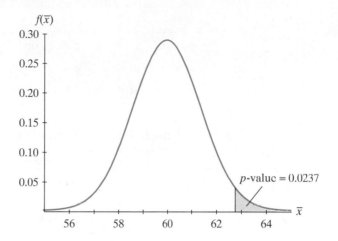

FIGURE 7.2-2: Illustration of p-value

If the alternative were the two-sided $H_1: \mu \neq 60$, then the p-value would have been double 0.0237; that is, then the p-value $= 2(0.0237) = 0.0474$ because we include both tails. ∎

When we sample from a normal distribution, the null hypothesis is generally of the form $H_0: \mu = \mu_0$. There are essentially three possibilities for the alternative hypothesis: (i) that μ has increased, or $H_1: \mu > \mu_0$; (ii) that μ has decreased, or $H_1: \mu < \mu_0$; and (iii) that μ has changed, but it is not known whether it has increased or decreased, which leads to the two-sided alternative hypothesis, or $H_1: \mu \neq \mu_0$.

To test $H_0: \mu = \mu_0$ against one of these three alternative hypotheses, a random sample is taken from the distribution and an observed sample mean, x, that is close to μ_0 supports H_0. The closeness of $\overline{x}$ to μ_0 is measured in terms of standard deviations of $\overline{X}$, $\sigma/\sqrt{n}$, when σ is known, a measure that is sometimes called the **standard error of the mean**. Thus, the test statistic could be defined by

$$Z = \frac{X - \mu_0}{\sqrt{\sigma^2/n}} = \frac{\overline{X} - \mu_0}{\sigma/\sqrt{n}}, \tag{7.2-1}$$

and the critical regions, at a significance level α, for the three respective alternative hypotheses would be (i) $z \geq z_\alpha$, (ii) $z \leq -z_\alpha$, and (iii) $|z| \geq z_{\alpha/2}$. In terms of $\overline{x}$ these three critical regions become (i) $\overline{x} \geq \mu_0 + z_\alpha(\sigma/\sqrt{n})$, (ii) $\overline{x} \leq \mu_0 - z_\alpha(\sigma/\sqrt{n})$, and (iii) $|\overline{x} - \mu_0| \geq z_{\alpha/2}(\sigma/\sqrt{n})$.

The three tests and critical regions are summarized in Table 7.2-1 on the next page. The underlying assumption is that the distribution is $N(\mu, \sigma^2)$ and σ^2 is known.

It is usually the case that the variance σ^2 is not known. Accordingly, we now take a more realistic position and assume that the variance is unknown. Suppose our null hypothesis is $H_0: \mu = \mu_0$ and the two-sided alternative hypothesis is $H_1: \mu \neq \mu_0$. Recall from Section 6.2, that, for a random sample $X_1, X_2, \ldots, X_n$ taken from a normal distribution $N(\mu, \sigma^2)$, a confidence interval for μ is based on

$$T = \frac{\overline{X} - \mu}{\sqrt{S^2/n}} = \frac{\overline{X} - \mu}{S/\sqrt{n}}.$$

TABLE 7.2-1: Tests of Hypotheses About One Mean, Variance Known

H_0	H_1	Critical Region
$\mu = \mu_0$	$\mu > \mu_0$	$z \geq z_\alpha$ or $\bar{x} \geq \mu_0 + z_\alpha \sigma/\sqrt{n}$
$\mu = \mu_0$	$\mu < \mu_0$	$z \leq -z_\alpha$ or $\bar{x} \leq \mu_0 - z_\alpha \sigma/\sqrt{n}$
$\mu = \mu_0$	$\mu \neq \mu_0$	$|z| \geq z_{\alpha/2}$ or $|\bar{x} - \mu_0| \geq z_{\alpha/2} \sigma/\sqrt{n}$

This suggests that T might be a good statistic to use for the test of H_0: $\mu = \mu_0$ with μ replaced by μ_0. In addition, it is the natural statistic to use if we replace σ^2/n by its unbiased estimator S^2/n in $(\bar{X} - \mu_0)/\sqrt{\sigma^2/n}$ in Equation 7.2-1. If $\mu = \mu_0$, we know that T has a t distribution with $n - 1$ degrees of freedom. Thus, with $\mu = \mu_0$,

$$P[|T| \geq t_{\alpha/2}(n - 1)] = P\left[\frac{|\bar{X} - \mu_0|}{S/\sqrt{n}} \geq t_{\alpha/2}(n - 1)\right] = \alpha.$$

Accordingly, if $\bar{x}$ and s are, respectively, the sample mean and sample standard deviation, then the rule that rejects H_0: $\mu = \mu_0$ and accepts H_1: $\mu \neq \mu_0$ if and only if

$$|t| = \frac{|\bar{x} - \mu_0|}{s/\sqrt{n}} \geq t_{\alpha/2}(n - 1)$$

provides a test of this hypothesis with significance level α. Note that this rule is equivalent to rejecting H_0: $\mu = \mu_0$ if μ_0 is not in the open $100(1 - \alpha)\%$ confidence interval

$$(\bar{x} - t_{\alpha/2}(n - 1)(s/\sqrt{n}), \bar{x} + t_{\alpha/2}(n - 1)(s/\sqrt{n})).$$

Table 7.2-2 summarizes tests of hypotheses for a single mean, along with the three possible alternative hypotheses, when the underlying distribution is $N(\mu, \sigma^2)$, σ^2 is unknown, $t = (\bar{x} - \mu_0)/(s/\sqrt{n})$, and $n \leq 30$. If $n > 30$, we use Table 7.2-1 for approximate tests, with σ replaced by s.

TABLE 7.2-2: Tests of Hypotheses for One Mean, Variance Unknown

H_0	H_1	Critical Region
$\mu = \mu_0$	$\mu > \mu_0$	$t \geq t_\alpha(n - 1)$ or $\bar{x} \geq \mu_0 + t_\alpha(n - 1)s/\sqrt{n}$
$\mu = \mu_0$	$\mu < \mu_0$	$t \leq -t_\alpha(n - 1)$ or $\bar{x} \leq \mu_0 - t_\alpha(n - 1)s/\sqrt{n}$
$\mu = \mu_0$	$\mu \neq \mu_0$	$|t| \geq t_{\alpha/2}(n - 1)$ or $|\bar{x} - \mu_0| \geq t_{\alpha/2}(n - 1)s/\sqrt{n}$

EXAMPLE 7.2-3   Let X (in millimeters) equal the growth in 15 days of a tumor induced in a mouse. Assume that the distribution of X is $N(\mu, \sigma^2)$. We shall test the null hypothesis H_0: $\mu = \mu_0 = 4.0$ millimeters against the two-sided alternative hypothesis H_1: $\mu \neq 4.0$. If we use $n = 9$ observations and a significance level of $\alpha = 0.10$, the critical region is

$$|t| = \frac{|\bar{x} - 4.0|}{s/\sqrt{9}} \geq t_{\alpha/2}(8) = 1.860.$$

If we are given that $n = 9$, $\bar{x} = 4.3$, and $s = 1.2$, we see that

$$t = \frac{4.3 - 4.0}{1.2/\sqrt{9}} = \frac{0.3}{0.4} = 0.75.$$

Thus,

$$|t| = |0.75| < 1.860,$$

and we accept (do not reject) H_0: $\mu = 4.0$ at the $\alpha = 10\%$ significance level. (See Figure 7.2-3.) The p-value is

$$p\text{-value} = P(|T| \geq 0.75) = 2P(T \geq 0.75).$$

With our t tables with eight degrees of freedom, we cannot find this p-value exactly. It is about 0.50, because

$$P(|T| \geq 0.706) = 2P(T \geq 0.706) = 0.50.$$

Minitab gives a p-value of 0.4747. (See Figure 7.2-3.) ∎

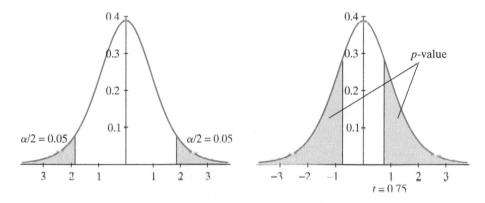

FIGURE 7.2-3: Test about mean of tumor growths

REMARK   In discussing the test of a statistical hypothesis, the word *accept H_0* might better be replaced by *do not reject H_0*. That is, if, in Example 7.2-3, $\bar{x}$ is close enough to 4.0 so that we accept $\mu = 4.0$, we do not want that acceptance to imply that μ is actually equal to 4.0. We want to say that the data do not deviate enough from $\mu = 4.0$ for us to reject that hypothesis; that is, we do not reject $\mu = 4.0$ with these observed data. With this understanding, we sometimes use *accept*, and sometimes *fail to reject* or *do not reject*, the null hypothesis. ∎

The next example illustrates the use of the t statistic with a one-sided alternative hypothesis.

EXAMPLE 7.2-4   In attempting to control the strength of the wastes discharged into a nearby river, a paper firm has taken a number of measures. Members of the firm believe that they have reduced the oxygen-consuming power of their wastes from a previous mean μ of 500 (measured in parts per million of permanganate). They plan to test

$H_0: \mu = 500$ against $H_1: \mu < 500$, using readings taken on $n = 25$ consecutive days. If these 25 values can be treated as a random sample, then the critical region, for a significance level of $\alpha = 0.01$, is

$$t = \frac{\bar{x} - 500}{s/\sqrt{25}} \leq -t_{0.01}(24) = -2.492.$$

The observed values of the sample mean and sample standard deviation were $\bar{x} = 308.8$ and $s = 115.15$. Since

$$t = \frac{308.8 - 500}{115.15/\sqrt{25}} = -8.30 < -2.492,$$

we clearly reject the null hypothesis and accept $H_1: \mu < 500$. Note, however, that although an improvement has been made, there still might exist the question of whether the improvement is adequate. The one-sided 99% confidence interval for μ, namely,

$$[0, 308.8 + 2.492(115.25/\sqrt{25}\,)] = [0, 366.191],$$

provides an upper bound for μ and may help the company answer this question. ∎

Oftentimes, there is interest in comparing the means of two different distributions or populations. We must consider two situations: that in which X and Y are dependent and that in which X and Y are independent. We consider the independent case in the next section.

If X and Y are dependent, let $W = X - Y$, and the hypothesis that $\mu_X = \mu_Y$ would be replaced with the hypothesis $H_0: \mu_W = 0$. For example, suppose that X and Y equal the resting pulse rate for a person before and after taking an eight-week program in aerobic dance. We would be interested in testing $H_0: \mu_W = 0$ (no change) against $H_1: \mu_W > 0$ (the aerobic dance program decreased the resting pulse rate). Because X and Y are measurements on the same person, X and Y are clearly dependent. If we can assume that the distribution of W is (approximately) $N(\mu_W, \sigma^2)$, then we can choose to use the appropriate t-test for a single mean from Table 7.2-2. This is often called a **paired t test**.

EXAMPLE 7.2-5   Twenty-four girls in the 9th and 10th grades were put on an ultraheavy rope-jumping program. Someone thought that such a program would increase their speed in the 40-yard dash. Let W equal the difference in time to run the 40-yard dash—the "before-program time" minus the "after-program time." Assume that the distribution of W is approximately $N(\mu_W, \sigma_W^2)$. We shall test the null hypothesis $H_0: \mu_W = 0$ against the alternative hypothesis $H_1: \mu_W > 0$. The test statistic and the critical region that has an $\alpha = 0.05$ significance level are given by

$$t = \frac{\bar{w} - 0}{s_w/\sqrt{24}} \geq t_{0.05}(23) = 1.714.$$

The following data give the difference in time that it took each girl to run the 40-yard dash, with positive numbers indicating a faster time after the program:

0.28	0.01	0.13	0.33	−0.03	0.07	−0.18	−0.14
−0.33	0.01	0.22	0.29	−0.08	0.23	0.08	0.04
−0.30	−0.08	0.09	0.70	0.33	−0.34	0.50	0.06

For these data, $\overline{w} = 0.079$ and $s_w = 0.255$. Thus, the observed value of the test statistic is

$$t = \frac{0.079 - 0}{0.255/\sqrt{24}} = 1.518.$$

Since $1.518 < 1.714$, the null hypothesis is not rejected. Note, however, that $t_{0.10}(23) = 1.319$ and $t = 1.518 > 1.319$. Hence, the null hypothesis would be rejected at an $\alpha = 0.10$ significance level. Another way of saying this is that

$$0.05 < p\text{-value} < 0.10.$$

It would be instructive for you to draw a figure illustrating this double inequality. ∎

There are two ways of viewing a statistical test. One of these is through the p-value of the test; this approach is becoming more popular and is included in most computer printouts, so we mention it again. After observing the test statistic, we can say that the p-value is the probability, under the hypothesis H_0, of the test statistic being as extreme (in the direction of rejection of H_0) as the observed one. That is, the p-value is the tail-end probability. As an illustration, say a golfer averages about 90, with a standard deviation of 3, and she takes some lessons to improve. To test her possible improvement, namely, H_0: $\mu = 90$, against H_1: $\mu < 90$, she plays $n = 16$ rounds of golf. Assume a normal distribution with $\sigma = 3$. If the golfer averaged $\overline{x} = 87.94$, then

$$p\text{-value} = P(\overline{X} \le 87.94) = P\left(\frac{\overline{X} - 90}{3/4} \le \frac{87.94 - 90}{3/4}\right) = 0.0027.$$

The fact that the p-value is less than 0.05 is equivalent to the fact that $\overline{x} < 88.77$, because $P(\overline{X} \le 88.77; \mu = 90) = 0.05$. Since $\overline{x} = 87.94$ is an observed value of a random variable, namely, X, it follows that the p-value, a function of x, is also an observed value of a random variable. That is, before the random experiment is performed, the probability that the p-value is less than α is equal to α when the null hypothesis is true. Many statisticians believe that the observed p-value provides an understandable measure of the truth of H_0: The smaller the p-value, the less they believe in H_0.

Three additional examples of the p-value are given in Examples 7.2-3, 7.2-4, and 7.2-5. In two-sided tests for means and proportions, the p-value is the probability of the extreme values in both directions. With the mouse data (Example 7.2-3), the p-value is

$$p\text{-value} = P(|T| \ge 0.75).$$

In Table VI in the appendix, we see that if T has a t distribution with eight degrees of freedom, then $P(T \ge 0.706) = 0.25$. Thus, $P(|T| \ge 0.706) = 0.50$ and the p-value will be a little smaller than 0.50. In fact, $P(|T| \ge 0.75) = 0.4747$ (a probability that was found with Minitab), which is not less than $\alpha = 0.10$; hence, we do not reject H_0 at that significance level. In the example concerned with waste (Example 7.2-4), the p-value is essentially zero, since $P(Z \le -8.30) \approx 0$, where Z is a standardized normal variable. Consequently, we reject H_0.

The other way of looking at tests of hypotheses is through the consideration of confidence intervals, particularly for two-sided alternatives and the corresponding tests. For example, with the mouse data (Example 7.2-3), a 90% confidence interval for the unknown mean is

$$4.3 \pm (1.86)(1.2)/\sqrt{9}, \qquad \text{or} \qquad [3.56, 5.04],$$

since $t_{0.05}(8) = 1.86$. Note that this confidence interval covers the hypothesized value $\mu = 4.0$ and we do not reject H_0: $\mu = 4.0$. If the confidence interval did not cover $\mu = 4.0$, then we would have rejected H_0: $\mu = 4.0$. Many statisticians believe that estimation is much more important than tests of hypotheses and accordingly approach statistical tests through confidence intervals. For one-sided tests, we use one-sided confidence intervals.

EXERCISES

7.2-1. Assume that IQ scores for a certain population are approximately $N(\mu, 100)$. To test H_0: $\mu = 110$ against the one-sided alternative hypothesis H_1: $\mu > 110$, we take a random sample of size $n = 16$ from this population and observe $\bar{x} = 113.5$.

(a) Do we accept or reject H_0 at the 5% significance level?

(b) Do we accept or reject H_0 at the 10% significance level?

(c) What is the p-value of this test?

7.2-2. Assume that the weight of cereal in a "12.6-ounce box" is $N(\mu, 0.2^2)$. The Food and Drug Association (FDA) allows only a small percentage of boxes to contain less than 12.6 ounces. We shall test the null hypothesis H_0: $\mu = 13$ against the alternative hypothesis H_1: $\mu < 13$.

(a) Use a random sample of $n = 25$ to define the test statistic and the critical region that has a significance level of $\alpha = 0.025$.

(b) If $\bar{x} = 12.9$, what is your conclusion?

(c) What is the p-value of this test?

7.2-3. Let X equal the Brinell hardness measurement of ductile iron subcritically annealed. Assume that the distribution of X is $N(\mu, 100)$. We shall test the null hypothesis H_0: $\mu = 170$ against the alternative hypothesis H_1: $\mu > 170$, using $n = 25$ observations of X.

(a) Define the test statistic and a critical region that has a significance level of $\alpha = 0.05$. Sketch a figure showing this critical region.

(b) A random sample of $n = 25$ observations of X yielded the following measurements:

170 167 174 179 179 156 163 156 187
156 183 179 174 179 170 156 187
179 183 174 187 167 159 170 179

Calculate the value of the test statistic and state your conclusion clearly.

(c) Give the approximate p-value of this test.

7.2-4. Let X equal the thickness of spearmint gum manufactured for vending machines. Assume that the distribution of X is $N(\mu, \sigma^2)$. The target thickness is 7.5 hundredths of an inch. We shall test the null hypothesis H_0: $\mu = 7.5$ against a two-sided alternative hypothesis, using 10 observations.

(a) Define the test statistic and critical region for an $\alpha = 0.05$ significance level. Sketch a figure illustrating this critical region.

(b) Calculate the value of the test statistic and state your decision clearly, using the following $n = 10$ thicknesses in hundredths of an inch for pieces of gum that were selected randomly from the production line:

7.65   7.60   7.65   7.70   7.55
7.55   7.40   7.40   7.50   7.50

(c) Is $\mu = 7.50$ contained in a 95% confidence interval for μ?

7.2-5. The mean birth weight in the United States is $\mu = 3315$ grams, with a standard deviation of $\sigma = 575$. Let X equal the birth weight in grams in Jerusalem. Assume that the distribution of X is $N(\mu, \sigma^2)$. We shall test the null hypothesis H_0: $\mu = 3315$ against the alternative hypothesis H_1: $\mu < 3315$, using a random sample of size $n = 30$.

(a) Define a critical region that has a significance level of $\alpha = 0.05$.

(b) If the random sample of $n = 30$ yielded $\bar{x} = 3189$ and $s = 488$, what would be your conclusion?

(c) What is the approximate p-value of your test?

7.2-6. Let X equal the forced vital capacity (FVC) in liters for a female college student. (The FVC is the amount of air that a student can force out of her lungs.) Assume that the distribution of X is approximately $N(\mu, \sigma^2)$. Suppose it is known that $\mu = 3.4$ liters. A volleyball coach claims that the FVC of volleyball players is greater than 3.4. She plans to test her claim with a random sample of size $n = 9$.

(a) Define the null hypothesis.

(b) Define the alternative (coach's) hypothesis.

(c) Define the test statistic.

(d) Define a critical region for which $\alpha = 0.05$. Draw a figure illustrating your critical region.

(e) Calculate the value of the test statistic given that the random sample yielded the following FVC's:

3.4 3.6 3.8 3.3 3.4 3.5 3.7 3.6 3.7

(f) What is your conclusion?

(g) What is the approximate p-value of this test?

7.2-7. Vitamin B_6 is one of the vitamins in a multiple vitamin pill manufactured by a pharmaceutical company. The pills are produced with a mean of 50 milligrams of vitamin B_6 per pill. The company believes that there is a deterioration of 1 milligram per month, so that after 3 months it expects that $\mu = 47$. A consumer group suspects that $\mu < 47$ after 3 months.

(a) Define a critical region to test H_0: $\mu = 47$ against H_1: $\mu < 47$ at an $\alpha = 0.05$ significance level based on a random sample of size $n = 20$.

(b) If the 20 pills yielded a mean of $\bar{x} = 46.94$ with a standard deviation of $s = 0.15$, what is your conclusion?

(c) What is the approximate p-value of this test?

7.2-8. Assume that the birth weight in grams of a baby born in the United States is $N(3315, 525^2)$ for boys and girls combined. Let X equal the weight of a baby girl who is born at home in Ottawa County, and assume that the distribution of X is $N(\mu_X, \sigma_X^2)$.

(a) Using 11 observations of X, give the test statistic and the critical region for testing

H_0: $\mu_X = 3315$ against the alternative hypothesis H_1: $\mu_X > 3315$ (home-born babies are heavier) if $\alpha = 0.01$.

(b) Calculate the value of the test statistic and state your conclusion, using the following weights:

3119   2657   3459   3629   3345   3629
3515   3856   3629   3345   3062

(c) What is the approximate p-value of the test?

7.2-9. Let Y equal the weight in grams of a baby boy who is born at home in Ottawa County, and assume that the distribution of Y is $N(\mu_Y, \sigma_Y^2)$. Use the weights:

4082   3686   4111   3686   3175   4139
3686   3430   3289   3657   4082

to answer the questions in Exercise 7.2-8.

7.2-10. A company that manufactures brackets for an automaker regularly selects brackets from the production line and performs a torque test. The goal is for mean torque to equal 125. Let X equal the torque and assume that X is $N(\mu, \sigma^2)$. We shall use a sample of size $n = 15$ to test H_0: $\mu = 125$ against a two-sided alternative hypothesis.

(a) Give the test statistic and a critical region with significance level $\alpha = 0.05$. Sketch a figure illustrating the critical region.

(b) Use the following observations to calculate the value of the test statistic and state your conclusion:

128 149 136 114 126 142 124 136
122 118 122 129 118 122 129

7.2-11. The ornamental ground cover *Vinca minor* is spreading rapidly through the Hope College Biology Field Station because it can outcompete the small, native woody vegetation. In an attempt to discover whether *Vinca minor* utilized natural chemical weapons to inhibit the growth of the native vegetation, Hope biology students conducted an experiment in which they treated 33 sunflower seedlings with extracts taken from *Vinca minor* roots for several weeks and then measured the heights of the seedlings. Let X equal the height of one of these seedlings and assume that the distribution of X is $N(\mu, \sigma^2)$. The observed growths (in cm) were

11.5	11.8	15.7	16.1	14.1	10.5	15.2	19.0	12.8	12.4	19.2
13.5	16.5	13.5	14.4	16.7	10.9	13.0	15.1	17.1	13.3	12.4
8.5	14.3	12.9	11.1	15.0	13.3	15.8	13.5	9.3	12.2	10.3

The students also planted some control sunflower seedlings that had a mean height of 15.7 cm. We shall test the null hypothesis H_0: $\mu = 15.7$ against the alternative hypothesis H_1: $\mu < 15.7$.

(a) Calculate the value of the test statistic and give limits for the p-value of this test.

(b) What is your conclusion?

(c) Find an approximate 98% one-sided confidence interval that gives an upper bound for μ.

7.2-12. In a mechanical testing lab, Plexiglass® strips are stretched to failure. Let X equal the change in

length in mm before breaking. Assume that the distribution of X is $N(\mu, \sigma^2)$. We shall test the null hypothesis H_0: $\mu = 5.70$ against the alternative hypothesis H_1: $\mu > 5.70$, using $n = 8$ observations of X.

(a) Define the test statistic and a critical region that has a significance level of $\alpha = 0.05$. Sketch a figure showing this critical region.

(b) A random sample of eight observations of X yielded the following data:

5.71 5.80 6.03 5.87 6.22 5.92 5.57 5.83

Calculate the value of the test statistic and state your conclusion clearly.

(c) Give the approximate value of or bounds for the p-value of this test.

7.2-13. A vendor of milk products produces and sells low-fat dry milk to a company that uses it to produce baby formula. In order to determine the fat content of the milk, both the company and the vendor take an observation from each lot and test

it for fat content in percent. Ten sets of paired test results are as follows:

Lot Number	Company Test Results (X)	Vendor Test Results (Y)
1	0.50	0.79
2	0.58	0.71
3	0.90	0.82
4	1.17	0.82
5	1.14	0.73
6	1.25	0.77
7	0.75	0.72
8	1.22	0.79
9	0.74	0.72
10	0.80	0.91

Let μ_D denote the mean of the difference $X - Y$. Test H_0: $\mu_D = 0$ against H_1: $\mu_D > 0$, using a paired t test with the differences. Let $\alpha = 0.05$.

7.2-14. To test whether a golf ball of brand A can be hit a greater distance off the tee than a golf ball of brand B, each of 17 golfers hit a ball of each brand, 8 hitting ball A before ball B and 9 hitting ball B before ball A. The results are as follows:

Golfer	Distance for Ball A	Distance for Ball B	Golfer	Distance for Ball A	Distance for Ball B
1	265	252	10	274	260
2	272	276	11	274	267
3	246	243	12	269	267
4	260	246	13	244	251
5	274	275	14	212	222
6	263	246	15	235	235
7	255	244	16	254	255
8	258	245	17	224	231
9	276	259			

Assume that the differences of the paired A distance and B distance are approximately normally distributed, and test the null hypothesis H_0: $\mu_D = 0$ against the alternative hypothesis H_1: $\mu_D > 0$, using a paired t test with the 17 differences. Let $\alpha = 0.05$.

7.2-15. The following data give the times of 22 male swimmers who entered the 50 yard freestyle race in the Conference Championship Meet (for these data, x is the swimmer's best time for the season and y is the swimmer's time in the meet).

(24.97, 23.98) (24.76, 23.63) (23.80, 23.61) (22.84, 23.58) (24.07, 23.57)
(22.93, 22.83) (23.41, 22.75) (22.10, 22.74) (23.08, 22.70) (23.59, 22.62)
(22.38, 22.39) (22.91, 22.34) (22.08, 21.95) (22.46, 21.73) (21.34, 21.68)
(22.76, 21.66) (21.82, 21.68) (21.80, 21.58) (22.26, 21.57) (21.36, 21.35)
(22.98, 23.17) (23.80, 22.53)

Let $d = x - y$, a swimmer's best time minus the swimmer's meet time. Assume that the distribution of D is $N(\mu_D, \sigma_D^2)$. Test the null hypothesis H_0: $\mu_D = 0$ against the alternative hypothesis H_1: $\mu_D > 0$. You may select the significance level. Interpret your conclusion.

7.2-16. A company that manufactures motors receives reels of 10,000 terminals per reel. Before using a reel of terminals, 20 terminals are randomly selected to be tested. The test is the amount of pressure needed to pull the terminal apart from its mate. This amount of pressure should continue to increase from test to test as the terminal is "roughed up." (Since this kind of testing is destructive testing, a terminal that is tested cannot be used in a motor.) Let W equal the difference of the pressures: "test No. 1 pressure" minus "test No. 2 pressure." Assume that the distribution of W is $N(\mu_W, \sigma_W^2)$. We shall test the null hypothesis H_0: $\mu_W = 0$ against the alternative hypothesis H_1: $\mu_W < 0$, using 20 pairs of observations.

(a) Give the test statistic and a critical region that has a significance level of $\alpha = 0.05$. Sketch a figure illustrating this critical region.

(b) Use the following data to calculate the value of the test statistic, and state your conclusion clearly:

Terminal	Test 1	Test 2	Terminal	Test 1	Test 2
1	2.5	3.8	11	7.3	8.2
2	4.0	3.9	12	7.2	6.6
3	5.2	4.7	13	5.9	6.8
4	4.9	6.0	14	7.5	6.6
5	5.2	5.7	15	7.1	7.5
6	6.0	5.7	16	7.2	7.5
7	5.2	5.0	17	6.1	7.3
8	6.6	6.2	18	6.3	7.1
9	6.7	7.3	19	6.5	7.2
10	6.6	6.5	20	6.5	6.7

(c) What would the conclusion be if $\alpha = 0.01$?

(d) What is the approximate p-value of this test?

7.3 TESTS OF THE EQUALITY OF TWO MEANS

Let independent random variables X and Y have normal distributions $N(\mu_X, \sigma_X^2)$ and $N(\mu_Y, \sigma_Y^2)$, respectively. There are times when we are interested in testing whether the distributions of X and Y are the same. So if the assumption of normality is valid, we would be interested in testing whether the two means are equal. (A test for the equality of the two variances is given in the next section.)

When X and Y are independent and normally distributed, we can test hypotheses about their means with the same t statistic that we used to construct a confidence interval for $\mu_X - \mu_Y$ in Section 6.3. Recall that the t statistic used to construct the confidence interval assumed that the variances of X and Y were equal. (That is why we shall consider a test for the equality of two variances in the next section.)

We begin with an example and then give a table that lists some hypotheses and critical regions. A botanist is interested in comparing the growth response of dwarf pea stems against two different levels of the hormone indoleacetic acid (IAA). Using 16-day-old pea plants, the botanist obtains 5-millimeter sections and floats these sections on solutions with different hormone concentrations to observe the effect of the hormone on the growth of the pea stem. Let X and Y denote, respectively, the independent growths that can be attributed to the hormone during the first 26 hours after sectioning for $(0.5)(10)^{-4}$ and 10^{-4} levels of concentration of IAA. The botanist would like to test the null hypothesis H_0: $\mu_X - \mu_Y = 0$ against the alternative hypothesis H_1: $\mu_X - \mu_Y < 0$. If we can assume that X and Y are independent and normally distributed with a common variance, and if we assume respective random samples of sizes n and m, then we can find a test based on the statistic

$$T = \frac{\overline{X} - \overline{Y}}{\sqrt{\{[(n - 1)S_X^2 + (m - 1)S_Y^2]/(n + m - 2)\}(1/n + 1/m)}} \qquad (7.3\text{-}1)$$

$$= \frac{\overline{X} - \overline{Y}}{S_P\sqrt{1/n + 1/m}},$$

where

$$S_P = \sqrt{\frac{(n - 1)S_X^2 + (m - 1)S_Y^2}{n + m - 2}}. \qquad (7.3\text{-}2)$$

Now, T has a t distribution with $r = n + m - 2$ degrees of freedom when H_0 is true and the variances are (approximately) equal. Thus, the hypothesis H_0 will be rejected in favor of H_1 if the observed value of T is less than $-t_\alpha(n + m - 2)$.

EXAMPLE 7.3-1 In the preceding discussion, the botanist measured the growths of pea stem segments, in millimeters, for $n = 11$ observations of X:

$$0.8 \quad 1.8 \quad 1.0 \quad 0.1 \quad 0.9 \quad 1.7 \quad 1.0 \quad 1.4 \quad 0.9 \quad 1.2 \quad 0.5$$

She did the same with $m = 13$ observations of Y:

$$1.0 \quad 0.8 \quad 1.6 \quad 2.6 \quad 1.3 \quad 1.1 \quad 2.4$$
$$1.8 \quad 2.5 \quad 1.4 \quad 1.9 \quad 2.0 \quad 1.2$$

For these data, $\overline{x} = 1.03$, $s_x^2 = 0.24$, $\overline{y} = 1.66$, and $s_y^2 = 0.35$. The critical region for testing $H_0: \mu_X - \mu_Y = 0$ against $H_1: \mu_X - \mu_Y < 0$ is $t \leq -t_{0.05}(22) = -1.717$, where t is the two-sample t found in Equation 7.3-1. Since

$$t = \frac{1.03 - 1.66}{\sqrt{\{[10(0.24) + 12(0.35)]/(11 + 13 - 2)\}(1/11 + 1/13)}}$$

$$= -2.81 < -1.717,$$

H_0 is clearly rejected at an $\alpha = 0.05$ significance level. Notice that the approximate p-value of this test is 0.005, because $-t_{0.005}(22) = -2.819$. (See Figure 7.3-1.) Notice also that the sample variances do not differ too much; thus, most statisticians would use this two-sample t test.

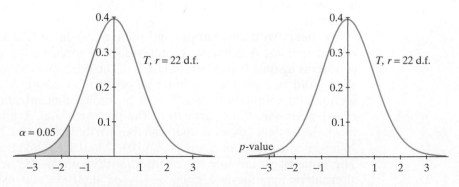

FIGURE 7.3-1: Critical region and p-value for pea stem growths

It is instructive to construct box-and-whisker diagrams to gain a visual comparison of the two samples. For these two sets of data, the five-number summaries (minimum, three quartiles, maximum) are

$$0.1 \quad 0.8 \quad 1.0 \quad 1.4 \quad 1.8$$

for the X sample and

$$0.8 \quad 1.15 \quad 1.6 \quad 2.2 \quad 2.6$$

for the Y sample. The two box plots are shown in Figure 7.3-2.   ■

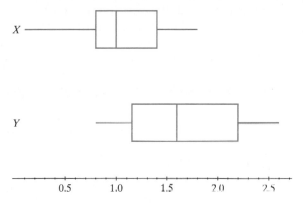

FIGURE 7.3-2: Box plots for pea stem growths

Assuming independent random samples of sizes n and m, let $\bar{x}$, $\bar{y}$, and s_p^2 represent the observed unbiased estimates of the respective parameters μ_X, μ_Y, and $\sigma_X^2 = \sigma_Y^2$ of two normal distributions with a common variance. Then α-level tests of certain hypotheses are given in Table 7.3-1 when $\sigma_X^2 = \sigma_Y^2$. If the common-variance assumption is violated, but not too badly, the test is satisfactory, but the significance levels are only approximate. The t statistic and s_p are given in Equations 7.3-1 and 7.3-2, respectively.

TABLE 7.3-1: Tests of Hypotheses for Equality of Two Means		
H_0	H_1	Critical Region
$\mu_X = \mu_Y$	$\mu_X > \mu_Y$	$t \geq t_\alpha(n + m - 2)$ or
		$\bar{x} - \bar{y} \geq t_\alpha(n + m - 2)s_p\sqrt{1/n + 1/m}$
$\mu_X = \mu_Y$	$\mu_X < \mu_Y$	$t \leq -t_\alpha(n + m - 2)$ or
		$\bar{x} - \bar{y} \leq -t_\alpha(n + m - 2)s_p\sqrt{1/n + 1/m}$
$\mu_X = \mu_Y$	$\mu_X \neq \mu_Y$	$|t| \geq t_{\alpha/2}(n + m - 2)$ or
		$|\bar{x} - \bar{y}| \geq t_{\alpha/2}(n + m - 2)s_p\sqrt{1/n + 1/m}$

REMARK   Again to emphasize the relationship between confidence intervals and tests of hypotheses, we note that each of the tests in Table 7.3-1 has a corresponding confidence interval. For example, the first one-sided test is equivalent to saying that we reject H_0: $\mu_X - \mu_Y = 0$ if zero is not in the one-sided confidence interval with lower bound

$$\bar{x} - \bar{y} - t_\alpha(n + m - 2)s_p\sqrt{1/n + 1/m}.$$ ■

EXAMPLE 7.3-2 A product is packaged by a machine with 24 filler heads numbered 1 to 24, with the odd numbered heads on one side of the machine and the even on the other side. Let X and Y equal the fill weights in grams when a package is filled by an odd-numbered head and an even-numbered head, respectively. Assume that the distributions of X and Y are $N(\mu_X, \sigma^2)$ and $N(\mu_Y, \sigma^2)$, respectively, and that X and Y are independent. We would like to test the null hypothesis $H_0: \mu_X - \mu_Y = 0$ against the alternative hypothesis $H_1: \mu_X - \mu_Y \neq 0$. To perform the test, after the machine has been set up and is running, we shall select one package at random from each filler head and weigh it. The test statistic is that given by Equation 7.3-1 with $n = m = 12$. At an $\alpha = 0.10$ significance level, the critical region is $|t| \geq t_{0.05}(22) = 1.717$.

For the $n = 12$ observations of X, namely,

$$
\begin{array}{cccccc}
1071 & 1076 & 1070 & 1083 & 1082 & 1067 \\
1078 & 1080 & 1075 & 1084 & 1075 & 1080
\end{array}
$$

$\bar{x} = 1076.75$ and $s_x^2 = 29.30$. For the $m = 12$ observations of Y, namely,

$$
\begin{array}{cccccc}
1074 & 1069 & 1075 & 1067 & 1068 & 1079 \\
1082 & 1064 & 1070 & 1073 & 1072 & 1075
\end{array}
$$

$\bar{y} = 1072.33$ and $s_y^2 = 26.24$. The calculated value of the test statistic is

$$
t = \frac{1076.75 - 1072.33}{\sqrt{\dfrac{11(29.30) + 11(26.24)}{22}\left(\dfrac{1}{12} + \dfrac{1}{12}\right)}} = 2.05.
$$

Since

$$
|t| = |2.05| = 2.05 > 1.717,
$$

the null hypothesis is rejected at an $\alpha = 0.10$ significance level. Note, however, that

$$
|t| = 2.05 < 2.074 = t_{0.025}(22),
$$

so that the null hypothesis would not be rejected at an $\alpha = 0.05$ significance level. That is, the p-value is between 0.05 and 0.10.

Again, it is instructive to construct box plots on the same graph for these two sets of data. The box plots in Figure 7.3-3 were constructed with the use of the five-number summary for the observations of X (1067, 1072, 1077, 1081.5, and 1084) and the five-number summary for the observations of Y (1064, 1068.25, 1072.5, 1075, and 1082). It looks like additional sampling would be advisable to test that the filler heads on the two sides of the machine are filling in a similar manner. If not, some corrective action needs to be taken. ∎

We would like to give two modifications of tests about two means. If we are able to assume that we know the variances of X and Y, then the appropriate test statistic to use for testing $H_0: \mu_X = \mu_Y$ is

$$
Z = \frac{\bar{X} - \bar{Y}}{\sqrt{\dfrac{\sigma_X^2}{n} + \dfrac{\sigma_Y^2}{m}}}, \tag{7.3-3}
$$

which has a standard normal distribution when the null hypothesis is true and, of course, when the populations are normally distributed. If the variances are unknown

FIGURE 7.3-3: Box plots for fill weights

and the sample sizes are large, replace σ_X^2 with S_X^2 and σ_Y^2 with S_Y^2 in Equation 7.3-3. The resulting statistic will have an approximate $N(0, 1)$ distribution.

EXAMPLE 7.3-3 The target thickness for Fruit Flavored Gum and for Fruit Flavored Bubble Gum is 6.7 hundredths of an inch. Let the independent random variables X and Y equal the respective thicknesses of these gums in hundredths of an inch, and assume that their distributions are $N(\mu_X, \sigma_X^2)$ and $N(\mu_Y, \sigma_Y^2)$, respectively. Because bubble gum has more elasticity than regular gum, it seems as if it would be harder to roll it out to the correct thickness. Thus, we shall test the null hypothesis H_0: $\mu_X = \mu_Y$ against the alternative hypothesis H_1: $\mu_X < \mu_Y$, using samples of sizes $n = 50$ and $m = 40$.

Because the variances are unknown and the sample sizes are large, the test statistic that is used is

$$Z = \frac{\overline{X} - \overline{Y}}{\sqrt{\dfrac{S_X^2}{50} + \dfrac{S_Y^2}{40}}}.$$

At an approximate significance level of $\alpha = 0.01$, the critical region is

$$z \le -z_{0.01} = -2.326.$$

The observed values of X were

6.85	6.60	6.70	6.75	6.75	6.90	6.85	6.90	6.70	6.85
6.60	6.70	6.75	6.70	6.70	6.70	6.55	6.60	6.95	6.95
6.80	6.80	6.70	6.75	6.60	6.70	6.65	6.55	6.55	6.60
6.60	6.70	6.80	6.75	6.60	6.75	6.50	6.75	6.70	6.65
6.70	6.70	6.55	6.65	6.60	6.65	6.60	6.65	6.80	6.60

for which $\overline{x} = 6.701$ and $s_x = 0.108$. The observed values of Y were

7.10	7.05	6.70	6.75	6.90	6.90	6.65	6.60	6.55	6.55
6.85	6.90	6.60	6.85	6.95	7.10	6.95	6.90	7.15	7.05
6.70	6.90	6.85	6.95	7.05	6.75	6.90	6.80	6.70	6.75
6.90	6.90	6.70	6.70	6.90	6.90	6.70	6.70	6.90	6.95

for which $\bar{y} = 6.841$ and $s_y = 0.155$. Since the calculated value of the test statistic is

$$z = \frac{6.701 - 6.841}{\sqrt{0.108^2/50 + 0.155^2/40}} = -4.848 < -2.326,$$

the null hypothesis is clearly rejected.

The box-and-whisker diagrams in Figure 7.3-4 were constructed with the use of the five-number summary of the observations of X (6.50, 6.60, 6.70, 6.75, and 6.95) and the five-number summary of the observations of Y (6.55, 6.70, 6.90, 6.94, and 7.15). This graphical display also confirms our conclusion. ∎

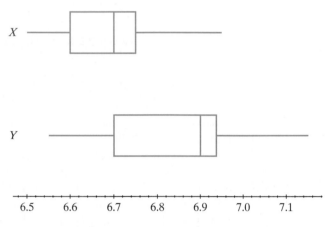

FIGURE 7.3-4: Box plots for gum thicknesses

REMARKS To have satisfactory tests, our assumptions must be satisfied reasonably well. As long as the underlying distributions are not highly skewed, the normal assumptions are not too critical, as $\overline{X}$ and $\overline{Y}$ have approximate normal distributions by the central limit theorem. As distributions become nonnormal and highly skewed, the sample mean and sample variance become more dependent, and that causes problems in using the Student's t as an approximating distribution for T. In these cases, some of the nonparametric methods described later could be used.

When the distributions are close to normal, but the variances seem to differ by a great deal, the t statistic should again be avoided, particularly if the sample sizes are also different. In that case, use Z or the modification produced by substituting the sample variances for the distribution variances. In the latter situation, if n and m are large enough, there is no problem. With small n and m, most statisticians would use Welch's suggestion (or other modifications of it); that is, they would use an approximating Student's t distribution with r degrees of freedom, where r is given by Equation 6.3-1. We actually give a test for the equality of variances that could be employed to decide whether to use T or a modification of Z. However, most statisticians do not place much confidence in this test of $\sigma_X^2 = \sigma_Y^2$ and would use a modification of Z (possibly Welch's) if they suspected that the variances differed greatly. ∎

EXERCISES

(In some of the exercises that follow, we must make assumptions such as the existence of normal distributions with equal variances.)

7.3-1. The botanist in Example 7.3-1 is really interested in testing for synergistic interaction. That is, given the two hormones gibberellin (GA_3) and indoleacetic acid (IAA), let X_1 and X_2 equal the growth responses (in millimeters) of dwarf pea stem segments to GA_3 and IAA, respectively and separately. Also, let $X = X_1 + X_2$ and let Y equal the growth response when both hormones are present. Assuming that X is $N(\mu_X, \sigma^2)$ and Y is $N(\mu_Y, \sigma^2)$, the botanist is interested in testing the hypothesis H_0: $\mu_X - \mu_Y$ against the alternative hypothesis of synergistic interaction H_1: $\mu_X < \mu_Y$.

 (a) Using $n = m = 10$ observations of X and Y, define the test statistic and the critical region. Sketch a figure of the t p.d.f. and show the critical region on your figure. Let $\alpha - 0.05$.

 (b) Given $n = 10$ observations of X, namely,

 2.1 2.6 2.6 3.4 2.1 1.7 2.6 2.6 2.2 1.2

and $m = 10$ observations of Y, namely,

 3.5 3.9 3.0 2.3 2.1 3.1 3.6 1.8 2.9 3.3

calculate the value of the test statistic and state your conclusion. Locate the test statistic on your figure.

 (c) Construct two box plots on the same figure. Does this confirm your conclusion?

7.3-2. Let X and Y denote the weights in grams of male and female gallinules, respectively. Assume that X is $N(\mu_X, \sigma_X^2)$ and Y is $N(\mu_Y, \sigma_Y^2)$.

 (a) Given $n = 16$ observations of X and $m = 13$ observations of Y, define a test statistic and a critical region for testing the null hypothesis H_0: $\mu_X = \mu_Y$ against the one-sided alternative hypothesis H_1: $\mu_X > \mu_Y$. Let $\alpha = 0.01$. (Assume that the variances are equal.)

 (b) Given that $\bar{x} = 415.16$, $s_x^2 = 1356.75$, $\bar{y} = 347.40$, and $s_y^2 = 692.21$, calculate the value of the test statistic and state your conclusion.

 (c) Although we assumed that $\sigma_X^2 = \sigma_Y^2$, let us say we suspect that that equality is not valid. Thus, use the test proposed by Welch.

7.3-3. Let X equal the weight in grams of a Low Fat Strawberry Kudo and Y the weight of a Low Fat Blueberry Kudo. Assume that the distributions of X and Y are $N(\mu_X, \sigma_X^2)$ and $N(\mu_Y, \sigma_Y^2)$, respectively. Let

$$21.7 \quad 21.0 \quad 21.2 \quad 20.7 \quad 20.4 \quad 21.9 \quad 20.2 \quad 21.6 \quad 20.6$$

be $n = 9$ observations of X, and let

$$21.5 \quad 20.5 \quad 20.3 \quad 21.6 \quad 21.7 \quad 21.3 \quad 23.0 \quad 21.3 \quad 18.9 \quad 20.0 \quad 20.4 \quad 20.8 \quad 20.3$$

be $m = 13$ observations of Y. Use these observations to answer the following questions:

 (a) Test the null hypothesis H_0: $\mu_X = \mu_Y$ against a two-sided alternative hypothesis. You may select the significance level. Assume that the variances are equal.

 (b) Construct and interpret box-and-whisker diagrams to support your conclusions.

7.3-4. Among the data collected for the World Health Organization air quality monitoring project is a measure of suspended particles, in $\mu g/m^3$. Let X and Y equal the concentration of suspended particles in $\mu g/m^3$ in the city centers (commercial districts) of Melbourne and Houston, respectively. Using $n = 13$ observations of X and $m = 16$ observations of Y, we shall test H_0: $\mu_X = \mu_Y$ against H_1: $\mu_X < \mu_Y$.

 (a) Define the test statistic and critical region, assuming that the variances are equal. Let $\alpha = 0.05$.

 (b) If $\bar{x} = 72.9$, $s_x = 25.6$, $\bar{y} = 81.7$, and $s_y = 28.3$, calculate the value of the test statistic and state your conclusion.

 (c) Give limits for the p-value of this test.

7.3-5. Some nurses in county public health conducted a survey of women who had received inadequate prenatal care. They used information from birth certificates to select mothers for the survey. The mothers selected were divided into two groups: 14 mothers who said they had five or fewer prenatal visits and 14 mothers who said they had six or more prenatal visits. Let X and Y equal the respective birth weights of the babies from these two sets of mothers, and assume that the distribution of X is $N(\mu_X, \sigma^2)$ and the distribution of Y is $N(\mu_Y, \sigma^2)$,

(a) Define the test statistic and critical region for testing $H_0: \mu_X - \mu_Y = 0$ against $H_1: \mu_X - \mu_Y < 0$. Let $\alpha = 0.05$.

(b) Given that the observations of X were

$$\begin{array}{ccccccc} 49 & 108 & 110 & 82 & 93 & 114 & 134 \\ 114 & 96 & 52 & 101 & 114 & 120 & 116 \end{array}$$

and the observations of Y were

$$\begin{array}{ccccccc} 133 & 108 & 93 & 119 & 119 & 98 & 106 \\ 131 & 87 & 153 & 116 & 129 & 97 & 110 \end{array}$$

calculate the value of the test statistic and state your conclusion.

(c) Approximate the p-value.

(d) Construct box plots on the same figure for these two sets of data. Do the box plots support your conclusion?

7.3-6. Let X and Y equal the forces required to pull stud No. 3 and stud No. 4 out of a window that has been manufactured for an automobile. Assume that the distributions of X and Y are $N(\mu_X, \sigma_X^2)$ and $N(\mu_Y, \sigma_Y^2)$, respectively.

(a) If $m = n = 10$ observations are selected randomly, define a test statistic and a critical region for testing $H_0: \mu_X - \mu_Y = 0$ against a two-sided alternative hypothesis. Let $\alpha = 0.05$. Assume that the variances are equal.

(b) Given $n = 10$ observations of X, namely,

111 120 139 136 138 149 143 145 111 123

and $m = 10$ observations of Y, namely,

152 155 133 134 119 155 142 146 157 149

calculate the value of the test statistic and state your conclusion clearly.

(c) What is the approximate p-value of this test?

(d) Construct box plots on the same figure for these two sets of data. Do the box plots confirm your decision in part (b)?

7.3-7. Let X and Y equal the number of milligrams of tar in filtered and nonfiltered cigarettes, respectively. Assume that the distributions of X and Y are $N(\mu_X, \sigma_X^2)$ and $N(\mu_Y, \sigma_Y^2)$, respectively. We shall test the null hypothesis $H_0: \mu_X - \mu_Y = 0$ against the alternative hypothesis $H_1: \mu_X - \mu_Y < 0$, using random samples of sizes $n = 9$ and $m = 11$ observations of X and Y, respectively.

(a) Define the test statistic and a critical region that has an $\alpha = 0.01$ significance level. Sketch a figure illustrating this critical region.

(b) Given $n = 9$ observations of X, namely,

0.9 1.1 0.1 0.7 0.4 0.9 0.8 1.0 0.4

and $m = 11$ observations of Y, namely,

1.5 0.9 1.6 0.5 1.4 1.9 1.0 1.2 1.3 1.6 2.1

calculate the value of the test statistic and state your conclusion clearly. Locate the value of the test statistic on your figure.

7.3-8. In Exercise 6.3-4, we let X and Y equal the blood volumes in milliliters for males who are paraplegics participating in vigorous physical activities and males who are able bodied participating in everyday activities. We assumed that X was $N(\mu_X, \sigma_X^2)$ and Y was $N(\mu_Y, \sigma_Y^2)$. Using $n = 7$ observations of X, namely,

1612 1352 1456 1222 1560 1456 1924

and $m = 10$ observations of Y, namely,

$$\begin{array}{ccccc} 1082 & 1300 & 1092 & 1040 & 910 \\ 1248 & 1092 & 1040 & 1092 & 1288 \end{array}$$

you were asked to find a 95% confidence interval for $\mu_X - \mu_Y$.

(a) Test the hypothesis that the means are equal against a two-sided alternative hypothesis. Let $\alpha = 0.05$.

(b) Did 0 belong to the 95% confidence interval in Exercise 6.3-4? Did you reject the null hypothesis in the current exercise? Are your answers to these questions compatible?

7.3-9. An office furniture manufacturer installed a new adhesive application process. To compare the new process with the old process, random samples were selected from the two processes and "pull tests" were performed to determine the number of pounds of pressure that were required to pull apart the glued parts. (This kind of test is an example of destructive testing.) Let X and Y denote the pounds of pressure needed for the new and old processes, respectively.

(a) On the basis of $n = m = 24$ observations, define the test statistic and the critical region for testing $H_0: \mu_X - \mu_Y = 0$ against $H_1: \mu_X - \mu_Y > 0$. Let $\alpha = 0.05$. State your assumptions.

(b) Let the following be $n = 24$ observations of X:

$$\begin{array}{cccccccc} 1250 & 1210 & 990 & 1310 & 1320 & 1200 & 1290 & 1360 \\ 1120 & 1360 & 1310 & 1110 & 1320 & 980 & 950 & 1430 \\ 960 & 1050 & 1310 & 1240 & 1420 & 1170 & 1470 & 1060 \end{array}$$

Let the following be $m = 24$ observations of Y:

1180 1360 1310 1190 920 1060 1440 1010
1310 980 1310 1030 960 800 1280 1080
930 1050 1010 1310 940 860 1450 1070

Use these observations of X and Y to calculate the value of the test statistic and state your conclusion clearly.

(c) What is the approximate p-value of this test?

(d) Construct two box plots on the same graph. Do the box plots support your conclusion?

7.3-10. Let X and Y denote the tarsus lengths of male and female grackles, respectively. Assume that X is $N(\mu_X, \sigma_X^2)$ and Y is $N(\mu_Y, \sigma_Y^2)$. Given that $n = 25$, $\bar{x} - 33.80$, $s_x^2 = 4.88$, $m = 29$, $\bar{y} = 31.66$, and $s_y^2 = 5.81$, test the null hypothesis $H_0: \mu_X = \mu_Y$ against $H_1: \mu_X > \mu_Y$ with $\alpha = 0.01$.

7.3-11. Weight checks are done on the scales for an automatic bagger for water softener pellets. Each bagger has two scales (say, south and north) that operate alternately to fill 80-lb bags of pellets. Let X and Y equal the weights of the bags of pellets from these two scales. Assume that the distributions of X and Y are $N(\mu_X, \sigma^2)$ and $N(\mu_Y, \sigma^2)$, respectively. Ten bags were selected randomly from each scale and weighed by hand on a scale with a "high" degree of accuracy, yielding the following weights:

X: 80.51 80.46 80.75 80.50 80.36
80.32 80.36 80.78 80.26 80.34
Y: 80.51 80.28 80.40 80.35 80.38
80.28 80.27 80.16 80.59 80.56

(a) Test $H_0: \mu_X = \mu_Y$ against $H_1: \mu_X \neq \mu_Y$. Assume that $\sigma_X^2 = \sigma_Y^2$. Give limits for the p-value of the test and state your conclusion.

(b) Draw box-and-whisker diagrams on the same graph. Does this figure confirm your answer?

7.3-12. Two different saws are used to cut columns for a gazebo that has a diameter of 16 feet. Suspecting that one saw is cutting columns shorter than the other saw, the company is interested in comparing the two saws. Let the lengths of the columns cut by the two saws be observations of random variables X and Y that are $N(\mu_X, \sigma_X^2)$ and $N(\mu_Y, \sigma_Y^2)$, respectively. We shall use the following observations of X and Y:

X: 8.02 8.10 8.04 8.04 8.00 8.11 8.07 8.02 8.04
Y: 8.04 8.04 8.10 8.06 8.08 8.10 8.07 8.08 8.06

(a) Test $H_0: \mu_X = \mu_Y$ against $H_1: \mu_X < \mu_Y$. Assume that $\sigma_X^2 = \sigma_Y^2$. Give limits for the p-value of the test and state your conclusion.

(b) Draw box-and-whisker diagrams on the same graph. Does this figure confirm your answer?

7.3-13. When a stream is turbid, it is not completely clear due to suspended solids in the water. The higher the turbidity, the less clear is the water. A stream was studied on 26 days, half during dry weather (say, observations of X) and the other half immediately after a significant rainfall (say, observations of Y). Assume that the distributions of X and Y are $N(\mu_X, \sigma^2)$ and $N(\mu_Y, \sigma^2)$, respectively. The following turbidities were recorded in units of NTUs (nephelometric turbidity units):

X: 2.9 14.9 1.0 12.6 9.4 7.6 3.6
3.1 2.7 4.8 3.4 7.1 7.2
Y: 7.8 4.2 2.4 12.9 17.3 10.4 5.9
4.9 5.1 8.4 10.8 23.4 9.7

(a) Test the null hypothesis $H_0: \mu_X = \mu_Y$ against $H_1: \mu_X < \mu_Y$. Give limits for the p-value and state your conclusion.

(b) Draw box-and-whisker diagrams on the same graph. Does this figure confirm your answer?

7.3-14. Plants convert carbon dioxide (CO_2) in the atmosphere, along with water and energy from sunlight, into the energy they need for growth and reproduction. Experiments were performed under normal atmospheric air conditions and in air with enriched CO_2 concentrations to determine the effect on plant growth. The plants were given the same amount of water and light for a four-week period. The following table gives the plant growths in grams:

Normal Air	4.67	4.21	2.18	3.91	4.09	5.24	2.94	4.71
	4.04	5.79	3.80	4.38				
Enriched Air	5.04	4.52	6.18	7.01	4.36	1.81	6.22	5.70

On the basis of these data, determine whether CO_2-enriched atmosphere increases plant growth.

7.3-15. Let X equal the fill weight in April and Y the fill weight in June for an 8-pound box of bleach. We shall test the null hypothesis H_0:

$\mu_X - \mu_Y = 0$ against the alternative hypothesis $H_1: \mu_X - \mu_Y > 0$ given that $n = 90$ observations of X yielded $\bar{x} = 8.10$ and $s_x = 0.117$ and

$m = 110$ observations of Y yielded $\bar{x} = 8.07$ and $s_y = 0.054$.

(a) What is your conclusion if $\alpha = 0.05$?
    HINT: Do the variances seem to be equal?

(b) What is the approximate p-value of this test?

7.3-16. Let X and Y denote the respective lengths of male and female green lynx spiders. Assume that the distributions of X and Y are $N(\mu_X, \sigma_X^2)$ and $N(\mu_Y, \sigma_Y^2)$, respectively. In Example 7.4-2, we will show that $\sigma_Y^2 > \sigma_X^2$. Thus, use the modification of Z to test the hypothesis H_0:

$\mu_X - \mu_Y = 0$ against the alternative hypothesis $H_1: \mu_X - \mu_Y < 0$.

(a) Define the test statistic and a critical region that has a significance level of $\alpha = 0.025$.

(b) Using the data given in Exercise 6.3-5 (see also Example 7.4-2), calculate the value of the test statistic and state your conclusion.

(c) Draw two box-and-whisker diagrams on the same figure. Does your figure confirm both the conclusion of this exercise and that of Example 7.4-2?

7.3-17. Some measurements in mm were made on a species of spiders named *Sosippus floridanus* that are native to Florida. (See Exercise 6.7-16.) There are 10 female spiders and 10 male spiders. The total lengths and the lengths of their front and their back legs are repeated here:

Female body lengths	Female front legs	Female back legs	Male body lengths	Male front legs	Male back legs
11.06	15.03	19.29	12.26	20.22	25.54
13.87	17.96	22.74	11.66	18.62	23.94
12.93	17.56	21.28	12.53	18.62	23.94
15.08	21.22	25.54	13.00	19.95	25.80
17.82	22.61	28.86	11.79	19.15	25.40
14.14	20.08	25.14	12.46	19.02	25.27
12.26	16.49	20.22	10.65	17.29	22.21
17.82	18.75	24.61	10.39	17.02	21.81
20.17	23.01	28.46	12.26	18.49	23.41
16.88	22.48	28.59	14.07	22.61	28.86

In this exercise, we shall compare the sizes of the female and male spiders. For each of these questions, construct box plots on the same graph to confirm your answers. Do not assume that the variances are equal.

(a) Test the null hypothesis that the lengths of female and male spiders are equal against the alternative hypothesis that female spiders are longer.

(b) Test the null hypothesis that the lengths of the front legs of the female and male spiders are equal against a two-sided alternative.

(c) Test the null hypothesis that the lengths of the back legs of the female and male spiders are equal against a two-sided alternative.

7.3-18. An ecology laboratory studied tree dispersion patterns for the sugar maple, whose seeds are dispersed by the wind, and the American beech, whose seeds are dispersed by mammals. In a plot of area 50 m by 50 m, they measured distances between like trees, yielding the following distances in meters for 19 American beech trees and 19 sugar maple trees:

American beech:	5.00	5.00	6.50	4.25	4.25	8.80	6.50
	7.15	6.15	2.70	2.70	11.40	9.70	
	6.10	9.35	2.85	4.50	4.50	6.50	
sugar maple:	6.00	4.00	6.00	6.45	5.00	5.00	5.50
	2.35	2.35	3.90	3.90	5.35	3.15	
	2.10	4.80	3.10	5.15	3.10	6.25	

(a) Test the null hypothesis that the means are equal against the one-sided alternative that the mean for the distances between beech trees is greater than that between maple trees.

(b) Construct two box plots to confirm your answer.

7.3-19. Students looked at the effect of a certain fertilizer on plant growth. The students tested this fertilizer on one group of plants (Group A) and did not give fertilizer to a second group (Group B). The growths of the plants, in mm, over six weeks were as follows:

Group A:	55	61	33	57	17	46	50	42	71	51	63
Group B:	31	27	12	44	9	25	34	53	33	21	32

(a) Test the null hypothesis that the mean growths are equal against the alternative that the fertilizer enhanced growth. Assume that the variances are equal.

(b) Construct box plots of the two sets of growths on the same graph. Does this confirm your answer to part (a)?

7.3-20. Ledolter and Hogg (see references) report that a new emergency procedure was developed to reduce the time that is required to fix a certain manufacturing problem. Past data under the old system were available ($n = 25$). The staff was trained under the new procedure, and the response times for the next 15 occurrences of this manufacturing problem were recorded. The data are as follows:

Old Procedure:	4.3	6.5	4.6	4.3	6.4	4.8	5.1	6.8	4.9
	4.5	5.1	7.3	3.3	5.0	4.6	7.0	5.1	3.8
	5.2	4.1	5.7	4.6	5.9	3.1	6.2		
New Procedure:	6.2	4.0	3.3	4.5	2.3	3.0	3.2	6.0	3.7
	4.5	5.3	4.0	5.4	4.3	3.8			

(a) Test the null hypothesis that the mean response times are equal against a one-sided alternative that the mean for the new procedure is less than that for the old procedure. You may assume that the variances are equal.

(b) Construct box plots on the same graph.

(c) What is your conclusion?

7.4 TESTS FOR VARIANCES

Tests for a single variance and for equality of two variances are given in this section. An example follows a brief discussion, and then formal rules are given for tests for a single variance.

A psychology professor claims that the variance of IQ scores for college students is equal to $\sigma^2 = 100$. To test this claim, it is decided to test the hypothesis H_0: $\sigma^2 = 100$ against a two-sided alternative hypothesis H_1: $\sigma^2 \neq 100$. A random sample of n students will be selected, and the test will be based on the observed unbiased estimate s^2 of the variance σ^2 of their IQ scores. The hypothesis H_0 will be rejected if s^2 differs "too much" from $\sigma^2 = 100$. That is, H_0 will be rejected if $s^2 \leq c_1$ or $s^2 \geq c_2$ for some constants $c_1 < 100$ and $c_2 > 100$. Generally, c_1 and c_2 are selected so that

$$P(S^2 \leq c_1) = \alpha/2 \quad \text{and} \quad P(S^2 \geq c_2) = \alpha/2.$$

If we may assume that IQ scores are normally distributed, then the distribution of $(n - 1)S^2/100$ is $\chi^2(n - 1)$ when H_0 is true. Thus, the chi-square probability table can be used to give us

$$P\left[\frac{(n - 1)S^2}{100} \leq \frac{(n - 1)c_1}{100}\right] = \alpha/2 \quad \text{and} \quad P\left[\frac{(n - 1)S^2}{100} \geq \frac{(n - 1)c_2}{100}\right] = \alpha/2.$$

Accordingly,

$$\frac{(n - 1)c_1}{100} = \chi^2_{1-\alpha/2}(n - 1) \quad \text{and} \quad \frac{(n - 1)c_2}{100} = \chi^2_{\alpha/2}(n - 1).$$

Hence,

$$c_1 = [100/(n-1)]\chi^2_{1-\alpha/2}(n-1) \quad \text{and} \quad c_2 = [100/(n-1)]\chi^2_{\alpha/2}(n-1).$$

Actually, it is easier to base this test on the value of the chi-square statistic

$$\chi^2 = \frac{(n-1)S^2}{100}.$$

We would reject the null hypothesis if

$$\chi^2 \le \chi^2_{1-\alpha/2}(n-1) \quad \text{or} \quad \chi^2 \ge \chi^2_{\alpha/2}(n-1).$$

EXAMPLE 7.4-1 Suppose that, in the preceding discussion, $n = 23$ and $\alpha = 0.05$. To test $H_0: \sigma^2 = 100$ against $H_1: \sigma^2 \ne 100$, we let $\chi^2 = (n-1)S^2/\sigma_0^2 = 22S^2/100$. Because

$$\chi^2_{0.975}(22) = 10.98 \quad \text{and} \quad \chi^2_{0.025}(22) = 36.78,$$

H_0 will be rejected if

$$\chi^2 = \frac{22s^2}{100} \le 10.98 \quad \text{or} \quad \chi^2 = \frac{22s^2}{100} \ge 36.78$$

or, equivalently, if

$$s^2 \le c_1 = \frac{100(10.98)}{22} = 49.91 \quad \text{or} \quad s^2 \ge c_2 = \frac{100(36.78)}{22} = 167.18.$$

(Exercise 7.4-6 gives some insight into why these critical values are so far apart.) Given that the observed value of the sample variance was $s^2 = 147.82$, the hypothesis $H_0: \sigma^2 = 100$ was not rejected. Note that the 95% confidence interval for σ^2, namely,

$$\left[\frac{(22)(147.82)}{36.78}, \frac{(22)(147.82)}{10.98} \right] = [88.42, 296.18],$$

contains $\sigma^2 = 100$. Also, the observed value of the chi-square test statistic is

$$\chi^2 = \frac{22(147.82)}{100} = 32.52.$$

Because

$$10.98 < 32.52 < 36.78,$$

we would again accept H_0, as expected. (See Figure 7.4-1.)

If $H_1: \sigma^2 > 100$ had been the alternative hypothesis, $H_0: \sigma^2 = 100$ would have been rejected if

$$\chi^2 = \frac{22s^2}{100} \ge \chi^2_{0.05}(22) = 33.92$$

or, equivalently,

$$s^2 \ge \frac{100\chi^2_{0.05}(22)}{22} = \frac{(100)(33.92)}{22} = 154.18.$$

Because

$$\chi^2 = 32.52 < 33.92 \quad \text{and} \quad s^2 = 147.82 < 154.18,$$

FIGURE 7.4-1: Test about the variance of IQ scores

H_0 would not be rejected in favor of this one-sided alternative hypothesis for $\alpha = 0.05$, although we suspect a slightly larger data set might lead to rejection. ■

H_0	H_1	Critical Region
$\sigma^2 = \sigma_0^2$	$\sigma^2 > \sigma_0^2$	$s^2 \geq \dfrac{\sigma_0^2 \chi_\alpha^2(n-1)}{n-1}$ or $\chi^2 \geq \chi_\alpha^2(n-1)$
$\sigma^2 = \sigma_0^2$	$\sigma^2 < \sigma_0^2$	$s^2 \leq \dfrac{\sigma_0^2 \chi_{1-\alpha}^2(n-1)}{n-1}$ or $\chi^2 \leq \chi_{1-\alpha}^2(n-1)$
$\sigma^2 = \sigma_0^2$	$\sigma^2 \neq \sigma_0^2$	$s^2 \leq \dfrac{\sigma_0^2 \chi_{1-\alpha/2}^2(n-1)}{n-1}$ or $s^2 \geq \dfrac{\sigma_0^2 \chi_{\alpha/2}^2(n-1)}{n-1}$ or $\chi^2 \leq \chi_{1-\alpha/2}^2(n-1)$ or $\chi^2 \geq \chi_{\alpha/2}^2(n-1)$

TABLE 7.4-1: Tests About the Variance

With normal distributions, Table 7.4-1 summarizes tests of hypotheses for a single variance. The critical region is given in terms of the sample variance. The critical region is also given in terms of the chi-square test statistic

$$\chi^2 = \frac{(n-1)s^2}{\sigma_0^2}.$$

We discussed the p-value for tests about proportions and means. We can also find p-values for tests about a variance. In Example 7.4-1, with the one-sided alternative $H_1 \colon \sigma^2 > 100$, the p-value is

$$p\text{-value} = P(W \geq 32.52),$$

where $W = 22S^2/100$ has a chi-square distribution with 22 degrees of freedom when H_0 is true. From Table IV in the appendix, we see that

$$P(W \geq 30.81) = 0.10 \qquad \text{and} \qquad P(W \geq 33.92) = 0.05.$$

Thus,

$$0.05 < p\text{-value} = P(W \geq 32.52) < 0.10,$$

and the null hypothesis would not be rejected at an $\alpha = 0.05$ significance level. To find the p-value for a two-sided alternative for tests about the variance, double the tail probability beyond the chi-square test statistic if the sample variance is larger than σ_0^2, or double the tail probability below the chi-square test statistic if the sample variance is below σ_0^2. So, in Example 7.4-1, since $s^2 = 147.82 > 100$, the p-value equals $2[P(W \geq 32.52)]$ and it follows that

$$0.10 < p\text{-value} < 0.20.$$

We now give a test for the equality of two variances when sampling occurs from normal populations. Let the independent random variables X and Y have respective distributions that are $N(\mu_X, \sigma_X^2)$ and $N(\mu_Y, \sigma_Y^2)$. To test the null hypothesis H_0: $\sigma_X^2/\sigma_Y^2 = 1$ (or, equivalently, $\sigma_X^2 = \sigma_Y^2$), take random samples of n observations of X and m observations of Y. Recall that $(n-1)S_X^2/\sigma_X^2$ and $(m-1)S_Y^2/\sigma_Y^2$ have independent chi-square distributions $\chi^2(n-1)$ and $\chi^2(m-1)$, respectively. Thus, when H_0 is true,

$$F = \frac{\dfrac{(n-1)S_X^2}{\sigma_X^2(n-1)}}{\dfrac{(m-1)S_Y^2}{\sigma_Y^2(m-1)}} = \frac{S_X^2}{S_Y^2}$$

has an F distribution with $r_1 = n-1$ and $r_2 = m-1$ degrees of freedom. This F statistic is our test statistic. When H_0 is true, we would expect the observed value of F to be close to 1.

With normal distributions, three possible alternative hypotheses, along with critical regions of size α, are summarized in Table 7.4-2. Recall that $1/F$, the reciprocal of F, has an F distribution with $m-1$ and $n-1$ degrees of freedom so all critical regions may be written in terms of right-tail rejection regions and the critical values can be selected easily from Table VII in the appendix.

TABLE 7.4-2: Tests of Hypotheses of the Equality of Variances

H_0	H_1	Critical Region
$\sigma_X^2 = \sigma_Y^2$	$\sigma_X^2 > \sigma_Y^2$	$\dfrac{s_x^2}{s_y^2} \geq F_\alpha(n-1, m-1)$
$\sigma_X^2 = \sigma_Y^2$	$\sigma_X^2 < \sigma_Y^2$	$\dfrac{s_y^2}{s_x^2} \geq F_\alpha(m-1, n-1)$
$\sigma_X^2 = \sigma_Y^2$	$\sigma_X^2 \neq \sigma_Y^2$	$\dfrac{s_x^2}{s_y^2} \geq F_{\alpha/2}(n-1, m-1)$ or
		$\dfrac{s_y^2}{s_x^2} \geq F_{\alpha/2}(m-1, n-1)$

EXAMPLE 7.4-2 A biologist who studies spiders believes that not only do female green lynx spiders tend to be longer than their male counterparts, but also the lengths of the female spiders seem to vary more than those of the male spiders. We shall test whether this

latter belief is true. Suppose that the distribution of the length X of male spiders is $N(\mu_X, \sigma_X^2)$, the distribution of the length Y of female spiders is $N(\mu_Y, \sigma_Y^2)$, and X and Y are independent. We shall test $H_0\colon \sigma_X^2/\sigma_Y^2 = 1$ (i.e., $\sigma_X^2 = \sigma_Y^2$) against the alternative hypothesis $H_1\colon \sigma_X^2/\sigma_Y^2 < 1$ (i.e., $\sigma_X^2 < \sigma_Y^2$). If we use $n = 30$ and $m = 30$ observations of X and Y, respectively, then a critical region that has a significance level of $\alpha = 0.01$ is

$$\frac{s_y^2}{s_x^2} \geq F_{0.01}(29, 29) \approx 2.42,$$

by interpolation from Table VII in the appendix (or 2.4234 with Minitab). In Exercise 6.3-5, $n = 30$ observations of X yielded $\bar{x} = 5.917$ and $s_x^2 = 0.4399$, while $m = 30$ observations of Y yielded $\bar{y} = 8.153$ and $s_y^2 = 1.4100$. Since

$$\frac{s_y^2}{s_x^2} = \frac{1.4100}{0.4399} = 3.2053 > 2.42,$$

the null hypothesis is rejected in favor of the biologist's belief. Using Minitab, we find that the p-value is equal to 0.0012, as illustrated in Figure 7.4-2. (See Exercise 7.3-16.) ∎

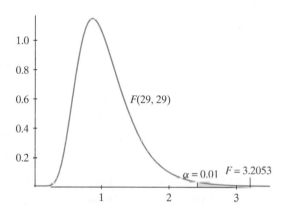

FIGURE 7.4-2: Test of variances for green lynx spiders

In Example 7.3-1, we used a t statistic for testing the equality of means which assumed that the variances were equal. In the next example, we shall test whether that assumption is valid.

EXAMPLE 7.4-3   For Example 7.3-1, given $n = 11$ observations of X and $m = 13$ observations of Y, where X is $N(\mu_X, \sigma_X^2)$ and Y is $N(\mu_Y, \sigma_Y^2)$, we shall test the null hypothesis $H_0\colon \sigma_X^2/\sigma_Y^2 = 1$ against a two-sided alternative hypothesis. At an $\alpha = 0.05$ significance level, H_0 is rejected if

$$s_x^2/s_y^2 \geq F_{0.025}(10, 12) = 3.37$$

or

$$s_y^2/s_x^2 \geq F_{0.025}(12, 10) = 3.62.$$

Using the data in Example 7.3-1, we obtain

$$s_x^2/s_y^2 = 0.24/0.35 = 0.686 \qquad \text{and} \qquad s_y^2/s_x^2 = 1.458,$$

so we do not reject H_0. Thus, the assumption of equal variances for the t statistic that was used in Example 7.3-1 seems to be valid. (See Figure 7.4-3, noting that $F_{0.975}(10, 12) = 1/F_{0.025}(12, 10) = 1/3.62 = 0.276$.) ∎

FIGURE 7.4-3: Test of variances for pea stem growths

EXERCISES

(In some of the exercises that follow, we must make assumptions such as the existence of normal distributions with the usual notation.)

7.4-1. Assume that the birth weight in grams of a baby born in the United States is $N(3315, 525^2)$ for boys and girls combined. Let X equal the weight of a baby girl who is born at home in Ottawa County, and assume that the distribution of X is $N(\mu_X, \sigma_X^2)$.

(a) Using 11 observations of X, give the test statistic and the critical region for testing H_0: $\sigma_X^2 = 525^2$ against the alternative hypothesis H_1: $\sigma_X^2 < 525^2$ (less variation of weights of home-born babies) if $\alpha = 0.05$.

(b) Calculate the value of the test statistic and state your conclusion, using the following weights:

3119	2657	3459	3629	3345	3629
3515	3856	3629	3345	3062	

(c) What is the approximate p-value of the test?

7.4-2. Let Y equal the weight in grams of a baby boy who is born at home in Ottawa County, and assume that the distribution of Y is $N(\mu_Y, \sigma_Y^2)$. The following weights are for 11 boys:

4082	3686	4111	3686	3175	4139
3686	3430	3289	3657	4082	

(a) Test H_0: $\sigma_Y^2 = 525^2$ against the alternative hypothesis H_1: $\sigma_Y^2 < 525^2$. Use $\alpha = 0.05$.

(b) Test the equality of the variances of the weights of girls (Exercise 7.4) and the weights of boys against a two-sided alternative.

(c) Test the equality of the means of the weights of girls and boys born at home in Ottawa County against a two-sided alternative.

7.4-3. Let X equal the number of pounds of butterfat produced by a Holstein cow during the 305-day milking period following the birth of a calf. We shall test the null hypothesis H_0: $\sigma^2 = 140^2$ against the alternative hypothesis H_1: $\sigma^2 > 140^2$.

(a) Give the test statistic and a critical region that has a significance level of $\alpha = 0.05$, assuming that there are $n = 25$ observations.

(b) Calculate the value of the test statistic and state your conclusion, using the following 25 observations of X:

425	710	661	664	732	714	934	761	744
653	725	657	421	573	535	602	537	405
874	791	721	849	567	468	975		

7.4-4. In May, the fill weights of 6-pound boxes of laundry soap had a mean of 6.13 pounds with a standard deviation of 0.095. The goal was to decrease the standard deviation. The company decided to adjust the filling machines and then test $H_0: \sigma = 0.095$ against $H_1: \sigma < 0.095$. In June, a random sample of size $n = 20$ yielded $\bar{x} = 6.10$ and $s = 0.065$.

   (a) At an $\alpha = 0.05$ significance level, was the company successful?

   (b) What is the approximate p-value of your test?

7.4-5. The mean birth weight in the United States is $\mu = 3315$ grams, with a standard deviation of $\sigma = 575$. Let X equal the birth weight in Rwanda. Assume that the distribution of X is $N(\mu, \sigma^2)$. We shall test the hypothesis $H_0: \sigma = 575$ against the alternative hypothesis $H_1: \sigma < 575$ at an $\alpha = 0.10$ significance level.

   (a) What is your decision if a random sample of size $n = 81$ yielded $\bar{x} = 2819$ and $s = 496$?

   (b) What is the approximate p-value of this test?

7.4-6. Let $X_1, X_2, \ldots, X_{23}$ be a random sample from a normal distribution that has variance $\sigma^2 = 100$. Let $S^2 = (1/22)\sum_{i=1}^{23}(X_i - \overline{X})^2$ be the sample variance. Of course, $22S^2/100$ is $\chi^2(22)$. Show that $\mathrm{Var}(S^2) = 10{,}000/11$. Thus, the standard deviation of S^2 is 30.15, and this helps explain the critical region $s^2 \le 49.91$ or $s^2 \ge 167.18$ in Example 7.4-1.

7.4-7. Let $X_1, X_2, \ldots, X_{19}$ be a random sample of size $n = 19$ from the normal distribution $N(\mu, \sigma^2)$.

   (a) Find a critical region, C, of size $\alpha = 0.05$ for testing $H_0: \sigma^2 = 30$ against $H_1: \sigma^2 = 80$.

   (b) Find the approximate value of β, the probability of a Type II error, for the critical region C of part (a).

7.4-8. Let X and Y equal the times in days required for maturation of Guardiola seeds from narrow-leaved and broad-leaved parents, respectively.

Assume that X is $N(\mu_X, \sigma_X^2)$ and Y is $N(\mu_Y, \sigma_Y^2)$ and that X and Y are independent. Test the hypothesis $H_0: \sigma_X^2/\sigma_Y^2 = 1$ against the alternative hypothesis $H_1: \sigma_X^2/\sigma_Y^2 > 1$ if a sample size $n = 13$ yielded $\bar{x} = 18.97$, $s_x^2 = 9.88$ and a sample of size $m = 9$ yielded $\bar{y} = 23.20$, $s_y^2 = 4.08$. Let $\alpha = 0.05$.

7.4-9. To measure air pollution in a home, let X and Y equal the amount of suspended particulate matter (in $\mu g/m^3$) measured during a 24-hour period in a home in which there is no smoker and a home in which there is a smoker, respectively. We shall test the null hypothesis $H_0: \sigma_X^2/\sigma_Y^2 = 1$ against the one-sided alternative hypothesis $H_1: \sigma_X^2/\sigma_Y^2 > 1$. If a random sample of size $n = 9$ yielded $\bar{x} = 93$ and $s_x = 12.9$ while a random sample of size $m = 11$ yielded $\bar{y} = 132$ and $s_y = 7.1$, define a critical region and give your conclusion if $\alpha = 0.05$. Now test $H_0: \mu_X = \mu_Y$ against $H_1: \mu_X < \mu_Y$ if $\alpha = 0.05$.

7.4-10. A random sample of $n = 7$ brand X light bulbs yielded $\bar{x} = 891$ hours and $s_x^2 = 9201$. A random sample of $m = 10$ brand Y light bulbs yielded $\bar{y} = 592$ hours and $s_y^2 = 4856$. Use these data to test $H_0: \sigma_X^2/\sigma_Y^2 = 1$ against $H_1: \sigma_X^2/\sigma_Y^2 > 1$. Let $\alpha = 0.05$ and state your assumptions.

7.4-11. In Section 7.3 for tests of the equality of two means, the assumption was often made that the variances were equal. In this exercise, we shall test $H_0: \sigma_X^2/\sigma_Y^2 = 1$ against the alternative hypothesis $H_1: \sigma_X^2/\sigma_Y^2 \neq 1$. Perform this test at the $\alpha = 0.05$ significance level for the following exercises:

   (a) Exercise 7.3-3.

   (b) Exercise 7.3-4.

   (c) Exercise 7.3-5.

   (d) Exercise 7.3-6.

   (e) Exercise 7.3-8.

   (f) Exercise 7.3-10.

   (g) Exercise 7.3-11.

7.5   ONE-FACTOR ANALYSIS OF VARIANCE

Frequently, experimenters want to compare more than two treatments: yields of several different corn hybrids, results due to three or more teaching techniques, or miles per gallon obtained from many different types of compact cars. Sometimes the different treatment distributions of the resulting observations are due to changing the level of a certain factor (e.g., different doses of a given drug). Thus, the consideration of the equality of the different means of the various distributions comes under the analysis of a **one-factor experiment**.

   In Section 7.3, we discussed how to compare the means of two normal distributions. More generally, let us now consider m normal distributions with unknown means

$\mu_1, \mu_2, \ldots, \mu_m$ and an unknown, but common, variance σ^2. One inference that we wish to consider is a test of the equality of the m means, namely, $H_0: \mu_1 = \mu_2 = \cdots = \mu_m = \mu$, with μ unspecified, against all possible alternative hypotheses H_1. In order to test this hypothesis, we shall take independent random samples from these distributions. Let $X_{i1}, X_{i2}, \ldots, X_{in_i}$ represent a random sample of size n_i from the normal distribution $N(\mu_i, \sigma^2)$, $i = 1, 2, \ldots, m$. In Table 7.5-1, we have indicated these random samples along with the row means (sample means), where, with $n = n_1 + n_2 + \cdots + n_m$,

$$\overline{X}_{..} = \frac{1}{n} \sum_{i=1}^{m} \sum_{j=1}^{n_i} X_{ij} \quad \text{and} \quad \overline{X}_{i\cdot} = \frac{1}{n_i} \sum_{j=1}^{n_i} X_{ij}, \quad i = 1, 2, \ldots, m.$$

TABLE 7.5-1: One-factor Random Samples

					Means
X_1:	X_{11}	X_{12}	$\cdots$	X_{1n_1}	$\overline{X}_{1\cdot}$
X_2:	X_{21}	X_{22}	$\cdots$	X_{2n_2}	$\overline{X}_{2\cdot}$
.	.	.	.	.	.
.	.	.	.	.	.
.	.	.	.	.	.
X_m:	X_{m1}	X_{m2}	$\cdots$	X_{mn_m}	$\overline{X}_{m\cdot}$
Grand Mean:					$\overline{X}_{..}$

The dot in the notation for the means, $\overline{X}_{..}$ and $\overline{X}_{i\cdot}$, indicates the index over which the average is taken. Here $\overline{X}_{..}$ is an average taken over both indices, while $\overline{X}_{i\cdot}$ is taken over just the index j.

To determine a critical region for a test of H_0, we shall first partition the sum of squares associated with the variance of the combined samples into two parts. This sum of squares is given by

$$SS(TO) = \sum_{i=1}^{m} \sum_{j=1}^{n_i} (X_{ij} - \overline{X}_{..})^2$$

$$= \sum_{i=1}^{m} \sum_{j=1}^{n_i} (X_{ij} - \overline{X}_{i\cdot} + \overline{X}_{i\cdot} - \overline{X}_{..})^2$$

$$= \sum_{i=1}^{m} \sum_{j=1}^{n_i} (X_{ij} - \overline{X}_{i\cdot})^2 + \sum_{i=1}^{m} \sum_{j=1}^{n_i} (\overline{X}_{i\cdot} - \overline{X}_{..})^2$$

$$+ 2 \sum_{i=1}^{m} \sum_{j=1}^{n_i} (X_{ij} - \overline{X}_{i\cdot})(\overline{X}_{i\cdot} - \overline{X}_{..}).$$

The last term of the right-hand member of this identity may be written as

$$2 \sum_{i=1}^{m} \left[(\overline{X}_{i\cdot} - \overline{X}_{..}) \sum_{j=1}^{n_i} (X_{ij} - \overline{X}_{i\cdot}) \right] = 2 \sum_{i=1}^{m} (\overline{X}_{i\cdot} - \overline{X}_{..})(n_i \overline{X}_{i\cdot} - n_i \overline{X}_{i\cdot}) = 0,$$

and the preceding term may be written as

$$\sum_{i=1}^{m} \sum_{j=1}^{n_i} (\overline{X}_{i\cdot} - \overline{X}_{\cdot\cdot})^2 = \sum_{i=1}^{m} n_i (\overline{X}_{i\cdot} - \overline{X}_{\cdot\cdot})^2.$$

Thus,

$$SS(TO) = \sum_{i=1}^{m} \sum_{j=1}^{n_i} (X_{ij} - \overline{X}_{i\cdot})^2 + \sum_{i=1}^{m} n_i (\overline{X}_{i\cdot} - \overline{X}_{\cdot\cdot})^2.$$

For notation, let

$$SS(TO) = \sum_{i=1}^{m} \sum_{j=1}^{n_i} (X_{ij} - \overline{X}_{\cdot\cdot})^2, \text{ the total sum of squares;}$$

$$SS(E) = \sum_{i=1}^{m} \sum_{j=1}^{n_i} (X_{ij} - \overline{X}_{i\cdot})^2, \quad \text{the sum of squares within treatments,}$$
groups, or classes, often called the error
sum of squares;

$$SS(T) = \sum_{i=1}^{m} n_i (\overline{X}_{i\cdot} - \overline{X}_{\cdot\cdot})^2, \quad \text{the sum of squares among the different}$$
treatments, groups, or classes, often called
the between-treatment sum of squares.

Hence,

$$SS(TO) = SS(E) + SS(T).$$

When H_0 is true, we may regard X_{ij}, $i = 1, 2, \ldots, m$, $j = 1, 2, \ldots, n_i$, as a random sample of size $n = n_1 + n_2 + \cdots + n_m$ from the normal distribution $N(\mu, \sigma^2)$. Then $SS(TO)/(n-1)$ is an unbiased estimator of σ^2 because $SS(TO)/\sigma^2$ is $\chi^2(n-1)$, so that $E[SS(TO)/\sigma^2] = n - 1$ and $E[SS(TO)/(n-1)] = \sigma^2$. An unbiased estimator of σ^2 based only on the sample from the ith distribution is

$$W_i = \frac{\sum_{j=1}^{n_i} (X_{ij} - \overline{X}_i)^2}{n_i - 1} \qquad \text{for } i = 1, 2, \ldots, m,$$

because $(n_i - 1)W_i/\sigma^2$ is $\chi^2(n_i - 1)$. Thus,

$$E\left[\frac{(n_i - 1)W_i}{\sigma^2}\right] = n_i - 1,$$

and so

$$E(W_i) = \sigma^2, \qquad i = 1, 2, \ldots, m.$$

It follows that the sum of m of these independent chi-square random variables, namely,

$$\sum_{i=1}^{m} \frac{(n_i - 1)W_i}{\sigma^2} = \frac{SS(E)}{\sigma^2},$$

is also chi-square with $(n_1 - 1) + (n_2 - 1) + \cdots + (n_m - 1) = n - m$ degrees of freedom. Hence, $SS(E)/(n - m)$ is an unbiased estimator of σ^2. We now have

$$\frac{SS(TO)}{\sigma^2} = \frac{SS(E)}{\sigma^2} + \frac{SS(T)}{\sigma^2},$$

where

$$\frac{SS(TO)}{\sigma^2} \text{ is } \chi^2(n-1) \quad \text{and} \quad \frac{SS(E)}{\sigma^2} \text{ is } \chi^2(n-m).$$

Because $SS(T) \geq 0$, there is a theorem (see subsequent remark) which states that $SS(E)$ and $SS(T)$ are independent and the distribution of $SS(T)/\sigma^2$ is $\chi^2(m-1)$.

REMARK   The sums of squares, $SS(T)$, $SS(E)$, and $SS(TO)$, are examples of **quadratic forms** in the variables X_{ij}, $i = 1, 2, \ldots, m$, $j = 1, 2, \ldots, n_i$. That is, each term in these sums of squares is of second degree in X_{ij}. Furthermore, the coefficients of the variables are real numbers, so these sums of squares are called **real quadratic forms**. The next theorem, stated without proof, is used in this chapter. [For a proof, see Hogg, McKean, and Craig, *Introduction to Mathematical Statistics*, 6th ed. (Upper Saddle River: Prentice Hall, 2005).] ∎

THEOREM
7.5-1

Let $Q = Q_1 + Q_2 + \cdots + Q_k$, where $Q, Q_1, \ldots, Q_k$ are $k + 1$ real quadratic forms in n mutually independent random variables normally distributed with the same variance σ^2. Let $Q/\sigma^2, Q_1/\sigma^2, \ldots, Q_{k-1}/\sigma^2$ have chi-square distributions with $r, r_1, \ldots, r_{k-1}$ degrees of freedom, respectively. If Q_k is nonnegative, then

(a) $Q_1, \ldots, Q_k$ are mutually independent, and hence,

(b) Q_k/σ^2 has a chi-square distribution with $r - (r_1 + \cdots + r_{k-1}) = r_k$ degrees of freedom.

Since, under H_0, $SS(T)/\sigma^2$ is $\chi^2(m-1)$, we have $E[SS(T)/\sigma^2] = m - 1$ and it follows that $E[SS(T)/(m-1)] = \sigma^2$. Now, the estimator of σ^2, namely, $SS(E)/(n-m)$, which is based on $SS(E)$, is always unbiased, whether H_0 is true or false. However, if the means $\mu_1, \mu_2, \ldots, \mu_m$ are not equal, the expected value of the estimator that is based on $SS(T)$ will be greater than σ^2. To make this last statement clear, we have

$$E[SS(T)] = E\left[\sum_{i=1}^{m} n_i (\overline{X}_{i\cdot} - \overline{X}_{\cdot\cdot})^2\right] = E\left[\sum_{i=1}^{m} n_i \overline{X}_{i\cdot}^2 - n\overline{X}_{\cdot\cdot}^2\right]$$

$$= \sum_{i=1}^{m} n_i \{\text{Var}(\overline{X}_{i\cdot}) + [E(\overline{X}_{i\cdot})]^2\} - n\{\text{Var}(\overline{X}_{\cdot\cdot}) + [E(\overline{X}_{\cdot\cdot})]^2\}$$

$$= \sum_{i=1}^{m} n_i \left\{\frac{\sigma^2}{n_i} + \mu_i^2\right\} - n\left\{\frac{\sigma^2}{n} + \overline{\mu}^2\right\}$$

$$= (m-1)\sigma^2 + \sum_{i=1}^{m} n_i (\mu_i - \overline{\mu})^2,$$

where $\overline{\mu} = (1/n) \sum_{i=1}^{m} n_i \mu_i$. If $\mu_1 = \mu_2 = \cdots = \mu_m = \mu$, then

$$E\left(\frac{SS(T)}{m-1}\right) = \sigma^2.$$

If the means are not all equal, then

$$E\left[\frac{SS(T)}{m-1}\right] = \sigma^2 + \sum_{i=1}^{m} n_i \frac{(\mu_i - \overline{\mu})^2}{m-1} > \sigma^2.$$

Exercise 7.5-4 also illustrates the fact that the estimator using SS(T) is usually greater than that using SS(E) when H_0 is false.

We can base our test of H_0 on the ratio of $SS(T)/(m-1)$ and $SS(E)/(n-m)$, both of which are unbiased estimators of σ^2, provided that $H_0: \mu_1 = \mu_2 = \cdots = \mu_m$ is true, so that, under H_0, the ratio would assume values near 1. However, in the case that the means $\mu_1, \mu_2, \ldots, \mu_m$ begin to differ, this ratio tends to become large, since $E[SS(T)/(m-1)]$ gets larger. Under H_0, the ratio

$$\frac{SS(T)/(m-1)}{SS(E)/(n-m)} = \frac{[SS(T)/\sigma^2]/(m-1)}{[SS(E)/\sigma^2]/(n-m)} = F$$

has an F distribution with $m-1$ and $n-m$ degrees of freedom because $SS(T)/\sigma^2$ and $SS(E)/\sigma^2$ are independent chi-square variables. We would reject H_0 if the observed value of F is too large, because this would indicate that we have a relatively large SS(T), suggesting that the means are unequal. Thus, the critical region is of the form $F \geq F_\alpha(m-1, n-m)$.

The information used for tests of the equality of several means is often summarized in an **analysis-of-variance table**, or **ANOVA** table, like that given in Table 7.5-2, where the mean square (MS) is the sum of squares (SS) divided by its degrees of freedom.

TABLE 7.5-2: Analysis-of-Variance Table				
Source	Sum of Squares (SS)	Degrees of Freedom	Mean Square (MS)	F Ratio
Treatment	SS(T)	$m-1$	$MS(T) = \dfrac{SS(T)}{m-1}$	$\dfrac{MS(T)}{MS(E)}$
Error	SS(E)	$n-m$	$MS(E) = \dfrac{SS(E)}{n-m}$	
Total	SS(TO)	$n-1$		

EXAMPLE 7.5-1   Let X_1, X_2, X_3, X_4 be independent random variables that have normal distributions $N(\mu_i, \sigma^2)$, $i = 1, 2, 3, 4$. We shall test

$$H_0: \mu_1 = \mu_2 = \mu_3 = \mu_4 = \mu$$

against all alternatives on the basis of a random sample of size $n_i = 3$ from each of the four distributions. A critical region of size $\alpha = 0.05$ is given by

$$F = \frac{SS(T)/(4-1)}{SS(E)/(12-4)} \geq 4.07 = F_{0.05}(3, 8).$$

The observed data are shown in Table 7.5-3. (Clearly, these data are not observations from normal distributions; they were selected to illustrate the calculations.)

TABLE 7.5-3: Illustrative Data

	Observations			$\overline{X}_{i\cdot}$
X_1:	13	8	9	10
X_2:	15	11	13	13
X_3:	8	12	7	9
X_4:	11	15	10	12
$\overline{X}_{\cdot\cdot}$				11

For the given data, the calculated SS(TO), SS(E), and SS(T) are

$$SS(TO) = (13 - 11)^2 + (8 - 11)^2 + \cdots + (15 - 11)^2 + (10 - 11)^2 = 80;$$
$$SS(E) = (13 - 10)^2 + (8 - 10)^2 + \cdots + (15 - 12)^2 + (10 - 12)^2 = 50;$$
$$SS(T) = 3[(10 - 11)^2 + (13 - 11)^2 + (9 - 11)^2 + (12 - 11)^2] = 30.$$

Note that since SS(TO) = SS(E) + SS(T), only two of the three values need to be calculated directly from the data. Here the computed value of F is

$$\frac{30/3}{50/8} = 1.6 < 4.07,$$

and H_0 is not rejected. The p-value is the probability, under H_0, of obtaining an F that is at least as large as this computed value of F. It is often given by computer programs.

The information for this example is summarized in Table 7.5-4. Again we note that (here and elsewhere) the F statistic is the ratio of two appropriate mean squares. ∎

TABLE 7.5-4: ANOVA Table for Illustrative Data

Source	Sum of Squares (SS)	Degrees of Freedom	Mean Square (MS)	F Ratio	p-value
Treatment	30	3	30/3	1.6	0.264
Error	50	8	50/8		
Total	80	11			

Formulas that sometimes simplify the calculations of SS(TO), SS(T), and SS(E) (and also reduce roundoff errors created by subtracting the averages from the observations) are

$$SS(TO) = \sum_{i=1}^{m} \sum_{j=1}^{n_i} X_{ij}^2 - \frac{1}{n}\left[\sum_{i=1}^{m} \sum_{j=1}^{n_i} X_{ij}\right]^2,$$

$$SS(T) = \sum_{i=1}^{m} \frac{1}{n_i}\left[\sum_{j=1}^{n_i} X_{ij}\right]^2 - \frac{1}{n}\left[\sum_{i=1}^{m} \sum_{j=1}^{n_i} X_{ij}\right]^2,$$

and

$$SS(E) = SS(TO) - SS(T).$$

It is interesting to note that in these formulas each square is divided by the number of observations in the sum being squared: X_{ij}^2 by 1, $\left(\sum_{j=1}^{n_i} X_{ij}\right)^2$ by n_i, and $\left(\sum_{i=1}^{m} \sum_{j=1}^{n_i} X_{ij}\right)^2$ by n. The preceding formulas are used in Example 7.5-2. Although they are useful, you are encouraged to use appropriate statistical packages on a computer to aid you with these calculations.

If the sample sizes are all at least equal to 7, insight can be gained by plotting box-and-whisker diagrams on the same figure, for each of the samples. This technique is also illustrated in Example 7.5-2.

EXAMPLE 7.5-2 A window that is manufactured for an automobile has five studs for attaching it. A company that manufactures these windows performs "pullout tests" to determine the force needed to pull a stud out of the window. Let X_i, $i = 1, 2, 3, 4, 5$, equal the force required at position i, and assume that the distribution of X_i is $N(\mu_i, \sigma^2)$. We shall test the null hypothesis $H_0: \mu_1 = \mu_2 = \mu_3 = \mu_4 = \mu_5$, using seven independent observations at each position. At an $\alpha = 0.01$ significance level, H_0 is rejected if the computed

$$F = \frac{SS(T)/(5-1)}{SS(E)/(35-5)} > 4.02 = F_{0.01}(4, 30).$$

TABLE 7.5-5: Pullout Test Data								$\sum_{j=1}^{7} x_{ij}$	$\sum_{j=1}^{7} x_{ij}^2$
	Observations								
X_1:	92	90	87	105	86	83	102	645	59,847
X_2:	100	108	98	110	114	97	94	721	74,609
X_3:	143	149	138	136	139	120	145	970	134,936
X_4:	147	144	160	149	152	131	134	1017	148,367
X_5:	142	155	119	134	133	146	152	981	138,415
Totals								4334	556,174

The observed data, along with certain sums are given in Table 7.5-5. For these data,

$$SS(TO) = 556{,}174 - \frac{1}{35}(4334)^2 = 19{,}500.97,$$

$$SS(T) = \frac{1}{7}(645^2 + 721^2 + 970^2 + 1017^2 + 981^2)$$

$$- \frac{1}{35}(4334)^2 = 16{,}672.11,$$

$$SS(E) = 19{,}500.97 - 16{,}672.11 = 2828.86.$$

Since the computed F is

$$F = \frac{16{,}672.11/4}{2828.86/30} = 44.20,$$

the null hypothesis is clearly rejected. This information obtained from the equations is summarized in Table 7.5-6.

TABLE 7.5-6: ANOVA Table for Pullout Tests				
Source	Sum of Squares (SS)	Degrees of Freedom	Mean Square (MS)	F
Treatment	16,672.11	4	4,168.03	44.20
Error	2,828.86	30	94.30	
Total	19,500.97	34		

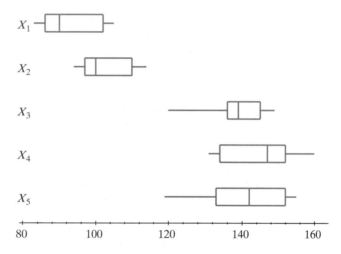

FIGURE 7.5-1: Box plots for pullout tests

But why is H_0 rejected? The box-and-whisker diagrams shown in Figure 7.5-1 help to answer this question. It looks like the forces required to pull out studs in positions 1 and 2 are similar, and those in positions 3, 4, and 5 are quite similar, but different from, positions 1 and 2. (See Exercise 7.5-12.) An examination of the window would confirm that this is the case. ∎

As with the two-sample t test, the F test works quite well even if the underlying distributions are nonnormal, unless they are highly skewed or the variances are quite different. In these latter cases, we might need to transform the observations to make the data more symmetric with about the same variances or to use certain nonparametric methods that are beyond the scope of this text.

EXERCISES

(In some of the exercises that follow, we must make assumptions, such as normal distributions with equal variances.)

7.5-1. Let μ_1, μ_2, μ_3 be, respectively, the means of three normal distributions with a common, but unknown, variance σ^2. In order to test, at the $\alpha = 0.05$ significance level, the hypothesis $H_0: \mu_1 = \mu_2 = \mu_3$ against all possible alternative hypotheses, we take a random sample of size 4 from each of these distributions. Determine whether we accept or reject H_0 if the observed values from the three distributions are, respectively, as follows:

$$X_1: \quad 5 \quad 9 \quad 6 \quad 8$$
$$X_2: \quad 11 \quad 13 \quad 10 \quad 12$$
$$X_3: \quad 10 \quad 6 \quad 9 \quad 9$$

7.5-2. Let μ_i be the average yield in bushels per acre of variety i of corn, $i = 1, 2, 3, 4$. In order to test the hypothesis $H_0: \mu_1 = \mu_2 = \mu_3 = \mu_4$ at the 5% significance level, four test plots for each of the four varieties of corn are planted. Determine whether we accept or reject H_0 if the yield in bushels per acre of the four varieties of corn are, respectively, as follows:

$$X_1: \quad 68.82 \quad 76.99 \quad 74.30 \quad 78.73$$
$$X_2: \quad 86.84 \quad 75.69 \quad 77.87 \quad 76.18$$
$$X_3: \quad 90.16 \quad 78.84 \quad 80.65 \quad 83.58$$
$$X_4: \quad 61.58 \quad 73.51 \quad 74.57 \quad 70.75$$

7.5-3. Four groups of three pigs each were fed individually four different feeds for a specified length of time to test the hypothesis $H_0: \mu_1 = \mu_2 = \mu_3 = \mu_4$, where μ_i, $i = 1, 2, 3, 4$, is the mean weight gain for each of the feeds. Determine whether the null hypothesis is accepted or rejected at a 5% significance level if the observed weight gains are, respectively, as follows:

$$X_1: \quad 194.11 \quad 182.80 \quad 187.43$$
$$X_2: \quad 216.06 \quad 203.50 \quad 216.88$$
$$X_3: \quad 178.10 \quad 189.20 \quad 181.33$$
$$X_4: \quad 197.11 \quad 202.68 \quad 209.18$$

7.5-4. For the following set of data, show that the computed $SS(E)/(n - m) = 1$ and $SS(T)/(m - 1) = 75$ (this suggests that the unbiased estimate of σ^2 based on $SS(T)$ is usually greater than σ^2 when the true means are unequal):

$$X_1: \quad 4 \quad 5 \quad 6$$
$$X_2: \quad 9 \quad 10 \quad 11$$
$$X_3: \quad 14 \quad 15 \quad 16$$

7.5-5. The female cuckoo lays her eggs in other birds' nests. The "foster parents" are usually deceived, probably because of the similarity in sizes of their own eggs and cuckoo eggs. Latter (see references) investigated this possible explanation and measured the lengths of cuckoo eggs (in millimeters) that were found in the nests of three species. Following are his results:

Hedge sparrow:	22.0	23.9	20.9	23.8	25.0
	24.0	21.7	23.8	22.8	23.1
	23.1	23.5	23.0	23.0	
Robin:	21.8	23.0	23.3	22.4	23.0
	23.0	23.0	22.4	23.9	22.3
	22.0	22.6	22.0	22.1	21.1
	23.0				
Wren:	19.8	22.1	21.5	20.9	22.0
	21.0	22.3	21.0	20.3	20.9
	22.0	20.0	20.8	21.2	21.0

(a) Construct an ANOVA table to test the equality of the three means.

(b) For each set of data, construct box-and-whisker diagrams on the same figure.

(c) Interpret your results.

7.5-6. Ledolter and Hogg (see references) report that a civil engineer wishes to compare the strengths of three different types of beams, one (A) made of steel and two (B and C) made of different and more expensive alloys. A certain deflection (in units of 0.001 inch) was measured for each beam when submitted to a given force; thus, a small deflection would indicate a beam of great strength. The order statistics for the three samples, of respective sizes $n_1 = 8$, $n_2 = 6$, and $n_3 = 6$, are as follows:

A:	79	82	83	84	85	86	86	87
B:	74	75	76	77	78	82		
C:	77	78	79	79	79	82		

(a) Use these data, $\alpha = 0.05$, and the F test to test the equality of the three means.

(b) For each set of data, construct box-and-whisker diagrams on the same figure and give an interpretation of your diagrams.

7.5-7. Montgomery (see references) examines the strengths of a synthetic fiber that may be affected by the percentage of cotton in the fiber. Five levels of this percentage are considered, with five observations taken at each level.

Percentage of Cotton	Tensile Strength in Pounds per Square Inch				
15	7	7	15	11	9
20	12	17	12	18	18
25	14	18	18	19	19
30	19	25	22	19	23
35	7	10	11	15	11

Use the F test, with $\alpha = 0.05$, to see if there are differences in the breaking strengths due to the percentages of cotton used.

7.5-8. Let X_1, X_2, X_3, X_4 equal the cholesterol level of a woman under the age of 50, a man under 50, a woman 50 or older, and a man 50 or older, respectively. Assume that the distribution of X_i is $N(\mu_i, \sigma^2)$, $i = 1, 2, 3, 4$. We shall test the null hypothesis $H_0: \mu_1 = \mu_2 = \mu_3 = \mu_4$, using seven observations of each X_i.

(a) Give a critical region for an $\alpha = 0.05$ significance level.

(b) Construct an ANOVA table and state your conclusion, using the following data:

X_1: 221 213 202 183 185 197 162
X_2: 271 192 189 209 227 236 142
X_3: 262 193 224 201 161 178 265
X_4: 192 253 248 278 232 267 289

(c) Give bounds on the p-value for this test.

(d) For each set of data, construct box-and-whisker diagrams on the same figure and give an interpretation of your diagram.

7.5-9. Let X_i, $i = 1, 2, 3, 4$, equal the distance that a golf ball travels when hit from a tee, where i denotes the index of the ith manufacturer. Assume that the distribution of X_i is $N(\mu_i, \sigma^2)$, $i = 1, 2, 3, 4$ when a ball is hit by a certain golfer. We shall test the null hypothesis $H_0: \mu_1 = \mu_2 = \mu_3 = \mu_4$, using three observations of each random variable.

(a) Give a critical region for an $\alpha = 0.05$ significance level.

(b) Construct an ANOVA table and state your conclusion, using the following data:

X_1: 240 221 265
X_2: 286 256 272
X_3: 259 245 232
X_4: 239 215 223

(c) What would your conclusion be if $\alpha = 0.025$?

(d) What is the approximate p-value of this test?

7.5-10. Different sizes of nails are packaged in "1-pound" boxes. Let X_i equal the weight of a box with nail size $(4i)C$, $i = 1, 2, 3, 4, 5$, where $4C, 8C, 12C, 16C$, and $20C$ are the sizes of the sinkers from smallest to largest. Assume that the distribution of X_i is $N(\mu_i, \sigma^2)$. To test the null hypothesis that the mean weights of "1-pound" boxes are all equal for different sizes of nails, we shall use random samples of size 7,

weighing the nails to the nearest hundredth of a pound.

(a) Give a critical region for an $\alpha = 0.05$ significance level.

(b) Construct an ANOVA table and state your conclusion, using the following data:

X_1: 1.03 1.04 1.07 1.03 1.08 1.06 1.07
X_2: 1.03 1.10 1.08 1.05 1.06 1.06 1.05
X_3: 1.03 1.08 1.06 1.02 1.04 1.04 1.07
X_4: 1.10 1.10 1.09 1.09 1.06 1.05 1.08
X_5: 1.04 1.06 1.07 1.06 1.05 1.07 1.05

(c) For each set of data, construct box-and-whisker diagrams on the same figure and give an interpretation of your diagrams.

7.5-11. The driver of a diesel-powered automobile decided to test the quality of three types of diesel fuel sold in the area. The test is to be based on miles per gallon (mpg). Make the usual assumptions, take $\alpha = 0.05$, and use the following data to test the null hypothesis that the three means are equal:

Brand A: 38.7   39.2   40.1   38.9
Brand B: 41.9   42.3   41.3
Brand C: 40.8   41.2   39.5   38.9   40.3

7.5-12. From the box-and-whisker diagrams in Figure 7.5-1, it looks like the means of X_1 and X_2 could be equal and also that the means of X_3, X_4, and X_5 could be equal but different from the first two.

(a) Using the data in Example 7.5-2, as well as a t test and an F test, test $H_0: \mu_1 = \mu_2$ against a two-sided alternative hypothesis. Let $\alpha = 0.05$. Do the F and t tests give the same result?

(b) Using the data in Example 7.5-2, test $H_0: \mu_3 = \mu_4 = \mu_5$. Let $\alpha = 0.05$.

7.5-13. For an aerosol product, there are three weights: the tare weight (container weight), the concentrate weight, and the propellant weight. Let X_1, X_2, X_3 denote the propellant weights on three different days. Assume that each of these independent random variables has a normal distribution with common variance and respective means μ_1, μ_2, and μ_3. We shall test the null hypothesis $H_0: \mu_1 = \mu_2 = \mu_3$, using nine observations of each of the random variables.

(a) Give a critical region for an $\alpha = 0.01$ significance level.

(b) Construct an ANOVA table and state your conclusion, using the following data:

X_1: 43.06 43.32 42.63 42.86 43.05
    42.87 42.94 42.80 42.36

X_2: 42.33 42.81 42.13 42.41 42.39
    42.10 42.42 41.42 42.52

X_3: 42.83 42.57 42.96 43.16 42.25
    42.24 42.20 41.97 42.61

(c) For each set of data, construct box-and-whisker diagrams on the same figure and give an interpretation of your diagrams.

7.5-14. A particular process puts a coating on a piece of glass so that it is sensitive to touch. Randomly throughout the day, pieces of glass are selected from the production line and the resistance is measured at 12 different locations on the glass. On each of three different days, December 6, December 7, and December 22, the following data give the means of the 12 measurements on each of 11 pieces of glass:

December 6:   175.05 177.44 181.94 176.51 182.12 164.34
       163.20 168.12 171.26 171.92 167.87

December 7:   175.93 176.62 171.39 173.90 178.34 172.90
       174.67 174.27 177.16 184.13 167.21

December 22: 167.27 161.48 161.86 173.83 170.75 172.90
       173.27 170.82 170.93 173.89 177.68

(a) Use these data to test whether the means on all three days are equal.

(b) Use box-and-whisker diagrams to confirm your answer.

7.5-15. Ledolter and Hogg (see references) report that an operator of a feedlot wants to compare the effectiveness of three different cattle feed supplements. He selects a random sample of 15 one-year old heifers from his lot of over 1000 and divides them into three groups at random. Each group gets a different feed supplement. Upon noting that one heifer in group A was lost due to an accident, the operator records the gains in weight over a 6-month period as follows:

Group A:	500	650	530	680	
Group B:	700	620	780	830	860
Group C:	500	520	400	580	410

(a) Test whether there are differences in the mean weight gains due to the three different feed supplements.

(b) Do box plots of the data confirm your answer in part (a)?

7.5-16. Ledolter and Hogg (see references) report the comparison of three workers with different experience who manufacture brake wheels for a magnetic brake. Worker A has four years of experience, worker B has seven years, and worker C has one year. The company is concerned about the product's quality, which is measured by the difference between the specified diameter and the actual diameter of the brake wheel. On a given day, the supervisor selects nine brake wheels at random from the output of each worker. The following data give the differences between specified and actual diameters in hundredths of an inch:

Worker A:	2.0	3.0	2.3	3.5	3.0	2.0	4.0	4.5	3.0
Worker B:	1.5	3.0	4.5	3.0	3.0	2.0	2.5	1.0	2.0
Worker C:	2.5	3.0	2.0	2.5	1.5	2.5	2.5	3.0	3.5

(a) Test whether there are statistically significant differences in the quality among the three different workers.

(b) Do box plots of the data confirm your answer in part (a)?

7.6 TWO-FACTOR ANALYSIS OF VARIANCE

The test of the equality of several means, considered in Section 7.5, is an example of a statistical inference method called the analysis of variance (ANOVA). This method derives its name from the fact that the quadratic form $SS(TO) = (n - 1)S^2$—the total sum of squares about the combined sample mean—is decomposed into its components and analyzed. In this section, other problems in the analysis of variance will be investigated; here we restrict our considerations to the two-factor case, but the reader can see how it can be extended to three-factor and other cases.

Consider a situation in which it is desirable to investigate the effects of two factors that influence an outcome of an experiment. For example, a teaching method (lecture, discussion, computer assisted, television, etc.) and the size of a class might influence a student's score on a standard test; or the type of car and the grade of gasoline used might change the number of miles per gallon. In this latter example, if the number of miles per gallon is not affected by the grade of gasoline, we would no doubt use the least expensive grade.

The first analysis-of-variance model that we discuss is referred to as a **two-way classification with one observation per cell**. Assume that there are two factors (attributes), one of which has a levels and the other b levels. There are thus $n = ab$ possible combinations, each of which determines a cell. Let us think of these cells as being arranged in a rows and b columns. Here we take one observation per cell, and we denote the observation in the ith row and jth column by X_{ij}. Assume further that X_{ij} is $N(\mu_{ij}, \sigma^2)$, $i = 1, 2, \ldots, a$, and $j = 1, 2, \ldots, b$; and the $n = ab$ random variables are independent. [The assumptions of normality and homogeneous (same) variances can be somewhat relaxed in applications, with little change in the significance levels of the resulting tests.] We shall assume that the means μ_{ij} are composed of a row effect, a column effect, and an overall effect in some additive way, namely, $\mu_{ij} = \mu + \alpha_i + \beta_j$, where $\sum_{i=1}^{a} \alpha_i = 0$ and $\sum_{j=1}^{b} \beta_j = 0$. The parameter α_i represents the ith row effect, and the parameter β_j represents the jth column effect.

REMARK   There is no loss in generality in assuming that

$$\sum_{i=1}^{a} \alpha_i = \sum_{j=1}^{b} \beta_j = 0.$$

To see this, let $\mu_{ij} = \mu' + \alpha_i' + \beta_j'$. Write

$$\overline{\alpha}' = \left(\frac{1}{a}\right)\sum_{i=1}^{a} \alpha_i' \text{ and } \overline{\beta}' = \left(\frac{1}{b}\right)\sum_{j=1}^{b} \beta_j'.$$

We have

$$\mu_{ij} = (\mu' + \overline{\alpha}' + \overline{\beta}') + (\alpha_i' - \overline{\alpha}') + (\beta_j' - \overline{\beta}') = \mu + \alpha_i + \beta_j,$$

where $\sum_{i=1}^{a} \alpha_i = 0$ and $\sum_{j=1}^{b} \beta_j = 0$. The reader is asked to find μ, α_i, and β_j for one display of μ_{ij} in Exercise 7.6-2. ∎

To test the hypothesis that there is no row effect, we would test $H_A: \alpha_1 = \alpha_2 = \cdots = \alpha_a = 0$, since $\sum_{i=1}^{a} \alpha_i = 0$. Similarly, to test that there is no column effect, we would test $H_B: \beta_1 = \beta_2 = \cdots = \beta_b = 0$, since $\sum_{j=1}^{b} \beta_j = 0$. To test these hypotheses, we shall again partition the total sum of squares into several components. Letting

$$\overline{X}_{i\cdot} = \frac{1}{b}\sum_{j=1}^{b} X_{ij}, \ \overline{X}_{\cdot j} = \frac{1}{a}\sum_{i=1}^{a} X_{ij}, \ \overline{X}_{\cdot\cdot} = \frac{1}{ab}\sum_{i=1}^{a}\sum_{j=1}^{b} X_{ij},$$

we have

$$SS(TO) = \sum_{i=1}^{a} \sum_{j=1}^{b} (X_{ij} - \overline{X}_{..})^2$$

$$= \sum_{i=1}^{a} \sum_{j=1}^{b} [(\overline{X}_{i\cdot} - \overline{X}_{..}) + (\overline{X}_{\cdot j} - \overline{X}_{..}) + (X_{ij} - \overline{X}_{i\cdot} - \overline{X}_{\cdot j} + \overline{X}_{..})]^2$$

$$= b \sum_{i=1}^{a} (\overline{X}_{i\cdot} - \overline{X}_{..})^2 + a \sum_{j=1}^{b} (\overline{X}_{\cdot j} - \overline{X}_{..})^2$$

$$+ \sum_{i=1}^{a} \sum_{j=1}^{b} (X_{ij} - \overline{X}_{i\cdot} - \overline{X}_{\cdot j} + \overline{X}_{..})^2$$

$$= SS(A) + SS(B) + SS(E),$$

where SS(A) is the sum of squares among levels of factor A, or among rows; SS(B) is the sum of squares among levels of factor B, or among columns; and SS(E) is the error or residual sum of squares. In Exercise 7.6-4, the reader is asked to show that the three cross-product terms in the square of the trinomial sum to zero. The distribution of the error sum of squares does not depend on the mean μ_{ij}, provided that the additive model is correct. Hence, its distribution is the same whether H_A or H_B is true or not, and thus SS(E) acts as a "measuring stick," as did SS(E) in Section 7.5. This can be seen more clearly by writing

$$SS(E) = \sum_{i=1}^{a} \sum_{j=1}^{b} (X_{ij} - \overline{X}_{i\cdot} - \overline{X}_{\cdot j} + \overline{X}_{..})^2$$

$$= \sum_{i=1}^{a} \sum_{j=1}^{b} [X_{ij} - (\overline{X}_{i\cdot} - \overline{X}_{..}) - (\overline{X}_{\cdot j} - \overline{X}_{..}) - \overline{X}]^2$$

and noting the similarity of the summand in the right-hand member to

$$X_{ij} - \mu_{ij} = X_{ij} - \alpha_i - \beta_j - \mu.$$

We now show that $SS(A)/\sigma^2$, $SS(B)/\sigma^2$, and $SS(E)/\sigma^2$ are independent chi-square variables, provided that both H_A and H_B are true—that is, when all the means μ_{ij} have a common value μ. To do this, we first note that $SS(TO)/\sigma^2$ is $\chi^2(ab - 1)$. In addition, from Section 7.5, we see that expressions such as $SS(A)/\sigma^2$ and $SS(B)/\sigma^2$ are chi-square variables, namely, $\chi^2(a - 1)$ and $\chi^2(b - 1)$, by replacing the n_i of Section 7.5 by a and b, respectively. Obviously, $SS(E) \geq 0$, and hence by Theorem 7.5-1, $SS(A)/\sigma^2$, $SS(B)/\sigma^2$, and $SS(E)/\sigma^2$ are independent chi-square variables with $a - 1$, $b - 1$, and $ab - 1 - (a - 1) - (b - 1) = (a - 1)(b - 1)$ degrees of freedom, respectively.

To test the hypothesis $H_A: \alpha_1 = \alpha_2 = \cdots = \alpha_a = 0$, we shall use the row sum of squares $SS(A)$ and the residual sum of squares $SS(E)$. When H_A is true, $SS(A)/\sigma^2$ and $SS(E)/\sigma^2$ are independent chi-square variables with $a - 1$ and $(a - 1)(b - 1)$ degrees of freedom, respectively. Thus, $SS(A)/(a - 1)$ and $SS(E)/[(a - 1)(b - 1)]$ are both unbiased estimators of σ^2 when H_A is true. However, $E[SS(A)/(a - 1)] > \sigma^2$ when H_A is not true, and hence we would reject H_A when

$$F_A = \frac{SS(A)/[\sigma^2(a - 1)]}{SS(E)/[\sigma^2(a - 1)(b - 1)]} = \frac{SS(A)/(a - 1)}{SS(E)/[(a - 1)(b - 1)]}$$

is "too large." Since F_A has an F distribution with $a - 1$ and $(a - 1)(b - 1)$ degrees of freedom when H_A is true, H_A is rejected if the observed value of $F_A \geq F_\alpha[a - 1, (a - 1)(b - 1)]$.

Similarly, the test of the hypothesis $H_B: \beta_1 = \beta_2 = \cdots = \beta_b = 0$ against all alternatives can be based on

$$F_B = \frac{\mathrm{SS(B)}/[\sigma^2(b - 1)]}{\mathrm{SS(E)}/[\sigma^2(a - 1)(b - 1)]} = \frac{\mathrm{SS(B)}/(b - 1)}{\mathrm{SS(E)}/[(a - 1)(b - 1)]},$$

which has an F distribution with $b - 1$ and $(a - 1)(b - 1)$ degrees of freedom, provided that H_B is true.

Table 7.6-1 is the ANOVA table that summarizes the information needed for these tests of hypotheses. The formulas for F_A and F_B show that each of them is a ratio of two mean squares.

TABLE 7.6-1: Two-Way ANOVA Table, One Observation per Cell

Source	Sum of Squares (SS)	Degrees of Freedom	Mean Square (MS)	F
Factor A (row)	SS(A)	$a - 1$	$\mathrm{MS(A)} = \dfrac{\mathrm{SS(A)}}{a - 1}$	$\dfrac{\mathrm{MS(A)}}{\mathrm{MS(E)}}$
Factor B (column)	SS(B)	$b - 1$	$\mathrm{MS(B)} = \dfrac{\mathrm{SS(B)}}{b - 1}$	$\dfrac{\mathrm{MS(B)}}{\mathrm{MS(E)}}$
Error	SS(E)	$(a - 1)(b - 1)$	$\mathrm{MS(E)} = \dfrac{\mathrm{SS(E)}}{(a - 1)(b - 1)}$	
Total	SS(TO)	$ab - 1$		

EXAMPLE 7.6-1 Each of three cars is driven with each of four different brands of gasoline. The number of miles per gallon driven for each of the $ab = (3)(4) = 12$ different combinations is recorded in Table 7.6-2.

TABLE 7.6-2: Gas Mileage Data

Car	Gasoline 1	2	3	4	$\overline{X}_{i\cdot}$
1	26	28	31	31	29
2	24	25	28	27	26
3	25	25	28	26	26
$\overline{X}_{\cdot j}$	25	26	29	28	27

We would like to test whether we can expect the same mileage for each of these four brands of gasoline. In our notation, we test the hypothesis

$$H_B: \beta_1 = \beta_2 = \beta_3 = \beta_4 = 0$$

against all alternatives. At a 1% significance level, we shall reject H_B if the computed F, namely,

$$\frac{SS(B)/(4-1)}{SS(E)/[(3-1)(4-1)]} \geq 9.78 = F_{0.01}(3,6).$$

We have

$$SS(B) = 3[(25-27)^2 + (26-27)^2 + (29-27)^2 + (28-27)^2] = 30;$$

$$SS(E) = (26 - 29 - 25 + 27)^2 + (24 - 26 - 25 + 27)^2 + \cdots$$

$$+ (26 - 26 - 28 + 27)^2 = 4.$$

Hence, the computed F is

$$\frac{30/3}{4/6} = 15 > 9.78,$$

and the hypothesis H_B is rejected. That is, the gasolines seem to give different performances (at least with these three cars).

The information for this example is summarized in Table 7.6-3. ■

Source	Sum of Squares (SS)	Degrees of Freedom	Mean Square (MS)	F	p-value
TABLE 7.6-3: ANOVA Table for Gas Mileage Data					
Row (A)	24	2	12	18	0.003
Column (B)	30	3	10	15	0.003
Error	4	6	2/3		
Total	58	11			

In a two-way classification problem, particular combinations of the two factors might interact differently from what is expected from the additive model. For instance, in Example 7.6-1 gasoline 3 seemed to be the best gasoline and car 1 the best car; however, it sometimes happens that the two best do not "mix" well and the joint performance is poor. That is, there might be a strange interaction between this combination of car and gasoline, and accordingly, the joint performance is not as good as expected. Sometimes it happens that we get good results from a combination of some of the poorer levels of each factor. This phenomenon is called interaction, and it frequently occurs in practice (e.g., in chemistry). In order to test for possible interaction, we shall consider a two-way classification problem in which $c > 1$ independent observations per cell are taken.

Assume that X_{ijk}, $i = 1, 2, \ldots, a$; $j = 1, 2, \ldots, b$; and $k = 1, 2, \ldots, c$, are $n = abc$ random variables that are mutually independent and have normal distributions with a common, but unknown, variance σ^2. The mean of each X_{ijk}, $k = 1, 2, \ldots, c$, is $\mu_{ij} = \mu + \alpha_i + \beta_j + \gamma_{ij}$, where $\sum_{i=1}^a \alpha_i = 0$, $\sum_{j=1}^b \beta_j = 0$, $\sum_{i=1}^a \gamma_{ij} = 0$, and $\sum_{j=1}^b \gamma_{ij} = 0$. The parameter γ_{ij} is called the **interaction** associated with cell (i, j). That is, the interaction between the ith level of one classification and the jth level of the other classification is γ_{ij}. In Exercise 7.6-6, the reader is asked to determine μ, α_i, β_j, and γ_{ij} for some given μ_{ij}.

To test the hypotheses that (a) the row effects are equal to zero, (b) the column effects are equal to zero, and (c) there is no interaction, we shall again partition the total sum of squares into several components. Letting

$$\overline{X}_{ij\cdot} = \frac{1}{c} \sum_{k=1}^{c} X_{ijk},$$

$$\overline{X}_{i\cdot\cdot} = \frac{1}{bc} \sum_{j=1}^{b} \sum_{k=1}^{c} X_{ijk},$$

$$\overline{X}_{\cdot j\cdot} = \frac{1}{ac} \sum_{i=1}^{a} \sum_{k=1}^{c} X_{ijk},$$

$$\overline{X}_{\cdots} = \frac{1}{abc} \sum_{i=1}^{a} \sum_{j=1}^{b} \sum_{k=1}^{c} X_{ijk},$$

we have

$$SS(TO) = \sum_{i=1}^{a} \sum_{j=1}^{b} \sum_{k=1}^{c} (X_{ijk} - \overline{X}_{\cdots})^2$$

$$= bc \sum_{i=1}^{a} (\overline{X}_{i\cdot\cdot} - \overline{X}_{\cdots})^2 + ac \sum_{j=1}^{b} (\overline{X}_{\cdot j\cdot} - \overline{X}_{\cdots})^2$$

$$+ c \sum_{i=1}^{a} \sum_{j=1}^{b} (\overline{X}_{ij\cdot} - \overline{X}_{i\cdot\cdot} - \overline{X}_{\cdot j\cdot} + \overline{X}_{\cdots})^2 + \sum_{i=1}^{a} \sum_{j=1}^{b} \sum_{k=1}^{c} (X_{ijk} - \overline{X}_{ij\cdot})^2$$

$$= SS(A) + SS(B) + SS(AB) + SS(E),$$

where $SS(A)$ is the row sum of squares, or the sum of squares among levels of factor A; $SS(B)$ is the column sum of squares, or the sum of squares among levels of factor B; $SS(AB)$ is the interaction sum of squares; and $SS(E)$ is the error sum of squares. Again, we can show that the cross-product terms sum to zero.

To consider the joint distribution of $SS(A), SS(B), SS(AB)$, and $SS(E)$, let us assume that all the means equal the same value μ. Of course, we know that $SS(TO)/\sigma^2$ is $\chi^2(abc - 1)$. Also, by letting the n_i of Section 7.5 equal bc and ac, respectively, we know that $SS(A)/\sigma^2$ and $SS(B)/\sigma^2$ are $\chi^2(a - 1)$ and $\chi^2(b - 1)$. Moreover,

$$\frac{\sum_{k=1}^{c} (X_{ijk} - \overline{X}_{ij\cdot})^2}{\sigma^2}$$

is $\chi^2(c - 1)$; hence, $SS(E)/\sigma^2$ is the sum of ab independent chi-square variables such as this and thus is $\chi^2[ab(c - 1)]$. Of course, $SS(AB) \geq 0$; so, according to Theorem 7.5-1, $SS(A)/\sigma^2$, $SS(B)/\sigma^2$, $SS(AB)/\sigma^2$, and $SS(E)/\sigma^2$ are mutually independent chi-square variables with $a - 1, b - 1, (a - 1)(b - 1)$, and $ab(c - 1)$ degrees of freedom, respectively.

To test the hypotheses concerning row, column, and interaction effects, we form F statistics in which the numerators are affected by deviations from the respective hypotheses whereas the denominator is a function of $SS(E)$, whose distribution depends only on the value of σ^2 and not on the values of the cell means. Hence, $SS(E)$ acts as our measuring stick here.

The statistic for testing the hypothesis

$$H_{AB}: \gamma_{ij} = 0, \; i = 1, 2, \ldots, a, \; j = 1, 2, \ldots, b,$$

against all alternatives is

$$F_{AB} = \frac{c \sum_{i=1}^{a} \sum_{j=1}^{b} (\overline{X}_{ij\cdot} - \overline{X}_{i\cdot\cdot} - \overline{X}_{\cdot j\cdot} + \overline{X}_{\cdots})^2 / [\sigma^2(a-1)(b-1)]}{\sum_{i=1}^{a} \sum_{j=1}^{b} \sum_{k=1}^{c} (X_{ijk} - \overline{X}_{ij\cdot})^2 / [\sigma^2 ab(c-1)]}$$

$$= \frac{SS(AB)/[(a-1)(b-1)]}{SS(E)/[ab(c-1)]},$$

which has an F distribution with $(a-1)(b-1)$ and $ab(c-1)$ degrees of freedom when H_{AB} is true. If the computed $F_{AB} \geq F_{\alpha}[(a-1)(b-1), ab(c-1)]$, we reject H_{AB} and say that there is a difference among the means, since there seems to be interaction. Most statisticians do *not* proceed to test row and column effects if H_{AB} is rejected.

The statistic for testing the hypothesis

$$H_A: \alpha_1 = \alpha_2 = \cdots = \alpha_a = 0$$

against all alternatives is

$$F_A = \frac{bc \sum_{i=1}^{a} (\overline{X}_{i\cdot\cdot} - \overline{X}_{\cdots})^2 / [\sigma^2(a-1)]}{\sum_{i=1}^{a} \sum_{j=1}^{b} \sum_{k=1}^{c} (X_{ijk} - \overline{X}_{ij\cdot})^2 / [\sigma^2 ab(c-1)]} = \frac{SS(A)/(a-1)}{SS(E)/[ab(c-1)]},$$

which has an F distribution with $a-1$ and $ab(c-1)$ degrees of freedom when H_A is true. The statistic for testing the hypothesis

$$H_B: \beta_1 = \beta_2 = \cdots = \beta_b = 0$$

against all alternatives is

$$F_B = \frac{ac \sum_{j=1}^{b} (\overline{X}_{\cdot j\cdot} - \overline{X}_{\cdots})^2 / [\sigma^2(b-1)]}{\sum_{i=1}^{a} \sum_{j=1}^{b} \sum_{k=1}^{c} (X_{ijk} - \overline{X}_{ij\cdot})^2 / [\sigma^2 ab(c-1)]} = \frac{SS(B)/(b-1)}{SS(E)/[ab(c-1)]},$$

which has an F distribution with $b-1$ and $ab(c-1)$ degrees of freedom when H_B is true. Each of these hypotheses is rejected if the observed value of F is greater than a given constant that is selected to yield the desired significance level.

On the next page, Table 7.6-4 gives the ANOVA table that summarizes the information needed for these tests of hypotheses.

EXAMPLE 7.6-2 Consider the following experiment: One hundred eight people were randomly divided into six groups with 18 people in each group. Each person was given sets of three

TABLE 7.6-4: Two-Way ANOVA Table, c Observations per Cell

Source	Sum of Squares (SS)	Degrees of Freedom	Mean Square (MS)	F
Factor A (row)	SS(A)	$a - 1$	$MS(A) = \dfrac{SS(A)}{a - 1}$	$\dfrac{MS(A)}{MS(E)}$
Factor B (column)	SS(B)	$b - 1$	$MS(B) = \dfrac{SS(B)}{b - 1}$	$\dfrac{MS(B)}{MS(E)}$
Factor AB (interaction)	SS(AB)	$(a - 1)(b - 1)$	$MS(AB) = \dfrac{SS(AB)}{(a - 1)(b - 1)}$	$\dfrac{MS(AB)}{MS(E)}$
Error	SS(E)	$ab(c - 1)$	$MS(E) = \dfrac{SS(E)}{ab(c - 1)}$	
Total	SS(TO)	$abc - 1$		

numbers to add. The three numbers were either in a "down array" or an "across array," representing the two levels of factor A. The levels of factor B are determined by the number of digits in the numbers to be added: one-digit, two-digit, or three-digit numbers. Table 7.6-5 illustrates this experiment with a sample problem for each cell; note, however, that an individual person works problems only of one of these types. Each person was placed in one of the six groups and was told to work as many problems as possible in 90 seconds. The measurement that was recorded was the average number of problems worked correctly in two trials.

TABLE 7.6-5: Illustration of Arrays for Numbers of Digits

	Number of Digits		
Type of Array	1	2	3
Down	5	25	259
	3	69	567
	8	37	130
Across	$5 + 3 + 8 =$	$25 + 69 + 37 =$	$259 + 567 + 130 =$

Whenever this many subjects are used, a computer becomes an invaluable tool. A computer program provided the summary shown in Table 7.6-6 of the sample means of the rows, the columns, and the six cells. Each cell mean is the average for 18 people.

Simply considering these means, we can see clearly that there is a column effect: It is not surprising that it is easier to add one-digit than three-digit numbers.

The most interesting feature of these results is that they show the possibility of interaction. The largest cell mean occurs for those adding one-digit numbers in an across array. Note, however, that for two- and three-digit numbers, the down arrays have larger means than the across arrays.

The computer provided the ANOVA table given in Table 7.6-7. The number of degrees of freedom for SS(E) is not in our F table in the appendix. However, the

TABLE 7.6-6: Cell, Row, and Column Means for Adding Numbers

Type of Array	Number of Digits 1	2	3	Row Means
Down	23.806	10.694	6.278	13.593
Across	26.056	6.750	3.944	12.250
Column means	24.931	8.722	5.111	

TABLE 7.6-7: ANOVA Table for Adding Numbers

Source	Sum of Squares	Degrees of Freedom	Mean Square	F	p-value
Factor A (array)	48.678	1	48.669	2.885	0.0925
Factor B (number of digits)	8022.73	2	4011.363	237.778	<0.0001
Interaction	185.92	2	92.961	5.510	0.0053
Error	1720.76	102	16.870		
Total	9978.08	107			

right-most column, obtained from the computer printout, provides the p-value of each test, namely, the probability of obtaining an F as large as or larger than the calculated F ratio. Note, for example, that, to test for interaction, $F = 5.51$ and the p-value is 0.006. Thus, the hypothesis of no interaction would be rejected at the $\alpha = 0.05$ or $\alpha = 0.01$ significance level, but it would not be rejected with $\alpha = 0.001$. ∎

EXERCISES

(In some of the exercises that follow, we must make assumptions, such as normal distributions with equal variances.)

7.6-1. For the data given in Example 7.6-1, test the hypothesis H_A: $\alpha_1 = \alpha_2 = \alpha_3 = 0$ against all alternatives at the 5% significance level.

7.6-2. With $a = 3$ and $b = 4$, find μ, α_i, and β_j if μ_{ij}, $i = 1, 2, 3$ and $j = 1, 2, 3, 4$, are given by

$$\begin{array}{cccc} 6 & 3 & 7 & 8 \\ 10 & 7 & 11 & 12 \\ 8 & 5 & 9 & 10 \end{array}$$

Note that in an "additive" model such as this one, one row (column) can be determined by adding a constant value to each of the elements of another row (column).

7.6-3. We wish to compare compressive strengths of concrete corresponding to $a = 3$ different drying methods (treatments). Concrete is mixed in batches that are just large enough to produce three cylinders. Although care is taken to achieve uniformity, we expect some variability among the $b = 5$ batches used to obtain the following compressive strengths (there is little reason to suspect interaction; hence, only one observation is taken in each cell):

Treatment	B_1	B_2	B_3	B_4	B_5
A_1	52	47	44	51	42
A_2	60	55	49	52	43
A_3	56	48	45	44	38

(a) Use the 5% significance level and test H_A: $\alpha_1 = \alpha_2 = \alpha_3 = 0$ against all alternatives.

(b) Use the 5% significance level and test H_B: $\beta_1 = \beta_2 = \beta_3 = \beta_4 = \beta_5 = 0$ against all alternatives. (See Ledolter and Hogg in references.)

7.6-4. Show that the cross-product terms formed from $(\overline{X}_{i\cdot} - \overline{X}_{\cdot\cdot})$, $(\overline{X}_{\cdot j} - \overline{X}_{\cdot\cdot})$, and $(X_{ij} - \overline{X}_{i\cdot} - \overline{X}_{\cdot j} + \overline{X}_{\cdot\cdot})$ sum to zero, $i = 1, 2, \ldots, a$ and $j = 1, 2, \ldots, b$.

HINT: For example, write

$$\sum_{i=1}^{a} \sum_{j=1}^{b} (\overline{X}_{\cdot j} - \overline{X}_{\cdot\cdot})(X_{ij} - \overline{X}_{i\cdot} - \overline{X}_{\cdot j} + \overline{X}_{\cdot\cdot})$$
$$= \sum_{j=1}^{b} (\overline{X}_{\cdot j} - \overline{X}_{\cdot\cdot}) \sum_{i=1}^{a} [(X_{ij} - \overline{X}_{\cdot j}) - (\overline{X}_{i\cdot} - \overline{X}_{\cdot\cdot})]$$

and sum each term in the inner summation, as grouped here, to get zero.

7.6-5. A psychology student was interested in testing how food consumption by rats would be affected by a particular drug. She used two levels of one attribute, namely, drug and placebo, and four levels of a second attribute, namely, male (M), castrated (C), female (F), and ovariectomized (O). For each cell, she observed five rats. The amount of food consumed in grams per 24 hours is listed in the following table:

	M	C	F	O
Drug	22.56	16.54	18.58	18.20
	25.02	24.64	15.44	14.56
	23.66	24.62	16.12	15.54
	17.22	19.06	16.88	16.82
	22.58	20.12	17.58	14.56
Placebo	25.64	22.50	17.82	19.74
	28.84	24.48	15.76	17.48
	26.00	25.52	12.96	16.46
	26.02	24.76	15.00	16.44
	23.24	20.62	19.54	15.70

(a) Use the 5% significance level and test H_{AB}: $\gamma_{ij} = 0, i = 1, 2, j = 1, 2, 3, 4$.

(b) Use the 5% significance level and test H_A: $\alpha_1 = \alpha_2 = 0$.

(c) Use the 5% significance level and test H_B: $\beta_1 = \beta_2 = \beta_3 = \beta_4 = 0$.

(d) How could you modify this model so that there are three attributes of classification, each with two levels?

7.6-6. With $a = 3$ and $b = 4$, find μ, α_i, β_j, and γ_{ij} if μ_{ij}, $i = 1, 2, 3$ and $j = 1, 2, 3, 4$ are given by

6	7	7	12
10	3	11	8
8	5	9	10

Note the difference between the layout here and that in Exercise 7.6-2. Does the interaction help explain the difference?

7.6-7. In order to test whether four brands of gasoline give equal performance in terms of mileage, each of three cars was driven with each of the four brands of gasoline. Then each of the $(3)(4) = 12$ possible combinations was repeated four times. The number of miles per gallon for each of the four repetitions in each cell is recorded in the following table:

Car	Brand of Gasoline			
	1	2	3	4
1	31.0 24.9	26.3 30.0	25.8 29.4	27.8 27.3
	26.2 28.8	25.2 31.6	24.5 24.8	28.2 30.4
2	30.6 29.5	25.5 26.8	26.6 23.7	28.1 27.1
	30.8 28.9	27.4 29.4	28.2 26.1	31.5 29.1
3	24.2 23.1	27.4 28.1	25.2 26.7	26.3 26.4
	26.8 27.4	26.4 26.9	27.7 28.1	27.9 28.8

Test the hypotheses H_{AB}: no interaction, H_A: no row effect, and H_B: no column effect, each at the 5% significance level.

7.6-8. The data in the following table could represent the outcomes of some experiment, but in actuality they were generated on a computer:

	B_1	B_2
A_1	14.552	13.980
	17.024	5.777
	10.132	7.638
	16.979	12.089
A_2	11.378	20.825
	12.578	17.333
	17.709	20.329
	11.037	25.277
A_3	12.345	22.223
	11.044	21.844
	9.458	16.176
	16.156	17.946

Recall that $\mu_{ij} = \mu + \alpha_i + \beta_j + \gamma_{ij}$ is the mean of cell (i, j), and in the simulation, we took $\mu = 15$,

$\alpha_1 = -3$, $\alpha_2 = 2$, $\alpha_3 = 1$, $\beta_1 = -1.5$, $\beta_2 = 1.5$, $\gamma_{11} = 4$, $\gamma_{12} = -4$, $\gamma_{21} = -2$, $\gamma_{22} = 2$, $\gamma_{31} = -2$, $\gamma_{32} = 2$, and $\sigma^2 = 9$. Use the data and the two-factor analysis-of-variance model with four repetitions per cell to test the hypotheses H_{AB}: no interaction, H_A: no row effect, and H_B: no column effect, each at the $\alpha = 0.05$ significance level.

7.6-9. Ledolter and Hogg (see references) report on an engineer in a textile mill who studies the effects of temperature and time on the brightness of a synthetic fabric in a process involving dye. (Brightness is measured on a 50-point scale.) Three observations were taken at each combination of temperature and time:

	Temperature		
Time (cycles)	350°F	375°F	400°F
40	38, 32, 30	37, 35, 40	36, 39, 43
50	40, 45, 36	39, 42, 46	39, 48, 47

Construct the ANOVA table, and conduct tests for the interaction first and then, if appropriate, the main effects. Use $\alpha = 0.05$ for each test.

7.6-10. There is another way of looking at Exercise 7.5-8, namely, as a two-factor analysis-of-variance problem with the levels of gender being female and male, the levels of age being less than 50 and at least 50, and the measurement for each subject being their cholesterol level. The data would then be set up as follows:

	Age	
Gender	<50	≥50
Female	221	262
	213	193
	202	224
	183	201
	185	161
	197	178
	162	265
Male	271	192
	192	253
	189	248
	209	278
	227	232
	236	267
	142	289

(a) Test H_{AB}: $\gamma_{ij} = 0$, $i = 1, 2$; $j = 1, 2$ (no interaction)

(b) Test H_A: $\alpha_1 = \alpha_2 = 0$ (no row effect).

(c) Test H_B: $\beta_1 = \beta_2 = 0$ (no column effect).

Use a 5% significance level for each test.

7.6-11. Ledolter and Hogg (see references) report that volunteers who had a smoking history classified as heavy, moderate, and nonsmoker were accepted until nine men were in each category. Three men in each category were randomly assigned to each of the following three stress tests: bicycle ergometer, treadmill, and step tests. The time until maximum oxygen uptake was recorded in minutes as follows:

	Test		
Smoking History	Bicycle	Treadmill	Step Test
Nonsmoker	12.8, 13.5, 11.2	16.2, 18.1, 17.8	22.6, 19.3, 18.9
Moderate	10.9, 11.1, 9.8	15.5, 13.8, 16.2	20.1, 21.0, 15.9
Heavy	8.7, 9.2, 7.5	14.7, 13.2, 8.1	16.2, 16.1, 17.8

(a) Analyze the results of this experiment. Obtain the ANOVA table and test for main effects and interactions.

(b) Use box plots to compare the data graphically.

7.7 TESTS CONCERNING REGRESSION AND CORRELATION

In Section 6.7, we considered the estimation of the parameters of a very simple regression curve, namely, a straight line. We can use confidence intervals for the parameters to test hypotheses about them. For example, with the same model as that in Section 6.7, we could test the hypothesis H_0: $\beta = \beta_0$ by using a t random variable that was used for a confidence interval with β replaced by β_0, namely,

$$T_1 = \frac{\widehat{\beta} - \beta_0}{\sqrt{\dfrac{n\widehat{\sigma^2}}{(n-2)\sum\limits_{i=1}^{n}(x_i - \bar{x})^2}}}.$$

The null hypothesis, along with three possible alternative hypotheses, is given in Table 7.7-1; these tests are equivalent to stating that we reject H_0 if β_0 is not in certain confidence intervals. For example, the first test is equivalent to rejecting H_0 if β_0 is not in the one-sided confidence interval with lower bound

$$\widehat{\beta} - t_\alpha(n-2)\sqrt{\dfrac{n\widehat{\sigma^2}}{(n-2)\sum\limits_{i=1}^{n}(x_i - \bar{x})^2}}.$$

Often we let $\beta_0 = 0$ and test the hypothesis H_0: $\beta = 0$. That is, we test the null hypothesis that the slope is equal to zero.

TABLE 7.7-1: Tests About the Slope of the Regression Line

H_0	H_1	Critical Region
$\beta = \beta_0$	$\beta > \beta_0$	$t_1 \geq t_\alpha(n-2)$
$\beta = \beta_0$	$\beta < \beta_0$	$t_1 \leq -t_\alpha(n-2)$
$\beta = \beta_0$	$\beta \neq \beta_0$	$|t_1| \geq t_{\alpha/2}(n-2)$

EXAMPLE 7.7-1 Let x equal a student's preliminary test score in a psychology course and y equal the same student's score on the final examination. With $n = 10$ students, we shall test H_0: $\beta = 0$ against H_1: $\beta \neq 0$. At the 0.01 significance level, the critical region is $|t_1| \geq t_{0.005}(8) = 3.355$. Using the data in Example 6.7-1, we find that the observed value of T_1 is

$$t_1 = \frac{0.742 - 0}{\sqrt{10(21.7709)/8(756.1)}} = \frac{0.742}{0.1897} = 3.911.$$

Thus, we reject H_0. ∎

We consider tests about the correlation coefficient ρ of a bivariate normal distribution. Let X and Y have a bivariate normal distribution. We know that if the correlation coefficient ρ is zero, then X and Y are independent random variables. Furthermore, the value of ρ gives a measure of the linear relationship between X and Y. We now give methods for using the sample correlation coefficient to test the hypothesis H_0: $\rho = 0$ and also to form a confidence interval for ρ.

Let $(X_1, Y_1), (X_2, Y_2), \ldots, (X_n, Y_n)$ denote a random sample from a bivariate normal distribution with parameters μ_X, μ_Y, σ_X^2, σ_Y^2, and ρ. That is, the n pairs of (X, Y) are independent, and each pair has the same bivariate normal distribution.

The **sample correlation coefficient** is

$$R = \frac{\frac{1}{n-1}\sum_{i=1}^{n}(X_i - \overline{X})(Y_i - \overline{Y})}{\sqrt{\frac{1}{n-1}\sum_{i=1}^{n}(X_i - \overline{X})^2}\sqrt{\frac{1}{n-1}\sum_{i=1}^{n}(Y_i - \overline{Y})^2}} = \frac{S_{XY}}{S_X S_Y}.$$

We note that

$$R\frac{S_Y}{S_X} = \frac{S_{XY}}{S_X^2} = \frac{\frac{1}{n-1}\sum_{i=1}^{n}(X_i - \overline{X})(Y_i - \overline{Y})}{\frac{1}{n-1}\sum_{i=1}^{n}(X_i - \overline{X})^2}$$

is exactly the solution that we obtained for $\widehat{\beta}$ in Section 6.7 when the X values were fixed at $X_1 = x_1, X_2 = x_2, \ldots, X_n = x_n$. Let us consider these values fixed temporarily so that we are considering conditional distributions, given $X_1 = x_1, \ldots, X_n = x_n$. Moreover, if $H_0\colon \rho = 0$ is true, then the distributions of $Y_1, Y_2, \ldots, Y_n$ are independent of $x_1, x_2, \ldots, x_n$ and thus $\beta = \rho\sigma_Y/\sigma_X = 0$. Under these conditions. the conditional distribution of

$$\widehat{\beta} = \frac{\sum_{i=1}^{n}(x_i - \overline{x})(Y_i - \overline{Y})}{\sum_{i=1}^{n}(x_i - \overline{x})^2}$$

is $N[0, \sigma_Y^2/(n-1)s_x^2]$ when $s_x^2 > 0$. Moreover, recall from Section 6.7 that the conditional distribution of

$$\frac{\sum_{i=1}^{n}[Y_i - \overline{Y} - (S_{XY}/s_x^2)(x_i - \overline{x})]^2}{\sigma_Y^2} = \frac{(n-1)S_Y^2(1 - R^2)}{\sigma_Y^2},$$

given that $X_1 = x_1, \ldots, X_n = x_n$, is $\chi^2(n-2)$ and is independent of $\widehat{\beta}$. (See Exercise 7.7-8.) Thus, when $\rho = 0$, the conditional distribution of

$$T = \frac{(RS_Y/s_x)/(\sigma_Y/\sqrt{n-1}\,s_x)}{\sqrt{[(n-1)S_Y^2(1-R^2)/\sigma_Y^2][1/(n-2)]}} = \frac{R\sqrt{n-2}}{\sqrt{1-R^2}}$$

is t with $n-2$ degrees of freedom. However, since the conditional distribution of T given that $X_1 = x_1, \ldots, X_n = x_n$, does not depend on $x_1, x_2, \ldots, x_n$, the unconditional distribution of T must be t with $n-2$ degrees of freedom, and T and $(X_1, X_2, \ldots, X_n)$ are independent when $\rho = 0$.

REMARK   It is interesting to note that in the discussion about the distribution of T, nothing was said about the distribution of $X_1, X_2, \ldots, X_n$. This means that if X and Y are independent and Y has a normal distribution, then T has a t distribution whatever the distribution of X. Obviously, the roles of X and Y can be reversed in all of this development. In particular, if X and Y are independent, then T and $Y_1, Y_2, \ldots, Y_n$ are also independent. ∎

Now T can be used to test H_0: $\rho = 0$. If the alternative hypothesis is H_1: $\rho > 0$, we would use the critical region defined by the observed $T \geq t_\alpha(n - 2)$, since large T implies large R. Obvious modifications would be made for the alternative hypotheses H_1: $\rho < 0$ and H_1: $\rho \neq 0$, the latter leading to a two-sided test.

Using the p.d.f. $h(t)$ of T, we can find the distribution function and p.d.f. of R when $-1 < r < 1$, provided that $\rho = 0$:

$$G(r) = P(R \leq r) = P\left(T \leq \frac{r\sqrt{n-2}}{\sqrt{1-r^2}}\right)$$

$$= \int_{-\infty}^{r\sqrt{n-2}/\sqrt{1-r^2}} h(t)\, dt$$

$$= \int_{-\infty}^{r\sqrt{n-2}/\sqrt{1-r^2}} \frac{\Gamma[(n-1)/2]}{\Gamma(1/2)\,\Gamma[(n-2)/2]} \frac{1}{\sqrt{n-2}}\left(1 + \frac{t^2}{n-2}\right)^{-(n-1)/2} dt.$$

The derivative of $G(r)$, with respect to r, is (see Appendix D.4)

$$g(r) = h\left(\frac{r\sqrt{n-2}}{\sqrt{1-r^2}}\right) \frac{d(r\sqrt{n-2}/\sqrt{1-r^2})}{dr},$$

which equals

$$g(r) = \frac{\Gamma[(n-1)/2]}{\Gamma(1/2)\,\Gamma[(n-2)/2]} (1 - r^2)^{(n-4)/2}, \qquad -1 < r < 1.$$

Thus, to test the hypothesis H_0: $\rho = 0$ against the alternative hypothesis H_1: $\rho \neq 0$ at a significance level α, select either a constant $r_{\alpha/2}(n - 2)$ or a constant $t_{\alpha/2}(n - 2)$ so that

$$\alpha = P(\,|R| \geq r_{\alpha/2}(n - 2); H_0) = P(\,|T| \geq t_{\alpha/2}(n - 2); H_0),$$

depending on the availability of R or T tables.

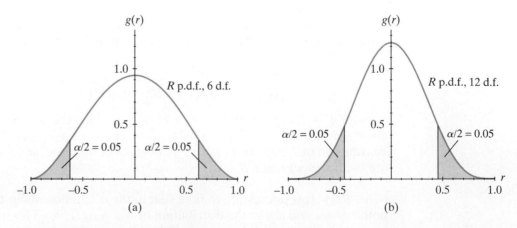

FIGURE 7.7-1: R p.d.f.'s when $n = 8$ and $n = 14$

It is interesting to graph the p.d.f. of R. Note in particular that if $n = 4$, $g(r) = 1/2$, $-1 < r < 1$, and if $n = 6$, $g(r) = (3/4)(1 - r^2)$, $-1 < r < 1$. The graphs of the p.d.f. of R when $n = 8$ and when $n = 14$ are given in Figure 7.7-1. Recall that this is the p.d.f. of R when $\rho = 0$. As n increases, R is more likely to equal values close to 0.

Table XI in the appendix lists selected values of the distribution function of R when $\rho = 0$. For example, if $n = 8$, then the number of degrees of freedom is 6 and $P(R \leq 0.7887) = 0.99$. Also, if $\alpha = 0.10$, then $r_{\alpha/2}(6) = r_{0.05}(6) = 0.6215$. [See Figure 7.7-1(a).]

It is also possible to obtain an approximate test of size α by using the fact that

$$W = \frac{1}{2} \ln \frac{1 + R}{1 - R}$$

has an approximate normal distribution with mean $(1/2) \ln[(1 + \rho)/(1 - \rho)]$ and variance $1/(n - 3)$. We accept this statement without proof. (See Exercise 7.7-10.) Thus, a test of $H_0: \rho = \rho_0$ can be based on the statistic

$$Z = \frac{\dfrac{1}{2} \ln \dfrac{1 + R}{1 - R} - \dfrac{1}{2} \ln \dfrac{1 + \rho_0}{1 - \rho_0}}{\sqrt{\dfrac{1}{n - 3}}},$$

which has a distribution that is approximately $N(0, 1)$.

EXAMPLE 7.7-2   We would like to test the hypothesis $H_0: \rho = 0$ against $H_1: \rho \neq 0$ at an $\alpha = 0.05$ significance level. A random sample of size 18 from a bivariate normal distribution yielded a sample correlation coefficient of $r = 0.35$. From Table XI in the appendix, since $0.35 < 0.4683$, H_0 is accepted (not rejected) at an $\alpha = 0.05$ significance level. Using the t distribution, we would reject H_0 if $|t| \geq 2.120 = t_{0.025}(16)$. Since

$$t = \frac{0.35 \sqrt{16}}{\sqrt{1 - (0.35)^2}} = 1.495,$$

H_0 is not rejected. If we had used the normal approximation for Z, H_0 would be rejected if $|z| \geq 1.96$. Because

$$z = \frac{(1/2) \ln[(1 + 0.35)/(1 - 0.35)] - 0}{\sqrt{1/(18 - 3)}} = 1.415,$$

H_0 is not rejected.   ∎

To develop an approximate $100(1 - \alpha)\%$ confidence interval for ρ, we use the normal approximation for the distribution of Z. Thus, we select a constant $c = z_{\alpha/2}$ from Table V in the appendix so that

$$P\left(-c \leq \frac{(1/2) \ln[(1 + R)/(1 - R)] - (1/2) \ln[(1 + \rho)/(1 - \rho)]}{\sqrt{1/(n - 3)}} \leq c\right) \approx 1 - \alpha.$$

After several algebraic manipulations, this formula becomes

$$P\left(\frac{1 + R - (1 - R) \exp(2c/\sqrt{n - 3})}{1 + R + (1 - R) \exp(2c/\sqrt{n - 3})} \leq \rho \leq\right.$$

$$\left.\frac{1 + R - (1 - R) \exp(-2c/\sqrt{n - 3})}{1 + R + (1 - R) \exp(-2c/\sqrt{n - 3})}\right) \approx 1 - \alpha.$$

EXAMPLE 7.7-3   Suppose that a random sample of size 12 from a bivariate normal distribution yielded a correlation coefficient of $r = 0.6$. An approximate 95% confidence interval for ρ would be

$$
\left[\frac{1 + 0.6 - (1 - 0.6)\exp\left(\dfrac{2(1.96)}{3}\right)}{1 + 0.6 + (1 - 0.6)\exp\left(\dfrac{2(1.96)}{3}\right)}, \frac{1 + 0.6 - (1 - 0.6)\exp\left(\dfrac{-2(1.96)}{3}\right)}{1 + 0.6 + (1 - 0.6)\exp\left(\dfrac{-2(1.96)}{3}\right)} \right]
$$

$$
= [0.040, 0.873].
$$

If the sample size had been $n = 39$ and $r = 0.6$, the approximate 95% confidence interval would have been $[0.351, 0.770]$. ∎

EXERCISES

(In some of the exercises that follow, we must make assumptions of normal distributions with the usual notation.)

7.7-1. For the data given in Exercise 6.7-1, use a t test to test $H_0: \beta = 0$ against $H_1: \beta > 0$ at the $\alpha = 0.025$ significance level.

7.7-2. For the data given in Exercise 6.7-2, use a t test to test $H_0: \beta = 0$ against $H_1: \beta > 0$ at the $\alpha = 0.025$ significance level.

7.7-3. For the data given in Exercise 6.7-3, use a t test to test $H_0: \beta = 0$ against $H_1: \beta > 0$ at the $\alpha = 0.01$ significance level.

7.7-4. For the candy data given in Exercise 6.7-11, test $H_0: \beta = 0$ against $H_1: \beta > 0$ at an $\alpha = 0.01$ significance level.

7.7-5. A random sample of size $n = 27$ from a bivariate normal distribution yielded a sample correlation coefficient of $r = -0.45$. Would the hypothesis $H_0: \rho = 0$ be rejected in favor of $H_1: \rho \neq 0$ at an $\alpha = 0.05$ significance level?

7.7-6. In bowling, it is often possible to score well in the first game and then bowl poorly in the second game, or vice versa. The following six pairs of numbers give the scores of the first and second games bowled by the same person on six consecutive Tuesday evenings:

Game 1:	170	190	200	183	187	178
Game 2:	197	178	150	176	205	153

Assume a bivariate normal distribution, and use these scores to test the hypothesis $H_0: \rho = 0$ against $H_1: \rho \neq 0$ at $\alpha = 0.10$.

7.7-7. A random sample of size 28 from a bivariate normal distribution yielded a sample correlation coefficient of $r = 0.65$. Find an approximate 90% confidence interval for ρ.

7.7-8. By squaring the binomial expression $[(Y_i - \overline{Y}) - (S_{xY}/s_x^2)(x_i - \overline{x})]$, show that

$$
\sum_{i=1}^{n} [(Y_i - \overline{Y}) - (S_{xY}/s_x^2)(x_i - \overline{x})]^2
$$

$$
= \sum_{i=1}^{n} (Y_i - \overline{Y})^2 - 2\left(\frac{S_{xY}}{s_x^2}\right)\sum_{i=1}^{n}(x_i - \overline{x})
$$

$$
(Y_i - \overline{Y}) + \frac{S_{xY}^2}{s_x^4}\sum_{i=1}^{n}(x_i - \overline{x})^2
$$

equals $(n - 1)S_Y^2(1 - R^2)$, where $X_1 = x_1$, $X_2 = x_2, \ldots, X_n = x_n$.

HINT: Replace $S_{xY} = \sum_{i=1}^{n}(x_i - \overline{x})(Y_i - \overline{Y})/(n - 1)$ by $R s_x s_Y$.

7.7-9. To help determine whether gallinules selected their mate on the basis of weight, 14 pairs of gallinules were captured and weighed. Test the null hypothesis $H_0: \rho = 0$ against a two-sided alternative at an $\alpha = 0.01$ significance level. Given that the male and female weights for the $n = 14$ pairs of birds yielded a sample correlation coefficient of $r = -0.252$, would H_0 be rejected?

7.7-10. In sampling from a bivariate normal distribution, it is true that the correlation coefficient R has an approximate normal distribution $N[\rho, (1 - \rho^2)^2/n]$ if the sample size n is large. Since, for large n, R is close to ρ, use two terms of the Taylor's expansion of $u(R)$ about ρ and determine that function $u(R)$ such that it has a variance which is (essentially) free of ρ. (The solution of this exercise explains why the transformation $(1/2)\ln[(1 + R)/(1 - R)]$ was suggested.)

7.7-11. Show that when $\rho = 0$,

(a) the points of inflection for the graph of the p.d.f. of R are at $r = \pm 1/\sqrt{n - 5}$ for $n \geq 7$;

(b) $E(R) = 0$;

(c) $\text{Var}(R) = 1/(n - 1), n \geq 3$.
   HINT: Note that $E(R^2) = E[1 - (1 - R^2)]$.

7.7-12. In a college health fitness program, let X equal the weight in kilograms of a female freshman at the beginning of the program and let Y equal her change in weight during the semester. We shall use the following data for $n = 16$ observations of (x, y) to test the null hypothesis $H_0: \rho = 0$ against a two-sided alternative hypothesis:

(61.4, −3.2) (62.9, 1.4) (58.7, 1.3) (49.3, 0.6)
(71.3, 0.2) (81.5, −2.2) (60.8, 0.9) (50.2, 0.2)
(60.3, 2.0) (54.6, 0.3) (51.1, 3.7) (53.3, 0.2)
(81.0, −0.5) (67.6, −0.8) (71.4, −0.1) (72.1, −0.1)

(a) What is the conclusion if $\alpha = 0.10$?
(b) What is the conclusion if $\alpha = 0.05$?

7.7-13. Let X and Y have a bivariate normal distribution with correlation coefficient ρ. To test $H_0: \rho = 0$ against $H_1: \rho \neq 0$, a random sample of n pairs of observations is selected. Suppose that the sample correlation coefficient is $r = 0.68$. Using a significance level of $\alpha = 0.05$, find the smallest value of the sample size n so that H_0 is rejected.

7.7-14. In Exercise 6.7-5, data are given for horsepower, the time it takes a car to go from 0 to 60, and the weight in pounds of a car, for 14 cars. Those data are repeated here:

Horsepower	0–60	Weight	Horsepower	0–60	Weight
230	8.1	3516	282	6.2	3627
225	7.8	3690	300	6.4	3892
375	4.7	2976	220	7.7	3377
322	6.6	4215	250	7.0	3625
190	8.4	3761	315	5.3	3230
150	8.4	2940	200	6.2	2657
178	7.2	2818	300	5.5	3518

(a) Let ρ be the correlation coefficient of horsepower and weight. Test $H_0: \rho = 0$ against $H_1: \rho \neq 0$.
(b) Let ρ be the correlation coefficient of horsepower and "0–60." Test $H_0: \rho = 0$ against $H_1: \rho < 0$.
(c) Let ρ be the correlation coefficient of weight and "0–60." Test $H_0: \rho = 0$ against $H_1: \rho \neq 0$.

7.7-15. The data in the following table, collected by Ledolter and Hogg (see references), for 2007 midsize model cars, provide the number of miles per gallon (MPG) for city and highway driving, as well as the curb weight of the cars:

Type	MPG City	MPG Hwy	Curb Weight
Ford Fusion V6 SE	20	28	3230
Chevrolet Sebring Sedan Base	24	32	3287
Toyota Camry Solara SE	24	34	3240
Honda Accord Sedan	20	29	3344
Audi A6 3.2	21	29	3825
BMW 5-series 525i Sedan	20	29	3450
Chrysler PT Cruiser Base	22	29	3076
Mercedes E-Class E350 Sedan	19	26	3740
Volkswagen Passat Sedan 2.0T	23	32	3305
Nissan Altima 2.5	26	35	3055
Kia Optima LX	24	34	3142

(a) Construct a scatter plot of city MPG against curb weight and find the least squares regression line.
(b) Let ρ be the correlation coefficient, and test the null hypothesis $H_0: \rho = 0$ against the alternative hypothesis $H_1: \rho < 0$. Give limits for the p-value of your test.
(c) Repeat parts (a) and (b) for highway MPG against curb weight.
(d) Construct a scatter plot of city MPG against highway MPG. What is the correlation coefficient of these two variables?

HISTORICAL COMMENTS   Most of the tests presented in this section result from the use of methods found in the theories of Neyman and Pearson (Egon, who is Karl's son). J. Neyman and E. Pearson formed a team, particularly in the 1920s and 1930s, which produced theoretical results that were important contributions to the area of testing statistical hypotheses. These results will be studied in Sections 10.1–10.3.

The other important item in this section is the analysis of variance (ANOVA). This is just the beginning of what is called the design of experiments, developed by R. A. Fisher. In our simple cases in this section, he shows how to test for the best levels of factors in the one-factor and two-factor cases. We study a few important generalizations in Section 11.3. The design of experiments was a huge contribution by Fisher.

Nonparametric Methods

8.1 CHI-SQUARE GOODNESS-OF-FIT TESTS

We now consider applications of the very important chi-square statistic, first proposed by Karl Pearson in 1900. As the reader will see, it is a very adaptable test statistic and can be used for many different types of tests. In particular, one application allows us to test the appropriateness of different probabilistic models and, in this sense, is a competitor of the Kolmogorov–Smirnov test (Section 8.7).

So that the reader can get some idea as to why Pearson first proposed his chi-square statistic, we begin with the binomial case. That is, let Y_1 be $b(n, p_1)$, where $0 < p_1 < 1$. According to the central limit theorem,

$$Z = \frac{Y_1 - np_1}{\sqrt{np_1(1 - p_1)}}$$

has a distribution that is approximately $N(0, 1)$ for large n, particularly when $np_1 \geq 5$ and $n(1 - p_1) \geq 5$. Thus, it is not surprising that $Q_1 = Z^2$ is approximately $\chi^2(1)$. If we let $Y_2 = n - Y_1$ and $p_2 = 1 - p_1$, we see that Q_1 may be written as

$$Q_1 = \frac{(Y_1 - np_1)^2}{np_1(1 - p_1)} = \frac{(Y_1 - np_1)^2}{np_1} + \frac{(Y_1 - np_1)^2}{n(1 - p_1)}.$$

Since

$$(Y_1 - np_1)^2 = (n - Y_1 - n[1 - p_1])^2 = (Y_2 - np_2)^2,$$

we have

$$Q_1 = \frac{(Y_1 - np_1)^2}{np_1} + \frac{(Y_2 - np_2)^2}{np_2}.$$

Let us now carefully consider each term in this last expression for Q_1. Of course, Y_1 is the number of "successes," and np_1 is the expected number of "successes"; that is, $E(Y_1) = np_1$. Likewise, Y_2 and np_2 are, respectively, the number and the expected number of "failures." So each numerator consists of the square of the difference of an observed number and an expected number. Note that Q_1 can be written as

$$Q_1 = \sum_{i=1}^{2} \frac{(Y_i - np_i)^2}{np_i}, \tag{8.1-1}$$

and we have seen intuitively that it has an approximate chi-square distribution with one degree of freedom. In a sense, Q_1 measure the "closeness" of the observed numbers to the corresponding expected numbers. For example, if the observed values of Y_1 and Y_2 equal their expected values, then the computed Q_1 is equal to $q_1 = 0$; but if they differ much from them, then the computed $Q_1 = q_1$ is relatively large.

To generalize, we let an experiment have k (instead of only two) mutually exclusive and exhaustive outcomes, say, $A_1, A_2, \ldots, A_k$. Let $p_i = P(A_i)$ and thus $\sum_{i=1}^{k} p_i = 1$. The experiment is repeated n independent times, and we let Y_i represent the number of times the experiment results in A_i, $i = 1, 2, \ldots, k$. This joint distribution of $Y_1, Y_2, \ldots, Y_{k-1}$ is a straightforward generalization of the binomial distribution, as follows.

In considering the joint p.m.f., we see that

$$f(y_1, y_2, \ldots, y_{k-1}) = P(Y_1 = y_1, Y_2 = y_2, \ldots, Y_{k-1} = y_{k-1}),$$

where $y_1, y_2, \ldots, y_{k-1}$ are nonnegative integers such that $y_1 + y_2 + \cdots + y_{k-1} \leq n$. Note that we do not need to consider Y_k, since, once the other $k - 1$ random variables are observed to equal $y_1, y_2, \ldots, y_{k-1}$, respectively, we know that

$$Y_k = n - y_1 - y_2 - \cdots - y_{k-1} = y_k, \quad \text{say.}$$

From the independence of the trials, the probability of each particular arrangement of $y_1 \ A_1$s, $y_2 \ A_2$s, $\ldots, y_k \ A_k$s is

$$p_1^{y_1} p_2^{y_2} \cdots p_k^{y_k}.$$

The number of such arrangements is the multinomial coefficient

$$\binom{n}{y_1, y_2, \ldots, y_k} = \frac{n!}{y_1! \, y_2! \cdots y_k!}.$$

Hence, the product of these two expressions gives the joint p.m.f. of $Y_1, Y_2, \ldots, Y_{k-1}$:

$$f(y_1, y_2, \ldots, y_{k-1}) = \frac{n!}{y_1! \, y_2! \cdots y_k!} p_1^{y_1} p_2^{y_2} \cdots p_k^{y_k}.$$

(Recall that $y_k = n - y_1 - y_2 - \cdots - y_{k-1}$.)

Pearson then constructed an expression similar to Q_1 (Equation 8.1-1), which involves Y_1 and $Y_2 = n - Y_1$, that we denote by Q_{k-1}, which involves $Y_1, Y_2, \ldots, Y_{k-1}$, and $Y_k = n - Y_1 - Y_2 - \cdots - Y_{k-1}$, namely,

$$Q_{k-1} = \sum_{i=1}^{k} \frac{(Y_i - np_i)^2}{np_i}.$$

He argued that Q_{k-1} has an approximate chi-square distribution with $k - 1$ degrees of freedom in much the same way we argued that Q_1 is approximately $\chi^2(1)$. We accept Pearson's conclusion, as the proof is beyond the level of this text.

Some writers suggest that n should be large enough so that $np_i \geq 5, i = 1, 2, \ldots, k$, to be certain that the approximating distribution is adequate. This is probably good advice for the beginner to follow, although we have seen the approximation work very well when $np_i \geq 1, i = 1, 2, \ldots, k$. The important thing to guard against is allowing some particular np_i to become so small that the corresponding term in Q_{k-1}, namely, $(Y_i - np_i)^2/np_i$, tends to dominate the others because of its small denominator. In any case, it is important to realize that Q_{k-1} has only an approximate chi-square distribution.

We shall now show how we can use the fact that Q_{k-1} is approximately $\chi^2(k-1)$ to test hypotheses about probabilities of various outcomes. Let an experiment have k mutually exclusive and exhaustive outcomes, $A_1, A_2, \ldots, A_k$. We would like to test whether $p_i = P(A_i)$ is equal to a known number $p_{i0}, i = 1, 2, \ldots, k$. That is, we shall test the hypothesis

$$H_0: p_i = p_{i0}, \qquad i = 1, 2, \ldots, k.$$

In order to test such a hypothesis, we shall take a sample of size n; that is, we repeat the experiment n independent times. We tend to favor H_0 if the observed number of times that A_i occurred, say, y_i, and the number of times A_i was expected to occur if H_0 were true, namely, np_{i0}, are approximately equal. That is, if

$$q_{k-1} = \sum_{i=1}^{k} \frac{(y_i - np_{i0})^2}{np_{i0}}$$

is "small," we tend to favor H_0. Since the distribution of Q_{k-1} is approximately $\chi^2(k-1)$, we shall reject H_0 if $q_{k-1} \geq \chi^2_\alpha(k-1)$, where α is the desired significance level of the test.

EXAMPLE 8.1-1   If persons are asked to record a string of random digits, such as

$$3 \quad 7 \quad 2 \quad 4 \quad 1 \quad 9 \quad 7 \quad 2 \quad 1 \quad 5 \quad 0 \quad 8 \ldots,$$

we usually find that they are reluctant to record the same or even the two closest numbers in adjacent positions. And yet, in true random-digit generation, the probability of the next digit being the same as the preceding one is $p_{10} = 1/10$, the probability of the next being only one away from the preceding (assuming that 0 is one away from 9) is $p_{20} = 2/10$, and the probability of all other possibilities is $p_{30} = 7/10$. We shall test one person's concept of a random sequence by asking her to record a string of 51 digits that seems to represent a random-digit generation. Thus, we shall test

$$H_0: p_1 = p_{10} = \frac{1}{10}, \quad p_2 = p_{20} = \frac{2}{10}, \quad p_3 = p_{30} = \frac{7}{10}.$$

The critical region for an $\alpha = 0.05$ significance level is $q_2 \geq \chi^2_{0.05}(2) = 5.991$. The sequence of digits was as follows:

$$
\begin{array}{cccccccccccccc}
5 & 8 & 3 & 1 & 9 & 4 & 6 & 7 & 9 & 2 & 6 & 3 & 0 \\
8 & 7 & 5 & 1 & 3 & 6 & 2 & 1 & 9 & 5 & 4 & 8 & 0 \\
3 & 7 & 1 & 4 & 6 & 0 & 4 & 3 & 8 & 2 & 7 & 3 & 9 \\
8 & 5 & 6 & 1 & 8 & 7 & 0 & 3 & 5 & 2 & 5 & 2 &
\end{array}
$$

We went through this listing and observed how many times the next digit was the same as or was one away from the preceding one:

	Frequency	Expected Number
Same	0	$50(1/10) = 5$
One away	8	$50(2/10) = 10$
Other	42	$50(7/10) = 35$
Total	50	50

The computed chi-square statistic is

$$
\frac{(0 - 5)^2}{5} + \frac{(8 - 10)^2}{10} + \frac{(42 - 35)^2}{35} = 6.8 > 5.991 = \chi^2_{0.05}(2).
$$

Thus, we would say that this string of 51 digits does not seem to be random. ∎

One major disadvantage in the use of the chi-square test is that it is a many-sided test. That is, the alternative hypothesis is very general, and it would be difficult to restrict alternatives to situations such as H_1: $p_1 > p_{10}$, $p_2 > p_{20}$, $p_3 < p_{30}$ (with $k = 3$). As a matter of fact, some statisticians would probably test H_0 against this particular alternative H_1 by using a linear function of Y_1, Y_2, and Y_3. However, that sort of discussion is beyond the scope of the book because it involves knowing more about the distributions of linear functions of the dependent random variables Y_1, Y_2, and Y_3. In any case, the student who truly recognizes that this chi-square statistic tests H_0: $p_i = p_{i0}$, $i = 1, 2, \ldots, k$, against all alternatives can usually appreciate the fact that it is more difficult to reject H_0 at a given significance level α when the chi-square statistic is used than it would be if some appropriate "one-sided" test statistic were available.

Many experiments yield a set of data, say, $x_1, x_2, \ldots, x_n$, and the experimenter is often interested in determining whether these data can be treated as the observed values of a random sample $X_1, X_2, \ldots, X_n$ from a given distribution. That is, would this proposed distribution be a reasonable probabilistic model for these sample items? To see how the chi-square test can help us answer questions of this sort, consider a very simple example.

EXAMPLE 8.1-2   Let X denote the number of heads that occur when four coins are tossed at random. Under the assumption that the four coins are independent and the probability of heads on each coin is $1/2$, X is $b(4, 1/2)$. One hundred repetitions of this experiment resulted in 0, 1, 2, 3, and 4 heads being observed on 7, 18, 40, 31, and 4 trials, respectively. Do these results support the assumptions? That is, is $b(4, 1/2)$ a reasonable model for the distribution of X? To answer this, we begin by letting $A_1 = \{0\}$, $A_2 = \{1\}$, $A_3 = \{2\}$, $A_4 = \{3\}$, and $A_5 = \{4\}$. If $p_{i0} = P(X \in A_i)$ when X is $b(4, 1/2)$, then

$$p_{10} = p_{50} = \binom{4}{0}\left(\frac{1}{2}\right)^4 = \frac{1}{16} = 0.0625,$$

$$p_{20} = p_{40} = \binom{4}{1}\left(\frac{1}{2}\right)^4 = \frac{4}{16} = 0.25,$$

$$p_{30} = \binom{4}{2}\left(\frac{1}{2}\right)^4 = \frac{6}{16} = 0.375.$$

At an approximate $\alpha = 0.05$ significance level, the null hypothesis

$$H_0: p_i = p_{i0}, \qquad i = 1, 2, \ldots, 5,$$

is rejected if the observed value of Q_4 is greater than $\chi^2_{0.05}(4) = 9.488$. If we use the 100 repetitions of this experiment that resulted in the observed values $y_1 = 7$, $y_2 = 18$, $y_3 = 40$, $y_4 = 31$, and $y_5 = 4$, of $Y_1, Y_2, \ldots, Y_5$, respectively, then the computed value of Q_4 is

$$q_4 = \frac{(7 - 6.25)^2}{6.25} + \frac{(18 - 25)^2}{25} + \frac{(40 - 37.5)^2}{37.5} + \frac{(31 - 25)^2}{25} + \frac{(4 - 6.25)^2}{6.25}$$

$$= 4.47.$$

Since $4.47 < 9.488$, the hypothesis is not rejected. That is, the data support the hypothesis that $b(4, 1/2)$ is a reasonable probabilistic model for X. Recall that the mean of a chi-square random variable is its number of degrees of freedom. In this example, the mean is 4 and the observed value of Q_4 is 4.47, just a little greater than the mean. ∎

Thus far, all the hypotheses H_0 tested with the chi-square statistic Q_{k-1} have been simple ones (i.e., completely specified—namely, in $H_0: p_i = p_{i0}$, $i = 1, 2, \ldots, k$, each p_{i0} has been known). This is not always the case, and it frequently happens that $p_{10}, p_{20}, \ldots, p_{k0}$ are functions of one or more unknown parameters. For example, suppose that the hypothesized model for X in Example 8.1-2 was $H_0: X$ is $b(4, p)$, $0 < p < 1$. Then

$$p_{i0} = P(X \in A_i) = \frac{4!}{(i-1)!(5-i)!}p^{i-1}(1-p)^{5-i}, \qquad i = 1, 2, \ldots, 5,$$

which is a function of the unknown parameter p. Of course, if $H_0: p_i = p_{i0}$, $i = 1, 2, \ldots, 5$, is true, then, for large n,

$$Q_4 = \sum_{i=1}^{5} \frac{(Y_i - np_{i0})^2}{np_{i0}}$$

still has an approximate chi-square distribution with four degrees of freedom. The difficulty is that when $Y_1, Y_2, \ldots, Y_5$ are observed to be equal to $y_1, y_2, \ldots, y_5$, Q_4 cannot be computed, since $p_{10}, p_{20}, \ldots, p_{50}$ (and hence Q_4) are functions of the unknown parameter p.

One way out of the difficulty would be to estimate p from the data and then carry out the computations with the use of this estimate. It is interesting to note the following: Say the estimation of p is carried out by minimizing Q_4 with respect to p, yielding $\widetilde{p}$. This $\widetilde{p}$ is sometimes called a **minimum chi-square estimator** of p. If, then, this $\widetilde{p}$ is used in Q_4, the statistic Q_4 still has an approximate chi-square distribution, but with only $4 - 1 = 3$ degrees of freedom. That is, the number of degrees of freedom of the approximating chi-square distribution is reduced by one for each parameter estimated by the minimum chi-square technique. We accept this result without proof (as it is a rather difficult one). Although we have considered it when p_{i0}, $i = 1, 2, \ldots, k$, is a function of only one parameter, it holds when there is more than one unknown parameter, say, d. Hence, in a more general situation, the test would be completed by computing Q_{k-1}, using Y_i and the estimated p_{i0}, $i = 1, 2, \ldots, k$, to obtain q_{k-1} (i.e., q_{k-1} is the minimized chi-square). This value q_{k-1} would then be compared with a critical value $\chi_\alpha^2(k - 1 - d)$. In our special case, the computed (minimized) chi-square q_4 would be compared with $\chi_\alpha^2(3)$.

There is still one trouble with all of this: It is usually very difficult to find minimum chi-square estimators. Hence, most statisticians usually use some reasonable method of estimating the parameters. (Maximum likelihood is satisfactory.) They then compute q_{k-1}, recognizing that it is somewhat larger than the minimized chi-square, and compare it with $\chi_\alpha^2(k - 1 - d)$. Note that this approach provides a slightly larger probability of rejecting H_0 than would the scheme in which the minimized chi-square were used because the computed q_{k-1} is larger than the minimum q_{k-1}.

EXAMPLE 8.1-3   Let X denote the number of alpha particles emitted by barium-133 in one-tenth of a second. The following 50 observations of X were taken with a Geiger counter in a fixed position:

$$
\begin{array}{cccccccccc}
7 & 4 & 3 & 6 & 4 & 4 & 5 & 3 & 5 & 3 \\
5 & 5 & 3 & 2 & 5 & 4 & 3 & 3 & 7 & 6 \\
6 & 4 & 3 & 11 & 9 & 6 & 7 & 4 & 5 & 4 \\
7 & 3 & 2 & 8 & 6 & 7 & 4 & 1 & 9 & 8 \\
4 & 8 & 9 & 3 & 9 & 7 & 7 & 9 & 3 & 10
\end{array}
$$

The experimenter is interested in determining whether X has a Poisson distribution. To test H_0: X is Poisson, we first estimate the mean of X—say, λ—with the sample mean, $\bar{x} = 5.4$, of these 50 observations. We then partition the set of outcomes for this experiment into the sets $A_1 = \{0, 1, 2, 3\}$, $A_2 = \{4\}$, $A_3 = \{5\}$, $A_4 = \{6\}$, $A_5 = \{7\}$, and $A_6 = \{8, 9, 10, \ldots\}$. (Note that we combined $\{0, 1, 2, 3\}$ into one set A_1 and $\{8, 9, 10, \ldots\}$ into another A_6 so that the expected number of outcomes for each set would be at least five when H_0 is true.) In Table 8.1-1, the data are grouped and the estimated probabilities specified by the hypothesis that X has a Poisson distribution with an estimated $\widehat{\lambda} = \bar{x} = 5.4$ are given. Since one parameter was estimated, Q_{6-1} has an approximate chi-square distribution with $r = 6 - 1 - 1 = 4$ degrees of freedom. Also, since

$$q_5 = \frac{[13 - 50(0.213)]^2}{50(0.213)} + \cdots + \frac{[10 - 50(0.178)]^2}{50(0.178)}$$

$$= 2.763 < 9.488 = \chi_{0.05}^2(4),$$

H_0 is not rejected at the 5% significance level. That is, with only these data, we are quite willing to accept the model that X has a Poisson distribution.   ∎

TABLE 8.1-1: Grouped Geiger Counter Data					
	Outcome				
	A_1	A_2	A_3	A_4	A_5	A_6
Frequency	13	9	6	5	7	10
Probability	0.213	0.160	0.173	0.156	0.120	0.178
Expected $(50p_i)$	10.65	8.00	8.65	7.80	6.00	8.90

Let us now consider the problem of testing a model for the distribution of a random variable W of the continuous type. That is, if $F(w)$ is the distribution function of W, we wish to test

$$H_0: F(w) = F_0(w),$$

where $F_0(w)$ is some known distribution function of the continuous type. In Section 8.7, we will use a Kolmogorov–Smirnov test for H_0. Also, recall that we have considered problems of this type in which we used q–q plots. In order to use the chi-square statistic, we must partition the set of possible values of W into k sets. One way this can be done is as follows: Partition the interval $[0, 1]$ into k sets with the points $b_0, b_1, b_2, \ldots, b_k$, where

$$0 = b_0 < b_1 < b_2 < \cdots < b_k - 1.$$

Let $u_i = F_0^{-1}(b_i)$, $i = 1, 2, \ldots, k - 1$; $A_1 = (-\infty, a_1]$, $A_i = (a_{i-1}, a_i]$ for $i = 2, 3, \ldots, k - 1$, and $A_k = (a_{k-1}, \infty)$; and $p_i = P(W \in A_i)$, $i = 1, 2, \ldots, k$. Let Y_i denote the number of times the observed value of W belongs to A_i, $i = 1, 2, \ldots, k$, in n independent repetitions of the experiment. Then $Y_1, Y_2, \ldots, Y_k$ have a multinomial distribution with parameters $n, p_1, p_2, \ldots, p_{k-1}$. Also, let $p_{i0} = P(W \in A_i)$ when the distribution function of W is $F_0(w)$. The hypothesis that we actually test is a modification of H_0, namely,

$$H_0': p_i = p_{i0}, \qquad i = 1, 2, \ldots, k.$$

This hypothesis is rejected if the observed value of the chi-square statistic

$$Q_{k-1} = \sum_{i=1}^{k} \frac{(Y_i - np_{i0})^2}{np_{i0}}$$

is at least as great as $\chi_\alpha^2(k - 1)$. If the hypothesis $H_0': p_i = p_{i0}$, $i = 1, 2, \ldots, k$, is not rejected, we do not reject the hypothesis $H_0: F(w) = F_0(w)$.

EXAMPLE 8.1-4   Example 3.3-3 gives 105 observations of the times between calls to 911. Also given is a histogram of these data, with the exponential p.d.f. with $\theta = 20$ superimposed. We shall now use a chi-square goodness-of-fit test to see whether or not this is an appropriate model for the data. That is, if X is equal to the time between calls to 911, we shall test the null hypothesis that the distribution of X is exponential with a mean of $\theta = 20$. Table 8.1-2 groups the data into nine classes and gives the probabilities and expected values of these classes. Using the frequencies and expected values, the chi-square goodness-of-fit statistic is

$$q_8 = \frac{(41 - 38.0520)^2}{38.0520} + \frac{(22 - 24.2655)^2}{24.2655} + \cdots + \frac{(2 - 2.8665)^2}{2.8665} = 4.6861.$$

TABLE 8.1-2: Summary of Times between Calls to 911			
Class	Frequency	Probability	Expected
$A_1 = [0, 9]$	41	0.3624	38.0520
$A_2 = (9, 18]$	22	0.2311	24.2655
$A_3 = (18, 27]$	11	0.1473	15.4665
$A_4 = (27, 36]$	10	0.0939	9.8595
$A_5 = (36, 45]$	9	0.0599	6.2895
$A_6 = (45, 54]$	5	0.0382	4.0110
$A_7 = (54, 63]$	2	0.0244	2.5620
$A_8 = (63, 72]$	3	0.0155	1.6275
$A_9 = (72, \infty)$	2	0.0273	2.8665

The p-value associated with this test is 0.7905, which means that it is an extremely good fit.

Note that we assumed that we knew $\theta = 20$. We could also have run this test letting $\theta = \bar{x}$, remembering that we then lose one degree of freedom. For this example, the outcome would be about the same. ∎

It is also true, in dealing with models of random variables of the continuous type, that we must frequently estimate unknown parameters. For example, let H_0 be that W is $N(\mu, \sigma^2)$, where μ and σ^2 are unknown. With a random sample $W_1, W_2, \ldots, W_n$, we first can estimate μ and σ^2, possibly with $\bar{w}$ and s_w^2. We partition the space $\{w: -\infty < w < \infty\}$ into k mutually disjoint sets $A_1, A_2, \ldots, A_k$. We then use the estimates of μ and σ^2—say, $\bar{w}$ and $s^2 = s_w^2$, respectively, to estimate

$$p_{i0} = \int_{A_i} \frac{1}{s\sqrt{2\pi}} \exp\left[-\frac{(w - \bar{w})^2}{2s^2}\right] dw,$$

$i = 1, 2, \ldots, k$. Using the observed frequencies $y_1, y_2, \ldots, y_k$ of $A_1, A_2, \ldots, A_k$, respectively, from the observed random sample $w_1, w_2, \ldots, w_n$, and $\widehat{p}_{10}, \widehat{p}_{20}, \ldots, \widehat{p}_{k0}$ estimated with $\bar{w}$ and $s^2 = s_w^2$, we compare the computed

$$q_{k-1} = \sum_{i=1}^{k} \frac{(y_i - n\widehat{p}_{i0})^2}{n\widehat{p}_{i0}}$$

with $\chi_\alpha^2(k - 1 - 2)$. This value q_{k-1} will again be somewhat larger than that which would be found using minimum chi-square estimation, and certain caution should be observed. Several exercises illustrate the procedure, in which one or more parameters must be estimated. Finally, note that the methods given in this section frequently are classified under the more general title of goodness-of-fit tests. In particular, then, the tests in this section would be **chi-square goodness-of-fit tests**.

EXERCISES

8.1-1. A 1-pound bag of candy-coated chocolate-covered peanuts contained 224 pieces of candy, each colored brown, orange, green, or yellow. Test the null hypothesis that the machine filling these bags treats the four colors of candy equally likely; that is, test

$$H_0: p_B = p_O = p_G = p_Y = \frac{1}{4}.$$

The observed values were 42 brown, 64 orange, 53 green, and 65 yellow candies. You may select the significance level or give an approximate p-value.

8.1-2. A particular brand of candy-coated chocolate comes in five different colors that we shall denote as $A_1 = \{\text{brown}\}$, $A_2 = \{\text{yellow}\}$, $A_3 = \{\text{orange}\}$, $A_4 = \{\text{green}\}$, and $A_5 = \{\text{coffee}\}$. Let p_i equal the probability that the color of a piece of candy selected at random belongs to A_i, $i = 1, 2, \ldots, 5$. Test the null hypothesis

$$H_0: p_1 = 0.4, \ p_2 = 0.2, \ p_3 = 0.2,$$
$$p_4 = 0.1, \ p_5 = 0.1,$$

using a random sample of $n = 580$ pieces of candy whose colors yielded the respective frequencies 224, 119, 130, 48, and 59. You may select the significance level or give an approximate p-value.

8.1-3. In the Michigan Daily Lottery, each weekday a three-digit integer is generated one digit at a time. Let p_i denote the probability of generating digit i, $i = 0, 1, \ldots, 9$. Let $\alpha = 0.05$, and use the following 50 digits to test $H_0: p_0 = p_1 = \cdots = p_9 = 1/10$:

1	6	9	9	3	8	5	0	6	7
4	7	5	9	4	6	5	6	4	4
4	8	0	9	3	2	1	5	4	5
7	3	2	1	4	6	7	1	3	4
4	8	8	6	1	6	1	2	8	8

8.1-4. In a biology laboratory, students use corn to test the Mendelian theory of inheritance. The theory claims that frequencies of the four categories "smooth and yellow," "wrinkled and yellow," "smooth and purple," and "wrinkled and purple" will occur in the ratio 9:3:3:1. If a student counted 124, 30, 43, and 11, respectively, for these four categories, would these data support the Mendelian theory? Let $\alpha = 0.05$.

8.1-5. Let X equal the number of female children in a three-child family. We shall use a chi-square goodness-of-fit statistic to test the null hypothesis that the distribution of X is $b(3, 0.5)$.

(a) Define the test statistic and critical region, using an $\alpha = 0.05$ significance level.

(b) Among students who were taking statistics, 52 came from families with three children. For these families, $x = 0, 1, 2$, and 3 for 5, 17, 24, and 6 families, respectively. Calculate the value of the test statistic and state your

conclusion, considering how the sample was selected.

8.1-6. It has been claimed that, for a penny minted in 1999 or earlier, the probability of observing heads upon spinning the penny is $p = 0.30$. Three students got together, and they would each spin a penny and record the number X of heads out of the three spins. They repeated this experiment $n = 200$ times, observing 0, 1, 2, and 3 heads 57, 95, 38, and 10 times, respectively. Use these data to test the hypotheses that X is $b(3, 0.30)$. Give limits for the p-value of this test. In addition, out of the 600 spins, calculate the number of heads occurring and then a 95% confidence interval for p.

8.1-7. In Example 2.6-3, it is claimed that X has a Poisson distribution, and 100 observations of X are given in Table 2.6-1. Use a chi-square goodness-of-fit statistic to test whether this claim is true. Since $\bar{x} = 5.59$, let the estimate of λ be 5.6 so that you can use the Poisson probability table in the appendix. Let $A_1 = \{x \le 1\}$ and $A_{11} = \{x \ge 11\}$. Use an $\alpha = 0.05$ significance level.

8.1-8. In Exercise 2.6-12, we are asked whether the data look like observations of a Poisson random variable with mean $\lambda = 3$. Use $\alpha = 0.05$ and a chi-square goodness-of-fit test to answer this question.

8.1-9. While testing a used tape for bad records, a computer operator counted the number of flaws per 100 feet of tape. Let X equal this random variable. Test the null hypothesis that X has a Poisson distribution with a mean of $\lambda = 2.4$, given that 40 observations of X yielded 5 zeros, 7 ones, 12 twos, 9 threes, 5 fours, 1 five, and 1 six. Let $\alpha = 0.05$.

HINT: Combine the last two sets into one; that is, the last set would be all x values ≥ 5.

8.1-10. Let X equal the distance between bad records on a used computer tape. Letting $\alpha = 0.05$ and taking $\bar{x} = 42.2$ as an estimate of θ, use the following 90 observations of X and 10 classes of equal probability to test the hypothesis that the distribution of X is exponential:

30	79	38	47	22	52	36	36	7	57
3	22	30	14	8	32	15	21	12	12
6	67	6	7	35	78	28	74	5	9
37	1	3	3	44	160	50	27	61	15
39	44	130	18	6	1	32	116	23	12
58	101	68	53	58	21	21	7	79	41
80	33	71	81	17	10	13	49	21	56
107	21	17	64	14	36	26	1	54	207
64	238	25	51	82	8	2	3	43	87

8.1-11. A sample of 100 2.2K-ohm resistors was tested, yielding the following measurements:

$$
\begin{array}{cccccccccc}
2.17 & 2.18 & 2.17 & 2.19 & 2.23 & 2.18 & 2.16 & 2.23 & 2.29 & 2.20 \\
2.24 & 2.13 & 2.18 & 2.22 & 2.21 & 2.22 & 2.23 & 2.18 & 2.23 & 2.25 \\
2.24 & 2.20 & 2.21 & 2.20 & 2.25 & 2.15 & 2.20 & 2.25 & 2.16 & 2.19 \\
2.20 & 2.18 & 2.18 & 2.19 & 2.22 & 2.19 & 2.20 & 2.19 & 2.22 & 2.21 \\
2.22 & 2.18 & 2.18 & 2.28 & 2.19 & 2.23 & 2.21 & 2.19 & 2.21 & 2.21 \\
2.19 & 2.18 & 2.21 & 2.17 & 2.20 & 2.18 & 2.18 & 2.21 & 2.22 & 2.18 \\
2.22 & 2.23 & 2.19 & 2.23 & 2.18 & 2.19 & 2.18 & 2.21 & 2.18 & 2.15 \\
2.20 & 2.23 & 2.20 & 2.20 & 2.23 & 2.20 & 2.19 & 2.22 & 2.17 & 2.20 \\
2.20 & 2.17 & 2.19 & 2.19 & 2.25 & 2.19 & 2.19 & 2.20 & 2.20 & 2.19 \\
2.21 & 2.14 & 2.24 & 2.21 & 2.19 & 2.23 & 2.18 & 2.22 & 2.23 & 2.19
\end{array}
$$

(a) Group these data into nine classes of equal length, with the class boundaries for the first class being 2.125–2.145.

(b) Use a chi-square goodness-of-fit statistic to test whether the data constitute observations of a normally distributed random variable, after grouping the first two classes and the last two classes to avoid small expected numbers. Estimate μ and σ with the mean and standard deviation, respectively, of the grouped data. Let $\alpha = 0.05$.

8.1-12. Let X equal the amount of butterfat (in pounds) produced by 90 cows during a 305-day milk production period following the birth of their first calf. Test the hypothesis that the distribution of X is $N(\mu, \sigma^2)$, using $k = 10$ classes of equal probability. You may take $\overline{x} = 511.633$ and $s_x = 87.576$ as estimates of μ and σ, respectively. The data are as follows:

$$
\begin{array}{cccccccccc}
486 & 537 & 513 & 583 & 453 & 510 & 570 & 500 & 458 & 555 \\
618 & 327 & 350 & 643 & 500 & 497 & 421 & 505 & 637 & 599 \\
392 & 574 & 492 & 635 & 460 & 696 & 593 & 422 & 499 & 524 \\
539 & 339 & 472 & 427 & 532 & 470 & 417 & 437 & 388 & 481 \\
537 & 489 & 418 & 434 & 466 & 464 & 544 & 475 & 608 & 444 \\
573 & 611 & 586 & 613 & 645 & 540 & 494 & 532 & 691 & 478 \\
513 & 583 & 457 & 612 & 628 & 516 & 452 & 501 & 453 & 643 \\
541 & 439 & 627 & 619 & 617 & 394 & 607 & 502 & 395 & 470 \\
531 & 526 & 496 & 561 & 491 & 380 & 345 & 274 & 672 & 509
\end{array}
$$

8.1-13. A rare type of heredity change causes the bacterium *E. coli* to become resistant to the drug streptomycin. This type of change, called *mutation*, can be detected by plating many bacteria on petri dishes containing an antibiotic medium. Any colonies that grow on this medium result from a single mutant cell. A sample of $n = 150$ petri dishes of streptomycin agar were each plated with 10^6 bacteria, and the numbers of colonies were counted on each dish. The observed results were that 92 dishes had 0 colonies, 46 had 1, 8 had 2, 3 had 3, and 1 dish had 4 colonies. Let X equal the number of colonies per dish. Test the hypothesis that X has a Poisson distribution. Use $\overline{x} = 0.5$ as an estimate of λ. Let $\alpha = 0.01$.

8.1-14. In Exercise 3.4-10, data are given for the distances that snails travel. These distances are repeated here for your convenience:

Distances for Infected Snail Group (ordered):

$$
\begin{array}{ccccccccccccc}
263 & 238 & 226 & 220 & 170 & 155 & 139 & 123 & 119 & 107 & 107 & 97 & 90 \\
90 & 90 & 79 & 75 & 74 & 71 & 66 & 60 & 55 & 47 & 47 & 47 & 45 \\
43 & 41 & 40 & 39 & 38 & 38 & 35 & 32 & 32 & 28 & 19 & 10 & 10
\end{array}
$$

Distances for Control Snail Group (ordered):

$$
\begin{array}{ccccccccccc}
314 & 300 & 274 & 246 & 190 & 186 & 185 & 182 & 180 & 141 & 132 \\
129 & 110 & 100 & 95 & 95 & 93 & 83 & 55 & 52 & 50 & 48 \\
48 & 44 & 40 & 32 & 30 & 25 & 24 & 18 & 7 & &
\end{array}
$$

(a) For the control snail group, test the hypothesis that the distances come from an exponential distribution. Use $\overline{x}$ as an estimate of θ. Group the data into 5 or 10 classes, with equal probabilities for each class. Thus, the expected values will be either 6.2 or 3.1, respectively.

(b) For the infected snail group, test the hypothesis that the distances come from a gamma distribution with $\alpha = 2$ and $\theta = 42$. Use 10 classes with equal probabilities so that the expected value of each class is 3.9. Use Minitab or some other computer program to calculate the boundaries of the classes.

8.1-15. In Exercise 3.1-8, data are given for the melting points for 50 metal alloy filaments. Here the data are repeated:

320	326	325	318	322	320	329	317	316	331
320	320	317	329	316	308	321	319	322	335
318	313	327	314	329	323	327	323	324	314
308	305	328	330	322	310	324	314	312	318
313	320	324	311	317	325	328	319	310	324

Test the hypothesis that these are observations of a normally distributed random variable. Note that you must estimate two parameters: μ and σ.

8.1-16. In Exercise 3.1-4, the weights of 100 laptop computers are given. Use 10 classes of equal probability to test the hypothesis that these weights come from a normal distribution. Note that you must estimate two parameters. Let $\alpha = 0.05$.

8.2 CONTINGENCY TABLES

In this section, we demonstrate the flexibility of the chi-square test. We first look at a method for testing whether two or more multinomial distributions are equal, sometimes called a *test for homogeneity*. Then we consider a *test for independence of attributes of classification*. Both of these lead to a similar test statistic.

Suppose that each of two independent experiments can end in one of the k mutually exclusive and exhaustive events $A_1, A_2, \ldots, A_k$. Let

$$p_{ij} = P(A_i), \qquad i = 1, 2, \ldots, k, \qquad j = 1, 2.$$

That is, $p_{11}, p_{21}, \ldots, p_{k1}$ are the probabilities of the events in the first experiment, and $p_{12}, p_{22}, \ldots, p_{k2}$ are those associated with the second experiment. Let the experiments be repeated n_1 and n_2 independent times, respectively. Also, let $Y_{11}, Y_{21}, \ldots, Y_{k1}$ be the frequencies of $A_1, A_2, \ldots, A_k$ associated with the n_1 independent trials of the first experiment. Similarly, let $Y_{12}, Y_{22}, \ldots, Y_{k2}$ be the respective frequencies associated with the n_2 trials of the second experiment. Of course, $\sum_{i=1}^{k} Y_{ij} = n_j, j = 1, 2$. From the sampling distribution theory corresponding to the basic chi-square test, we know that each of

$$\sum_{i=1}^{k} \frac{(Y_{ij} - n_j p_{ij})^2}{n_j p_{ij}}, \qquad j = 1, 2,$$

has an approximate chi-square distribution with $k - 1$ degrees of freedom. Since the two experiments are independent (and thus the two chi-square statistics are independent), the sum

$$\sum_{j=1}^{2} \sum_{i=1}^{k} \frac{(Y_{ij} - n_j p_{ij})^2}{n_j p_{ij}}$$

is approximately chi-square with $k - 1 + k - 1 = 2k - 2$ degrees of freedom.

Usually the $p_{ij}, i = 1, 2, \ldots, k, j = 1, 2$, are unknown, and frequently we wish to test the hypothesis

$$H_0: p_{11} = p_{12}, p_{21} = p_{22}, \ldots, p_{k1} = p_{k2};$$

that is, H_0 is the hypothesis that the corresponding probabilities associated with the two independent experiments are equal. Under H_0, we can estimate the unknown

$$p_{i1} = p_{i2}, \qquad i = 1, 2, \ldots, k,$$

by using the relative frequency $(Y_{i1} + Y_{i2})/(n_1 + n_2)$, $i = 1, 2, \ldots, k$. That is, if H_0 is true, we can say that the two experiments are actually parts of a larger one in which $Y_{i1} + Y_{i2}$ is the frequency of the event A_i, $i = 1, 2, \ldots, k$. Note that we have to estimate only the $k - 1$ probabilities $p_{i1} = p_{i2}$, using

$$\frac{Y_{i1} + Y_{i2}}{n_1 + n_2}, \qquad i = 1, 2, \ldots, k - 1,$$

since the sum of the k probabilities must equal 1. That is, the estimator of $p_{k1} = p_{k2}$ is

$$1 - \frac{Y_{11} + Y_{12}}{n_1 + n_2} - \cdots - \frac{Y_{k-1,1} + Y_{k-1,2}}{n_1 + n_2} = \frac{Y_{k1} + Y_{k2}}{n_1 + n_2}.$$

Substituting these estimators, we find that

$$Q = \sum_{j=1}^{2} \sum_{i=1}^{k} \frac{[Y_{ij} - n_j(Y_{i1} + Y_{i2})/(n_1 + n_2)]^2}{n_j(Y_{i1} + Y_{i2})/(n_1 + n_2)}$$

has an approximate chi-square distribution with $2k - 2 - (k - 1) = k - 1$ degrees of freedom. Here $k - 1$ is subtracted from $2k - 2$, because that is the number of estimated parameters. The critical region for testing H_0 is of the form

$$q \geq \chi_\alpha^2(k - 1).$$

EXAMPLE 8.2-1 To test two methods of instruction, 50 students are selected at random from each of two groups. At the end of the instruction period, each student is assigned a grade (A, B, C, D, or F) by an evaluating team. The data are recorded as follows:

	Grade					
	A	B	C	D	F	Totals
Group I	8	13	16	10	3	50
Group II	4	9	14	16	7	50

Accordingly, if the hypothesis H_0 that the corresponding probabilities are equal is true, then the respective estimates of the probabilities are

$$\frac{8 + 4}{100} = 0.12,\ 0.22,\ 0.30,\ 0.26,\ \frac{3 + 7}{100} = 0.10.$$

Thus, the estimates of $n_1 p_{i1} = n_2 p_{i2}$ are 6, 11, 15, 13, and 5, respectively. Hence, the computed value of Q is

$$q = \frac{(8 - 6)^2}{6} + \frac{(13 - 11)^2}{11} + \frac{(16 - 15)^2}{15} + \frac{(10 - 13)^2}{13} + \frac{(3 - 5)^2}{5}$$

$$+ \frac{(4 - 6)^2}{6} + \frac{(9 - 11)^2}{11} + \frac{(14 - 15)^2}{15} + \frac{(16 - 13)^2}{13} + \frac{(7 - 5)^2}{5}$$

$$= \frac{4}{6} + \frac{4}{11} + \frac{1}{15} + \frac{9}{13} + \frac{4}{5} + \frac{4}{6} + \frac{4}{11} + \frac{1}{15} + \frac{9}{13} + \frac{4}{5} = 5.18.$$

Now, under H_0, Q has an approximate chi-square distribution with $k - 1 = 4$ degrees of freedom, so the $\alpha = 0.05$ critical region is $q \geq 9.488 = \chi^2_{0.05}(4)$. Here $q = 5.18 < 9.488$, and hence H_0 is not rejected at the 5% significance level. Furthermore, the p-value for $q = 5.18$ is 0.268, which is greater than most significance levels. Thus, with these data, we cannot say that there is a difference between the two methods of instruction. ◼

It is fairly obvious how this procedure can be extended to testing the equality of h independent multinomial distributions. That is, let

$$p_{ij} = P(A_i), \qquad i = 1, 2, \ldots, k, \qquad j = 1, 2, \ldots, h,$$

and test

$$H_0: \; p_{i1} = p_{i2} = \cdots = p_{ih} = p_i, \qquad i = 1, 2, \ldots, k.$$

Repeat the jth experiment n_j independent times, and let $Y_{1j}, Y_{2j}, \ldots, Y_{kj}$ denote the frequencies of the respective events $A_1, A_2, \ldots, A_k$. Now,

$$Q = \sum_{j=1}^{h} \sum_{i=1}^{k} \frac{(Y_{ij} - n_j p_{ij})^2}{n_j p_{ij}}$$

has an approximate chi-square distribution with $h(k - 1)$ degrees of freedom. Under H_0, we must estimate $k - 1$ probabilities, using

$$\widehat{p}_i = \frac{\displaystyle\sum_{j=1}^{h} Y_{ij}}{\displaystyle\sum_{j=1}^{h} n_j}, \qquad i = 1, 2, \ldots, k - 1,$$

because the estimate of p_k follows from $\widehat{p}_k = 1 - \widehat{p}_1 - \widehat{p}_2 \quad \cdots \quad - \widehat{p}_{k-1}$. We use these estimates to obtain

$$Q = \sum_{j=1}^{h} \sum_{i=1}^{k} \frac{(Y_{ij} - n_j \widehat{p}_i)^2}{n_j \widehat{p}_i},$$

which has an approximate chi-square distribution, with its degrees of freedom given by $h(k - 1) - (k - 1) = (h - 1)(k - 1)$.

Let us see how we can use the preceding procedures to test the equality of two or more independent distributions that are not necessarily multinomial. Suppose first that we are given random variables U and V with distribution functions $F(u)$ and $G(v)$, respectively. It is sometimes of interest to test the hypothesis $H_0: F(x) = G(x)$ for all x. Previously, we considered tests of $\mu_U = \mu_V$, $\sigma_U^2 = \sigma_V^2$. In Section 8.5, we will look at the two-sample Wilcoxon test. Now we shall assume only that the distributions are independent and of the continuous type.

We are interested in testing the hypothesis $H_0: F(x) = G(x)$ for all x. This hypothesis will be replaced by another one. Partition the real line into k mutually disjoint sets $A_1, A_2, \ldots, A_k$. Let

$$p_{i1} = P(U \in A_i), \qquad i = 1, 2, \ldots, k,$$

and

$$p_{i2} = P(V \in A_i), \qquad i = 1, 2, \ldots, k.$$

We observe that if $F(x) = G(x)$ for all x, then $p_{i1} = p_{i2}$, $i = 1, 2, \ldots, k$. We replace the hypothesis H_0: $F(x) = G(x)$ with the less restrictive hypothesis H_0': $p_{i1} = p_{i2}$, $i = 1, 2, \ldots, k$. That is, we are now essentially interested in testing the equality of two multinomial distributions.

Let n_1 and n_2 denote the number of independent observations of U and V, respectively. For $i = 1, 2, \ldots, k$, let Y_{ij} denote the number of these observations of U and V, $j = 1, 2$, respectively, that fall into a set A_i. At this point, we proceed to make the test of H_0' as described earlier. Of course, if H_0' is rejected at the (approximate) significance level α, then H_0 is rejected with the same probability. However, if H_0' is true, H_0 is not necessarily true. Thus, if H_0' is accepted, it is probably better to say that we do not reject H_0 than to say that H_0 is accepted.

In applications, the question of how to select $A_1, A_2, \ldots, A_k$ is frequently raised. Obviously, there is no single choice for k or for the dividing marks of the partition. But it is interesting to observe that the combined sample can be used in this selection without upsetting the approximate distribution of Q. For example, suppose that $n_1 = n_2 = 20$. Then we could easily select the dividing marks of the partition so that $k = 4$ and one-fourth of the combined sample falls into each of the four sets.

EXAMPLE 8.2-2 Select, at random, 20 cars of each of two comparable major-brand models. All 40 cars are submitted to accelerated life testing; that is, they are driven many miles over very poor roads in a short time, and their failure times (in weeks) are recorded as follows:

Brand U:	25	31	20	42	39	19	35	36	44	26
	38	31	29	41	43	36	28	31	25	38
Brand V:	28	17	33	25	31	21	16	19	31	27
	23	19	25	22	29	32	24	20	34	26

If we use 23.5, 28.5, and 34.5 as dividing marks, we note that exactly one-fourth of the 40 cars fall into each of the resulting four sets. Thus, the data can be summarized as follows:

	A_1	A_2	A_3	A_4	Totals
Brand U	2	4	4	10	20
Brand V	8	6	6	0	20

The estimate of each p_i is $10/40 = 1/4$, which, multiplied by $n_j = 20$, gives 5. Hence, the computed Q is

$$q = \frac{(2-5)^2}{5} + \frac{(4-5)^2}{5} + \frac{(4-5)^2}{5} + \frac{(10-5)^2}{5} + \frac{(8-5)^2}{5}$$
$$+ \frac{(6-5)^2}{5} + \frac{(6-5)^2}{5} + \frac{(0-5)^2}{5}$$
$$= \frac{72}{5} = 14.4 > 7.815 = \chi^2_{0.05}(3).$$

Also, the p-value is 0.0028. Thus, it seems that the two brands of cars have different distributions for the length of life under accelerated life testing. Brand U seems better than brand V. ∎

Again, it should be clear how this approach can be extended to more than two distributions, and this extension will be illustrated in the exercises.

Now let us suppose that a random experiment results in an outcome that can be classified by two different attributes, such as height and weight. Assume that the first attribute is assigned to one and only one of k mutually exclusive and exhaustive events—say, $A_1, A_2, \ldots, A_k$—and the second attribute falls into one and only one of h mutually exclusive and exhaustive events—say, $B_1, B_2, \ldots, B_h$. Let the probability of $A_i \cap B_j$ be defined by

$$p_{ij} = P(A_i \cap B_j), \qquad i = 1, 2, \ldots, k, \qquad j = 1, 2, \ldots, h.$$

The random experiment is to be repeated n independent times, and Y_{ij} will denote the frequency of the event $A_i \cap B_j$. Since there are kh such events as $A_i \cap B_j$, the random variable

$$Q_{kh-1} = \sum_{j=1}^{h} \sum_{i=1}^{k} \frac{(Y_{ij} - np_{ij})^2}{np_{ij}}$$

has an approximate chi-square distribution with $kh - 1$ degrees of freedom, provided that n is large.

Suppose that we wish to test the hypothesis of the independence of the A and B attributes, namely,

$$H_0: P(A_i \cap B_j) = P(A_i)P(B_j), \qquad i = 1, 2, \ldots, k, \qquad j = 1, 2, \ldots, h.$$

Let us denote $P(A_i)$ by $p_{i\cdot}$ and $P(B_j)$ by $p_{\cdot j}$; that is,

$$p_{i\cdot} = \sum_{j=1}^{h} p_{ij} = P(A_i) \qquad \text{and} \qquad p_{\cdot j} = \sum_{i=1}^{k} p_{ij} = P(B_j).$$

Of course,

$$1 = \sum_{j=1}^{h} \sum_{i=1}^{k} p_{ij} = \sum_{j=1}^{h} p_{\cdot j} = \sum_{i=1}^{k} p_{i\cdot}.$$

Then the hypothesis can be formulated as

$$H_0: p_{ij} = p_{i\cdot}p_{\cdot j}, \qquad i = 1, 2, \ldots, k, \qquad j = 1, 2, \ldots, h.$$

To test H_0, we can use Q_{kh-1} with p_{ij} replaced by $p_{i\cdot}p_{\cdot j}$. But if $p_{i\cdot}$, $i = 1, 2, \ldots, k$, and $p_{\cdot j}$, $j = 1, 2, \ldots, h$, are unknown, as they usually are in applications, we cannot compute Q_{kh-1} once the frequencies are observed. In such a case, we estimate these unknown parameters by

$$\widehat{p}_{i\cdot} = \frac{y_{i\cdot}}{n}, \qquad \text{where} \quad y_{i\cdot} = \sum_{j=1}^{h} y_{ij}$$

is the observed frequency of A_i, $i = 1, 2, \ldots, k$; and

$$\widehat{p}_{\cdot j} = \frac{y_{\cdot j}}{n}, \qquad \text{where} \quad y_{\cdot j} = \sum_{i=1}^{k} y_{ij}$$

is the observed frequency of B_j, $j = 1, 2, \ldots, h$. Since $\sum_{i=1}^{k} p_{i\cdot} = \sum_{j=1}^{h} p_{\cdot j} = 1$, we actually estimate only $k - 1 + h - 1 = k + h - 2$ parameters. So if these estimates are used in Q_{kh-1}, with $p_{ij} = p_{i\cdot}p_{\cdot j}$, then, according to the rule stated earlier, the random variable

$$Q = \sum_{j=1}^{h} \sum_{i=1}^{k} \frac{[Y_{ij} - n(Y_{i\cdot}/n)(Y_{\cdot j}/n)]^2}{n(Y_{i\cdot}/n)(Y_{\cdot j}/n)}$$

has an approximate chi-square distribution with $kh - 1 - (k + h - 2) = (k - 1)(h - 1)$ degrees of freedom, provided that H_0 is true. The hypothesis H_0 is rejected if the computed value of this statistic exceeds $\chi^2_\alpha[(k - 1)(h - 1)]$.

EXAMPLE 8.2-3 The 400 undergraduate students in a random sample at the University of Iowa were classified according to the college in which the students were enrolled and according to their gender. The results are recorded in Table 8.2-1, called a $k \times h$ **contingency table**, where, in this case, $k = 2$ and $h = 5$. (Do not be concerned about the numbers in parentheses at this point.) Incidentally, these data do actually reflect the composition of the undergraduate colleges at Iowa, but they were modified a little to make the computations easier in this example.

We desire to test the null hypothesis H_0: $p_{ij} = p_{i\cdot}p_{\cdot j}$, $i = 1, 2$ and $j = 1, 2, 3, 4, 5$, that the college in which a student enrolls is independent of the gender of that student. Under H_0, estimates of the probabilities are

$$\widehat{p}_{1\cdot} = \frac{190}{400} = 0.475 \qquad \text{and} \qquad \widehat{p}_{2\cdot} = \frac{210}{400} = 0.525$$

and

$$\widehat{p}_{\cdot 1} = \frac{35}{400} = 0.0875, \ \widehat{p}_{\cdot 2} = 0.05, \ \widehat{p}_{\cdot 3} = 0.8, \ \widehat{p}_{\cdot 4} = 0.0375, \ \widehat{p}_{\cdot 5} = 0.025.$$

The expected numbers $n(y_{i\cdot}/n)(y_{\cdot j}/n)$ are computed as follows:

$$400(0.475)(0.0875) = 16.625,$$
$$400(0.525)(0.0875) = 18.375,$$
$$400(0.475)(0.05) = 9.5,$$

and so on. These are the values recorded in parentheses in Table 8.2-1. The computed chi-square statistic is

$$q = \frac{(21 - 16.625)^2}{16.625} + \frac{(14 - 18.375)^2}{18.375} + \cdots + \frac{(4 - 5.25)^2}{5.25}$$
$$= 1.15 + 1.04 + 4.45 + 4.02 + 0.32 + 0.29 + 3.69$$
$$+ 3.34 + 0.33 + 0.30 = 18.93.$$

TABLE 8.2-1: Undergraduates at the University of Iowa						
	College					
Gender	Business	Engineering	Liberal Arts	Nursing	Pharmacy	Totals
Male	21	16	145	2	6	190
	(16.625)	(9.5)	(152)	(7.125)	(4.75)	
Female	14	4	175	13	4	210
	(18.375)	(10.5)	(168)	(7.875)	(5.25)	
Totals	35	20	320	15	10	400

Since the number of degrees of freedom equals $(k-1)(h-1) = 4$, this $q = 18.93 > 13.28 = \chi^2_{0.01}(4)$, and we reject H_0 at the $\alpha = 0.01$ significance level. Moreover, since the first two terms of q come from the business college, the next two from engineering, and so on, it is clear that the enrollments in engineering and nursing are more highly dependent on gender than in the other colleges, because they have contributed the most to the value of the chi-square statistic. It is also interesting to note that one expected number is less than 5, namely 4.75. However, as the associated term in q does not contribute an unusual amount to the chi-square value, it does not concern us. ■

It is fairly obvious how to extend the preceding testing procedure to more than two attributes. For example, if the third attribute falls into one and only one of m mutually exclusive and exhaustive events—say, $C_1, C_2, \ldots, C_m$—then we test the independence of the three attributes by using

$$Q = \sum_{r=1}^{m} \sum_{j=1}^{h} \sum_{i=1}^{k} \frac{[Y_{ijr} - n(Y_{i\cdot\cdot}/n)(Y_{\cdot j\cdot}/n)(Y_{\cdot\cdot r}/n)]^2}{n(Y_{i\cdot\cdot}/n)(Y_{\cdot j\cdot}/n)(Y_{\cdot\cdot r}/n)},$$

where Y_{ijr}, $Y_{i\cdot\cdot}$, $Y_{\cdot j\cdot}$, and $Y_{\cdot\cdot r}$ are the respective observed frequencies of the events $A_i \cap B_j \cap C_r$, A_i, B_j, and C_r in n independent trials of the experiment. If n is large and if the three attributes are independent, then Q has an approximate chi-square distribution with $khm - 1 - (k-1) - (h-1) - (m-1) = khm - k - h - m + 2$ degrees of freedom.

Rather than explore this extension further, it is more instructive to note some interesting uses of contingency tables.

EXAMPLE 8.2-4 Say we observed 30 values $x_1, x_2, \ldots, x_{30}$ that are claimed to be the values of a random sample. That is, the corresponding random variables $X_1, X_2, \ldots, X_{30}$ were supposed to be mutually independent and each of these random variables is supposed to have the same distribution. Say, however, by looking at the 30 values, we detect an upward trend which indicates that there might have been some dependence and/or the random variables did not actually have the same distribution. One simple way to test whether they could be thought of as being observed values of a random sample is the following: Mark each x high (H) or low (L), depending on whether it is above or below the sample median. Then divide the x values into three groups: $x_1, \ldots, x_{10}$; $x_{11}, \ldots, x_{20}$; and $x_{21}, \ldots, x_{30}$. Certainly, if the observations are those of a random sample, we would expect five H's and five L's in each group. That is, the attribute classified as H or L should be independent of the group number. The summary of these data provides a 3×2 contingency table. For example, say the 30 values are

5.6	8.2	7.8	4.8	5.5	8.1	6.7	7.7	9.3	6.9
8.2	10.1	7.5	6.9	11.1	9.2	8.7	10.3	10.7	10.0
9.2	11.6	10.3	11.7	9.9	10.6	10.0	11.4	10.9	11.1

The median can be taken to be the average of the two middle observations in magnitude, namely, 9.2 and 9.3. Marking each item H or L after comparing it with this median, we obtain the following 3×2 contingency table:

Group	L	H	Totals
1	9	1	10
2	5	5	10
3	1	9	10
Totals	15	15	30

Here each $n(y_{i\cdot}/n)(y_{\cdot j}/n) = 30(10/30)(15/30) = 5$, so that the computed value of Q is

$$q = \frac{(9-5)^2}{5} + \frac{(1-5)^2}{5} + \frac{(5-5)^2}{5} + \frac{(5-5)^2}{5} + \frac{(1-5)^2}{5} + \frac{(9-5)^2}{5}$$

$$= 12.8 > 5.991 = \chi_{0.05}^2(2),$$

since in this instance $(k-1)(h-1) = 2$ degrees of freedom. (The p-value is 0.0017.) Hence, we reject the conjecture that these 30 values could be the observations of a random sample. Obviously, modifications could be made to this scheme: dividing the sample into more (or fewer) than three groups and rating items differently, such as low (L), middle (M), and high (H). ∎

It cannot be emphasized enough that the chi-square statistic can be used fairly effectively in almost any situation in which there should be independence. For example, suppose that we have a group of workers who have essentially the same qualifications (training, experience, etc.). Many believe that the salary and gender of the workers should be independent attributes, yet there have been several claims in special cases that there is a dependence—or discrimination—in attributes associated with such a problem.

EXAMPLE 8.2-5   Two groups of workers have the same qualifications for a particular type of work. Their experience in salaries is summarized by the following 2×5 contingency table, in which the upper bound of each salary range is not included in that listing:

	\multicolumn{5}{c	}{Salary (Thousands of Dollars)}	
Group	27–29	29–31	31–33	33–35	35 and over	Totals
1	6	11	16	14	13	60
2	5	9	8	6	2	30
Totals	11	20	24	20	15	90

To test whether the group assignment and the salaries seem to be independent with these data at the $\alpha = 0.05$ significance level, we compute

$$q = \frac{[6 - 90(60/90)(11/90)]^2}{90(60/90)(11/90)} + \cdots + \frac{[2 - 90(30/90)(15/90)]^2}{90(30/90)(15/90)}$$

$$= 4.752 < 9.488 = \chi_{0.05}^2(4).$$

Also, the p-value is 0.313. Hence, with these limited data, group assignment and salaries seem to be independent. ∎

Before turning to the exercises, note that we could have thought of the last two examples in this section as testing the equality of two or more multinomial distributions. In Example 8.2-4 the three groups define three binomial distributions, and in Example 8.2-5 the two groups define two multinomial distributions. What would have happened if we had used the computations outlined earlier in the section? It is interesting to note that we obtain exactly the same value of chi-square and in each case the number of degrees of freedom is equal to $(k-1)(h-1)$. Hence, it makes no difference whether we think of it as a test of independence or a test of the equality of several multinomial distributions. Our advice is to use the terminology that seems most natural for the particular situation.

EXERCISES

8.2-1. We wish to see if two groups of nurses distribute their time in six different categories about the same way. That is, the hypothesis under consideration is $H_0: p_{i1} = p_{i2}, i = 1, 2, \ldots, 6$. To test this hypothesis, nurses are observed at random throughout several days, each observation resulting in a mark in one of the six categories. A summary of the results is given by the following frequency table:

	1	2	3	4	5	6	Totals
Group I	95	36	71	21	45	32	300
Group II	53	26	43	18	32	28	200

Category (spanning columns 1–6)

Use a chi-square test with $\alpha = 0.05$.

8.2-2. Suppose that a third group of nurses was observed along with groups I and II of Exercise 8.2-1, resulting in the respective frequencies 130, 75, 136, 33, 61, and 65. Test $H_0: p_{i1} = p_{i2} = p_{i3}, i = 1, 2, \ldots, 6$, at the $\alpha = 0.025$ significance level.

8.2-3. Each of two comparable classes of 15 students responded to two different methods of instructions, giving the following scores on a standardized test:

Class U: 91 42 39 62 55 82 67 44
51 77 61 52 76 41 59

Class V: 80 71 55 67 61 93 49 78
57 88 79 81 63 51 75

Use a chi-square test with $\alpha = 0.05$ to test the equality of the distributions of test scores by dividing the combined sample into three equal parts (low, middle, high).

8.2-4. Suppose that a third class (W) of 15 students was observed along with classes U and V of Exercise 8.2-3, resulting in scores of

91 73 67 83 59 98 87 69
78 80 65 94 82 74 85

Again, use a chi-square test with $\alpha = 0.05$ to test the equality of the three distributions by dividing the combined sample into three equal parts.

8.2-5. In the following contingency table, 1015 individuals are classified by gender and by whether they favor, oppose, or have no opinion on a complete ban on smoking in public places:

Gender	Favor	Oppose	No Opinion	Totals
Male	262	231	10	503
Female	302	205	5	512
Totals	564	436	15	1015

Smoking in Public Places (spanning Favor, Oppose, No Opinion)

Test the null hypothesis that gender and opinion on smoking in public places are independent. Give the approximate p-value of this test.

8.2-6. A random survey of 100 students asked each student to select the most preferred form of recreational activity from five choices. Following are the results of the survey:

Gender	Basketball	Baseball Softball	Swimming	Jogging Running	Tennis	Totals
Male	21	5	9	12	13	60
Female	9	3	1	15	12	40
Totals	30	8	10	27	25	100

Recreational Choice (spanning columns)

Test whether the choice is independent of the gender of the respondent. Approximate the p-value of the test. Would we reject the null hypothesis at $\alpha = 0.05$?

8.2-7. One hundred students in a random sample were classified as follows by gender and by the kind of instrument (including voice) that they played:

Gender	Piano	Woodwind	Brass	String	Vocal	Totals
Male	4	11	15	6	9	45
Female	7	18	6	6	18	55
Totals	11	29	21	12	27	100

Test whether the selection of instrument is independent of the gender of the respondent. Approximate the p-value of this test.

8.2-8. A student who uses a certain college's recreational facilities was interested in whether there is a difference between the facilities used by men and those used by women. Use $\alpha = 0.05$ and the following data to test the null hypothesis that facility and gender are independent attributes:

Gender	Racquetball Court	Track	Totals
Male	51	30	81
Female	43	48	91
Totals	94	78	172

8.2-9. A survey of high school girls classified them by two attributes: whether or not they participated in sports and whether or not they had one or more older brothers. Use the following data to test the null hypothesis that these two attributes of classification are independent:

Older Brother(s)	Participated in Sports Yes	No	Totals
Yes	12	8	20
No	13	27	40
Totals	25	35	60

Approximate the p-value of this test. Do we reject the null hypothesis if $\alpha = 0.05$?

8.2-10. A random sample of 50 women who were tested for cholesterol was classified according to age and cholesterol level and grouped into the following contingency table.

Age	Cholesterol Level < 180	180–210	> 210	Totals
< 50	5	11	9	25
≥ 50	4	3	18	25
Totals	9	14	27	50

Test the null hypothesis H_0: Age and cholesterol level are independent attributes of classification. What is your conclusion if $\alpha = 0.01$?

8.2-11. Although high school grades and testing scores, such as SAT or ACT, can be used to predict first-year college grade point average (GPA), many educators claim that a more important factor influencing GPA is the living conditions of students. In particular, it is claimed that the roommate of the student will have a great influence on his or her grades. To test this hypothesis, suppose we selected at random 200 students and classified each according to the following two attributes:

(a) Ranking of the student's roommate on a scale from 1 to 5, with 1 denoting a person who was difficult to live with and discouraged scholarship, and 5 signifying a person who was congenial and encouraged scholarship.

(b) The student's first-year GPA.

Say this classification gives the following 5 × 4 contingency table:

Rank of Roommate	Under 2.00	2.00–2.69	2.70–3.19	3.20–4.00	Totals
			Grade-Point Average		
1	8	9	10	4	31
2	5	11	15	11	42
3	6	7	20	14	47
4	3	5	22	23	53
5	1	3	11	12	27
Totals	23	35	78	64	200

Compute the chi-square statistic used to test the independence of the two attributes, and compare it with the critical value associated with $\alpha = 0.05$.

8.2-12. (*How Americans Watch TV: A Nation of Grazers.*) A random sample of 395 people were classified by their age and by whether or not they change channels during programs when watching television. Use the following data to test the null hypothesis H_0: Changing channels and age are independent attributes:

Change?	18–24	25–34	35–49	50–64	Totals
			Age		
Yes	60	54	46	41	201
No	40	44	53	57	194
Totals	100	98	99	98	395

(a) Give a significance level α for which H_0 is rejected.

(b) Give a value of the significance level α for which H_0 is not rejected.

8.2-13. A study was conducted to determine the media credibility for reporting news. Those surveyed were asked to give their age, gender, education, and the most credible medium. The results of the survey are as follows:

Age	Newspaper	Television	Radio	Totals
		Most Credible Medium		
Under 35	30	68	10	108
35–54	61	79	20	160
Over 54	98	43	21	162
Totals	189	190	51	430

Gender	Newspaper	Television	Radio	Totals
		Most Credible Medium		
Male	92	108	19	219
Female	97	81	32	210
Totals	189	189	51	429

Education	Newspaper	Television	Radio	Totals
		Most Credible Medium		
Grade School	45	22	6	73
High School	94	115	30	239
College	49	52	13	114
Totals	188	189	49	426

(a) Test whether media credibility and age are independent.

(b) Test whether media credibility and gender are independent.

(c) Test whether media credibility and education are independent.

(d) Give the approximate *p*-value for each test.

8.2-14. In a psychology experiment, 140 students were divided into majors emphasizing left-hemisphere brain skills (e.g., philosophy, physics, and mathematics) and majors emphasizing right-hemisphere skills (e.g., art, music, theater, and dance). They were also classified into one of three groups on the basis of hand posture (right noninverted, left inverted, and left noninverted). The data are as follows:

	LH	RH
RN	89	29
LI	5	4
LN	5	8

Do these data show sufficient evidence to reject the claim that the choice of college major is independent of hand posture? Let $\alpha = 0.025$.

8.2-15. Ledolter and Hogg (see references) report the starting salaries of $n = 200$ engineers classified as being in the lower 25% (A_1), second 25% (A_2), third 25% (A_3), and upper 25% (A_4) of their class. In addition, these 200 engineers were classified according to the color of their eyes. Here are the summarized results:

	Salary				
Eyes	A_1	A_2	A_3	A_4	Totals
Blue	22	17	21	20	80
Brown	14	20	20	16	70
Other	14	13	9	14	50
Totals	50	50	50	50	200

Are color of eyes and starting salary independent attributes. Use $\alpha = 0.05$.

8.2-16. In the Michigan Daily Lottery, one three-digit integer is generated six days a week. Twenty weeks of lottery numbers have been classified as follows by the day of the week and the magnitude of the number:

	Magnitude of Number		
Days	000–499	500–999	Totals
Monday and Tuesday	22	18	40
Wednesday and Thursday	19	21	40
Friday and Saturday	13	27	40
Totals	54	66	120

Test whether these attributes of classification are independent at the 10% significance level.

8.3 ORDER STATISTICS

Order statistics are the observations of the random sample, arranged, or ordered, in magnitude from the smallest to the largest. In recent years, the importance of order statistics has increased owing to the more frequent use of nonparametric inferences and robust procedures. However, order statistics have always been prominent because, among other things, they are needed to determine rather simple statistics, such as the sample median, the sample range, and the empirical distribution function. Recall that in Section 3.2 we discussed observed order statistics in connection with descriptive and exploratory statistical methods. We will consider certain interesting aspects about their distributions in this section.

In most of our discussions about order statistics, we will assume that the n independent observations come from a continuous-type distribution. This means, among other things, that the probability of any two observations being equal is zero. That is, the probability is 1 that the observations can be ordered from smallest to largest without having two equal values. Of course, in practice, we do frequently observe *ties*; but if the probability of a tie is small, the distribution theory that follows will hold approximately. Thus, in the discussion here, we are assuming that the probability of a tie is zero.

EXAMPLE 8.3-1 The values $x_1 = 0.62$, $x_2 = 0.98$, $x_3 = 0.31$, $x_4 = 0.81$, and $x_5 = 0.53$ are the $n = 5$ observed values of five independent trials of an experiment with p.d.f. $f(x) = 2x$, $0 < x < 1$. The observed order statistics are

$$y_1 = 0.31 < y_2 = 0.53 < y_3 = 0.62 < y_4 = 0.81 < y_5 = 0.98.$$

Recall that the middle observation in the ordered arrangement, here $y_3 = 0.62$, is called the *sample median* and the difference of the largest and the smallest, here

$$y_5 - y_1 = 0.98 - 0.31 = 0.67,$$

is called the *sample range*.   ∎

If $X_1, X_2, \ldots, X_n$ are observations of a random sample of size n from a continuous-type distribution, we let the random variables

$$Y_1 < Y_2 < \cdots < Y_n$$

denote the order statistics of that sample. That is,

$$Y_1 = \text{smallest of } X_1, X_2, \ldots, X_n,$$
$$Y_2 = \text{second smallest of } X_1, X_2, \ldots, X_n,$$

$$\vdots$$

$$Y_n = \text{largest of } X_1, X_2, \ldots, X_n.$$

There is a very simple procedure for determining the distribution function of the rth order statistic, Y_r. This procedure depends on the binomial distribution and is illustrated in Example 8.3-2.

EXAMPLE 8.3-2   Let $Y_1 < Y_2 < Y_3 < Y_4 < Y_5$ be the order statistics associated with n independent observations X_1, X_2, X_3, X_4, X_5, each from the distribution with p.d.f. $f(x) = 2x$, $0 < x < 1$. Consider $P(Y_4 < 1/2)$. For the event $\{Y_4 < 1/2\}$ to occur, at least four of the random variables X_1, X_2, X_3, X_4, X_5 must be less than $1/2$, because Y_4 is the fourth smallest among the five observations. Thus, if the event $\{X_i < 1/2\}$, $i = 1, 2, \ldots, 5$, is called "success," we must have at least four successes in the five mutually independent trials, each of which has probability of success

$$P\left(X_i \leq \frac{1}{2}\right) = \int_0^{1/2} 2x\, dx = \left(\frac{1}{2}\right)^2 = \frac{1}{4}.$$

Hence,

$$P\left(Y_4 \leq \frac{1}{2}\right) = \binom{5}{4}\left(\frac{1}{4}\right)^4\left(\frac{3}{4}\right) + \left(\frac{1}{4}\right)^5 = 0.0156.$$

In general, if $0 < y < 1$, then the distribution function of Y_4 is

$$G(y) = P(Y_4 < y) = \binom{5}{4}(y^2)^4(1 - y^2) + (y^2)^5,$$

since this represents the probability of at least four "successes" in five independent trials, each of which has probability of success

$$P(X_i < y) = \int_0^y 2x\, dx = y^2.$$

For $0 < y < 1$, the p.d.f. of Y_4 is therefore

$$g(y) = G'(y) = \binom{5}{4} 4(y^2)^3(2y)(1 - y^2) + \binom{5}{4}(y^2)^4(-2y) + 5(y^2)^4(2y)$$

$$= \frac{5!}{3!\,1!}(y^2)^3(1 - y^2)(2y), \qquad 0 < y < 1.$$

Note that in this example the distribution function of each X is $F(x) = x^2$ when $0 < x < 1$. Thus,

$$g(y) = \frac{5!}{3!\,1!}[F(y)]^3[1 - F(y)]f(y), \qquad 0 < y < 1. \qquad \blacksquare$$

The preceding example should make the following generalization easier to read: Let $Y_1 < Y_2 < \cdots < Y_n$ be the order statistics of n independent observations from a distribution of the continuous type with distribution function $F(x)$ and p.d.f. $F'(x) = f(x)$, where $0 < F(x) < 1$ for $a < x < b$ and $F(a) = 0$, $F(b) = 1$. (It is possible that $a = -\infty$ and/or $b = +\infty$.) The event that the rth order statistic Y_r is at most y, $\{Y_r \leq y\}$, can occur if and only if at least r of the n observations are less than or equal to y. That is, here the probability of "success" on each trial is $F(y)$, and we must have at least r successes. Thus,

$$G_r(y) = P(Y_r \leq y) = \sum_{k=r}^{n} \binom{n}{k}[F(y)]^k[1 - F(y)]^{n-k}.$$

Rewriting this slightly, we have

$$G_r(y) = \sum_{k=r}^{n-1} \binom{n}{k}[F(y)]^k[1 - F(y)]^{n-k} + [F(y)]^n.$$

Hence, the p.d.f. of Y_r is

$$g_r(y) = G'_r(y) = \sum_{k=r}^{n-1} \binom{n}{k}(k)[F(y)]^{k-1}f(y)[1 - F(y)]^{n-k}$$

$$+ \sum_{k=r}^{n-1} \binom{n}{k}[F(y)]^k(n - k)[1 - F(y)]^{n-k-1}[-f(y)]$$

$$+ n[F(y)]^{n-1}f(y). \qquad (8.3\text{-}1)$$

However, since

$$\binom{n}{k}k = \frac{n!}{(k - 1)!\,(n - k)!} \qquad \text{and} \qquad \binom{n}{k}(n - k) = \frac{n!}{k!\,(n - k - 1)!},$$

it follows that the p.d.f. of Y_r is

$$g_r(y) = \frac{n!}{(r - 1)!\,(n - r)!}[F(y)]^{r-1}[1 - F(y)]^{n-r}f(y), \qquad a < y < b,$$

which is the first term of the first summation in $g_r(y) = G'_r(y)$, Equation 8.3-1. The remaining terms in $g_r(y) = G'_r(y)$ sum to zero because the second term of the first summation (when $k = r + 1$) equals the negative of the first term in the second summation (when $k = r$), and so on. Finally, the last term of the second summation equals the negative of $n[F(y)]^{n-1}f(y)$. To see this clearly, the student is urged to write out a number of terms in these summations. (See Exercise 8.3-4.)

It is worth noting that the p.d.f. of the smallest order statistic is

$$g_1(y) = n[1 - F(y)]^{n-1}f(y), \qquad a < y < b,$$

and the p.d.f. of the largest order statistic is

$$g_n(y) = n[F(y)]^{n-1}f(y), \qquad a < y < b.$$

REMARK   There is one quite satisfactory way to construct heuristically the expression for the p.d.f. of Y_r. To do this, we must recall the multinomial probability and then consider the probability element $g_r(y)(\Delta y)$ of Y_r. If the length Δy is *very* small, $g_r(y)(\Delta y)$ represents approximately the probability

$$P(y < Y_r \le y + \Delta y).$$

Thus, we want the probability, $g_r(y)(\Delta y)$, that $(r - 1)$ items fall less than y, that $(n - r)$ items are greater than $y + \Delta y$, and that one item falls between y and $y + \Delta y$. Recall that the probabilities on a single trial are

$$P(X \le y) = F(y),$$
$$P(X > y + \Delta y) = 1 - F(y + \Delta y) \approx 1 - F(y),$$
$$P(y < X \le y + \Delta y) \approx f(y)(\Delta y).$$

Thus, the multinomial probability is approximately

$$g_r(y)(\Delta y) - \frac{n!}{(r - 1)!\,1!\,(n - r)!}[F(y)]^{r-1}[1 - F(y)]^{n-r}[f(y)(\Delta y)].$$

If we divide both sides by the length Δy, the formula for $g_r(y)$ results.   ∎

EXAMPLE 8.3-3   Returning to Example 8.3-2, we shall now graph the p.d.f.s of the order statistics $Y_1 < Y_2 < Y_3 < Y_4 < Y_5$ when sampling from a distribution with p.d.f. $f(x) = 2x, 0 < x < 1$. These graphs are given in Figure 8.3-1 on the next page. The respective p.d.f.'s and their means are as follows:

$$g_1(y) = 10y(1 - y^2)^4; \qquad \mu_1 = \frac{256}{693},$$

$$g_2(y) = 40y^3(1 - y^2)^3; \qquad \mu_2 = \frac{128}{231},$$

$$g_3(y) = 60y^5(1 - y^2)^2; \qquad \mu_3 = \frac{160}{231},$$

$$g_4(y) = 40y^7(1 - y^2); \qquad \mu_4 = \frac{80}{99},$$

$$g_5(y) = 10y^9; \qquad \mu_5 = \frac{10}{11}.$$ ∎

FIGURE 8.3-1: p.d.f.'s of order statistics, $f(x) = 2x, 0 < x < 1$

EXAMPLE 8.3-4   Let $Y_1 < Y_2 < Y_3 < Y_4$ be the order statistics of a random sample X_1, X_2, X_3, X_4 from a uniform distribution with p.d.f. $f(x; \theta) = 1/\theta, 0 < x \leq \theta$. The likelihood function is

$$f(x; \theta) = \left(\frac{1}{\theta}\right)^4, \qquad 0 < x_i \leq \theta, \ i = 1, 2, 3, 4,$$

and equals zero if $\theta < x_i$. To maximize $L(\theta)$, we must make θ as small as possible; hence,

$$\widehat{\theta} = \max(X_i) = Y_4$$

because θ cannot be less than any X_i. Since $F(x; \theta) = x/\theta, 0 < x \leq \theta$, the p.d.f. of Y_4 is

$$g_4(y_4) = \frac{4!}{3!1!}\left(\frac{y_4}{\theta}\right)^3\left(\frac{1}{\theta}\right) = 4\frac{y_4^3}{\theta^4}, \qquad 0 < y_4 \leq \theta.$$

Accordingly,

$$E(Y_4) = \int_0^\theta y_4 \cdot 4\frac{y_4^3}{\theta^4}\,dy_4 = \frac{4}{5}\theta$$

and Y_4 is a biased estimator of θ. However, $5Y_4/4$ is unbiased. So, since, with $0 < c < 1$, we have

$$\int_{c\theta}^\theta 4\,y_4^3/\theta^4\,dy_4 = 1 - c^4,$$

it follows that

$$1 - c^4 = P(c\theta < Y_4 < \theta) = P(Y_4 < \theta < Y_4/c).$$

Then, with an observed Y_4 of y_4, the interval from y_4 to y_4/c would be a $100(1 - c^4)$ percent confidence interval for θ. Say $c = (0.05)^{1/4} = 0.47$; then the interval from y_4 to $2.11y_4$ is a 95% confidence interval for θ. ■

Recall that in Theorem 5.1-2 we proved that if X has a distribution function $F(x)$ of the continuous type, then $F(X)$ has a uniform distribution on the interval from 0 to 1. If $Y_1 < Y_2 < \cdots < Y_n$ are the order statistics of n independent observations $X_1, X_2, \ldots, X_n$, then

$$F(Y_1) < F(Y_2) < \cdots < F(Y_n),$$

because F is a nondecreasing function and the probability of an equality is again zero. Note that this ordering could be looked upon as an ordering of the mutually independent random variables $F(X_1)$, $F(X_2)$,...,$F(X_n)$, each of which is $U(0,1)$. That is,

$$W_1 = F(Y_1) < W_2 = F(Y_2) < \cdots < W_n = F(Y_n)$$

can be thought of as the order statistics of n independent observations from that uniform distribution. Since the distribution function of $U(0,1)$ is $G(w) = w$, $0 < w < 1$, the p.d.f. of the rth order statistic, $W_r = F(Y_r)$, is

$$h_r(w) = \frac{n!}{(r-1)!(n-r)!} w^{r-1}(1-w)^{n-r}, \qquad 0 < w < 1.$$

Of course, the mean, $E(W_r) = E[F(Y_r)]$ of $W_r = F(Y_r)$, is given by the integral

$$E(W_r) = \int_0^1 w \frac{n!}{(r-1)!(n-r)!} w^{r-1}(1-w)^{n-r}\, dw.$$

This integral can be evaluated by integrating by parts several times, but it is easier to obtain the answer if we rewrite the integration as follows:

$$E(W_r) = \left(\frac{r}{n+1}\right) \int_0^1 \frac{(n+1)!}{r!(n-r)!} w^r (1-w)^{n-r}\, dw.$$

The integrand in this last expression can be thought of as the p.d.f. of the $(r+1)$st order statistic of $n+1$ independent observations from a $U(0,1)$ distribution. This is a beta p.d.f. with $\alpha = r + 1$ and $\beta = n - r + 1$; hence, the integral must equal 1, and it follows that

$$E(W_r) = \frac{r}{n+1}, \qquad r = 1, 2, \ldots, n.$$

There is an extremely interesting interpretation of $W_r = F(Y_r)$. Note that $F(Y_r)$ is the cumulated probability up to and including Y_r or, equivalently, the area under $f(x) = F'(x)$ but less than Y_r. Consequently, $F(Y_r)$ can be treated as a random area. Since $F(Y_{r-1})$ is also a random area, $F(Y_r) - F(Y_{r-1})$ is the random area under $f(x)$ between Y_{r-1} and Y_r. The expected value of the random area between any two adjacent order statistics is then

$$E[F(Y_r) - F(Y_{r-1})] = E[F(Y_r)] - E[F(Y_{r-1})]$$
$$= \frac{r}{n+1} - \frac{r-1}{n+1} = \frac{1}{n+1}.$$

Also, it is easy to show (see Exercise 8.3-6) that

$$E[F(Y_1)] = \frac{1}{n+1} \qquad \text{and} \qquad E[1 - F(Y_n)] = \frac{1}{n+1}.$$

That is, the order statistics $Y_1 < Y_2 < \cdots < Y_n$ partition the support of X into $n+1$ parts and thus create $n+1$ areas under $f(x)$ and above the x-axis. On the average, each of the $n+1$ areas equals $1/(n+1)$.

If we recall that the $(100p)$th percentile π_p is such that the area under $f(x)$ to the left of π_p is p, then the preceding discussion suggests that we let Y_r be an estimator of π_p, where $p = r/(n+1)$. For this reason, we define the **(100p)th percentile of the sample** as Y_r, where $r = (n+1)p$. In case $(n+1)p$ is not an integer, we use a weighted average (or an average) of the two adjacent order statistics Y_r and Y_{r+1}, where r is the greatest integer $\lfloor (n + 1)p \rfloor$ in $(n + 1)p$. In particular, the sample median is

$$\tilde{m} = \begin{cases} Y_{(n+1)/2}, & \text{when } n \text{ is odd,} \\ \dfrac{Y_{n/2} + Y_{(n/2)+1}}{2}, & \text{when } n \text{ is even.} \end{cases}$$

EXAMPLE 8.3-5 Let X equal the weight of soap in a "1000-gram" bottle. A random sample of $n = 12$ observations of X yielded the following weights, which have been ordered:

$$\begin{array}{cccccc} 1013 & 1019 & 1021 & 1024 & 1026 & 1028 \\ 1033 & 1035 & 1039 & 1040 & 1043 & 1047 \end{array}$$

Since $n = 12$ is even, the sample median is

$$\tilde{m} = \frac{y_6 + y_7}{2} = \frac{1028 + 1033}{2} = 1030.5.$$

The location of the 25th percentile (or first quartile) is

$$(n + 1)(0.25) = (12 + 1)(0.25) = 3.25.$$

Thus, using a weighted average, we find that the first quartile is

$$\tilde{q}_1 = y_3 + (0.25)(y_4 - y_3) = (0.75)y_3 + (0.25)y_4$$
$$= (0.75)(1021) + (0.25)(1024) = 1021.75.$$

Similarly, the 75th percentile (or third quartile) is

$$\tilde{q}_3 = y_9 + (0.75)(y_{10} - y_9) = (0.25)y_9 + (0.75)y_{10}$$
$$= (0.25)(1039) + (0.75)(1040) = 1039.75,$$

because $(12 + 1)(0.75) = 9.75$. Since $(12 + 1)(0.60) = 7.8$, the 60th percentile is

$$\tilde{\pi}_{0.60} = (0.2)y_7 + (0.8)y_8 = (0.2)(1033) + (0.8)(1035) = 1034.6. \qquad \blacksquare$$

EXERCISES

8.3-1. Some biology students were interested in analyzing the amount of time that bees spend gathering nectar in flower patches. Thirty-nine bees visited a high-density flower patch and spent the following times (in seconds) gathering nectar:

$$\begin{array}{cccccccccc} 235 & 210 & 95 & 146 & 195 & 840 & 185 & 610 & 680 & 990 \\ 146 & 404 & 119 & 47 & 9 & 4 & 10 & 169 & 270 & 95 \\ 329 & 151 & 211 & 127 & 154 & 35 & 225 & 140 & 158 & 116 \\ 46 & 113 & 149 & 420 & 120 & 45 & 10 & 18 & 105 \end{array}$$

(a) Find the order statistics.
(b) Find the median and 80th percentile of the sample.

(c) Determine the first and third quartiles (i.e., 25th and 75th percentiles) of the sample.

8.3-2. Let X equal the forced vital capacity (the volume of air a person can expel from his or her lungs) of a male freshman. Seventeen observations of X, which have been ordered, are

$$\begin{array}{cccccccc} 3.7 & 3.8 & 4.0 & 4.3 & 4.7 & 4.8 & 4.9 & 5.0 \\ 5.2 & 5.4 & 5.6 & 5.6 & 5.6 & 5.7 & 6.2 & 6.8 & 7.6 \end{array}$$

(a) Find the median, the first quartile, and the third quartile.
(b) Find the 35th and 65th percentiles.

8.3-3. Let $Y_1 < Y_2 < Y_3 < Y_4 < Y_5$ be the order statistics of five independent observations from an exponential distribution that has a mean of $\theta = 3$.

(a) Find the p.d.f. of the sample median Y_3.
(b) Compute the probability that Y_4 is less than 5.
(c) Determine $P(1 < Y_1)$.

8.3-4. In the expression for $g_r(y) = G'_r(y)$ in Equation 8.3-1, let $n = 6$, and $r = 3$, and write out the summations, showing that the "telescoping" suggested in the text is achieved.

8.3-5. Let $Y_1 < Y_2 < \cdots < Y_8$ be the order statistics of eight independent observations from a continuous-type distribution with 70th percentile $\pi_{0.7} = 27.3$.

(a) Determine $P(Y_7 < 27.3)$.
(b) Find $P(Y_5 < 27.3 < Y_8)$.

8.3-6. Let $W_1 < W_2 < \cdots < W_n$ be the order statistics of n independent observations from a $U(0,1)$ distribution.

(a) Find the p.d.f. of W_1 and that of W_n.
(b) Use the results of (a) to verify that $E(W_1) = 1/(n + 1)$ and $E(W_n) = n/(n + 1)$.
(c) Show that the p.d.f. of W_r is beta.

8.3-7. Let $Y_1 < Y_2 < \cdots < Y_{19}$ be the order statistics of $n = 19$ independent observations from the exponential distribution with mean θ.

(a) What is the p.d.f. of Y_1?
(b) Using integration, find the value of $E[F(Y_1)]$, where F is the distribution function of the exponential distribution.

8.3-8. Let $W_1 < W_2 < \cdots < W_n$ be the order statistics of n independent observations from a $U(0,1)$ distribution.

(a) Show that $E(W_r^2) = \dfrac{r(r + 1)}{(n + 1)(n + 2)}$, using a technique similar to that used in determining that $E(W_r) = r/(n + 1)$.
(b) Find the variance of W_r.

8.3-9. Let $X_1, X_2, \ldots, X_{10}$ be a random sample of size $n = 10$ from a distribution with p.d.f. $f(x; \theta) = e^{-(x-\theta)}$, $\theta \le x < \infty$.

(a) Show that $Y_1 = \min(X_i)$ is the maximum likelihood estimator of θ.
(b) Find the p.d.f. of Y_1 and show that $E(Y_1) = \theta + 1/10$, so that $Y_1 - 1/10$ is an unbiased estimator of θ.
(c) Compute $P(\theta \le Y_1 \le \theta + c)$ and use the result to construct a 95% confidence interval for θ.

8.3-10. Let $Y_1 < Y_2 < \cdots < Y_n$ be the order statistics of a random sample of size n from an exponential distribution with mean θ. Let $1 < m < n$. Then

$$W = \left[\frac{2}{\theta}\right]\left[\sum_{i=1}^{m} Y_i + (n - m)Y_m\right]$$

has a chi-square distribution with $2m$ degrees of freedom. (This statistic could be used in a life-testing experiment in which only the first m "deaths" are observed.)

(a) Use W to construct a $100(1 - \alpha)\%$ confidence interval for θ.
(b) The following observed order statistics were simulated from an exponential distribution with $\theta = 10$:

$$0.043 \quad 2.836 \quad 4.962 \quad 5.297 \quad 6.730 \quad 7.338 \quad 8.888 \quad 13.259 \quad 20.558 \quad 36.852$$

Use the first (i) four, (ii) five, (iii) six, and (iv) seven order statistics to construct 90% confidence intervals for θ. Describe these intervals.

REMARK   It is possible to simulate only the first m out of n order statistics. [See Karian and Tanis (Questions and Comments 10.1-2) or Lurie and Hartley in the references.] ∎

8.3-11. Let $Y_1 < Y_2 < \cdots < Y_n$ be the order statistics of a random sample of size n from a distribution with p.d.f. $f(x) = e^{-x}, 0 < x < \infty$.

(a) Find the p.d.f. of Y_r.
(b) Determine the p.d.f. of $U = e^{-Y_r}$.

8.3-12. Use the heuristic argument to show that the joint p.d.f. of the two order statistics $Y_i < Y_j$ is

$$g(y_i, y_j) = \frac{n!}{(i - 1)!(j - i - 1)!(n - j)!} \times$$
$$[F(y_i)]^{i-1}[F(y_j) - F(y_i)]^{j-i-1} \times$$
$$[1 - F(y_j)]^{n-j}f(y_i)f(y_j),$$
$$-\infty < y_i < y_j < \infty.$$

8.3-13. Use the result of Exercise 8.3-12.

(a) Find the joint p.d.f. of Y_1 and Y_n, the first and the nth order statistics of a random sample of size n from the $U(0,1)$ distribution.

(b) Find the joint and the marginal p.d.f.s of $W_1 = Y_1/Y_n$ and $W_2 = Y_n$.

(c) Are W_1 and W_2 independent?

(d) Use simulation to confirm your theoretical results.

8.3-14. Let $Y_1 < Y_2 < Y_3$ be the order statistics of a random sample X_1, X_2, X_3 of size 3 from the uniform distribution $U(\theta - 1/2, \theta + 1/2)$. That is, $Y_1 = \min\{X_1, X_2, X_3\}$, $Y_2 = \text{median}\{X_1, X_2, X_3\}$, and $Y_3 = \max\{X_1, X_2, X_3\}$. Three possible estimators of θ are the sample mean,

$$W_1 = \overline{X} = \frac{1}{3}\sum_{i=1}^{3} X_i,$$

the sample median, $W_2 = Y_2$, and the midrange, $W_3 = (Y_1 + Y_3)/2$.

(a) Simulate 100 samples of size 3 from $U(\theta - 1/2, \theta + 1/2)$ for a particular value of θ—for example, $\theta = 1/2$. For each sample, calculate the values of W_1, W_2, and W_3.

(b) Compare the values of the sample means and sample variances of the 100 observations of each of W_1, W_2, and W_3. Which of these statistics seems to be the best estimator of θ?

(c) Verify that $E(W_1) = E(W_2) = E(W_3) = 1/2$ and that $\text{Var}(W_1) = 1/36$, $\text{Var}(W_2) = 1/20$, and $\text{Var}(W_3) = 1/40$ when $\theta = 1/2$.

8.4 DISTRIBUTION-FREE CONFIDENCE INTERVALS FOR PERCENTILES

In Section 8.3, we defined sample percentiles in terms of order statistics and noted that sample percentiles can be used to estimate corresponding distribution percentiles. In this section, we use order statistics to construct confidence intervals for unknown distribution percentiles. Since little is assumed about the underlying distribution (except that it is of the continuous type) in the construction of these confidence intervals, they are often called **distribution-free confidence intervals**.

If $Y_1 < Y_2 < Y_3 < Y_4 < Y_5$ are the order statistics of a random sample of size $n = 5$ from a continuous-type distribution, then the sample median Y_3 could be thought of as an estimator of the distribution median $\pi_{0.5}$. We shall let $m = \pi_{0.5}$. We could simply use the sample median Y_3 as an estimator of the distribution median m. However, we are certain that all of us recognize that, with only a sample of size 5, we would be quite lucky if the observed $Y_3 = y_3$ were very close to m. Thus, we now describe how a confidence interval can be constructed for m.

Instead of simply using Y_3 as an estimator of m, let us also compute the probability that the random interval (Y_1, Y_5) includes m. That is, let us determine $P(Y_1 < m < Y_5)$. Doing this is easy if we say that we have success if an individual observation—say, X—is less than m; then the probability of success on one of the independent trials is $P(X < m) = 0.5$. In order for the first order statistic Y_1 to be less than m and the last order statistic Y_5 to be greater than m, we must have at least one success, but not five successes. That is,

$$P(Y_1 < m < Y_5) = \sum_{k=1}^{4} \binom{5}{k}\left(\frac{1}{2}\right)^k\left(\frac{1}{2}\right)^{5-k}$$

$$= 1 - \left(\frac{1}{2}\right)^5 - \left(\frac{1}{2}\right)^5 = \frac{15}{16}.$$

So the probability that the random interval (Y_1, Y_5) includes m is $15/16 \approx 0.94$. Suppose now that this random sample is actually taken and the order statistics are observed to equal $y_1 < y_2 < y_3 < y_4 < y_5$, respectively. Then (y_1, y_5) is a 94% confidence interval for m.

It is interesting to note what happens as the sample size increases. Let $Y_1 < Y_2 < \cdots < Y_n$ be the order statistics of a random sample of size n from a distribution of the continuous type. Then $P(Y_1 < m < Y_n)$ is the probability that there is at least one "success" but not n successes, where the probability of success on each trial is

$P(X < m) = 0.5$. Consequently,

$$P(Y_1 < m < Y_n) = \sum_{k=1}^{n-1} \binom{n}{k}\left(\frac{1}{2}\right)^k \left(\frac{1}{2}\right)^{n-k}$$

$$= 1 - \left(\frac{1}{2}\right)^n - \left(\frac{1}{2}\right)^n = 1 - \left(\frac{1}{2}\right)^{n-1}.$$

This probability increases as n increases, so that the corresponding confidence interval (y_1, y_n) would have the very large confidence coefficient $1 - (1/2)^{n-1}$. Unfortunately, the interval (y_1, y_n) tends to get wider as n increases; thus, we are not "pinning down" m very well. However, if we used the interval (y_2, y_{n-1}) or (y_3, y_{n-2}), we would obtain shorter intervals, but also smaller confidence coefficients. Let us investigate this possibility further.

With the order statistics $Y_1 < Y_2 < \cdots < Y_n$ associated with a random sample of size n from a continuous-type distribution, consider $P(Y_i < m < Y_j)$, where $i < j$. For example, we might want

$$P(Y_2 < m < Y_{n-1}) \qquad \text{or} \qquad P(Y_3 < m < Y_{n-2}).$$

On each of the n independent trials, we say that we have success if that X is less than m; thus, the probability of success on each trial is $P(X < m) = 0.5$. Consequently, to have the ith order statistic Y_i less than m and the jth order statistic greater than m, we must have at least i successes, but fewer than j successes (or else $Y_j < m$). That is,

$$P(Y_i < m < Y_j) = \sum_{k=i}^{j-1} \binom{n}{k}\left(\frac{1}{2}\right)^k \left(\frac{1}{2}\right)^{n-k} = 1 - \alpha.$$

For particular values of n, i, and j, this probability—say, $1 - \alpha$—which is the sum of probabilities from a binomial distribution, can be calculated directly or approximated by an area under the normal p.d.f., provided that n is large enough. The observed interval (y_i, y_j) could then serve as a $100(1 - a)\%$ confidence interval for the unknown distribution median.

EXAMPLE 8.4-1   The lengths in centimeters of $n = 9$ fish of a particular species (*nezumia*) captured off the New England coast were 32.5, 27.6, 29.3, 30.1, 15.5, 21.7, 22.8, 21.2, and 19.0. Thus, the observed order statistics are

$$15.5 < 19.0 < 21.2 < 21.7 < 22.8 < 27.6 < 29.3 < 30.1 < 32.5.$$

Before the sample is drawn, we know that

$$P(Y_2 < m < Y_8) = \sum_{k=2}^{7} \binom{9}{k}\left(\frac{1}{2}\right)^k \left(\frac{1}{2}\right)^{9-k}$$

$$= 0.9805 - 0.0195 = 0.9610,$$

from Table II in the appendix. Thus, the confidence interval ($y_2 = 19.0$, $y_8 = 30.1$) for m, the median of the lengths of all fish of this species, has a 96.1% confidence coefficient. ∎

So that the student need not compute many of these probabilities, Table 8.4-1 on the next page lists the necessary information for constructing confidence intervals

n	$(i, n + 1 - i)$	$P(Y_i < m < Y_{n+1-i})$	n	$(i, n + 1 - i)$	$P(Y_i < m < Y_{n+1-i})$
5	$(1,5)$	0.9376	13	$(3,11)$	0.9776
6	$(1,6)$	0.9688	14	$(4,11)$	0.9426
7	$(1,7)$	0.9844	15	$(4,12)$	0.9648
8	$(2,7)$	0.9296	16	$(5,12)$	0.9232
9	$(2,8)$	0.9610	17	$(5,13)$	0.9510
10	$(2,9)$	0.9786	18	$(5,14)$	0.9692
11	$(3,9)$	0.9346	19	$(6,14)$	0.9364
12	$(3,10)$	0.9614	20	$(6,15)$	0.9586

TABLE 8.4-1: Information for Confidence Intervals for m

of the form (y_i, y_{n+1-i}) for the unknown m for sample sizes $n = 5, 6, \cdots, 20$. The subscript i is selected so that the confidence coefficient $P(Y_i < m < Y_{n+1-i})$ is greater than 90% and as close to 95% as possible.

For sample sizes larger than 20, we approximate those binomial probabilities with areas under the normal curve. To illustrate how good these approximations are, we compute the probability corresponding to $n = 16$ in Table 8.4-1. Here, using Table II, we have

$$1 - \alpha = P(Y_5 < m < Y_{12}) = \sum_{k=5}^{11} \binom{16}{k} \left(\frac{1}{2}\right)^k \left(\frac{1}{2}\right)^{16-k}$$
$$= P(W = 5, 6, \ldots, 11)$$
$$= 0.9616 - 0.0384 = 0.9232,$$

where W is $b(16, 1/2)$. The normal approximation gives

$$1 - \alpha = P(4.5 < W < 11.5) = P\left(\frac{4.5 - 8}{2} < \frac{W - 8}{2} < \frac{11.5 - 8}{2}\right),$$

because W has mean $np = 8$ and variance $np(1 - p) = 4$. The standardized variable $Z = (W - 8)/2$ has an approximate normal distribution. Thus,

$$1 - \alpha \approx \Phi\left(\frac{3.5}{2}\right) - \Phi\left(\frac{-3.5}{2}\right) = \Phi(1.75) - \Phi(-1.75)$$
$$= 0.9599 - 0.0401 = 0.9198.$$

This value compares very favorably with the probability 0.9232 recorded in Table 8.4-1. (Note that Minitab or some other computer program can also be used.)

The argument used to find a confidence interval for the median m of a distribution of the continuous type can be applied to any percentile π_p. In this case, we say that we have success on a single trial if that X is less than π_p. Thus, the probability of success on each of the independent trials is $P(X < \pi_p) = p$. Accordingly, with $i < j$, $1 - \alpha = P(Y_i < \pi_p < Y_j)$ is the probability that we have at least i successes, but fewer than j successes. Hence,

$$1 - \alpha = P(Y_i < \pi_p < Y_j) = \sum_{k=i}^{j-1} \binom{n}{k} p^k (1 - p)^{n-k}.$$

Once the sample is observed and the order statistics determined, the known interval (y_i, y_j) could serve as a $100(1 - \alpha)\%$ confidence interval for the unknown distribution percentile π_p.

EXAMPLE 8.4-2   Let the following numbers represent the order statistics of the $n = 27$ observations obtained in a random sample from a certain population of incomes (measured in hundreds of dollars):

$$
\begin{array}{cccccc}
261 & 280 & 292 & 305 & 329 & 364 \\
269 & 283 & 293 & 313 & 341 & 391 \\
271 & 284 & 296 & 321 & 343 & 417 \\
274 & 286 & 300 & 322 & 356 & 476 \\
279 & 287 & 304 & & &
\end{array}
$$

Say we are interested in estimating the 25th percentile, $\pi_{0.25}$, of the population. Since $(n + 1)p = 28(1/4) = 7$, the seventh order statistic, namely, $y_7 = 283$, would be a point estimate of $\pi_{0.25}$. To find a confidence interval for $\pi_{0.25}$, let us move down and up a few order statistics from y_7—say, to y_4 and y_{10}. What is the confidence coefficient associated with the interval (y_4, y_{10})? Before the sample was drawn, we had

$$1 - \alpha = P(Y_4 < \pi_{0.25} < Y_{10}) = \sum_{k=4}^{9} \binom{27}{k}(0.25)^k(0.75)^{27-k} - 0.8201,$$

from Minitab. For the normal approximation, we use W, which is $b(27, 1/4)$ with mean $27/4 = 6.75$ and variance $81/16$. Hence,

$$1 - \alpha = P(4 \le W \le 9) = P(3.5 < W < 9.5)$$

$$\approx \Phi\left(\frac{9.5 - 6.75}{9/4}\right) - \Phi\left(\frac{3.5 - 6.75}{9/4}\right)$$

$$= \Phi\left(\frac{11}{9}\right) - \Phi\left(-\frac{13}{9}\right) = 0.8149.$$

Thus, $(y_4 = 274, y_{10} = 287)$ is an 82.01% (or approximate 81.49%) confidence interval for $\pi_{0.25}$. Note that we could choose other intervals, such as $(y_3 = 271, y_{11} = 292)$, and these would have different confidence coefficients. The persons involved in the study must select the desired confidence coefficient, and then the appropriate order statistics are taken, usually quite symmetrically about the $(n + 1)p$th order statistic.   ■

When the number of observations is large, it is important to be able to determine the order statistics rather easily. As illustrated in the next example, a stem-and-leaf diagram, as introduced in Section 3.2, can be helpful in determining the needed order statistics.

EXAMPLE 8.4-3   The measurements of butterfat produced by $n = 90$ cows during a 305-day milk production period following their first calf are summarized in Table 8.4-2, in which each leaf consists of two digits. From this display, it is quite easy to see that $y_8 = 392$. It takes a little more work to show that $y_{38} = 494$ and $y_{53} = 526$ creates an interval

TABLE 8.4-2: Ordered Stem-and-Leaf Diagram of Butterfat Production

Stems	Leaves									Depths
2s	74									1
2•										1
3*										1
3t	27	39								3
3f	45	50								5
3s										5
3•	80	88	92	94	95					10
4*	17	18								12
4t	21	22	27	34	37	39				18
4f	44	52	53	53	57	58				24
4s	60	64	66	70	70	72	75	78		32
4•	81	86	89	91	92	94	96	97	99	41
5*	00	00	01	02	05	09	10	13	13	16	(10)
5t	24	26	31	32	32	37	37	39		39
5f	40	41	44	55						31
5s	61	70	73	74						27
5•	83	83	86	93	99					23
6*	07	08	11	12	13	17	18	19		18
6t	27	28	35	37						10
6f	43	43	45							6
6s	72									3
6•	91	96								2

$(494, 526)$ which serves as a confidence interval for the unknown median m of all butterfat production for the given breed of cows. Its confidence coefficient is

$$P(Y_{38} < m < Y_{53}) = \sum_{k=38}^{52} \binom{90}{k} \left(\frac{1}{2}\right)^k \left(\frac{1}{2}\right)^{90-k}$$

$$\approx \Phi\left(\frac{52.5 - 45}{\sqrt{22.5}}\right) - \Phi\left(\frac{37.5 - 45}{\sqrt{22.5}}\right)$$

$$= \Phi(1.58) - \Phi(-1.58) = 0.8858.$$

Similarly, $(y_{17} = 437, y_{29} = 470)$ is a confidence interval for the first quartile, $\pi_{0.25}$, with confidence coefficient

$$P(Y_{17} < \pi_{0.25} < Y_{29}) \approx \Phi\left(\frac{28.5 - 22.5}{\sqrt{16.875}}\right) - \Phi\left(\frac{16.5 - 22.5}{\sqrt{16.875}}\right)$$

$$= \Phi(1.46) - \Phi(-1.46) = 0.8558.$$

Minitab gives the exact confidence coefficients of 0.8867 and 0.8569, respectively. ∎

It is interesting to compare the length of a confidence interval for the mean μ obtained with $\bar{x} \pm t_{\alpha/2}(n - 1)(s/\sqrt{n})$ against the length of a $100(1 - \alpha)\%$ confidence interval for the median m obtained with the distribution-free techniques of this section. Usually, if the sample arises from a distribution that does not deviate too much from the normal, the confidence interval based upon $\bar{x}$ is much shorter. After all, we assume much more when we create that confidence interval. With the distribution-free method, all we assume is that the distribution is of the continuous type. So if the distribution is highly skewed or heavy tailed so that outliers could exist, a distribution-free technique is safer and much more robust. Moreover, the distribution-free technique provides a way to get confidence intervals for various percentiles, and investigators are often interested in such intervals.

EXERCISES

8.4-1. Let $Y_1 < Y_2 < Y_3 < Y_4 < Y_5 < Y_6$ be the order statistics of a random sample of size $n = 6$ from a distribution of the continuous type having $(100p)$th percentile π_p. Compute

(a) $P(Y_2 < \pi_{0.5} < Y_5)$.

(b) $P(Y_1 < \pi_{0.25} < Y_4)$.

(c) $P(Y_4 < \pi_{0.9} < Y_6)$.

8.4-2. For $n = 12$ year 2007 model sedans whose horsepower is between 290 and 390, the following measurements give the time in seconds for the car to go from 0 to 60 mph:

6.0 6.3 5.0 6.0 5.7 5.9 6.8 5.5 5.4 4.8 5.4 5.8

(a) Find a 96.14% confidence interval for the median, m.

(b) The interval (y_1, y_7) could serve as a confidence interval for $\pi_{0.3}$. Find it and give its confidence coefficient.

8.4-3. A sample of $n = 9$ electrochromic mirrors was used to measure the following low-end reflectivity percentages:

7.12 7.22 6.78 6.31 5.99 6.58 7.80 7.40 7.05

(a) Find the endpoints for an approximate 95% confidence interval for the median, m.

(b) The interval (y_3, y_7) could serve as a confidence interval for m. Find it and give its confidence coefficient.

8.4-4. Let m denote the median weight of "80-pound" bags of water softener pellets. Use the following random sample of $n = 14$ weights to find an approximate 95% confidence interval for m:

80.51 80.28 80.40 80.35 80.38 80.28 80.27
80.16 80.59 80.56 80.32 80.27 80.53 80.32

(a) Find a 94.26% confidence interval for m.

(b) The interval (y_6, y_{12}) could serve as a confidence interval for $\pi_{0.6}$. What is its confidence coefficient?

8.4-5. Use the following weights of an observed random sample of $n = 11$ pieces of candy to find

an approximate 95% confidence interval for the median weight, m:

2.76 2.96 2.67 2.97 2.77 2.72
2.74 2.84 2.61 2.82 2.72

Also, give the exact confidence coefficient.

8.4-6. A company manufactures mints that have a label weight of 20.4 grams. The company regularly samples from the production line and weighs the selected mints. During two mornings of production it sampled 81 mints, obtaining the following weights.

21.8 21.7 21.7 21.6 21.3 21.6 21.5 21.3 21.2
21.0 21.6 21.6 21.6 21.5 21.4 21.8 21.7 21.6
21.6 21.3 21.9 21.9 21.6 21.0 20.7 21.8 21.7
21.7 21.4 20.9 22.0 21.3 21.2 21.0 21.0 21.9
21.7 21.5 21.5 21.1 21.3 21.3 21.2 21.0 20.8
21.6 21.6 21.5 21.5 21.2 21.5 21.4 21.4 21.3
21.2 21.8 21.7 21.7 21.6 20.5 21.8 21.7 21.5
21.4 21.4 21.9 21.8 21.7 21.4 21.3 20.9 21.9
20.7 21.1 20.8 20.6 20.6 22.0 22.0 21.7 21.6

(a) Construct an ordered stem-and-leaf display using stems of $20f$, $20s$, $20\bullet$, $21*$, ..., $22*$.

(b) Find (i) the three quartiles, (ii) the 60th percentile, and (iii) the 15th percentile.

(c) Find approximate 95% confidence intervals for (i) $\pi_{0.25}$, (ii) $m = \pi_{0.5}$, and (iii) $\pi_{0.75}$.

8.4-7. A biologist who studies spiders selected a random sample of 20 male green lynx spiders (a spider that does not weave a web, but chases and leaps on its prey) and measured the lengths (in millimeters) of one of the front legs of the 20 spiders. Use the following measurements to construct a confidence interval for m that has a confidence coefficient about equal to 0.95:

15.10 13.55 15.75 20.00 15.45
13.60 16.45 14.05 16.95 19.05
16.40 17.05 15.25 16.65 16.25
17.75 15.40 16.80 17.55 19.05

8.4-8. The biologist of Exercise 8.4-7 also selected a random sample of 20 female green lynx spiders and measured the length (again in millimeters) of one of their front legs. Use the following data to construct a confidence interval for m that has a confidence coefficient about equal to 0.95:

15.85	18.00	11.45	15.60	16.10
18.80	12.85	15.15	13.30	16.65
16.25	16.15	15.25	12.10	16.20
14.80	14.60	17.05	14.15	15.85

8.4-9. Here are the weights (in grams) of 25 indicator housings used on gauges (see Exercise 3.2-10):

102.0	106.3	106.6	108.8	107.7
106.1	105.9	106.7	106.8	110.2
101.7	106.6	106.3	110.2	109.9
102.0	105.8	109.1	106.7	107.3
102.0	106.8	110.0	107.9	109.3

(a) List the observations in order of magnitude.

(b) Give point estimates of $\pi_{0.25}$, m, and $\pi_{0.75}$.

(c) Find the following confidence intervals and, from Table II in the appendix, state the associated confidence coefficient:

   (i) (y_3, y_{10}), a confidence interval for $\pi_{0.25}$.

   (ii) (y_9, y_{17}), a confidence interval for the median m.

   (iii) (y_{16}, y_{23}), a confidence interval for $\pi_{0.75}$.

(d) Use $\bar{x} \pm t_{\alpha/2}(24)(s/\sqrt{25})$ to find a confidence interval for μ, whose confidence coefficient corresponds to that of (d), part (ii). Compare these two confidence intervals of the middles.

8.4-10. Let X equal the weight in grams of a miniature candy bar. A random sample of candy bar weights yielded the following observations of X:

22.4	23.7	22.4	23.6	24.0	24.5	23.4	23.9
22.8	22.5	23.4	23.0	22.8	22.6	24.0	23.3

(a) Give a point estimate of the median $m = \pi_{0.50}$.

(b) Find a confidence interval for m with confidence level between 90% and 95%. Give the value of the exact confidence coefficient.

8.4-11. Let X equal the amount of fluoride in a certain brand of toothpaste. The specifications are 0.85–1.10 mg/g. Table 3.1-3 lists 100 such measurements.

(a) Give a point estimate of the median $m = \pi_{0.50}$.

(b) Find an approximate 95% confidence interval for the median m. If possible, use a computer to find the exact confidence level.

(c) Give a point estimate for the first quartile.

(d) Find an approximate 95% confidence interval for the first quartile and, if possible, give the exact confidence coefficient.

(e) Give a point estimate for the third quartile.

(f) Find an approximate 95% confidence interval for the third quartile and, if possible, give the exact confidence coefficient.

8.4-12. When placed in solutions of varying ionic strength, paramecia grow blisters in order to counteract the flow of water. The following 60 measurements in microns are blister lengths:

7.42	5.73	3.80	5.20	11.66	8.51	6.31	8.49
10.31	6.92	7.36	5.92	6.74	8.93	9.61	11.38
12.78	11.43	6.57	13.50	10.58	8.03	10.07	8.71
10.09	11.16	7.22	10.10	6.32	10.30	10.75	11.51
11.55	11.41	9.40	4.74	6.52	12.10	6.01	5.73
7.57	7.80	6.84	6.95	8.93	8.92	5.51	6.71
10.40	13.44	9.33	8.57	7.08	8.11	13.34	6.58
8.82	7.70	12.22	7.46				

(a) Construct an ordered stem-and-leaf diagram.

(b) Give a point estimate of the median $m = \pi_{0.50}$.

(c) Find an approximate 95% confidence interval for m.

(d) Give a point estimate for the 40th percentile, $\pi_{0.40}$.

(e) Find an approximate 90% confidence interval for $\pi_{0.40}$.

8.4-13. Using the weights of Verica's 39 gold coins given in Example 3.2-4, find approximate 95% confidence intervals for $\pi_{0.25}$, $\pi_{0.5}$, and $\pi_{0.75}$.

8.4-14. Let $Y_1 < Y_2 < \cdots < Y_8$ be the order statistics of eight independent observations from a continuous-type distribution with 70th percentile $\pi_{0.7} = 27.3$.

(a) Determine $P(Y_7 < 27.3)$.

(b) Find $P(Y_5 < 27.3 < Y_8)$.

8.4-15. Let X equal the weight in grams of peanuts in a "14-gram" snack package. The order statistics of a sample of size $n = 16$ are

13.9	14.4	14.6	14.7	14.7	15.2	15.2	15.2
15.3	15.4	15.4	15.5	15.6	15.6	15.9	16.4

(a) Find the 30th percentile of this sample.

(b) Find $P(Y_2 < \pi_{0.30} < Y_8)$, where Y_2 and Y_8 are the respective second- and eighth-order statistics of 16 independent observations.

8.5 THE WILCOXON TESTS

As mentioned earlier in the text, at times it is clear that the normal assumptions are not met and that other procedures, sometimes referred to as **nonparametric** or **distribution-free** methods, should be considered. For example, suppose some hypothesis, say, H_0: $m = m_0$, against H_1: $m \neq m_0$, is made about the unknown median, m, of a continuous-type distribution. From the data, we could construct a $100(1 - \alpha)$ percent confidence interval for m, and if m_0 is not in that interval, we would reject H_0 at the α significance level.

Now let X be a continuous-type random variable and let m denote the median of X. To test the hypothesis H_0: $m = m_0$ against an appropriate alternative hypothesis, we could also use a **sign test**. That is, if $X_1, X_2, \ldots, X_n$ denote the observations of a random sample from this distribution, and if we let Y equal the number of negative differences among $X_1 - m_0, X_2 - m_0, \ldots, X_n - m_0$, then Y has the binomial distribution $b(n, 1/2)$ under H_0 and is the test statistic for the sign test. If Y is too large or too small, we reject H_0: $m = m_0$.

EXAMPLE 8.5-1 Let X denote the length of time in seconds between two calls entering a college switchboard. Let m be the unique median of this continuous-type distribution. We test the null hypothesis H_0: $m = 6.2$ against the alternative hypothesis H_1: $m < 6.2$. Table II in the appendix tells us that if Y is the number of lengths of time in a random sample of size 20 that are less than 6.2, then the critical region $C = \{y : y \geq 14\}$ has a significance level of $\alpha = 0.0577$. A random sample of size 20 yielded the following data:

6.8	5.7	6.9	5.3	4.1	9.8	1.7	7.0
2.1	19.0	18.9	16.9	10.4	44.1	2.9	2.4
4.8	18.9	4.8	7.9				

Since $y = 9$, the null hypothesis is not rejected.   ■

The sign test can also be used to test the hypothesis that two dependent continuous-type random variables X and Y are such that $p = P(X > Y) = 1/2$. To test the hypothesis H_0: $p = 1/2$ against an appropriate alternative hypothesis, consider the independent pairs $(X_1, Y_1), (X_2, Y_2), \ldots, (X_n, Y_n)$. Let W denote the number of pairs for which $X_i - Y_i > 0$. When H_0 is true, W is $b(n, 1/2)$, and the test can be based upon the statistic W. For example, say X is the length of the right foot of a person and Y the length of the corresponding left foot. Thus, there is a natural pairing, and here H_0: $p = P(X > Y) - 1/2$ suggests that either foot of a particular individual is equally likely to be longer.

One major objection to the sign test is that it does not take into account the magnitude of the differences $X_1 - m_0, \ldots, X_n - m_0$. We now discuss a **test of Wilcoxon** that does take into account the magnitude of the differences $|X_i - m_0|$, $i = 1, 2, \ldots, n$. However, in addition to assuming that the random variable X is of the continuous type, we must also assume that the p.d.f. of X is symmetric about the median in order to find the distribution of this new statistic. Because of the continuity assumption, we assume, in the discussion which follows, that no two observations are equal and that no observation is equal to the median.

We are interested in testing the hypothesis H_0: $m = m_0$, where m_0 is some given constant. With our random sample $X_1, X_2, \ldots, X_n$, we rank the absolute values $|X_1 - m_0|, |X_2 - m_0|, \ldots, |X_n - m_0|$ in ascending order according to magnitude.

That is, for $i = 1, 2, \ldots, n$, we let R_i denote the rank of $|X_i - m_0|$ among $|X_1 - m_0|, |X_2 - m_0|, \ldots, |X_n - m_0|$. Note that $R_1, R_2, \ldots, R_n$ is a permutation of the first n positive integers, $1, 2, \ldots, n$. Now, with each R_i, we associate the sign of the difference $X_i - m_0$; that is, if $X_i - m_0 > 0$, we use R_i, but if $X_i - m_0 < 0$, we use $-R_i$. The Wilcoxon statistic W is the sum of these n signed ranks.

EXAMPLE 8.5-2

Suppose the lengths of $n = 10$ sunfish are

$$x_i: \ 5.0 \ \ 3.9 \ \ 5.2 \ \ 5.5 \ \ 2.8 \ \ 6.1 \ \ 6.4 \ \ 2.6 \ \ 1.7 \ \ 4.3$$

We shall test H_0: $m = 3.7$ against the alternative hypothesis H_1: $m > 3.7$. Thus, we have

$$
\begin{array}{rrrrrrrrrrr}
x_i - m_0: & 1.3, & 0.2, & 1.5, & 1.8, & -0.9, & 2.4, & 2.7, & -1.1, & -2.0, & 0.6 \\
|x_i - m_0|: & 1.3, & 0.2, & 1.5, & 1.8, & 0.9, & 2.4, & 2.7, & 1.1, & 2.0, & 0.6 \\
\text{Ranks:} & 5, & 1, & 6, & 7, & 3, & 9, & 10, & 4, & 8, & 2 \\
\text{Signed Ranks:} & 5, & 1, & 6, & 7, & -3, & 9, & 10, & -4, & -8, & 2 \\
\end{array}
$$

Therefore, the Wilcoxon statistic is equal to

$$W = 5 + 1 + 6 + 7 - 3 + 9 + 10 - 4 - 8 + 2 = 25.$$

Incidentally, the positive answer seems reasonable because the number of the 10 lengths that are less than 3.7 is 3, which is the statistic used in the sign test. ∎

If the hypothesis H_0: $m = m_0$ is true, about one-half of the differences would be negative and thus about one-half of the signs would be negative. Hence, it seems that the hypothesis H_0: $m = m_0$ is supported if the observed value of W is close to zero. If the alternative hypothesis is H_1: $m > m_0$, we would reject H_0 if the observed $W = w$ is too large, since, in this case, the larger deviations $|X_i - m_0|$ would usually be associated with observations for which $x_i - m_0 > 0$. That is, the critical region would be of the form $\{w: w \geq c_1\}$. If the alternative hypothesis is H_1: $m < m_0$, the critical region would be of the form $\{w: w \leq c_2\}$. Of course, the critical region would be of the form $\{w: w \leq c_3 \text{ or } w \geq c_4\}$ for a two-sided alternative hypothesis H_1: $m \neq m_0$. In order to find the values of c_1, c_2, c_3, and c_4 that yield desired significance levels, it is necessary to determine the distribution of W under H_0. Accordingly, we consider certain characteristics of this distribution.

When H_0: $m = m_0$ is true,

$$P(X_i < m_0) = P(X_i > m_0) = \frac{1}{2}, \qquad i = 1, 2, \ldots, n.$$

Hence, the probability is $1/2$ that a negative sign is associated with the rank R_i of $|X_i - m_0|$. Moreover, the assignments of these n signs are independent because $X_1, X_2, \ldots, X_n$ are mutually independent. In addition, W is a sum that contains the integers $1, 2, \ldots, n$, each with a positive or negative sign. Since the underlying distribution is symmetric, it seems intuitively obvious that W has the same distribution as the random variable

$$V = \sum_{i=1}^{n} V_i,$$

where $V_1, V_2, \ldots, V_n$ are independent and

$$P(V_i = i) = P(V_i = -i) = \frac{1}{2}, \qquad i = 1, 2, \ldots, n.$$

That is, V is a sum that contains the integers $1, 2, \ldots, n$, and these integers receive their algebraic signs by independent assignments.

Since W and V have the same distribution, their means and variances are equal, and we can easily find those of V. Now, the mean of V_i is

$$E(V_i) = -i\left(\frac{1}{2}\right) + i\left(\frac{1}{2}\right) = 0;$$

thus,

$$E(W) = E(V) = \sum_{i=1}^{n} E(V_i) = 0.$$

The variance of V_i is

$$\mathrm{Var}(V_i) = E(V_i^2) = (-i)^2\left(\frac{1}{2}\right) + (i)^2\left(\frac{1}{2}\right) = i^2.$$

Hence,

$$\mathrm{Var}(W) = \mathrm{Var}(V) = \sum_{i=1}^{n} \mathrm{Var}(V_i) = \sum_{i=1}^{n} i^2 = \frac{n(n+1)(2n+1)}{6}.$$

We shall not try to find the distribution of W in general, since that p.m.f. does not have a convenient expression. However, we demonstrate how we could find the distribution of W (or V) with enough patience and computer support. Recall that the moment-generating function of V_i is

$$M_i(t) = e^{t(-i)}\left(\frac{1}{2}\right) + e^{t(+i)}\left(\frac{1}{2}\right) = \frac{e^{-it} + e^{it}}{2}, \qquad i = 1, 2, \ldots, n.$$

Let $n = 2$; then the moment-generating function of $V_1 + V_2$ is

$$M(t) = E[e^{t(V_1+V_2)}].$$

From the independence of V_1 and V_2, we obtain

$$M(t) = E(e^{tV_1})E(e^{tV_2})$$

$$= \left(\frac{e^{-t} + e^{t}}{2}\right)\left(\frac{e^{-2t} + e^{2t}}{2}\right)$$

$$= \frac{e^{-3t} + e^{-t} + e^{t} + e^{3t}}{4}.$$

This means that each of the points $-3, -1, 1, 3$ in the support of $V_1 + V_2$ has probability $1/4$.

Next let $n = 3$; then the moment-generating function of $V_1 + V_2 + V_3$ is

$$M(t) = E[e^{t(V_1+V_2+V_3)}]$$

$$= E[e^{t(V_1+V_2)}]E(e^{tV_3})$$

$$= \left(\frac{e^{-3t} + e^{-t} + e^{t} + e^{3t}}{4}\right)\left(\frac{e^{-3t} + e^{3t}}{2}\right)$$

$$= \frac{e^{-6t} + e^{-4t} + e^{-2t} + 2e^{0} + e^{-2t} + e^{-4t} + e^{-6t}}{8}.$$

Thus, the points $-6, -4, -2, 0, 2, 4,$ and 6 in the support of $V_1 + V_2 + V_3$ have the respective probabilities $1/8, 1/8, 1/8, 2/8, 1/8, 1/8,$ and $1/8$. Obviously, this procedure can be continued for $n = 4, 5, 6, \ldots$, but it is rather tedious. Fortunately, however, even though $V_1, V_2, \ldots, V_n$ are not identically distributed random variables, the sum V of them still has an approximate normal distribution. To obtain this normal approximation for V (or W), a more general form of the central limit theorem, due to Liapounov, can be used which allows us to say that the standardized random variable

$$Z = \frac{W - 0}{\sqrt{n(n+1)(2n+1)/6}}$$

is approximately $N(0, 1)$ when H_0 is true. We accept this theorem without proof; therefore we can use this normal distribution to approximate probabilities such as $P(W \geq c; H_0) \approx P(Z \geq z_\alpha; H_0)$ when the sample size n is sufficiently large. The next example illustrates this approximation.

EXAMPLE 8.5-3 The moment-generating function of W or of V is given by

$$M(t) = \prod_{i=1}^{n} \frac{e^{-it} + e^{it}}{2}.$$

Using a computer algebra system (CAS) such as *Maple*, we can expand $M(t)$ and find the coefficients of e^{kt}, which is equal to $P(W = k)$. In Figure 8.5-1, we have drawn a probability histogram for the distribution of W along with the approximating $N[0, n(n+1)(2n+1)/6]$ p.d.f. for $n = 4$ (a poor approximation) and for $n = 10$. It is important to note that the widths of the rectangles in the probability histogram are equal to 2, so the "half-unit correction for continuity" now becomes a unit correction. ∎

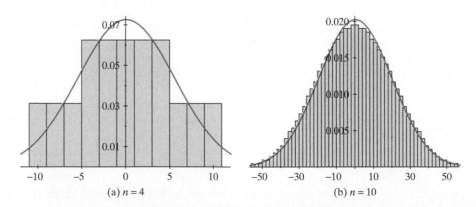

(a) $n = 4$ (b) $n = 10$

FIGURE 8.5-1: The Wilcoxon distribution

EXAMPLE 8.5-4 Let m be the median of a symmetric distribution of the continuous type. To test the hypothesis $H_0: m = 160$ against the alternative hypothesis $H_1: m > 160$, we take a random sample of size $n = 16$. For an approximate significance level of $\alpha = 0.05$, H_0 is rejected if the computed $W = w$ is such that

$$z = \frac{w}{\sqrt{16(17)(33)/6}} \geq 1.645,$$

TABLE 8.5-1: Ordered Absolute Differences from 160

0.1	0.3	0.5	1.2	1.7	2.7	3.4	5.2
1	2	3	4	5	6	7	8
7.9	9.8	12.2	12.4	14.5	16.9	17.8	24.4
9	10	11	12	13	14	15	16

or

$$w \geq 1.645 \sqrt{\frac{16(17)(33)}{6}} = 63.626.$$

Say the observed values of a random sample are 176.9, 158.3, 152.1, 158.8, 172.4, 169.8, 159.7, 162.7, 156.6, 174.5, 184.4, 165.2, 147.8, 177.8, 160.1, and 160.5. In Table 8.5-1, the magnitudes of the differences $|x_i - 160|$ have been ordered and ranked. Those differences $x_i - 160$ which were negative have been underlined, and the ranks are under the ordered values. For this set of data,

$$w = 1 - 2 + 3 - 4 - 5 + 6 + \cdots + 16 = 60.$$

Since $60 < 63.626$, H_0 is not rejected at the 0.05 significance level. It is interesting to note that H_0 would have been rejected at $\alpha = 0.10$, since, with a unit correction made for continuity, the approximate p-value is,

$$p\text{-value} = P(W \geq 60)$$

$$= P\left(\frac{W - 0}{\sqrt{(16)(17)(33)/6}} \geq \frac{59 - 0}{\sqrt{(16)(17)(33)/6}} \right)$$

$$\approx P(Z \geq 1.525) = 0.0636.$$

(*Maple* produces a p-value equal to $4{,}251/65{,}536 = 0.0649$.) Such a p-value would indicate that the data are too few to reject H_0, but if the pattern continues, we shall most certainly reject H_0 with a larger sample size.   ■

Although theoretically we could ignore the possibilities that $x_i = m_0$ for some i and that $|x_i - m_0| = |x_j - m_0|$ for some $i \neq j$, these situations do occur in applications. Usually, in practice, if $x_i = m_0$ for some i, that observation is deleted and the test is performed with a reduced sample size. If the absolute values of the differences from m_0 of two or more observations are equal, each observation is assigned the average of the corresponding ranks. The change this causes in the distribution of W is not very great; thus, we continue using the same normal approximation.

We now give an example that has some tied observations. This example also illustrates that the Wilcoxon rank sum test can be used to test hypotheses about the median of paired data.

EXAMPLE 8.5-5   We consider some paired data for percentage of body fat measured at the beginning and the end of a semester. Let m equal the median of the differences, $x - y$. We shall use the Wilcoxon statistic to test the null hypothesis H_0: $m = 0$ against the alternative hypothesis H_1: $m > 0$ with the differences given shortly. Since there are

$n = 25$ nonzero differences, we reject H_0 if

$$z = \frac{w - 0}{\sqrt{(25)(26)(51)/6}} \geq 1.645$$

or, equivalently, if

$$w \geq 1.645 \sqrt{\frac{(25)(26)(51)}{6}} = 122.27$$

at an approximate $\alpha = 0.05$ significance level. The 26 differences are

$$
\begin{array}{ccccccccc}
1.8 & -3.1 & 0.1 & 1.1 & 0.6 & -5.1 & 9.2 & 0.2 & 0.4 \\
0.0 & 1.9 & -0.4 & -1.5 & 1.4 & -1.0 & 2.2 & 0.8 & -0.4 \\
2.0 & -5.8 & -3.4 & -2.3 & 3.0 & 2.7 & 0.2 & 3.2 \\
\end{array}
$$

Table 8.5-2 lists the ordered nonzero absolute values, with those that were originally negative underlined. The rank is under each observation. Note that in the case of ties, the average of the ranks of the tied measurements is given.

TABLE 8.5-2: Ordered Absolute Values, Changes in Percentage of Body Fat

0.1	0.2	0.2	0.4	0.4	0.4	0.6	0.8	1.0	1.1	1.4	1.5	1.8
1	2.5	2.5	5	5	5	7	8	9	10	11	12	13
1.9	2.0	2.2	2.3	2.7	3.0	3.1	3.2	3.4	5.1	5.8	9.2	
14	15	16	17	18	19	20	21	22	23	24	25	

The value of the Wilcoxon statistic is

$$w = 1 + 2.5 + 2.5 + 5 - 5 - 5 + \cdots + 25 = 51.$$

Since $51 < 122.27$, we fail to reject the null hypothesis. The approximate p-value of this test is

$$p\text{-value} = P(W \geq 51) = P(W \geq 50)$$

$$\approx P\left(Z \geq \frac{50 - 0}{\sqrt{(25)(26)(51)/6}} \right) = P(Z \geq 0.673) = 0.2505. \qquad \blacksquare$$

Another method due to Wilcoxon for testing the equality of two distributions of the continuous type uses the magnitudes of the observations. For this test, it is assumed that the respective distribution functions F and G have similar shapes. To proceed with the test, place the combined sample of $X_1, X_2, \ldots, X_{n_1}$ and $Y_1, Y_2, \ldots, Y_{n_2}$ in increasing order of magnitude. Assign the ranks $1, 2, 3, \ldots, n_1 + n_2$ to the ordered values. In the case of ties, assign the average of the ranks associated with the tied values. Let W equal the sum of the ranks of $Y_1, Y_2, \ldots, Y_{n_2}$. If the distribution of Y is shifted to the right of that of X, the values of Y would tend to be larger than the values of X and W would usually be larger than expected when $F(z) = G(z)$. If m_X and m_Y are the respective medians, the critical region for testing $H_0 : m_X = m_Y$ against $H_1 : m_X < m_Y$ would be of the form $w \geq c$. Similarly, if the alternative hypothesis is $m_X > m_Y$, the critical region would be of the form $w \leq c$.

We shall not derive the distribution of W. However, if n_1 and n_2 are both greater than 7, a normal approximation can be used. With $F(z) = G(z)$, the mean and variance of W are

$$\mu_W = \frac{n_2(n_1 + n_2 + 1)}{2}$$

and

$$\text{Var}(W) = \frac{n_1 n_2 (n_1 + n_2 + 1)}{12},$$

and the statistic

$$Z = \frac{W - n_2(n_1 + n_2 + 1)/2}{\sqrt{n_1 n_2 (n_1 + n_2 + 1)/12}}$$

is approximately $N(0, 1)$.

EXAMPLE 8.5-6   The weights of of the contents of $n_1 = 8$ and $n_2 = 8$ tins of cinnamon packaged by companies A and B, respectively, selected at random, yielded the following observations of X and Y:

$$X: \quad 117.1 \quad 121.3 \quad 127.8 \quad 121.9 \quad 117.4 \quad 124.5 \quad 119.5 \quad 115.1,$$
$$Y: \quad 123.5 \quad 125.3 \quad 126.5 \quad 127.9 \quad 122.1 \quad 125.6 \quad 129.8 \quad 117.2.$$

The critical region for testing $H_0: m_X = m_Y$ against $H_1: m_X < m_Y$ is of the form $w \geq c$. Since $n_1 = n_2 = 8$, at an approximate $\alpha = 0.05$ significance level H_0 is rejected if

$$z = \frac{w - 8(8 + 8 + 1)/2}{\sqrt{[(8)(8)(8 + 8 + 1)]/12}} \geq 1.645,$$

or

$$w \geq 1.645 \sqrt{\frac{(8)(8)(17)}{12}} + 4(17) = 83.66.$$

To calculate the value of W, it is sometimes helpful to construct a **back-to-back stem-and-leaf display**. In such a display, the stems are put in the center and the leaves go to the left and the right. (See Table 8.5-3.)

TABLE 8.5-3: Back-to-Back Stem-and-Leaf Diagram of Weights of Cinnamon				
X leaves		Stems	Y leaves	
	51	11f		
74	71	11s	72	
	95	11•		
19	13	12*		
		12t	21	35
	45	12f	53	56
	78	12s	65	79
		12•	98	

Multiply numbers by 10^{-1}

TABLE 8.5-4: Combined Ordered Samples							
115.1	117.1	_117.2_	117.4	119.5	121.3	121.9	_122.1_
1	2	3	4	5	6	7	8
_123.5_	124.5	_125.3_	_125.6_	_126.5_	127.8	_127.9_	129.8
9	10	11	12	13	14	15	16

Reading from this two-sided stem-and-leaf display, we show the combined sample in Table 8.5-4, with the Company B (Y) weights underlined. The ranks are given beneath the values.

From Table 8.5-4, the computed W is

$$w = 3 + 8 + 9 + 11 + 12 + 13 + 15 + 16 = 87 > 83.66.$$

Thus, H_0 is rejected. Finally, making a half-unit correction for continuity, we see that the p-value of this test is

$$p\text{-value} = P(W \geq 87)$$
$$= P\left(\frac{W - 68}{\sqrt{90.667}} \geq \frac{86.5 - 68}{\sqrt{90.667}}\right)$$
$$\approx P(Z \geq 1.943) = 0.0260. \qquad \blacksquare$$

REMARKS Shortly after Wilcoxon proposed his two-sample test, Mann and Whitney suggested a test based on the estimate of the probability $P(X < Y)$. In this test, they let U equal the number of times that $X_i < Y_j$, $i = 1, 2, \ldots, n_1$ and $j = 1, 2, \ldots, n_2$. Using the data in Example 8.5-6, we find that the computed U is $u = 51$ among all $n_1 n_2 = (8)(8) = 64$ pairs of (X, Y). Thus, the estimate of $P(X < Y)$ is $51/64$ or, in general, $u/n_1 n_2$. At the time of the Mann–Whitney suggestion, it was noted that U was just a linear function of Wilcoxon's W and hence really provided the same test. That relationship is

$$U = W - \frac{n_2(n_2 + 1)}{2},$$

which in our special case is

$$51 = 87 - \frac{8(9)}{2} = 87 - 36.$$

Thus, we often read about the test of Mann, Whitney, and Wilcoxon. From this observation, this test could be thought of as one testing $H_0: P(X < Y) = 1/2$ against the alternative $H_1: P(X < Y) > 1/2$ with critical region of the form $w \geq c$.

Note that the two-sample Wilcoxon test is much less sensitive to extreme values than is the Student's t test based on $\overline{X} - \overline{Y}$. Therefore, if there is considerable skewness or contamination, these proposed distribution-free tests are much safer. In particular, that of Wilcoxon is quite good and does not lose too much in case the distributions are close to normal ones. It is important to note that the one-sample Wilcoxon test requires symmetry of the underlying distribution, but the two-sample Wilcoxon test does not and thus can be used for skewed distributions.

From theoretical developments beyond the scope of this text, the two Wilcoxon tests are strong competitors of the usual one- and two-sample tests based upon normal assumptions, so the Wilcoxon tests should be considered if those assumptions are questioned.

Computer programs, including Minitab, will calculate the value of the Wilcoxon or Mann–Whitney statistic. However, it is instructive to do these tests by hand so that you can see what is being calculated! ∎

EXERCISES

8.5-1. It is claimed that the median weight m of certain loads of candy is 40,000 pounds.

(a) Use the following 13 observations and the Wilcoxon statistic to test the null hypothesis H_0: $m = 40,000$ against the one-sided alternative hypothesis H_1: $m < 40,000$ at an approximate significance level of $\alpha = 0.05$:

$$41,195 \quad 39,485 \quad 41,229 \quad 36,840 \quad 38,050 \quad 40,890 \quad 38,345$$
$$34,930 \quad 39,245 \quad 31,031 \quad 40,780 \quad 38,050 \quad 30,906$$

(b) What is the approximate p-value of this test?

(c) Use the sign test to test the same hypothesis.

(d) Calculate the p-value from the sign test and compare it with the p-value obtained from the Wilcoxon test.

8.5-2. A course in economics was taught to two groups of students, one in a classroom situation and the other by TV. There were 24 students in each group. The students were first paired according to cumulative grade point averages and background in economics, and then assigned to the courses by a flip of a coin. (The procedure was repeated 24 times.) At the end of the course each class was given the same final examination. Use the Wilcoxon test to test the hypothesis that the two methods of teaching are equally effective against a two-sided alternative. The differences in the final scores for each pair of students were as follows (the TV student's score was subtracted from the corresponding classroom student's score):

14	-4	-6	-2	-1	18
6	12	8	-4	13	7
2	6	21	7	-2	11
-3	-14	-2	17	-4	-5

8.5-3. Let X equal the weight (in grams) of a Hershey's grape-flavored Jolly Rancher. Denote the median of X by m. We shall test H_0: $m = 5.900$ against H_1: $m > 5.900$. A random sample of size $n = 25$ yielded the following ordered data:

$$5.625 \quad 5.665 \quad 5.697 \quad 5.837 \quad 5.863 \quad 5.870 \quad 5.878 \quad 5.884 \quad 5.908$$
$$5.967 \quad 6.019 \quad 6.020 \quad 6.029 \quad 6.032 \quad 6.037 \quad 6.045 \quad 6.049$$
$$6.050 \quad 6.079 \quad 6.116 \quad 6.159 \quad 6.186 \quad 6.199 \quad 6.307 \quad 6.387$$

(a) Use the sign test to test the hypothesis.

(b) Use the Wilcoxon test statistic to test the hypothesis.

(c) Use a t test to test the hypothesis.

(d) Write a short comparison of the three tests.

8.5-4. The outcomes on $n = 10$ simulations of a Cauchy random variable were -1.9415, 0.5901, -5.9848, -0.0790, -0.7757, -1.0962, 9.3820, -74.0216, -3.0678, and 3.8545. For the Cauchy distribution, the mean does not exist, but for this one, the median is believed to equal zero. Use the Wilcoxon test and these data to test H_0: $m = 0$ against the alternative hypothesis H_1: $m \neq 0$. Let $\alpha \approx 0.05$.

8.5-5. Let x equal a student's GPA in the fall semester and y the same student's GPA in the spring semester. Let m equal the median of the differences, $x - y$. We shall test the null hypothesis H_0: $m = 0$ against an appropriate alternative hypothesis that you select on the basis of your past experience. Use a Wilcoxon test and the following 15 observations of paired data to test H_0:

x	y	x	y
2.88	3.22	3.98	3.76
3.67	3.49	4.00	3.96
2.76	2.54	3.39	3.52
2.34	2.17	2.59	2.36
2.46	2.53	2.78	2.62
3.20	2.98	2.85	3.06
3.17	2.98	3.25	3.16
2.90	2.84		

8.5-6. Let m equal the median of the posttest grip strengths in the right arms of male freshmen in a study of health dynamics. We shall use observations on $n = 15$ such students to test the null hypothesis $H_0: m = 50$ against the alternative hypothesis $H_1: m > 50$.

(a) Using the Wilcoxon statistic, define a critical region that has an approximate significance level of $\alpha = 0.05$.

(b) Given the observed values

58.0 52.5 46.0 57.5 52.0 45.5 65.5 71.0
57.0 54.0 48.0 58.0 35.5 44.0 53.0,

what is your conclusion?

(c) What is the p-value of this test?

8.5-7. Let X equal the weight in pounds of a "1-pound" bag of carrots. Let m equal the median weight of a population of these bags. Test the null hypothesis $H_0: m = 1.14$ against the alternative hypothesis $H_1: m > 1.14$.

(a) With a sample of size $n = 14$, use the Wilcoxon statistic to define a critical region. Use $\alpha \approx 0.10$.

(b) What would be your conclusion if the observed weights were

1.12 1.13 1.19 1.25 1.06 1.31 1.12
1.23 1.29 1.17 1.20 1.11 1.18 1.23

(c) What is the p-value of your test?

8.5-8. A pharmaceutical company is interested in testing the effect of humidity on the weight of pills that are sold in aluminum packaging. Let X and Y denote the respective weights of pills and their packaging, when the packaging is good and when it is defective, after the pill has spent 1 week in a chamber containing 100% humidity and heated to $30°C$.

(a) Use the Wilcoxon test to test $H_0: m_X = m_Y$ against $H_0: m_X - m_Y < 0$ on the following random samples of $n_1 = 12$ observations of X and $n_2 = 12$ observations of Y:

X: 0.7565   0.7720   0.7776   0.7750   0.7494   0.7615
        0.7741   0.7701   0.7712   0.7719   0.7546   0.7719

Y: 0.7870   0.7750   0.7720   0.7876   0.7795   0.7972
        0.7815   0.7811   0.7731   0.7613   0.7816   0.7851

What is the p-value?

(b) Construct and interpret a q–q plot of these data. HINT: This is a q–q plot of the empirical distribution of X against that of Y.

8.5-9. Let us compare the failure times of a certain type of light bulb produced by two different manufacturers, X and Y, by testing 10 bulbs selected at random from each of the outputs. The data, in hundreds of hours used before failure, are

X: 5.6 4.6 6.8 4.9 6.1 5.3 4.5 5.8 5.4 4.7
Y: 7.2 8.1 5.1 7.3 6.9 7.8 5.9 6.7 6.5 7.1

(a) Use the Wilcoxon test to test the equality of medians of the failure times at the approximate 5% significance level. What is the p-value?

(b) Construct and interpret a q–q plot of these data. HINT: This is a q–q plot of the empirical distribution of X against that of Y.

8.5-10. Let X and Y denote the heights of blue spruce trees, measured in centimeters, growing in two large fields. We shall compare these heights by measuring 12 trees selected at random from each of the fields. Take $\alpha \approx 0.05$, and use the statistic W—the sum of the ranks of the observations of Y in the combined sample—to test the hypothesis $H_0: m_X = m_Y$ against the alternative hypothesis

$H_1: m_X < m_Y$ on the basis of the following $n_1 = 12$ observations of X and $n_2 = 12$ observations of Y:

X: 90.4   77.2   75.9   83.2   84.0   90.2
      87.6   67.4   77.6   69.3   83.3   72.7

Y: 92.7   78.9   82.5   88.6   95.0   94.4
      73.1   88.3   90.4   86.5   84.7   87.5

8.5-11. Let X and Y equal the sizes of grocery orders from, respectively, a south-side and a north-side food store of the same chain. We shall test the null hypothesis $H_0: m_X = m_Y$ against a two-sided alternative, using the following ordered observations:

X: 5.13   8.22   11.81   13.77   15.36
     23.71   31.39   34.65   40.17   75.58

Y: 4.42   6.47   7.12   10.50   12.12
     12.57   21.29   33.14   62.84   72.05

(a) Use the Wilcoxon test when $\alpha = 0.05$. What is the p-value of this two-sided test?

(b) Construct a q–q plot and interpret it. HINT: This is a q–q plot of the empirical distribution of X against that of Y.

8.5-12. A charter bus line has 48-passenger and 38-passenger buses. Let m_{48} and m_{38} denote the median number of miles traveled per day by the respective buses. With $\alpha = 0.05$, use the Wilcoxon statistic to test $H_0: m_{48} = m_{38}$ against the one-sided alternative $H_1: m_{48} > m_{38}$. Use the following data, which give the numbers of miles traveled per day for respective random samples of sizes 9 and 11:

48-passenger buses:	331	308	300	414	253	
	323	452	396	104		
38-passenger buses:	248	393	260	355	279	184
	386	450	432	196	197	

8.5-13. A company manufactures and packages soap powder in 6-pound boxes. The quality assurance department was interested in comparing the fill weights of packages from the East and West lines. Taking random samples from the two lines, the department obtained the following weights:

East line (X): 6.06 6.04 6.11 6.06 6.06
6.07 6.06 6.08 6.05 6.09

West line (Y): 6.08 6.03 6.04 6.07 6.11
6.08 6.08 6.10 6.06 6.04

(a) Let m_X and m_Y denote the median weights for the East and West lines, respectively. Test $H_0: m_X = m_Y$ against a two-sided alternative hypothesis, using the Wilcoxon test with $\alpha \approx 0.05$. Find the p-value of this two-sided test.

(b) Construct and interpret a q–q plot of these data.

8.5-14. In Exercise 7.3-19, data are given that show the effect of a certain fertilizer on plant growth. The growths of the plants in mm over six weeks are repeated here, where Group A received fertilizer and Group B did not:

Group A:	55	61	33	57	17	46	50	42	71	51	63
Group B:	31	27	12	44	9	25	34	53	33	21	32

We shall test the hypothesis that fertilizer enhanced the growth of the plants.

(a) Construct a back-to-back stem-and-leaf display in which the stems are put down the center of the diagram and the Group A leaves go to the left while the Group B leaves go to the right.

(b) Calculate the value of the Wilcoxon statistic and give your conclusion.

(c) How does this result compare with that using the t test in Exercise 7.3-19?

8.5-15. With $\alpha = 0.05$, use the Wilcoxon statistic to test $H_0: m_X = m_Y$ against a two-sided alternative. Use the following observations of X and Y, which have been ordered for your convenience:

X:	−2.3864	−2.2171	−1.9148	−1.9097	−1.4883
	−1.2007	−1.1077	−0.3601	0.4325	1.0598
	1.3035	1.5241	1.7133	1.7656	2.4912
Y:	−1.7613	−0.9391	−0.7437	−0.5530	−0.2469
	0.0647	0.2031	0.3219	0.3579	0.6431
	0.6557	0.6724	0.6762	0.9041	1.3571

8.5-16. Data were collected during a step-direction experiment in the biomechanics laboratory of Hope College. The goal of the study is to establish differences in stepping responses between healthy young and healthy older adults. In one part of the experiment, the subjects are told in what direction they should take a step. Then, when given a signal, the subject takes a step in that direction, and the time it takes for them to lift their foot to take the step is measured. The direction is repeated a few times throughout the testing, and for each subject, a mean of all the "liftoff" times in a certain direction is calculated. The mean liftoff times (in thousandths of a second) for the anterior direction, ordered for your convenience, are as follows:

Young Subjects

397 433 450 468 485 488 498 504 561 565 569 576 577 579 581 586 696

Older Subjects

463 538 549 573 588 590 594 626 627 653 674 728 818 835 863 888 936

(a) Construct a back-to-back stem-and-leaf display. Use stems $3\bullet, 4*, \ldots, 9*$.

(b) Use the Wilcoxon statistic to test the null hypothesis that the response times are equal against the alternative that the times for the young subjects are less than that for the older subjects.

(c) What outcome does a t test give?

8.5-17. Some measurements in mm were made on a species of spiders named *Sosippus floridanus* that are native to Florida. (See Exercises 6.7-16 and 7.3-17.) There are 10 female spiders and 10 male spiders. The body lengths and the lengths of their front and their back legs are repeated here:

Female body lengths	Female front legs	Female back legs	Male body lengths	Male front legs	Male back legs
11.06	15.03	19.29	12.26	21.22	25.54
13.87	17.96	22.74	11.66	18.62	23.94
12.93	17.56	21.28	12.53	18.62	23.94
15.08	21.22	25.54	13.00	19.95	25.80
17.82	22.61	28.86	11.79	19.15	25.40
14.14	20.08	25.14	12.46	19.02	25.27
12.26	16.49	20.22	10.65	17.29	22.21
17.82	18.75	24.61	10.39	17.02	21.81
20.17	23.01	28.46	12.26	18.49	23.41
16.88	22.48	28.59	14.07	22.61	28.86

In this exercise, we shall use the Wilcoxon statistic to compare the sizes of the female and male spiders. For each of the following instructions, construct back-to-back stem-and-leaf displays:

(a) Test the null hypothesis that the lengths of female and male spiders are equal against the alternative hypothesis that female spiders are longer.

(b) Test the null hypothesis that the lengths of the front legs of the female and male spiders are equal against a two-sided alternative.

(c) Test the null hypothesis that the lengths of the back legs of the female and male spiders are equal against a two-sided alternative.

8.5-18. In Exercise 7.3-14, growth data are given for plants in normal air and for plants in CO_2-enriched air. Those data are repeated here:

Normal Air (X)	4.67	4.21	2.18	3.91	4.09	5.24	2.94	4.71
	4.04	5.79	3.80	4.38				
Enriched Air (Y)	5.04	4.52	6.18	7.01	4.36	1.81	6.22	5.70

In this exercise, we shall test test the null hypothesis that the medians are equal, namely, $H_0: m_X = m_Y$, against the alternative hypothesis $H_1: m_X < m_Y$. You may select the significance level. However, give the approximate p-value or state clearly why you arrived at a particular conclusion, for each of the tests. Show your work.

(a) What is your conclusion from the Wilcoxon test?

(b) What was your conclusion from the t test in Exercise 7.3-14?

(c) Write a comparison of these two tests.

8.5-19. In Exercise 7.3-20, we noted that Ledolter and Hogg reported the times needed for an old procedure and for a new procedure. The data are repeated here:

Old Procedure:	4.3	6.5	4.6	4.3	6.4	4.8	5.1	6.8	4.9
	4.5	5.1	7.3	3.3	5.0	4.6	7.0	5.1	3.8
	5.2	4.1	5.7	4.6	5.9	3.1	6.2		
New Procedure:	6.2	4.0	3.3	4.5	2.3	3.0	3.2	6.0	3.7
	4.5	5.3	4.0	5.4	4.3	3.8			

(a) Construct a back-to-back stem-and-leaf display.

(b) Use the Wilcoxon test to test the null hypothesis that the median response times are equal against a one-sided alternative that the median for the new procedure is less than that for the old procedure. Let $\alpha = 0.01$.

(c) What is the approximate p-value of your test?

(d) What was your conclusion from the t test in Exercise 7.3-20?

(e) How do the two tests compare?

8.6 RUN TEST AND TEST FOR RANDOMNESS

Under the assumption that the random variables X and Y are of the continuous type and have distribution functions $F(x)$ and $G(y)$, respectively, we describe another test of the hypothesis $H_0: F(z) = G(z)$. This new test can also be used to test for randomness. In either application, we need the concept of runs, which we now define.

Suppose that we have n_1 observations of the random variable X and n_2 observations of the random variable Y. The combination of two sets of independent observations into one collection of $n_1 + n_2$ observations, placed in ascending order of magnitude, might yield an ordered arrangement

$$\underline{y\,y\,y}\;\underline{x\,x}\;\underline{y}\;\underline{x}\;\underline{y}\;\underline{x\,x}\;\underline{y\,y},$$

where x denotes an observation of X and y an observation of Y. We have underlined groups of successive values of X and Y. Each underlined group is called a **run**. Thus, we have a run of three values of Y, followed by a run of two values of X, followed by a run of one value of Y, and so on. In this example, there are seven runs.

We give two more examples to show what might be indicated by the number of runs. If the five x's and seven y's had the ordering

$$\underline{x\,x\,x\,x}\;\underline{y}\;\underline{x}\;\underline{y\,y\,y\,y\,y\,y},$$

we might suspect that $F(z) \geq G(z)$. Note that there are four runs in this ordering. By contrast, the ordered arrangement

$$\underline{y\,y\,y}\;\underline{x\,x}\;\underline{y}\;\underline{x\,x\,x}\;\underline{y\,y\,y}$$

might suggest that the medians of the two distributions are equal, but that the spread of the Y distribution is greater than the spread of the X distribution—for example, that $\sigma_Y > \sigma_X$. All three examples suggest that the hypothesis $F(z) = G(z)$ should be rejected if the number of runs is too small, where a small number of runs could be caused by differences in the location or in the spread of the two distributions.

Let the random variable R equal the number of runs in the combined ordered sample of n_1 observations of X and n_2 observations of Y. We shall find the distribution of R when $F(z) = G(z)$ and then describe a test of the hypothesis $H_0: F(z) = G(z)$.

Under H_0, all permutations of the n_1 observations of X and n_2 observations of Y have equal probabilities. We can select the n_1 positions for the n_1 values of X in

$$\binom{n_1 + n_2}{n_1}$$

ways, the probability of each arrangement being

$$\frac{1}{\dbinom{n_1 + n_2}{n_1}}.$$

To find $P(R = r)$, we must determine the number of permutations that yield r runs.

First suppose that $r = 2k$, where k is a positive integer. In this case, the n_1 ordered values of X and the n_2 ordered values of Y must each be separated into k runs. We can form k runs of the n_1 values of X by inserting $k - 1$ dividers into the $n_1 - 1$ spaces between the values of X, with no more than one divider per space. This can be done in

$$\binom{n_1 - 1}{k - 1}$$

ways. (See Exercise 8.6-4.) Similarly, k runs of the n_2 values of Y can be formed in

$$\binom{n_2 - 1}{k - 1}$$

ways. These two sets of runs can be placed together to form $r = 2k$ runs, of which

$$\binom{n_1 - 1}{k - 1}\binom{n_2 - 1}{k - 1}$$

begin with a run of x's and

$$\binom{n_2 - 1}{k - 1}\binom{n_1 - 1}{k - 1}$$

begin with a run of y's. Thus,

$$P(R = 2k) = \frac{2\binom{n_1 - 1}{k - 1}\binom{n_2 - 1}{k - 1}}{\binom{n_1 + n_2}{n_2}}, \tag{8.6-1}$$

where $2k$ is an element of the space of R.

When $r = 2k + 1$, it is possible to have $k + 1$ runs of the ordered values of X and k runs of the ordered values of Y or k runs of Xs and $k + 1$ runs of Ys. We can form $k + 1$ runs of the n_1 values of X by inserting k dividers into the $n_1 - 1$ spaces between the values of X, with no more than one divider per space, in

$$\binom{n_1 - 1}{k}$$

ways. Similarly, k runs of n_2 values of Y can be done in

$$\binom{n_2 - 1}{k - 1}$$

ways. These two sets of runs can be placed together to form $2k + 1$ runs in

$$\binom{n_1 - 1}{k}\binom{n_2 - 1}{k - 1}$$

ways. In addition, $k + 1$ runs of the n_2 values of Y and k runs of the n_1 values of X can be placed together to form

$$\binom{n_2 - 1}{k}\binom{n_1 - 1}{k - 1}$$

sets of $2k + 1$ runs. Hence,

$$P(R = 2k + 1) = \frac{\binom{n_1 - 1}{k}\binom{n_2 - 1}{k - 1} + \binom{n_1 - 1}{k - 1}\binom{n_2 - 1}{k}}{\binom{n_1 + n_2}{n_2}}, \qquad (8.6\text{-}2)$$

for $2k + 1$ in the space of R.

A test based on the number of runs can be used to test the hypothesis $H_0: F(z) = G(z)$. The hypothesis is rejected if the observed number of runs r is too small. That is, the critical region is of the form $r \leq c$, where the constant c is determined by using the p.d.f. of R to yield the desired significance level. The **run test** is sensitive to both differences in location and differences in spread of the two distributions.

EXAMPLE 8.6-1 Let X and Y equal the percentages of body fat for freshman women and men, respectively, with distribution functions $F(x)$ and $G(y)$. We shall use the run test to test the hypothesis $H_0: F(z) = G(z)$ against the alternative hypothesis $H_1: F(z) < G(z)$. (That is, the alternative hypothesis is that the X distribution is to the right of the Y distribution.) Ten observations of both X and Y that have been ordered are

$$
\begin{array}{lccccc}
X: & 16.6 & 16.7 & 18.5 & 19.2 & 21.5 \\
   & 22.4 & 22.6 & 23.2 & 24.2 & 26.3 \\
Y: & 9.4 & 9.7 & 11.3 & 11.8 & 13.3 \\
   & 15.6 & 16.1 & 16.5 & 18.2 & 21.7
\end{array}
$$

The critical region is of the form $r \leq c$. To determine the value of c, we use Formulas 8.6-1 and 8.6-2 with $n_1 = n_2 = 10$. Table I in the appendix is very useful for evaluating these probabilities. We have

$$P(R = 2) = \frac{2}{184{,}756}; \quad P(R = 3) = \frac{18}{184{,}756};$$

$$P(R = 4) = \frac{162}{184{,}756}; \quad P(R = 5) = \frac{648}{184{,}756};$$

$$P(R = 6) = \frac{2592}{184{,}756}; \quad P(R = 7) = \frac{6048}{184{,}756}.$$

The sum of these six probabilities is $9470/184{,}756 = 0.051$, so we can take for our critical region $C = \{r: r \leq 7\}$ with a significance level of $\alpha = 0.051$. To determine the number of runs, we order the combined samples and underline adjacent x and y values:

$$
\begin{array}{cccccccccc}
9.4 & 9.7 & 11.3 & 11.8 & 13.3 & 15.6 & 16.1 & 16.5 & 16.6 & 16.7 \\
18.2 & 18.5 & 19.2 & 21.5 & 21.7 & 22.4 & 22.6 & 23.2 & 24.2 & 26.3
\end{array}
$$

We see that the number of runs is $r = 6$, so we reject the null hypothesis. Note that the p-value of this test is

$$p\text{-value} = P(R \leq 6) = \frac{3422}{184{,}756} = 0.0185.$$

When n_1 and n_2 are large (say, each is at least equal to 10), R can be approximated with a normally distributed random variable. That is, it can be shown that

$$\mu_R = E(R) = \frac{2n_1 n_2}{n_1 + n_2} + 1,$$

$$\mathrm{Var}(R) = \frac{(\mu_R - 1)(\mu_R - 2)}{n_1 + n_2 - 1} = \frac{2n_1 n_2(2n_1 n_2 - n_1 - n_2)}{(n_1 + n_2)^2(n_1 + n_2 - 1)},$$

and

$$Z = \frac{R - \mu_R}{\sqrt{\mathrm{Var}(R)}}$$

is approximately $N(0, 1)$. The critical region for testing the null hypothesis H_0: $F(z) = G(z)$ is of the form $z \leq -z_\alpha$, where α is the desired significance level.

EXAMPLE 8.6-2 We use the normal approximation to calculate the significance level and the p-value for Example 8.6-1. With $n_1 = n_2 = 10$,

$$\mu_R = \frac{2(10)(10)}{10 + 10} + 1 = 11; \qquad \sigma_R^2 = \frac{(11 - 1)(11 - 2)}{19} = \frac{90}{19}.$$

With the critical region $C = \{r: \ r \leq 7\}$, the approximate significance level, with a half-unit correction for continuity, is

$$\alpha = P(R \leq 7) = P\left(\frac{R - 11}{\sqrt{90/19}} \leq \frac{7.5 - 11}{\sqrt{90/19}}\right)$$

$$\approx P(Z \leq -1.608) = 0.0539.$$

Note that this value compares very favorably with $\alpha = 0.051$ given in Example 8.6-1. Since $r = 6$, the approximate p-value, under the assumption of a normal approximation, is

$$p\text{-value} = P(R \leq 6)$$

$$\approx P\left(Z \leq \frac{6.5 - 11}{\sqrt{90/19}}\right) = P(Z \leq -2.068) = 0.0193,$$

which is close to the p-value given in Example 8.6-1. ∎

Applications of the run test include tests for randomness. Analyses of runs can also be useful in quality-control studies. To illustrate these applications, let $x_1, x_2, \ldots, x_k$ be the observed values of a random variable X, where the subscripts now designate the order in which the outcomes were observed and the observations are not arranged in order of magnitude. In a quality-control situation, the observations may be made systematically—for example, every hour. Assume that k is even. The median divides the k numbers into a lower and an upper half. Replace each observation by L if it falls below the median and by U if it falls above the median. Then, on the one hand, a sequence such as

$$\text{U U U L U L L L}$$

might suggest a trend toward decreasing values of X. If the existence of a trend is the alternative hypothesis to randomness, the critical region would be of the form $r \leq c$.

On the other hand, if we have a sequence such as

U L U L U L U L,

we would suspect a cyclic effect and would reject the hypothesis of randomness if r were too large. To test for both trend and cyclic effects, the critical region for testing the hypothesis of randomness is of the form $r \leq c_1$ or $r \geq c_2$.

If the sample size k is odd, the number of observations in the "upper half" and "lower half" will differ by one. In this case, we will always put the extra observation in the upper group and, of course, $n_2 = n_1 + 1$. If the median is equal to a value that is tied with other values, we will again put the tied values in the upper group and then perform the test in which n_1 and n_2 are not equal to each other.

EXAMPLE 8.6-3 We shall use a sample of size $k = 14$ to test for both trend and cyclic effects. To determine the critical region for rejecting the hypothesis of randomness, we use the p.m.f. of R with $n_1 = n_2 = 7$. Since

$$P(R = 2) = P(R = 14) = \frac{2}{3432},$$

$$P(R = 3) = P(R = 13) = \frac{12}{3432},$$

$$P(R = 4) = P(R = 12) = \frac{72}{3432},$$

the critical region $\{r: \ r \leq 4 \text{ or } r \geq 12\}$ would yield a test at a significance level of $\alpha = 172/3432 = 0.05$. The 14 observations are

81.4 76.3 85.6 76.4 88.4 80.2 85.6
81.6 78.3 82.8 88.1 85.4 87.7 86.6

The median of these outcomes is $(84.6 + 85.4)/2 = 85.0$. Replacing each outcome with L if it falls below 85.0 and U if it falls above 85.0 yields the sequence

L L U L U L U L U L L L U U U U.

Since $r = 8$, the hypothesis of randomness is not rejected. ∎

EXERCISES

8.6-1. Let the total lengths of the male and female trident lynx spiders be denoted by X and Y, respectively, with corresponding distribution functions $F(x)$ and $G(y)$. Measurement of the lengths, in millimeters, of eight male and eight female spiders yielded the following observations of X and Y:

X: 5.40 5.55 6.00 5.00 5.70 5.20 5.45 4.95
Y: 6.20 6.25 5.75 5.85 6.55 6.05 5.50 6.65

(a) Use these data to test the hypothesis $H_0: F(z) = G(z)$ against the alternative $H_1: F(z) > G(z)$. Let $\alpha \approx 0.10$, and use a run test.

(b) Use the two-sample Wilcoxon test to test $H_0: m_X = m_Y$ against the alternative $H_1: m_X < m_Y$ with $\alpha \approx 0.10$.

8.6-2. Let X and Y denote the times in hours per week that students in two different schools watch television. Let $F(x)$ and $G(y)$ denote the respective distribution functions. To test the hypothesis $H_0: F(z) = G(z)$, a random sample of eight students was selected from each school. Use the following observations to test this hypothesis:

X: 16.75 19.25 22.00 20.50 22.50 15.50 17.25 20.75
Y: 24.75 21.50 19.75 17.50 22.75 23.50 13.00 19.00

8.6-3. A parade consists of six bands and 12 floats. How many different parade lineups are possible if the parade begins and ends with a band and there is at least 1 float between each pair of bands?

8.6-4. List the $\binom{5}{3} = 10$ ways in which three dividers can be inserted between six given values of x, with no more than one divider per space.

8.6-6. Let $\alpha \approx 0.05$, and use the run test to test $H_0: F(z) = G(z)$ on the following observations of X and Y, which have been ordered for your convenience:

$$
\begin{array}{rrrrr}
X: & -2.0482 & -1.5748 & -0.8797 & -0.7170 & -0.4907 \\
 & -0.2051 & 0.1651 & 0.2893 & 0.3186 & 0.3550 \\
 & 0.4056 & 0.6975 & 0.7113 & 0.7377 & 1.7356 \\
\end{array}
$$

$$
\begin{array}{rrrrr}
Y: & -1.2311 & -1.0228 & -0.8836 & -0.6684 & -0.6157 \\
 & -0.5755 & -0.1019 & -0.0297 & 0.3781 & 0.7400 \\
 & 0.8479 & 1.0901 & 1.1397 & 1.1748 & 1.2921 \\
\end{array}
$$

NOTE: These data were simulated on a computer, using the standard normal distribution for X and a $U(-1.5, 1.5)$ for Y.

8.6-7. Use the following sample of size $k = 14$ to test the hypothesis of randomness against the alternative hypothesis of a trend effect at an $\alpha = 0.025$ significance level:

12.4	14.2	11.7	14.0	12.7	15.7	12.8
14.1	17.9	18.4	17.5	20.2	20.8	20.3

8.6-8. Use the following sample of size $k = 16$ to test the hypothesis of randomness against the alternative

8.6-5. Given the observations of X and Y listed in Exercise 8.5-15, use the run test to test $H_0: F(z) = G(z)$.

(a) Show that if $C = \{r: r \leq 11\}$, then the significance level is $\alpha \approx 0.0473$.

(b) Calculate the value of r and state your conclusion.

(c) What is the p-value of this test?

hypothesis of a cyclic effect:

12.4	31.8	22.2	24.5	17.9	24.6	15.7	27.3
22.7	26.0	14.5	22.0	21.8	31.9	11.5	28.3

What is your conclusion if
(a) $\alpha = 0.0317$?
(b) $\alpha = 0.10$?

8.6-9. A manufacturer of powdered soap checks the weights of soap in the company's 6-pound boxes periodically throughout the day. At each of 22 times, four boxes are selected at random, and the average of the weights of the soap in the boxes is recorded. Use the following 22 average weights to test the hypothesis of randomness against the alternative hypothesis of a trend at an approximate significance level of $\alpha = 0.025$:

6.050	6.038	6.003	6.015	6.025	6.063	6.033	6.010
5.995	6.020	6.060	6.060	6.065	6.050	6.043	6.040
6.045	6.065	6.055	6.060	6.060	6.070		

8.6-10. Each hour, a manufacturer of mints selects four mints at random from the production line and finds the average weight in grams. For 1 week, the following average weights were observed:

21.2	21.7	21.3	21.4	21.8	21.9	21.6	21.7	21.3	21.9
21.3	22.0	21.3	21.5	21.6	21.3	21.6	21.9	21.3	21.6
21.9	22.0	21.9	21.4	21.4	21.3	21.7	21.6	21.5	21.7
21.3	21.7	21.0	21.3	21.3	21.6	20.9	21.4		

Use these weights to test the hypothesis of randomness against the alternative hypothesis of a cyclic effect.

8.6-11. It is claimed that X is $U(0, 1)$. A random sample of 14 observations of X yielded the following data:

0.15	0.67	0.05	0.47	0.29	0.23	0.10
0.01	0.96	0.92	0.51	0.73	0.91	0.82

(a) Test $H_0: m = 0.5$ against the two-sided alternative hypothesis $H_1: m \neq 0.5$, using the sign test with significance level $\alpha = 0.0574$.

(b) Use a run test with the preceding observations of X to test for both trend and cyclic effects at the approximate significance level $\alpha = 0.05$.

8.6-12. On an introductory statistics test, 27 form A and 29 form B tests were used. The order in which the tests were returned is as follows:

A B B A B A B A A B B A B B
A B B A A A B A B A A A B B
B A B A B A B A A A B A B B
A B A A B A B B A B A B B B

(a) Use a run test to test the null hypothesis that this is a random sequence against the alternative hypothesis that there is a cyclic effect. Use the normal approximation and an $\alpha = 0.01$ significance level.

(a) On the basis of your conclusion, what advice would you give to the professor who gave this test?

8.6-13. In Exercise 11.1-4, the U.S. birthrates are given for 1960–2002. In order, they are

23.7 23.3 22.4 21.7 21.1 19.4 18.4 17.8 17.6
17.9 18.4 17.2 15.6 14.8 14.8 14.6 14.6 15.1
15.0 15.6 15.9 15.8 15.9 15.5 15.5 15.8 15.6
15.7 15.9 16.3 16.7 16.2 15.8 15.4 15.0 14.6
14.4 14.2 14.3 14.2 14.4 14.1 13.9

Use the run test to test whether this is a random sequence against the alternative that there is a trend effect. Give the p-value of your test.

8.6-14. A window is manufactured that will be inserted into a car. To attach the window, five studs are located in the frame in locations A, B, C, D, and E. Periodically, a window is selected randomly from the production line and a standard "stud pull-out test" is performed that measures the force required to pull a stud out of the window. For each window, the tests are always performed in

the order A, B, C, D, and E. Seventy observations that resulted from 14 windows were

140 159 138 102 84 126 147 126 103 92
149 155 135 120 94 149 143 109 101 86
144 154 120 105 97 149 151 140 103 99
157 140 120 96 87 146 137 120 93 89
149 154 139 100 84 148 142 130 98 81
112 135 109 84 87 112 135 109 84 87
126 118 135 90 77 125 126 131 78 75

For these data, the median is equal to 120. Use the run test to test whether this is a random sequence against the alternative hypothesis of a cyclic effect. Give the approximate p-value of the test. Interpret your conclusion. (See Exercise 11.1-6.)

8.6-15. In Exercises 7.3-19 and 8.5-14, the effect of a certain fertilizer was tested. For those data, calculate the number of runs and compare this number with the expected number of runs. Does the run test detect any fertilizer effect?

8.6-16. In Exercise 8.5-16, reaction times are given for young and older subjects. Use the run test to test whether the reaction times are equal or whether the young react more quickly. Compare the p-value for this test with the result for the Wilcoxon test.

8.6-17. In Exercise 8.5-17, several measurements are given for spiders. Does the run test detect any differences between females and males for either body lengths or leg lengths?

8.6-18. In Exercise 8.5-18, plants growths are compared for normal and enriched air. Compare the number of runs with the mean number of runs. Would the run test detect any difference?

8.6-19. In Exercises 7.3-20 and 8.5-19, data are given comparing an old and a new procedure. How does the result of the run test compare with the tests performed in those exercises?

8.7 KOLMOGOROV–SMIRNOV GOODNESS-OF-FIT TEST

In this section, we discuss a test that considers the goodness of fit between a hypothesized distribution function and an empirical distribution function. The empirical distribution function is given here in terms of its order statistics. Let $y_1 < y_2 < \cdots < y_n$ be the observed values of the order statistics of a random sample $x_1, x_2, \ldots, x_n$ of size n. When no two observations are equal, the empirical distribution function is defined by

$$F_n(x) = \begin{cases} 0, & x < y_1, \\ k/n, & y_k \le x < y_{k+1}, \quad k = 1, 2, \ldots, n-1, \\ 1, & y_n \le x. \end{cases}$$

In this case, the empirical distribution function has a jump of magnitude $1/n$ occurring at each observation. If n_k observations are equal to x_k, a jump of magnitude n_k/n

occurs at x_k. In any case, $F_n(x)$ is simply the fraction of sample observations that are less than or equal to x.

Suppose that a random sample of size n is taken from a distribution of the continuous type that has the distribution function $F(x)$. How can we measure the "closeness" of $F(x)$ and the empirical distribution function $F_n(x)$? How does the sample size affect this closeness? We give some theoretical results to help answer these questions and then give a test for goodness of fit.

Let $X_1, X_2, \ldots, X_n$ denote a random sample of size n from a distribution of the continuous type with the distribution function $F(x)$. Consider a fixed value of x. Then $W = F_n(x)$, the value of the empirical distribution function at x, can be thought of as a random variable that takes on the values $0, 1/n, 2/n, \ldots, 1$. Now, $nW = k$ if, and only if, exactly k observations are less than or equal to x (say, success) and $n - k$ observations are greater than x. The probability that an observation is less than or equal to x is given by $F(x)$. That is, the probability of success is $F(x)$. Because of the independence of the random variables $X_1, X_2, \ldots, X_n$, the probability of k successes is given by the binomial distribution, namely,

$$P(nW = k) = P\left(W = \frac{k}{n}\right)$$

$$= \binom{n}{k}[F(x)]^k[1 - F(x)]^{n-k}, \qquad k = 0, 1, 2, \ldots, n.$$

Since nW has a binomial distribution with $p = F(x)$, the mean and variance of nW are, respectively, given by

$$E(nW) = nF(x) \qquad \text{and} \qquad \text{Var}(nW) = n[F(x)][1 - F(x)].$$

Hence, the mean and the variance of $W = F_n(x)$ are

$$E[F_n(x)] = E(W) = F(x)$$

and

$$\text{Var}[F_n(x)] = \text{Var}(W) = \frac{F(x)[1 - F(x)]}{n}.$$

Since the variance of $F_n(x)$ approaches zero as n becomes large, $F_n(x)$ and its mean $F(x)$ tend to be closer with large n. As a matter of fact, there is a theorem by Glivenko, the proof of which is beyond the level of this book, which states that, with probability 1, $F_n(x)$ converges to $F(x)$ uniformly in x as $n \to \infty$.

Because of the convergence of the empirical distribution function to the theoretical distribution function, it makes sense to construct a goodness-of-fit test based on the closeness of the empirical and a hypothesized distribution function, say, $F_n(x)$ and $F_0(x)$, respectively. We shall use the Kolmogorov–Smirnov statistic defined by

$$D_n = \sup_x[\,|F_n(x) - F_0(x)|\,].$$

That is, D_n is the least upper bound of all pointwise differences $|F_n(x) - F_0(x)|$.

The exact distribution of the statistic D_n can be derived. We shall not do so, but we give some values of the distribution function of D_n, namely, $P(D_n \le d)$, in Table VIII in the appendix, that will be used for goodness-of-fit tests. We also point out that the distribution of D_n does not depend on the particular function $F_0(x)$ of the continuous type. [This is essentially due to the fact that $Y = F_0(X)$ has a uniform distribution $U(0, 1)$.] Thus, D_n can be thought of as a distribution-free statistic.

We are interested in using the Kolmogorov–Smirnov statistic D_n to test the hypothesis H_0: $F(x) = F_0(x)$ against all alternatives H_1: $F(x) \neq F_0(x)$, where $F_0(x)$ is some specified distribution function. Intuitively, we accept H_0 if the empirical distribution function $F_n(x)$ is sufficiently close to $F_0(x)$—that is, if the value of D_n is sufficiently small. The hypothesis H_0 is rejected if the observed value of D_n is greater than the critical value selected from Table VIII in the appendix, where this critical value depends upon the desired significance level and sample size.

We illustrate the use of the Kolmogorov–Smirnov statistic by two examples.

EXAMPLE 8.7-1   We shall test the hypothesis H_0: $F(x) = F_0(x)$ against H_1: $F(x) \neq F_0(x)$, where

$$F_0(x) = \begin{cases} 0, & x < 0, \\ x, & 0 \leq x < 1, \\ 1, & 1 \leq x. \end{cases}$$

That is, the null hypothesis is that X is $U(0,1)$. If the test is based on a sample of size $n = 10$ and if $\alpha = 0.10$, then the critical region is $C = \{d_{10}: d_{10} \geq 0.37\}$, where d_{10} is the observed value of the Kolmogorov–Smirnov statistic D_{10}. Suppose that the observed values of the random sample are 0.62, 0.36, 0.23, 0.76, 0.65, 0.09, 0.55, 0.26, 0.38, and 0.24. In Figure 8.7-1, we have plotted the empirical and hypothesized distribution functions for $0 \leq x \leq 1$. We see that $d_{10} = F_{10}(0.65) - F_0(0.65) = 0.25$; hence, H_0 is not rejected. Note that the more carefully you plot the empirical and theoretical distribution functions, the more you can minimize the number of points at which you should calculate the possible value of the Kolmogorov–Smirnov goodness-of-fit statistic.   ∎

FIGURE 8.7-1: H_0: X is $U(0, 1)$

REMARK   If you have a computer program such as Minitab available, you can have it plot the empirical and theoretical distribution functions for you. Minitab will also give the value of the Kolmogorov–Smirnov goodness-of-fit statistic when you are testing for normality.   ∎

EXAMPLE 8.7-2   For a Poisson process with a mean rate of arrivals $\lambda = 1/\theta$, the random variable W, which denotes the waiting time until the αth arrival, has a gamma distribution. The p.d.f. of W is

$$f(w) = \frac{w^{\alpha-1}e^{-w/\theta}}{\Gamma(\alpha)\theta^\alpha}, \qquad 0 \le w < \infty.$$

A Geiger counter was set up to record the waiting time W in seconds required to observe $\alpha = 100$ alpha particle emissions of barium 133. It is claimed that the number of counts per second has a Poisson distribution with $\lambda = 14.7$ and hence, that $\theta = 0.068$. We shall test the hypothesis

$$H_0: F(w) = \int_{-\infty}^{w} f(t)\,dt,$$

where $f(t)$ is the gamma p.d.f. with $\theta = 0.068$ and $\alpha = 100$. On the basis of 25 observations, H_0 is rejected if $d_{25} \ge 0.24$ for $\alpha = 0.10$. The empirical and theoretical distribution functions are depicted in Figure 8.7-2 for 25 observations. For these data (Exercise 3.5-6), $d_{25} = 0.123$ and hence H_0 is not rejected. ■

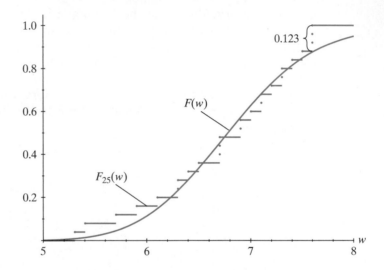

FIGURE 8.7-2: H_0: X has a gamma distribution

Note that we have been assuming that $F(x)$ is a continuous function. That is, we have considered only random variables of the continuous type. This procedure may also be applied in the discrete case; however, the true significance level will then be at most α. That is, the resulting test will be conservative.

Another application of the Kolmogorov–Smirnov statistic is in forming a confidence band for an unknown distribution function $F(x)$. To form a confidence band based on a sample of size n, select a number d such that

$$P(D_n \ge d) = \alpha.$$

Then

$$1 - \alpha = P[\sup_x |F_n(x) - F(x)| \le d]$$

$$= P[\,|F_n(x) - F(x)| \le d \text{ for all } x]$$

$$= P[F_n(x) - d \le F(x) \le F_n(x) + d \text{ for all } x].$$

Let

$$F_L(x) = \begin{cases} 0, & F_n(x) - d \leq 0, \\ F_n(x) - d, & F_n(x) - d > 0, \end{cases}$$

and

$$F_U(x) = \begin{cases} F_n(x) + d, & F_n(x) + d < 1, \\ 1, & F_n(x) + d \geq 0. \end{cases}$$

The two-step functions $F_L(x)$ and $F_U(x)$ yield a $100(1 - \alpha)\%$ confidence band for the unknown distribution function $F(x)$.

EXAMPLE 8.7-3 A random sample of size $n = 15$ from an unknown distribution yielded the sample values 3.88, 3.97, 4.03, 2.49, 3.18, 3.08, 2.91, 3.43, 2.41, 1.57, 3.78, 3.25, 1.29, 2.57, and 3.40. Now,

$$P(D_{15} \geq 0.30) = 0.10.$$

A 90% confidence band for the unknown distribution function $F(x)$ is depicted in Figure 8.7-3. ■

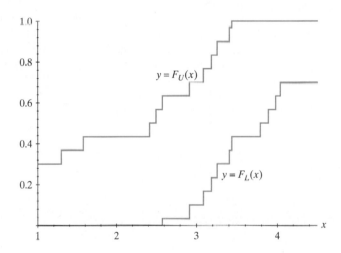

FIGURE 8.7-3: A 90% confidence band for an unknown distribution function

EXERCISES

8.7-1. Five ordered observations of X are 0.40, 0.51, 0.53, 0.62, 0.74. Use the Kolmogorov–Smirnov statistic to test the hypothesis that X is $U(0,1)$. Let $\alpha = 0.20$.

8.7-2. Select 10 sets of 10 random numbers from Table IX in the appendix. For each set of 10 random numbers, calculate the value of the Kolmogorov–Smirnov statistic d_{10} with $F_0(x) = x$, $0 \leq x \leq 1$. Is it true that about 20% of the observations of D_{10} are greater than 0.32? (Compare your results with those of other students, provided that they selected different random numbers.)

8.7-3. A doctor of obstetrics used an ultrasound examination between the 16th and 25th weeks of a patient's pregnancy to measure the widest diameter of the fetal head, in millimeters. Let X equal this diameter. Use the following 10 observations of X to test the hypothesis that the distribution of X is $N(60, 100)$:

56   65   47   57   62   48   68   75   79   49

8.7-4. In Exercise 8.5-4, the outcomes of 10 simulations of a Cauchy random variable are given.

(a) Sketch the empirical distribution function and superimpose the Cauchy distribution function.

(b) Use the Kolmogorov–Smirnov statistic to test whether the 10 observations represent a random sample from a Cauchy distribution. Let $\alpha = 0.10$.

8.7-5. In Exercise 3.6-9, diameters (in mm) of 30 individual grains of soil were given as follows:

1.24 1.36 1.28 1.31 1.35 1.20 1.39 1.35 1.41 1.31
1.28 1.26 1.37 1.49 1.32 1.40 1.33 1.28 1.25 1.39
1.38 1.34 1.40 1.27 1.33 1.36 1.43 1.33 1.29 1.34

(a) Calculate the sample mean and the sample variance for these data.

(b) Use a Kolmogorov–Smirnov goodness-of-fit statistic to test the null hypothesis that these data are observations of a normally distributed random variable.

8.7-6. Use the following 15 observations of X to construct a 90% confidence band for the unknown distribution function $F(x)$:

20.2 85.4 59.9 72.7 88.0 33.7 87.1 99.5 93.8
18.4 60.6 98.9 90.9 86.9 74.2

8.7-7. While testing a computer tape for bad records, the computer operator counted the number of flaws per 100 feet. Let X equal this number, and test, at a significance level of $\alpha \approx 0.10$, the hypothesis that the distribution of X is Poisson with a mean of $\lambda = 2.4$, using the following 40 observations of X:

x	Frequency
0	5
1	7
2	12
3	9
4	5
5	1
6	1
	40

8.7-8. The computer operator in Exercise 8.7-7 also measured the distances between flaws. These distances, reported in Exercise 8.1-10, in which a chi-square goodness-of-fit test was used, are repeated here:

30	79	38	47	22	52	36	36	7	57
3	22	30	14	8	32	15	21	12	12
6	67	6	7	35	78	28	74	5	9
37	1	3	3	44	160	50	27	61	15
39	44	130	18	6	1	32	116	23	12
58	101	68	53	58	21	21	7	79	41
80	33	71	81	17	10	13	49	21	56
107	21	17	64	14	36	26	1	54	207
64	238	25	51	82	8	2	3	43	87

Now use a Kolmogorov–Smirnov goodness-of-fit statistic to test the null hypothesis that we are observing observations of an exponential random variable. Are your conclusions consistent with each other?

8.7-9. In Exercise 8.6-14, 70 observations of a random variable X are listed that give the force required to pull a stud out of a window.

(a) Test the null hypothesis that the distribution of X is $N(117.5, 252)$. Use the Kolmogorov–Smirnov statistic with a significance level of $\alpha = 0.20$. What is your conclusion?

(b) Carefully analyzing the graphs of the empirical and theoretical distribution functions or by constructing a histogram, do you agree with the conclusion in part (a)? Why or why not?

(c) Construct and interpret a q–q plot for these data.

8.7-10. Let X equal the number of chocolate chips in a chocolate-chip cookie. Let $\alpha \approx 0.10$, and use the Kolmogorov–Smirnov goodness-of-fit statistic and the following data to test the hypothesis that the distribution of X is Poisson with a mean of $\lambda = 5.6$:

x	Frequency
0	0
1	0
2	2
3	8
4	7
5	13
6	13
7	10
8	4
9	4
10	1
	62

8.7-11. In Example 3.3-3, 105 times between calls to 911 are listed. Also given is a graph of the empirical distribution function with an exponential distribution function superimposed. Does this graph and the Kolmogorov–Smirnov goodness-of-fit statistic support the suggestion that the 105 times are observations of an exponential random variable? Why?

8.7-12. In Exercise 8.4-15, the weights in grams of "14-gram" snack packages of peanuts were given as follows:

13.9 14.4 14.6 14.7 14.7 15.2 15.2 15.2
15.3 15.4 15.4 15.5 15.6 15.6 15.9 16.4

Test the null hypothesis that these weights are observations of a normally distributed random variable with mean $\mu = 15.3$ and standard deviation $\sigma = 0.6$. Use a 10% significance level.

8.7-13. The March 29, 1992, issue of *Parade Magazine* reported the results of a study by Runzheimer International that gathered data from around the world on the cost in U.S. dollars per gallon for the least expensive gasoline available. The study compared each cost in December 1990 (before the Gulf War) with the cost in December 1991. Let X equal the difference—the 1991 cost minus the 1990 cost. Assume that the following $n = 18$

differences represent a random sample of observations of X:

$$
\begin{array}{cccccc}
-0.28 & -0.40 & 0.06 & 0.04 & -0.29 & 0.38 \\
-0.08 & -0.02 & -0.22 & -0.28 & 0.11 & 0.44 \\
-0.17 & -0.39 & -0.36 & -0.18 & -0.39 & -0.31
\end{array}
$$

Test whether the times are observations of a shifted (two-parameter) exponential distribution. Estimate δ and θ with their unbiased estimators. That is, let y_1 equal the minimum value of the X observations. Then unbiased estimates of θ and δ are

$$\widehat{\theta} = [n/(n-1)][\bar{x} - y_1] \quad \text{and} \quad \widehat{\delta} = y_1 - \widehat{\theta}/n.$$

8.8   RESAMPLING METHODS

Sampling and resampling methods have become more useful in recent years due to the power of computers. These methods are even used in introductory courses to convince students that statistics have distributions—that is, that statistics are random variables with distributions. At this stage in the book, the reader should be convinced that this is true, although we did use some sampling in Section 5.6 to help sell the idea that the sample mean has an approximate normal distribution.

Resampling methods, however, are used for more than showing that statistics have certain distributions. Rather, they are needed in finding approximate distributions of certain statistics that are used to make statistical inferences. We already know a great deal about the distribution of $\overline{X}$, and resampling methods are not needed for $\overline{X}$. In particular, $\overline{X}$ has an approximate normal distribution with mean μ and standard deviation $\sigma/\sqrt{n}$. Of course, if the latter is unknown, we can estimate it by $s/\sqrt{n}$ and note that $(\overline{X} - \mu)/(s/\sqrt{n})$ has an approximate $N(0,1)$ distribution, provided that the sample size is large enough and the underlying distribution is not too badly skewed with a long, heavy tail.

We know something about the distribution of S^2 *if the random sample arises from a normal distribution* or one fairly close to it. However, the statistic S^2 is not very robust, in that its distribution changes a great deal as the underlying distribution changes. It is not like $\overline{X}$, which always has an approximate normal distribution, provided that the mean μ and variance σ^2 of the underlying distribution exist. So what do we do about distributions of statistics like the sample variance S^2, whose distribution depends so much on having a given underlying distribution? We use resampling methods that essentially substitute computation for theory. We need to have some idea about the distributions of these various estimators to find confidence intervals for the corresponding parameters.

Let us now explain resampling. Suppose that we need to find the distribution of some statistic, such as S^2, but we do not believe that we are sampling from a normal distribution. We observe the values of $X_1, X_2, \ldots, X_n$ to be $x_1, x_2, \ldots, x_n$. Actually, if we know nothing about the underlying distribution, then the empirical distribution found by placing the weight $1/n$ on each x_i is the best estimate of that distribution. Therefore, to get some idea about the distribution of S^2, let us take a random sample of size n from this empirical distribution; then we are sampling from the n values with replacement. We compute S^2 for that sample; say it is s_1^2. We then do it again, getting s_2^2. And again, we compute s_3^2. We continue to do this a large number of times, say, N, where N might be 1000, 2000, or even 10,000. Once we have these

N values of S^2, we can construct a histogram, a stem-and-leaf display, or a q–q plot—anything to help us get some information about the distribution of S^2 when the sample arises from this empirical distribution, which is an estimate of the real underlying distribution. Clearly, we must use the computer for all of this sampling. We illustrate the resampling procedure by using, not S^2, but a statistic called the *trimmed mean*.

Although we usually do not know the underlying distribution, we state that, in this illustration, it is of the Cauchy type, because there are certain basic ideas we want to review or introduce for the first time. The p.d.f. of the Cauchy is

$$f(x) = \frac{1}{\pi(1 + x^2)}, \qquad -\infty < x < \infty.$$

The distribution function is

$$F(x) = \int_{-\infty}^{x} \frac{1}{\pi(1 + w^2)} \, dw = \frac{1}{\pi} \arctan x + \frac{1}{2}, \qquad -\infty < x < \infty.$$

If we want to generate some X values that have this distribution, we let Y have the uniform distribution $U(0,1)$ and define X by

$$Y = F(X) = \frac{1}{\pi} \arctan X + \frac{1}{2}$$

or, equivalently,

$$X = \tan\left[\pi\left(Y - \frac{1}{2}\right)\right].$$

We can generate 40 values of Y on the computer and then calculate the 40 values of X. Let us now add $\theta = 5$ to each X value to create a sample from a Cauchy distribution with a median of 5. That is, we have a random sample of 40 W values, where $W = X + 5$. We will consider some statistics used to estimate the median, θ, of this distribution. Of course, usually the value of the median is unknown, but here we know that it is equal to $\theta = 5$, and our statistics are estimates of this known number. These 40 values of W are as follows, after ordering:

−7.34	−5.92	−2.98	0.19	0.77	0.95	2.86	3.17	3.76	4.20
4.20	4.27	4.31	4.42	4.60	4.73	4.84	4.87	4.90	4.96
4.98	5.00	5.09	5.09	5.14	5.22	5.23	5.42	5.50	5.83
5.94	5.95	6.00	6.01	6.24	6.82	9.62	10.03	18.27	93.62

It is interesting to observe that many of these 40 values are between 3 and 7 and hence, are close to $\theta = 5$; it is almost as if they had arisen from a normal distribution with mean $\mu = 5$ and $\sigma^2 = 1$. But then we note the outliers; these very large or small values occur because of the heavy and long tails of the Cauchy distribution and suggest that the sample mean $\overline{X}$ is not a very good estimator of the middle. And it is not in this sample because $\bar{x} = 6.67$. In a more theoretical course, it can be shown that, due to the fact that the mean μ and the variance σ^2 do not exist for a Cauchy distribution, $\overline{X}$ is not any better than a single observation X_i in estimating the median θ. The sample median $\widetilde{m}$ is a much better estimate of θ, as it is not influenced by the outliers. Here the median equals 4.97, which is fairly close to 5. Actually, the maximum likelihood estimator found by maximizing

$$L(\theta) = \prod_{i=1}^{40} \frac{1}{\pi[1 + (x_i - \theta)^2]}$$

is extremely good, but requires difficult numerical methods to compute. Then advanced theory shows that, in the case of a Cauchy distribution, a **trimmed mean**, found by ordering the sample, discarding the smallest and largest $3/8 = 37.5\%$ of the sample, and averaging the middle 25%, is almost as good as the maximum likelihood estimator, but is much easier to compute. This trimmed mean is usually denoted by $\overline{X}_{0.375}$; we use $\overline{X}_t$ for brevity, and here $\bar{x}_t = 4.96$. For this sample, it is not quite as good as the median; but, for most samples, it is better. Trimmed means are often very useful and many times are used with a smaller trimming percentage. For example, in sporting events such as skating and diving, often the smallest and largest of the judges' scores are discarded.

For this Cauchy example, let us resample from the empirical distribution created by placing the "probability" 1/40 on each of our 40 observations. With each of these samples, we find our trimmed mean $\overline{X}_t$. That is, we order the observations of each resample and average the middle 25% of the order statistics—namely, the middle 10 order statistics. We do this $N = 1000$ times, thus obtaining $N = 1000$ values of $\overline{X}_t$. These values are summarized with the histogram in Figure 8.8-1(a).

From this resampling procedure, which is called **bootstrapping**, we have some idea about the distribution if the sample arises from the empirical distribution and, hopefully, from the underlying distribution, which is approximated by the empirical distribution. While the distribution of the sample mean $\overline{X}$ is not normal if the sample arises from a Cauchy-type distribution, the approximate distribution of $\overline{X}_t$ is normal. From the histogram of trimmed mean values in Figure 8.8-1(a), that looks to be the case. This observation is supported by the q–q plot in Figure 8.8-1(b) of the quantiles of a standard normal distribution versus those of the 1000 $\bar{x}_t$ values: The plot is very close to being a straight line.

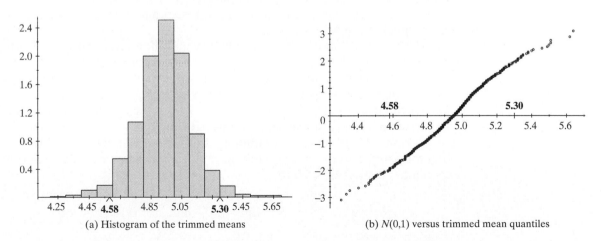

(a) Histogram of the trimmed means   (b) $N(0,1)$ versus trimmed mean quantiles

FIGURE 8.8-1: $N = 1000$ observations of trimmed means

How do we find a confidence interval for θ? Recall that the middle of the distribution of $\overline{X}_t - \theta$ is zero. So a guess at θ would be the amount needed to move the histogram of $\overline{X}_t$ values over so that zero is more or less in the middle of the translated histogram. We recognize that this histogram was generated from the original sample $X_1, X_2, \ldots, X_{40}$ and thus is really only an estimate of the distribution of $\overline{X}_t$.

We could get a point estimate of θ by moving it over until its median (or mean) is at zero. Clearly, however, some error is incurred in doing so—and we really want some bounds for θ as given by a confidence interval.

To find that confidence interval, let us proceed as follows: In the $N = 1000$ resampled values of $\overline{X}_t$, we find two points—say, c and d—such that about 25 values are less than c and about 25 are greater than d. That is, c and d are about on the respective 2.5th and 97.5th percentiles of the empirical distribution of these $N = 1000$ resampled $\overline{X}_t$ values. Thus, θ should be big enough so that over 2.5 percent of the $\overline{X}_t$ values are less than c and small enough so that over 2.5 percent of the $\overline{X}_t$ values are greater than d. This requires that $c < \theta$ and $\theta < d$; thus, $[c, d]$ serves as an approximate 95% confidence interval for θ as found by the **percentile method**. With our bootstrapped distribution of $N = 1000$ $\overline{X}_t$ values, this 95% confidence interval for θ runs from 4.58 to 5.30, and these two points are marked on the histogram and the q–q plot. Clearly, we could change this percentage to other values, such as 90%.

This percentile method, associated with the bootstrap method, is a nonparametric procedure, as we make no assumptions about the underlying distribution. It is interesting to compare the answer it produces with that obtained by using the order statistics $Y_1 < Y_2 < \cdots < Y_{40}$. If the sample arises from a continuous-type distribution, then, with the use of a calculator or computer, we have, when θ is the median,

$$P(Y_{14} < \theta < Y_{27}) = \sum_{k=14}^{26} \binom{40}{k} \left(\frac{1}{2}\right)^{40} = 0.9615.$$

(See Section 8.4.) Since, in our illustration, $Y_{14} = 4.42$ and $Y_{27} = 5.23$, the interval [4.42, 5.23] is an approximate 96% confidence interval for θ. Of course, $\theta = 5$ is included in each of the two confidence intervals. In this case, the bootstrap confidence interval is a little more symmetric about $\theta = 5$ and somewhat shorter, but it did require much more work.

We have now illustrated bootstrapping, which allows us to substitute computation for theory to make statistical inferences about characteristics of the underlying distribution. This method is becoming more important as we encounter complicated data sets that clearly do not satisfy certain underlying assumptions. For example, consider the distribution of $T = (\overline{X} - \mu)/(S/\sqrt{n})$ when the random sample arises from an exponential distribution that has p.d.f. $f(x) = e^{-x}$, $0 < x < \infty$, with mean $\mu = 1$. First we will *not* use resampling, but we will simulate the distribution of T when the sample size $n = 16$ by taking $N = 1000$ random samples from this known exponential distribution. Here

$$F(x) = \int_0^x e^{-w}\, dw = 1 - e^{-x}, \qquad 0 < x < \infty.$$

So $Y = F(X)$ means

$$X = -\ln(1 - Y)$$

and X has that given exponential distribution with $\mu = 1$, provided that Y has the uniform distribution $U(0,1)$. With the computer, we select $n = 16$ values of Y, determine the corresponding $n = 16$ values of X, and, finally, compute the value of $T = (\overline{X} - 1)/(S/\sqrt{16})$—say, T_1. We repeat this process over and over again, obtaining not only T_1, but also the values of $T_2, T_3, \ldots, T_{1000}$. We have done this and display the histogram of the 1000 T values in Figure 8.8-2(a). Moreover the q–q plot with quantiles of $N(0,1)$ on the y-axis is displayed in Figure 8.8-2(b). Both the histogram and the q–q plot show that the distribution of T in this case is skewed to the left.

(a) 1000 observations of T (b) $N(0,1)$ quantiles versus T quantiles

FIGURE 8.8-2: T observations from an exponential distribution

In the preceding illustration, we knew the underlying distribution. Let us now sample from the exponential distribution with mean $\mu = 1$, but add a value θ to each X. Thus, we will try to estimate the new mean $\theta + 1$. The authors know the value of θ, but the readers do not at this time. The observed 16 values of this random sample are

11.9776	9.3889	9.9798	13.4676	9.2895	10.1242	9.5798	9.3148
9.0605	9.1680	11.0394	9.1083	10.3720	9.0523	13.2969	10.5852

At this point we are trying to find a confidence interval for $\mu = \theta + 1$, and we pretend that we do not know that the underlying distribution is exponential. Actually, this is the case in practice: We do not know the underlying distribution. So we use the empirical distribution as the best guess of the underlying distribution; it is found by placing the weight 1/16 on each of the observations. The mean of this empirical distribution is $\bar{x} = 10.3003$. Therefore, we obtain some idea about the distribution of T by now simulating

$$T = \frac{\overline{X} - 10.3003}{S/\sqrt{16}}$$

with $N = 1000$ random samples from the empirical distribution.

We obtain $t_1, t_2, \ldots, t_{1000}$, and these values are used to construct a histogram, shown in Figure 8.8-3(a), and a q–q plot, illustrated in Figure 8.8-3(b). These two figures look somewhat like those in Figure 8.8-2. Moreover, the 0.025th and 0.975th quantiles of the 1000 t-values are $c = -3.1384$ and $d = 1.8167$, respectively.

Now we have some idea about the 2.5th and 97.5th percentiles of the T distribution. Hence, as a very rough approximation, we can write

$$P\left(-3.1384 \le \frac{\overline{X} - \mu}{S/\sqrt{16}} \le 1.8167\right) \approx 0.95.$$

(a) 1000 observations of T (b) $N(0,1)$ quantiles versus T quantiles

FIGURE 8.8-3: T observations from an empirical distribution

This formula leads to the rough approximate 95 percent confidence interval

$$[\bar{x} - 1.8167s/\sqrt{16},\ \bar{x} - (-3.1384)s/\sqrt{16}]$$

once the $\bar{x}$ and s of the *original* sample are substituted. With $\bar{x} = 10.3003$ and $s = 1.4544$, we have

$$[10.3003 - 1.8167(1.4544)/4,\ 10.3003 + 3.1384(1.4544)/4] = [9.6397, 11.4414]$$

as a 95% approximate confidence interval for $\mu = \theta + 1$. Note that, because we added $\theta = 9$ to each x-value, the interval does cover $\theta + 1 = 10$.

It is easy to see how this procedure gets its name, because it is like "pulling yourself up by your own bootstraps," with the empirical distribution acting as the bootstraps.

EXERCISES

8.8-1. If time and computing facilities are available, consider the following 40 losses, due to wind-related catastrophes, that were recorded to the nearest $1 million (these data include only those losses of $2 million or more, and, for convenience, they have been ordered and recorded in millions of dollars):

2	2	2	2	2	2	2	2	2	2
2	2	3	3	3	3	4	4	4	5
5	5	5	6	6	6	6	8	8	9
15	17	22	23	24	24	25	27	32	43

To illustrate bootstrapping, take resamples of size $n = 40$ as many as $N = 100$ times, computing $T = (\overline{X} - 5)/(S/\sqrt{40})$ each time. Here the value 5 is the median of the original sample. Construct a histogram of the bootstrapped values of T.

8.8-2. Consider the following 16 observed values, rounded to the nearest tenth, from the exponential distribution that was given in this section:

12.0	9.4	10.0	13.5	9.3	10.1	9.6	9.3
9.1	9.2	11.0	9.1	10.4	9.1	13.3	10.6

(a) Take resamples of size $n = 16$ from these observations about $N = 200$ times, and compute s^2 each time. Construct a histogram of these 200 bootstrapped values of S^2.

(b) Simulate $N = 200$ random samples of size $n = 16$ from an exponential distribution with θ equal to the mean of the data in part (a) minus 9. For each sample, calculate the value of s^2. Construct a histogram of these 200 values of S^2.

(c) Construct a q–q plot of the two sets of sample variances and compare these two empirical distributions of S^2.

8.8-3. Let $X_1, X_2, \ldots, X_{21}$ and $Y_1, Y_2, \ldots, X_{21}$ be random samples of sizes $n = 21$ and $m = 21$ from independent $N(0,1)$ distributions. Then $F = S_X^2/S_Y^2$ has an F distribution with 20 and 20 degrees of freedom.

(a) Illustrate this situation empirically by simulating 100 observations of F.
(i) Plot a relative frequency histogram with the $F(20, 20)$ p.d.f. superimposed.
(ii) Construct a q–q plot of the quantiles of $F(20, 20)$ versus the order statistics of your simulated data. Is the plot linear?

(b) Consider the following 21 observations of the $N(0,1)$ random variable X:

0.1616	−0.8593	0.3105	0.3932	−0.2357	0.9697	1.3633
−0.4166	0.7540	−1.0570	−0.1287	−0.6172	0.3208	0.9637
0.2494	1.1907	−2.4699	−0.1931	1.2274	−1.2826	−1.1532

Consider also the following 21 observations of the $N(0,1)$ random variable Y:

0.4419	−0.2313	0.9233	−0.1203	1.7659	−0.2022	0.9036
−0.4996	−0.8778	−0.8574	2.7574	1.1033	0.7066	1.3595
−0.0056	−0.5545	−0.1491	−0.9774	−0.0868	1.7462	−0.2636

Sampling with replacement, resample with a sample of size 21 from each of these sets of observations. Calculate the value of $w = s_x^2/s_y^2$. Repeat in order to simulate 100 observations of W from these two empirical distributions. Use the same graphical comparisons that you used in part (a) to see if the 100 observations represent observations from an approximate $F(20, 20)$ distribution.

(c) Consider the following 21 observations of the exponential random variable X with mean 1:

0.6958	1.6394	0.2464	1.5827	0.0201	0.4544	0.8427
0.6385	0.1307	1.0223	1.3423	1.6653	0.0081	5.2150
0.5453	0.08440	1.2346	0.5721	1.5167	0.4843	0.9145

Consider also the following 21 observations of the exponential random variable Y with mean 1:

1.1921	0.3708	0.0874	0.5696	0.1192	0.0164	1.6482
0.2453	0.4522	3.2312	1.4745	0.8870	2.8097	0.8533
0.1466	0.9494	0.0485	4.4379	1.1244	0.2624	1.3655

Sampling with replacement, resample with a sample of size 21 from each of these sets of observations. Calculate the value of $w = s_x^2/s_y^2$. Repeat in order to simulate 100 observations of W from these two empirical distributions. Use the same graphical comparisons that you used in part (a) to see if the 100 observations represent observations from an approximate $F(20, 20)$ distribution.

8.8-4. The following 54 pairs of data give, for Old Faithful geyser, the duration in minutes of an eruption and the time in minutes until the next eruption:

(2.500, 72)	(4.467, 88)	(2.333, 62)	(5.000, 87)	(1.683, 57)	(4.500, 94)
(4.500, 91)	(2.083, 51)	(4.367, 98)	(1.583, 59)	(4.500, 93)	(4.550, 86)
(1.733, 70)	(2.150, 63)	(4.400, 91)	(3.983, 82)	(1.767, 58)	(4.317, 97)
(1.917, 59)	(4.583, 90)	(1.833, 58)	(4.767, 98)	(1.917, 55)	(4.433, 107)
(1.750, 61)	(4.583, 82)	(3.767, 91)	(1.833, 65)	(4.817, 97)	(1.900, 52)
(4.517, 94)	(2.000, 60)	(4.650, 84)	(1.817, 63)	(4.917, 91)	(4.000, 83)
(4.317, 84)	(2.133, 71)	(4.783, 83)	(4.217, 70)	(4.733, 81)	(2.000, 60)
(4.717, 91)	(1.917, 51)	(4.233, 85)	(1.567, 55)	(4.567, 98)	(2.133, 49)
(4.500, 85)	(1.717, 65)	(4.783, 102)	(1.850, 56)	(4.583, 86)	(1.733, 62)

(a) Calculate the correlation coefficient, and construct a scatterplot, of these data.
(b) To estimate the distribution of the correlation coefficient, R, resample 500 samples of size 54 from the empirical distribution, and for each sample, calculate the value of R.
(c) Construct a histogram of these 500 observations of R.
(d) Simulate 500 samples of size 54 from a bivariate normal distribution with correlation coefficient equal to the correlation coefficient of the geyser data. For each sample of 54, calculate the correlation coefficient.
(e) Construct a histogram of the 500 observations of the correlation coefficient.
(f) Construct a q–q plot of the 500 observations of R from the bivariate normal distribution of part (d) versus the 500 observations in part (b). Do the two distributions of R appear to be about equal?

8.8-5. Consider the following 12 pairs of data, which are values of a random sample of size 12 from a bivariate normal distribution with correlation coefficient $\rho = -0.8$:

$$
\begin{array}{llll}
(-0.8118, 0.5205) & (-1.7527, 1.2374) & (-0.0639, 1.1181) & (-0.5526, 0.7047) \\
(-0.6387, -0.3772) & (0.1743, 0.2298) & (1.7126, -2.2842) & (0.3114, 0.0846) \\
(-0.6270, 1.0964) & (0.5996, -1.1564) & (-0.9370, 0.6896) & (0.6739, -0.0470)
\end{array}
$$

(a) To estimate the distribution of the correlation coefficient, R, resample 500 samples of size 12 from the empirical distribution, and for each sample, calculate the value of R.

(b) Construct a histogram of these 500 observations of R.

(c) Simulate 500 samples of size 12 from a bivariate normal distribution with correlation coefficient equal to the correlation coefficient of the given data. For each sample of 12, calculate the correlation coefficient.

(d) Construct a histogram of the 500 observations of the correlation coefficient given in part (c).

(e) Construct a q–q plot of the 500 observations of R from the bivariate normal distribution of part (c) versus the 500 observations found in part (a). Do the two distributions of R appear to be about equal?

8.8-6. Consider the following 20 pairs of data, which are paired data from two exponential distributions, each having a mean of 5:

$$
\begin{array}{llll}
(5.4341, 8.4902) & (33.2097, 4.7063) & (0.4034, 1.8961) & (1.4137, 0.2996) \\
(17.9365, 3.1350) & (4.4867, 6.2089) & (11.5107, 10.9784) & (8.2473, 19.6554) \\
(1.9995, 3.6339) & (1.8965, 1.7850) & (1.7116, 1.1545) & (4.4594, 1.2344) \\
(0.4036, 0.7260) & (3.0578, 19.0489) & (21.4049, 4.6495) & (3.8845, 13.7945) \\
(5.9536, 9.2438) & (11.3942, 1.7863) & (5.4813, 4.3356) & (7.0590, 1.15834)
\end{array}
$$

(a) Calculate the correlation coefficient, and construct a scatterplot, of these data.

(b) To estimate the distribution of the correlation coefficient, R, resample 500 samples of size 20 from the empirical distribution, and for each sample, calculate the value of R.

(c) Construct a histogram of these 500 observations of R.

(d) Simulate 500 samples of size 20 from a bivariate normal distribution with correlation coefficient 0. For each sample of 20, calculate the correlation coefficient.

(e) Construct a histogram of the 500 observations of the correlation coefficient.

(f) Construct a q–q plot of the 500 observations of R from the bivariate normal distribution of part (d) versus the 500 observations in part (b). Do the two distributions of R appear to be about equal?

8.8-7. Consider the following 20 pairs of data, which are paired data from two Cauchy distributions:

$$
\begin{array}{llll}
(-6.0141, 3.1634) & (-1.0420, 16.6134) & (0.0894, 0.1775) & (-4.0680, -0.4082) \\
(2.4561, -3.0045) & (-3.7596, -1.0007) & (0.1678, -1.1852) & (-0.5504, -0.9796) \\
(1.4399, -0.1211) & (-0.2510, -0.7397) & (-5.3430, 0.4839) & (-0.8826, -2.9297) \\
(1.4219, 1.7424) & (-5.7973, 0.1788) & (1.0064, -0.2806) & (-2.1313, -0.5802) \\
(5.8900, 0.8383) & (-2.2183, -0.4968) & (1.1598, 1.2758) & (-1.0269, -15.7753)
\end{array}
$$

(a) To estimate the distribution of the correlation coefficient, R, resample 500 samples of size 20 from the empirical distribution, and for each sample, calculate the value of R.

(b) Construct a histogram of these 500 observations of R.

(c) Simulate 500 samples of size 20 from a bivariate normal distribution with correlation coefficient 0. For each sample of 20, calculate the correlation coefficient.

(d) Construct a histogram of the 500 observations of the correlation coefficient.

(e) Construct a q–q plot of the 500 observations of R from the bivariate normal distribution of part (c) versus the 500 observations found in part (a). Do the two distributions of R appear to be about equal?

HISTORICAL COMMENTS   From the 1940s to well into the 1960s, most statisticians thought that nonparametric methods referred to distribution-free inferences. For example, under the null hypothesis, the one-sample Wilcoxon test does not depend upon the underlying distribution as long as it is symmetric and of the continuous type. Also, the two-sample Wilcoxon test of $F(x) = G(x)$ does not depend upon the distributions $F(x) = G(x)$ as long as they are of the continuous type. In later years, the term *nonparametric methods* came to mean something that is possibly broader than distribution-free methods, such as the bootstrap technique.

The name *bootstrap* and the resulting technique were first used by Brad Efron of Stanford University. Efron knew that the expression "to pull oneself up by his or her own bootstraps" seems to come from *The Surprising Adventures of Baron Munchausen* by Rudolph Erich Raspe. The Baron had fallen from the sky and found himself in a hole 9 fathoms deep and had no idea how to get out. He comments as follows: "Looking down I observed that I had on a pair of boots with exceptionally sturdy straps. Grasping them firmly, I pulled with all my might. Soon I had hoisted myself to the top and stepped out on terra firma without further ado."

Of course, in statistical *bootstrapping*, statisticians pull themselves up by their bootstraps (the empirical distributions) by recognizing that the empirical distribution is the best estimate of the underlying distribution without a lot of other assumptions. So they use the empirical distribution as if it is the underlying distribution to find approximate distributions of statistics of interest.

Chi-square tests were the invention of Karl Pearson, except that he had it wrong in the case in which parameters are estimated. When R. A. Fisher was a brash young man, he told his senior Pearson that he should reduce the number of degrees of freedom of the chi-square distribution by 1 for every parameter that was estimated. Pearson never believed this (of course, Fisher was correct), and, as editor of the very prestigious journal *Biometrika*, Pearson blocked Fisher in his later professional life from publishing in that journal. Fisher was disappointed, and the two men battled during their lifetimes; however, later Fisher saw this conflict to be to his advantage, as it made him consider applied journals in which to publish, and thus he became a better, more well rounded scientist.

Bayesian Methods

9.1 SUBJECTIVE PROBABILITY

Subjective probability is the foundation for Bayesian methods. Thus, we devote this short section to understanding how a Bayesian would assign probabilities to certain events. In a thorough study of Bayesian methods, there would be greater discussion about how a Bayesian would assess the prior probabilities.

Consider an event A, and let us say that we believe that $P(A) = 2/5$. This means that the odds *against* the event A are

$$O(A) = \frac{1 - P(A)}{P(A)} = \frac{1 - 2/5}{2/5} = \frac{3}{2},$$

or 3 to 2. Likewise the odds *for A* are

$$\frac{P(A)}{1 - P(A)} = \frac{2/5}{1 - 2/5} = \frac{2}{3},$$

or 2 to 3. Moreover, if a person really believes that $P(A) = 2/5$ and is willing to bet, he or she would take either side of this bet:

1. Win 3 dollars if A occurs and lose 2 dollars if A does not occur or
2. Win 2 dollars if A does not occur and lose 3 dollars if A does occur.

If this is not the case, then that person should review his or her subjective probability $P(A)$.

Another way of thinking about this subjective probability, $P(A)$, is to say that it is equal to the amount you are willing to pay to receive \$1 [the bet of $P(A)$ plus the winnings of $\{1 - P(A)\}$] if A actually occurs. That is, you lose $P(A)$ if A does not occur, and you win $1 - P(A)$ if A occurs. So if you believe that $P(A) = 2/5$, you would be willing to pay 2/5 of a dollar to play "a game" in which you win 3/5 (plus your bet of 2/5) if A occurs. Note that $(2/5)/(3/5) = 2/3$, or 2 to 3, are the odds in favor of A. Or $(3/5)/(2/5) = 3/2$, or 3 to 2, are the odds against A. Again, you must be willing to take either side of the bet.

This is really much like two children dividing a candy bar as equally as possible: One divides it, and the other gets to choose which of the two parts seems most desirable—that is, the larger. Accordingly, the child dividing the candy bar tries extremely hard to cut it as equally as possible. Clearly, this is exactly what the person selecting the subjective probability does, as he or she must be willing to take either side of the bet with the odds established.

Now, every person has his or her own personal probability of almost any statement A, which could be something like "There must be some type of life on the planet Mars." Since we often think of A as something on which we might bet, let us consider the statement A that Iowa will beat Michigan in their football game this year. A man who studies the strengths of those two teams might offer the subjective probability $P(A) = 1/3$, or odds against Iowa of

$$O(A) = \frac{1 - 1/3}{1/3} = 2 = \frac{2}{1},$$

(i.e., 2 to 1) and be willing to take either side of the bet. Suppose his opponent in this gamble is a strong Michigan fan who he thinks might accept 3 to 1 odds. Then he might offer those odds to that individual, hoping that she will bet on Michigan. In that case, he would have quite a satisfactory bet: He wins three units if Iowa wins and loses one unit if Michigan wins. Does he dare do that if he really believes that $P(A) = 1/3$? His opponent also knows a lot about these two teams and actually thinks that $P(A) = 1/3$ is about right; so she takes the 3 to 1 bet, but chooses Iowa, as she likes those 3 to 1 odds even though she is a Michigan fan. So the first person has a bad bet according to his subjective probability $P(A) = 1/3$.

Recall, however, that each person has his or her own subjective probability and that the Michigan fan might really believe that $P(A) = 1/5$, or odds of $(1 - 1/5)/(1/5)$—that is, 4 to 1 against Iowa. Thus, she would be glad to bet three units to one unit; that is, she would be happy to give three units if Iowa wins and accept one unit if Michigan wins. In this case, each player would have a satisfactory bet, consistent with their subjective probabilities. We are certain that this is why bookmakers (bookies) tend to stay in business and often make a great deal of money. They offer odds, and others think that they are wrong and thus bet. Later we shall point out how bookies (and racetracks, casinos, and even state lotteries) can "cover themselves." Accordingly, it is a good rule, in our opinion, never to bet against people who are professional gamblers. How can casinos afford to have such fancy buildings? The persons who bet against them pay for those nice facilities. (Incidentally, we have no objection to, say, a friendly poker game in which none of the participants could "be hurt" by the stakes involved. Even one of the authors has participated in friendly games and wagers; still, the reader must be warned about playing against statisticians, as well as professional gamblers.)

Let us now explain how a racetrack can always make money; of course, the owners have certain fixed costs that must be covered, so they obviously must make some money. To keep it simple, suppose there are only three horses, A, B, and C, and we will worry only about bets to "win," not "place" or "show." The following table shows the amount of money bet on each horse to win and the "subjective probability" determined by all the bettors:

Horse	Amount	Probability
A	$50,000	0.50
B	$40,000	0.40
C	$10,000	0.10
Total	$100,000	1.00

In a sense, these are not really subjective probabilities as we have been describing them, but some sort of degrees of belief in the various horses as determined by all the bettors.

If we used those probabilities and the associated odds to determine fair payoffs (track bets are made in $2 units), a $2 bet on A would yield $4 if A wins: the $2 bet plus $2 more. A $2 bet on B would yield $5 if B wins: Recall that the odds are $(1 - 0.4)/0.4 = 1.5$ against B, so the track puts up $3 against the bettor's $2. Finally, a $2 bet on C would yield $20 if C wins: the $2 bet plus $18 more, as the odds are $(1 - 0.1)/0.1 = 9$ to 1 against C.

If the track would make these payoffs, what would be the total payoff in case A wins, in case B wins, and in case C wins? There are 25,000 $2 bets in that $50,000 bet on A; so if A wins, then

$$25,000 \times \$4 = \$100,000$$

would be the total payoff. There are 20,000 $2 bets in that $40,000 bet on B; so if B wins, then

$$20,000 \times \$5 = \$100,000$$

would be the total payoff. Similarly, for C winning, the total payoff is

$$5000 \times \$20 = \$100,000.$$

That is, with these probabilities, the track would pay off as much as the total amount bet. The track could never stay in business in that situation, so it must take something off the top, and this is now explained.

A typical percentage of the track's take is 17%, so we will use that figure. In our example, this amounts to $17,000, and thus there is only $83,000 to distribute. Accordingly, the payoffs are adjusted by the factor $83/100 = 0.83$. That is, instead of payoffs of $4, $5, and $20, they are, respectively,

$$4(0.83) = 3.32,$$
$$5(0.83) = 4.15,$$
$$20(0.83) = 16.60,$$

and the three possible total payoffs are, respectively,

$$25,000(3.32) = 83,000,$$
$$20,000(4.15) = 83,000,$$
$$5000(16.60) = 83,000.$$

Hence, this track is always assured of making the money necessary to cover all expenses, provided that it has enough bettors.

Other gambling establishments cover themselves by having the probabilities slightly favor the "house." Sometimes the take is nothing like 17%, but the probabilities of the bettor winning are slightly under 0.50. For example, in American roulette, the probability of winning by betting on red is $18/38 = 0.474$, because 18 out of the 38 equally likely places favor red. Actually, the game that provides a probability closest to one-half is "craps," if you bet on the caster. There, the probability of the caster winning is 0.493, which is extremely close to 0.5. Sometimes the side bets in craps are not so favorable; be careful making those. With such slim odds favoring the house, how does a casino make money? Mainly because it has lots of bettors. And, of course, some games (e.g., the slots) are not quite as favorable as roulette or craps. (As a side note, we mention that, in state lotteries, the expected payoff for a $1 bet is usually about 45 or 50 cents; so these are bad bets, although states claim they do good things with their winnings.)

In addition to the preceding, many bookies have two sets of odds, such as 2 to 1 if you bet on Iowa and 3 to 1 if you bet on Michigan. That is, on the one hand, they will pay $3 (2 + 1 = 3) for a $1 bet on Iowa if Iowa wins, while on the other hand, they will pay $4 (1 + 3 = 4) for a $3 bet on Michigan if Michigan wins. Clearly, if they can get the right number of bettors on each side, they have created a position such that they cannot lose. Suppose 10,000 people bet on Iowa (you might not need 10,000 people, as some will bet more than the minimum) and 7500 bet on Michigan. Then the bookies take in

$$(10{,}000)(\$1) + (7500)(\$3) = \$32{,}500.$$

If Iowa wins, they pay off $2 + 1 = 3$ for each $1 bet, or a total of

$$(10{,}000)(\$3) = \$30{,}000.$$

If Michigan wins, they pay off $1 + 3 = 4$ for each $3 bet, or a total of

$$(7500)(\$4) = \$30{,}000.$$

In either case, they make money, namely,

$$\$32{,}500 - \$30{,}000 = \$2500.$$

Clearly, they must balance the bettors or change the odds (which they do) to have a sure win. That is, just like the racetrack, they change the odds, depending upon the amount of money bet on each side. So if the bookies can get enough bettors, they can make lots of money, and they generally do! Hence, do not bet against professional gamblers unless you achieve a certain amount of enjoyment from betting; but, even then, make certain that the stakes are such that you can afford possible losses. [For a further discussion of betting, see Berry (1996).]

Now, not all betting is bad. For example, most people are betting when they buy insurance. The insurance companies do exactly the same thing that casinos do, as they know the probabilities and the corresponding premiums favor them somewhat. Is this bad? Certainly not, and one of the authors is associated with a Department of Statistics and Actuarial Science, has helped teach many actuaries over the years, and even has a daughter who is an actuary. However, to understand the value of buying insurance, the reader must understand the utility of money.

Most of us would say that $4 are worth twice as much as $2. Also, $100 are worth 20 times as much as $5. But to most individuals, is $2,000,000 worth twice as much as $1,000,000? For very rich people, it is. However, suppose a person has retired and has exactly $1,000,000. A very rich person offers that person the following bet: Toss a coin fairly, and if a head occurs, the rich person gives the retiree $2,000,000; but if a tail results, the retiree gives the rich person $1,000,000. That is, the retiree gains $2,000,000 with probability 0.5 and loses $1,000,000 with probability 0.5. The expected gain is

$$(\$2,000,000)(0.5) + (-\$1,000,000)(0.5) = \$500,000.$$

Seemingly, this is a great bet, but any reasonable person with only $1,000,000 would turn it down because that first $1,000,000 is worth a lot more than the second $1,000,000 or a third $1,000,000. Clearly, with the bet described, he could have zero or $3,000,000, each with probability 0.5, but the retiree would rather have $1,000,000 with certainty. While the utility function $u(x)$ of money for this individual is just about $u(x) = x$ for relatively small values of x, it starts to bend downward for larger values of x. Where this bending occurs is also subjective; Bill Gates would immediately accept the proposed bet, because his utility function is $u(x) = x$ into the millions, but Bob Hogg would not take the bet, because his utility function bends downward long before $1,000,000.

So why do we buy insurance? To keep it simple, suppose a man wishes to buy term life insurance valued at $100,000 for one year, possibly to protect a young family—say, a wife and two children—in case he dies. Say his probability of dying in that year is 0.002. What is a fair premium for this insurance if it will cost the company $100,000 with probability 0.002 and zero with probability 0.998? The answer is

$$(\$100,000)(0.002) + (\$0)(0.998) = \$200.$$

However, the company asks for a loading factor of, say, 15%, or $30. Thus, the charge is $230, which seemingly is not a fair bet. However, if the insured does die, the widow with two children have nothing without the insurance and $100,000 with the insurance. Under those circumstances, most men would buy the insurance because the value of the policy is worth much more than the $230 premium in their minds, and hence the bet is more than fair to them *personally*. Note, then, that insurance companies, like casinos, have nice facilities, but we believe in that type of betting.

So it is smart to insure valuable items, such as one's life, liability associated with car accidents, home, and so on. Things of smaller value should not be insured; that is, you carry your own insurance. The cutoff value differs with each individual, although we find that the rule of not insuring anything with a value under one or two months' salary is a good one; most of our utility functions are $u(x) = x$ in that range, and there is no sense paying a "load." Bill Gates does not need to insure for collision on a new car valued at $40,000 (he carries liability, however, as most states require it); Bob Hogg does. However, once the value of the car drops to under $10,000, even Bob Hogg drops the collision insurance. But note that these are personal decisions that depend upon the subjective probabilities and a person's utility of money.

Let us now say that the reader is willing to accept the subjective probability $P(A)$ as the fair price for event A, given that you will receive one unit [win $1 - P(A)$] in case A occurs and, of course, lose $P(A)$ if it does not occur. Then it turns out that all rules (definitions and theorems) in Section 1.2 follow for subjective probabilities. We do not give proofs of them all, but only one of them, leaving some of the others as exercises. The following proof was given to us in a personal communication from George Woodworth of the University of Iowa.

THEOREM 9.1-1	If A_1 and A_2 are mutually exclusive, then

$$P(A_1 \cup A_2) = P(A_1) + P(A_2).$$

Proof. Suppose a person thinks that a fair price for A_1 is $p_1 = P(A_1)$ and a fair price for A_2 is $p_2 = P(A_2)$. However, that person believes that the fair price for $A_1 \cup A_2$ is p_3, which differs from $p_1 + p_2$. Say $p_3 < p_1 + p_2$, and let the difference be $d = (p_1 + p_2) - p_3$. A gambler offers the person the price $p_3 + d/4$ for $A_1 \cup A_2$. The person takes the offer because it is better than p_3. The gambler sells A_1 at a discount price of $p_1 - d/4$ and sells A_2 at a discount price of $p_2 - d/4$ to the person. Being a rational person, and given the prices p_1, p_2, and p_3, the person believes that all three of these deals are satisfactory. However, the person received $p_3 + d/4$ and paid $p_1 + p_2 - d/2$. Thus, before any bets are paid off, the person has

$$p_3 + \frac{d}{4} - \left(p_1 + p_2 - \frac{d}{2} \right) = p_3 - p_1 - p_2 + \frac{3d}{4} = -\frac{d}{4}.$$

That is, the person is down $d/4$ before any bets are settled. Now,

1. Suppose A_1 happens: The gambler has $A_1 \cup A_2$ and the person has A_1; so they exchange units and the person is still down $d/4$. The same thing occurs if A_2 happens.
2. Suppose neither A_1 nor A_2 happens; then the gambler and that person receive zero, and the person is still down $d/4$.
3. Of course, A_1 and A_2 cannot occur together, since they are mutually exclusive.

Thus, we see that it is bad for the person to assign

$$p_3 = P(A_1 \cup A_2) < p_1 + p_2 = P(A_1) + P(A_2),$$

because the gambler can put the person in a position to lose $(p_1 + p_2 - p_3)/4$ no matter what happens. This kind of bet is sometimes referred to as a *Dutch book*.

The argument when $p_3 > p_1 + p_2$ is similar and also can lead to a Dutch book; it is left as an exercise. Hence, p_3 must equal $p_1 + p_2$ to avoid a Dutch book; that is, $P(A_1 \cup A_2) = P(A_1) + P(A_2)$. □

EXERCISES

9.1-1. Think of some event A (it could or could not be a sporting outcome) that might or might not happen on your college campus. Get several students to assign their personal probabilities to A. If they differ and the students are not opposed to betting, could you find any friendly (small-stakes) bets among them? For example, if two students, S_1 and S_2, assigned $P_1(A) = 0.5$ and $P_2(A) = 0.3$, respectively, they might agree that their odds of $(1 - 0.4)/0.4 = 1.5$ to 1 would be reasonable as they are between odds associated with their respective probabilities. That is, S_1 would place one unit on A to 1.5 units on A' by S_2. According to their subjective probabilities, this would be a satisfactory bet for both.

9.1-2. As in Exercise 9.1-1, consider some event A around your campus. Let one student, S_1, determine $P_1(A)$, the subjective probability of A. Then, on the basis of $P_1(A)$, let a second student, S_2, take either side of a friendly bet established by these odds. Try this several times, with students exchanging roles.

9.1-3. In American roulette, suppose that 3800 consecutive $5 bets were placed on red. Show that the casino's expected gain is $1000. Determine the standard deviation associated with the casino's gain.

9.1-4. On a certain fight involving boxer A and boxer B, A is the favorite over B. A bookie finds that $30,000 is bet on A and only $5000 is bet on B. Determine two sets of odds, one for those betting on A and one for those betting on B.

HINT: Note that the bets suggest a "subjective probability" of 5000/35,000 = 1/7 for B, or odds of 6 to 1 against B. Find odds on either side of the 6 to 1 for the two possible odds. Make certain that your choice (it is not unique) will be such that the bookie always makes money.

9.1-5. The following amounts are bet on horses A, B, C, D, and E to win:

Horse	Amount
A	$600,000
B	$200,000
C	$100,000
D	$75,000
E	$25,000
Total	$1,000,000

Suppose the track wants to take 20% off the top, or $200,000. Determine the payoff for winning with a $2 bet on each of the five horses.

9.1-6. In the proof of Theorem 9.1-1, we considered the case in which $p_3 < p_1 + p_2$. Now say the person believes that $p_3 > p_1 + p_2$. Create a Dutch book for him.

HINT: Let $d = p_3 - (p_1 + p_2)$. The gambler buys A_1 from the person at a premium price of $p_1 + d/4$ and A_2 for $p_2 + d/4$. Then the gambler sells $A_1 \cup A_2$ to the person at a discount of $p_3 - d/4$. All those deals are good for the person, who believes that p_1, p_2, and p_3 are correct, with $p_3 > p_1 + p_2$. Show that the person has a Dutch book.

9.1-7. Show that $P(S) = 1$.

HINT: Suppose a person thinks $P(S) = p \neq 1$. Consider two cases: $p > 1$ and $p < 1$. In the first case, say $d = p - 1$ and the gambler sells the person S at a discount price $1 + d/2$. Of course, S happens and the gambler pays the person one unit, but the person is down $1 + d/2 - 1 = d/2$, so he has a Dutch book. Do the other case.

9.1-8. Show that $P(A') = 1 - P(A)$.

HINT: Note that $A' \cup A = S$, and use the result of Theorem 9.1-1 and Exercise 9.1-7.

9.2 BAYESIAN ESTIMATION

We now describe another approach to estimation that is used by a group of statisticians who call themselves Bayesians. To understand their approach fully would require more text than we can allocate to this topic, but let us begin this brief introduction by considering a simple application of the theorem of the Reverend Thomas Bayes. (See Section 1.6.)

EXAMPLE 9.2-1 Suppose we know that we are going to select an observation from a Poisson distribution with mean λ equal to 2 or 4. Moreover, prior to performing the experiment, we believe that $\lambda = 2$ has about four times as much chance of being the parameter as does $\lambda = 4$; that is, the prior probabilities are $P(\lambda = 2) = 0.8$ and $P(\lambda = 4) = 0.2$. The experiment is now performed and we observe that $x = 6$. At this point, our intuition tells us that $\lambda = 2$ seems less likely than before, as the observation $x = 6$ is much more probable with $\lambda = 4$ than with $\lambda = 2$, because, in an obvious notation,

$$P(X = 6 \mid \lambda = 2) = 0.995 - 0.983 = 0.012$$

and

$$P(X = 6 \mid \lambda = 4) = 0.889 - 0.785 = 0.104,$$

from Table III in the appendix. Our intuition can be supported by computing the conditional probability of $\lambda = 2$, given that $X = 6$:

$$P(\lambda = 2 \mid X = 6) = \frac{P(\lambda = 2, X = 6)}{P(X = 6)}$$

$$= \frac{P(\lambda = 2)P(X = 6 \mid \lambda = 2)}{P(\lambda = 2)P(X = 6 \mid \lambda = 2) + P(\lambda = 4)P(X = 6 \mid \lambda = 4)}$$

$$= \frac{(0.8)(0.012)}{(0.8)(0.012) + (0.2)(0.104)} = 0.316.$$

This conditional probability is called the posterior probability of $\lambda = 2$, given the single data point (here, $x = 6$). In a similar fashion, the posterior probability of $\lambda = 4$ is found to be 0.684. Thus, we see that the probability of $\lambda = 2$ has decreased from 0.8 (the prior probability) to 0.316 (the posterior probability) with the observation of $x = 6$. ∎

In a more practical application, the parameter, say, θ can take many more than two values as in Example 9.2-1. Somehow Bayesians must assign prior probabilities to this total parameter space through a prior p.d.f. $h(\theta)$. They have developed procedures for assessing these prior probabilities, and we simply cannot do justice to these methods here. Somehow $h(\theta)$ reflects the prior weights that the Bayesian wants to assign to the various possible values of θ. In some instances, if $h(\theta)$ is a constant and thus θ has the uniform prior distribution, we say that the Bayesian has a *noninformative* prior. If, in fact, some knowledge of θ exists in advance of experimentation, noninformative priors should be avoided if at all possible.

Also, in more practical examples, we usually take several observations, not just one. That is, we take a random sample, and there is frequently a good statistic, say, Y, for the parameter θ. Suppose we are considering a continuous case and the p.d.f. of Y, say, $g(y; \theta)$, can be thought of as the conditional p.d.f. of Y, given θ. [Henceforth in this section, we write $g(y; \theta) = g(y \mid \theta)$.] Thus, we can treat

$$g(y \mid \theta)h(\theta) = k(y, \theta)$$

as the joint p.d.f. of the statistic Y and the parameter. Of course, the marginal p.d.f. of Y is

$$k_1(y) = \int_{-\infty}^{\infty} h(\theta)g(y \mid \theta)\, d\theta.$$

Consequently,

$$\frac{k(y, \theta)}{k_1(y)} = \frac{g(y \mid \theta)h(\theta)}{k_1(y)} = k(\theta \mid y)$$

would serve as the conditional p.d.f. of the parameter, given that $Y = y$. This formula is essentially Bayes's theorem, and $k(\theta \mid y)$ is called the *posterior p.d.f. of* θ, given that $Y = y$.

Bayesians believe that everything which needs to be known about the parameter is summarized in this posterior p.d.f. $k(\theta \mid y)$. Suppose, for example, that they were pressed into making a point estimate of the parameter θ. They would note that they would be guessing the value of a random variable, here θ, given its p.d.f. $k(\theta \mid y)$. There are many ways that this could be done: The mean, the median, or the mode of that distribution would be reasonable guesses. However, in the final analysis, the best guess would clearly depend upon the penalties for various errors created by incorrect guesses. For instance, if we were penalized by taking the square of the error between the guess, say, $w(y)$, and the real value of the parameter θ, clearly we would use the conditional mean

$$w(y) = \int_{-\infty}^{\infty} \theta k(\theta \mid y)\, d\theta$$

as our Bayes estimate of θ. The reason is that, in general, if Z is a random variable, then the function of b, $E[(Z - b)^2]$, is minimized by $b = E(Z)$. (See Example 2.2-4.) Likewise, if the penalty (loss) function is the absolute value of the error, $|\theta - w(y)|$, then we use the median of the distribution, because with any random variable Z, $E[|Z - b|]$ is minimized when b equals the median of the distribution of Z. (See Exercise 2.2-10.)

EXAMPLE 9.2-2   Suppose that Y has a binomial distribution with parameters n and $p = \theta$. Then the p.m.f. of Y, given θ, is

$$g(y \mid \theta) = \binom{n}{y} \theta^y (1 - \theta)^{n-y}, \qquad y = 0, 1, 2, \ldots, n.$$

Let us take the prior p.d.f. of the parameter to be the beta p.d.f:

$$h(\theta) = \frac{\Gamma(\alpha + \beta)}{\Gamma(\alpha)\Gamma(\beta)} \theta^{\alpha-1}(1 - \theta)^{\beta-1}, \qquad 0 < \theta < 1.$$

Such a prior p.d.f. provides a Bayesian a great deal of flexibility through the selection of the parameters α and β. Thus, the joint probabilities can be described by a product of a binomial p.m.f. with parameters n and θ and this beta p.d.f., namely,

$$k(y, \theta) = \binom{n}{y} \frac{\Gamma(\alpha + \beta)}{\Gamma(\alpha)\Gamma(\beta)} \theta^{y+\alpha-1}(1 - \theta)^{n-y+\beta-1},$$

on the support given by $y = 0, 1, 2, \ldots, n$ and $0 < \theta < 1$. We find

$$k_1(y) = \int_0^1 k(y, \theta)\, d\theta$$

$$= \binom{n}{y} \frac{\Gamma(\alpha + \beta)}{\Gamma(\alpha)\Gamma(\beta)} \frac{\Gamma(\alpha + y)\Gamma(n + \beta - y)}{\Gamma(n + \alpha + \beta)}$$

on the support $y = 0, 1, 2, \ldots, n$ by comparing the integral with one involving a beta p.d.f. with parameters $y + \alpha$ and $n - y + \beta$. Therefore,

$$k(\theta \mid y) = \frac{k(y, \theta)}{k_1(y)}$$

$$= \frac{\Gamma(n + \alpha + \beta)}{\Gamma(\alpha + y)\Gamma(n + \beta - y)} \theta^{y+\alpha-1}(1 - \theta)^{n-y+\beta-1}, \qquad 0 < \theta < 1,$$

which is a beta p.d.f. with parameters $y + \alpha$ and $n - y + \beta$. With the square error loss function, we must minimize, with respect to $w(y)$, the integral

$$\int_0^1 [\theta - w(y)]^2 k(\theta \mid y)\, d\theta$$

to obtain the Bayes solution. But, as noted earlier, if Z is a random variable with a second moment, then $E[(Z - b)^2]$ is minimized by $b = E(Z)$. In the preceding integration, θ is like the Z with p.d.f. $k(\theta\,|\,y)$, and $w(y)$ is like the b, so the minimization is accomplished by taking

$$w(y) = E(\theta\,|\,y) = \frac{\alpha + y}{\alpha + \beta + n},$$

which is the mean of the beta distribution with parameters $y + \alpha$ and $n - y + \beta$. (See Exercise 5.2-8.) It is instructive to note that this Bayes solution can be written as

$$w(y) = \left(\frac{n}{\alpha + \beta + n}\right)\left(\frac{y}{n}\right) + \left(\frac{\alpha + \beta}{\alpha + \beta + n}\right)\left(\frac{\alpha}{\alpha + \beta}\right),$$

which is a weighted average of the maximum likelihood estimate y/n of θ and the mean $\alpha/(\alpha + \beta)$ of the prior p.d.f. of the parameter. Moreover, the respective weights are $n/(\alpha + \beta + n)$ and $(\alpha + \beta)/(\alpha + \beta + n)$. Thus, we see that α and β should be selected so that not only is $\alpha/(\alpha + \beta)$ the desired prior mean, but also the sum $\alpha + \beta$ plays a role corresponding to a sample size. That is, if we want our prior opinion to have as much weight as a sample size of 20, we would take $\alpha + \beta = 20$. So if our prior mean is 3/4, we select $\alpha = 15$ and $\beta = 5$. That is, the prior p.d.f. of θ is beta(15, 5). If we observe $n = 40$ and $y = 28$, then the posterior p.d.f. is beta($28 + 15 = 43, 12 + 5 = 17$). The prior and posterior p.d.f.'s are shown in Figure 9.2-1. ∎

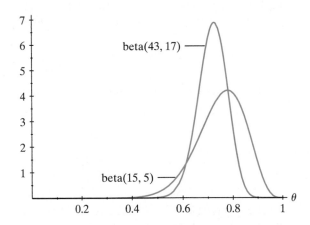

FIGURE 9.2-1: Beta prior and posterior p.d.f.'s

In Example 9.2-2, it is quite convenient to note that it is not really necessary to determine $k_1(y)$ to find $k(\theta\,|\,y)$. If we divide $k(y, \theta)$ by $k_1(y)$, we get the product of a factor that depends on y but does *not* depend on θ—say, $c(y)$—and we have

$$\theta^{y+\alpha-1}(1 - \theta)^{n-y+\beta-1}.$$

That is,

$$k(\theta\,|\,y) = c(y)\,\theta^{y+\alpha-1}(1 - \theta)^{n-y+\beta-1}, \qquad 0 < \theta < 1.$$

However, $c(y)$ must be that "constant" needed to make $k(\theta\,|\,y)$ a p.d.f., namely,

$$c(y) = \frac{\Gamma(n + \alpha + \beta)}{\Gamma(y + \alpha)\Gamma(n - y + \beta)}.$$

Accordingly, Bayesians frequently write that $k(\theta \mid y)$ is proportional to $k(y, \theta) = g(y \mid \theta)h(\theta)$; that is,

$$k(\theta \mid y) \propto g(y \mid \theta) \, h(\theta).$$

Then, to actually form the p.d.f. $k(\theta \mid y)$, they simply find a "constant" (which is, of course, actually some function of y) such that the expression integrates to 1.

EXAMPLE 9.2-3   Suppose that $Y = \overline{X}$ is the mean of a random sample of size n that arises from the normal distribution $N(\theta, \sigma^2)$, where σ^2 is known. Then $g(y \mid \theta)$ is $N(\theta, \sigma^2/n)$. Suppose further that we are able to assign prior weights to θ through a prior p.d.f. $h(\theta)$ that is $N(\theta_0, \sigma_0^2)$. Then we have

$$k(\theta \mid y) \propto \frac{1}{\sqrt{2\pi}\,(\sigma/\sqrt{n})} \frac{1}{\sqrt{2\pi}\sigma_0} \exp\left[-\frac{(y - \theta)^2}{2(\sigma^2/n)} - \frac{(\theta - \theta_0)^2}{2\sigma_0^2} \right].$$

If we eliminate all constant factors (including factors involving y only), then

$$k(\theta \mid y) \propto \exp\left[-\frac{(\sigma_0^2 + \sigma^2/n)\theta^2 - 2(y\sigma_0^2 + \theta_0\sigma^2/n)\theta}{2(\sigma^2/n)\sigma_0^2} \right].$$

This expression can be simplified by completing the square, to read (after eliminating factors not involving θ)

$$k(\theta \mid y) \propto \exp\left\{ -\frac{[\theta - (y\sigma_0^2 + \theta_0\sigma^2/n)/(\sigma_0^2 + \sigma^2/n)]^2}{[2(\sigma^2/n)\sigma_0^2]/[\sigma_0^2 + (\sigma^2/n)]} \right\}.$$

That is, the posterior p.d.f. of the parameter is obviously normal with mean

$$\frac{y\sigma_0^2 + \theta_0\sigma^2/n}{\sigma_0^2 + \sigma^2/n} = \left(\frac{\sigma_0^2}{\sigma_0^2 + \sigma^2/n} \right)y + \left(\frac{\sigma^2/n}{\sigma_0^2 + \sigma^2/n} \right)\theta_0$$

and variance $(\sigma^2/n)\sigma_0^2/(\sigma_0^2 + \sigma^2/n)$. If the square error loss function is used, then this posterior mean is the Bayes solution. Again, note that it is a weighted average of the maximum likelihood estimate $y = \overline{x}$ and the prior mean θ_0. The Bayes solution $w(y)$ will always be a value between the prior judgment and the usual estimate. Note also, here and in Example 9.2-2, that the Bayes solution gets closer to the maximum likelihood estimate as n increases. Thus, the Bayesian procedures permit the decision maker to enter his or her prior opinions into the solution in a very formal way so that the influence of those prior notions will be less and less as n increases.   ■

In Bayesian statistics, all the information is contained in the posterior p.d.f. $k(\theta \mid y)$. In Examples 9.2-2 and 9.2-3, we found Bayesian point estimates with the use of the squared error loss function. Note that if the loss function is the absolute value of the error, $|w(y) - \theta|$, then the Bayes solution would be the median of the posterior distribution of the parameter, which is given by $k(\theta \mid y)$. Hence, the Bayes solution changes—as it should—with different loss functions.

Finally, if an interval estimate of θ is desired, we would find two functions of y—say, $u(y)$ and $v(y)$—such that

$$\int_{u(y)}^{v(y)} k(\theta \mid y) \, d\theta = 1 - \alpha,$$

where α is small—say, $\alpha = 0.05$. Then the observed interval from $u(y)$ to $v(y)$ would serve as an interval estimate for the parameter in the sense that the posterior probability of the parameter's being in that interval is $1 - \alpha$. In Example 9.2-3, where the posterior p.d.f. of the parameter was normal, the interval

$$\frac{y\sigma_0^2 + \theta_0\sigma^2/n}{\sigma_0^2 + \sigma^2/n} \pm 1.96\sqrt{\frac{(\sigma^2/n)\sigma_0^2}{\sigma_0^2 + \sigma^2/n}}$$

serves as an interval estimate for θ with posterior probability of 0.95.

In closing this short section on Bayesian estimation, note that we could have begun with the sample observations $X_1, X_2, \ldots, X_n$, rather than some statistic Y. Then, in our discussion, we would replace $g(y \mid \theta)$ by the likelihood function

$$L(\theta) = f(x_1 \mid \theta)f(x_2 \mid \theta) \cdots f(x_n \mid \theta),$$

which is the joint p.d.f. of $X_1, X_2, \ldots, X_n$, given θ. Thus, we find that

$$k(\theta \mid x_1, x_2, \ldots, x_n) \propto h(\theta)f(x_1 \mid \theta)f(x_2 \mid \theta) \cdots f(x_n \mid \theta) = h(\theta)L(\theta).$$

Now, $k(\theta \mid x_1, x_2, \ldots, x_n)$ contains all the information about θ, given the data. Thus, depending on the loss function, we would choose our Bayesian estimate of θ as some characteristic of this posterior distribution, such as the mean or the median. It is interesting to observe that if the loss function is zero for some small neighborhood about the true parameter θ and is some large positive constant otherwise, then the Bayesian estimate, $w(x_1, x_2, \ldots, x_n)$, is essentially the mode of this conditional p.d.f., $k(\theta \mid x_1, x_2, \ldots, x_n)$. The reason for this is that we want to take the estimate so that it has as much posterior probability as possible in a small neighborhood around it. Finally, note that if $h(\theta)$ is a constant (a noninformative prior), then this Bayesian estimate using the mode is exactly the same as the maximum likelihood estimate. More generally, if $h(\theta)$ is not a constant, then the Bayesian estimate using the mode can be thought of as a weighted maximum likelihood estimate in which the weights reflect prior opinion about θ. That is, that value of θ which maximizes $h(\theta)L(\theta)$ is the mode of the posterior distribution of the parameter given the data and can be used as the Bayesian estimate associated with the appropriate loss function.

EXAMPLE 9.2-4 Let us consider again Example 9.2-2, but now say that $X_1, X_2, \ldots, X_n$ is a random sample from the Bernoulli distribution with p.m.f.

$$f(x \mid \theta) = \theta^x(1 - \theta)^{1-x}, \qquad x = 0, 1.$$

With the same prior p.d.f. of θ, the joint distribution of $X_1, X_2, \ldots, X_n$ and θ is given by

$$\frac{\Gamma(\alpha + \beta)}{\Gamma(\alpha)\Gamma(\beta)} \theta^{\alpha-1}(1 - \theta)^{\beta-1}\theta^{\sum_{i=1}^n x_i}(1 - \theta)^{n-\sum_{i=1}^n x_i}, \qquad 0 < \theta < 1, \ x_i = 0, 1.$$

Of course, the posterior p.d.f. of θ, given that $X_1 = x_1, X_2 = x_2, \ldots, X_n = x_n$ is such that

$$k(\theta \mid x_1, x_2, \ldots, x_n) \propto \theta^{\sum_{i=1}^n x_i + \alpha - 1}(1 - \theta)^{n-\sum_{i=1}^n x_i + \beta - 1}, \qquad 0 < \theta < 1,$$

which is beta with $\alpha^* = \sum x_i + \alpha$, $\beta^* = n - \sum x_i + \beta$. The conditional mean of θ is

$$\frac{\sum_{i=1}^n x_i + \alpha}{n + \alpha + \beta} = \left(\frac{n}{n + \alpha + \beta}\right)\left(\frac{\sum_{i=1}^n x_i}{n}\right) + \left(\frac{\alpha + \beta}{n + \alpha + \beta}\right)\left(\frac{\alpha}{\alpha + \beta}\right),$$

which, with $y = \sum x_i$, is exactly the same result as that of Example 9.2-2. ∎

REMARK   In Section 6.5, there was a remark about improving the confidence intervals for probability $p = \theta$ if Y is $b(n,p)$. There it was suggested that we should use $\tilde{p} = (Y + 2)/(n + 4)$ instead of $\hat{p} = Y/n$. Note that $\tilde{p}$ is the Bayes estimator when the prior p.d.f. is beta with $\alpha = \beta = 2$. Since it brings the estimator Y/n closer to the prior mean $\alpha/(\alpha + \beta)$—here, $1/2 - \tilde{p}$ is often called a *shrinkage estimator*. To get an interval estimate for p, the Bayesian takes the posterior p.d.f. $k(p \mid y)$, which is beta, and finds $u(y)$ and $v(y)$ so that

$$\int_{u(y)}^{v(y)} k(p \mid y)\, dp = 1 - \alpha.$$

Beta tables or the computer are needed to do this, but it would be an interesting project to compare such intervals for various values of n, p, and y with those found by

$$\tilde{p} \pm z_{\alpha/2}\sqrt{\tilde{p}(1 - \tilde{p})/n}.$$ ∎

EXERCISES

9.2-1. Let Y be the sum of the observations of a random sample from a Poisson distribution with mean θ. Let the prior p.d.f. of θ be gamma with parameters α and β.

(a) Find the posterior p.d.f. of θ, given that $Y = y$.

(b) If the loss function is $[w(y) - \theta]^2$, find the Bayesian point estimate $w(y)$.

(c) Show that the $w(y)$ found in part (b) is a weighted average of the maximum likelihood estimate y/n and the prior mean $\alpha\beta$, with respective weights of $n/(n + 1/\beta)$ and $(1/\beta)/(n + 1/\beta)$.

9.2-2. Let $X_1, X_2, \ldots, X_n$ be a random sample from a gamma distribution with known α and with $\theta = 1/\tau$. Say τ has a prior p.d.f. that is gamma with parameters α_0 and θ_0, so that the prior mean is $\alpha_0\theta_0$.

(a) Find the posterior p.d.f. of τ, given that $X_1 = x_1, X_2 = x_2, \ldots, X_n = x_n$.

(b) Find the mean of the posterior distribution found in part (a), and write it as a function of the sample mean $\overline{X}$ and $\alpha_0\theta_0$.

(c) Explain how you would find a 95% interval estimate of τ if $n = 10$, $\alpha = 3$, $\alpha_0 = 10$, and $\theta_0 = 2$.

9.2-3. In Example 9.2-2, take $n = 30$, $\alpha = 15$, and $\beta = 5$.

(a) Using the squared error loss, compute the expected loss (risk function) associated with the Bayes solution $w(Y)$.

(b) The risk function associated with the usual estimator Y/n is, of course, $\theta(1 - \theta)/30$. Find those values of θ for which the risk function in part (a) is less than $\theta(1 - \theta)/30$. In particular, if the prior mean $\alpha/(\alpha + \beta) = 3/4$ is a reasonable guess, then the risk function in part (a) is the better of the two (i.e., is smaller in a neighborhood of $\theta = 3/4$) for what values of θ?

9.2-4. Consider a random sample $X_1, X_2, \ldots, X_n$ from a distribution with p.d.f.

$$f(x \mid \theta) = 3\theta x^2 e^{-\theta x^3}, \qquad 0 < x < \infty.$$

Let θ have a prior p.d.f. that is gamma with $\alpha = 4$ and the usual $\theta = 1/4$. Find the conditional mean of θ, given that $X_1 = x_1, X_2 = x_2, \ldots, X_n = x_n$.

9.2-5. In Example 9.2-3, suppose the loss function $|\theta - w(Y)|$ is used. What is the Bayesian estimator $w(Y)$?

9.2-6. Let Y be the largest order statistic of a random sample of size n from a distribution with p.d.f.

$f(x \mid \theta) = 1/\theta$, $0 < x < \theta$. Say θ has the prior p.d.f.

$$h(\theta) = \beta \alpha^\beta / \theta^{\beta+1}, \qquad \alpha < \theta < \infty,$$

where $\alpha > 0$, $\beta > 0$.

(a) If $w(Y)$ is the Bayesian estimator of θ and $[\theta - w(Y)]^2$ is the loss function, find $w(Y)$.

(b) If $n = 4$, $\alpha = 1$, and $\beta = 2$, find the Bayesian estimator $w(Y)$ if the loss function is $|\theta - w(Y)|$.

9.2-7. Refer to Example 9.2-3. Suppose we select $\sigma_0^2 = d\sigma^2$, where σ^2 is known in that example. What value do we assign to d so that the variance of the posterior p.d.f. of the parameter is two-thirds of the variance of $Y = \overline{X}$, namely, σ^2/n?

9.3 MORE BAYESIAN CONCEPTS

Let $X_1, X_2, \ldots, X_n$ be a random sample from a distribution with p.d.f. (p.m.f.) $f(x \mid \theta)$, and let $h(\theta)$ be the prior p.d.f. Then the distribution associated with the marginal p.d.f. of $X_1, X_2, \ldots, X_n$, namely,

$$k_1(x_1, x_2, \ldots, x_n) = \int_{-\infty}^{\infty} f(x_1 \mid \theta) f(x_2 \mid \theta) \cdots f(x_n \mid \theta) h(\theta) \, d\theta,$$

is called the **predictive distribution** because it provides the best description of the probabilities on $X_1, X_2, \ldots, X_n$. Often this creates some interesting distributions. For example, suppose there is only one X with the normal p.d.f.

$$f(x \mid \theta) = \frac{\sqrt{\theta}}{\sqrt{2\pi}} e^{-(\theta x^2)/2}, \qquad -\infty < x < \infty.$$

Here, $\theta = 1/\sigma^2$, the inverse of the variance, is called the **precision** of X. Say this precision has the gamma p.d.f.

$$h(\theta) = \frac{1}{\Gamma(\alpha)\beta^\alpha} \theta^{\alpha-1} e^{-\theta/\beta}, \qquad 0 < \theta < \infty.$$

Then the predictive p.d.f. is

$$k_1(x) = \int_0^\infty \frac{\theta^{\alpha+1/2-1} e^{-(x^2/2+1/\beta)\theta}}{\Gamma(\alpha)\beta^\alpha \sqrt{2\pi}} \, d\theta$$

$$= \frac{\Gamma(\alpha + 1/2)}{\Gamma(\alpha)\beta^\alpha \sqrt{2\pi}} \frac{1}{(1/\beta + x^2/2)^{\alpha+1/2}}, \qquad -\infty < x < \infty.$$

Note that if $\alpha = r/2$ and $\beta = 2/r$, where r is a positive integer, then

$$k_1(x) \propto \frac{1}{(1 + x^2/r)^{(r+1)/2}}, \qquad -\infty < x < \infty,$$

which is a t p.d.f. with r degrees of freedom. So if the inverse of the variance—or precision θ—of a normal distribution varies as a gamma random variable, a generalization of a t distribution has been created that has heavier tails than the normal distribution. This **mixture** of normals (different from a mixed distribution) is attained by weighing with the gamma distribution in a process often called **compounding**.

Another illustration of compounding is given in the next example.

EXAMPLE 9.3-1 Suppose X has a gamma distribution with the two parameters k and θ^{-1}. (That is, the usual α is replaced by k and θ by its reciprocal.) Say $h(\theta)$ is gamma with parameters

α and β, so that

$$k_1(x) = \int_0^\infty \frac{\theta^k x^{k-1} e^{-\theta x}}{\Gamma(k)} \frac{1}{\Gamma(\alpha)\beta^\alpha} \theta^{\alpha-1} e^{-\theta/\beta} \, d\theta$$

$$= \int_0^\infty \frac{x^{k-1}\theta^{k+\alpha-1}e^{-\theta(x+1/\beta)}}{\Gamma(k)\Gamma(\alpha)\beta^\alpha} d\theta$$

$$= \frac{\Gamma(k+\alpha)x^{k-1}}{\Gamma(k)\Gamma(\alpha)\beta^\alpha} \frac{1}{(x+1/\beta)^{k+\alpha}}$$

$$= \frac{\Gamma(k+x)\beta^k x^{k-1}}{\Gamma(k)\Gamma(\alpha)(1+\beta x)^{k+\alpha}}, \qquad 0 < x < \infty.$$

Of course, this is a generalization of the F distribution, which we obtain by letting $\alpha = r_2/2$, $k = r_1/2$, and $\beta = r_1/r_2$. ∎

Note how well the prior $h(\theta)$ "fits" with $f(x \mid \theta)$ or $f(x_1 \mid \theta)f(x_2 \mid \theta)\cdots f(x_n \mid \theta)$ in all of our examples, and the posterior distribution is of exactly the same form as the prior. In Example 9.2-2, both the prior and the posterior were beta. In Example 9.2-3, both the prior and posterior were normal. In Example 9.3-1, both the prior and the posterior (if we had found it) were gamma. When this type of pairing occurs, we say that the class of prior p.d.f.'s (p.m.f.'s) is a **conjugate family of priors**. Obviously, this makes the mathematics easier, and usually the parameters in the prior distribution give us enough flexibility to obtain good fits.

EXAMPLE 9.3-2

(Berry, 1996) This example deals with *predictive probabilities*, and it concerns the breakage of glass panels in high-rise buildings. One such case involved 39 panels, and of the 39 panels that broke, it was known that 3 broke due to nickel sulfide (NiS) stones found in them. Loss of evidence prevented the causes of breakage of the other 36 panels from being known. So the court wanted to know whether the manufacturer of the panels or the builder was at fault for the breakage of these 36 panels.

From expert testimony, it was thought that usually about 5% breakage is caused by NiS stones. That is, if this value of p is selected from a beta distribution, we have

$$\frac{\alpha}{\alpha+\beta} = 0.05. \tag{9.3-1}$$

Moreover, the expert thought that if two panels from the same lot break and one breakage was caused by NiS stones, then, due to the pervasive nature of the manufacturing process, the probability of the second panel's breaking due to NiS stones increases to about 95%. Thus, the posterior estimate of p (see Example 9.2-2) with one "success" after one trial is

$$\frac{\alpha+1}{\alpha+\beta+1} = 0.95. \tag{9.3-2}$$

Solving Equations 9.3-1 and 9.3-2 for α and β, we obtain

$$\alpha = \frac{1}{360} \qquad \text{and} \qquad \beta = \frac{19}{360}.$$

Now updating the posterior probability with 3 "successes" out of 3 trials, we obtain the posterior estimate of p:

$$\frac{\alpha + 3}{\alpha + \beta + 3} = \frac{1/360 + 3}{20/360 + 3} = \frac{1081}{1100} = 0.983.$$

Of course, the court that heard the case wanted to know the expert's opinion about the probability that all of the remaining 36 panels broke because of NiS stones. Using updated probabilities after the third break, then the fourth, and so on, we obtain the product

$$\left(\frac{1/360 + 3}{20/360 + 3}\right)\left(\frac{1/360 + 4}{20/360 + 4}\right)\left(\frac{1/360 + 5}{20/360 + 5}\right)\cdots\left(\frac{1/360 + 38}{20/360 + 38}\right) = 0.8664.$$

That is, the expert held that the probability that all 36 breakages were caused by NiS stones was about 87%, which is the needed value in the court's decision. ∎

We now look at a situation in which we have two unknown parameters; we will use, for convenience, what is called a **noninformative prior**, which usually puts uniform distributions on the parameters. Let us begin with a random sample $X_1, X_2, \ldots, X_n$ from the normal distribution $N(\theta_1, \theta_2)$, and suppose we have little prior knowledge about θ_1 and θ_2. We then use the noninformative prior that θ_1 and $\ln \theta_2$ are uniform and independent; that is,

$$h_1(\theta_1)h_2(\theta_2) \propto \frac{1}{\theta_2}, \qquad -\infty < \theta_1 < \infty, \; 0 < \theta_2 < \infty.$$

Of course, we immediately note that we cannot find a constant c such that c/θ_2 is a joint p.d.f. on that support. That is, this noninformative prior p.d.f. is not a p.d.f. at all; hence, it is called an **improper** prior. However, we use it anyway, because it will be satisfactory when multiplied by the joint p.d.f. of $X_1, X_2, \ldots, X_n$. We have the product

$$\left(\frac{1}{\theta_2}\right)\left(\frac{1}{\sqrt{2\pi\theta_2}}\right)^n \exp\left[-\sum_{i=1}^n \frac{(x_i - \theta_1)^2}{2\theta_2}\right].$$

Thus,

$$k_{12}(\theta_1, \theta_2 \mid x_1, x_2, \ldots, x_n) \propto \left(\frac{1}{\theta_2}\right)^{\frac{n}{2} + 1} \exp\left[-\frac{1}{2}\{(n - 1)s^2 + n(\bar{x} - \theta_1)^2\}/\theta_2\right]$$

since $\sum_{i=1}^n (x_i - \theta_1)^2 = (n - 1)s^2 + n(\bar{x} - \theta_1)^2 = D$. It then follows that

$$k_1(\theta_1 \mid x_1, x_2, \ldots, x_n) \propto \int_0^\infty k_{12}(\theta_1, \theta_2 \mid x_1, x_2, \ldots, x_n)\, d\theta_2.$$

Changing variables by letting $z = 1/\theta_2$, we obtain

$$k_1(\theta_1 \mid x_1, x_2, \ldots, x_n) \propto \int_0^\infty \frac{z^{n/2+1}}{z^2} e^{(-1/2)Dz}\, dz$$

$$\propto D^{-n/2} = \left[(n - 1)s^2 + n(\bar{x} - \theta_1)^2\right]^{-n/2}.$$

To get this p.d.f. in a more familiar form, let $t = (\theta_1 - \bar{x})/(s/\sqrt{n})$, with Jacobian $s/\sqrt{n}$, to yield

$$k(t \mid x_1, x_2, \ldots, x_n) \propto \frac{1}{[1 + t^2/(n-1)]^{[(n-1)+1]/2}}, \qquad -\infty < t < \infty.$$

That is, the conditional p.d.f. of t, given $x_1, x_2, \ldots, x_n$, is Student's t with $n - 1$ degrees of freedom. Thus, a $(1 - \alpha)$ probability interval for θ_1 is given by

$$-t_{\alpha/2}(n-1) < \frac{\theta_1 - \bar{x}}{s/\sqrt{n}} < t_{\alpha/2}(n-1),$$

or

$$\bar{x} - t_{\alpha/2}(n-1)\, s/\sqrt{n} < \theta_1 < \bar{x} + t_{\alpha/2}(n-1)\, s/\sqrt{n},$$

which is the same as the standard $100(1 - \alpha)\%$ confidence interval for θ_1.

The reason we get the same answer in this case is that we use a noninformative prior. Bayesians do not like to use a noninformative prior if they really know something about the parameters. For example, say they believe that the precision $1/\theta_2$ has a gamma distribution with parameters α and β instead of the noninformative prior. Then finding the conditional p.d.f. of θ_1 becomes a much more difficult integration. However, it can be done, but we leave it to a more advanced course. (See Hogg, McKean, and Craig, 2005.)

EXAMPLE 9.3-3   (Johnson and Albert, 1999) The data in this example, a sample of $n = 13$ measurements of the National Oceanographic and Atmospheric Administration (NOAA)/Environmental Protection Agency (EPA) ultraviolet (UV) index taken in Los Angeles, were collected from archival data of every Sunday in October during the years 1995–1997 in a database maintained by NOAA. The 13 UV readings are

$$7, 6, 5, 5, 3, 6, 5, 5, 3, 5, 5, 4, 4,$$

and, although they are integer values, we assume that they are taken from a $N(\mu, \sigma^2)$ distribution.

The Bayesian analysis, using a noninformative prior in the preceding discussion, implies that, with $\mu = \theta_1$,

$$\frac{\mu - 4.846}{0.317}, \qquad \text{where} \qquad x = 4.846 \qquad \text{and} \qquad \frac{s}{\sqrt{n}} = 0.317,$$

has a posterior t distribution with $n - 1 = 12$ degrees of freedom. For example, a posterior 95% probability interval for μ is

$$(4.846 - [t_{0.025}(12)][0.317], \; 4.846 + [t_{0.025}(12)][0.317]) = (4.155, 5.537). \quad \blacksquare$$

EXAMPLE 9.3-4   Tsutakawa et. al. (1985) discuss the rates of death from stomach cancer in males aged 45–61 in the 20 largest cities in Missouri. Suppose p_i, $i = 1, 2, \ldots, 20$, are 20 respective probabilities of death due to stomach cancer. On one model, we can assume that $p_1, p_2, \ldots, p_{20}$ are taken independently from a beta distribution with parameters α and β. If y_i is the number of "successes" out of n_i observations in city i, then the posterior mean of p_i is

$$\widehat{p_i}\left(\frac{n_i}{n_i + \alpha + \beta}\right) + \left(\frac{\alpha}{\alpha + \beta}\right)\left(\frac{\alpha + \beta}{n_i + \alpha + \beta}\right), \quad i = 1, 2, \ldots, 20,$$

where $\widehat{p_i} = y_i/n_i$. Of course, the parameters α and β are unknown, but we have assumed that $p_1, p_2, \ldots, p_{20}$ arose from a similar distribution for these cities in Missouri; that is, we assume that our prior knowledge concerning the proportions is *exchangeable*. So it would be reasonable to estimate $\alpha/(\alpha + \beta)$, the prior mean of a proportion, with the formula

$$\bar{y} = \frac{y_1 + y_2 + \cdots + y_{20}}{n_1 + n_2 + \cdots + n_{20}} = \frac{71}{71,478} = 0.000993,$$

for the data given in Table 9.3-1. Thus, the posterior estimate of p_i is found by *shrinking* $\widehat{p_i}$ towards the pooled estimate of the mean $\alpha/(\alpha + \beta)$—namely, $\bar{y}$. That is, the posterior estimate is

$$\widehat{p_i}\left(\frac{n_i}{n_i + \alpha + \beta}\right) + \bar{y}\left(\frac{\alpha + \beta}{n_i + \alpha + \beta}\right).$$

The only question remaining is how much weight should be given to the prior, represented by $\alpha + \beta$, relative to $n_1, n_2, \ldots, n_{20}$. Considering the sizes of the samples from the various cities, we selected $\alpha + \beta = 3000$ (which means that the prior is worth about a sample of 3000), which resulted in the posterior probabilities given in Table 9.3-1. Note how this type of shrinkage tends to pull the posterior estimates much closer to the average, particularly those associated with small sample sizes. Baseball fans might try this type of shrinkage in predicting some of the final batting averages of the better batters about a quarter of the way through the season. ∎

			TABLE 9.3-1: Cancer Mortality Rates				
y_i	n_i	$\widehat{p_i}$	posterior estimate	y_i	n_i	$\widehat{p_i}$	posterior estimate
0	1083	0	0.00073	0	855	0	0.00077
2	3461	0.00058	0.00077	0	657	0	0.00081
1	1208	0.00083	0.00095	1	1025	0.00098	0.00099
0	527	0	0.00084	2	1668	0.00120	0.00107
1	583	0.00172	0.00111	3	582	0.00515	0.00167
0	917	0	0.00076	1	857	0.00117	0.00103
1	680	0.00147	0.00108	1	917	0.00109	0.00102
54	53637	0.00101	0.00101	0	874	0	0.00077
0	395	0	0.00088	1	581	0.00172	0.00111
3	588	0.00510	0.00167	0	383	0	0.00088

It is clear that difficult integration caused Bayesians great problems until very recent times, in which advances in computer methods "solved" many of these problems. As a simple illustration, suppose the p.d.f. of a good statistic Y is $f(y \mid \theta)$ and the prior p.d.f. $h(\theta)$ is such that

$$k(\theta \mid y) = \frac{f(y \mid \theta) h(\theta)}{\int_{-\infty}^{\infty} f(y \mid \theta) h(\theta) \, d\theta}$$

is not a nice p.d.f. with which to deal. In particular, say that we have a squared error loss and we wish to determine $E(\theta \mid y)$, namely,

$$\delta(y) = \frac{\int_{-\infty}^{\infty} \theta f(y \mid \theta) h(\theta) \, d\theta}{\int_{-\infty}^{\infty} f(y \mid \theta) h(\theta) \, d\theta},$$

but cannot do it easily. Let $f(y \mid \theta) = w(\theta)$. Then we wish to evaluate the ratio

$$\frac{E[\theta w(\theta)]}{E[w(\theta)]},$$

where y is given and the expected values are taken with respect to θ. To do so, we simply generate a number of θ values, say, $\theta_1, \theta_2, \ldots, \theta_m$ (where m is large), from the distribution given by $h(\theta)$. Then we estimate the numerator and denominator of the desired ratio by

$$\sum_{i=1}^{m} \frac{\theta_i w(\theta_i)}{m} \qquad \text{and} \qquad \sum_{i-1}^{m} \frac{w(\theta_i)}{m},$$

respectively, to obtain

$$\tau = \frac{\sum_{i=1}^{m} \theta_i w(\theta_i)/m}{\sum_{i=1}^{m} w(\theta_i)/m}.$$

In addition to this simple Monte Carlo procedure, there are additional ones that are extremely useful in Bayesian inferences. Two of these are the *Gibbs sampler* and the *Markov chain Monte Carlo (MCMC)*. The latter is used in *hierarchical Bayes models* in which the prior has another parameter that has its own prior. That is, we have

$$f(y \mid \theta), \qquad h(\theta \mid \tau), \qquad \text{and} \qquad g(\tau).$$

Hence,

$$k(\theta, \tau \mid y) = \frac{f(y \mid \theta) h(\theta \mid \tau) g(\tau)}{\int_{-\infty}^{\infty} \int_{-\infty}^{\infty} f(y \mid \theta) h(\theta \mid \tau) g(\tau) \, d\theta \, d\tau}$$

and

$$k_1(\theta \mid y) = \int_{-\infty}^{\infty} k(\theta, \tau \mid y) \, d\tau.$$

Thus, a Bayes estimator, for a squared error loss, is

$$\int_{-\infty}^{\infty} \theta k_1(\theta \mid y) \, d\theta.$$

Using the Gibbs sampler, we can generate a stream of values $(\theta_1, \tau_1), (\theta_2, \tau_2), \ldots$ that allows us to estimate $k(\theta, \tau \mid y)$ and $\int_{-\infty}^{\infty} \theta k_1(\theta \mid y) \, d\theta$. These procedures are the MCMC procedures. (For additional references, see Hogg, McKean, and Craig, 2005.)

EXERCISES

9.3-1. Let X have a Poisson distribution with parameter θ. Let θ be $\Gamma(\alpha, \beta)$. Show that the marginal p.m.f. of X (the compound distribution) is

$$k_1(x) = \frac{\Gamma(\alpha + x)\beta^x}{\Gamma(\alpha)x!(1 + \beta)^{\alpha+x}}, \quad x = 0, 1, 2, 3, \ldots,$$

which is a generalization of the negative binomial distribution.

9.3-2. Suppose X is $b(n, \theta)$ and θ is beta(α, β). Show that the marginal p.d.f. of X (the compound distribution) is

$$k_1(x) = \frac{n!\,\Gamma(\alpha + \beta)\,\Gamma(x + \alpha)\,\Gamma(n - x + \beta)}{x!\,(n - x)!\,\Gamma(\alpha)\,\Gamma(\beta)\,\Gamma(n + \alpha + \beta)},$$
$$x = 0, 1, 2, \ldots, n.$$

9.3-3. Let X have the geometric p.m.f. $\theta(1 - \theta)^{x-1}$, $x = 1, 2, 3, \ldots$, where θ is beta with parameters α and β. Show that the compound p.m.f. is

$$\frac{\Gamma(\alpha + \beta)\,\Gamma(\alpha + 1)\,\Gamma(\beta + x - 1)}{\Gamma(\alpha)\,\Gamma(\beta)\,\Gamma(\alpha + \beta + x)},$$
$$x = 1, 2, 3, \ldots.$$

With $\alpha = 1$, this is one form of **Zipf's law**,

$$\frac{\beta}{(\beta + x)(\beta + x - 1)}, \quad x = 1, 2, 3, \ldots.$$

9.3-4. Let X have the p.d.f.

$$f(x \mid \theta) = \theta \tau x^{\tau-1} e^{-\theta x^\tau}, \quad 0 < x < \infty,$$

where the distribution of θ is $\Gamma(\alpha, \beta)$. Find the compound distribution of X, which is called the **Burr distribution**.

9.3-5. Let $X_1, X_2, \ldots, X_n$ be a random sample from a gamma distribution with $\alpha = 1, \theta$. Let $h(\theta) \propto 1/\theta$, $0 < \theta < \infty$, be an improper noninformative prior.

(a) Find the posterior p.d.f. of θ.

(b) Change variables by letting $z = 1/\theta$, and show that the posterior distribution of Z is $\Gamma(n, 1/y)$, where $y = \sum_{i=1}^{n} x_i$.

(c) Use $2yz$ to obtain a $(1 - \alpha)$ probability interval for z and, of course, for θ.

9.3-6. Let X_1, X_2 be a random sample from the Cauchy distribution with p.d.f.

$$f(x \mid \theta_1, \theta_2) = \frac{1}{\pi} \frac{\theta_2}{\theta_2^2 + (x - \theta_1)^2},$$
$$-\infty < x < \infty, \ -\infty < \theta_1 < \infty, \ 0 < \theta_2 < \infty.$$

Consider the noninformative prior $h(\theta_1, \theta_2) \propto 1$ on that support. Obtain the posterior p.d.f. (except for constants) of θ_1, θ_2 if $x_1 = 3$ and $x_2 = 7$. For estimates, find θ_1, θ_2 that maximizes this posterior p.d.f.; that is, find the mode of that posterior. (This might require some reasonable "trial and error" or an advanced method of maximizing a function of two variables.)

HISTORICAL COMMENTS   The Neo-Bayesian movement in America really started with J. Savage in the 1950s. Initially, Bayesians were limited in their work because it was extremely difficult to compute certain distributions, such as the conditional one, $k(\theta \mid x_1, x_2, \ldots, x_n)$. However, towards the end of the 1970s, computers were becoming more useful and thus computing was much easier. In particular, the Bayesians developed Gibbs sampling and Markov chain Monte Carlo (MCMC). It is our opinion that the Bayesians will continue to expand and Bayes's methods will be a major approach to statistical inferences, possibly even dominating professional applications. This is difficult for two fairly classical statisticians (as we are) to admit, but, in all fairness, we cannot ignore the strong trend toward Bayesian methods.

Some Theory

10.1 SUFFICIENT STATISTICS

We first define a sufficient statistic $Y = u(X_1, X_2, \ldots, X_n)$ for a parameter, using a statement that, in most books, is given as a necessary and sufficient condition for sufficiency, namely, the well-known Fisher–Neyman factorization theorem. We do this because we find that readers at the introductory level can apply such a definition easily. However, using this definition, we shall note, by examples, its implications, one of which is also sometimes used as the definition of sufficiency. The understanding of Example 10.1-3 is most important in an appreciation of the value of sufficient statistics.

DEFINITION 10.1-1

(Factorization Theorem) Let $X_1, X_2, \ldots, X_n$ denote random variables with joint p.d.f. or p.m.f. $f(x_1, x_2, \ldots, x_n; \theta)$, which depends on the parameter θ. The statistic $Y = u(X_1, X_2, \ldots, X_n)$ is sufficient for θ if and only if

$$f(x_1, x_2, \ldots, x_n; \theta) = \phi[u(x_1, x_2, \ldots, x_n); \theta]h(x_1, x_2, \ldots, x_n),$$

where ϕ depends on $x_1, x_2, \ldots, x_n$ only through $u(x_1, \ldots, x_n)$ and $h(x_1, \ldots, x_n)$ does not depend on θ.

Let us consider several important examples and consequences of this definition. We first note, however, that in all instances in this book the random variables $X_1, X_2, \ldots, X_n$ will be of a random sample, and hence their joint p.d.f. or p.m.f. will be of the form

$$f(x_1; \theta)f(x_2; \theta) \cdots f(x_n; \theta).$$

EXAMPLE 10.1-1 Let $X_1, X_2, \ldots, X_n$ denote a random sample from a Poisson distribution with parameter $\lambda > 0$. Then

$$f(x_1; \lambda)f(x_2; \lambda) \cdots f(x_n; \lambda) = \frac{\lambda^{\Sigma x_i} e^{-n\lambda}}{x_1! x_2! \cdots x_n!} = (\lambda^{n\bar{x}} e^{-n\lambda})\left(\frac{1}{x_1! x_2! \cdots x_n!}\right),$$

where $\bar{x} = (1/n) \sum_{i=1}^{n} x_i$. Thus, from the factorization theorem (Definition 10.1-1), it is clear that the sample mean $\overline{X}$ is a sufficient statistic for λ. It can easily be shown that the maximum likelihood estimator for λ is also $\overline{X}$, so here the maximum likelihood estimator is a function of the sufficient statistic. ∎

In Example 10.1-1, if we replace $n\bar{x}$ by $\sum_{i=1}^{n} x_i$, it is quite obvious that the sum $\sum_{i=1}^{n} X_i$ is also a sufficient statistic for λ. This certainly agrees with our intuition, because if we know one of the statistics $\overline{X}$ and $\sum_{i=1}^{n} X_i$, we can easily find the other. If we generalize this idea, we see that if Y is sufficient for a parameter θ, then every single-valued function of Y not involving θ, but with a single-valued inverse, is also a sufficient statistic for θ. Again, the reason is that knowing either Y or that function of Y, we know the other. More formally, if $W = v(Y) = v[u(X_1, X_2, \ldots, X_n)]$ is that function and $Y = v^{-1}(W)$ is the single-valued inverse, then the factorization theorem can be written as

$$f(x_1, x_2, \ldots, x_n; \theta) = \phi[v^{-1}\{v[u(x_1, x_2, \ldots, x_n)]\}; \theta] h(x_1, x_2, \ldots, x_n).$$

The first factor of the righthand member of this equation depends on $x_1, x_2, \ldots, x_n$ through $v[u(x_1, x_2, \ldots, x_n)]$, so $W = v[u(X_1, X_2, \ldots, X_n)]$ is a sufficient statistic for θ. We illustrate this fact and the factorization theorem with an underlying distribution of the continuous type.

EXAMPLE 10.1-2 Let $X_1, X_2, \ldots, X_n$ be a random sample from $N(\mu, 1)$, $-\infty < \mu < \infty$. The joint p.d.f. of these random variables is

$$\frac{1}{(2\pi)^{n/2}} \exp\left[-\frac{1}{2}\sum_{i=1}^{n}(x_i - \mu)^2\right]$$

$$= \frac{1}{(2\pi)^{n/2}} \exp\left[-\frac{1}{2}\sum_{i=1}^{n}[(x_i - \bar{x}) + (\bar{x} - \mu)]^2\right]$$

$$= \left\{\exp\left[-\frac{n}{2}(\bar{x} - \mu)^2\right]\right\}\left\{\frac{1}{(2\pi)^{n/2}} \exp\left[-\frac{1}{2}\sum_{i=1}^{n}(x_i - \bar{x})^2\right]\right\}.$$

From the factorization theorem, we see that $\overline{X}$ is sufficient for μ. Now, $\overline{X}^3$ is also sufficient for μ, because knowing $\overline{X}^3$ is equivalent to having knowledge of the value of $\overline{X}$. However, $\overline{X}^2$ does not have this property, and it is not sufficient for μ. ∎

One extremely important consequence of the sufficiency of a statistic Y is that the conditional probability of any given event A in the support of $X_1, X_2, \ldots, X_n$, given that $Y = y$, does not depend on θ. This consequence is sometimes used as the definition of sufficiency and is illustrated in the next example.

EXAMPLE 10.1-3 Let $X_1, X_2, \ldots, X_n$ be a random sample from a distribution with p.m.f.

$$f(x; p) = p^x(1 - p)^{1-x}, \qquad x = 0, 1,$$

where the parameter p is between 0 and 1. We know that

$$Y = X_1 + X_2 + \cdots + X_n$$

is $b(n, p)$ and Y is sufficient for p because the joint p.m.f. of $X_1, X_2, \ldots, X_n$ is

$$p^{x_1}(1 - p)^{1-x_1} \cdots p^{x_n}(1 - p)^{1-x_n} = [p^{\Sigma x_i}(1 - p)^{n-\Sigma x_i}](1),$$

where $\phi(y; p) = p^y(1 - p)^{n-y}$ and $h(x_1, x_2, \ldots, x_n) = 1$. What, then, is the conditional probability $P(X_1 = x_1, \ldots, X_n = x_n \mid Y = y)$, where $y = 0, 1, \ldots, n - 1$, or n? Unless the sum of the nonnegative integers $x_1, x_2, \ldots, x_n$ equals y, this conditional probability is obviously equal to zero, which does not depend on p. Hence, it is interesting to consider the solution only when $y = x_1 + \cdots + x_n$. From the definition of conditional probability, we have

$$P(X_1 = x_1, \ldots, X_n = x_n \mid Y = y) = \frac{P(X_1 = x_1, \ldots, X_n = x_n)}{P(Y = y)}$$

$$= \frac{p^{x_1}(1 - p)^{1-x_1} \cdots p^{x_n}(1 - p)^{1-x_n}}{\binom{n}{y}p^y(1 - p)^{n-y}}$$

$$= \frac{1}{\binom{n}{y}},$$

where $y = x_1 + \cdots + x_n$. Since y equals the number of ones in the collection $x_1, x_2, \ldots, x_n$, this answer is only the probability of selecting a particular arrangement, namely, $x_1, x_2, \ldots, x_n$, of y ones and $n - y$ zeros, and does not depend on the parameter p. That is, given that the sufficient statistic $Y = y$, the conditional probability of $X_1 = x_1, X_2 = x_2, \ldots, X_n = x_n$ does not depend on the parameter p. ∎

It is interesting to observe that the underlying p.d.f. or p.m.f. in Examples 10.1-1, 10.1-2, and 10.1-3 can be written in the exponential form

$$f(x; \theta) = \exp[K(x)p(\theta) + S(x) + q(\theta)],$$

where the support is free of θ. That is, we have, respectively,

$$\frac{e^{-\lambda}\lambda^x}{x!} = \exp\{x \ln \lambda - \ln x! - \lambda\}, \qquad x = 0, 1, 2, \ldots,$$

$$\frac{1}{\sqrt{2\pi}} e^{-(x-\mu)^2/2} = \exp\left\{x\mu - \frac{x^2}{2} - \frac{\mu^2}{2} - \frac{1}{2}\ln(2\pi)\right\}, \qquad -\infty < x < \infty,$$

and

$$p^x(1 - p)^{1-x} = \exp\left\{x \ln\left(\frac{p}{1 - p}\right) + \ln(1 - p)\right\}, \qquad x = 0, 1.$$

In each of these examples, the sum $\sum_{i=1}^{n} X_i$ of the observations of the random sample was a sufficient statistic for the parameter. This idea is generalized by Theorem 10.1-1.

THEOREM 10.1-1

Let $X_1, X_2, \ldots, X_n$ be a random sample from a distribution with a p.d.f. or p.m.f. of the exponential form

$$f(x; \theta) = \exp[K(x)p(\theta) + S(x) + q(\theta)]$$

on a support free of θ. Then the statistic $\sum_{i=1}^{n} K(X_i)$ is sufficient for θ.

Proof. The joint p.d.f. (p.m.f.) of $X_1, X_2, \ldots, X_n$ is

$$\exp\left[p(\theta) \sum_{i=1}^{n} K(x_i) + \sum_{i=1}^{n} S(x_i) + nq(\theta)\right]$$

$$= \left\{\exp\left[p(\theta) \sum_{i=1}^{n} K(x_i) + nq(\theta)\right]\right\}\left\{\exp\left[\sum_{i=1}^{n} S(x_i)\right]\right\}.$$

In accordance with the factorization theorem, the statistic $\sum_{i=1}^{n} K(X_i)$ is sufficient for θ. $\qquad\square$

In many cases, Theorem 10.1-1 permits the student to find a sufficient statistic for a parameter with very little effort, as shown in the next example.

EXAMPLE 10.1-4

Let $X_1, X_2, \ldots, X_n$ be a random sample from an exponential distribution with p.d.f.

$$f(x; \theta) = \frac{1}{\theta} e^{-x/\theta} = \exp\left[x\left(-\frac{1}{\theta}\right) - \ln\theta\right], \qquad 0 < x < \infty,$$

provided that $0 < \theta < \infty$. Here, $K(x) = x$. Thus, $\sum_{i=1}^{n} X_i$ is sufficient for θ; of course, $\overline{X} = \sum_{i=1}^{n} X_i/n$ is also sufficient. ∎

Note that if there is a sufficient statistic for the parameter under consideration and if the maximum likelihood estimator of this parameter is unique, then the maximum likelihood estimator is a function of the sufficient statistic. To see this heuristically, consider the following: If a sufficient statistic exists, then the likelihood function is

$$L(\theta) = f(x_1, x_2, \ldots, x_n; \theta) = \phi[u(x_1, x_2, \ldots, x_n); \theta]\, h(x_1, x_2, \ldots, x_n).$$

Since $h(x_1, x_2, \ldots, x_n)$ does not depend on θ, we maximize $L(\theta)$ by maximizing $\phi[u(x_1, x_2, \ldots, x_n); \theta]$. But ϕ is a function of $x_1, x_2, \ldots, x_n$ only through the statistic $u(x_1, x_2, \ldots, x_n)$. Thus, if there is a unique value of θ that maximizes ϕ, then it must be a function of $u(x_1, x_2, \ldots, x_n)$. That is, $\widehat{\theta}$ is a function of the sufficient statistic $u(X_1, X_2, \ldots, X_n)$. This fact was alluded to in Example 10.1-1, but it could be checked with the use of other examples and exercises.

In many cases, we have two (or more) parameters—say, θ_1 and θ_2. All of the preceding concepts can be extended to these situations. For example, Definition 10.1-1 (the factorization theorem) becomes the following in the case of two parameters: If

$$f(x_1, \ldots, x_n; \theta_1, \theta_2) = \phi[u_1(x_1, \ldots, x_n), u_2(x_1, \ldots, x_n); \theta_1, \theta_2] h(x_1, \ldots, x_n),$$

where ϕ depends on $x_1, x_2, \ldots, x_n$ only through $u_1(x_1, \ldots, x_n)$, $u_2(x_1, \ldots, x_n)$, and $h(x_1, x_2, \ldots, x_n)$ does not depend upon θ_1 or θ_2, then $Y_1 = u_1(X_1, X_2, \ldots, X_n)$ and $Y_2 = u_2(X_1, X_2, \ldots, X_n)$ are **joint sufficient statistics** for θ_1 and θ_2.

EXAMPLE 10.1-5 Let $X_1, X_2, \ldots, X_n$ denote a random sample from a normal distribution $N(\theta_1, \theta_2)$. That is, $\theta_1 = \mu$ and $\theta_2 = \sigma^2$. Then

$$\prod_{i=1}^{n} f(x_i; \theta_1, \theta_2) = \left(\frac{1}{\sqrt{2\pi\theta_2}} \right)^n \exp\left[-\sum_{i=1}^{n} (x_i - \theta_1)^2 \bigg/ 2\theta_2 \right]$$

$$= \exp\left[\left(-\frac{1}{2\theta_2} \right) \sum_{i=1}^{n} x_i^2 + \left(\frac{\theta_1}{\theta_2} \right) \sum_{i=1}^{n} x_i - \frac{n\theta_1^2}{2\theta_2} - n \ln \sqrt{2\pi\theta_2} \right] \cdot (1).$$

Thus,

$$Y_1 = \sum_{i=1}^{n} X_i^2 \quad \text{and} \quad Y_2 = \sum_{i=1}^{n} X_i$$

are joint sufficient statistics for θ_1 and θ_2. Of course, the single-valued functions of Y_1 and Y_2, namely,

$$\overline{X} = \frac{Y_2}{n} \quad \text{and} \quad S^2 = \frac{Y_1 - Y_2^2/n}{n-1},$$

are also joint sufficient statistics for θ_1 and θ_2. ∎

Actually, we can see from Definition 10.1-1 and Example 10.1-5 that if we can write the p.d.f. in the exponential form, it is easy to find the joint sufficient statistics. In that example,

$$f(x; \theta_1, \theta_2) = \exp\left(\frac{-1}{2\theta_2} x^2 + \frac{\theta_1}{\theta_2} x - \frac{\theta_1^2}{2\theta_2} - \ln \sqrt{2\pi\theta_2} \right);$$

so

$$Y_1 = \sum_{i=1}^{n} X_i^2 \quad \text{and} \quad Y_2 = \sum_{i=1}^{n} X_i$$

are joint sufficient statistics for θ_1 and θ_2. A much more complicated illustration is given if we take a random sample $(X_1, Y_1), (X_2, Y_2), \ldots, (X_n, Y_n)$ from a bivariate normal distribution with parameters $\theta_1 = \mu_X, \theta_2 = \mu_Y, \theta_3 = \sigma_X^2, \theta_4 = \sigma_Y^2,$ and $\theta_5 = \rho$. In Exercise 10.1-3, we write the bivariate normal p.d.f. $f(x, y; \theta_1, \theta_2, \theta_3, \theta_4, \theta_5)$ in exponential form and see that $Z_1 = \sum_{i=1}^{n} X_i^2$, $Z_2 = \sum_{i=1}^{n} Y_i^2$, $Z_3 = \sum_{i=1}^{n} X_i Y_i$, $Z_4 = \sum_{i=1}^{n} X_i$, and $Z_5 = \sum_{i=1}^{n} Y_i$ are joint sufficient statistics for $\theta_1, \theta_2, \theta_3, \theta_4,$ and θ_5. Of course, the single-valued functions

$$\overline{X} = \frac{Z_4}{n}, \qquad \overline{Y} = \frac{Z_5}{n}, \qquad S_X^2 = \frac{Z_1 - Z_4^2/n}{n-1},$$

$$S_Y^2 = \frac{Z_2 - Z_5^2/n}{n-1}, \qquad R = \frac{(Z_3 - Z_4 Z_5/n)/(n-1)}{S_X S_Y},$$

are also joint sufficient statistics for those parameters.

The important point to stress for cases in which sufficient statistics exist is that once the sufficient statistics are given, there is no additional information about the parameters left in the remaining (conditional) distribution. That is, all statistical inferences should be based upon the sufficient statistics. To help convince the reader of this in point estimation, we state and prove the well-known **Rao–Blackwell theorem**.

THEOREM 10.1-2

Let $X_1, X_2, \ldots, X_n$ be a random sample from a distribution with p.d.f. or p.m.f. $f(x; \theta)$, $\theta \in \Omega$. Let $Y_1 = u_1(X_1, X_2, \ldots, X_n)$ be a sufficient statistic for θ, and let $Y_2 = u_2(X_1, X_2, \ldots, X_n)$ be an unbiased estimator of θ, where Y_2 is not a function of Y_1 alone. Then $E(Y_2 | y_1) = u(y_1)$ defines a statistic $u(Y_1)$, a function of the sufficient statistic Y_1, which is an unbiased estimator of θ, and its variance is less than that of Y_2.

Proof. Let $g(y_1, y_2; \theta)$ be the joint p.d.f. or p.m.f. of Y_1 and Y_2. Let $g_1(y_1; \theta)$ be the marginal of Y_1; thus,

$$\frac{g(y_1, y_2; \theta)}{g_1(y_1; \theta)} = h(y_2 | y_1)$$

is the conditional p.d.f. or p.m.f. of Y_2, given that $Y_1 = y_1$. This equation does not depend upon θ, since Y_1 is a sufficient statistic for θ. Of course, in the continuous case,

$$u(y_1) = \int_{y_2} y_2 h(y_2 | y_1)\, dy_2 = \int_{y_2} y_2 \frac{g(y_1, y_2; \theta)}{g_1(y_1; \theta)}\, dy_2$$

and

$$E[u(Y_1)] = \int_{y_1} \left(\int_{y_2} y_2 \frac{g(y_1, y_2; \theta)}{g_1(y_1; \theta)}\, dy_2 \right) g_1(y_1; \theta)\, dy_1$$

$$= \int_{y_1} \int_{y_2} y_2\, g(y_1, y_2; \theta)\, dy_2\, dy_1 = \theta,$$

because Y_2 is an unbiased estimator of θ. Thus, $u(Y_1)$ is also an unbiased estimator of θ.

Now, consider

$$\mathrm{Var}(Y_2) = E[(Y_2 - \theta)^2] = E[\{Y_2 - u(Y_1) + u(Y_1) - \theta\}^2]$$
$$= E[\{Y_2 - u(Y_1)\}^2] + E[\{u(Y_1) - \theta\}^2]$$
$$+ 2E[\{Y_2 - u(Y_1)\}\{u(Y_1) - \theta\}].$$

But the latter expectation is equal to

$$\int_{y_1} [u(y_1) - \theta] \left\{ \int_{y_2} [y_2 - u(y_1)] h(y_2 | y_1)\, dy_2 \right\} g(y_1; \theta)\, dy_1 = 0,$$

because $u(y_1)$ is the mean $E(Y_2 | y_1)$ of Y_2 in the conditional distribution given by $h(y_2 | y_1)$. Thus,

$$\mathrm{Var}(Y_2) = E[\{Y_2 - u(Y_1)\}^2] + \mathrm{Var}[u(Y_1)].$$

However, $E[\{(Y_2 - u(Y_1)\}^2 \geq 0$, as it is the expected value of a positive expression. Therefore,

$$\text{Var}(Y_2) \geq \text{Var}[u(Y_1)]. \qquad \square$$

The importance of this theorem is that it shows that for every other unbiased estimator of θ, we can always find an unbiased estimator based on the sufficient statistic which has a smaller variance than the first unbiased estimator. Hence, in that sense, the one based upon the sufficient statistic is better than the first one. More important, we might as well begin our search for an unbiased estimator with the smallest variance by considering only those unbiased estimators based upon the sufficient statistics. Moreover, in an advanced course we show that if the underlying distribution is described by a p.d.f. or p.m.f. of the exponential form, then, if an unbiased estimator exists, there is only one function of the sufficient statistic that is unbiased. That is, that unbiased estimator is unique. (See Hogg, McKean, and Craig, 2005.)

There is one other useful result involving a sufficient statistic Y for a parameter θ, particularly with a p.d.f. of the exponential form. It is that if another statistic Z has a distribution that is free of θ, then Y and Z are independent. This is the reason $Z = (n - 1)S^2$ is independent of $Y = \overline{X}$ when the sample arises from a distribution that is $N(\theta, \sigma^2)$. The sample mean is the sufficient statistic for θ, and

$$Z = (n - 1)S^2 = \sum_{i=1}^{n}(X_i - \overline{X})^2$$

has a distribution that is free of θ. To see this, we note that the m.g.f. of Z, namely, $E(e^{tZ})$, is

$$\int_{-\infty}^{\infty}\int_{-\infty}^{\infty}\cdots\int_{-\infty}^{\infty}\exp\left[t\sum_{i=1}^{n}(x_i - \overline{x})^2\right]\left(\frac{1}{\sqrt{2\pi}\sigma}\right)^n\exp\left[-\frac{\sum(x_i - \theta)^2}{2\sigma^2}\right]dx_1 dx_2\ldots dx_n.$$

Changing variables by letting $x_i - \theta = w_i$, $i = 1, 2, \ldots, n$, the preceding expression becomes

$$\int_{-\infty}^{\infty}\int_{-\infty}^{\infty}\cdots\int_{-\infty}^{\infty}\exp\left[t\sum_{i=1}^{n}(w_i - \overline{w})^2\right]\left(\frac{1}{\sqrt{2\pi}\sigma}\right)^n\exp\left[-\frac{\sum w_i^2}{2\sigma^2}\right]dw_1 dw_2\ldots dw_n,$$

which is free of θ.

An outline of the proof of this result is given by noting that

$$\int_{y}[h(z\,|\,y) - g_2(z)]g_1(y; \theta)\,dy = g_2(z) - g_2(z) = 0$$

for all $\theta \in \Omega$. However, because $h(z\,|\,y)$ is free of θ due to the hypothesis of sufficiency, $h(z\,|\,y) - g_2(z)$ is free of θ, since Z has a distribution that is free of θ. Because $N(\theta, \sigma^2)$ is of the exponential form, and thus $Y = \overline{X}$ has a p.d.f. $g_1(y\,|\,\theta)$ that is called a **complete kernel**, $h(z\,|\,y) - g_2(z)$ must be equal to zero. That is,

$$h(z\,|\,y) = g_2(z),$$

which means that Z and Y are independent. This proves the independence of $\overline{X}$ and S^2, which was stated in Theorem 5.5-2.

EXAMPLE 10.1-6 Let $X_1, X_2, \ldots, X_n$ be a random sample from a gamma distribution with α (given) and $\theta > 0$. Now, $Y = \sum_{i=1}^{n} X_i$ is a sufficient statistic for θ, since the gamma p.d.f. is of the exponential form. Clearly, then,

$$Z = \frac{\sum_{i=1}^{n} a_i X_i}{\sum_{i=1}^{n} X_i},$$

where not all constants $a_1, a_2, \ldots, a_n$ are equal, has a distribution that is free of the spread parameter θ because the moment-generating function of Z, namely,

$$E(e^{tZ}) = \int_0^\infty \int_0^\infty \cdots \int_0^\infty \frac{e^{t\Sigma a_i X_i / \Sigma X_i}}{[\Gamma(\alpha)]^n \theta^{n\alpha}} (x_1 x_2 \cdots x_n)^{\alpha-1} e^{-\Sigma x_i/\theta} \, dx_1 dx_2 \ldots dx_n,$$

does not depend upon θ, as is seen by the transformation $w_i = x_i/\theta$, $i = 1, 2, \ldots, n$. So Y and Z are independent statistics. ∎

This special case of the independence of Y and Z concerning one sufficient statistic Y and one parameter θ was first observed by Hogg (1953) and then generalized to several sufficient statistics for more than one parameter by Basu (1955) and is usually called **Basu's theorem**.

Due to these results, sufficient statistics are extremely important and future estimation problems are based upon them when they exist.

EXERCISES

10.1-1. Let $X_1, X_2, \ldots, X_n$ be a random sample from $N(0, \sigma^2)$.

(a) Find a sufficient statistic Y for σ^2.

(b) Show that the maximum likelihood estimator for σ^2 is a function of Y.

(c) Is the maximum likelihood estimator for σ^2 unbiased?

10.1-2. Let $X_1, X_2, \ldots, X_n$ be a random sample from a Poisson distribution with mean $\lambda > 0$. Find the conditional probability $P(X_1 = x_1, \ldots, X_n = x_n | Y = y)$, where $Y = X_1 + \cdots + X_n$ and the nonnegative integers $x_1, x_2, \ldots, x_n$ sum to y. Note that this probability does not depend on λ.

10.1-3. Write the bivariate normal p.d.f. $f(x, y; \theta_1, \theta_2, \theta_3, \theta_4, \theta_5)$ in exponential form, and show that $Z_1 = \sum_{i=1}^{n} X_i^2$, $Z_2 = \sum_{i=1}^{n} Y_i^2$, $Z_3 = \sum_{i=1}^{n} X_i Y_i$, $Z_4 = \sum_{i=1}^{n} X_i$, and $Z_5 = \sum_{i=1}^{n} Y_i$ are joint sufficient statistics for $\theta_1, \theta_2, \theta_3, \theta_4$, and θ_5.

10.1-4. Let $X_1, X_2, \ldots, X_n$ be a random sample from a distribution with p.d.f. $f(x; \theta) = \theta x^{\theta-1}$, $0 < x < 1$, where $0 < \theta$.

(a) Find the sufficient statistic Y for θ.

(b) Show that the maximum likelihood estimator $\widehat{\theta}$ is a function of Y.

(c) Argue that $\widehat{\theta}$ is also sufficient for θ.

10.1-5. Let $X_1, X_2, \ldots, X_n$ be a random sample from a gamma distribution with $\alpha = 1$ and $1/\theta > 0$. Show that $Y = \sum_{i=1}^{n} X_i$ is a sufficient statistic, Y has a gamma distribution with parameters n and $1/\theta$, and $(n-1)/Y$ is an unbiased estimator of θ.

10.1-6. Let $X_1, X_2, \ldots, X_n$ be a random sample from a gamma distribution with known parameter α and unknown parameter $\theta > 0$.

(a) Show that $Y = \sum_{i=1}^{n} X_i$ is a sufficient statistic for θ.

(b) Show that the maximum likelihood estimator of θ is a function of the sufficient statistic and is an unbiased estimator of θ.

10.1-7. Let $X_1, X_2, \ldots, X_n$ be a random sample from the distribution with p.m.f. $f(x; p) = p(1-p)^{x-1}$, $x = 1, 2, 3, \ldots$, where $0 < p < 1$.

(a) Show that $Y = \sum_{i=1}^{n} X_i$ is a sufficient statistic for p.

(b) Find a function of $Y = \sum_{i=1}^{n} X_i$ that is an unbiased estimator of $\theta = 1/p$.

10.1-8. Let $X_1, X_2, \ldots, X_n$ be a random sample from $N(0, \theta)$, where $\sigma^2 = \theta > 0$ is unknown. Argue that the sufficient statistic $Y = \sum_{i=1}^{n} X_i^2$ for θ and $Z = \sum_{i=1}^{n} a_i X_i / \sum_{i=1}^{n} X_i$ are independent.

HINT: Let $x_i = \theta w_i$, $i = 1, 2, \ldots, n$, in the multivariate integral representing $E[e^{tZ}]$.

10.1-9. Let $X_1, X_2, \ldots, X_n$ be a random sample from $N(\theta_1, \theta_2)$. Show that the sufficient statistics $Y_1 = \overline{X}$ and $Y_2 = S^2$ are independent of the statistic

$$Z = \sum_{i=1}^{n-1} \frac{(X_{i+1} - X_i)^2}{S^2}$$

because Z has a distribution that is free of θ_1 and θ_2.

HINT: Let $w_i = (x_i - \theta_1)/\sqrt{\theta_2}$, $i = 1, 2, \ldots, n$, in the multivariate integral representing $E[e^{tZ}]$.

10.2 POWER OF A STATISTICAL TEST

In Chapter 7, we gave several tests of fairly common statistical hypotheses in such a way that we described the significance level α and the p-values of each. Of course, those tests were based on good (sufficient) statistics of the parameters, when the latter exist. In this section, we consider the probability of making the other type of error: accepting the null hypothesis H_0 when the alternative hypothesis H_1 is true. This consideration leads to ways to find most powerful tests of the null hypothesis H_0 against the alternative hypothesis H_1.

The first example introduces a new concept using a test about p, the probability of success. The sample size is kept small so that Table II in the appendix can be used to find probabilities. The application is one that you can actually perform.

EXAMPLE 10.2-1 Assume that when given a name tag, a person puts it on either the right or left side. Let p equal the probability that the name tag is placed on the right side. We shall test the null hypothesis H_0: $p = 1/2$ against the composite alternative hypothesis H_1: $p < 1/2$. (Included with the null hypothesis are those values of p which are greater than 1/2; that is, we could think of H_0 as H_0: $p \geq 1/2$.) We shall give name tags to a random sample of $n = 20$ people, denoting the placements of their name tags with Bernoulli random variables, $X_1, X_2, \ldots, X_{20}$, where $X_i = 1$ if a person places the name tag on the right and $X_i = 0$ if a person places the name tag on the left. For our test statistic, we can then use $Y = \sum_{i=1}^{20} X_i$, which has the binomial distribution $b(20, p)$. Say the critical region is defined by $C = \{y : y \leq 6\}$ or, equivalently, by $\{(x_1, x_2, \ldots, x_{20}) : \sum_{i=1}^{20} x_i \leq 6\}$. Since Y is $b(20, 1/2)$ if $p = 1/2$, the significance level of the corresponding test is

$$\alpha = P\left(Y \leq 6; p = \frac{1}{2}\right) = \sum_{y=0}^{6} \binom{20}{y}\left(\frac{1}{2}\right)^{20} = 0.0577,$$

from Table II in the appendix. Of course, the probability β of a Type II error has different values, with different values of p selected from the composite alternative hypothesis H_1: $p < 1/2$. For example, with $p = 1/4$,

$$\beta = P\left(7 \leq Y \leq 20; p = \frac{1}{4}\right) = \sum_{y=7}^{20} \binom{20}{y}\left(\frac{1}{4}\right)^{y}\left(\frac{3}{4}\right)^{20-y} = 0.2142,$$

whereas with $p = 1/10$,

$$\beta = P\left(7 \leq Y \leq 20; p = \frac{1}{10}\right) = \sum_{y=7}^{20} \binom{20}{y}\left(\frac{1}{10}\right)^{y}\left(\frac{9}{10}\right)^{20-y} = 0.0024.$$

Instead of considering the probability β of accepting H_0 when H_1 is true, we could compute the probability K of rejecting H_0 when H_1 is true. After all, β and $K = 1 - \beta$ provide the same information. Since K is a function of p, we denote this explicitly by writing $K(p)$. The probability

$$K(p) = \sum_{y=0}^{6} \binom{20}{y} p^y (1 - p)^{20-y}, \qquad 0 < p \le \frac{1}{2},$$

is called the **power function of the test**. Thus, $\alpha = K(1/2) = 0.0577$, $1 - K(1/4) = 0.2142$, and $1 - K(1/10) = 0.0024$. The value of the power function at a specified p is called the **power** of the test at that point. For instance, $K(1/4) = 0.7858$ and $K(1/10) = 0.9976$ are the powers at $p = 1/4$ and $p = 1/10$, respectively. An acceptable power function is a power function that assumes small values when H_0 is true and larger values when p differs much from $p = 1/2$. (See Figure 10.2-1 for a graph of this power function.) ∎

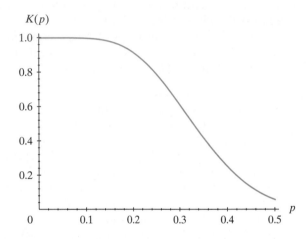

FIGURE 10.2-1: Power function $K(p) = P(Y \le 6; p)$, where Y is $b(20, p)$

In Example 10.2-1, we introduced the new concept of the power function of a test. We now show how the sample size can be selected so as to create a test with appropriate power.

EXAMPLE 10.2-2   Let $X_1, X_2, \ldots, X_n$ be a random sample of size n from the normal distribution $N(\mu, 100)$, which we can suppose is a possible distribution of scores of students in a statistics course that uses a new method of teaching (e.g., computer-related materials). We wish to decide between $H_0: \mu = 60$ (the *no-change* hypothesis because, let us say, this was the mean by the previous method of teaching) and the research worker's hypothesis $H_1: \mu > 60$. Let us consider a sample of size $n = 25$. Of course, the sample mean $\overline{X}$ is the maximum likelihood estimator of μ; thus, it seems reasonable to base our decision on this statistic. Initially, we use the rule to reject H_0 and accept H_1 if and only if $\overline{x} \ge 62$. What are the consequences of this test? These are summarized in the power function of the test.

We first find the probability of rejecting $H_0: \mu = 60$ for various values of $\mu \ge 60$. The probability of rejecting H_0 is given by

$$K(\mu) = P(\overline{X} \ge 62; \mu),$$

because this test calls for the rejection of $H_0: \mu = 60$ when $\overline{x} \ge 62$. When the new process has the general mean μ, $\overline{X}$ has the normal distribution $N(\mu, 100/25 = 4)$.

Accordingly,

$$K(\mu) = P\left(\frac{\overline{X} - \mu}{2} \geq \frac{62 - \mu}{2}; \mu\right)$$

$$= 1 - \Phi\left(\frac{62 - \mu}{2}\right), \qquad 60 \leq \mu,$$

is the probability of rejecting H_0: $\mu = 60$ by using this particular test. Several values of $K(\mu)$ are given in Table 10.2-1. Figure 10.2-2 depicts the graph of the function $K(\mu)$.

TABLE 10.2-1: Values of the Power Function

μ	$K(\mu)$
60	0.1587
61	0.3085
62	0.5000
63	0.6915
64	0.8413
65	0.9332
66	0.9772

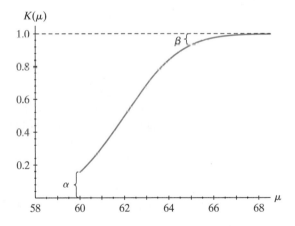

FIGURE 10.2-2: Power function $K(\mu) = 1 - \Phi([62 - \mu]/2)$

The probability $K(\mu)$ of rejecting H_0: $\mu = 60$ is called the *power function* of the test. At the value μ_1 of the parameter, $K(\mu_1)$ is the power at μ_1. The power at $\mu = 60$ is $K(60) = 0.1587$, and this is the probability of rejecting H_0: $\mu = 60$ when H_0 is true. That is, $K(60) = 0.1587 = \alpha$ is the probability of a Type I error and is called the *significance level* of the test.

The power at $\mu = 65$ is $K(65) = 0.9332$, and this is the probability of making the correct decision (namely, rejecting H_0: $\mu = 60$ when $\mu = 65$). Hence, we are pleased that here it is large. When $\mu = 65$, $1 - K(65) = 0.0668$ is the probability of not rejecting H_0: $\mu = 60$ when $\mu = 65$; that is, it is the probability of a Type II error and is denoted by $\beta = 0.0668$. These α and β values are displayed in Figure 10.2-2. Clearly, the probability $\beta = 1 - K(\mu_1)$ of a Type II error depends on which value—say,

μ_1, is taken in the alternative hypothesis H_1: $\mu > 60$. Thus, while $\beta = 0.0668$ when $\mu = 65$, β is equal to $1 - K(63) = 0.3085$ when $\mu = 63$.

Frequently, statisticians like to have the significance level α smaller than 0.1587—say, around 0.05 or less—because it is a probability of an error, namely, a Type I error. Thus, if we would like $\alpha = 0.05$, then, with $n = 25$, we can no longer use the critical region $\bar{x} \geq 62$; rather, we use $\bar{x} \geq c$, where c is selected such that

$$K(60) = P(\overline{X} \geq c; \mu = 60) = 0.05.$$

However, when $\mu = 60$, $\overline{X}$ is $N(60, 4)$, and it follows that

$$K(60) = P\left(\frac{\overline{X} - 60}{2} \geq \frac{c - 60}{2}; \mu = 60\right)$$

$$= 1 - \Phi\left(\frac{c - 60}{2}\right) = 0.05.$$

From Table Va in the appendix, we have

$$\frac{c - 60}{2} = 1.645 = z_{0.05} \qquad \text{and} \qquad c = 60 + 3.29 = 63.29.$$

Although this change reduces α from 0.1587 to 0.05, it increases β at $\mu = 65$ from 0.0668 to

$$\beta = 1 - P(\overline{X} \geq 63.29; \mu = 65)$$

$$= 1 - P\left(\frac{\overline{X} - 65}{2} \geq \frac{63.29 - 65}{2}; \mu = 65\right)$$

$$= \Phi(-0.855) = 0.1963.$$

In general, without changing the sample size or the type of test of the hypothesis, a decrease in α causes an increase in β, and a decrease in β causes an increase in α. *Both* probabilities α and β of the two types of errors can be decreased only by increasing the sample size or, in some way, constructing a better test of the hypothesis.

For example, if $n = 100$ and we desire a test with significance level $\alpha = 0.05$, then, since $\overline{X}$ is $N(\mu, 100/100 = 1)$,

$$\alpha = P(\overline{X} \geq c; \mu = 60) = 0.05$$

means that

$$P\left(\frac{\overline{X} - 60}{1} \geq \frac{c - 60}{1}; \mu = 60\right) = 0.05$$

and $c - 60 = 1.645$. Thus, $c = 61.645$. The power function is

$$K(\mu) = P(\overline{X} \geq 61.645; \mu)$$

$$= P\left(\frac{\overline{X} - \mu}{1} \geq \frac{61.645 - \mu}{1}; \mu\right) = 1 - \Phi(61.645 - \mu).$$

In particular, this means that at $\mu = 65$,

$$\beta = 1 - K(\mu) = \Phi(61.645 - 65) = \Phi(-3.355) \approx 0;$$

so, with $n = 100$, both α and β have decreased from their respective original values of 0.1587 and 0.0668 when $n = 25$.

Rather than guess at the value of n, an ideal power function determines the sample size. Let us use a critical region of the form $\bar{x} \geq c$. Further, suppose that we want $\alpha = 0.025$ and, when $\mu = 65$, $\beta = 0.05$. Thus, since $\overline{X}$ is $N(\mu, 100/n)$, it follows that

$$0.025 = P(\overline{X} \geq c; \mu = 60) = 1 - \Phi\left(\frac{c - 60}{10/\sqrt{n}}\right)$$

and

$$0.05 = 1 - P(\overline{X} \geq c; \mu = 65) = \Phi\left(\frac{c - 65}{10/\sqrt{n}}\right).$$

That is,

$$\frac{c - 60}{10/\sqrt{n}} = 1.96 \qquad \text{and} \qquad \frac{c - 65}{10/\sqrt{n}} = -1.645.$$

Solving these equations simultaneously for c and $10/\sqrt{n}$, we obtain

$$c = 60 + 1.96\,\frac{5}{3.605} = 62.718;$$

$$\frac{10}{\sqrt{n}} = \frac{5}{3.605}.$$

Hence,

$$\sqrt{n} = 7.21 \qquad \text{and} \qquad n = 51.98.$$

Since n must be an integer, we would use $n = 52$ and obtain $\alpha \approx 0.025$ and $\beta \approx 0.05$. ∎

The next example is an extension of Example 10.2-1.

EXAMPLE 10.2-3   To test H_0: $p = 1/2$ against H_1: $p < 1/2$, we take a random sample of Bernoulli trials, $X_1, X_2, \ldots, X_n$, and use for our test statistic $Y = \sum_{i=1}^{n} X_i$, which has a binomial distribution $b(n, p)$. Let the critical region be defined by $C = \{y: y \leq c\}$. The power function for this test is defined by $K(p) = P(Y \leq c; p)$. We shall find the values of n and c so that $K(1/2) \approx 0.05$ and $K(1/4) \approx 0.90$. That is, we would like the significance level to be $\alpha = K(1/2) = 0.05$ and the power at $p = 1/4$ to equal 0.90. We proceed as follows: Since

$$0.05 = P\left(Y \leq c; p = \frac{1}{2}\right) = P\left(\frac{Y - n/2}{\sqrt{n(1/2)(1/2)}} \leq \frac{c - n/2}{\sqrt{n(1/2)(1/2)}}\right),$$

it follows that

$$(c - n/2)/\sqrt{n/4} \approx -1.645;$$

and since

$$0.90 = P\left(Y \leq c; p = \frac{1}{4}\right) = P\left(\frac{Y - n/4}{\sqrt{n(1/4)(3/4)}} \leq \frac{c - n/4}{\sqrt{n(1/4)(3/4)}}\right),$$

it follows that

$$(c - n/4)/\sqrt{3n/16} \approx 1.282.$$

Therefore,

$$\frac{n}{4} \approx 1.645\sqrt{\frac{n}{4}} + 1.282\sqrt{\frac{3n}{16}} \qquad \text{and} \qquad \sqrt{n} \approx 4(1.378) = 5.512.$$

Thus, n is about equal to 31, and from either of the first two approximate equalities, we find that c is about equal to 10.9. Using $n = 31$ and $c = 10.9$ means that $K(1/2) = 0.05$ and $K(1/4) = 0.90$ are only approximate. In fact, since Y must be an integer, we could let $c = 10.5$. Then, with $n = 31$,

$$\alpha = K\left(\frac{1}{2}\right) = P\left(Y \le 10.5; p = \frac{1}{2}\right) \approx 0.0362;$$

$$K\left(\frac{1}{4}\right) = P\left(Y \le 10.5; p = \frac{1}{4}\right) \approx 0.8729.$$

Or we could let $c = 11.5$ and $n = 31$, in which case

$$\alpha = K\left(\frac{1}{2}\right) = P\left(Y \le 11.5; p = \frac{1}{2}\right) \approx 0.0558;$$

$$K\left(\frac{1}{4}\right) = P\left(Y \le 11.5; p = \frac{1}{4}\right) \approx 0.9235. \qquad \blacksquare$$

EXERCISES

10.2-1. A certain size of bag is designed to hold 25 pounds of potatoes. A farmer fills such bags in the field. Assume that the weight X of potatoes in a bag is $N(\mu, 9)$. We shall test the null hypothesis $H_0: \mu = 25$ against the alternative hypothesis $H_1: \mu < 25$. Let X_1, X_2, X_3, X_4 be a random sample of size 4 from this distribution, and let the critical region C for this test be defined by $\bar{x} \le 22.5$, where $\bar{x}$ is the observed value of $\overline{X}$.

(a) What is the power function $K(\mu)$ of this test? In particular, what is the significance level $\alpha = K(25)$ for your test?

(b) If the random sample of four bags of potatoes yielded the values $x_1 = 21.24$, $x_2 = 24.81$, $x_3 = 23.62$, and $x_4 = 26.82$, would your test lead you to accept or reject H_0?

(c) What is the p-value associated with $\bar{x}$ in part (b)?

10.2-2. Let X equal the number of milliliters in a bottle that has a label volume of 350 milliliters. Assume that the distribution of X is $N(\mu, 4)$. To test the null hypothesis $H_0: \mu = 355$ against the alternative hypothesis $H_1: \mu < 355$, let the critical region be defined by $C = \{\bar{x}: \bar{x} \le 354.05\}$, where

$\bar{x}$ is the sample mean of the contents of a random sample of $n = 12$ bottles.

(a) Find the power function $K(\mu)$ for this test.

(b) What is the (approximate) significance level of the test?

(c) Find the values of $K(354.05)$ and $K(353.1)$, and sketch the graph of the power function.

(d) Use the following 12 observations to state your conclusion from this test:

350	353	354	356	353	352
354	355	357	353	354	355

(e) What is the approximate p-value of the test?

10.2-3. Assume that SAT mathematics scores of students who attend small liberal arts colleges are $N(\mu, 8100)$. We shall test $H_0: \mu = 530$ against the alternative hypothesis $H_1: \mu < 530$. Given a random sample of size $n = 36$ SAT mathematics scores, let the critical region be defined by $C = \{\bar{x}: \bar{x} \le 510.77\}$, where $\bar{x}$ is the observed mean of the sample.

(a) Find the power function, $K(\mu)$, for this test.

(b) What is the value of the significance level of the test?

(c) What is the value of $K(510.77)$?

(d) Sketch the graph of the power function.

(e) What is the p-value associated with (i) $\bar{x} = 507.35$; (ii) $\bar{x} = 497.45$?

10.2-4. Let X be $N(\mu, 100)$. To test H_0: $\mu = 80$ against H_1: $\mu > 80$, let the critical region be defined by $C = \{(x_1, x_2, \ldots, x_{25}): \bar{x} \geq 83\}$, where $\bar{x}$ is the sample mean of a random sample of size $n = 25$ from this distribution.

(a) What is the power function $K(\mu)$ for this test?

(b) What is the significance level of the test?

(c) What are the values of $K(80)$, $K(83)$, and $K(86)$?

(d) Sketch the graph of the power function.

(e) What is the p-value corresponding to $\bar{x} = 83.41$?

10.2-5. Let X equal the yield of alfalfa in tons per acre per year. Assume that X is $N(1.5, 0.09)$. It is hoped that a new fertilizer will increase the average yield. We shall test the null hypothesis H_0: $\mu = 1.5$ against the alternative hypothesis H_1: $\mu > 1.5$. Assume that the variance continues to equal $\sigma^2 = 0.09$ with the new fertilizer. Using $\bar{X}$, the mean of a random sample of size n, as the test statistic, reject H_0 if $\bar{x} \geq c$. Find n and c so that the power function $K(\mu) = P(\bar{X} \geq c : \mu)$ is such that $\alpha = K(1.5) = 0.05$ and $K(1.7) = 0.95$.

10.2-6. Let X equal the number of pounds of butterfat produced by a Holstein cow during the 305-day milking period following the birth of a calf. Assume that the distribution of X is $N(\mu, 140^2)$. To test the null hypothesis H_0: $\mu = 715$ against the alternative hypothesis H_1: $\mu < 715$, let the critical region be defined by $C = \{\bar{x}: \bar{x} \leq 668.94\}$, where $\bar{x}$ is the sample mean of $n = 25$ butterfat weights from 25 cows selected at random.

(a) Find the power function $K(\mu)$ for this test.

(b) What is the significance level of the test?

(c) What are the values of $K(668.94)$ and $K(622.88)$?

(d) Sketch a graph of the power function.

(e) What conclusion do you draw from the following 25 observations of X?

425 710 661 664 732 714 934 761 744

653 725 657 421 573 535 602 537 405

874 791 721 849 567 468 975

(f) What is the approximate p-value of the test?

10.2-7. In Exercise 10.2-6, let $C = \{\bar{x}: \bar{x} \leq c\}$ be the critical region. Find values for n and c so that

the significance level of this test is $\alpha = 0.05$ and the power at $\mu = 650$ is 0.90.

10.2-8. Let X have a Bernoulli distribution with p.m.f.

$$f(x; p) = p^x(1 - p)^{1-x}, \quad x = 0, 1, \quad 0 \leq p \leq 1.$$

We would like to test the null hypothesis H_0: $p \leq 0.4$ against the alternative hypothesis H_1: $p > 0.4$. For the test statistic, use $Y = \sum_{i=1}^{n} X_i$, where $X_1, X_2, \ldots, X_n$ is a random sample of size n from this Bernoulli distribution. Let the critical region be of the form $C = \{y: y \geq c\}$.

(a) Let $n = 100$. On the same set of axes, sketch the graphs of the power functions corresponding to the three critical regions, $C_1 = \{y: y \geq 40\}$, $C_2 = \{y: y \geq 50\}$, and $C_3 = \{y: y \geq 60\}$. Use the normal approximation to compute the probabilities.

(b) Let $C = \{y: y \geq 0.45n\}$. On the same set of axes, sketch the graphs of the power functions corresponding to the three samples of sizes 10, 100, and 1000.

10.2-9. Let p denote the probability that, for a particular tennis player, the first serve is good. Since $p = 0.40$, this player decided to take lessons in order to increase p. When the lessons are completed, the hypothesis H_0: $p = 0.40$ will be tested against H_1: $p > 0.40$ on the basis of $n = 25$ trials. Let y equal the number of first serves that are good, and let the critical region be defined by $C = \{y: y \geq 14\}$.

(a) Find the power function $K(p)$ for this test.

(b) What is the value of the significance level, $\alpha = K(0.40)$? Use Table II in the appendix.

(c) Evaluate $K(p)$ at $p = 0.45, 0.50, 0.60, 0.70, 0.80,$ and 0.90. Use Table II.

(d) Sketch the graph of the power function.

(e) If $y = 15$ following the lessons, would H_0 be rejected?

(f) What is the p-value associated with $y = 15$?

10.2-10. Let $X_1, X_2, \ldots, X_8$ be a random sample of size $n = 8$ from a Poisson distribution with mean λ. Reject the simple null hypothesis H_0: $\lambda = 0.5$, and accept H_1: $\lambda > 0.5$, if the observed sum $\sum_{i=1}^{8} x_i \geq 8$.

(a) Compute the significance level α of the test.

(b) Find the power function $K(\lambda)$ of the test as a sum of Poisson probabilities.

(c) Using Table III in the appendix, determine $K(0.75)$, $K(1)$, and $K(1.25)$.

10.2-11. Let p equal the fraction defective of a certain manufactured item. To test H_0: $p = 1/26$ against

H_1: $p > 1/26$, we inspect n items selected at random and let Y be the number of defective items in this sample. We reject H_0 if the observed $y \geq c$. Find n and c so that $\alpha = K(1/26) \approx 0.05$ and $K(1/10) \approx 0.90$, where $K(p) = P(Y \geq c; p)$.

HINT: Use either the normal or Poisson approximation to help solve this exercise.

10.2-12. Let X_1, X_2, X_3 be a random sample of size $n = 3$ from an exponential distribution with mean $\theta > 0$. Reject the simple null hypothesis H_0: $\theta = 2$,

and accept the composite alternative hypothesis H_1: $\theta < 2$, if the observed sum $\sum_{i=1}^{3} x_i \leq 2$.

(a) What is the power function $K(\theta)$, written as an integral?

(b) Using integration by parts, define the power function as a summation.

(c) With the help of Table III in the appendix, determine $\alpha = K(2)$, $K(1)$, $K(1/2)$, and $K(1/4)$.

10.3   BEST CRITICAL REGIONS

In this section, we consider the properties a satisfactory test (or critical region) should possess. To introduce our investigation, we begin with a nonstatistical example.

EXAMPLE 10.3-1   Say that you have α dollars with which to buy books. Further, suppose that you are not interested in the books themselves, but only in filling as much of your bookshelves as possible. How do you decide which books to buy? Does the following approach seem reasonable? First of all, take all the available free books. Then start choosing those books for which the cost of filling an inch of bookshelf is smallest. That is, choose those books for which the ratio c/w is a minimum, where w is the width of the book in inches and c is the cost of the book. Continue choosing books this way until you have spent the α dollars.   ∎

To see how Example 10.3-1 provides the background for selecting a good critical region of size α, let us consider a test of the simple hypothesis H_0: $\theta = \theta_0$ against a simple alternative hypothesis H_1: $\theta = \theta_1$. In this discussion, we assume that the random variables $X_1, X_2, \ldots, X_n$ under consideration have a joint p.m.f. of the discrete type, which we here denote by $L(\theta; x_1, x_2, \ldots, x_n)$. That is,

$$P(X_1 = x_1, X_2 = x_2, \ldots, X_n = x_n) = L(\theta; x_1, x_2, \ldots, x_n).$$

A critical region C of size α is a set of points $(x_1, x_2, \ldots, x_n)$ with probability α when $\theta = \theta_0$. For a good test, this set C of points should have a large probability when $\theta = \theta_1$, because, under H_1: $\theta = \theta_1$, we wish to reject H_0: $\theta = \theta_0$. Accordingly, the first point we would place in the critical region C is the one with the smallest ratio

$$\frac{L(\theta_0; x_1, x_2, \ldots, x_n)}{L(\theta_1; x_1, x_2, \ldots, x_n)}.$$

That is, the "cost" in terms of probability under H_0: $\theta = \theta_0$ is small compared with the probability that we can "buy" if $\theta = \theta_1$. The next point to add to C would be the one with the next-smallest ratio. We would continue to add points to C in this manner until the probability of C, under H_0: $\theta = \theta_0$, equals α. In this way, for the given significance level α, we have achieved the region C with the largest probability when H_1: $\theta = \theta_1$ is true. We now formalize this discussion by defining a best critical region and proving the well-known Neyman–Pearson lemma.

DEFINITION 10.3-1   Consider the test of the simple null hypothesis H_0: $\theta = \theta_0$ against the simple alternative hypothesis H_1: $\theta = \theta_1$. Let C be a critical region of size α; that is,

$\alpha = P(C; \theta_0)$. Then C is a **best critical region of size** α if, for every other critical region D of size $\alpha = P(D; \theta_0)$, we have

$$P(C; \theta_1) \geq P(D; \theta_1).$$

That is, when $H_1: \theta = \theta_1$ is true, the probability of rejecting $H_0: \theta = \theta_0$ with the use of the critical region C is at least as great as the corresponding probability with the use of any other critical region D of size α.

Thus, a best critical region of size α is the critical region that has the greatest power among all critical regions of size α. The Neyman–Pearson lemma gives sufficient conditions for a best critical region of size α.

THEOREM
10.3-1

(Neyman–Pearson Lemma) Let $X_1, X_2, \ldots, X_n$ be a random sample of size n from a distribution with p.d.f. or p.m.f. $f(x; \theta)$, where θ_0 and θ_1 are two possible values of θ. Denote the joint p.d.f. or p.m.f. of $X_1, X_2, \ldots, X_n$ by the likelihood function

$$L(\theta) = L(\theta; x_1, x_2, \ldots, x_n) = f(x_1; \theta)f(x_2; \theta) \cdots f(x_n; \theta).$$

If there exist a positive constant k and a subset C of the sample space such that

(a) $P[(X_1, X_2, \ldots, X_n) \in C; \theta_0] = \alpha$,

(b) $\dfrac{L(\theta_0)}{L(\theta_1)} \leq k$ for $(x_1, x_2, \ldots, x_n) \in C$, and

(c) $\dfrac{L(\theta_0)}{L(\theta_1)} \geq k$ for $(x_1, x_2, \ldots, x_n) \in C'$,

then C is a best critical region of size α for testing the simple null hypothesis $H_0: \theta = \theta_0$ against the simple alternative hypothesis $H_1: \theta = \theta_1$.

Proof. We prove the theorem when the random variables are of the continuous type; for discrete-type random variables, replace the integral signs by summation signs. To simplify the exposition, we shall use the following notation:

$$\int_B L(\theta) = \int \cdots \int_B L(\theta; x_1, x_2, \ldots, x_n) \, dx_1 \, dx_2 \cdots dx_n.$$

Assume that there exists another critical region of size α—say, D—such that, in this new notation,

$$\alpha = \int_C L(\theta_0) = \int_D L(\theta_0).$$

Then we have

$$0 = \int_C L(\theta_0) - \int_D L(\theta_0)$$

$$= \int_{C \cap D'} L(\theta_0) + \int_{C \cap D} L(\theta_0) - \int_{C \cap D} L(\theta_0) - \int_{C' \cap D} L(\theta_0).$$

Hence,

$$0 = \int_{C \cap D'} L(\theta_0) - \int_{C' \cap D} L(\theta_0).$$

By hypothesis (b), $kL(\theta_1) \geq L(\theta_0)$ at each point in C and therefore in $C \cap D'$; thus,

$$k \int_{C \cap D'} L(\theta_1) \geq \int_{C \cap D'} L(\theta_0).$$

By hypothesis (c), $kL(\theta_1) \leq L(\theta_0)$ at each point in C' and therefore in $C' \cap D$; thus, we obtain

$$k \int_{C' \cap D} L(\theta_1) \leq \int_{C' \cap D} L(\theta_0).$$

Consequently,

$$0 = \int_{C \cap D'} L(\theta_0) - \int_{C' \cap D} L(\theta_0) \leq (k) \left\{ \int_{C \cap D'} L(\theta_1) - \int_{C' \cap D} L(\theta_1) \right\}.$$

That is,

$$0 \leq (k) \left\{ \int_{C \cap D'} L(\theta_1) + \int_{C \cap D} L(\theta_1) - \int_{C \cap D} L(\theta_1) - \int_{C' \cap D} L(\theta_1) \right\}$$

or, equivalently,

$$0 \leq (k) \left\{ \int_{C} L(\theta_1) - \int_{D} L(\theta_1) \right\}.$$

Thus,

$$\int_{C} L(\theta_1) \geq \int_{D} L(\theta_1);$$

that is, $P(C; \theta_1) \geq P(D; \theta_1)$. Since that is true for every critical region D of size α, C is a best critical region of size α. $\qquad\square$

For a realistic application of the Neyman–Pearson lemma, consider the next example, in which the test is based on a random sample from a normal distribution.

EXAMPLE 10.3-2 Let $X_1, X_2, \ldots, X_n$ be a random sample from the normal distribution $N(\mu, 36)$. We shall find the best critical region for testing the simple hypothesis H_0: $\mu = 50$ against the simple alternative hypothesis H_1: $\mu = 55$. Using the ratio of the likelihood functions, namely, $L(50)/L(55)$, we shall find those points in the sample space for which this ratio is less than or equal to some constant k. That is, we shall solve the following inequality:

$$\frac{L(50)}{L(55)} = \frac{(72\pi)^{-n/2} \exp\left[-\left(\dfrac{1}{72}\right)\sum\limits_{i=1}^{n}(x_i - 50)^2\right]}{(72\pi)^{-n/2} \exp\left[-\left(\dfrac{1}{72}\right)\sum\limits_{i=1}^{n}(x_i - 55)^2\right]}$$

$$= \exp\left[-\left(\frac{1}{72}\right)\left(10\sum_{i=1}^{n}x_i + n50^2 - n55^2\right)\right] \le k.$$

If we take the natural logarithm of each member of the inequality, we find that

$$-10\sum_{i=1}^{n}x_i - n50^2 + n55^2 \le (72)\ln k.$$

Thus,

$$\frac{1}{n}\sum_{i=1}^{n}x_i \ge -\frac{1}{10n}[n50^2 - n55^2 + (72)\ln k]$$

or, equivalently,

$$\bar{x} \ge c,$$

where $c = -(1/10n)[n50^2 - n55^2 + (72)\ln k]$. Hence, $L(50)/L(55) \le k$ is equivalent to $\bar{x} \ge c$. According to the Neyman–Pearson lemma, a best critical region is

$$C = \{(x_1, x_2, \ldots, x_n): \bar{x} \ge c\},$$

where c is selected so that the size of the critical region is α. Say $n = 16$ and $c = 53$. Then, since $\bar{X}$ is $N(50, 36/16)$ under H_0, we have

$$\alpha = P(\bar{X} \ge 53; \mu = 50)$$

$$= P\left(\frac{\bar{X} - 50}{6/4} \ge \frac{3}{6/4}; \mu = 50\right) = 1 - \Phi(2) = 0.0228. \qquad \blacksquare$$

This last example illustrates what is often true, namely, that the inequality

$$L(\theta_0)/L(\theta_1) \le k$$

can be expressed in terms of a function $u(x_1, x_2, \ldots, x_n)$, say,

$$u(x_1, \ldots, x_n) \le c_1$$

or

$$u(x_1, \ldots, x_n) \ge c_2,$$

where c_1 or c_2 is selected so that the size of the critical region is α. Thus, the test can be based on the statistic $u(x_1, \ldots, x_n)$. As an example, if we want α to be a given value—say, 0.05—we could then choose our c_1 or c_2. In Example 10.3-2, with

$\alpha = 0.05$, we want

$$0.05 = P(\overline{X} \geq c; \mu = 50)$$

$$= P\left(\frac{\overline{X} - 50}{6/4} \geq \frac{c - 50}{6/4}; \mu = 50\right) = 1 - \Phi\left(\frac{c - 50}{6/4}\right).$$

Hence, it must be true that $(c - 50)/(3/2) = 1.645$, or, equivalently,

$$c = 50 + \frac{3}{2}(1.645) \approx 52.47.$$

EXAMPLE 10.3-3 Let $X_1, X_2, \ldots, X_n$ denote a random sample of size n from a Poisson distribution with mean λ. A best critical region for testing $H_0: \lambda = 2$ against $H_1: \lambda = 5$ is given by

$$\frac{L(2)}{L(5)} = \frac{2^{\Sigma x_i} e^{-2n}}{x_1! x_2! \cdots x_n!} \frac{x_1! x_2! \cdots x_n!}{5^{\Sigma x_i} e^{-5n}} \leq k.$$

This inequality can be written as

$$\left(\frac{2}{5}\right)^{\Sigma x_i} e^{3n} \leq k, \qquad \text{or} \qquad (\Sigma x_i) \ln\left(\frac{2}{5}\right) + 3n \leq \ln k.$$

Since $\ln(2/5) < 0$, the latter inequality is the same as

$$\sum_{i=1}^{n} x_i \geq \frac{\ln k - 3n}{\ln(2/5)} = c.$$

If $n = 4$ and $c = 13$, then

$$\alpha = P\left(\sum_{i=1}^{4} X_i \geq 13; \lambda = 2\right) = 1 - 0.936 = 0.064,$$

from Table III in the appendix, since $\sum_{i=1}^{4} X_i$ has a Poisson distribution with mean 8 when $\lambda = 2$. ∎

When $H_0: \theta = \theta_0$ and $H_1: \theta = \theta_1$ are both simple hypotheses, a critical region of size α is a best critical region if the probability of rejecting H_0 when H_1 is true is a maximum compared with all other critical regions of size α. The test using the best critical region is called a **most powerful test**, because it has the greatest value of the power function at $\theta = \theta_1$ compared with that of other tests with significance level α. If H_1 is a composite hypothesis, the power of a test depends on each simple alternative in H_1.

DEFINITION 10.3-2 A test defined by a critical region C of size α is a **uniformly most powerful test** if it is a most powerful test against each simple alternative in H_1. The critical region C is called a **uniformly most powerful critical region of size** α.

Let us consider again Example 10.3-2 when the alternative hypothesis is composite.

EXAMPLE 10.3-4 Let $X_1, \ldots, X_n$ be a random sample from $N(\mu, 36)$. We have seen that, in testing $H_0: \mu = 50$ against $H_1: \mu = 55$, a best critical region C is defined by

$C = \{(x_1, x_2, \ldots, x_n) : \bar{x} \geq c\}$, where c is selected so that the significance level is α. Now consider testing $H_0 : \mu = 50$ against the one-sided composite alternative hypothesis $H_1 : \mu > 50$. For each simple hypothesis in H_1—say, $\mu = \mu_1$—the quotient of the likelihood functions is

$$\frac{L(50)}{L(\mu_1)} = \frac{(72\pi)^{-n/2} \exp\left[-\left(\dfrac{1}{72}\right)\sum_{i=1}^{n}(x_i - 50)^2\right]}{(72\pi)^{-n/2} \exp\left[-\left(\dfrac{1}{72}\right)\sum_{i=1}^{n}(x_i - \mu_1)^2\right]}$$

$$= \exp\left[-\frac{1}{72}\left\{2(\mu_1 - 50)\sum_{i=1}^{n} x_i + n(50^2 - \mu_1^2)\right\}\right].$$

Now, $L(50)/L(\mu_1) \leq k$ if and only if

$$\bar{x} \geq \frac{(-72)\ln(k)}{2n(\mu_1 - 50)} + \frac{50 + \mu_1}{2} = c.$$

Thus, the best critical region of size α for testing $H_0 : \mu = 50$ against $H_1 : \mu = \mu_1$, where $\mu_1 > 50$, is given by $C = \{(x_1, x_2, \ldots, x_n) : \bar{x} \geq c\}$, where c is selected such that $P(\bar{X} \geq c; H_0 : \mu = 50) = \alpha$. Note that the same value of c can be used for each $\mu_1 > 50$, but (of course) k does not remain the same. Since the critical region C defines a test that is most powerful against each simple alternative $\mu_1 > 50$, this is a uniformly most powerful test, and C is a uniformly most powerful critical region of size α. Again, if $\alpha = 0.05$, then $c \approx 52.47$. ∎

EXAMPLE 10.3-5 Let Y have the binomial distribution $b(n, p)$. To find a uniformly most powerful test of the simple null hypothesis $H_0 : p = p_0$ against the one-sided alternative hypothesis $H_1 : p > p_0$, consider, with $p_1 > p_0$,

$$\frac{L(p_0)}{L(p_1)} = \frac{\binom{n}{y} p_0^y (1 - p_0)^{n-y}}{\binom{n}{y} p_1^y (1 - p_1)^{n-y}} \leq k.$$

This is equivalent to

$$\left[\frac{p_0(1 - p_1)}{p_1(1 - p_0)}\right]^y \left[\frac{1 - p_0}{1 - p_1}\right]^n \leq k$$

and

$$y \ln\left[\frac{p_0(1 - p_1)}{p_1(1 - p_0)}\right] \leq \ln k - n \ln\left[\frac{1 - p_0}{1 - p_1}\right].$$

Since $p_0 < p_1$, we have $p_0(1 - p_1) < p_1(1 - p_0)$, and it follows that $\ln[p_0(1 - p_1)/p_1(1 - p_0)] < 0$. Thus,

$$\frac{y}{n} \geq \frac{\ln k - n \ln[(1 - p_0)/(1 - p_1)]}{n \ln[p_0(1 - p_1)/p_1(1 - p_0)]} = c$$

for each $p_1 > p_0$.

It is interesting to note that if the alternative hypothesis is the one-sided $H_1: p < p_0$, then a uniformly most powerful test is of the form $(y/n) \leq c$. Thus, the tests of $H_0: p = p_0$ against the one-sided alternatives given in Table 7.1-1 are uniformly most powerful. ∎

Exercise 10.3-5 will demonstrate that uniformly most powerful tests do not always exist; in particular, they usually do not exist when the composite alternative hypothesis is two sided.

REMARK   We close this section with one easy, but important, observation: If a sufficient statistic $Y = u(X_1, X_2, \ldots, X_n)$ exists for θ, then, by the factorization theorem,

$$\frac{L(\theta_0)}{L(\theta_1)} = \frac{\phi[u(x_1, x_2, \ldots, x_n); \theta_0] \, h(x_1, x_2, \ldots, x_n)}{\phi[u(x_1, x_2, \ldots, x_n); \theta_1] \, h(x_1, x_2, \ldots, x_n)}$$

$$= \frac{\phi[u(x_1, x_2, \ldots, x_n); \theta_0)}{\phi[u(x_1, x_2, \ldots, x_n); \theta_1]}.$$

Thus, $L(\theta_0)/L(\theta_1) \leq k$ provides a critical region that is a function of the observations $x_1, x_2, \ldots x_n$ only through the observed value of the sufficient statistic $y = u(x_1, x_2, \ldots, x_n)$. Hence, best critical and uniformly most powerful critical regions are based upon sufficient statistics when they exist. ∎

EXERCISES

10.3-1. Let $X_1, X_2, \ldots, X_n$ be a random sample from a normal distribution $N(\mu, 64)$.

(a) Show that $C = \{(x_1, x_2, \ldots, x_n): \bar{x} \leq c\}$ is a best critical region for testing $H_0: \mu = 80$ against $H_1: \mu = 76$.

(b) Find n and c so that $\alpha \approx 0.05$ and $\beta \approx 0.05$.

10.3-2. Let $X_1, X_2, \ldots, X_n$ be a random sample from $N(0, \sigma^2)$.

(a) Show that a best critical region for testing $H_0: \sigma^2 = 4$ against $H_1: \sigma^2 = 16$ is given by $C = \{(x_1, x_2, \ldots, x_n): \sum_{i=1}^{n} x_i^2 \geq c\}$.

(b) If $n = 15$, find the value of c so that $\alpha = 0.05$. HINT: Recall that $\sum_{i=1}^{n} X_i^2 / \sigma^2$ is $\chi^2(n)$.

(c) If $n = 15$ and c is the value found in part (b), find the approximate value of $\beta = P(\sum_{i=1}^{n} X_i^2 < c; \sigma^2 = 16)$.

10.3-3. Let X have an exponential distribution with a mean of θ; that is, the p.d.f. of X is $f(x; \theta) = (1/\theta)e^{-x/\theta}, 0 < x < \infty$. Let $X_1, X_2, \ldots, X_n$ be a random sample from this distribution.

(a) Show that a best critical region for testing $H_0: \theta = 3$ against $H_1: \theta = 5$ can be based on the statistic $\sum_{i=1}^{n} X_i$.

(b) If $n = 12$, use the fact that $(2/\theta)\sum_{i=1}^{12} X_i$ is $\chi^2(24)$ to find a best critical region of size $\alpha = 0.10$.

(c) If $n = 12$, find a best critical region of size $\alpha = 0.10$ for testing $H_0: \theta = 3$ against $H_1: \theta = 7$.

(d) If $H_1: \theta > 3$, is the common region found in parts (b) and (c) a uniformly most powerful critical region of size $\alpha = 0.10$?

10.3-4. Let $X_1, X_2, \ldots, X_n$ be a random sample of Bernoulli trials $b(1, p)$.

(a) Show that a best critical region for testing $H_0: p = 0.9$ against $H_1: p = 0.8$ can be based on the statistic $Y = \sum_{i=1}^{n} X_i$, which is $b(n, p)$.

(b) If $C = \{(x_1, x_2, \ldots, x_n): \sum_{i=1}^{n} x_i \leq (0.85)n\}$ and $Y = \sum_{i=1}^{n} X_i$, find the value of n such that $\alpha = P[Y \leq n(0.85); p = 0.9] \approx 0.10$.
   HINT: Use the normal approximation for the binomial distribution.

(c) What is the approximate value of $\beta = P[Y > n(0.85); p = 0.8]$ for the test given in part (b)?

(d) Is the test of part (b) a uniformly most powerful test when the alternative hypothesis is $H_1: p < 0.9$?

10.3-5. Let $X_1, X_2, \ldots, X_n$ be a random sample from the normal distribution $N(\mu, 36)$.

(a) Show that a uniformly most powerful critical region for testing $H_0: \mu = 50$ against $H_1: \mu < 50$ is given by $C_2 = \{\bar{x}: \bar{x} \leq c\}$.

(b) With this result and that of Example 10.3-4, argue that a uniformly most powerful test for testing $H_0: \mu = 50$ against $H_1: \mu \neq 50$ does not exist.

10.3-6. Let $X_1, X_2, \ldots, X_n$ be a random sample from the normal distribution $N(\mu, 9)$. To test the hypothesis $H_0: \mu = 80$ against $H_1: \mu \neq 80$, consider the following three critical regions: $C_1 = \{\bar{x}: \bar{x} \geq c_1\}$, $C_2 = \{\bar{x}: \bar{x} \leq c_2\}$, and $C_3 = \{\bar{x}: |\bar{x} - 80| \geq c_3\}$.

(a) If $n = 16$, find the values of c_1, c_2, c_3 such that the size of each critical region is 0.05.

That is, find c_1, c_2, c_3 such that

$$0.05 = P(\overline{X} \in C_1; \mu = 80)$$
$$= P(\overline{X} \in C_2; \mu = 80)$$
$$= P(\overline{X} \in C_3; \mu = 80).$$

(b) On the same graph paper, sketch the power functions for these three critical regions.

10.3-7. Let $X_1, X_2, \ldots, X_{10}$ be a random sample of size 10 from a Poisson distribution with mean μ.

(a) Show that a uniformly most powerful critical region for testing $H_0: \mu = 0.5$ against $H_1: \mu > 0.5$ can be defined with the use of the statistic $\sum_{i=1}^{10} X_i$.

(b) What is a uniformly most powerful critical region of size $\alpha = 0.068$? Recall that $\sum_{i=1}^{10} X_i$ has a Poisson distribution with mean 10μ.

(c) Sketch the power function of this test.

10.4   LIKELIHOOD RATIO TESTS

In this section, we consider a general test-construction method that is applicable when both the null and alternative hypotheses—say, H_0 and H_1—are composite. We continue to assume that the functional form of the p.d.f. is known, but that it depends on an unknown parameter or unknown parameters. That is, we assume that the p.d.f. of X is $f(x; \theta)$, where θ represents one or more unknown parameters. We let Ω denote the total parameter space—that is, the set of all possible values of the parameter θ given by either H_0 or H_1. These hypotheses will be stated as

$$H_0: \theta \in \omega, \qquad H_1: \theta \in \omega',$$

where ω is a subset of Ω and ω' is the complement of ω with respect to Ω. The test will be constructed with the use of a ratio of likelihood functions that have been maximized in ω and Ω, respectively. In a sense, this is a natural generalization of the ratio appearing in the Neyman–Pearson lemma when the two hypotheses were simple.

DEFINITION 10.4-1

The **likelihood ratio** is the quotient

$$\lambda = \frac{L(\widehat{\omega})}{L(\widehat{\Omega})},$$

where $L(\widehat{\omega})$ is the maximum of the likelihood function with respect to θ when $\theta \in \omega$ and $L(\widehat{\Omega})$ is the maximum of the likelihood function with respect to θ when $\theta \in \Omega$.

Because λ is the quotient of nonnegative functions, $\lambda \geq 0$. In addition, since $\omega \subset \Omega$, it follows that $L(\widehat{\omega}) \leq L(\widehat{\Omega})$ and hence $\lambda \leq 1$. Thus, $0 \leq \lambda \leq 1$. If the maximum of L in ω is much smaller than that in Ω, it would seem that the data $x_1, x_2, \ldots, x_n$ do not support the hypothesis $H_0: \theta \in \omega$. That is, a small value of the ratio $\lambda = L(\widehat{\omega})/L(\widehat{\Omega})$ would lead to the rejection of H_0. In contrast, a value of the ratio λ that is close to 1 would support the null hypothesis H_0. This reasoning leads us to the next definition.

DEFINITION
10.4-2

To test $H_0: \theta \in \omega$ against $H_1: \theta \in \omega'$, the **critical region for the likelihood ratio test** is the set of points in the sample space for which

$$\lambda = \frac{L(\widehat{\omega})}{L(\widehat{\Omega})} \leq k,$$

where $0 < k < 1$ and k is selected so that the test has a desired significance level α.

The next example illustrates these definitions.

EXAMPLE 10.4-1 Assume that the weight X in ounces of a "10-pound" bag of sugar is $N(\mu, 5)$. We shall test the hypothesis $H_0: \mu = 162$ against the alternative hypothesis $H_1: \mu \neq 162$. Thus, $\Omega = \{\mu: -\infty < \mu < \infty\}$ and $\omega = \{162\}$. To find the likelihood ratio, we need $L(\widehat{\omega})$ and $L(\widehat{\Omega})$. When H_0 is true, μ can take on only one value, namely, $\mu = 162$. Hence, $L(\widehat{\omega}) = L(162)$. To find $L(\widehat{\Omega})$, we must find the value of μ that maximizes $L(\mu)$. Recall that $\widehat{\mu} = \bar{x}$ is the maximum likelihood estimate of μ. Then $L(\widehat{\Omega}) = L(\bar{x})$, and the likelihood ratio $\lambda = L(\widehat{\omega})/L(\widehat{\Omega})$ is given by

$$\lambda = \frac{(10\pi)^{-n/2} \exp\left[-\left(\frac{1}{10}\right) \sum_{i=1}^{n} (x_i - 162)^2\right]}{(10\pi)^{-n/2} \exp\left[-\left(\frac{1}{10}\right) \sum_{i=1}^{n} (x_i - \bar{x})^2\right]}$$

$$= \frac{\exp\left[-\left(\frac{1}{10}\right) \sum_{i=1}^{n} (x_i - \bar{x})^2 - \left(\frac{n}{10}\right)(\bar{x} - 162)^2\right]}{\exp\left[-\left(\frac{1}{10}\right) \sum_{i=1}^{n} (x_i - \bar{x})^2\right]}$$

$$= \exp\left[-\frac{n}{10}(\bar{x} - 162)^2\right].$$

On the one hand, a value of $\bar{x}$ close to 162 would tend to support H_0, and in that case λ is close to 1. On the other hand, an $\bar{x}$ that differs from 162 by too much would tend to support H_1. (See Figure 10.4-1 for the graph of this likelihood ratio when $n = 5$.)

A critical region for a likelihood ratio test is given by $\lambda \leq k$, where k is selected so that the significance level of the test is α. Using this criterion and simplifying the inequality as we do when we use the Neyman–Pearson lemma, we find that $\lambda \leq k$ is equivalent to each of the following inequalities:

$$-\left(\frac{n}{10}\right)(\bar{x} - 162)^2 \leq \ln k,$$

$$(\bar{x} - 162)^2 \geq -\left(\frac{10}{n}\right) \ln k,$$

$$\frac{|\bar{x} - 162|}{\sigma/\sqrt{n}} \geq \frac{\sqrt{-(10/n) \ln k}}{\sigma/\sqrt{n}} = c.$$

Since $Z = (\overline{X} - 162)/(\sigma/\sqrt{n})$ is $N(0,1)$ when $H_0\colon \mu = 162$ is true, let $c = z_{\alpha/2}$. Thus, the critical region is

$$C = \left\{\overline{x}\colon \frac{|\overline{x} - 162|}{\sigma/\sqrt{n}} \geq z_{\alpha/2}\right\}.$$

To illustrate, if $\alpha = 0.05$, then $z_{0.025} = 1.96$. ∎

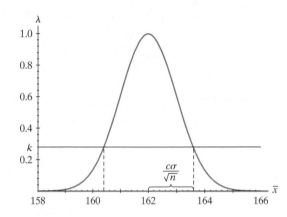

FIGURE 10.4-1: The likelihood ratio for testing $H_0\colon \mu = 162$

As illustrated in Example 10.4-1, the inequality $\lambda \leq k$ can often be expressed in terms of a statistic whose distribution is known. Also, note that although the likelihood ratio test is an intuitive test, it leads to the same critical region as that given by the Neyman–Pearson lemma when H_0 and H_1 are both simple hypotheses.

Suppose now that the random sample $X_1, X_2, \ldots, X_n$ arises from the normal population $N(\mu, \sigma^2)$. where both μ and σ^2 are unknown. Let us consider the likelihood ratio test of the null hypothesis $H_0\colon \mu = \mu_0$ against the two-sided alternative hypothesis $H_1\colon \mu \neq \mu_0$. For this test,

$$\omega = \{(\mu, \sigma^2)\colon \mu = \mu_0,\ 0 < \sigma^2 < \infty\}$$

and

$$\Omega = \{(\mu, \sigma^2)\colon -\infty < \mu < \infty,\ 0 < \sigma^2 < \infty\}.$$

If $(\mu, \sigma^2) \in \Omega$, then the observed maximum likelihood estimates are $\widehat{\mu} = \overline{x}$ and $\widehat{\sigma^2} = (1/n)\sum_{i=1}^{n}(x_i - \overline{x})^2$. Thus,

$$L(\widehat{\Omega}) = \left[\frac{1}{2\pi\left(\dfrac{1}{n}\right)\sum_{i=1}^{n}(x_i - \overline{x})^2}\right]^{n/2} \exp\left[-\frac{\sum_{i=1}^{n}(x_i - \overline{x})^2}{\left(\dfrac{2}{n}\right)\sum_{i=1}^{n}(x_i - \overline{x})^2}\right]$$

$$= \left[\frac{ne^{-1}}{2\pi\sum_{i=1}^{n}(x_i - \overline{x})^2}\right]^{n/2}.$$

Similarly, if $(\mu, \sigma^2) \in \omega$, then the observed maximum likelihood estimates are $\widehat{\mu} = \mu_0$ and $\widehat{\sigma^2} = (1/n)\sum_{i=1}^{n}(x_i - \mu_0)^2$. Hence,

$$L(\widehat{\omega}) = \left[\frac{1}{2\pi\left(\frac{1}{n}\right)\sum_{i=1}^{n}(x_i - \mu_0)^2}\right]^{n/2} \exp\left[-\frac{\sum_{i=1}^{n}(x_i - \mu_0)^2}{\left(\frac{2}{n}\right)\sum_{i=1}^{n}(x_i - \mu_0)^2}\right]$$

$$= \left[\frac{ne^{-1}}{2\pi\sum_{i=1}^{n}(x_i - \mu_0)^2}\right]^{n/2} .$$

The likelihood ratio $\lambda = L(\widehat{\omega})/L(\widehat{\Omega})$ for this test is

$$\lambda = \frac{\left[\dfrac{ne^{-1}}{2\pi\sum_{i=1}^{n}(x_i - \mu_0)^2}\right]^{n/2}}{\left[\dfrac{ne^{-1}}{2\pi\sum_{i=1}^{n}(x_i - \overline{x})^2}\right]^{n/2}}$$

$$= \left[\frac{\sum_{i=1}^{n}(x_i - \overline{x})^2}{\sum_{i=1}^{n}(x_i - \mu_0)^2}\right]^{n/2} .$$

However, note that

$$\sum_{i=1}^{n}(x_i - \mu_0)^2 = \sum_{i=1}^{n}(x_i - \overline{x} + \overline{x} - \mu_0)^2$$

$$= \sum_{i=1}^{n}(x_i - \overline{x})^2 + n(\overline{x} - \mu_0)^2.$$

If this substitution is made in the denominator of λ, we have

$$\lambda = \left[\frac{\sum_{i=1}^{n}(x_i - \overline{x})^2}{\sum_{i=1}^{n}(x_i - \overline{x})^2 + n(\overline{x} - \mu_0)^2}\right]^{n/2}$$

$$= \left[\frac{1}{1 + \dfrac{n(\overline{x} - \mu_0)^2}{\sum_{i=1}^{n}(x_i - \overline{x})^2}}\right]^{n/2} .$$

Note that λ is close to 1 when $\bar{x}$ is close to μ_0 and λ is small when $\bar{x}$ and μ_0 differ by a great deal. The likelihood ratio test, given by the inequality $\lambda \leq k$, is the same as

$$\frac{1}{1 + \dfrac{n(\bar{x} - \mu_0)^2}{\displaystyle\sum_{i=1}^{n}(x_i - \bar{x})^2}} \leq k^{2/n}$$

or, equivalently,

$$\frac{n(\bar{x} - \mu_0)^2}{\dfrac{\displaystyle\sum_{i=1}^{n}(x_i - \bar{x})^2}{n-1}} \geq (n-1)(k^{-2/n} - 1).$$

When H_0 is true, $\sqrt{n}(\bar{X} - \mu_0)/\sigma$ is $N(0,1)$ and $\sum_{i=1}^{n}(X_i - \bar{X})^2/\sigma^2$ has an independent chi-square distribution $\chi^2(n-1)$. Hence, under H_0,

$$T = \frac{\sqrt{n}(\bar{X} - \mu_0)/\sigma}{\sqrt{\displaystyle\sum_{i=1}^{n}(X_i - \bar{X})^2/[\sigma^2(n-1)]}}$$

$$= \frac{\sqrt{n}(\bar{X} - \mu_0)}{\sqrt{\displaystyle\sum_{i=1}^{n}(X_i - \bar{X})^2/(n-1)}} = \frac{\bar{X} - \mu_0}{S/\sqrt{n}}$$

has a t distribution with $r = n - 1$ degrees of freedom. In accordance with the likelihood ratio test criterion, H_0 is rejected if the observed

$$T^2 \geq (n-1)(k^{-2/n} - 1).$$

That is, we reject $H_0: \mu = \mu_0$ and accept $H_1: \mu \neq \mu_0$ if the observed $|T| \geq t_{\alpha/2}(n-1)$.

Note that this test is exactly the same as that listed in Table 7.2-2 for testing $H_0: \mu = \mu_0$ against $H_1: \mu \neq \mu_0$. That is, the test listed there is a likelihood ratio test. As a matter of fact, all six of the tests given in Tables 7.2-2 and 7.3-1 are likelihood ratio tests. Thus, the examples and exercises associated with those tables are illustrations of the use of such tests.

The final development of this section concerns a test about the variance of a normal population. Let $X_1, X_2, \ldots, X_n$ be a random sample from $N(\mu, \sigma^2)$, where μ and σ^2 are unknown. We wish to test $H_0: \sigma^2 = \sigma_0^2$ against $H_1: \sigma^2 \neq \sigma_0^2$. For this purpose, we have

$$\omega = \{(\mu, \sigma^2): -\infty < \mu < \infty, \sigma^2 = \sigma_0^2\}$$

and

$$\Omega = \{(\mu, \sigma^2): -\infty < \mu < \infty, 0 < \sigma^2 < \infty\}.$$

As in the test concerning the mean, we obtain

$$L(\widehat{\Omega}) = \left[\frac{ne^{-1}}{2\pi \displaystyle\sum_{i=1}^{n}(x_i - \bar{x})^2}\right]^{n/2}.$$

If $(\mu, \sigma^2) \in \omega$, then $\hat{\mu} = \bar{x}$ and $\hat{\sigma^2} = \sigma_0^2$; thus,

$$L(\hat{\omega}) = \left(\frac{1}{2\pi\sigma_0^2}\right)^{n/2} \exp\left[-\frac{\sum_{i=1}^{n}(x_i - \bar{x})^2}{2\sigma_0^2}\right].$$

Accordingly, the likelihood ratio test $\lambda = L(\hat{\omega})/L(\hat{\Omega})$ is

$$\lambda = \left(\frac{w}{n}\right)^{n/2} \exp\left(-\frac{w}{2} + \frac{n}{2}\right) \leq k,$$

where $w = \sum_{i=1}^{n}(x_i - \bar{x})^2/\sigma_0^2$. Solving this inequality for w, we obtain a solution of the form $w \leq c_1$ or $w \geq c_2$, where the constants c_1 and c_2 are appropriate functions of the constants k and n so as to achieve the desired significance level α. Since $W = \sum_{i=1}^{n}(X_i - \bar{X})^2/\sigma_0^2$ is $\chi^2(n-1)$ if $H_0: \sigma^2 = \sigma_0^2$ is true, most statisticians modify this test slightly by taking $c_1 = \chi^2_{1-\alpha/2}(n-1)$ and $c_2 = \chi^2_{\alpha/2}(n-1)$. That is, this test and the others listed in Table 7.4-1 are either likelihood ratio tests or slight modifications of likelihood ratio tests. As a matter of fact, most tests involving normal assumptions are likelihood ratio tests or modifications of them; included are tests involving regression and analysis of variance.

REMARK   Note that likelihood ratio tests are based on sufficient statistics when they exist, as was also true of best critical and uniformly most powerful critical regions. ∎

EXERCISES

10.4-1. In Example 10.4-1, suppose that $n = 20$ and $\bar{x} = 161.1$.

(a) Is H_0 accepted if $\alpha = 0.10$?
(b) Is H_0 accepted if $\alpha = 0.05$?
(c) What is the p-value of this test?

10.4-2. Assume that the weight X in ounces of a "10-ounce" box of cornflakes is $N(\mu, 0.03)$. Let $X_1, X_2, \ldots, X_n$ be a random sample from this distribution.

(a) To test the hypothesis $H_0: \mu \geq 10.35$ against the alternative hypothesis $H_1: \mu < 10.35$, what is the critical region of size $\alpha = 0.05$ specified by the likelihood ratio test criterion?
    HINT: Note that if $\mu \geq 10.35$ and $\bar{x} < 10.35$, then $\hat{\mu} = 10.35$.
(b) If a random sample of n = 50 boxes yielded a sample mean of $\bar{x} = 10.31$, is H_0 rejected?
    HINT: Find the critical value z_α when H_0 is true by taking $\mu = 10.35$, which is the extreme value in $\mu \geq 10.35$.
(c) What is the p-value of this test?

10.4-3. Let $X_1, X_2, \ldots, X_n$ be a random sample from the normal distribution $N(\mu, 100)$.

(a) To test $H_0: \mu = 230$ against $H_1: \mu > 230$, what is the critical region specified by the likelihood ratio test criterion?
(b) Is this test uniformly most powerful?
(c) If a random sample of $n = 16$ yielded $\bar{x} = 232.6$, is H_0 accepted at a significance level of $\alpha = 0.10$?
(d) What is the p-value of this test?

10.4-4. Let $X_1, X_2, \ldots, X_n$ be a random sample from the normal distribution $N(\mu, 225)$.

(a) To test $H_0: \mu = 59$ against $H_1: \mu \neq 59$, what is the critical region of size $\alpha = 0.05$ specified by the likelihood ratio test criterion?
(b) If a sample of size $n = 100$ yielded $\bar{x} = 56.13$, is H_0 accepted?
(c) What is the p-value of this test? Note that H_1 is a two-sided alternative.

10.4-5. It is desired to test the hypothesis $H_0: \mu = 30$ against the alternative hypothesis $H_1: \mu \neq 30$, where μ is the mean of a normal distribution and σ^2 is unknown. If a random sample of size $n = 9$ has $\bar{x} = 32.8$ and $s = 4$, is H_0 accepted at an $\alpha = 0.05$ significance level? What is the approximate p-value of this test?

10.4-6. To test H_0: $\mu = 335$ against H_1: $\mu < 335$ under normal assumptions, a random sample of size 17 yielded $\bar{x} = 324.8$ and $s = 40$. Is H_0 accepted at an $\alpha = 0.10$ significance level?

10.4-7. Let X have a normal distribution in which μ and σ^2 are both unknown. It is desired to test H_0: $\mu = 1.80$ against H_1: $\mu > 1.80$ at an $\alpha = 0.10$ significance level. If a random sample of size $n = 121$ yielded $\bar{x} = 1.84$ and $s = 0.20$, is H_0 accepted or rejected? What is the p-value of this test?

10.4-8. Let $X_1, X_2, \ldots, X_n$ be a random sample from an exponential distribution with mean θ. Show that the likelihood ratio test of H_0: $\theta = \theta_0$ against

H_1: $\theta \neq \theta_0$ has a critical region of the form $\sum_{i=1}^n x_i \leq c_1$ or $\sum_{i=1}^n x_i \geq c_2$. How would you modify this test so that chi-square tables can be used easily?

10.4-9. Let independent random samples of sizes n and m be taken respectively from two normal distributions with unknown means μ_X and μ_Y and unknown variances σ_X^2 and σ_Y^2.

(a) Show that when $\sigma_X^2 = \sigma_Y^2$, the likelihood ratio for testing H_0: $\mu_X = \mu_Y$ against H_1: $\mu_X \neq \mu_Y$ is a function of the usual two-sample t statistic.

(b) Show that the likelihood ratio for testing H_0: $\sigma_X^2 = \sigma_Y^2$ against H_1: $\sigma_X^2 \neq \sigma_Y^2$ is a function of the usual two-sample F statistic.

The tests described in parts (a) and (b) are listed in Tables 7.3-1 and 7.4-2, respectively, with the second modified slightly.

10.5   CHEBYSHEV'S INEQUALITY AND CONVERGENCE IN PROBABILITY

In this section, we use Chebyshev's inequality to show, in another sense, that the sample mean, $\bar{x}$, is a good statistic to use to estimate a population mean μ; the relative frequency of success in n Bernoulli trials, y/n, is a good statistic for estimating p; and the empirical distribution function $F_n(x)$ can be used to estimate the theoretical distribution function $F(x)$. We examine the effect of the sample size n on these estimates.

We begin by showing that Chebyshev's inequality gives added significance to the standard deviation in terms of bounding certain probabilities. The inequality is valid for *all* distributions for which the standard deviation exists. The proof is given for the discrete case, but it holds for the continuous case with integrals replacing summations.

THEOREM 10.5-1

(Chebyshev's Inequality) If the random variable X has a mean μ and variance σ^2, then, for every $k \geq 1$,

$$P(|X - \mu| \geq k\sigma) \leq \frac{1}{k^2}.$$

Proof. Let $f(x)$ denote the p.m.f. of X. Then

$$\sigma^2 = E[(X - \mu)^2] = \sum_{x \in S} (x - \mu)^2 f(x)$$

$$= \sum_{x \in A} (x - \mu)^2 f(x) + \sum_{x \in A'} (x - \mu)^2 f(x), \qquad (10.5\text{-}1)$$

where

$$A = \{x : |x - \mu| \geq k\sigma\}.$$

The second term in the right-hand member of Equation 10.5-1 is the sum of nonnegative numbers and thus is greater than or equal to zero. Hence,

$$\sigma^2 \geq \sum_{x \in A} (x - \mu)^2 f(x).$$

However, in A, $|x - \mu| \geq k\sigma$; so

$$\sigma^2 \geq \sum_{x \in A} (k\sigma)^2 f(x) = k^2 \sigma^2 \sum_{x \in A} f(x).$$

But the latter summation equals $P(X \in A)$; thus,

$$\sigma^2 \geq k^2 \sigma^2 P(X \in A) = k^2 \sigma^2 P(|X - \mu| \geq k\sigma).$$

That is,

$$P(|X - \mu| \geq k\sigma) \leq \frac{1}{k^2}. \qquad \square$$

COROLLARY 10.5-1 If $\varepsilon = k\sigma$, then

$$P(|X - \mu| \geq \varepsilon) \leq \frac{\sigma^2}{\varepsilon^2}.$$

In words, Chebyshev's inequality states that the probability that X differs from its mean by at least k standard deviations is less than or equal to $1/k^2$. It follows that the probability that X differs from its mean by less than k standard deviations is at least $1 - 1/k^2$. That is,

$$P(|X - \mu| < k\sigma) \geq 1 - \frac{1}{k^2}.$$

From the corollary, it also follows that

$$P(|X - \mu| < \varepsilon) \geq 1 - \frac{\sigma^2}{\varepsilon^2}.$$

Thus, Chebyshev's inequality can be used as a bound for certain probabilities. However, in many instances, the bound is not very close to the true probability.

EXAMPLE 10.5-1 If it is known that X has a mean of 25 and a variance of 16, then, since $\sigma = 4$, a lower bound for $P(17 < X < 33)$ is given by

$$P(17 < X < 33) = P(|X - 25| < 8)$$

$$= P(|X - 25| < 2\sigma) \geq 1 - \frac{1}{4} = 0.75$$

and an upper bound for $P(|X - 25| \geq 12)$ is found to be

$$P(|X - 25| \geq 12) = P(|X - \mu| \geq 3\sigma) \leq \frac{1}{9}. \qquad \blacksquare$$

Note that the results of the last example hold for any distribution with mean 25 and standard deviation 4. But, even stronger, the probability that any random variable X differs from its mean by 3 or more standard deviations is at most $1/9$, which may be seen by letting $k = 3$ in the theorem. Also, the probability that any random variable X differs from its mean by less than 2 standard deviations is at least 3/4, which may be seen by letting $k = 2$.

The following consideration partially indicates the value of Chebyshev's inequality in theoretical discussions: If Y is the number of successes in n Bernoulli trials with probability p of success on each trial, then Y is $b(n, p)$. Furthermore, Y/n gives the relative frequency of success, and when p is unknown, Y/n can be used as an estimate of its mean p. To gain some insight into the closeness of Y/n to p, we shall use Chebyshev's inequality. With $\varepsilon > 0$, we note from Corollary 10.5-1 that, since $\text{Var}(Y/n) = pq/n$, it follows that

$$P\left(\left|\frac{Y}{n} - p\right| \geq \varepsilon\right) \leq \frac{pq/n}{\varepsilon^2}$$

or, equivalently,

$$P\left(\left|\frac{Y}{n} - p\right| < \varepsilon\right) \geq 1 - \frac{pq}{n\varepsilon^2}. \tag{10.5-2}$$

On the one hand, when p is completely unknown, we can use the fact that $pq = p(1 - p)$ is a maximum when $p = 1/2$ in order to find a lower bound for the probability in Equation 10.5-2. That is,

$$1 - \frac{pq}{n\varepsilon^2} \geq 1 - \frac{(1/2)(1/2)}{n\varepsilon^2}.$$

For example, if $\varepsilon = 0.05$ and $n = 400$, then

$$P\left(\left|\frac{Y}{400} - p\right| < 0.05\right) \geq 1 - \frac{(1/2)(1/2)}{400(0.0025)} = 0.75.$$

On the other hand, if it is known that p is close to $1/10$, we would have

$$P\left(\left|\frac{Y}{400} - p\right| < 0.05\right) \geq 1 - \frac{(0.1)(0.9)}{400(0.0025)} = 0.91.$$

Note that Chebyshev's inequality is applicable to all distributions with a finite variance, and thus the bound is not always a tight one; that is, the bound is not necessarily close to the true probability.

In general, however, it should be noted that, with fixed $\varepsilon > 0$ and $0 < p < 1$, we have

$$\lim_{n \to \infty} P\left(\left|\frac{Y}{400} - p\right| < \varepsilon\right) \geq \lim_{n \to \infty}\left(1 - \frac{pq}{n\varepsilon^2}\right) = 1.$$

But since the probability of every event is less than or equal to 1, it must be that

$$\lim_{n \to \infty} P\left(\left|\frac{Y}{400} - p\right| < \varepsilon\right) = 1.$$

That is, the probability that the relative frequency Y/n is within ε of p is close to 1 when n is large enough. This is one form of the **law of large numbers**, and we say that Y/n **converges in probability** to p.

This theoretical result has something to contribute to our understanding of the properties of the **empirical distribution function**, $F_n(x)$. For a fixed x, $F_n(x)$ is the proportion of sample observations that are less than or equal to x. That is, if "success" is an observed X being less than or equal to the fixed x, then $F_n(x)$ is the relative frequency of success. The *law of large numbers* states that $F_n(x)$ converges, in that probabilistic sense, to the true probability $P(X \leq x)$ of success, namely, $p = F(x)$. Thus, $F_n(x)$ does approach $F(x)$ in this sense and hence provides an estimate of the distribution function. (We used this idea of the empirical distribution approximating the theoretical distribution function in Section 8.7.)

A more general form of the law of large numbers is found by considering the mean $\overline{X}$ of a random sample from a distribution with mean μ and variance σ^2. This form of the law is more general because the relative frequency Y/n can be thought of as $\overline{X}$ when the sample arises from a Bernoulli distribution. To derive it, we note that

$$E(\overline{X}) = \mu \qquad \text{and} \qquad \text{Var}(\overline{X}) = \frac{\sigma^2}{n}.$$

Thus, from Corollary 10.5-1, for every $\varepsilon > 0$, we have

$$P[\,|\overline{X} - \mu| \geq \varepsilon\,] \leq \frac{\sigma^2/n}{\varepsilon^2} = \frac{\sigma^2}{n\varepsilon^2}.$$

Since probability is nonnegative, it follows that

$$\lim_{n\to\infty} P(|\overline{X} - \mu| \geq \varepsilon) \leq \lim_{n\to\infty} \frac{\sigma^2}{\varepsilon^2 n} = 0,$$

which implies that

$$\lim_{n\to\infty} P(|\overline{X} - \mu| \geq \varepsilon) = 0,$$

or, equivalently,

$$\lim_{n\to\infty} P(|\overline{X} - \mu| < \varepsilon) = 1.$$

The preceding discussion shows that the probability associated with the distribution of $\overline{X}$ becomes concentrated in an arbitrarily small interval centered at μ as n increases. This is a more general form of the law of large numbers, and we say that $\overline{X}$ *converges in probability* to μ.

Although Chebyshev's inequality is quite useful in theoretical discussions, it also shows that in a collection of numbers—say, $x_1, x_2, \ldots, x_n$—a certain proportion of them must be within $k\sqrt{v}$ of $\bar{x}$, where $\bar{x}$ and v are the respective mean and variance of the empirical distribution defined by the numbers. That is,

$$\bar{x} = \frac{1}{n}\sum_{i=1}^{n} x_i \qquad \text{and} \qquad v = \frac{1}{n}\sum_{i=1}^{n}(x_i - \bar{x})^2.$$

To show this, think of the empirical distribution as defining the probability distribution; thus, $\bar{x}$ and v are, respectively, the mean and the variance of this distribution. Hence, for every $k \geq 1$, we have

$$\frac{\mathcal{N}\{x_i \colon |x_i - \bar{x}| \geq k\sqrt{v}\}}{n} \leq \frac{1}{k^2}$$

or, equivalently,

$$\frac{\mathcal{N}\{x_i \colon |x_i - \bar{x}| < k\sqrt{v}\}}{n} \geq 1 - \frac{1}{k^2}.$$

In words, the second inequality says, "The proportion of the numbers $x_1, x_2, \ldots, x_n$ that lie within $k\sqrt{v}$ of the mean $\bar{x}$ is at least $1 - 1/k^2$."

EXERCISES

10.5-1. If X is a random variable with mean 33 and variance 16, use Chebyshev's inequality to find

(a) A lower bound for $P(23 < X < 43)$.

(b) An upper bound for $P(|X - 33| \geq 14)$.

10.5-2. If $E(X) = 17$ and $E(X^2) = 298$, use Chebyshev's inequality to determine

(a) A lower bound for $P(10 < X < 24)$.

(b) An upper bound for $P(|X - 17| \geq 16)$.

10.5-3. Let X denote the outcome when a fair die is rolled. Then $\mu = 7/2$ and $\sigma^2 = 35/12$. Note that the maximum deviation of X from μ equals 5/2. Express this deviation in terms of number of standard deviations; that is, find k, where $k\sigma = 5/2$. Determine a lower bound for $P(|X - 3.5| < 2.5)$.

10.5-4. If Y is $b(n, 0.5)$, give a lower bound for $P(|Y/n - 0.5| < 0.08)$ when

(a) $n = 100$.

(b) $n = 500$.

(c) $n = 1000$.

10.5-5. If Y is $b(n, 0.25)$, give a lower bound for $P(|Y/n - 0.25| < 0.05)$ when

(a) $n = 100$.

(b) $n = 500$.

(c) $n = 1000$.

10.5-6. Let $\overline{X}$ be the mean of a random sample of size $n = 15$ from a distribution with mean $\mu = 80$ and variance $\sigma^2 = 60$. Use Chebyshev's inequality to find a lower bound for $P(75 < \overline{X} < 85)$.

10.5-7. The characteristics of the empirical distribution of test scores of 900 students are $\bar{x} = 83$ and

$v = 36$, respectively. At least how many students received test scores between 71 and 95?

10.5-8. Suppose that W is a random variable with mean $\mu = 0$ and variance $\sigma^2 = 1$. Then, for $k \geq 1$, Chebyshev's inequality becomes

$$P(|W| < k) \geq 1 - \frac{1}{k^2}.$$

Suppose further that W is a continuous random variable with a symmetric p.d.f. $f(w)$ and distribution function $F(w)$. Finally, suppose that there exists a continuous and symmetric random variable, W, such that **equality** holds in "Chebyshev's inequality." That is, suppose that

$$P(|W - 0| < k) = 1 - \frac{1}{k^2}.$$

Then the distribution function satisfies

$$F(w) - F(-w) = 1 - \frac{1}{w^2}, \qquad w \geq 1.$$

Also, the symmetry assumption implies that

$$F(-w) = 1 - F(w).$$

(a) Show that the p.d.f. of W is

$$f(w) = \begin{cases} \dfrac{1}{|w|^3}, & |w| > 1, \\ 0, & |w| \leq 1. \end{cases}$$

(b) Find the mean and the variance of W and interpret your results.

(c) Graph the distribution function of W.

10.6   LIMITING MOMENT-GENERATING FUNCTIONS

We would like to begin this section by showing that the binomial distribution can be approximated by the Poisson distribution when n is sufficiently large and p is fairly small. Of course, we proved this in Section 2.6 by showing that, under these conditions, the binomial p.m.f. is close to that of the Poisson. Here, however, we show that the binomial moment-generating function is close to that of the Poisson distribution. We do so by taking the limit of a moment-generating function.

Consider the moment-generating function of Y, which is $b(n, p)$. We shall take the limit of this function as $n \to \infty$ such that $np = \lambda$ is a constant; thus, $p \to 0$. The moment-generating function of Y is

$$M(t) = (1 - p + pe^t)^n.$$

Because $p = \lambda/n$, we have

$$M(t) = \left[1 - \frac{\lambda}{n} + \frac{\lambda}{n} e^t \right]^n$$

$$= \left[1 + \frac{\lambda(e^t - 1)}{n} \right]^n.$$

Since

$$\lim_{n \to \infty} \left(1 + \frac{b}{n} \right)^n = e^b,$$

we have

$$\lim_{n \to \infty} M(t) = e^{\lambda(e^t - 1)},$$

which exists for all real t. But this is the moment-generating function of a Poisson random variable with mean λ. Hence, this Poisson distribution seems like a reasonable approximation to the binomial distribution when n is large and p is small. That approximation is usually found to be fairly successful if $n \geq 20$ and $p \leq 0.05$ and is found to be very successful if $n \geq 100$ and $p \leq 0.10$, but it is not bad if these bounds are violated somewhat. That is, the approximation could be used in other situations, too; we only want to stress that it becomes better with larger n and smaller p.

The preceding result illustrates the theorem we now state without proof.

THEOREM 10.6-1   If a sequence of moment-generating functions approaches a certain m.g.f., say, $M(t)$, then the limit of the corresponding distributions must be the distribution corresponding to $M(t)$.

REMARK   This theorem certainly appeals to one's intuition! In a more advanced course, the theorem is proven, and there the existence of the moment-generating function is not even needed, for we would use the characteristic function $\phi(t) = E(e^{itX})$ instead. ∎

The next example illustrates graphically the convergence of the binomial moment-generating functions to that of a Poisson distribution.

EXAMPLE 10.6-1   Consider the moment-generating function for the Poisson distribution with $\lambda = 5$ and those for three binomial distributions for which $np = 5$, namely, $b(50, 1/10)$, $b(100, 1/20)$, and $b(200, 1/40)$. These four moment-generating functions are, respectively,

$$M(t) = e^{5(e^t - 1)},$$

$$M(t) = (0.9 + 0.1e^t)^{50},$$

$$M(t) = (0.95 + 0.05e^t)^{100},$$

$$M(t) = (0.975 + 0.025e^t)^{200}.$$

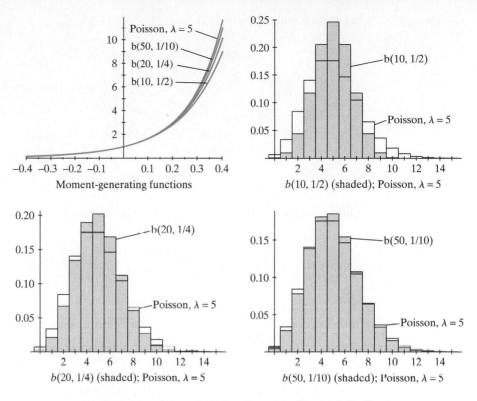

FIGURE 10.6-1: Poisson approximation to the binomial distribution

The graphs of the moment-generating functions with $n = 10, 20,$ and 50 are shown in Figure 10.6-1. Although the proof and the figure show the convergence of the binomial moment-generating functions to that of the Poisson distribution, the last three graphs in Figure 10.6-1 show more clearly how the Poisson distribution can be used to approximate binomial probabilities with large n and small p. ∎

The next example gives a numerical approximation.

EXAMPLE 10.6-2   Let Y be $b(50, 1/25)$. Then

$$P(Y \leq 1) = \left(\frac{24}{25}\right)^{50} + 50\left(\frac{1}{25}\right)\left(\frac{24}{25}\right)^{49} = 0.400.$$

Since $\lambda = np = 2$, the Poisson approximation is

$$P(Y \leq 1) \approx 0.406,$$

from Table III in the appendix. ∎

Theorem 10.6-1 is used to prove the central limit theorem. To help in understanding this proof, let us first consider a different problem: that of the limiting distribution of the mean $\overline{X}$ of a random sample $X_1, X_2, \ldots, X_n$ from a distribution with mean μ. If the distribution has moment-generating function $M(t)$, then the moment-generating function of $\overline{X}$ is $[M(t/n)]^n$. But, by Taylor's expansion, there exists a number t_1

between 0 and t/n such that

$$M\left(\frac{t}{n}\right) = M(0) + M'(t_1)\frac{t}{n}$$

$$= 1 + \frac{\mu t}{n} + \frac{[M'(t_1) - M'(0)]t}{n},$$

because $M(0) = 1$ and $M'(0) = \mu$. Since $M'(t)$ is continuous at $t = 0$ and since $t_1 \to 0$ as $n \to \infty$, it follows that

$$\lim_{n\to\infty}[M'(t_1) - M'(0)] = 0.$$

Thus, using a result from advanced calculus, we obtain

$$\lim_{n\to\infty}\left[M\left(\frac{t}{n}\right)\right]^n = \lim_{n\to\infty}\left\{1 + \frac{\mu t}{n} + \frac{[M'(t_1) - M'(0)]t}{n}\right\}^n$$

$$= e^{\mu t}$$

for all real t. But this limit is the moment-generating function of a degenerate distribution with all of the probability on μ. Accordingly, $\overline{X}$ has this limiting distribution, indicating that $\overline{X}$ converges to μ in a certain sense. This is one form of the law of large numbers.

We have seen that, in some probability sense, $\overline{X}$ converges to μ in the limit, or, equivalently, $\overline{X} - \mu$ converges to zero. Let us multiply the difference $\overline{X} - \mu$ by some function of n so that the result will not converge to zero. In our search for such a function, it is natural to consider

$$W = \frac{\overline{X} - \mu}{\sigma/\sqrt{n}} = \frac{\sqrt{n}(\overline{X} - \mu)}{\sigma} = \frac{Y - n\mu}{\sqrt{n}\,\sigma},$$

where Y is the sum of the observations of the random sample. The reason for this is that, by the remark after the proof of Theorem 3.6-1, W is a standardized random variable. That is, W has mean 0 and variance 1 for each positive integer n. We are now ready to prove the central limit theorem, which is stated in Section 5.6.

Proof. (of the Central Limit Theorem):
We first consider

$$E[\exp(tW)] = E\left\{\exp\left[\left(\frac{t}{\sqrt{n}\sigma}\right)\left(\sum_{i=1}^{n}X_i - n\mu\right)\right]\right\}$$

$$= E\left\{\exp\left[\left(\frac{t}{\sqrt{n}}\right)\left(\frac{X_1 - \mu}{\sigma}\right)\right]\cdots\exp\left[\left(\frac{t}{\sqrt{n}}\right)\left(\frac{X_n - \mu}{\sigma}\right)\right]\right\}$$

$$= E\left\{\exp\left[\left(\frac{t}{\sqrt{n}}\right)\left(\frac{X_1 - \mu}{\sigma}\right)\right]\right\}\cdots E\left\{\exp\left[\left(\frac{t}{\sqrt{n}}\right)\left(\frac{X_n - \mu}{\sigma}\right)\right]\right\},$$

which follows from the independence of $X_1, X_2, \ldots, X_n$. Then

$$E[\exp(tW)] = \left[m\left(\frac{t}{\sqrt{n}} \right) \right]^n, \qquad -h < \frac{t}{\sqrt{n}} < h,$$

where

$$m(t) = E\left\{ \exp\left[t\left(\frac{X_i - \mu}{\sigma} \right) \right] \right\}, \qquad -h < t < h,$$

is the common moment-generating function of each

$$Y_i = \frac{X_i - \mu}{\sigma}, \qquad i = 1, 2, \ldots, n.$$

Since $E(Y_i) = 0$ and $E(Y_i^2) = 1$, it must be that

$$m(0) = 1, \qquad m'(0) = E\left(\frac{X_i - \mu}{\sigma} \right) = 0, \qquad m''(0) = E\left[\left(\frac{X_i - \mu}{\sigma} \right)^2 \right] = 1.$$

Hence, using Taylor's formula with a remainder, we know that there exists a number t_1 between 0 and t such that

$$m(t) = m(0) + m'(0)t + \frac{m''(t_1)t^2}{2} = 1 + \frac{m''(t_1)t^2}{2}.$$

Adding and subtracting $t^2/2$, we have

$$m(t) = 1 + \frac{t^2}{2} + \frac{[m''(t_1) - 1]t^2}{2}.$$

Using this expression of $m(t)$ in $E[\exp(tW)]$, we can represent the moment-generating function of W by

$$E[\exp(tW)] = \left\{ 1 + \frac{1}{2}\left(\frac{t}{\sqrt{n}} \right)^2 + \frac{1}{2}[m''(t_1) - 1]\left(\frac{t}{\sqrt{n}} \right)^2 \right\}^n$$

$$= \left\{ 1 + \frac{t^2}{2n} + \frac{[m''(t_1) - 1]t^2}{2n} \right\}^n, \qquad -\sqrt{n}\,h < t < \sqrt{n}\,h,$$

where now t_1 is between 0 and $t/\sqrt{n}$. Since $m''(t)$ is continuous at $t = 0$ and $t_1 \to 0$ as $n \to \infty$, we have

$$\lim_{n \to \infty} [m''(t_1) - 1] = 1 - 1 = 0.$$

Thus, using a result from advanced calculus, we obtain

$$\lim_{n \to \infty} E[\exp(tW)] = \lim_{n \to \infty} \left\{ 1 + \frac{t^2}{2n} + \frac{[m''(t_1) - 1]t^2}{2n} \right\}^n$$

$$= \lim_{n \to \infty} \left\{ 1 + \frac{t^2}{2n} \right\}^n = e^{t^2/2}.$$

for all real t. We know that $e^{t^2/2}$ is the m.g.f. of the standard normal distribution, $N(0,1)$. It then follows that the limiting distribution of

$$W = \frac{\overline{X} - \mu}{\sigma/\sqrt{n}} = \frac{\sum_{i=1}^{n} X_i - n\mu}{\sqrt{n}\,\sigma}$$

is $N(0,1)$. This completes the proof of the central limit theorem. $\square$

Examples of the use of the central limit theorem as an approximating distribution were given in Sections 5.6 and 5.7.

To help appreciate the proof of the central limit theorem, the next example graphically illustrates the convergence of the moment-generating functions for two distributions.

EXAMPLE 10.6-3 Let $X_1, X_2, \ldots, X_n$ be a random sample of size n from an exponential distribution with $\theta = 2$. The moment-generating function of $(\overline{X} - \theta)/(\theta/\sqrt{n})$ is

$$M_n(t) = \frac{e^{-t\sqrt{n}}}{(1 - t/\sqrt{n}\,)^n}.$$

The central limit theorem says that, as n increases, this moment-generating function approaches that of the standard normal distribution, namely,

$$M(t) = e^{t^2/2}.$$

The moment-generating functions for $M(t)$ and $M_n(t), n = 5, 15, 50$, are shown in Figure 10.6-2(a). [See also Figure 5.6-2, in which samples were taken from a $\chi^2(1)$ distribution, and recall that the exponential distribution with $\theta = 2$ is $\chi^2(2)$.]

In Example 5.6-6, a U-shaped distribution was considered for which the p.d.f. is $f(x) = (3/2)x^2, -1 < x < 1$. For this distribution, $\mu = 0$ and $\sigma^2 = 3/5$. Its moment-generating function, for $t \neq 0$, is

$$M(t) = \left(\frac{3}{2}\right)\frac{e^t t^2 - 2e^t t + 2e^t - e^{-t}t^2 - 2e^{-t}t - 2e^{-t}}{t^3}.$$

Of course, $M(0) = 1$. The moment-generating function of

$$W_n = \frac{\overline{X} - 0}{\sqrt{3/(5n)}}$$

is

$$E[e^{tW_n}] = \left\{E\left[\exp\left(\sqrt{\frac{5}{3n}}\,t\right)\right]\right\}^n = \left[M\left(\sqrt{\frac{5}{3n}}\,t\right)\right]^n.$$

The graphs of these moment-generating functions when $n = 2, 5, 10$ and the graph of the moment-generating function for the standard normal distribution are shown in Figure 10.6-2(b). Note how much more quickly these moment-generating functions converge compared with those for the exponential distribution. ■

(a) Sampling from $\chi^2(2)$ (b) Sampling from U-shaped distribution

FIGURE 10.6-2: Convergence of moment-generating functions

EXERCISES

10.6-1. Let Y be the number of defectives in a box of 50 articles taken from the output of a machine. Each article is defective with probability 0.01. Find the probability that $Y = 0, 1, 2,$ or 3.

  (a) By using the binomial distribution.
  (b) By using the Poisson approximation.

10.6-2. The probability that a certain type of inoculation takes effect is 0.995. Use the Poisson distribution to approximate the probability that at most 2 out of 400 people given the inoculation find that it has not taken effect.

  HINT: Let $p = 1 - 0.995 = 0.005$.

10.6-3. Let S^2 be the sample variance of a random sample of size n from $N(\mu, \sigma)$. Show that the limit, as

$n \to \infty$, of the moment-generating function of S^2 is $e^{\sigma^2 t}$. Thus, in the limit, the distribution of S^2 is degenerate with probability 1 at σ^2.

10.6-4. Let Y be $\chi^2(n)$. Use the central limit theorem to demonstrate that $W = (Y - n)/\sqrt{2n}$ has a limiting distribution function that is $N(0, 1)$.

  HINT: Think of Y as being the sum of a random sample from a certain distribution.

10.6-5. Let Y have a Poisson distribution with mean $3n$. Use the central limit theorem to show that $W = (Y - 3n)/\sqrt{3n}$ has a limiting distribution that is $N(0, 1)$.

10.7 ASYMPTOTIC DISTRIBUTIONS OF MAXIMUM LIKELIHOOD ESTIMATORS

Let us consider a distribution of the continuous type with p.d.f. $f(x; \theta)$ such that the parameter θ is not involved in the support of the distribution. Moreover, we want $f(x; \theta)$ to possess a number of mathematical properties that we do not list here. However, in particular, we want to be able to find the maximum likelihood estimator $\widehat{\theta}$ by solving

$$\frac{\partial[\ln L(\theta)]}{\partial \theta} = 0,$$

where here we use a partial-derivative sign because $L(\theta)$ involves $x_1, x_2, \ldots, x_n$, too.
  That is,

$$\frac{\partial[\ln L(\widehat{\theta})]}{\partial \theta} = 0,$$

where now, with $\widehat{\theta}$ in this expression, $L(\widehat{\theta}) = f(X_1; \widehat{\theta}) f(X_2; \widehat{\theta}) \cdots f(X_n; \widehat{\theta})$. We can approximate the left-hand member of this latter equation by a linear function found from the first two terms of a Taylor's series expanded about θ, namely,

$$\frac{\partial[\ln L(\theta)]}{\partial\theta} + (\widehat{\theta} - \theta)\frac{\partial^2[\ln L(\theta)]}{\partial\theta^2} \approx 0,$$

when $L(\theta) = f(X_1; \theta)f(X_2; \theta)\cdots f(X_n; \theta)$.

Obviously, this approximation is good enough only if $\widehat{\theta}$ is close to θ, and an adequate mathematical proof involves those conditions, which we have not given here. (See Hogg, McKean, and Craig, 2005.) But a heuristic argument can be made by solving for $\widehat{\theta} - \theta$ to obtain

$$\widehat{\theta} - \theta = \frac{\dfrac{\partial[\ln L(\theta)]}{\partial\theta}}{-\dfrac{\partial^2[\ln L(\theta)]}{\partial\theta^2}}. \tag{10.7-1}$$

Recall that

$$\ln L(\theta) = \ln f(X_1; \theta) + \ln f(X_2; \theta) + \cdots + \ln f(X_n; \theta)$$

and

$$\frac{\partial \ln L(\theta)}{\partial\theta} = \sum_{i=1}^{n} \frac{\partial[\ln f(X_i; \theta)]}{\partial\theta}, \tag{10.7-2}$$

which is the numerator in Equation 10.7-1. However, Equation 10.7-2 gives the sum of the n independent and identically distributed random variables

$$Y_i = \frac{\partial[\ln f(X_i; \theta)]}{\partial\theta}, \qquad i = 1, 2, \ldots, n,$$

and thus, by the central limit theorem, has an approximate normal distribution with mean (in the continuous case) equal to

$$
\begin{aligned}
\int_{-\infty}^{\infty} \frac{\partial[\ln f(x; \theta)]}{\partial\theta} f(x; \theta)\, dx &= \int_{-\infty}^{\infty} \frac{\partial[f(x; \theta)]}{\partial\theta} \frac{f(x; \theta)}{f(x; \theta)}\, dx \\
&= \int_{-\infty}^{\infty} \frac{\partial[f(x; \theta)]}{\partial\theta}\, dx \\
&= \frac{\partial}{\partial\theta}\left[\int_{-\infty}^{\infty} f(x; \theta)\, dx\right] \\
&= \frac{\partial}{\partial\theta}[1] \\
&= 0.
\end{aligned}
$$

Clearly, we need a certain mathematical condition that makes it permissible to interchange the operations of integration and differentiation in those last steps. Of course, the integral of $f(x; \theta)$ is equal to 1 because it is a p.d.f.

Since we now know that the mean of each Y is

$$\int_{-\infty}^{\infty} \frac{\partial[\ln f(x; \theta)]}{\partial\theta} f(x; \theta)\, dx = 0,$$

let us take derivatives of each member of this equation with respect to θ, obtaining

$$\int_{-\infty}^{\infty} \left\{ \frac{\partial^2[\ln f(x;\theta)]}{\partial\theta^2} f(x;\theta) + \frac{\partial[\ln f(x;\theta)]}{\partial\theta} \frac{\partial[f(x;\theta)]}{\partial\theta} \right\} dx = 0.$$

However,

$$\frac{\partial[f(x;\theta)]}{\partial\theta} = \frac{\partial[\ln f(x;\theta)]}{\partial\theta} f(x;\theta);$$

so

$$\int_{-\infty}^{\infty} \left\{ \frac{\partial[\ln f(x;\theta)]}{\partial\theta} \right\}^2 f(x;\theta)\, dx = - \int_{-\infty}^{\infty} \frac{\partial^2[\ln f(x;\theta)]}{\partial\theta^2} f(x;\theta)\, dx.$$

Since $E(Y) = 0$, this last expression provides the variance of $Y = \partial[\ln f(X;\theta)]/\partial\theta$. Then the variance of the sum in Equation 10.7-2 is n times this value, namely,

$$-nE\left\{ \frac{\partial^2[\ln f(X;\theta)]}{\partial\theta^2} \right\}.$$

Let us rewrite Equation 10.7-1 as

$$\frac{\sqrt{n}\,(\hat{\theta} - \theta)}{\dfrac{1}{\sqrt{-E\{\partial^2[\ln f(X;\theta)]/\partial\theta^2\}}}} = \frac{\dfrac{\partial[\ln L(\theta)]/\partial\theta}{\sqrt{-nE\{\partial^2[\ln f(X;\theta)]/\partial\theta^2\}}}}{\dfrac{-\dfrac{1}{n}\dfrac{\partial^2[\ln L(\theta)]}{\partial\theta^2}}{E\{-\partial^2[\ln f(X;\theta)]/\partial\theta^2\}}}. \qquad (10.7\text{-}3)$$

The numerator of the right-hand member of Equation 10.7-3 has an approximate $N(0,1)$ distribution, and the aforementioned unstated mathematical conditions require, in some sense, that

$$\frac{1}{n}\frac{\partial^2[\ln L(\theta)]}{\partial\theta^2} \qquad \text{converge to} \qquad E\{ \partial^2[\ln f(X;\theta)]/\partial\theta^2\}.$$

Accordingly, the ratios given in Equation 10.7-3 must be approximately $N(0,1)$. That is, $\hat{\theta}$ has an approximate normal distribution with mean θ and standard deviation

$$\frac{1}{\sqrt{-nE\{\partial^2[\ln f(X;\theta)]/\partial\theta^2\}}}.$$

EXAMPLE 10.7-1 (continuation of Example 6.1-1). With the underlying exponential p.d.f.

$$f(x;\theta) = \frac{1}{\theta} e^{-x/\theta}, \qquad 0 < x < \infty, \qquad \theta \in \Omega = \{\theta : 0 < \theta < \infty\},$$

$\overline{X}$ is the maximum likelihood estimator. Since

$$\ln f(x;\theta) = -\ln\theta - \frac{x}{\theta}$$

and

$$\frac{\partial[\ln f(x;\theta)]}{\partial\theta} = -\frac{1}{\theta} + \frac{x}{\theta^2} \qquad \text{and} \qquad \frac{\partial^2[\ln f(x;\theta)]}{\partial\theta} = \frac{1}{\theta^2} - \frac{2x}{\theta^3},$$

we have

$$-E\left[\frac{1}{\theta^2} - \frac{2X}{\theta^3}\right] = -\frac{1}{\theta^2} + \frac{2\theta}{\theta^3} = \frac{1}{\theta^2},$$

because $E(X) = \theta$. That is, $\overline{X}$ has an approximate normal distribution with mean θ and standard deviation $\theta/\sqrt{n}$. Thus, the random interval $\overline{X} \pm 1.96(\theta/\sqrt{n})$ has an approximate probability of 0.95 that it covers θ. Substituting the observed $\overline{x}$ for θ, as well as for $\overline{X}$, we say that $\overline{x} \pm 1.96\,\overline{x}/\sqrt{n}$ is an approximate 95% confidence interval for θ.   ■

While the development of the preceding result used a continuous-type distribution, the result holds for the discrete type also, as long as the support does not involve the parameter. This is illustrated in the next example.

EXAMPLE 10.7-2   (continuation of Exercise 6.1-3). If the random sample arises from a Poisson distribution with p.m.f.

$$f(x; \lambda) = \frac{\lambda^x e^{-\lambda}}{x!}, \qquad x = 0, 1, 2, \ldots; \qquad \lambda \in \Omega = \{\lambda : 0 < \lambda < \infty\},$$

then the maximum likelihood estimator for λ is $\widehat{\lambda} = \overline{X}$. Now,

$$\ln f(x; \lambda) = x \ln \lambda - \lambda - \ln x!.$$

Also,

$$\frac{\partial[\ln f(x; \lambda)]}{\partial \lambda} = \frac{x}{\lambda} - 1 \qquad \text{and} \qquad \frac{\partial^2[\ln f(x; \lambda)]}{\partial \lambda^2} = -\frac{x}{\lambda^2}.$$

Thus,

$$-E\left(-\frac{X}{\lambda^2}\right) = \frac{\lambda}{\lambda^2} = \frac{1}{\lambda},$$

and $\widehat{\lambda} = \overline{X}$ has an approximate normal distribution with mean λ and standard deviation $\sqrt{\lambda/n}$. Finally, $\overline{x} \pm 1.645\sqrt{\overline{x}/n}$ serves as an approximate 90% confidence interval for λ. With the data in Exercise 6.1-3, $\overline{x} = 2.225$ and it follows that this interval ranges from 1.887 to 2.563.   ■

It is interesting that there is another theorem which is somewhat related to the preceding result in that the variance of $\widehat{\theta}$ serves as a lower bound for the variance of every unbiased estimator of θ. Thus, we know that if a certain unbiased estimator has a variance equal to that lower bound, we cannot find a better one, and hence that estimator is the best in the sense of being the minimum-variance unbiased estimator. So, in the limit, the maximum likelihood estimator is this type of best estimator.

We describe this **Rao–Cramér inequality** here without proof. Let $X_1, X_2, \ldots, X_n$ be a random sample from a distribution of the continuous type with p.d.f. $f(x; \theta)$, $\theta \in \Omega = \{\theta : c < \theta < d\}$, where the support of X does not depend upon θ, so that we can differentiate, with respect to θ, under integral signs like that in the following integral:

$$\int_{-\infty}^{\infty} f(x; \theta)\, dx = 1.$$

If $Y = u(X_1, X_2, \ldots, X_n)$ is an unbiased estimator of θ, then

$$\text{Var}(Y) \geq \frac{1}{n \int_{-\infty}^{\infty} \{[\partial \ln f(x; \theta)/\partial\theta]\}^2 f(x; \theta)\, dx}$$

$$= \frac{-1}{n \int_{-\infty}^{\infty} [\partial^2 \ln f(x; \theta)/\partial\theta^2]\, f(x; \theta)\, dx}.$$

Note that the integrals in the denominators are, respectively, the expectations

$$E\left\{\left[\frac{\partial \ln f(X; \theta)}{\partial\theta}\right]^2\right\} \qquad \text{and} \qquad E\left[\frac{\partial^2 \ln f(X; \theta)}{\partial\theta^2}\right];$$

sometimes one is easier to compute than the other. Note also that although the Rao–Cramér lower bound has been stated only for a continuous-type distribution, it is also true for a discrete-type distribution, with summations replacing integrals.

We have computed this lower bound for each of two distributions: exponential and Poisson. Those respective lower bounds were θ^2/n and λ/n. (See Examples 10.7-1 and 10.7-2.) Since, in each case, the variance of $\overline{X}$ equals the lower bound, then $\overline{X}$ is the minimum-variance unbiased estimator.

Let us consider another example.

EXAMPLE 10.7-3 (continuation of Exercise 6.1-7). Let the p.d.f. of X be given by

$$f(x; \theta) = \theta x^{\theta-1}, \qquad 0 < x < 1, \qquad \theta \in \Omega = \{\theta: 0 < \theta < \infty\}.$$

We then have

$$\ln f(x; \theta) = \ln \theta + (\theta - 1)\ln x,$$

$$\frac{\partial \ln f(x; \theta)}{\partial\theta} = \frac{1}{\theta} + \ln x,$$

and

$$\frac{\partial^2 \ln f(x; \theta)}{\partial\theta^2} = -\frac{1}{\theta^2}.$$

Since $E(-1/\theta^2) = -1/\theta^2$, the lower bound of the variance of every unbiased estimator of θ is θ^2/n. Moreover, the maximum likelihood estimator $\widehat{\theta} = -n/\ln \prod_{i=1}^{n} X_i$ has an approximate normal distribution with mean θ and variance θ^2/n. Thus, in a limiting sense, $\widehat{\theta}$ is the minimum variance unbiased estimator of θ. ∎

To measure the value of estimators, their variances are compared with the Rao–Cramér lower bound. The ratio of the Rao–Cramér lower bound to the actual variance of any unbiased estimator is called the **efficiency** of that estimator. An estimator with an efficiency of, say, 50%, means that $1/0.5 = 2$ times as many sample observations are needed to do as well in estimation as can be done with the minimum-variance unbiased estimator (the 100% efficient estimator).

EXERCISES

10.7-1. Let $X_1, X_2, \ldots, X_n$ be a random sample from $N(\theta, \sigma^2)$, where σ^2 is known.

(a) Show that $Y = (X_1 + X_2)/2$ is an unbiased estimator of θ.

(b) Find the Rao–Cramér lower bound for the variance of an unbiased estimator of θ for a general n.

(c) What is the efficiency of Y in part (a)?

10.7-2. Let $X_1, X_2, \ldots, X_n$ denote a random sample from $b(1, p)$. We know that $\overline{X}$ is an unbiased estimator of p and that $\mathrm{Var}(\overline{X}) = p(1 - p)/n$. (See Exercise 6.1-12.)

(a) Find the Rao–Cramér lower bound for the variance of every unbiased estimator of p.

(b) What is the efficiency of $\overline{X}$ as an estimator of p?

10.7-3. (Continuation of Exercise 6.1-2.) In sampling from a normal distribution with known mean μ, $\widehat{\theta} = \sum_{i=1}^{n}(X_i - \mu)^2/n$ is the maximum likelihood estimator of $\theta = \sigma^2$.

(a) Determine the Rao–Cramér lower bound.

(b) What is the approximate distribution of $\widehat{\theta}$?

(c) What is the exact distribution of $n\widehat{\theta}/\theta$, where $\theta = \sigma^2$?

10.7-4. Find the Rao–Cramér lower bound, and thus the asymptotic variance of the maximum likelihood estimator $\widehat{\theta}$, if the random sample $X_1, X_2, \ldots, X_n$ is taken from each of the distributions having the following p.d.f.'s:

(a) $f(x; \theta) = (1/\theta^2)\, x\, e^{-x/\theta}$,
$$0 < x < \infty, \qquad 0 < \theta < \infty.$$

(b) $f(x; \theta) = (1/2\theta^3)\, x^2\, e^{-x/\theta}$,
$$0 < x < \infty, \qquad 0 < \theta < \infty.$$

(c) $f(x; \theta) = (1/\theta)\, x^{(1-\theta)/\theta}$,
$$0 < x < 1, \qquad 0 < \theta < \infty.$$

HISTORICAL COMMENTS   This chapter on theory briefly explains the best tests of statistical hypotheses, the importance of the use of sufficient statistics in estimation and testing, and certain limiting distributions, including the proof of the central limit theorem. We mention again some of the statistical giants who contributed to these theories: R. A. Fisher, J. Neyman, and E. Pearson. Of course, we cannot forget those early pioneers associated with the central limit theorem: de Moivre, Laplace, and Gauss. Many fine statisticians have worked on generalizations and extensions of their results since the 1930s.

Quality Improvement through Statistical Methods

11.1 TIME SEQUENCES

Thus far, we have treated $x_1, x_2, \ldots, x_n$ as if they were observations from some random experiment. That is, each one of these observations has the same distribution with mean μ and variance σ^2, and we therefore might use $\bar{x}$ and s^2 as respective estimates of μ and σ^2. In practice, however, we often find that $x_1, x_2, \ldots, x_n$ are observed in *time sequence*, and that there is frequently a change in the distribution as we continue to observe the x values. Consequently, we do not in fact want to use $\bar{x}$ and s^2 as estimates of those characteristics. To see what is actually going on in these cases, we should plot the data in the order collected. In a **time sequence**, observations are recorded in the order in which they were collected as ordered pairs, where the usual x-coordinate denotes the time t (in order of collection, which could be days, weeks, years, etc.) and the usual y-coordinate records the observation x. When measurements are taken sequentially and plotted in this time sequence, it turns out that trends, cycles, or major changes in processes (like government interventions) are often observed. As an illustration, consider the three time sequences in Figures 11.1-1(a), 11.1-1(b), and 11.1-1(c).

Figure 11.1-1(a) might be a record of a company's sales for the last 12 years. Despite occasional drops in sales from one year to the next, there has been an upward trend. Suppose, however, that only the last two points had been recorded for the stockholders. Since sales went down during that time, the stockholders might believe that there are some difficulties and that changes should be made. If they look at the entire sequence, however, they clearly see the long-term improvement, and only variability about that upward trend accounts for the last drop. If, by contrast, several successive drops were to appear, then some action might be needed.

The points in Figure 11.1-1(b) might also represent sales, but now plotted by quarters. Clearly, in this business, there is a cyclic effect, with increases from the first to second quarters and from the second to the third, but decreases from the third to

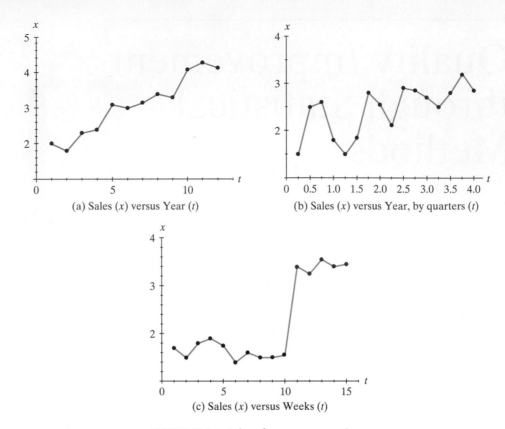

(a) Sales (x) versus Year (t)

(b) Sales (x) versus Year, by quarters (t)

(c) Sales (x) versus Weeks (t)

FIGURE 11.1-1: Sales of company over time

the fourth and the fourth to the first quarter of the next year. Again we note that there seems to be an overall upward trend.

Figure 11.1-1(c) could be the plot of some measure of success (possibly sales) of the company. Clearly, something happened after the tenth week. Perhaps management did not like the past level of performance and instituted a new program (possibly one based on statistical quality control). Alternatively, there might have been some outside intervention (like war) that increased the sales of this particular product.

We have listed three possible scenarios associated with the three time sequences shown. There are others, but in any case, we clearly see that plotting the points in a time sequence often helps explore the variability of the outcomes associated with a particular system. That is, time is frequently an important factor in dealing with variability, possibly when purchasing stock. The following example is a simple one based on one student's experience.

EXAMPLE 11.1-1   A student monitored some stock that she purchased in March for $5 per share. She recorded the value of a share every two weeks, yielding the following paired data, (t, x), with t equal to the time in weeks after the purchase and x the value of a share at that time:

$$(0, 5.00) \quad (2, 5.00) \quad (4, 5.25) \quad (6, 5.25) \quad (8, 5.50)$$
$$(10, 5.50) \quad (12, 5.25) \quad (14, 5.75) \quad (16, 6.25) \quad (18, 6.50)$$
$$(20, 6.25) \quad (22, 7.25) \quad (24, 7.50)$$

A time sequence of these data is shown in Figure 11.1-2.   ∎

FIGURE 11.1-2: Value of a share of stock plotted biweekly

Sometimes, however, time is not an important factor in the response variable, and yet time-sequence plots still help us understand the concept of variation. Suppose that a process or an individual performs at some average level, sometimes doing better and sometimes doing worse than that average. For example, suppose a certain worker's output is plotted by the month, as in Figure 11.1-3.

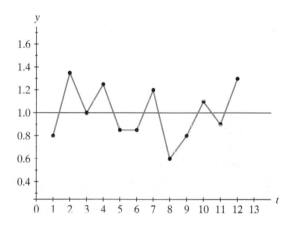

FIGURE 11.1-3: Average output of worker (y) versus month (t)

Let us say that the manager of the company decides to do the following: On good months (like the second month), the worker is given a "pat on the back" (maybe a bonus or even being named "worker of the month"). What happens? Usually the performance goes down the next month. Recall that the worker is averaging below the performance of her high month; that is, she was performing above average when she was made worker of the month. So there is a tendency to do worse the month following a very good performance. (Statisticians sometimes call this "regression toward the mean.")

This phenomenon explains what is often called the "sophomore jinx" in sports. A player might have an unusually good first year, actually one far above his average. So he tends to do worse the next year. Fans will say it's the sophomore jinx; he probably got overconfident and didn't work as hard. Although there might be some truth to this, it could also be that he worked just as hard as before, but still went down because his first year was over his average. So, often the "rookie of the year" will do worse the second year because not many players win that award with an average year. That is, he was performing far above his average when he won the award.

As regards the worker described by Figure 11.1-3, on bad months (such as the eighth month) the manager could call in the worker and have a serious talk with her or, in some way, penalize her. What happens? Usually, she does better the next month (again regression toward the mean). So the boss gets the idea that it is better to get tough because the workers seem to do better after these little talks. Of course, like the "pat on the back" technique, this strategy does not really work, because the workers keep varying about their average. They usually do better after a very poor performance and worse after an excellent performance.

If the manager does not like the average level of performance, then he or she must do something to change the system. Most of the time, individual workers can do very little to change the system; major decisions regarding improvements must be made by the manager or even in the boardroom. After these decisions, the workers must be given "road maps" directing them to use the improvements. Most workers (as well as students) want to do better and would like to take pride in their performance. Managers (as well as teachers) should provide workers (students) the opportunity to improve their self-esteem.

The next example could help some of you analyze performances in events that are repeated over time.

EXAMPLE 11.1-2 Consider the following 20 quiz scores of a student in a two-semester calculus course (each quiz was graded on a 20-point scale, and the quizzes are given in the order in which they were taken):

$$18 \quad 20 \quad 16 \quad 12 \quad 15 \quad 15 \quad 10 \quad 13 \quad 14 \quad 9$$
$$12 \quad 10 \quad 8 \quad 11 \quad 12 \quad 11 \quad 10 \quad 11 \quad 8 \quad 6$$

We could calculate the average of all 20 quiz scores, which is 12.05. We could calculate the average of the first 10 and the second 10. A graphic display of what has happened to this student over the course of the two semesters is shown in Figure 11.1-4. Clearly, the student was not making a great effort throughout the year in this calculus class. If interested, students can correct trends like this, but they must make changes in their behavior to do so. By continuing the past pattern, nothing will change and the downward trend will persist. What improvements could be made to change this trend? Each student must answer this in his or her own way. ∎

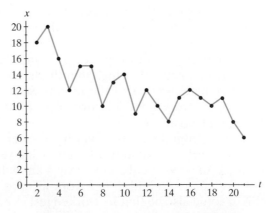

FIGURE 11.1-4: Calculus scores over time

The next example illustrates how, in a production process, tracking something over time can be beneficial.

EXAMPLE 11.1-3 Containers of a prewash concentrate are selected regularly from the production line, and the concentrate is weighed. The target weight is 330.5–336.5 grams. We now construct a time-sequence plot (see Figure 11.1-5) for the following 60 weights that were observed in order from left to right, line by line:

337.9	338.1	337.7	338.3	337.3	338.1	337.3	338.0	337.9	338.9
337.6	338.2	338.1	338.6	338.0	338.4	334.4	334.3	334.5	333.9
335.1	334.0	335.6	334.1	335.0	334.8	335.2	335.2	335.2	334.0
334.5	334.6	335.4	335.9	335.6	335.8	334.5	335.4	335.0	335.5
335.1	335.0	334.1	334.6	334.4	334.7	333.8	335.0	334.0	334.3
334.0	333.7	334.7	333.5	334.8	333.9	335.7	335.1	335.1	334.6

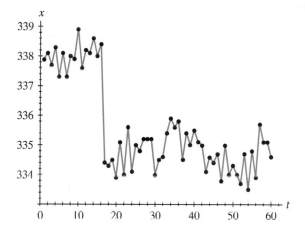

FIGURE 11.1-5: Prewash concentrate fill weights

It is pretty clear that during the beginning of production the fill weights were slightly above the target weight. This is good news for the customer, as long as the containers do not overflow. However, if the trend had continued, the company would have lost money. After adjustments were made, the process seemed to come under control, with the weights reflecting the normal variability for such a process. ∎

EXERCISES

11.1-1. The following data give the percentage of solids in a product:

69.8	71.3	65.6	66.3	70.1	71.9	69.6	71.9	71.1	71.7
71.9	69.8	66.8	68.3	64.4	64.2	65.1	63.7	66.2	61.9
66.1	62.9	66.9	67.3	63.3	63.4	67.2	67.4	65.5	66.2
67.5	67.3	66.9	66.5	65.5	63.9	64.6	62.3	66.2	67.2
66.0	69.8	69.7	71.0	69.8	66.0	70.3	65.5	67.0	66.8
67.6	68.6	66.5	66.2	70.4	68.1	64.3	65.2	68.0	65.1

The specifications call for the percentage to be between 60.0% and 72.0%. The data were observed in order from left to right, line by line.

(a) Construct a time sequence of these percentages.

(b) Interpret your output.

11.1-2. Twenty-four "3-pound" bags of apples were filled and weighed automatically by scale number 5. The weights are as follows, in the time order shown:

3.26 3.62 3.39 3.12 3.53 3.30 3.10 3.26
3.19 3.22 3.14 3.39 3.31 3.21 3.49 3.41
3.02 3.17 3.20 3.12 3.42 3.36 3.21 3.26

(a) Construct and interpret a time sequence of these weights.

(b) Compare scale number 5 with scale number 6 in the next exercise.

11.1-3. Twenty-four "3-pound" bags of apples were filled and weighed automatically by scale number 6. The weights are as follows, in the time order shown:

3.22 2.97 3.00 3.08 3.04 3.09 3.04 3.21
3.00 3.05 3.11 3.00 3.10 3.13 3.08 3.22
3.12 3.04 3.03 3.02 3.10 3.09 3.18 3.00

(a) Construct and interpret a time sequence of these weights.

(b) Compare scale number 6 with scale number 5 in the last exercise.

11.1-4. The year, x, and the birthrate, y (number of births per 1000 population) for 1960–2002 are as follows:

(1960, 23.7) (1961, 23.3) (1962, 22.4) (1963, 21.7)
(1964, 21.1) (1965, 19.4) (1966, 18.4) (1967, 17.8)
(1968, 17.6) (1969, 17.9) (1970, 18.4) (1971, 17.2)
(1972, 15.6) (1973, 14.8) (1974, 14.8) (1975, 14.6)
(1976, 14.6) (1977, 15.1) (1978, 15.0) (1979, 15.6)
(1980, 15.9) (1981, 15.8) (1982, 15.9) (1983, 15.5)
(1984, 15.5) (1985, 15.8) (1986, 15.6) (1987, 15.7)
(1988, 15.9) (1989, 16.3) (1990, 16.7) (1991, 16.2)
(1992, 15.8) (1993, 15.4) (1994, 15.0) (1995, 14.6)
(1996, 14.4) (1997, 14.2) (1998, 14.3) (1999, 14.2)
(2000, 14.4) (2001, 14.1) (2002, 13.9)

(a) Construct a time sequence of these birthrates.

(b) Interpret your output.

11.1-5. Dish drop viscosity has a specification of 120–230 cps (centipoise). One hundred observations were selected and measured, yielding the following viscosities (read across, line by line, in the time order shown):

158 147 158 159 169 151 166 151 143 169
153 174 151 164 185 168 140 180 176 154
160 187 145 164 158 169 153 149 144 157
156 183 157 140 162 158 160 180 154 160
164 168 154 158 164 159 153 170 158 170
150 161 169 166 154 157 138 155 134 165
161 172 156 145 153 143 152 152 156 163
179 157 135 172 143 154 165 145 152 145
171 189 144 154 147 187 147 159 167 151
153 168 148 188 152 165 155 140 157 176

(a) Construct a time sequence of the viscosities.

(b) If there seems to be no significant time factor, construct a histogram of the viscosities.

(c) Use the two graphs to interpret these measurements.

11.1-6. A manufacturer of car windows has studs in the windows for attaching them to the car. A standard "stud pullout test" measures the force required to pull each of the five studs out of a window. Because of the shape of the window, the two studs on one side are attached in one way and the three studs on the other side in a different way. The target mean is 123. Seventy observations, going around each of 14 windows in a clockwise direction, were as follows:

140 159 138 102   84 126 147 126 103   92
149 155 135 120   94 149 143 109 101   86
144 154 120 105   97 149 151 140 103   99
157 140 120   96   87 146 137 120   93   89
149 154 139 100   84 148 142 130   98   81
112 135 109   84   87 112 135 109   84   87
118 135   90   77 125 126 131   78   75 126

(a) Construct a time sequence of these forces.

(b) Construct a histogram of the forces.

(c) Give an interpretation.

11.1-7. The amount of fluoride in a certain brand of toothpaste has a specification of 0.85–1.10 mg/g. One hundred tubes of toothpaste were selected randomly from the production line, and the amounts of fluoride were measured, yielding the following data:

0.98	0.92	0.89	0.90	0.94	0.99	0.86	0.85	1.06	1.01
1.03	0.85	0.95	0.90	1.03	0.87	1.02	0.88	0.92	0.88
0.88	0.90	0.98	0.96	0.98	0.93	0.98	0.92	1.00	0.95
0.88	0.90	1.01	0.98	0.85	0.91	0.95	1.01	0.88	0.89
0.99	0.95	0.90	0.88	0.92	0.89	0.90	0.95	0.93	0.96
0.93	0.91	0.92	0.86	0.87	0.91	0.89	0.93	0.93	0.95
0.92	0.88	0.87	0.98	0.98	0.91	0.93	1.00	0.90	0.93
0.89	0.97	0.98	0.91	0.88	0.89	1.00	0.93	0.92	0.97
0.97	0.91	0.85	0.92	0.87	0.86	0.91	0.92	0.95	0.97
0.88	1.05	0.91	0.89	0.92	0.94	0.90	1.00	0.90	0.93

(a) Construct a time sequence of these measurements.

(b) Construct a histogram of the measurements.

(c) Construct a box-plot of the measurements.

11.1-8. Use the data in Exercise 3.1-11 for this exercise.

(a) Construct and interpret a time sequence of the durations of eruptions of Old Faithful geyser. Note the starting times in making your interpretation.

(b) Construct and interpret a time sequence of the times between eruptions of Old Faithful geyser. Note the starting times in making your interpretation.

11.1-9. Ledolter and Hogg (see references) report the following $n = 52$ weekly sales of thermostat replacement parts (read across):

206	245	185	169	162	177	207	216	193	230	212	192	162
189	244	209	207	211	210	173	194	234	156	206	188	162
172	210	205	244	218	182	206	211	273	248	262	258	233
255	303	282	291	280	255	312	296	307	281	308	280	345

(a) Construct a time sequence for these data and discuss your findings.

(b) What do you think will happen in the next month?

(c) What do you think will happen in the next two months?

11.2 STATISTICAL QUALITY CONTROL

Statistical methods can be used in many scientific fields, such as medical research, engineering, chemistry, and psychology. Often, it is necessary to compare two ways of doing something—say, the old way and a possible new way. We collect data on each way, quite possibly in a laboratory situation, and try to decide whether the new way is actually better than the old. Needless to say, it would be terrible to change to the new way at great expense, only to find out that it is really not any better than the old. That is, suppose the lab results indicate, by some statistical method, that the new is seemingly better than the old. Can we actually extrapolate those outcomes in the lab to the situations in the real world? Clearly, statisticians cannot make these decisions, but they should be made by some professional who knows both statistics and the specialty in question very well. The statistical analysis might provide helpful guidelines, but we still need the expert to make the final decision.

However, even before investigating possible changes in any process, it is extremely important to determine exactly what the process in question is doing at the present time. Often, people in charge of an organization do not understand the capabilities of many of its processes. Simply measuring what is going on frequently leads to improvements. In many cases measurement is easy, such as determining the diameter of a bolt, but sometimes it is extremely difficult, as in evaluating good teaching or many other service activities. But if at all possible, we encourage those involved to begin to "listen" to their processes; that is, they should measure what is going on in their organization. These measurements alone often are the beginning of desirable improvements. While most of our remarks in this chapter concern measurements made in manufacturing, service industries frequently find them just as useful.

At one time, some manufacturing plants would make parts to be used in the construction of some piece of equipment. Say a particular line in the plant, making a certain part, might produce several hundreds of them each day. These items would then be sent on to an inspection cage, where they would be checked for goodness, often several days or even weeks later. Occasionally, the inspectors would discover many defectives among the items made, say, two weeks ago. There was little that could be done at that point except scrap or rework the defective parts, both expensive outcomes.

In the 1920s, W. A. Shewhart, who was working for AT&T Bell Labs, recognized that this was an undesirable situation and suggested that, with some suitable frequency, a sample of these parts should be taken as they were being made. If the sample indicated that the items were satisfactory, the manufacturing process would continue. But if the sampled parts were not satisfactory, corrections should be made then so that things became satisfactory. This idea led to what are commonly called *Shewhart control charts*—the basis of what was called *statistical quality control* in those early days; today it is often referred to as *statistical process control.*

Shewhart control charts consist of calculated values of a statistic, say, $\bar{x}$, plotted in sequence. That is, in making products, every so often (each hour, each day, or each week, depending upon how many items are being produced) a sample of size n of them is taken, and they are measured, resulting in the observations $x_1, x_2, \ldots, x_n$. The average $\bar{x}$ and the standard deviation s are computed. This is done k times, and the k values of $\bar{x}$ and s are averaged, resulting in $\bar{\bar{x}}$ and $\bar{s}$, respectively; usually, k is equal to some number between 10 and 30.

The central limit theorem states that if the true mean μ and standard deviation σ of the process were known, then almost all of the $\bar{x}$ values would plot between $\mu - 3\sigma/\sqrt{n}$ and $\mu + 3\sigma/\sqrt{n}$, unless the system has actually changed. However, suppose we know neither μ nor σ, and thus μ is estimated by $\bar{\bar{x}}$ and $3\sigma/\sqrt{n}$ by $A_3\bar{s}$, where $\bar{\bar{x}}$ and $\bar{s}$ are the respective means of the k observations of $\bar{x}$ and s, and where A_3 is a factor depending upon n that can be found in books on statistical quality control. A few values of A_3 (and some other constants that will be used later) are given in Table 11.2-1 for typical values of n.

The estimates of $\mu \pm 3\sigma/\sqrt{n}$ are called the *upper control limit* (UCL), $\bar{\bar{x}} + A_3\bar{s}$, and the *lower control limit* (LCL), $\bar{\bar{x}} - A_3\bar{s}$, and $\bar{\bar{x}}$ provides the estimate of the centerline. A typical plot is given in Figure 11.2-1. Here, in the 13th sampling period, $\bar{x}$ is outside the control limits, indicating that the process has changed and that some investigation and action are needed to correct this change, which seems like an upward shift in the process.

Note that there is a control chart for the s values, too. From sampling distribution theory, values of B_3 and B_4 have been determined and are given in Table 11.2-1, so we know that almost all the s values should be between $B_3\bar{s}$ and $B_4\bar{s}$ if there is no

n	A_3	B_3	B_4	A_2	D_3	D_4
TABLE 11.2-1: Some Constants Used with Control Charts						
4	1.63	0	2.27	0.73	0	2.28
5	1.43	0	2.09	0.58	0	2.11
6	1.29	0.03	1.97	0.48	0	2.00
8	1.10	0.185	1.815	0.37	0.14	1.86
10	0.98	0.28	1.72	0.31	0.22	1.78
20	0.68	0.51	1.49	0.18	0.41	1.59

FIGURE 11.2-1: Typical control chart

change in the underlying distribution. So again, if an individual s value is outside these control limits, some action should be taken, as it seems as if there has been a change in the variation of the underlying distribution.

Often, when these charts are first constructed after $k = 10$ to 30 sampling periods, many points fall outside the control limits. A team consisting of workers, the manager of the process, the supervisor, an engineer, and even a statistician should try to find the reasons that this has occurred, and the situation should be corrected. After this is done and the points plot within the control limits, the process is "in statistical control." However, being in statistical control is not a guarantee of satisfaction with the products. Since $A_3\bar{s}$ is an estimate of $3\sigma/\sqrt{n}$, it follows that $\sqrt{n}A_3\bar{s}$ is an estimate of 3σ, and with an underlying distribution close to a normal one, almost all items would be between $\bar{\bar{x}} \pm \sqrt{n}A_3\bar{s}$. If these limits are too wide, then corrections must be made again.

If the variation is under control (i.e., if $\bar{x}$ and s are within their control limits), we say that the variations seen in $\bar{x}$ and s are due to common causes. If products made under such a system with these existing common causes are satisfactory, then production continues. If either $\bar{x}$ or s, however, is outside the control limits, that is an indication that some special causes are at work, and they must be corrected. That is, a team should investigate the problem and some action should be taken.

EXAMPLE 11.2-1   A company produces a storage console. Twice a day, nine critical characteristics are tested on five consoles that are selected randomly from the production line. One of these characteristics is the time it takes the lower storage component door to open completely. Table 11.2-2 lists the opening times in seconds for the consoles that were

TABLE 11.2-2: Console Opening Times							
Group	x_1 x_2 x_3 x_4 x_5					$\bar{x}$	s	R

Group	x_1	x_2	x_3	x_4	x_5	$\bar{x}$	s	R
1	1.2	1.8	1.7	1.3	1.4	1.480	0.259	0.60
2	1.5	1.2	1.0	1.0	1.8	1.300	0.346	0.80
3	0.9	1.6	1.0	1.0	1.0	1.100	0.283	0.70
4	1.3	0.9	0.9	1.2	1.0	1.060	0.182	0.40
5	0.7	0.8	0.9	0.6	0.8	0.760	0.114	0.30
6	1.2	0.9	1.1	1.0	1.0	1.040	0.104	0.30
7	1.1	0.9	1.1	1.0	1.4	1.100	0.187	0.50
8	1.4	0.9	0.9	1.1	1.0	1.060	0.207	0.50
9	1.3	1.4	1.1	1.5	1.6	1.380	0.192	0.50
10	1.6	1.5	1.4	1.3	1.5	1.460	0.114	0.30

$$\bar{\bar{x}} = 1.174 \quad \bar{s} = 0.200 \quad \bar{R} = 0.49$$

tested during one week. Also included in the table are the sample means, the sample standard deviations, and the ranges.

The upper control limit (UCL) and the lower control limit (LCL) for $\bar{x}$ are found using A_3 in Table 11.2-1 with $n = 5$ as follows:

$$\text{UCL} = \bar{\bar{x}} + A_3\bar{s} = 1.174 + 1.43(0.20) = 1.460$$

and

$$\text{LCL} = \bar{\bar{x}} - A_3\bar{s} = 1.174 - 1.43(0.20) = 0.888.$$

These control limits and the sample means are plotted on the $\bar{x}$ chart in Figure 11.2-2. There should be some concern about the fifth sampling period; thus, there should be an investigation to determine why that particular $\bar{x}$ is below the LCL.

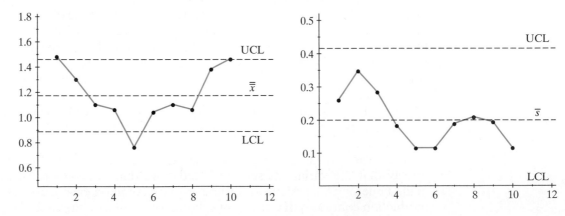

FIGURE 11.2-2: The $\bar{x}$ chart and s chart

The UCL and LCL for s are found using B_3 and B_4 in Table 11.2-1 with $n = 5$ as follows:

$$\text{UCL} = B_4\bar{s} = 2.09(0.200) = 0.418$$

and

$$\text{LCL} = B_3\bar{s} = 0(0.200) = 0.$$

These control limits and the sample standard deviations are plotted on the s chart in Figure 11.2-2.

Almost all of the observations should lie between $\bar{\bar{x}} \pm \sqrt{n}\, A_3 \bar{s}$, namely,

$$1.174 + \sqrt{5}\,(1.43)(0.20) = 1.814$$

and

$$1.174 - \sqrt{5}\,(1.43)(0.20) = 0.535.$$

This situation is illustrated in Figure 11.2-3, in which all 50 observations do fall within these control limits. ◼

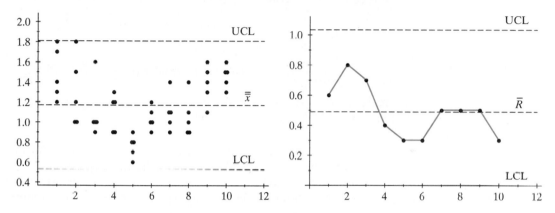

FIGURE 11.2-3: Plot of 50 console opening times and R chart

In most books on statistical quality control, there is an alternative way of constructing the limits on an $\bar{x}$ chart. For each sample, we compute the range, R, which is the absolute value of the difference of the extremes of the sample. This computation is much easier than that for calculating s. After k samples are taken, we compute the average of these R values, obtaining $\bar{R}$ as well as $\bar{\bar{x}}$. The statistic $A_2 \bar{R}$ serves as an estimate of $3\sigma/\sqrt{n}$, where A_2 is found in Table 11.2-1. Thus, the estimates of $\mu \pm 3\sigma/\sqrt{n}$, namely, $\bar{\bar{x}} \pm A_2 \bar{R}$, can be used as the UCL and LCL of an $\bar{x}$ chart.

In addition, $\sqrt{n}A_2 \bar{R}$ is an estimate of 3σ; so, with an underlying distribution that is close to a normal one, we find that almost all observations are within the limits $\bar{\bar{x}} \pm \sqrt{n}A_2 \bar{R}$.

Moreover, an R chart can be constructed with centerline $\bar{R}$ and control limits equal to $D_3\bar{R}$ and $D_4\bar{R}$, where D_3 and D_4 are given in Table 11.2-1 and were determined so that almost all R values should be between the control limits if there is no change in the underlying distribution. Thus, a value of R falling outside those limits would indicate a change in the spread of the underlying distribution, and some corrective action should be considered.

The use of R, rather than s, is illustrated in the next example.

EXAMPLE 11.2-2   Using the data in Example 11.2-1, we compute UCL and LCL for an $\bar{x}$ chart. We use $\bar{\bar{x}} \pm A_2 \bar{R}$ as follows:

$$\text{UCL} = \bar{\bar{x}} + A_2 \bar{R} = 1.174 + 0.58(0.49) = 1.458$$

and

$$\text{LCL} = \bar{\bar{x}} - A_2 \bar{R} = 1.174 - 0.58(0.49) = 0.890.$$

Note that these values are very close to the limits that we found for the $\bar{x}$ chart in Figure 11.2-2 using $\bar{\bar{x}} \pm A_3 \bar{s}$. In addition, almost all of the observations should lie within the limits $\bar{\bar{x}} + \sqrt{n}A_2 \bar{R}$, which are

$$\text{UCL} = 1.174 + \sqrt{5}(0.58)(0.49) = 1.809$$

and

$$\text{LCL} = 1.174 - \sqrt{5}(0.58)(0.49) = 0.539.$$

Note that these are almost the same as the limits found in Example 11.2-1 and plotted in Figure 11.2-3.

An R chart can be constructed with centerline $\bar{R} = 0.49$ and control limits given by

$$\text{UCL} = D_4 \bar{R} = 2.11(0.49) = 1.034$$

and

$$\text{LCL} = D_3 \bar{R} = 0(0.49) = 0.$$

Figure 11.2-3 illustrates this control chart for the range, and we see that its pattern is similar to that of the s chart in Figure 11.2-2. ■

There are two other Shewhart control charts: the p and c charts. The central limit theorem, which provided a justification for the three-sigma limits in the $\bar{x}$ chart, also justifies the control limits in the p chart. Suppose the number of defectives among n items that are selected randomly—say, D—has a binomial distribution $b(n,p)$. Then the limits $p \pm 3\sqrt{p(1-p)/n}$ should include almost all of the D/n values. However, p must be approximated by observing k values of D—say, $D_1, D_2, \ldots, D_k$—and computing what is called $\bar{p}$ in the statistical quality control literature, namely,

$$\bar{p} = \frac{D_1 + D_2 + \cdots + D_k}{kn}.$$

Thus, the LCL and UCL for the fraction defective, D/n, are respectively given by

$$\text{LCL} = \bar{p} - 3\sqrt{\bar{p}(1-\bar{p})/n}$$

and

$$\text{UCL} = \bar{p} + 3\sqrt{\bar{p}(1-\bar{p})/n}.$$

If the process is in control, almost all D/n values are between LCL and UCL. Still, this may not be satisfactory and improvements might be needed to decrease $\bar{p}$. If it is satisfactory, however, let the process continue under these common causes of variation until a point, D/n, outside the control limits would indicate that some special cause has changed the variation. (Incidentally, if D/n is below the LCL, this might very well indicate that some type of change for the better has been made, and we want to find out why. In general, outlying statistics can often suggest that good (as well as bad) breakthroughs have been made.)

The next example gives the results of a simple experiment that you can easily duplicate.

EXAMPLE 11.2-3   Let D_i equal the number of yellow candies in a 1.69-ounce bag. Because the number of pieces of candy varies slightly from bag to bag, we shall use an average value for n when we construct the control limits. Table 11.2-3 lists, for 20 packages, the number

TABLE 11.2-3: Data on Yellow Candies							
Package	n_i	D_i	D_i/n_i	Package	n_i	D_i	D_i/n_i
1	56	8	0.14	11	57	10	0.18
2	55	13	0.24	12	59	8	0.14
3	58	12	0.21	13	54	10	0.19
4	56	13	0.23	14	55	11	0.20
5	57	14	0.25	15	56	12	0.21
6	54	5	0.09	16	57	11	0.19
7	56	14	0.25	17	54	6	0.11
8	57	15	0.26	18	58	7	0.12
9	54	11	0.20	19	58	12	0.21
10	55	13	0.24	20	58	14	0.24

of pieces of candy in the package, the number of yellow ones, and the proportion of yellow ones.

For these data,

$$\sum_{i=1}^{20} n_i = 1124 \qquad \text{and} \qquad \sum_{i=1}^{20} D_i = 219.$$

It follows that

$$\bar{p} = \frac{219}{1124} = 0.195 \qquad \text{and} \qquad \bar{n} = \frac{1124}{20} \approx 56.$$

Thus, the LCL and UCL are respectively given by

$$\text{LCL} = \bar{p} - 3\sqrt{\bar{p}(1 - \bar{p})/56} = 0.195 - 3\sqrt{0.195(0.805)/56} = 0.036$$

and

$$\text{UCL} = \bar{p} + 3\sqrt{\bar{p}(1 - \bar{p})/56} = 0.195 + 3\sqrt{0.195(0.805)/56} = 0.354.$$

The control chart for p is depicted in Figure 11.2-4 on the next page. (For your information the "true" value for p is 0.20.) ∎

Consider the following explanation of the c chart: Suppose the number of flaws, say, C, on some product has a Poisson distribution with parameter λ. If λ is sufficiently large, as in Example 5.7-5, we consider approximating the discrete Poisson distribution with the continuous $N(\lambda, \lambda)$ distribution. Thus, the interval from $\lambda - 3\sqrt{\lambda}$ to $\lambda + 3\sqrt{\lambda}$ contains virtually all of the C values. Since λ is unknown, however, it must be approximated by $\bar{c}$, the average of the k values, $c_1, c_2, \ldots, c_k$. Hence, the two control limits for C are computed as

$$\text{LCL} = \bar{c} - 3\sqrt{\bar{c}} \qquad \text{and} \qquad \text{UCL} = \bar{c} + 3\sqrt{\bar{c}}.$$

The remarks made about the $\bar{x}$ and $\bar{p}$ charts apply to the c chart as well, but we must remember that each c value is the number of flaws on one manufactured item, not an average $\bar{x}$ or a fraction defective D/n.

FIGURE 11.2-4: The *p* chart

STATISTICAL COMMENTS   As with control charts, in observing time-sequence plots, do not read too much into a short sequence of points. For example, after a pep talk by the coach of a women's golf team, three successive decreasing scores (remember, in golf, the lower the score, the better it is) might indicate that the coach had some influence. But even if the process has not changed, the probability of three decreasing scores (ruling out ties) is 1/6, since there are 3! = 6 equally likely ways of arranging three points. Often, businesspeople are worse offenders because they might think that two increasing sales points indicate improvement. We know that with no change in the system, the probability of that is 1/2. Now, maybe 4 points that successively increase after an intervention would be cause for claiming improvement, because the probability of such an event is 1/4! = 1/24 if no change has actually been made. This probability is small enough to justify a celebration.

EXERCISES

11.2-1. It is important to control the viscosity of liquid dishwasher soap so that it flows out of the container but does not run out too rapidly. Thus, samples are taken randomly throughout the day and the viscosity is measured. Use the following 20 sets of five observations for this exercise:

Observations					$\bar{x}$	s	R
158	147	158	159	169	158.20	7.79	22
151	166	151	143	169	156.00	11.05	26
153	174	151	164	185	165.40	14.33	34
168	140	180	176	154	163.60	16.52	40
160	187	145	164	158	162.80	15.29	42
169	153	149	144	157	154.40	9.48	25
156	183	157	140	162	159.60	15.47	43
158	160	180	154	160	162.40	10.14	26
164	168	154	158	164	161.60	5.55	14
159	153	170	158	170	162.00	7.65	17
150	161	169	166	154	160.00	7.97	19
157	138	155	134	165	149.80	13.22	31
161	172	156	145	153	157.40	10.01	27

Observations					$\bar{x}$	s	R
143	152	152	156	163	153.20	7.26	20
179	157	135	172	143	157.20	18.63	44
154	165	145	152	145	152.20	8.23	20
171	189	144	154	147	161.00	18.83	45
187	147	159	167	151	162.20	15.85	40
153	168	148	188	152	161.80	16.50	40
165	155	140	157	176	158.60	13.28	36

(a) Calculate the values of $\bar{\bar{x}}$, $\bar{s}$, and $\bar{R}$.
(b) Use the values of A_3 and $\bar{s}$ to construct an $\bar{x}$ chart.
(c) Construct an s chart.
(d) Use the values of A_2 and $\bar{R}$ to construct an $\bar{x}$ chart.
(e) Construct an R chart.
(f) Do the charts indicate that viscosity is in statistical control?

11.2-2. It is necessary to control the percentage of solids in a product, so samples are taken randomly throughout the day and the percentage of solids is measured. Use the following 20 sets of five observations for this exercise:

Observations					$\bar{x}$	s	R
69.8	71.3	65.6	66.3	70.1	68.62	2.51	5.7
71.9	69.6	71.9	71.1	71.7	71.24	0.97	2.3
71.9	69.8	66.8	68.3	64.4	68.24	2.86	7.5
64.2	65.1	63.7	66.2	61.9	64.22	1.61	4.3
66.1	62.9	66.9	67.3	63.3	65.30	2.06	4.4
63.4	67.2	67.4	65.5	66.2	65.94	1.61	4.0
67.5	67.3	66.9	66.5	65.5	66.74	0.79	2.0
63.9	64.6	62.3	66.2	67.2	64.84	1.92	4.9
66.0	69.8	69.7	71.0	69.8	69.26	1.90	5.0
66.0	70.3	65.5	67.0	66.8	67.12	1.88	4.8
67.6	68.6	66.5	66.2	70.4	67.86	1.71	4.2
68.1	64.3	65.2	68.0	65.1	66.14	1.78	3.8
64.5	66.6	65.2	69.3	62.0	65.52	2.69	7.3
67.1	68.3	64.0	64.9	68.2	66.50	1.96	4.3
67.1	63.8	71.4	67.5	63.7	66.70	3.17	7.7
60.7	63.5	62.9	67.0	69.6	64.74	3.53	8.9
71.0	68.6	68.1	67.4	71.7	69.36	1.88	4.3
69.5	61.5	63.7	66.3	68.6	65.92	3.34	8.0
66.7	75.2	79.0	75.3	79.2	75.08	5.07	12.5
77.3	67.2	69.3	67.9	65.6	69.46	4.58	11.7

(a) Calculate the values of $\bar{\bar{x}}$, $\bar{s}$, and $\bar{R}$.
(b) Use the values of A_3 and $\bar{s}$ to construct an $\bar{x}$ chart.
(c) Construct an s chart.
(d) Use the values of A_2 and $\bar{R}$ to construct an $\bar{x}$ chart.
(e) Construct an R chart.
(f) Do the charts indicate that the percentage of solids in this product is in statistical control?

11.2-3. It is important to control the net weight of a packaged item; thus, items are selected randomly throughout the day from the production line and their weights are recorded. Use the following 20 sets of five weights (in grams) for this exercise (note that a weight recorded here is the actual weight minus 330):

Observations					$\bar{x}$	s	R
7.97	8.10	7.73	8.26	7.30	7.872	0.3740	0.96
8.11	7.26	7.99	7.88	8.88	8.024	0.5800	1.62
7.60	8.23	8.07	8.51	8.05	8.092	0.3309	0.91
8.44	4.35	4.33	4.48	3.89	5.098	1.8815	4.55
5.11	4.05	5.62	4.13	5.01	4.784	0.6750	1.57
4.79	5.25	5.19	5.23	3.97	4.886	0.5458	1.28
4.47	4.58	5.35	5.86	5.61	5.174	0.6205	1.39
5.82	4.51	5.38	5.01	5.54	5.252	0.5077	1.31
5.06	4.98	4.13	4.58	4.35	4.620	0.3993	0.93
4.74	3.77	5.05	4.03	4.29	4.376	0.5199	1.28
4.05	3.71	4.73	3.51	4.76	4.152	0.5748	1.25
3.94	5.72	5.07	5.09	4.61	4.886	0.6599	1.78
4.63	3.79	4.69	5.13	4.66	4.580	0.4867	1.34
4.30	4.07	4.39	4.63	4.47	4.372	0.2079	0.56
4.05	4.14	4.01	3.95	4.05	4.040	0.0693	0.19
4.20	4.50	5.32	4.42	5.24	4.736	0.5094	1.12
4.54	5.23	4.32	4.66	3.86	4.522	0.4999	1.37
5.02	4.10	5.08	4.94	5.18	4.864	0.4360	1.08
4.80	4.73	4.82	4.69	4.27	4.662	0.2253	0.55
4.55	4.76	4.45	4.85	4.02	4.526	0.3249	0.83

(a) Calculate the values of $\bar{\bar{x}}$, $\bar{s}$, and $\bar{R}$.
(b) Use the values of A_3 and $\bar{s}$ to construct an $\bar{x}$ chart.
(c) Construct an s chart.
(d) Use the values of A_2 and $\bar{R}$ to construct an $\bar{x}$ chart.
(e) Construct an R chart.
(f) Do the charts indicate that these fill weights are in statistical control?

11.2-4. The following data give fill weights (in grams) of "4-ounce bags" of candy as packed by a vertical form, fill, and seal machine:

118.8	116.6	116.2	113.8	116.9	118.1	116.2	115.6	113.5	119.2
115.2	117.0	113.6	115.1	113.5	117.0	113.6	115.5	113.5	117.0
119.0	117.5	119.2	118.1	121.2	117.2	113.5	119.7	118.4	116.6
119.0	113.6	118.8	118.1	117.7	116.0	117.1	119.1	116.8	118.7
115.9	116.8	118.7	115.9	116.8	117.2	115.5	120.0	117.6	117.8
117.0	116.2	119.9	118.9	118.2	118.8	118.9	116.5	116.8	116.6
119.3	116.2	116.4	114.5	114.8	121.1	118.3	113.6	116.7	118.0
118.3	117.3	115.2	114.6	114.2	117.2	115.8	113.6	121.6	120.8
117.2	117.0	115.4	118.2	119.4	117.5	113.5	119.1	116.3	118.1
119.7	118.5	119.6	119.2	116.7	117.3	115.6	115.0	117.5	114.1
118.4	118.3	117.4	117.2	116.3	116.5	116.7	117.8	118.8	118.0
115.4	113.5	118.2	116.5	117.4	113.6	113.5	119.0	116.7	117.0
117.5	117.9	117.2	118.7	117.6	118.2	120.5	116.4	114.9	116.8

116.7	116.5	117.8	116.4	116.9		118.4	116.8	120.8	118.2	119.3
118.9	118.0	120.1	114.9	119.5		119.0	115.5	115.8	115.8	115.4
120.8	118.3	115.9	117.7	114.8		116.6	118.1	114.0	113.4	113.4
116.0	121.8	119.3	117.7	118.3		118.4	116.8	118.0	117.8	116.1
119.1	115.9	118.8	117.9	118.5						

(a) Taking the data in consecutive sets of five (row by row; 118.8, 116.6, 116.2, 113.8, 116.9 is the first set), construct $\bar{x}$, R and $\bar{x}$, s charts.

(b) Describe what these charts show. Is the machine performing satisfactorily?

11.2-5. To give some indication of how the values in Table 11.2-1 are calculated, values of A_3 are found in this exercise. Let $X_1, X_2, \ldots, X_n$ be a random sample of size n from the normal distribution $N(\mu, \sigma^2)$. Let S^2 equal the sample variance of this random sample.

(a) Use the fact that $Y = (n - 1)S^2/\sigma^2$ has a distribution that is $\chi^2(n - 1)$ to show that $E[S^2] = \sigma^2$.

(b) Using the $\chi^2(n - 1)$ p.d.f., find the value of $E(\sqrt{Y})$.

(c) Show that

$$E\left[\frac{\sqrt{n - 1}\,\Gamma\left(\dfrac{n - 1}{2}\right)}{\sqrt{2}\,\Gamma\left(\dfrac{n}{2}\right)}\, S\right] = \sigma.$$

(d) Verify that

$$\frac{3}{\sqrt{n}}\left[\frac{\sqrt{n - 1}\,\Gamma\left(\dfrac{n - 1}{2}\right)}{\sqrt{2}\,\Gamma\left(\dfrac{n}{2}\right)}\right] = A_3,$$

found in Table 11.2-1 for $n = 5$ and $n = 6$. Thus, $A_3\bar{s}$ approximates $3\sigma/\sqrt{n}$.

11.2-6. A company has been producing bolts that are about $\bar{p} = 0.02$ defective, and this is satisfactory. To monitor the quality of the process, 100 bolts are selected at random each hour and the number of defective bolts counted. With $\bar{p} = 0.02$, compute the UCL and LCL of the $\bar{p}$ chart. Then suppose that, over the next 24 hours, the following numbers of defective bolts are observed:

4 1 1 0 5 2 1 3 4 3 1 0
0 4 1 1 6 2 0 0 2 8 7 5

Would any action have been required during this time?

11.2-7. In the past, $n = 50$ fuses are tested each hour and $\bar{p} = 0.03$ have been found defective. Calculate the UCL and LCL. After a production error, say the true p shifts to $p = 0.05$.

(a) What is the probability that the next observation exceeds the UCL?

(b) What is the probability that at least one of the next five observations exceeds the UCL? HINT: Assume independence and compute the probability that none of the next five observations exceeds the UCL.

11.2-8. In a woolen mill, 100-yard pieces are inspected. In the last 20 observations, the following numbers of flaws were found:

2 4 0 1 0 3 4 1 1 2 4 0 0 1 0 3 2 3 5 0

(a) Compute the control limits of the c chart and draw this control chart.

(b) Is the process in statistical control?

11.2-9. Suppose we find that the number of blemishes in 50-foot tin strips averages about $\bar{c} = 1.4$. Calculate the control limits. Say the process has gone out of control and this average has increased to 3.

(a) What is the probability that the next observation will exceed the UCL?

(b) What is the probability that at least 1 of the next 10 observations will exceed the UCL?

11.2-10. Snee (see references) has measured the thickness of the "ears" of paint cans. (The "ear" of a paint can is the tab that secures the lid of the can.) At periodic intervals, samples of five paint cans are taken from a hopper that collects the production from two machines, and the thickness of each ear is measured. The results (in inches × 1000) of 30 such samples are as follows:

Observations					$\bar{x}$	s	R
29	36	39	34	34	34.4	3.64692	10
29	29	28	32	31	29.8	1.64317	4
34	34	39	38	37	36.4	2.30217	5
35	37	33	38	41	36.8	3.03315	8
30	29	31	38	29	31.4	3.78153	9

Observations	$\bar{x}$	s	R
34 31 37 39 36	35.4	3.04959	8
30 35 33 40 36	34.8	3.70135	10
28 28 31 34 30	30.2	2.48998	6
32 36 38 38 35	35.8	2.48998	6
35 30 37 35 31	33.6	2.96648	7
35 30 35 38 35	34.6	2.88097	8
38 34 35 35 31	34.6	2.50998	7
34 35 33 30 34	33.2	1.92354	5
40 35 34 33 35	35.4	2.70185	7
34 35 38 35 30	34.4	2.88097	8
35 30 35 29 37	33.2	3.49285	8
40 31 38 35 31	35.0	4.06202	9
35 36 30 33 32	33.2	2.38747	6
35 34 35 30 36	34.0	2.34521	6
35 35 31 38 36	35.0	2.54951	7
32 36 36 32 36	34.4	2.19089	4
36 37 32 34 34	34.6	1.94936	5
29 34 33 37 35	33.6	2.96648	8
36 36 35 37 37	36.2	0.83666	2
36 30 35 33 31	33.0	2.54951	6
35 30 29 38 35	33.4	3.78153	9
35 36 30 34 36	34.2	2.48998	6
35 30 36 29 35	33.0	3.24037	7
38 36 35 31 31	34.2	3.11448	7
30 34 40 28 30	32.4	4.77493	12

(a) Calculate the values of $\bar{\bar{x}}$, $\bar{s}$, and $\bar{R}$.

(b) Use the values of A_3 and $\bar{s}$ to construct an $\bar{x}$ chart.

(c) Construct an s chart.

(d) Use the values of A_2 and $\bar{R}$ to construct an $\bar{x}$ chart.

(e) Construct an R chart.

(f) Do the charts indicate that these fill weights are in statistical control?

11.2-11. Ledolter and Hogg (see references) report that, in the production of stainless steel pipes, the number of defects per 100 feet should be controlled. From 15 randomly selected pipes of length 100 feet, the following data on the number of defects were observed:

$$6\ 10\ 8\ 1\ 7\ 9\ 7\ 4\ 5\ 10\ 3\ 4\ 9\ 8\ 5$$

(a) Compute the control limits of the c chart and draw this control chart.

(b) Is the process in statistical control?

11.3 GENERAL FACTORIAL AND 2^k FACTORIAL DESIGNS

In Section 7.6, we studied two-factor experiments in which the A factor is performed at a levels and the B factor has b levels. Without replications, we need ab level combinations, and with c replications with each of these combinations, we need a total of abc experiments.

Let us consider a situation with three factors—say, A, B, and C, with a, b, and c levels, respectively. Here there are a total of abc level combinations, and if, at each of these combinations, we have d replications, there is a need for $abcd$ experiments. Once these experiments are run, in some random order, and the data collected, there are computer programs available to calculate the entries in the ANOVA table, as in Table 11.3-1.

The main effects (A, B, and C) and the two-factor interactions (AB, AC, and BC) have the same interpretations as in the two-factor ANOVA. The three-factor interaction represents that part of the model for the means μ_{ijh}; $i = 1, 2, \ldots, a$; $j = 1, 2, \ldots, b$; $h = 1, 2, \ldots, c$ that cannot be explained by a model including only the main effects and two-factor interactions. In particular, if, for each fixed h, the "plane" created by μ_{ijh} is "parallel" to the "plane" created by every other fixed h, then the three-factor interaction is equal to zero. Usually, higher order interactions tend to be small.

Source	SS	d.f.	MS	F
TABLE 11.3-1: ANOVA Table				
A	SS(A)	$a - 1$	MS(A)	MS(A)/MS(E)
B	SS(B)	$b - 1$	MS(B)	MS(B)/MS(E)
C	SS(C)	$c - 1$	MS(C)	MS(C)/MS(E)
AB	SS(AB)	$(a - 1)(b - 1)$	MS(AB)	MS(AB)/MS(E)
AC	SS(AC)	$(a - 1)(c - 1)$	MS(AC)	MS(AC)/MS(E)
BC	SS(BC)	$(b - 1)(c - 1)$	MS(BC)	MS(BC)/MS(E)
ABC	SS(ABC)	$(a - 1)(b - 1)(c - 1)$	MS(ABC)	MS(ABC)/MS(E)
Error	SS(E)	$abc(d - 1)$	MS(E)	
Total	SS(TO)	$abcd - 1$		

In the testing sequence, we test the three-factor interaction first by checking to see whether or not

$$\text{MS(ABC)/MS(E)} \geq F_\alpha[(a - 1)(b - 1)(c - 1), abc(d - 1)].$$

If this inequality holds, the ABC interaction is significant at the α level. We would then not continue testing the two-factor interactions and the main effects with those F values, but analyze the data otherwise. For example, for each fixed h, we could look at a two-factor ANOVA for factors A and B. Of course, if the inequality does not hold, we next check the two-factor interactions with the appropriate F values. If these are not significant, we check the main effects, A, B, and C.

Factorial analyses with three or more factors require many experiments, particularly if each factor has several levels. Often, in the health, social, and physical sciences experimenters want to consider several factors (maybe as many as 10, 20, or even hundreds), and they cannot afford to run that many experiments. This is particularly true with preliminary or screening investigations, in which they want to detect the factors that seem most important. In these cases, they often consider factorial experiments such that each of k factors is run at just two levels, frequently without replication. We consider only this situation, although the reader should recognize that it has many variations. In particular, there are methods for investigating only *fractions of these 2^k designs*. The reader interested in more information should refer to a good book on the design of experiments, such as that by Box, Hunter, and Hunter (see references). Many statisticians in industry believe that these statistical methods are the most useful in improving product and process designs. Hence, this is clearly an extremely important topic, as many industries are greatly concerned about the quality of their products.

In factorial experiments in which each of the k factors is considered at only two levels, those levels are selected at some reasonable low and high values. That is, with the help of someone in the field, the typical range of each factor is considered. For instance, if we are considering baking temperatures in the range from $300°$ to $375°$, a representative low is selected—say, $320°$—and a representative high is selected—say, $355°$. There is no formula for these selections, and someone familiar with the experiment would help make them. Often, it happens that only two different types of a material (e.g., fabric) are considered, and one is called low and the other high.

Thus, we select a low and a high for each factor and code them as -1 and $+1$ or, more simply, $-$ and $+$, respectively. We give three 2^k designs, for $k = 2, 3$, and 4, in standard order in Tables 11.3-2, 11.3-3, and 11.3-4, respectively. From these three

TABLE 11.3-2: 2^2 Design			
2^2 Design			
Run	A	B	Observation
1	−	−	X_1
2	+	−	X_2
3	−	+	X_3
4	+	+	X_4

TABLE 11.3-3: 2^3 Design				
2^3 Design				
Run	A	B	C	Observation
1	−	−	−	X_1
2	+	−	−	X_2
3	−	+	−	X_3
4	+	+	−	X_4
5	−	−	+	X_5
6	+	−	+	X_6
7	−	+	+	X_7
8	+	+	+	X_8

tables, we can easily note what is meant by standard order. The A column starts with a minus sign and then the sign alternates. The B column begins with two minus signs and then the signs alternate in blocks of two. The C column has 4 minus signs and then 4 plus signs and so on. The D column starts with 8 minus signs and then 8 plus signs. It is easy to extend this idea to 2^k designs, where $k \geq 5$. To illustrate, under the E column in a 2^5 design, we have 16 minus signs followed by 16 plus signs, which together account for the 32 experiments.

To be absolutely certain what these runs mean, consider run number 12 in Table 11.3-4: A is set at its high level, B at its high, C at its low, and D at its high level. The value X_{12} is the random observation resulting from this one combination of these four settings. It must be emphasized that the runs are not necessarily performed in the order $1, 2, 3, \ldots, 2^k$; in fact, they should be performed in a random order if at all possible. That is, in a 2^3 design, we might perform the experiment in the order 3, 2, 8, 6, 5, 1, 4, 7 if this, in fact, was a random selection of a permutation of the first eight positive integers.

Once all 2^k experiments have been run, it is possible to consider the total sum of squares

$$\sum_{i=1}^{2^k} (X_i - \overline{X})^2$$

and decompose it very easily into $2^k - 1$ parts, which represent the respective measurements (estimators) of the k main effects, $\binom{k}{2}$ two-factor interactions, $\binom{k}{3}$ three-factor interactions, and so on, until we have the one k-factor interaction. We

TABLE 11.3-4: 2^4 Design

		2^4 Design			
Run	A	B	C	D	Observation
1	−	−	−	−	X_1
2	+	−	−	−	X_2
3	−	+	−	−	X_3
4	+	+	−	−	X_4
5	−	−	+	−	X_5
6	+	−	+	−	X_6
7	−	+	+	−	X_7
8	+	+	+	−	X_8
9	−	−	−	+	X_9
10	+	−	−	+	X_{10}
11	−	+	−	+	X_{11}
12	+	+	−	+	X_{12}
13	−	−	+	+	X_{13}
14	+	−	+	+	X_{14}
15	−	+	+	+	X_{15}
16	+	+	+	+	X_{16}

TABLE 11.3-5: 2^3 Design Decomposition

	2^3 Design							
Run	A	B	C	AB	AC	BC	ABC	Observation
1	−	−	−	+	+	+	−	X_1
2	+	−	−	−	−	+	+	X_2
3	−	+	−	−	+	−	+	X_3
4	+	+	−	+	−	−	−	X_4
5	−	−	+	+	−	−	+	X_5
6	+	−	+	−	+	−	−	X_6
7	−	+	+	−	−	+	−	X_7
8	+	+	+	+	+	+	+	X_8

illustrate this decomposition with the 2^3 design in Table 11.3-5. Note that column AB is found by formally multiplying the elements of column A by the corresponding ones in B. Likewise, AC is found by multiplying the elements of column A by the corresponding ones in column C, and so on, until column ABC is the product of the corresponding elements of columns A, B, and C. Next, we construct seven linear forms, using these seven columns of signs with the corresponding observations. The resulting measures (estimates) of the main effects (A, B, C), the two-factor interactions (AB, AC, BC), and the three-factor interaction (ABC) are then found by dividing the linear forms by $2^k = 2^3 = 8$. (Some statisticians divide by $2^{k-1} = 2^{3-1} = 4$.) These measures are denoted by

$$[A] = (-X_1 + X_2 - X_3 + X_4 - X_5 + X_6 - X_7 + X_8)/8,$$
$$[B] = (-X_1 - X_2 + X_3 + X_4 - X_5 - X_6 + X_7 + X_8)/8,$$
$$[C] = (-X_1 - X_2 - X_3 - X_4 + X_5 + X_6 + X_7 + X_8)/8,$$
$$[AB] = (+X_1 - X_2 - X_3 + X_4 + X_5 - X_6 - X_7 + X_8)/8,$$
$$[AC] = (+X_1 - X_2 + X_3 - X_4 - X_5 + X_6 - X_7 + X_8)/8,$$
$$[BC] = (+X_1 + X_2 - X_3 - X_4 - X_5 - X_6 + X_7 + X_8)/8,$$
$$[ABC] = (-X_1 + X_2 + X_3 - X_4 + X_5 - X_6 - X_7 + X_8)/8.$$

With assumptions of normality, mutual independence, and common variance σ^2, under the overall null hypothesis of the equality of all the means, each of these measures has a normal distribution with mean zero and variance $\sigma^2/8$ (in general, $\sigma^2/2^k$). This implies that the square of each measure divided by $\sigma^2/8$ is $\chi^2(1)$. Moreover, it can be shown (see Exercise 11.3-2) that

$$\sum_{i=1}^{8} (X_i - \overline{X})^2$$
$$= 8([A]^2 + [B]^2 + [C]^2 + [AB]^2 + [AC]^2 + [BC]^2 + [ABC]^2).$$

So, by Theorem 7.5-1, the terms on the right-hand side, divided by σ^2, are mutually independent random variables, each being $\chi^2(1)$. While it requires a little more theory, it follows that the linear forms [A], [B], [C], [AB], [AC], [BC], and [ABC] are mutually independent $N(0, \sigma^2/8)$ random variables.

Since we have assumed that we have not run any replications, how can we obtain an estimate of σ^2 to see if any of the main effects or interactions are significant? To help us, we fall back on the use of a q–q plot because, under the overall null hypothesis, those seven measures are mutually independently, normally distributed variables with the same mean and variance. Thus, a q–q plot of the normal percentiles against the corresponding ordered values of the measures should be about on a straight line if, in fact, the null hypothesis is true. If one of these points is "out of line," we might believe that the overall null hypothesis is not true and that the effect associated with the factor represented by that point is significant. It is possible that two or three points might be out of line; then all corresponding effects (main or interaction) should be investigated. Clearly, this is not a formal test, but it has been extremely successful in practice.

As an illustration, we use the data from an experiment designed to evaluate the effects of laundering on a certain fire-retardant treatment for fabrics. These data, somewhat modified, were taken from *Experimental Statistics, National Bureau of Standards Handbook 91*, by Mary G. Natrella (Washington, DC: U.S. Government Printing Office, 1963). Factor A is the type of fabric (sateen or monk's cloth), factor B corresponds to two different fire-retardant treatments, and factor C describes the laundering conditions (no laundering, after one laundering). The observations are inches burned, measured on a standard-size fabric after a flame test. They are as follows, in standard order:

$$x_1 = 41.0, \quad x_2 = 30.5, \quad x_3 = 47.5, \quad x_4 = 27.0,$$
$$x_5 = 39.5, \quad x_6 = 26.5, \quad x_7 = 48.0, \quad x_8 = 27.5.$$

Thus, the measures of the effects are

$$[A] = (-41.0 + 30.5 - 47.5 + 27.0 - 39.5 + 26.5 - 48.0 + 27.5)/8 = -8.06,$$
$$[B] = (-41.0 - 30.5 + 47.5 + 27.0 - 39.5 - 26.5 + 48.0 + 27.5)/8 = 1.56,$$
$$[C] = (-41.0 - 30.5 - 47.5 - 27.0 + 39.5 + 26.5 + 48.0 + 27.5)/8 = 0.56,$$
$$[AB] = (+41.0 - 30.5 - 47.5 + 27.0 + 39.5 - 26.5 - 48.0 + 27.5)/8 = -2.19,$$
$$[AC] = (+41.0 - 30.5 + 47.5 - 27.0 - 39.5 + 26.5 - 48.0 + 27.5)/8 = -0.31,$$
$$[BC] = (+41.0 + 30.5 - 47.5 - 27.0 - 39.5 - 26.5 + 48.0 + 27.5)/8 = 0.81,$$
$$[ABC] = (-41.0 + 30.5 + 47.5 - 27.0 + 39.5 - 26.5 - 48.0 + 27.5)/8 = 0.31.$$

In Table 11.3-6, we order these seven measures, determine their percentiles, and find the corresponding percentiles of the standard normal distribution.

TABLE 11.3-6: Seven Measures Ordered			
Identity of Effect	Ordered Effect	Percentile	Percentile from $N(0,1)$
[A]	−8.06	12.5	−1.15
[AB]	−2.19	25.0	−0.67
[AC]	−0.31	37.5	−0.32
[ABC]	0.31	50.0	0.00
[C]	0.56	62.5	0.32
[BC]	0.81	75.0	0.67
[B]	1.56	87.5	1.15

The q–q plot is given in Figure 11.3-1. Each point has been identified with its effect. A straight line fits six of those points reasonably well, but the point associated with $[A] = -8.06$ is far from this straight line. Hence, the main effect of factor A (the type of fabric) seems to be significant. It is interesting to note that the laundering factor, C, does not seem to be a significant factor.

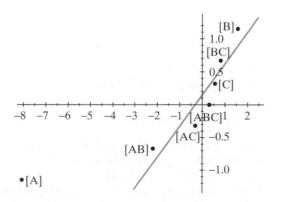

FIGURE 11.3-1: A q–q plot of normal percentiles versus estimated effects

EXERCISES

11.3-1. Write out a 2^2 design, displaying the A, B, and AB columns for the four runs.

(a) If X_1, X_2, X_3, and X_4 are the four observations for the respective runs in standard order, write out the three linear forms, [A], [B], and [AB], that measure the two main effects and the interaction. These linear forms should include the divisor $2^2 = 4$.

(b) Show that $\sum_{i=1}^{4}(X_i - \overline{X})^2 = 4([A]^2 + [B]^2 + [AB]^2)$.

(c) Under the null hypothesis that all the means are equal and with the usual assumptions (normality, mutual independence, and common variance), what can you say about the distributions of the expressions in (b) after each is divided by σ^2?

11.3-2. Show that, in a 2^3 design,

$$\sum_{i=1}^{8}(X_i - \overline{X})^2 = 8([A]^2 + [B]^2 + [C]^2 + [AB]^2 + [AC]^2 + [BC]^2 + [ABC]^2).$$

HINT: Since both the right and the left members of this equation are symmetric in the variables $X_1, X_2, \ldots, X_8$, it is necessary to show only that the corresponding coefficients of $X_1 X_i, i = 1, 2, \ldots, 8$, are the same in each member of the equation. Of course, recall that $\overline{X} = (X_1 + X_2 + \cdots + X_8)/8$.

11.3-3. Show that the unbiased estimator of the variance σ^2 from a sample of size $n = 2$ is one-half of the square of the difference of the two observations. Thus, show that, if a 2^k design is replicated, say, with X_{i1} and $X_{i2}, i = 1, 2, \ldots, 2^k$, then the estimate of the common σ^2 is

$$\frac{1}{2^{k+1}} \sum_{i=1}^{2^k}(X_{i1} - X_{i2})^2 = MS(E).$$

Under the usual assumptions, this equation implies that each of $2^k[A]^2/MS(E)$, $2^k[B]^2/MS(E)$, $2^k[AB]^2/MS(E)$, and so on has an $F(1, 2^k)$ distribution under the null hypothesis. This approach, of course, would provide tests for the significance of the various effects, including interactions.

11.3-4. Ledolter and Hogg (see references) note that percent yields from a certain chemical reaction for changing temperature (factor A), reaction time (factor B), and concentration (factor C) are $x_1 = 79.7$, $x_2 = 74.3$, $x_3 = 76.7$, $x_4 = 70.0$, $x_5 = 84.0$, $x_6 = 81.3$, $x_7 = 87.3$, and $x_8 = 73.7$, in standard order with a 2^3 design.

(a) Estimate the main effects, the three two-factor interactions, and the three-factor interaction.

(b) Construct an appropriate q–q plot to see if any of these effects seem to be significantly larger than the others.

11.3-5. Box, Hunter, and Hunter (see references) studied the effects of catalyst charge (10 pounds $= -1$, 20 pounds $= +1$), temperature (220°C $= -1$, 240°C $= +1$), pressure (50 psi $= -1$, 80 psi $= +1$), and concentration (10% $= -1$, 12% $= +1$) on percent conversion (X) of a certain chemical. The results of a 2^4 design, in standard order, are

$$x_1 = 71, \quad x_2 = 61, \quad x_3 = 90, \quad x_4 = 82, \quad x_5 = 68, \quad x_6 = 61,$$
$$x_7 = 87, \quad x_8 = 80, \quad x_9 = 61, \quad x_{10} = 50, \quad x_{11} = 89,$$
$$x_{12} = 83, \quad x_{13} = 59, \quad x_{14} = 51, \quad x_{15} = 85, \quad x_{16} = 78.$$

(a) Estimate the main effects and the two-, three-, and four-factor interactions.

(b) Construct an appropriate q–q plot and assess the significance of the various effects.

11.4 UNDERSTANDING VARIATION

Really understanding variation, which is the key element of probability and statistics, is extremely important in so many phases of life. Yet, as frequently happens in studying a textbook like this, we are so involved in the mathematics of probability, distributions, sampling distributions, and even descriptive statistics that we miss the real meanings of some of these concepts, particularly as far as the role they play in day-to-day activities.

To illustrate, variability explains what is commonly called the "sophomore jinx": Some freshmen (rookies) have great first years (above their averages), and then most of them do worse the second year. The same explanation can be applied to movies that have sequels; often the original was outstanding, but even with the same cast and same director, the sequel is usually worse than the original. This phenomenon is sometimes called *regression to the mean*.

We also note that, on the one hand, workers who do very well (above their averages) and who thus might be given some reward usually do worse during the next work period. On the other hand, those who have bad periods (below average) and are reprimanded frequently do better the next. Yet it is wrong to think that a policy of reprimand is better than that of a pat on the back. Both situations can be explained by understanding the worker's pattern of variation. If the employer really wants to improve the outcome of the process, he or she must consider ways of improving the *level* (average). The workers can sometimes make small adjustments, but they need road maps to make the major ones. Thus, it is really the responsibility of management to improve the process substantially. That is, management must realize that working harder with techniques that have failed will not improve the situation much. Changes must be made, and often data and the resulting statistical analysis can suggest changes for the better.

It is disturbing to see the following situation: Suppose that the manager has created a team (say, of 10) of outstanding workers, and it is time to give raises. The workers are ranked from 1 to 10. It is always true that 10 workers, no matter how good or how bad, will get the ranks $1, 2, 3, \ldots, 10$. Then say they get raises according to their ranks, which might have been determined simply by a random process for this particular period. That is, in the next period, their ranks might be entirely different. How does the small raise make the one with lowest rank feel if, in fact, all 10 are members of an outstanding team and there might be little or no difference among the persons ranked 1 and 10? Not good, and he or she is less likely to help the ones with higher ranks in a future period. That is, such a strategy does not promote teamwork. As a matter of fact, there is always this danger in any kind of reward, such as the "worker of the month." Others are not likely to help that person the next month, because they want to be the worker of the month and receive the corresponding bonus.

A better way of rewarding a good team would be to give them essentially the same raises. Clearly, a good manager will continuously monitor the team's performance and supply appropriate feedback to the workers. Of course, if one worker is consistently on the high side, then he or she should be considered for promotion or a substantial raise. By contrast, if a worker is on the low end most of the time, some help for this worker is in order: maybe additional training or even a transfer to a different department to which the worker's talents are more suitable. It is important not to demean the worker, because each of us needs to take pride in our accomplishments. If at all possible, firing the worker should be avoided, possibly by finding him or her some other suitable job in the company.

In one state, legislators appropriated enough money for 2% raises at a state university one year. The administrators thought that that percentage was not adequate, so they had the "brilliant" idea to give one-third of the faculty 6% raises and zero to the other two-thirds. Thus, in a department of 18, 6 received raises and 12 did not. Who ranked those 18? Was there much difference between the 6th and 7th? Or between the 5th and 10th for that matter? So the person who received a raise could have people close to him or her in the ranking who got nothing. The administrators' idea created a terrible morale problem. It certainly would have been better to give every faculty member an approximate 2% raise and blame it on the legislators. Of course, we must recognize differences among people, and accordingly, they must be treated differently, but certainly not on the basis of mere chance variation.

Many of the ideas in this area are those of W. Edwards Deming, an esteemed statistician who went to Japan after World War II and taught the Japanese how to make quality products. For his work there, he was awarded the Emperor's Medal, and the Japanese established the Deming Prize to be awarded each year to the company or individual contributing the most to quality improvement. One of the things Deming stressed was the need for "profound knowledge," and a major item in that concept is understanding variation and statistical theory.

In making quality products, you want to reduce variation as much as possible and move levels closer to their targets. Deming believed that barriers between departments, between management and workers, and among workers must be broken down to improve communication and the ability to work as a team. The lines of communication, all the way from suppliers to customers, must be open to help reduce variation and improve products and services. For example, he argued that a company should not buy only on price tag, but should have a few reliable suppliers (and possibly only one) for a single part because that will reduce variation. That is, many different suppliers would obviously increase variation. Moreover, he argued that you should become partners—friends if you like—with your suppliers. You learn to trust each other; that way, you can use methods like "just in time," in which your inventory can be kept reasonably small, to keep costs down.

If each of us thinks about these ideas, we might become obsessed by understanding variation, and that might make a big difference in our everyday lives. For example, once we have selected good suppliers, we continue to go to the same barber, the same service station, the same clothier, the same bank, and on and on. We like to buy items, even if a little more expensive, from places that will give us good service if something goes wrong. If this does not happen, then, of course, we must consider changing suppliers.

Clearly, listening to the customer can help improve the quality of our products and services. We can then meet—or even exceed—the expectations of our customers. We should continually try to improve by reducing variation and moving the process to a better level. Although the customer—as well as the supplier—must be part of the total team, more often management must continue to look for better ways to do things, ways that the customer never would have imagined. For example, in the early days of automobiles, not many owners would have thought of driving on pneumatic tires. Harvey Firestone did and, in this way, exceeded the expectations of those early customers.

Deming also preached constancy of purpose. If management's ideas tend to change too much, employees really do not know what to do and cannot do their best. That is, they get mixed up, thereby increasing variation. It is better for them to receive a constant signal from their employer—a signal that changes only if research dictates ways of improving.

More training and education for everyone associated with a company also decreases variation, by teaching how to make the product more uniform. Workers must know that they do not have to be afraid to make suggestions to improve the process. Often, being on a team will make it easier for workers to speak up without fear of reprisal.

Many of us remember playing a game called "telephone": One whispers a message to the next person, who whispers the message to the next, and so on. The message at the end is compared with the original message, and it is usually much different from the original. This is like trying to hit a target (say, at zero) with a random variable X_1 and then, starting with X_1 as the center and adding on another error X_2, and so on, creating the sum of the errors, $X_1 + X_2 + \cdots + X_n$, which has an ever-increasing variance with independent errors. Deming would say, "We are off to the Milky Way."

Yet we actually do this in business and industry by letting worker train worker. Once errors are introduced, they will stay there, and others will add on and on. Incidentally, we might say the same thing about too many layers of management. To decrease variation, wouldn't it be better to have a master instructor train each worker or have fewer layers of management?

Deming also noted that requiring quotas does not help the quality of a product or service. A manager or worker who has a quota of 100 items per day will often ship out 90 good and 10 bad items just to make the quota. Clearly, it would reduce variation and satisfy the customer better if only the 90 good ones were shipped.

All this discussion leads to the point that a final inspection of products often does not really improve quality. With such a procedure, you can only eliminate the bad items and send on the good ones. *Improvements* in the design of the products and manufacturing processes are needed. If these are done well, often with the help of statistical methods, that final inspection can be eliminated. That is, improvements should be made continuously in the manufacturing process, rather than trying to correct things at a final inspection by weeding out the bad items.

Once you recognize that there is variation in almost every process, you begin to think like a statistician. Often, you want to reduce that variation and center the variation that remains at the right level, because then "the doors will fit better." But while we want to reduce undesirable variation, we do not want to reduce *all* variation: Most of us do not want to eat, say, Chinese food every night, even though we enjoy it from time to time. A certain amount of variation is "the spice of life."

This desire to understand variation is the key to statistical thinking. As a matter of fact, many interested in quality improvement believe that statistical thinking recognizes that variation is present in everything we do. To provide better products and services we must strive to reduce undesirable variation as much as possible. They become obsessed with determining whether the variation is due to special or common causes and then correcting or controlling them, as the case may be.

In a sense, statistical thinking can affect everything about us, including our relationships with others. All of us have our "ups and downs," and we must seriously think about how to deal with others, recognizing this variation.

It was not until the 1980s that Americans paid much attention to Deming's ideas. At that time Deming started to give short courses on his quality philosophy, the highlight of which was his famous "red beads" experiment. He brought out a box with over 1000 beads, about 20 percent of which were red (defective), and a paddle that held exactly 50 beads. His audience did not know the exact proportion of defectives. He selected about six "willing workers" from the audience who had to dip the paddle into the box and remove a set of beads. Deming gave detailed instructions on how to do this, emphasizing the "correct" way to scooping with just the right amount of "shaking" and the appropriate "angle" of the paddle. Two inspectors would count the number of defectives (i.e., the number of reds) and report the results to a secretary. The first worker, say, Mike, happened to have 9 defectives. "We can do better than that," Deming would declare, and he advised a little more shaking before drawing out the paddle. The next worker, Joe, had 13 defectives. Deming would "give him hell" because, in his words, the paddle came out at the wrong angle. This would continue until Mary ended up with 6 defectives. "That's more like it," Deming would praise, and, pointing to the next worker, he would say, "Tim, you saw how she did this." Tim might end up with 12 defectives and would get the question, "Didn't you learn anything from Mary?" Occasionally, someone would get 16 defectives and be fired, although Deming was known to encourage managers to retrain, not fire, those who needed help. This process would continue for well over an hour, with Deming playing the role of the tough boss encouraging the participants to "work harder and reduce the number of defects."

The "red beads" experiment was a good and also very entertaining illustration, and it communicated a very important point. Members of the audience, as well as the willing participants, knew that the only way to reduce the number of defectives, on average, was to improve the process, and this meant removing a number of red beads from the box. It is the managers and the engineers who can facilitate improvements, usually by redesigning and changing the underlying processes. Workers cannot make major improvements to the system just by trying harder; if the system is bad, such good efforts will not help. Workers may have good ideas for improvements, but it is the "owner" of the process (the one responsible for the process and in charge of changes) who can affect improvements. Deming also realized that many valuable suggestions will not come to the surface if people are afraid to speak up and question the status quo. "Driving out fear" was another principle that Deming communicated to his audience.

It is the authors' opinion that quality improvement must begin with the individual who should create a personal vision, goals, and principles by which to live. Hopefully, one of those individuals is the chief of the organization, for then things go much smoother. Bob Galvin, who was the CEO of Motorola during the 1980s, believes that "quality improvement is a daily, personal, priority obligation," and his company made great improvements in the quality of its products during that period. Much of this progress was due to understanding variation and statistical methods. Even Motorola's program was called Six Sigma, in which *sigma* referred to the standard deviation, a name that is a delight to a statistician.

Although three-sigma limits have proved to be an effective guide, many companies are now considering six-sigma limits. The Six Sigma program has its statistical underpinnings in the following idea: There are specifications in manufacturing items that are usually set by the engineers—or possibly by customers. These specifications are usually given in terms of a target value and upper and lower specification limits—USL and LSL, respectively. The mean μ of the values resulting from the process is, hopefully, close to the target value. The "specs" are 6σ away from the target, where, as mentioned, σ is the standard deviation associated with the process. However, the mean μ is often dynamic, and Six Sigma companies try to keep it at least 4.5σ from the closest spec. If the values of the items are distributed normally and the nearest spec is 4.5σ from μ, there are only 3.4 defectives per million. This is the very worthwhile goal of the Six Sigma companies, and they use many of Deming's ideas, some of which are outlined earlier, in their attempts to achieve that goal.

While an understanding of quality improvement ideas and basic statistical methods, such as those in this chapter, is extremely important to the Six Sigma programs, possibly the major factor is the attitude of the CEOs and other important administrators. There were many total quality management programs in the 1980s, but they were not as successful as Six Sigma, for now each CEO is demanding "Show me the money!" That is, companies hire a Six Sigma expert to come in for four one-week periods, about one month apart, for a fee of about $15,000 per person. If the company has 20 participants, this cost is $300,000 plus the four weeks of time "lost" to the work process by each trainee. That is a great deal of money. However, each of these individuals (sometimes a pair) has a project associated with some process that has not been very efficient. They work on these projects, using statistical methods, during the "off weeks" and report to the Six Sigma expert during the training weeks to get advice. Often, these projects, if successful, will save millions of dollars for a company, so the expenses are well worth the benefits. As a matter of fact, if a participant's project saves the company at least $1 million (possibly the required amount differs from company to company), then the participant earns a "six-sigma Black Belt." So the CEO does see the money saved, and thus the bottom line looks extremely good to him or her.

Each of us should use as much statistical thinking as we know to help us build trust with the persons with whom we deal directly or indirectly. Let us not do what those administrators of that university did when they did not understand the variability associated with ranking people. That action destroyed any trust that was there and created a mean-spirited competition among the members of that faculty. Why should anyone help others when their gain only hurts that individual? While statistical thinking cannot solve all problems, it most certainly can help in many situations. We hope you use what statistics you know to benefit you the rest of your life.

There is a huge demand for statistical scientists in many fields to make sense out of the increasing volumes of data that scientists discover every day. Statistics is needed to turn quality data into useful information upon which decisions can be made. The striking thing about any data set is that there is variation: simply put, not all of the data points lie within a pattern. It is the statistician's job to find that pattern and describe the variation about it. Done properly, this clearly helps the decision maker significantly.

One observation about variation should be noted before any major adjustments or decisions are made. Frequently, persons in charge jump at conclusions too quickly; that is, major decisions are often made after too few observations. For example, we know that if X_1, X_2, and X_3 are independent and identically distributed continuous-type observations, then

$$P(X_1 < X_2 < X_3) = \frac{1}{3!} = \frac{1}{6}.$$

Yet if this event occurred and these observations were taken on, say, sales and plotted in time sequence, the fact that the three points plotted were "going up" might suggest to management that "the company is on a roll." In some cases, only two increasing points might cause such a reaction. If there is no change in the system, two or three increasing points have respective probabilities of 1/2 and 1/6, and those probabilities do not warrant that sort of reaction. If, however, we observe four or five such points, with respective probabilities of 1/24 and 1/120, then an appropriate reaction is in order.

An interesting question to ask is why statisticians treat $1/20 = 0.05$ as the value at which the probability of an event is considered small and often suggests some type of action. Possibly it was because Ronald A. Fisher suggested that 1 out of 20 seemed small enough. Obviously, the value of 0.05 is not written in stone, but statisticians seem to look for differences of two or three standard deviations as a guide for action. Since many estimators have approximate normal distributions, such differences do have small probabilities.

For instance, suppose a candidate believes that he or she has at least 50% of the votes. Yet, in a poll of $n = 400$, only 160 favor that candidate; thus, the standardized value

$$\frac{160 - 400(1/2)}{\sqrt{400(1/2)(1/2)}} = -4$$

suggests that the candidate does not in fact have 50% of the votes. Depending upon the financial situation, the candidate must change his or her approach or possibly even consider dropping from the race. However, we note that this simple statistical analysis can be a guide in the decision process.

Although the reader has now studied enough statistics to appreciate the importance of understanding variation, there are many more useful statistical techniques to be studied if the reader is so inclined. For example, there are courses in regression and time series in which we learn how to predict future observations. Or a study of the design of experiments can help an investigator select the most efficient levels of the various factors. After all, if we have 10 factors and run each even at only two levels, we have created $2^{10} = 1024$ runs. Can we perform just a fraction of these runs without losing too much information? Additional study of multivariate analysis can lead to interesting problems in classification. Say a doctor takes several measurements on a patient and then classifies the patient's condition as one of many possible diseases. There are errors of misclassification, and statisticians can help reduce the probabilities of those errors. Doctors—and statisticians—can make mistakes, and second or third opinions should be requested if there is some doubt.

The computer has opened the door to a wide variety of new statistical techniques, and researchers, computer scientists, and statisticians are working together to reduce huge amounts of data into nuggets of quality information on the basis of which important decisions can be made. Statistics is an exciting field that finds many useful applications in the social, health, and physical sciences. The authors have found statistics to be a great profession; we hope a few of you find it that way, too. In any case, we hope that statistical thinking will make you more aware of the need of understanding variation, which can be a great influence on your daily life.

HISTORICAL COMMENTS Quality improvement made a substantial change in manufacturing beginning in the 1920s, with Walter A. Shewart's control charts. In fairness, it should be noted that the British started a similar program about the same time. Statistical quality control, as described in Section 11.2, really had a huge influence during World War II, with many universities giving short courses in the subject. These courses continued after the war, but the development of the importance of total quality improvement lagged behind. W. Edwards Deming complained that the Japanese used his quality ideas beginning in the 1950s, but Americans did not adopt them until 1980. That year NBC televised a program entitled "If Japan can, why can't we?" and Deming was the "star" of that broadcast. He related that the next day his phone "started ringing off the hook." Various companies requested that he spend one day with them to get them started on the right path. According to Deming, they all wanted "instant pudding," and he noted that he had asked the Japanese to give him five years to make the improvements he pioneered. Actually, using his philosophy, many of these companies did achieve substantial results in quality sooner than that. However it was after the NBC program that Deming started his famous four-day courses, and he taught his last one in December of 1993, about 10 days before his death at the age of 93.

Many of these quality efforts in the 1970s and 1980s used the name "Total Quality Management" or, later, "Continuous Process Improvements." However, it was Motorola's Six Sigma program, which started in the late 1980s and has continued for over 20 years since then, that has had the biggest impact. In addition to Motorola, GE and Allied, as well as a number of smaller companies, have used this system. In our opinion, Six Sigma is leading development in the quality improvement effort.

References

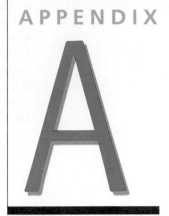

Aspin, A. A., "Tables for Use in Comparisons Whose Accuracy Involves Two Variances, Separately Estimated," *Biometrika*, **36** (1949), pp. 290–296.

Agaresti, A., and B.A. Coull, "Approximate is Better than 'Exact' for Interval Estimation of Binomial Proportions," *Amer. Statist.*, **52**, 2(1998), pp. 119–126.

Barnett, A., "How Numbers Can Trick You," *Technology Review*, 1994.

Basu, D., "On Statistics Independent of a Complete Sufficient Statistic," *Sankhya*, 15, 377 (1955).

Bernstein, P. L., *Against the Gods: The Remarkable Story of Risk*. New York: John Wiley & Sons, Inc., 1996.

Berry, D. A., *Statistics: A Bayesian Perspective*. Belmont, CA: Duxbury Press, An Imprint of Wadsworth Publishing Co., 1996.

Box, George E. P., J. Stuart Hunter, and William G. Hunter, *Statistics for Experimenters: Design, Innovation, and Discovery*, 2d ed., New York: John Wiley & Sons, Inc., 2005.

Box, G. E. P., and M. E. Muller, "A Note on the Generation of Random Normal Deviates," *Ann. Math. Statist.*, **29** (1958), p. 610.

Crisman, R., "Shortest Confidence Interval for the Standard Deviation of a Normal Distribution," *J. Undergrad. Math.*, **7**, 2 (1975), p. 57.

Douglas, Andrea, Courtney Fitzgerald, and Scott Mihalik, "Expected Areas of Randomly Generated Triangles of Fixed Perimeter," *The ΠME Journal*, Vol. 11, Num. 7, Fall, 2002, pp. 365–371.

Guenther, William C., "Shortest Confidence Intervals," *Amer. Statist.*, **23**, 1(1969), p. 22.

Hogg, R. V., "Testing the Equality of Means of Rectangular Populations," *Ann. Math. Statist.*, **24** (1953), p. 691.

Hogg, R. V., J. W. McKean, and A. T. Craig, *Introduction to Mathematical Statistics*, 6th ed. Upper Saddle River, NJ: Prentice Hall, 2005.

Hogg, R. V., and A. T. Craig, "On the Decomposition of Certain Chi-Square Variables," *Ann. Math. Statist.*, **29** (1958), p. 608.

Johnson, V. E., and J. H. Albert, *Ordinal Data Modeling*, New York: Springer-Verlag, 1999.

Karian, Z. A., and E. A. Tanis, *Probability & Statistics: Explorations with MAPLE*, 2d ed. Upper Saddle River, NJ: Prentice Hall, 1999.

Keating, Jerome P., and David W. Scott, "Ask Dr. STATS," *Stats, The Magazine for Students of Statistics*, **25**, Spring, 1999, pp. 16–22.

Kinney, J., "Mathematica As an Aid in Teaching Probability and Statistics," *Proceedings of the Statistical Computing Section*, (1998), American Statistical Association, pp. 25–32.

Latter, O. H., "The Cuckoo's Egg," *Biometrika*, 1: pp. 164–176, 1901.

Ledolter, J., and R. V. Hogg, *Applied Statistics for Engineers and Scientists*, 3d ed. Upper Saddle River, NJ: Prentice Hall, 2010.

Lurie, D., and H. O. Hartley, "Machine-Generation of Order Statistics for Monte Carlo Computations," *The American Statistician*, February, 1972, pp. 26–27.

Montgomery, D. C., *Design and Analysis of Experiments*, 2d ed. New York: John Wiley & Sons, Inc., 1984.

Natrella, M. G., *Experimental Statistics, National Bureau of Standards Handbook 91*. Washington, DC: U.S. Government Printing Office, 1963.

Nicol, Sherrie J., "Who's Picking Up the Pieces?" *Primus*, Volume IV, Number 2, June 1994, pp. 182–184.

Pearson, K., "On the Criterion That a Given System of Deviations from the Probable in the Case of a Correlated System of Variables Is Such That It Can Be Reasonably Supposed to Have Arisen from Random Sampling," *Phil. Mag.*, Series 5, **50** (1900), p. 157.

Putz, John, "The Golden Section and the Piano Sonatas of Mozart," *Mathematics Magazine*, Volume 68, No. 4, October, 1995, pp. 275–282.

Quain, J. R., "Going Mainstream," *PC Magazine*, February, 1994.

Rafter, J. A., M. L. Abell, and J. P. Braselton, *Statistics with Maple*. Amsterdam and Boston: Academic Press, An imprint of Elsevier Science (USA), 2003.

Raspe, Rudolph Erich, *The Surprising Adventures of Baron Munchausen*, IndyPublish.com, 2001.

Snee, R. D., "Graphical Analysis of Process Variation Studies," *Journal of Quality Technology* 15: April 1983, pp. 76–88.

Snee, R. D., L. B. Hare, and J. R. Trout, *Experiments in Industry*, Milwaukee: American Society of Quality Control, 1985.

Stigler, S. M., *The History of Statistics: The Measurement of Uncertainty Before 1900*. Cambridge, MA: Harvard University Press, 1986.

Tanis, E. A., "Maple Integrated Into the Instruction of Probability and Statistics," *Proceedings of the Statistical Computing Section* (1998), American Statistical Association, pp. 19–24.

Tanis, E. A., and R. V. Hogg, *A Brief Course in Mathematical Statistics*. Upper Saddle River, NJ: Prentice Hall, 2008.

Tate, R. F., and G. W. Klett, "Optimum Confidence Intervals for the Variance of a Normal Distribution," *J. Am. Statist. Assoc.*, **54** (1959), p. 674.

Tsutakawa, R. K., G. L. Shoop, and C. J. Marienfeld, "Empirical Bayes estimation of cancer mortality rates," *Statistics in Medicine*, **4** (1985), pp. 201–212.

Tukey, John W., *Exploratory Data Analysis*. Reading, MA: Addison-Wesley Publishing Company, 1977.

Velleman, P. F., and D. C. Hoaglin, *Applications, Basics, and Computing of Exploratory Data Analysis*, Boston: Duxbury Press, 1981.

Wilcoxon, F., "Individual Comparisons by Ranking Methods," *Biometrics Bull.*, **1** (1945), p. 80.

Zerger, Monte, "Mean Meets Variance," *Primus*, Volume IV, Number 2, June 1994, pp. 106–108.

Tables

TABLE I: Binomial Coefficients

$$\binom{n}{r} = \frac{n!}{r!(n-r)!} = \binom{n}{n-r}$$

n	$\binom{n}{0}$	$\binom{n}{1}$	$\binom{n}{2}$	$\binom{n}{3}$	$\binom{n}{4}$	$\binom{n}{5}$	$\binom{n}{6}$	$\binom{n}{7}$	$\binom{n}{8}$	$\binom{n}{9}$	$\binom{n}{10}$	$\binom{n}{11}$	$\binom{n}{12}$	$\binom{n}{13}$
0	1													
1	1	1												
2	1	2	1											
3	1	3	3	1										
4	1	4	6	4	1									
5	1	5	10	10	5	1								
6	1	6	15	20	15	6	1							
7	1	7	21	35	35	21	7	1						
8	1	8	28	56	70	56	28	8	1					
9	1	9	36	84	126	126	84	36	9	1				
10	1	10	45	120	210	252	210	120	45	10	1			
11	1	11	55	165	330	462	462	330	165	55	11	1		
12	1	12	66	220	495	792	924	792	495	220	66	12	1	
13	1	13	78	286	715	1,287	1,716	1,716	1,287	715	286	78	13	1
14	1	14	91	364	1,001	2,002	3,003	3,432	3,003	2,002	1,001	364	91	14
15	1	15	105	455	1,365	3,003	5,005	6,435	6,435	5,005	3,003	1,365	455	105
16	1	16	120	560	1,820	4,368	8,008	11,440	12,870	11,440	8,008	4,368	1,820	560
17	1	17	136	680	2,380	6,188	12,376	19,448	24,310	24,310	19,448	12,376	6,188	2,380
18	1	18	153	816	3,060	8,568	18,564	31,824	43,758	48,620	43,758	31,824	18,564	8,568
19	1	19	171	969	3,876	11,628	27,132	50,388	75,582	92,378	92,378	75,582	50,388	27,132
20	1	20	190	1,140	4,845	15,504	38,760	77,520	125,970	167,960	184,756	167,960	125,970	77,520
21	1	21	210	1,330	5,985	20,349	54,264	116,280	203,490	293,930	352,716	352,716	293,930	203,490
22	1	22	231	1,540	7,315	26,334	74,613	170,544	319,770	497,420	646,646	705,432	646,646	497,420
23	1	23	253	1,771	8,855	33,649	100,947	245,157	490,314	817,190	1,144,066	1,352,078	1,352,078	1,144,066
24	1	24	276	2,024	10,626	42,504	134,596	346,104	735,471	1,307,504	1,961,256	2,496,144	2,704,156	2,496,144
25	1	25	300	2,300	12,650	53,130	177,100	480,700	1,081,575	2,042,975	3,268,760	4,457,400	5,200,300	5,200,300
26	1	26	325	2,600	14,950	65,780	230,230	657,800	1,562,275	3,124,550	5,311,735	7,726,160	9,657,700	10,400,600

For $r > 13$ you may use the identity $\binom{n}{r} = \binom{n}{n-r}$.

TABLE II: The Binomial Distribution

$$F(x) = P(X \le x) = \sum_{k=0}^{x} \frac{n!}{k!(n-k)!} p^k (1-p)^{n-k}$$

n	x	p 0.05	0.10	0.15	0.20	0.25	0.30	0.35	0.40	0.45	0.50
2	0	0.9025	0.8100	0.7225	0.6400	0.5625	0.4900	0.4225	0.3600	0.3025	0.2500
	1	0.9975	0.9900	0.9775	0.9600	0.9375	0.9100	0.8775	0.8400	0.7975	0.7500
	2	1.0000	1.0000	1.0000	1.0000	1.0000	1.0000	1.0000	1.0000	1.0000	1.0000
3	0	0.8574	0.7290	0.6141	0.5120	0.4219	0.3430	0.2746	0.2160	0.1664	0.1250
	1	0.9928	0.9720	0.9392	0.8960	0.8438	0.7840	0.7182	0.6480	0.5748	0.5000
	2	0.9999	0.9990	0.9966	0.9920	0.9844	0.9730	0.9571	0.9360	0.9089	0.8750
	3	1.0000	1.0000	1.0000	1.0000	1.0000	1.0000	1.0000	1.0000	1.0000	1.0000
4	0	0.8145	0.6561	0.5220	0.4096	0.3164	0.2401	0.1785	0.1296	0.0915	0.0625
	1	0.9860	0.9477	0.8905	0.8192	0.7383	0.6517	0.5630	0.4752	0.3910	0.3125
	2	0.9995	0.9963	0.9880	0.9728	0.9492	0.9163	0.8735	0.8208	0.7585	0.6875
	3	1.0000	0.9999	0.9995	0.9984	0.9961	0.9919	0.9850	0.9744	0.9590	0.9375
	4	1.0000	1.0000	1.0000	1.0000	1.0000	1.0000	1.0000	1.0000	1.0000	1.0000
5	0	0.7738	0.5905	0.4437	0.3277	0.2373	0.1681	0.1160	0.0778	0.0503	0.0312
	1	0.9774	0.9185	0.8352	0.7373	0.6328	0.5282	0.4284	0.3370	0.2562	0.1875
	2	0.9988	0.9914	0.9734	0.9421	0.8965	0.8369	0.7648	0.6826	0.5931	0.5000
	3	1.0000	0.9995	0.9978	0.9933	0.9844	0.9692	0.9460	0.9130	0.8688	0.8125
	4	1.0000	1.0000	0.9999	0.9997	0.9990	0.9976	0.9947	0.9898	0.9815	0.9688
	5	1.0000	1.0000	1.0000	1.0000	1.0000	1.0000	1.0000	1.0000	1.0000	1.0000
6	0	0.7351	0.5314	0.3771	0.2621	0.1780	0.1176	0.0754	0.0467	0.0277	0.0156
	1	0.9672	0.8857	0.7765	0.6553	0.5339	0.4202	0.3191	0.2333	0.1636	0.1094
	2	0.9978	0.9842	0.9527	0.9011	0.8306	0.7443	0.6471	0.5443	0.4415	0.3438
	3	0.9999	0.9987	0.9941	0.9830	0.9624	0.9295	0.8826	0.8208	0.7447	0.6562
	4	1.0000	0.9999	0.9996	0.9984	0.9954	0.9891	0.9777	0.9590	0.9308	0.8906
	5	1.0000	1.0000	1.0000	0.9999	0.9998	0.9993	0.9982	0.9959	0.9917	0.9844
	6	1.0000	1.0000	1.0000	1.0000	1.0000	1.0000	1.0000	1.0000	1.0000	1.0000
7	0	0.6983	0.4783	0.3206	0.2097	0.1335	0.0824	0.0490	0.0280	0.0152	0.0078
	1	0.9556	0.8503	0.7166	0.5767	0.4449	0.3294	0.2338	0.1586	0.1024	0.0625
	2	0.9962	0.9743	0.9262	0.8520	0.7564	0.6471	0.5323	0.4199	0.3164	0.2266
	3	0.9998	0.9973	0.9879	0.9667	0.9294	0.8740	0.8002	0.7102	0.6083	0.5000
	4	1.0000	0.9998	0.9988	0.9953	0.9871	0.9712	0.9444	0.9037	0.8471	0.7734
	5	1.0000	1.0000	0.9999	0.9996	0.9987	0.9962	0.9910	0.9812	0.9643	0.9375

TABLE II: *continued*

n	x	0.05	0.10	0.15	0.20	0.25	0.30	0.35	0.40	0.45	0.50
						p					
	6	1.0000	1.0000	1.0000	1.0000	0.9999	0.9998	0.9994	0.9984	0.9963	0.9922
	7	1.0000	1.0000	1.0000	1.0000	1.0000	1.0000	1.0000	1.0000	1.0000	1.0000
8	0	0.6634	0.4305	0.2725	0.1678	0.1001	0.0576	0.0319	0.0168	0.0084	0.0039
	1	0.9428	0.8131	0.6572	0.5033	0.3671	0.2553	0.1691	0.1064	0.0632	0.0352
	2	0.9942	0.9619	0.8948	0.7969	0.6785	0.5518	0.4278	0.3154	0.2201	0.1445
	3	0.9996	0.9950	0.9786	0.9437	0.8862	0.8059	0.7064	0.5941	0.4770	0.3633
	4	1.0000	0.9996	0.9971	0.9896	0.9727	0.9420	0.8939	0.8263	0.7396	0.6367
	5	1.0000	1.0000	0.9998	0.9988	0.9958	0.9887	0.9747	0.9502	0.9115	0.8555
	6	1.0000	1.0000	1.0000	0.9999	0.9996	0.9987	0.9964	0.9915	0.9819	0.9648
	7	1.0000	1.0000	1.0000	1.0000	1.0000	0.9999	0.9998	0.9993	0.9983	0.9961
	8	1.0000	1.0000	1.0000	1.0000	1.0000	1.0000	1.0000	1.0000	1.0000	1.0000
9	0	0.6302	0.3874	0.2316	0.1342	0.0751	0.0404	0.0207	0.0101	0.0046	0.0020
	1	0.9288	0.7748	0.5995	0.4362	0.3003	0.1960	0.1211	0.0705	0.0385	0.0195
	2	0.9916	0.9470	0.8591	0.7382	0.6007	0.4628	0.3373	0.2318	0.1495	0.0898
	3	0.9994	0.9917	0.9661	0.9144	0.8343	0.7297	0.6089	0.4826	0.3614	0.2539
	4	1.0000	0.9991	0.9944	0.9804	0.9511	0.9012	0.8283	0.7334	0.6214	0.5000
	5	1.0000	0.9999	0.9994	0.9969	0.9900	0.9747	0.9464	0.9006	0.8342	0.7461
	6	1.0000	1.0000	1.0000	0.9997	0.9987	0.9957	0.9888	0.9750	0.9502	0.9102
	7	1.0000	1.0000	1.0000	1.0000	0.9999	0.9996	0.9986	0.9962	0.9909	0.9805
	8	1.0000	1.0000	1.0000	1.0000	1.0000	1.0000	0.9999	0.9997	0.9992	0.9980
	9	1.0000	1.0000	1.0000	1.0000	1.0000	1.0000	1.0000	1.0000	1.0000	1.0000
10	0	0.5987	0.3487	0.1969	0.1074	0.0563	0.0282	0.0135	0.0060	0.0025	0.0010
	1	0.9139	0.7361	0.5443	0.3758	0.2440	0.1493	0.0860	0.0464	0.0233	0.0107
	2	0.9885	0.9298	0.8202	0.6778	0.5256	0.3828	0.2616	0.1673	0.0996	0.0547
	3	0.9990	0.9872	0.9500	0.8791	0.7759	0.6496	0.5138	0.3823	0.2660	0.1719
	4	0.9999	0.9984	0.9901	0.9672	0.9219	0.8497	0.7515	0.6331	0.5044	0.3770
	5	1.0000	0.9999	0.9986	0.9936	0.9803	0.9527	0.9051	0.8338	0.7384	0.6230
	6	1.0000	1.0000	0.9999	0.9991	0.9965	0.9894	0.9740	0.9452	0.8980	0.8281
	7	1.0000	1.0000	1.0000	0.9999	0.9996	0.9984	0.9952	0.9877	0.9726	0.9453
	8	1.0000	1.0000	1.0000	1.0000	1.0000	0.9999	0.9995	0.9983	0.9955	0.9893
	9	1.0000	1.0000	1.0000	1.0000	1.0000	1.0000	1.0000	0.9999	0.9997	0.9990
	10	1.0000	1.0000	1.0000	1.0000	1.0000	1.0000	1.0000	1.0000	1.0000	1.0000
11	0	0.5688	0.3138	0.1673	0.0859	0.0422	0.0198	0.0088	0.0036	0.0014	0.0005
	1	0.8981	0.6974	0.4922	0.3221	0.1971	0.1130	0.0606	0.0302	0.0139	0.0059
	2	0.9848	0.9104	0.7788	0.6174	0.4552	0.3127	0.2001	0.1189	0.0652	0.0327
	3	0.9984	0.9815	0.9306	0.8389	0.7133	0.5696	0.4256	0.2963	0.1911	0.1133
	4	0.9999	0.9972	0.9841	0.9496	0.8854	0.7897	0.6683	0.5328	0.3971	0.2744
	5	1.0000	0.9997	0.9973	0.9883	0.9657	0.9218	0.8513	0.7535	0.6331	0.5000
	6	1.0000	1.0000	0.9997	0.9980	0.9924	0.9784	0.9499	0.9006	0.8262	0.7256
	7	1.0000	1.0000	1.0000	0.9998	0.9988	0.9957	0.9878	0.9707	0.9390	0.8867
	8	1.0000	1.0000	1.0000	1.0000	0.9999	0.9994	0.9980	0.9941	0.9852	0.9673
	9	1.0000	1.0000	1.0000	1.0000	1.0000	1.0000	0.9998	0.9993	0.9978	0.9941

TABLE II: *continued*

n	x	0.05	0.10	0.15	0.20	0.25	0.30	0.35	0.40	0.45	0.50
						p					
	10	1.0000	1.0000	1.0000	1.0000	1.0000	1.0000	1.0000	1.0000	0.9998	0.9995
	11	1.0000	1.0000	1.0000	1.0000	1.0000	1.0000	1.0000	1.0000	1.0000	1.0000
12	0	0.5404	0.2824	0.1422	0.0687	0.0317	0.0138	0.0057	0.0022	0.0008	0.0002
	1	0.8816	0.6590	0.4435	0.2749	0.1584	0.0850	0.0424	0.0196	0.0083	0.0032
	2	0.9804	0.8891	0.7358	0.5583	0.3907	0.2528	0.1513	0.0834	0.0421	0.0193
	3	0.9978	0.9744	0.9078	0.7946	0.6488	0.4925	0.3467	0.2253	0.1345	0.0730
	4	0.9998	0.9957	0.9761	0.9274	0.8424	0.7237	0.5833	0.4382	0.3044	0.1938
	5	1.0000	0.9995	0.9954	0.9806	0.9456	0.8822	0.7873	0.6652	0.5269	0.3872
	6	1.0000	0.9999	0.9993	0.9961	0.9857	0.9614	0.9154	0.8418	0.7393	0.6128
	7	1.0000	1.0000	0.9999	0.9994	0.9972	0.9905	0.9745	0.9427	0.8883	0.8062
	8	1.0000	1.0000	1.0000	0.9999	0.9996	0.9983	0.9944	0.9847	0.9644	0.9270
	9	1.0000	1.0000	1.0000	1.0000	1.0000	0.9998	0.9992	0.9972	0.9921	0.9807
	10	1.0000	1.0000	1.0000	1.0000	1.0000	1.0000	0.9999	0.9997	0.9989	0.9968
	11	1.0000	1.0000	1.0000	1.0000	1.0000	1.0000	1.0000	1.0000	0.9999	0.9998
	12	1.0000	1.0000	1.0000	1.0000	1.0000	1.0000	1.0000	1.0000	1.0000	1.0000
13	0	0.5133	0.2542	0.1209	0.0550	0.0238	0.0097	0.0037	0.0013	0.0004	0.0001
	1	0.8646	0.6213	0.3983	0.2336	0.1267	0.0637	0.0296	0.0126	0.0049	0.0017
	2	0.9755	0.8661	0.6920	0.5017	0.3326	0.2025	0.1132	0.0579	0.0269	0.0112
	3	0.9969	0.9658	0.8820	0.7473	0.5843	0.4206	0.2783	0.1686	0.0929	0.0461
	4	0.9997	0.9935	0.9658	0.9009	0.7940	0.6543	0.5005	0.3530	0.2279	0.1334
	5	1.0000	0.9991	0.9924	0.9700	0.9198	0.8346	0.7159	0.5744	0.4268	0.2905
	6	1.0000	0.9999	0.9987	0.9930	0.9757	0.9376	0.8705	0.7712	0.6437	0.5000
	7	1.0000	1.0000	0.9998	0.9988	0.9944	0.9818	0.9538	0.9023	0.8212	0.7095
	8	1.0000	1.0000	1.0000	0.9998	0.9990	0.9960	0.9874	0.9679	0.9302	0.8666
	9	1.0000	1.0000	1.0000	1.0000	0.9999	0.9993	0.9975	0.9922	0.9797	0.9539
	10	1.0000	1.0000	1.0000	1.0000	1.0000	0.9999	0.9997	0.9987	0.9959	0.9888
	11	1.0000	1.0000	1.0000	1.0000	1.0000	1.0000	1.0000	0.9999	0.9995	0.9983
	12	1.0000	1.0000	1.0000	1.0000	1.0000	1.0000	1.0000	1.0000	1.0000	0.9999
	13	1.0000	1.0000	1.0000	1.0000	1.0000	1.0000	1.0000	1.0000	1.0000	1.0000
14	0	0.4877	0.2288	0.1028	0.0440	0.0178	0.0068	0.0024	0.0008	0.0002	0.0001
	1	0.8470	0.5846	0.3567	0.1979	0.1010	0.0475	0.0205	0.0081	0.0029	0.0009
	2	0.9699	0.8416	0.6479	0.4481	0.2811	0.1608	0.0839	0.0398	0.0170	0.0065
	3	0.9958	0.9559	0.8535	0.6982	0.5213	0.3552	0.2205	0.1243	0.0632	0.0287
	4	0.9996	0.9908	0.9533	0.8702	0.7415	0.5842	0.4227	0.2793	0.1672	0.0898
	5	1.0000	0.9985	0.9885	0.9561	0.8883	0.7805	0.6405	0.4859	0.3373	0.2120
	6	1.0000	0.9998	0.9978	0.9884	0.9617	0.9067	0.8164	0.6925	0.5461	0.3953
	7	1.0000	1.0000	0.9997	0.9976	0.9897	0.9685	0.9247	0.8499	0.7414	0.6047
	8	1.0000	1.0000	1.0000	0.9996	0.9978	0.9917	0.9757	0.9417	0.8811	0.7880
	9	1.0000	1.0000	1.0000	1.0000	0.9997	0.9983	0.9940	0.9825	0.9574	0.9102
	10	1.0000	1.0000	1.0000	1.0000	1.0000	0.9998	0.9989	0.9961	0.9886	0.9713

TABLE II: *continued*

n	x	0.05	0.10	0.15	0.20	0.25	0.30	0.35	0.40	0.45	0.50
							p				
	11	1.0000	1.0000	1.0000	1.0000	1.0000	1.0000	0.9999	0.9994	0.9978	0.9935
	12	1.0000	1.0000	1.0000	1.0000	1.0000	1.0000	1.0000	0.9999	0.9997	0.9991
	13	1.0000	1.0000	1.0000	1.0000	1.0000	1.0000	1.0000	1.0000	1.0000	0.9999
	14	1.0000	1.0000	1.0000	1.0000	1.0000	1.0000	1.0000	1.0000	1.0000	1.0000
15	0	0.4633	0.2059	0.0874	0.0352	0.0134	0.0047	0.0016	0.0005	0.0001	0.0000
	1	0.8290	0.5490	0.3186	0.1671	0.0802	0.0353	0.0142	0.0052	0.0017	0.0005
	2	0.9638	0.8159	0.6042	0.3980	0.2361	0.1268	0.0617	0.0271	0.0107	0.0037
	3	0.9945	0.9444	0.8227	0.6482	0.4613	0.2969	0.1727	0.0905	0.0424	0.0176
	4	0.9994	0.9873	0.9383	0.8358	0.6865	0.5155	0.3519	0.2173	0.1204	0.0592
	5	0.9999	0.9978	0.9832	0.9389	0.8516	0.7216	0.5643	0.4032	0.2608	0.1509
	6	1.0000	0.9997	0.9964	0.9819	0.9434	0.8689	0.7548	0.6098	0.4522	0.3036
	7	1.0000	1.0000	0.9994	0.9958	0.9827	0.9500	0.8868	0.7869	0.6535	0.5000
	8	1.0000	1.0000	0.9999	0.9992	0.9958	0.9848	0.9578	0.9050	0.8182	0.6964
	9	1.0000	1.0000	1.0000	0.9999	0.9992	0.9963	0.9876	0.9662	0.9231	0.8491
	10	1.0000	1.0000	1.0000	1.0000	0.9999	0.9993	0.9972	0.9907	0.9745	0.9408
	11	1.0000	1.0000	1.0000	1.0000	1.0000	0.9999	0.9995	0.9981	0.9937	0.9824
	12	1.0000	1.0000	1.0000	1.0000	1.0000	1.0000	0.9999	0.9987	0.9989	0.9963
	13	1.0000	1.0000	1.0000	1.0000	1.0000	1.0000	1.0000	1.0000	0.9999	0.9995
	14	1.0000	1.0000	1.0000	1.0000	1.0000	1.0000	1.0000	1.0000	1.0000	1.0000
	15	1.0000	1.0000	1.0000	1.0000	1.0000	1.0000	1.0000	1.0000	1.0000	1.0000
16	0	0.4401	0.1853	0.0743	0.0281	0.0100	0.0033	0.0010	0.0003	0.0001	0.0000
	1	0.8108	0.5147	0.2839	0.1407	0.0635	0.0261	0.0098	0.0033	0.0010	0.0003
	2	0.9571	0.7892	0.5614	0.3518	0.1971	0.0994	0.0451	0.0183	0.0066	0.0021
	3	0.9930	0.9316	0.7899	0.5981	0.4050	0.2459	0.1339	0,0651	0.0281	0.0106
	4	0.9991	0.9830	0.9209	0.7982	0.6302	0.4499	0.2892	0.1666	0.0853	0.0384
	5	0.9999	0.9967	0.9765	0.9183	0.8103	0.6598	0.4900	0.3288	0.1976	0.1051
	6	1.0000	0.9995	0.9944	0.9733	0.9204	0.8247	0.6881	0.5272	0.3660	0.2272
	7	1.0000	0.9999	0.9989	0.9930	0.9729	0.9256	0.8406	0.7161	0.5629	0.4018
	8	1.0000	1.0000	0.9998	0.9985	0.9925	0.9743	0.9329	0.8577	0.7441	0.5982
	9	1.0000	1.0000	1.0000	0.9998	0.9984	0.9929	0.9771	0.9417	0.8759	0.7728
	10	1.0000	1.0000	1.0000	1.0000	0.9997	0.9984	0.9938	0.9809	0.9514	0.8949
	11	1.0000	1.0000	1.0000	1.0000	1.0000	0.9997	0.9987	0.9951	0.9851	0.9616
	12	1.0000	1.0000	1.0000	1.0000	1.0000	1.0000	0.9998	0.9991	0.9965	0.9894
	13	1.0000	1.0000	1.0000	1.0000	1.0000	1.0000	1.0000	0.9999	0.9994	0.9979
	14	1.0000	1.0000	1.0000	1.0000	1.0000	1.0000	1.0000	1.0000	0.9999	0.9997
	15	1.0000	1.0000	1.0000	1.0000	1.0000	1.0000	1.0000	1.0000	1.0000	1.0000
	16	1.0000	1.0000	1.0000	1.0000	1.0000	1.0000	1.0000	1.0000	1.0000	1.0000
20	0	0.3585	0.1216	0.0388	0.0115	0.0032	0.0008	0.0002	0.0000	0.0000	0.0000
	1	0.7358	0.3917	0.1756	0.0692	0.0243	0.0076	0.0021	0.0005	0.0001	0.0000
	2	0.9245	0.6769	0.4049	0.2061	0.0913	0.0355	0.0121	0.0036	0.0009	0.0002
	3	0.9841	0.8670	0.6477	0.4114	0.2252	0.1071	0.0444	0.0160	0.0049	0.0013
	4	0.9974	0.9568	0.8298	0.6296	0.4148	0.2375	0.1182	0.0510	0.0189	0.0059

TABLE II: *continued*

n	x	0.05	0.10	0.15	0.20	0.25	0.30	0.35	0.40	0.45	0.50
	5	0.9997	0.9887	0.9327	0.8042	0.6172	0.4164	0.2454	0.1256	0.0553	0.0207
	6	1.0000	0.9976	0.9781	0.9133	0.7858	0.6080	0.4166	0.2500	0.1299	0.0577
	7	1.0000	0.9996	0.9941	0.9679	0.8982	0.7723	0.6010	0.4159	0.2520	0.1316
	8	1.0000	0.9999	0.9987	0.9900	0.9591	0.8867	0.7624	0.5956	0.4143	0.2517
	9	1.0000	1.0000	0.9998	0.9974	0.9861	0.9520	0.8782	0.7553	0.5914	0.4119
	10	1.0000	1.0000	1.0000	0.9994	0.9961	0.9829	0.9468	0.8725	0.7507	0.5881
	11	1.0000	1.0000	1.0000	0.9999	0.9991	0.9949	0.9804	0.9435	0.8692	0.7483
	12	1.0000	1.0000	1.0000	1.0000	0.9998	0.9987	0.9940	0.9790	0.9420	0.8684
	13	1.0000	1.0000	1.0000	1.0000	1.0000	0.9997	0.9985	0.9935	0.9786	0.9423
	14	1.0000	1.0000	1.0000	1.0000	1.0000	1.0000	0.9997	0.9984	0.9936	0.9793
	15	1.0000	1.0000	1.0000	1.0000	1.0000	1.0000	1.0000	0.9997	0.9985	0.9941
	16	1.0000	1.0000	1.0000	1.0000	1.0000	1.0000	1.0000	1.0000	0.9997	0.9987
	17	1.0000	1.0000	1.0000	1.0000	1.0000	1.0000	1.0000	1.0000	1.0000	0.9998
	18	1.0000	1.0000	1.0000	1.0000	1.0000	1.0000	1.0000	1.0000	1.0000	1.0000
	19	1.0000	1.0000	1.0000	1.0000	1.0000	1.0000	1.0000	1.0000	1.0000	1.0000
	20	1.0000	1.0000	1.0000	1.0000	1.0000	1.0000	1.0000	1.0000	1.0000	1.0000
25	0	0.2774	0.0718	0.0172	0.0038	0.0008	0.0001	0.0000	0.0000	0.0000	0.0000
	1	0.6424	0.2712	0.0931	0.0274	0.0070	0.0016	0.0003	0.0001	0.0000	0.0000
	2	0.8729	0.5371	0.2537	0.0982	0.0321	0.0090	0.0021	0.0004	0.0001	0.0000
	3	0.9659	0.7636	0.4711	0.2340	0.0962	0.0332	0.0097	0.0024	0.0005	0.0001
	4	0.9928	0.9020	0.6821	0.4207	0.2137	0.0905	0.0320	0.0095	0.0023	0.0005
	5	0.9988	0.9666	0.8385	0.6167	0.3783	0.1935	0.0826	0.0294	0.0086	0.0020
	6	0.9998	0.9905	0.9305	0.7800	0.5611	0.3407	0.1734	0.0736	0.0258	0.0073
	7	1.0000	0.9977	0.9745	0.8909	0.7265	0.5118	0.3061	0.1536	0.0639	0.0216
	8	1.0000	0.9995	0.9920	0.9532	0.8506	0.6769	0.4668	0.2735	0.1340	0.0539
	9	1.0000	0.9999	0.9979	0.9827	0.9287	0.8106	0.6303	0.4246	0.2424	0.1148
	10	1.0000	1.0000	0.9995	0.9944	0.9703	0.9022	0.7712	0.5858	0.3843	0.2122
	11	1.0000	1.0000	0.9999	0.9985	0.9893	0.9558	0.8746	0.7323	0.5426	0.3450
	12	1.0000	1.0000	1.0000	0.9996	0.9966	0.9825	0.9396	0.8462	0.6937	0.5000
	13	1.0000	1.0000	1.0000	0.9999	0.9991	0.9940	0.9745	0.9222	0.8173	0.6550
	14	1.0000	1.0000	1.0000	1.0000	0.9998	0.9982	0.9907	0.9656	0.9040	0.7878
	15	1.0000	1.0000	1.0000	1.0000	1.0000	0.9995	0.9971	0.9868	0.9560	0.8852
	16	1.0000	1.0000	1.0000	1.0000	1.0000	0.9999	0.9992	0.9957	0.9826	0.9461
	17	1.0000	1.0000	1.0000	1.0000	1.0000	1.0000	0.9998	0.9988	0.9942	0.9784
	18	1.0000	1.0000	1.0000	1.0000	1.0000	1.0000	1.0000	0.9997	0.9984	0.9927
	19	1.0000	1.0000	1.0000	1.0000	1.0000	1.0000	1.0000	0.9999	0.9996	0.9980
	20	1.0000	1.0000	1.0000	1.0000	1.0000	1.0000	1.0000	1.0000	0.9999	0.9995
	21	1.0000	1.0000	1.0000	1.0000	1.0000	1.0000	1.0000	1.0000	1.0000	0.9999
	22	1.0000	1.0000	1.0000	1.0000	1.0000	1.0000	1.0000	1.0000	1.0000	1.0000
	23	1.0000	1.0000	1.0000	1.0000	1.0000	1.0000	1.0000	1.0000	1.0000	1.0000
	24	1.0000	1.0000	1.0000	1.0000	1.0000	1.0000	1.0000	1.0000	1.0000	1.0000
	25	1.0000	1.0000	1.0000	1.0000	1.0000	1.0000	1.0000	1.0000	1.0000	1.0000

TABLE III: The Poisson Distribution

$$F(x) = P(X \leq x) = \sum_{k=0}^{x} \frac{\lambda^k e^{-\lambda}}{k!}$$

x	$\lambda = E(X)$									
	0.1	0.2	0.3	0.4	0.5	0.6	0.7	0.8	0.9	1.0
0	0.905	0.819	0.741	0.670	0.607	0.549	0.497	0.449	0.407	0.368
1	0.995	0.982	0.963	0.938	0.910	0.878	0.844	0.809	0.772	0.736
2	1.000	0.999	0.996	0.992	0.986	0.977	0.966	0.953	0.937	0.920
3	1.000	1.000	1.000	0.999	0.998	0.997	0.994	0.991	0.987	0.981
4	1.000	1.000	1.000	1.000	1.000	1.000	0.999	0.999	0.998	0.996
5	1.000	1.000	1.000	1.000	1.000	1.000	1.000	1.000	1.000	0.999
6	1.000	1.000	1.000	1.000	1.000	1.000	1.000	1.000	1.000	1.000

x	1.1	1.2	1.3	1.4	1.5	1.6	1.7	1.8	1.9	2.0
0	0.333	0.301	0.273	0.247	0.223	0.202	0.183	0.165	0.150	0.135
1	0.699	0.663	0.627	0.592	0.558	0.525	0.493	0.463	0.434	0.406
2	0.900	0.879	0.857	0.833	0.809	0.783	0.757	0.731	0.704	0.677
3	0.974	0.966	0.957	0.946	0.934	0.921	0.907	0.891	0.875	0.857
4	0.995	0.992	0.989	0.986	0.981	0.976	0.970	0.964	0.956	0.947
5	0.999	0.998	0.998	0.997	0.996	0.994	0.992	0.990	0.987	0.983
6	1.000	1.000	1.000	0.999	0.999	0.999	0.998	0.997	0.997	0.995
7	1.000	1.000	1.000	1.000	1.000	1.000	1.000	0.999	0.999	0.999
8	1.000	1.000	1.000	1.000	1.000	1.000	1.000	1.000	1.000	1.000

x	2.2	2.4	2.6	2.8	3.0	3.2	3.4	3.6	3.8	4.0
0	0.111	0.091	0.074	0.061	0.050	0.041	0.033	0.027	0.022	0.018
1	0.355	0.308	0.267	0.231	0.199	0.171	0.147	0.126	0.107	0.092
2	0.623	0.570	0.518	0.469	0.423	0.380	0.340	0.303	0.269	0.238
3	0.819	0.779	0.736	0.692	0.647	0.603	0.558	0.515	0.473	0.433
4	0.928	0.904	0.877	0.848	0.815	0.781	0.744	0.706	0.668	0.629
5	0.975	0.964	0.951	0.935	0.916	0.895	0.871	0.844	0.816	0.785
6	0.993	0.988	0.983	0.976	0.966	0.955	0.942	0.927	0.909	0.889
7	0.998	0.997	0.995	0.992	0.988	0.983	0.977	0.969	0.960	0.949
8	1.000	0.999	0.999	0.998	0.996	0.994	0.992	0.988	0.984	0.979
9	1.000	1.000	1.000	0.999	0.999	0.998	0.997	0.996	0.994	0.992
10	1.000	1.000	1.000	1.000	1.000	1.000	0.999	0.999	0.998	0.997
11	1.000	1.000	1.000	1.000	1.000	1.000	1.000	1.000	0.999	0.999
12	1.000	1.000	1.000	1.000	1.000	1.000	1.000	1.000	1.000	1.000

TABLE III: *continued*

x	4.2	4.4	4.6	4.8	5.0	5.2	5.4	5.6	5.8	6.0
0	0.015	0.012	0.010	0.008	0.007	0.006	0.005	0.004	0.003	0.002
1	0.078	0.066	0.056	0.048	0.040	0.034	0.029	0.024	0.021	0.017
2	0.210	0.185	0.163	0.143	0.125	0.109	0.095	0.082	0.072	0.062
3	0.395	0.359	0.326	0.294	0.265	0.238	0.213	0.191	0.170	0.151
4	0.590	0.551	0.513	0.476	0.440	0.406	0.373	0.342	0.313	0.285
5	0.753	0.720	0.686	0.651	0.616	0.581	0.546	0.512	0.478	0.446
6	0.867	0.844	0.818	0.791	0.762	0.732	0.702	0.670	0.638	0.606
7	0.936	0.921	0.905	0.887	0.867	0.845	0.822	0.797	0.771	0.744
8	0.972	0.964	0.955	0.944	0.932	0.918	0.903	0.886	0.867	0.847
9	0.989	0.985	0.980	0.975	0.968	0.960	0.951	0.941	0.929	0.916
10	0.996	0.994	0.992	0.990	0.986	0.982	0.977	0.972	0.965	0.957
11	0.999	0.998	0.997	0.996	0.995	0.993	0.990	0.988	0.984	0.980
12	1.000	0.999	0.999	0.999	0.998	0.997	0.996	0.995	0.993	0.991
13	1.000	1.000	1.000	1.000	0.999	0.999	0.999	0.998	0.997	0.996
14	1.000	1.000	1.000	1.000	1.000	1.000	0.999	0.999	0.999	0.999
15	1.000	1.000	1.000	1.000	1.000	1.000	1.000	1.000	1.000	0.999
16	1.000	1.000	1.000	1.000	1.000	1.000	1.000	1.000	1.000	1.000

x	6.5	7.0	7.5	8.0	8.5	9.0	9.5	10.0	10.5	11.0
0	0.002	0.001	0.001	0.000	0.000	0.000	0.000	0.000	0.000	0.000
1	0.011	0.007	0.005	0.003	0.002	0.001	0.001	0.000	0.000	0.000
2	0.043	0.030	0.020	0.014	0.009	0.006	0.004	0.003	0.002	0.001
3	0.112	0.082	0.059	0.042	0.030	0.021	0.015	0.010	0.007	0.005
4	0.224	0.173	0.132	0.100	0.074	0.055	0.040	0.029	0.021	0.015
5	0.369	0.301	0.241	0.191	0.150	0.116	0.089	0.067	0.050	0.038
6	0.527	0.450	0.378	0.313	0.256	0.207	0.165	0.130	0.102	0.079
7	0.673	0.599	0.525	0.453	0.386	0.324	0.269	0.220	0.179	0.143
8	0.792	0.729	0.662	0.593	0.523	0.456	0.392	0.333	0.279	0.232
9	0.877	0.830	0.776	0.717	0.653	0.587	0.522	0.458	0.397	0.341
10	0.933	0.901	0.862	0.816	0.763	0.706	0.645	0.583	0.521	0.460
11	0.966	0.947	0.921	0.888	0.849	0.803	0.752	0.697	0.639	0.579
12	0.984	0.973	0.957	0.936	0.909	0.876	0.836	0.792	0.742	0.689
13	0.993	0.987	0.978	0.966	0.949	0.926	0.898	0.864	0.825	0.781
14	0.997	0.994	0.990	0.983	0.973	0.959	0.940	0.917	0.888	0.854
15	0.999	0.998	0.995	0.992	0.986	0.978	0.967	0.951	0.932	0.907
16	1.000	0.999	0.998	0.996	0.993	0.989	0.982	0.973	0.960	0.944
17	1.000	1.000	0.999	0.998	0.997	0.995	0.991	0.986	0.978	0.968
18	1.000	1.000	1.000	0.999	0.999	0.998	0.096	0.993	0.988	0.982
19	1.000	1.000	1.000	1.000	0.999	0.999	0.998	0.997	0.994	0.991
20	1.000	1.000	1.000	1.000	1.000	1.000	0.999	0.998	0.997	0.995
21	1.000	1.000	1.000	1.000	1.000	1.000	1.000	0.999	0.999	0.998
22	1.000	1.000	1.000	1.000	1.000	1.000	1.000	1.000	0.999	0.999
23	1.000	1.000	1.000	1.000	1.000	1.000	1.000	1.000	1.000	1.000

TABLE III: *continued*

x	11.5	12.0	12.5	13.0	13.5	14.0	14.5	15.0	15.5	16.0
0	0.000	0.000	0.000	0.000	0.000	0.000	0.000	0.000	0.000	0.000
1	0.000	0.000	0.000	0.000	0.000	0.000	0.000	0.000	0.000	0.000
2	0.001	0.001	0.000	0.000	0.000	0.000	0.000	0.000	0.000	0.000
3	0.003	0.002	0.002	0.001	0.001	0.000	0.000	0.000	0.000	0.000
4	0.011	0.008	0.005	0.004	0.003	0.002	0.001	0.001	0.001	0.000
5	0.028	0.020	0.015	0.011	0.008	0.006	0.004	0.003	0.002	0.001
6	0.060	0.046	0.035	0.026	0.019	0.014	0.010	0.008	0.006	0.004
7	0.114	0.090	0.070	0.054	0.041	0.032	0.024	0.018	0.013	0.010
8	0.191	0.155	0.125	0.100	0.079	0.062	0.048	0.037	0.029	0.022
9	0.289	0.242	0.201	0.166	0.135	0.109	0.088	0.070	0.055	0.043
10	0.402	0.347	0.297	0.252	0.211	0.176	0.145	0.118	0.096	0.077
11	0.520	0.462	0.406	0.353	0.304	0.260	0.220	0.185	0.154	0.127
12	0.633	0.576	0.519	0.463	0.409	0.358	0.311	0.268	0.228	0.193
13	0.733	0.682	0.629	0.573	0.518	0.464	0.413	0.363	0.317	0.275
14	0.815	0.772	0.725	0.675	0.623	0.570	0.518	0.466	0.415	0.368
15	0.878	0.844	0.806	0.764	0.718	0.669	0.619	0.568	0.517	0.467
16	0.924	0.899	0.869	0.835	0.798	0.756	0.711	0.664	0.615	0.566
17	0.954	0.937	0.916	0.890	0.861	0.827	0.790	0.749	0.705	0.659
18	0.974	0.963	0.948	0.930	0.908	0.883	0.853	0.819	0.782	0.742
19	0.986	0.979	0.969	0.957	0.942	0.923	0.901	0.875	0.846	0.812
20	0.992	0.988	0.983	0.975	0.965	0.952	0.936	0.917	0.894	0.868
21	0.996	0.994	0.991	0.986	0.980	0.971	0.960	0.947	0.930	0.911
22	0.999	0.997	0.995	0.992	0.989	0.983	0.976	0.967	0.956	0.942
23	0.999	0.999	0.998	0.996	0.994	0.991	0.986	0.981	0.973	0.963
24	1.000	0.999	0.999	0.998	0.997	0.995	0.992	0.989	0.984	0.978
25	1.000	1.000	0.999	0.999	0.998	0.997	0.996	0.994	0.991	0.987
26	1.000	1.000	1.000	1.000	0.999	0.999	0.998	0.997	0.995	0.993
27	1.000	1.000	1.000	1.000	1.000	0.999	0.999	0.998	0.997	0.996
28	1.000	1.000	1.000	1.000	1.000	1.000	0.999	0.999	0.999	0.998
29	1.000	1.000	1.000	1.000	1.000	1.000	1.000	1.000	0.999	0.999
30	1.000	1.000	1.000	1.000	1.000	1.000	1.000	1.000	1.000	0.999
31	1.000	1.000	1.000	1.000	1.000	1.000	1.000	1.000	1.000	1.000
32	1.000	1.000	1.000	1.000	1.000	1.000	1.000	1.000	1.000	1.000
33	1.000	1.000	1.000	1.000	1.000	1.000	1.000	1.000	1.000	1.000
34	1.000	1.000	1.000	1.000	1.000	1.000	1.000	1.000	1.000	1.000
35	1.000	1.000	1.000	1.000	1.000	1.000	1.000	1.000	1.000	1.000

TABLE IV: The Chi-Square Distribution

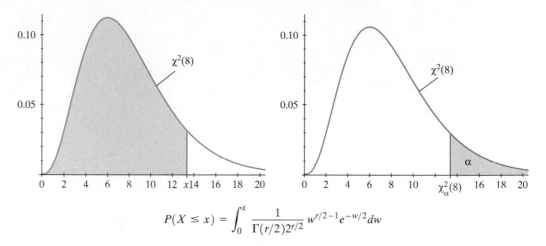

$$P(X \leq x) = \int_0^x \frac{1}{\Gamma(r/2)2^{r/2}} w^{r/2-1}e^{-w/2}dw$$

	$P(X \leq r)$							
	0.010	0.025	0.050	0.100	0.900	0.950	0.975	0.990
r	$\chi^2_{0.99}(r)$	$\chi^2_{0.975}(r)$	$\chi^2_{0.95}(r)$	$\chi^2_{0.90}(r)$	$\chi^2_{0.10}(r)$	$\chi^2_{0.05}(r)$	$\chi^2_{0.025}(r)$	$\chi^2_{0.01}(r)$
1	0.000	0.001	0.004	0.016	2.706	3.841	5.024	6.635
2	0.020	0.051	0.103	0.211	4.605	5.991	7.378	9.210
3	0.115	0.216	0.352	0.584	6.251	7.815	9.348	11.34
4	0.297	0.484	0.711	1.064	7.779	9.488	11.14	13.28
5	0.554	0.831	1.145	1.610	9.236	11.07	12.83	15.09
6	0.872	1.237	1.635	2.204	10.64	12.59	14.45	16.81
7	1.239	1.690	2.167	2.833	12.02	14.07	16.01	18.48
8	1.646	2.180	2.733	3.490	13.36	15.51	17.54	20.09
9	2.088	2.700	3.325	4.168	14.68	16.92	19.02	21.67
10	2.558	3.247	3.940	4.865	15.99	18.31	20.48	23.21
11	3.053	3.816	4.575	5.578	17.28	19.68	21.92	24.72
12	3.571	4.404	5.226	6.304	18.55	21.03	23.34	26.22
13	4.107	5.009	5.892	7.042	19.81	22.36	24.74	27.69
14	4.660	5.629	6.571	7.790	21.06	23.68	26.12	29.14
15	5.229	6.262	7.261	8.547	22.31	25.00	27.49	30.58
16	5.812	6.908	7.962	9.312	23.54	26.30	28.84	32.00
17	6.408	7.564	8.672	10.08	24.77	27.59	30.19	33.41
18	7.015	8.231	9.390	10.86	25.99	28.87	31.53	34.80
19	7.633	8.907	10.12	11.65	27.20	30.14	32.85	36.19
20	8.260	9.591	10.85	12.44	28.41	31.41	34.17	37.57
21	8.897	10.28	11.59	13.24	29.62	32.67	35.48	38.93
22	9.542	10.98	12.34	14.04	30.81	33.92	36.78	40.29
23	10.20	11.69	13.09	14.85	32.01	35.17	38.08	41.64
24	10.86	12.40	13.85	15.66	33.20	36.42	39.36	42.98
25	11.52	13.12	14.61	16.47	34.38	37.65	40.65	44.31
26	12.20	13.84	15.38	17.29	35.56	38.88	41.92	45.64
27	12.88	14.57	16.15	18.11	36.74	40.11	43.19	46.96
28	13.56	15.31	16.93	18.94	37.92	41.34	44.46	48.28
29	14.26	16.05	17.71	19.77	39.09	42.56	45.72	49.59
30	14.95	16.79	18.49	20.60	40.26	43.77	46.98	50.89
40	22.16	24.43	26.51	29.05	51.80	55.76	59.34	63.69
50	29.71	32.36	34.76	37.69	63.17	67.50	71.42	76.15
60	37.48	40.48	43.19	46.46	74.40	79.08	83.30	88.38
70	45.44	48.76	51.74	55.33	85.53	90.53	95.02	100.4
80	53.34	57.15	60.39	64.28	96.58	101.9	106.6	112.3

This table is abridged and adapted from Table III in *Biometrika Tables for Statisticians*, edited by E.S.Pearson and H.O.Hartley.

TABLE Va: The Normal Distribution

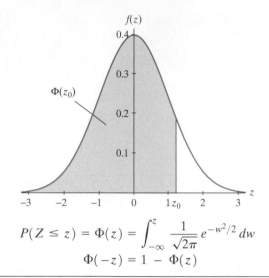

$$P(Z \leq z) = \Phi(z) = \int_{-\infty}^{z} \frac{1}{\sqrt{2\pi}} e^{-w^2/2} \, dw$$

$$\Phi(-z) = 1 - \Phi(z)$$

z	0.00	0.01	0.02	0.03	0.04	0.05	0.06	0.07	0.08	0.09
0.0	0.5000	0.5040	0.5080	0.5120	0.5160	0.5199	0.5239	0.5279	0.5319	0.5359
0.1	0.5398	0.5438	0.5478	0.5517	0.5557	0.5596	0.5636	0.5675	0.5714	0.5753
0.2	0.5793	0.5832	0.5871	0.5910	0.5948	0.5987	0.6026	0.6064	0.6103	0.6141
0.3	0.6179	0.6217	0.6255	0.6293	0.6331	0.6368	0.6406	0.6443	0.6480	0.6517
0.4	0.6554	0.6591	0.6628	0.6664	0.6700	0.6736	0.6772	0.6808	0.6844	0.6879
0.5	0.6915	0.6950	0.6985	0.7019	0.7054	0.7088	0.7123	0.7157	0.7190	0.7224
0.6	0.7257	0.7291	0.7324	0.7357	0.7389	0.7422	0.7454	0.7486	0.7517	0.7549
0.7	0.7580	0.7611	0.7642	0.7673	0.7703	0.7734	0.7764	0.7794	0.7823	0.7852
0.8	0.7881	0.7910	0.7939	0.7967	0.7995	0.8023	0.8051	0.8078	0.8106	0.8133
0.9	0.8159	0.8186	0.8212	0.8238	0.8264	0.8289	0.8315	0.8340	0.8365	0.8389
1.0	0.8413	0.8438	0.8461	0.8485	0.8508	0.8531	0.8554	0.8577	0.8599	0.8621
1.1	0.8643	0.8665	0.8686	0.8708	0.8729	0.8749	0.8770	0.8790	0.8810	0.8830
1.2	0.8849	0.8869	0.8888	0.8907	0.8925	0.8944	0.8962	0.8980	0.8997	0.9015
1.3	0.9032	0.9049	0.9066	0.9082	0.9099	0.9115	0.9131	0.9147	0.9162	0.9177
1.4	0.9192	0.9207	0.9222	0.9236	0.9251	0.9265	0.9279	0.9292	0.9306	0.9319
1.5	0.9332	0.9345	0.9357	0.9370	0.9382	0.9394	0.9406	0.9418	0.9429	0.9441
1.6	0.9452	0.9463	0.9474	0.9484	0.9495	0.9505	0.9515	0.9525	0.9535	0.9545
1.7	0.9554	0.9564	0.9573	0.9582	0.9591	0.9599	0.9608	0.9616	0.9625	0.9633
1.8	0.9641	0.9649	0.9656	0.9664	0.9671	0.9678	0.9686	0.9693	0.9699	0.9706
1.9	0.9713	0.9719	0.9726	0.9732	0.9738	0.9744	0.9750	0.9756	0.9761	0.9767
2.0	0.9772	0.9778	0.9783	0.9788	0.9793	0.9798	0.9803	0.9808	0.9812	0.9817
2.1	0.9821	0.9826	0.9830	0.9834	0.9838	0.9842	0.9846	0.9850	0.9854	0.9857
2.2	0.9861	0.9864	0.9868	0.9871	0.9875	0.9878	0.9881	0.9884	0.9887	0.9890
2.3	0.9893	0.9896	0.9898	0.9901	0.9904	0.9906	0.9909	0.9911	0.9913	0.9916
2.4	0.9918	0.9920	0.9922	0.9925	0.9927	0.9929	0.9931	0.9932	0.9934	0.9936
2.5	0.9938	0.9940	0.9941	0.9943	0.9945	0.9946	0.9948	0.9949	0.9951	0.9952
2.6	0.9953	0.9955	0.9956	0.9957	0.9959	0.9960	0.9961	0.9962	0.9963	0.9964
2.7	0.9965	0.9966	0.9967	0.9968	0.9969	0.9970	0.9971	0.9972	0.9973	0.9974
2.8	0.9974	0.9975	0.9976	0.9977	0.9977	0.9978	0.9979	0.9979	0.9980	0.9981
2.9	0.9981	0.9982	0.9982	0.9983	0.9984	0.9984	0.9985	0.9985	0.9986	0.9986
3.0	0.9987	0.9987	0.9987	0.9988	0.9988	0.9989	0.9989	0.9989	0.9990	0.9990

α	0.400	0.300	0.200	0.100	0.050	0.025	0.020	0.010	0.005	0.001
z_α	0.253	0.524	0.842	1.282	1.645	1.960	2.054	2.326	2.576	3.090
$z_{\alpha/2}$	0.842	1.036	1.282	1.645	1.960	2.240	2.326	2.576	2.807	3.291

TABLE Vb: The Normal Distribution

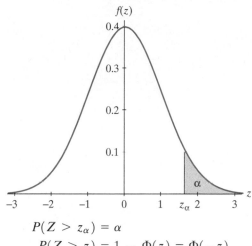

$$P(Z > z_\alpha) = \alpha$$
$$P(Z > z) = 1 - \Phi(z) = \Phi(-z)$$

z_α	0.00	0.01	0.02	0.03	0.04	0.05	0.06	0.07	0.08	0.09
0.0	0.5000	0.4960	0.4920	0.4880	0.4840	0.4801	0.4761	0.4721	0.4681	0.4641
0.1	0.4602	0.4562	0.4522	0.4483	0.4443	0.4404	0.4364	0.4325	0.4286	0.4247
0.2	0.4207	0.4168	0.4129	0.4090	0.4052	0.4013	0.3974	0.3936	0.3897	0.3859
0.3	0.3821	0.3783	0.3745	0.3707	0.3669	0.3632	0.3594	0.3557	0.3520	0.3483
0.4	0.3446	0.3409	0.3372	0.3336	0.3300	0.3264	0.3228	0.3192	0.3156	0.3121
0.5	0.3085	0.3050	0.3015	0.2981	0.2946	0.2912	0.2877	0.2843	0.2810	0.2776
0.6	0.2743	0.2709	0.2676	0.2643	0.2611	0.2578	0.2546	0.2514	0.2483	0.2451
0.7	0.2420	0.2389	0.2358	0.2327	0.2296	0.2266	0.2236	0.2206	0.2177	0.2148
0.8	0.2119	0.2090	0.2061	0.2033	0.2005	0.1977	0.1949	0.1922	0.1894	0.1867
0.9	0.1841	0.1814	0.1788	0.1762	0.1736	0.1711	0.1685	0.1660	0.1635	0.1611
1.0	0.1587	0.1562	0.1539	0.1515	0.1492	0.1469	0.1446	0.1423	0.1401	0.1379
1.1	0.1357	0.1335	0.1314	0.1292	0.1271	0.1251	0.1230	0.1210	0.1190	0.1170
1.2	0.1151	0.1131	0.1112	0.1093	0.1075	0.1056	0.1038	0.1020	0.1003	0.0985
1.3	0.0968	0.0951	0.0934	0.0918	0.0901	0.0885	0.0869	0.0853	0.0838	0.0823
1.4	0.0808	0.0793	0.0778	0.0764	0.0749	0.0735	0.0721	0.0708	0.0694	0.0681
1.5	0.0668	0.0655	0.0643	0.0630	0.0618	0.0606	0.0594	0.0582	0.0571	0.0559
1.6	0.0548	0.0537	0.0526	0.0516	0.0505	0.0495	0.0485	0.0475	0.0465	0.0455
1.7	0.0446	0.0436	0.0427	0.0418	0.0409	0.0401	0.0392	0.0384	0.0375	0.0367
1.8	0.0359	0.0351	0.0344	0.0336	0.0329	0.0322	0.0314	0.0307	0.0301	0.0294
1.9	0.0287	0.0281	0.0274	0.0268	0.0262	0.0256	0.0250	0.0244	0.0239	0.0233
2.0	0.0228	0.0222	0.0217	0.0212	0.0207	0.0202	0.0197	0.0192	0.0188	0.0183
2.1	0.0179	0.0174	0.0170	0.0166	0.0162	0.0158	0.0154	0.0150	0.0146	0.0143
2.2	0.0139	0.0136	0.0132	0.0129	0.0125	0.0122	0.0119	0.0116	0.0113	0.0110
2.3	0.0107	0.0104	0.0102	0.0099	0.0096	0.0094	0.0091	0.0089	0.0087	0.0084
2.4	0.0082	0.0080	0.0078	0.0075	0.0073	0.0071	0.0069	0.0068	0.0066	0.0064
2.5	0.0062	0.0060	0.0059	0.0057	0.0055	0.0054	0.0052	0.0051	0.0049	0.0048
2.6	0.0047	0.0045	0.0044	0.0043	0.0041	0.0040	0.0039	0.0038	0.0037	0.0036
2.7	0.0035	0.0034	0.0033	0.0032	0.0031	0.0030	0.0029	0.0028	0.0027	0.0026
2.8	0.0026	0.0025	0.0024	0.0023	0.0023	0.0022	0.0021	0.0021	0.0020	0.0019
2.9	0.0019	0.0018	0.0018	0.0017	0.0016	0.0016	0.0015	0.0015	0.0014	0.0014
3.0	0.0013	0.0013	0.0013	0.0012	0.0012	0.0011	0.0011	0.0011	0.0010	0.0010
3.1	0.0010	0.0009	0.0009	0.0009	0.0008	0.0008	0.0008	0.0008	0.0007	0.0007
3.2	0.0007	0.0007	0.0006	0.0006	0.0006	0.0006	0.0006	0.0005	0.0005	0.0005
3.3	0.0005	0.0005	0.0005	0.0004	0.0004	0.0004	0.0004	0.0004	0.0004	0.0003
3.4	0.0003	0.0003	0.0003	0.0003	0.0003	0.0003	0.0003	0.0003	0.0003	0.0002

TABLE VI: The *t* Distribution

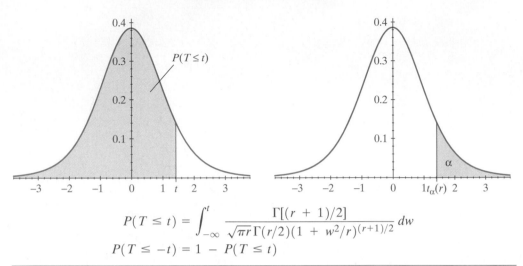

$$P(T \le t) = \int_{-\infty}^{t} \frac{\Gamma[(r+1)/2]}{\sqrt{\pi r}\,\Gamma(r/2)(1+w^2/r)^{(r+1)/2}}\, dw$$

$$P(T \le -t) = 1 - P(T \le t)$$

	$P(T \le t)$						
	0.60	0.75	0.90	0.95	0.975	0.99	0.995
r	$t_{0.40}(r)$	$t_{0.25}(r)$	$t_{0.10}(r)$	$t_{0.05}(r)$	$t_{0.025}(r)$	$t_{0.01}(r)$	$t_{0.005}(r)$
1	0.325	1.000	3.078	6.314	12.706	31.821	63.657
2	0.289	0.816	1.886	2.920	4.303	6.965	9.925
3	0.277	0.765	1.638	2.353	3.182	4.541	5.841
4	0.271	0.741	1.533	2.132	2.776	3.747	4.604
5	0.267	0.727	1.476	2.015	2.571	3.365	4.032
6	0.265	0.718	1.440	1.943	2.447	3.143	3.707
7	0.263	0.711	1.415	1.895	2.365	2.998	3.499
8	0.262	0.706	1.397	1.860	2.306	2.896	3.355
9	0.261	0.703	1.383	1.833	2.262	2.821	3.250
10	0.260	0.700	1.372	1.812	2.228	2.764	3.169
11	0.260	0.697	1.363	1.796	2.201	2.718	3.106
12	0.259	0.695	1.356	1.782	2.179	2.681	3.055
13	0.259	0.694	1.350	1.771	2.160	2.650	3.012
14	0.258	0.692	1.345	1.761	2.145	2.624	2.997
15	0.258	0.691	1.341	1.753	2.131	2.602	2.947
16	0.258	0.690	1.337	1.746	2.120	2.583	2.921
17	0.257	0.689	1.333	1.740	2.110	2.567	2.898
18	0.257	0.688	1.330	1.734	2.101	2.552	2.878
19	0.257	0.688	1.328	1.729	2.093	2.539	2.861
20	0.257	0.687	1.325	1.725	2.086	2.528	2.845
21	0.257	0.686	1.323	1.721	2.080	2.518	2.831
22	0.256	0.686	1.321	1.717	2.074	2.508	2.819
23	0.256	0.685	1.319	1.714	2.069	2.500	2.807
24	0.256	0.685	1.318	1.711	2.064	2.492	2.797
25	0.256	0.684	1.316	1.708	2.060	2.485	2.787
26	0.256	0.684	1.315	1.706	2.056	2.479	2.779
27	0.256	0.684	1.314	1.703	2.052	2.473	2.771
28	0.256	0.683	1.313	1.701	2.048	2.467	2.763
29	0.256	0.683	1.311	1.699	2.045	2.462	2.756
30	0.256	0.683	1.310	1.697	2.042	2.457	2.750
∞	0.253	0.674	1.282	1.645	1.960	2.326	2.576

This table is taken from Table III of Fisher and Yates: *Statistical Tables for Biological, Agricultrual, and Medical Research*, published by Longman Group Ltd., London (previously published by Oliver and Boyd, Edinburgh).

TABLE VII: The *F* Distribution

$$P(F \leq f) = \int_0^f \frac{\Gamma[(r_1 + r_2)/2](r_1/r_2)^{r_1/2}w^{r_1/2-1}}{\Gamma(r_1/2)\Gamma(r_2/2)(1 + r_1w/r_2)^{(r_1+r_2)/2}} \, dw$$

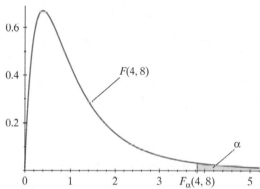

TABLE VII: *continued*

$$P(F \le f) = \int_0^f \frac{\Gamma[(r_1 + r_2)/2](r_1/r_2)^{r_1/2} w^{r_1/2-1}}{\Gamma(r_1/2)\Gamma(r_2/2)(1 + r_1w/r_2)^{(r_1+r_2)/2}}\, dw$$

α	$P(F \le f)$	Den. d.f. r_2	\multicolumn{10}{c}{Numerator Degrees of Freedom, r_1}

α	$P(F \le f)$	r_2	1	2	3	4	5	6	7	8	9	10
0.05	0.95	1	161.4	199.5	215.7	224.6	230.2	234.0	236.8	238.9	240.5	241.9
0.025	0.975		647.79	799.50	864.16	899.58	921.85	937.11	948.22	956.66	963.28	968.63
0.01	0.99		4052	4999.5	5403	5625	5764	5859	5928	5981	6022	6056
0.05	0.95	2	18.51	19.00	19.16	19.25	19.30	19.33	19.35	19.37	19.38	19.40
0.025	0.975		38.51	39.00	39.17	39.25	39.30	39.33	39.36	39.37	39.39	39.40
0.01	0.99		98.50	99.00	99.17	99.25	99.30	99.33	99.36	99.37	99.39	99.40
0.05	0.95	3	10.13	9.55	9.28	9.12	9.01	8.94	8.89	8.85	8.81	8.79
0.025	0.975		17.44	16.04	15.44	15.10	14.88	14.73	14.62	14.54	14.47	14.42
0.01	0.99		34.12	30.82	29.46	28.71	28.24	27.91	27.67	27.49	27.35	27.23
0.05	0.95	4	7.71	6.94	6.59	6.39	6.26	6.16	6.09	6.04	6.00	5.96
0.025	0.975		12.22	10.65	9.98	9.60	9.36	9.20	9.07	8.98	8.90	8.84
0.01	0.99		21.20	18.00	16.69	15.98	15.52	15.21	14.98	14.80	14.66	14.55
0.05	0.95	5	6.61	5.79	5.41	5.19	5.05	4.95	4.88	4.82	4.77	4.74
0.025	0.975		10.01	8.43	7.76	7.39	7.15	6.98	6.85	6.76	6.68	6.62
0.01	0.99		16.26	13.27	12.06	11.39	10.97	10.67	10.46	10.29	10.16	10.05
0.05	0.95	6	5.99	5.14	4.76	4.53	4.39	4.28	4.21	4.15	4.10	4.06
0.025	0.975		8.81	7.26	6.60	6.23	5.99	5.82	5.70	5.60	5.52	5.46
0.01	0.99		13.75	10.92	9.78	9.15	8.75	8.47	8.26	8.10	7.98	7.87
0.05	0.95	7	5.59	4.74	4.35	4.12	3.97	3.87	3.79	3.73	3.68	3.64
0.025	0.975		8.07	6.54	5.89	5.52	5.29	5.12	4.99	4.90	4.82	4.76
0.01	0.99		12.25	9.55	8.45	7.85	7.46	7.19	6.99	6.84	6.72	6.62
0.05	0.95	8	5.32	4.46	4.07	3.84	3.69	3.58	3.50	3.44	3.39	3.35
0.025	0.975		7.57	6.06	5.42	5.05	4.82	4.65	4.53	4.43	4.36	4.30
0.01	0.99		11.26	8.65	7.59	7.01	6.63	6.37	6.18	6.03	5.91	5.81
0.05	0.95	9	5.12	4.26	3.86	3.63	3.48	3.37	3.29	3.23	3.18	3.14
0.025	0.975		7.21	5.71	5.08	4.72	4.48	4.32	4.20	4.10	4.03	3.96
0.01	0.99		10.56	8.02	6.99	6.42	6.06	5.80	5.61	5.47	5.35	5.26
0.05	0.95	10	4.96	4.10	3.71	3.48	3.33	3.22	3.14	3.07	3.02	2.98
0.025	0.975		6.94	5.46	4.83	4.47	4.24	4.07	3.95	3.85	3.78	3.72
0.01	0.99		10.04	7.56	6.55	5.99	5.64	5.39	5.20	5.06	4.94	4.85

TABLE VII: *continued*

$$P(F \le f) = \int_0^f \frac{\Gamma[(r_1 + r_2)/2](r_1/r_2)^{r_1/2}w^{r_1/2-1}}{\Gamma(r_1/2)\Gamma(r_2/2)(1 + r_1w/r_2)^{(r_1+r_2)/2}}\,dw$$

Den. d.f. r_2	$P(F \le f)$	α	Numerator Degrees of Freedom, r_1									
			1	2	3	4	5	6	7	8	9	10
12	0.95	0.05	4.75	3.89	3.49	3.26	3.11	3.00	2.91	2.85	2.80	2.75
	0.975	0.025	6.55	5.10	4.47	4.12	3.89	3.73	3.61	3.51	3.44	3.37
	0.99	0.01	9.33	6.93	5.95	5.41	5.06	4.82	4.64	4.50	4.39	4.30
15	0.95	0.05	4.54	3.68	3.29	3.06	2.90	2.79	2.71	2.64	2.59	2.54
	0.975	0.025	6.20	4.77	4.15	3.80	3.58	3.41	3.29	3.20	3.12	3.06
	0.99	0.01	8.68	6.36	5.42	4.89	4.56	4.32	4.14	4.00	3.89	3.80
20	0.95	0.05	4.35	3.49	3.10	2.87	2.71	2.60	2.51	2.45	2.39	2.35
	0.975	0.025	5.87	4.46	3.86	3.51	3.29	3.13	3.01	2.91	2.84	2.77
	0.99	0.01	8.10	5.85	4.94	4.43	4.10	3.87	3.70	3.56	3.46	3.37
24	0.95	0.05	4.26	3.40	3.01	2.78	2.62	2.51	2.42	2.36	2.30	2.25
	0.975	0.025	5.72	4.32	3.72	3.38	3.15	2.99	2.87	2.78	2.70	2.64
	0.99	0.01	7.82	5.61	4.72	4.22	3.90	3.67	3.50	3.36	3.26	3.17
30	0.95	0.05	4.17	3.32	2.92	2.69	2.53	2.42	2.33	2.27	2.21	2.16
	0.975	0.025	5.57	4.18	3.59	3.25	3.03	2.87	2.75	2.65	2.57	2.51
	0.99	0.01	7.56	5.39	4.51	4.02	3.70	3.47	3.30	3.17	3.07	2.98
40	0.95	0.05	4.08	3.23	2.84	2.61	2.45	2.34	2.25	2.18	2.12	2.08
	0.975	0.025	5.42	4.05	3.46	3.13	2.90	2.74	2.62	2.53	2.45	2.39
	0.99	0.01	7.31	5.18	4.31	3.83	3.51	3.29	3.12	2.99	2.89	2.80
60	0.95	0.05	4.00	3.15	2.76	2.53	2.37	2.25	2.17	2.10	2.04	1.99
	0.975	0.025	5.29	3.93	3.34	3.01	2.79	2.63	2.51	2.41	2.33	2.27
	0.99	0.01	7.08	4.98	4.13	3.65	3.34	3.12	2.95	2.82	2.72	2.63
120	0.95	0.05	3.92	3.07	2.68	2.45	2.29	2.17	2.09	2.02	1.96	1.91
	0.975	0.025	5.15	3.80	3.23	2.89	2.67	2.52	2.39	2.30	2.22	2.16
	0.99	0.01	6.85	4.79	3.95	3.48	3.17	2.96	2.79	2.66	2.56	2.47
∞	0.95	0.05	3.84	3.00	2.60	2.37	2.21	2.10	2.01	1.94	1.88	1.83
	0.975	0.025	5.02	3.69	3.12	2.79	2.57	2.41	2.29	2.19	2.11	2.05
	0.99	0.01	6.63	4.61	3.78	3.32	3.02	2.80	2.64	2.51	2.41	2.32

TABLE VII: *continued*

$$P(F \le f) = \int_0^f \frac{\Gamma[(r_1 + r_2)/2](r_1/r_2)^{r_1/2} w^{r_1/2-1}}{\Gamma(r_1/2)\Gamma(r_2/2)(1 + r_1 w/r_2)^{(r_1+r_2)/2}} \, dw$$

α	$P(F \le f)$	Den. d.f. r_2	Numerator Degrees of Freedom, r_1								
			12	15	20	24	30	40	60	120	∞
0.05	0.95	1	243.9	245.9	248.0	249.1	250.1	251.1	252.2	253.3	254.3
0.025	0.975		976.71	984.87	993.10	997.25	1001.4	1005.6	1009.8	1014.0	1018.3
0.01	0.99		6106	6157	6209	6235	6261	6287	6313	6339	6366
0.05	0.95	2	19.41	19.43	19.45	19.45	19.46	19.47	19.48	19.49	19.50
0.025	0.975		39.42	39.43	39.45	39.46	39.47	39.47	39.48	39.49	39.50
0.01	0.99		99.42	99.43	99.45	99.46	99.47	99.47	99.48	99.49	99.50
0.05	0.95	3	8.74	8.70	8.66	8.64	8.62	8.59	8.57	8.55	8.53
0.025	0.975		14.34	14.25	14.17	14.12	14.08	14.04	13.99	13.95	13.90
0.01	0.99		27.05	26.87	26.69	26.60	26.50	26.41	26.32	26.22	26.13
0.05	0.95	4	5.91	5.86	5.80	5.77	5.75	5.72	5.69	5.66	5.63
0.025	0.975		8.75	8.66	8.56	8.51	8.46	8.41	8.36	8.31	8.26
0.01	0.99		14.37	14.20	14.02	13.93	13.84	13.75	13.65	13.56	13.46
0.05	0.95	5	4.68	4.62	4.56	4.53	4.50	4.46	4.43	4.40	4.36
0.025	0.975		6.52	6.43	6.33	6.28	6.23	6.18	6.12	6.07	6.02
0.01	0.99		9.89	9.72	9.55	9.47	9.38	9.29	9.20	9.11	9.02
0.05	0.95	6	4.00	3.94	3.87	3.84	3.81	3.77	3.74	3.70	3.67
0.025	0.975		5.37	5.27	5.17	5.12	5.07	5.01	4.96	4.90	4.85
0.01	0.99		7.72	7.56	7.40	7.31	7.23	7.14	7.06	6.97	6.88
0.05	0.95	7	3.57	3.51	3.41	3.41	3.38	3.34	3.30	3.27	3.23
0.025	0.975		4.67	4.57	4.47	4.42	4.36	4.31	4.25	4.20	4.14
0.01	0.99		6.47	6.31	6.16	6.07	5.99	5.91	5.82	5.74	5.65
0.05	0.95	8	3.28	3.22	3.15	3.12	3.08	3.04	3.01	2.97	2.93
0.025	0.975		4.20	4.10	4.00	3.95	3.89	3.84	3.78	3.73	3.67
0.01	0.99		5.67	5.52	5.36	5.28	5.20	5.12	5.03	4.95	4.86
0.05	0.95	9	3.07	3.01	2.94	2.90	2.86	2.83	2.79	2.75	2.71
0.025	0.975		3.87	3.77	3.67	3.61	3.56	3.51	3.45	3.39	3.33
0.01	0.99		5.11	4.96	4.81	4.73	4.65	4.57	4.48	4.40	4.31

TABLE VII: *continued*

$$P(F \le f) = \int_0^f \frac{\Gamma[(r_1 + r_2)/2](r_1/r_2)^{r_1/2}w^{r_1/2 - 1}}{\Gamma(r_1/2)\Gamma(r_2/2)(1 + r_1w/r_2)^{(r_1+r_2)/2}} \, dw$$

Den. d.f. r_2	$P(F \le f)$	α	Numerator Degrees of Freedom, r_1								
			12	15	20	24	30	40	60	120	∞
10	0.95	0.05	2.91	2.85	2.77	2.74	2.70	2.66	2.62	2.58	2.54
	0.975	0.025	3.62	3.52	3.42	3.37	3.31	3.26	3.20	3.14	3.08
	0.99	0.01	4.71	4.56	4.41	4.33	4.25	4.17	4.08	4.00	3.91
12	0.95	0.05	2.69	2.62	2.54	2.51	2.47	2.43	2.38	2.34	2.30
	0.975	0.025	3.28	3.18	3.07	3.02	2.96	2.91	2.85	2.79	2.72
	0.99	0.01	4.16	4.01	3.86	3.78	3.70	3.62	3.54	3.45	3.36
15	0.95	0.05	2.48	2.40	2.33	2.29	2.25	2.20	2.16	2.11	2.07
	0.975	0.025	2.96	2.86	2.76	2.70	2.64	2.59	2.52	2.46	2.40
	0.99	0.01	3.67	3.52	3.37	3.29	3.21	3.13	3.05	2.96	2.87
20	0.95	0.05	2.28	2.20	2.12	2.08	2.04	1.99	1.95	1.90	1.84
	0.975	0.025	2.68	2.57	2.46	2.41	2.35	2.29	2.22	2.16	2.09
	0.99	0.01	3.23	3.09	2.94	2.86	2.78	2.69	2.61	2.52	2.42
24	0.95	0.05	2.18	2.11	2.03	1.98	1.94	1.89	1.84	1.79	1.73
	0.975	0.025	2.54	2.44	2.33	2.27	2.21	2.15	2.08	2.01	1.94
	0.99	0.01	3.03	2.89	2.74	2.66	2.58	2.49	2.40	2.31	2.21
30	0.95	0.05	2.09	2.01	1.93	1.89	1.84	1.79	1.74	1.68	1.62
	0.975	0.025	2.41	2.31	2.20	2.14	2.07	2.01	1.94	1.87	1.79
	0.99	0.01	2.84	2.70	2.55	2.47	2.39	2.30	2.21	2.11	2.01
40	0.95	0.05	2.00	1.92	1.84	1.79	1.74	1.69	1.64	1.58	1.51
	0.975	0.025	2.29	2.18	2.07	2.01	1.94	1.88	1.80	1.72	1.64
	0.99	0.01	2.66	2.52	2.37	2.29	2.20	2.11	2.02	1.92	1.80
60	0.95	0.05	1.92	1.84	1.75	1.70	1.65	1.59	1.53	1.47	1.39
	0.975	0.025	2.17	2.06	1.94	1.88	1.82	1.74	1.67	1.58	1.48
	0.99	0.01	2.50	2.35	2.20	2.12	2.03	1.94	1.84	1.73	1.60
120	0.95	0.05	1.83	1.75	1.66	1.61	1.55	1.50	1.43	1.35	1.25
	0.975	0.025	2.05	1.95	1.82	1.76	1.69	1.61	1.53	1.43	1.31
	0.99	0.01	2.34	2.19	2.03	1.95	1.86	1.76	1.66	1.53	1.38
∞	0.95	0.05	1.75	1.67	1.57	1.52	1.46	1.39	1.32	1.22	1.00
	0.975	0.025	1.94	1.83	1.71	1.64	1.57	1.48	1.39	1.27	1.00
	0.99	0.01	2.18	2.04	1.88	1.79	1.70	1.59	1.47	1.32	1.00

TABLE VIII: Kolmogorov-Smirnov Acceptance Limits

$$D_n = \sup_x [\,|F_n(x) - F_0(x)|\,]$$
$$\alpha = 1 - P(D_n \leq d)$$

		α		
n	0.20	0.10	0.05	0.01
1	0.90	0.95	0.98	0.99
2	0.68	0.78	0.84	0.93
3	0.56	0.64	0.71	0.83
4	0.49	0.56	0.62	0.73
5	0.45	0.51	0.56	0.67
6	0.41	0.47	0.52	0.62
7	0.38	0.44	0.49	0.58
8	0.36	0.41	0.46	0.54
9	0.34	0.39	0.43	0.51
10	0.32	0.37	0.41	0.49
11	0.31	0.35	0.39	0.47
12	0.30	0.34	0.38	0.45
13	0.28	0.32	0.36	0.43
14	0.27	0.31	0.35	0.42
15	0.27	0.30	0.34	0.40
16	0.26	0.30	0.33	0.39
17	0.25	0.29	0.32	0.38
18	0.24	0.28	0.31	0.37
19	0.24	0.27	0.30	0.36
20	0.23	0.26	0.29	0.35
25	0.21	0.24	0.26	0.32
30	0.19	0.22	0.24	0.29
35	0.18	0.21	0.23	0.27
40	0.17	0.19	0.21	0.25
45	0.16	0.18	0.20	0.24
Large n	$\dfrac{1.07}{\sqrt{n}}$	$\dfrac{1.22}{\sqrt{n}}$	$\dfrac{1.36}{\sqrt{n}}$	$\dfrac{1.63}{\sqrt{n}}$

TABLE IX: Random Numbers on the Interval $(0, 1)$

3407	1440	6960	8675	5649	5793	1514
5044	9859	4658	7779	7986	0520	6697
0045	4999	4930	7408	7551	3124	0527
7536	1448	7843	4801	3147	3071	4749
7653	4231	1233	4409	0609	6448	2900
6157	1144	4779	0951	3757	9562	2354
6593	8668	4871	0946	3155	3941	9662
3187	7434	0315	4418	1569	1101	0043
4780	1071	6814	2733	7968	8541	1003
9414	6170	2581	1398	2429	4763	9192
1948	2360	7244	9682	5418	0596	4971
1843	0914	9705	7861	6861	7865	7293
4944	8903	0460	0188	0530	7790	9118
3882	3195	8287	3298	9532	9066	8225
6596	9009	2055	4081	4842	7852	5915
4793	2503	2906	6807	2028	1075	7175
2112	0232	5334	1443	7306	6418	9639
0743	1083	8071	9779	5973	1141	4393
8856	5352	3384	8891	9189	1680	3192
8027	4975	2346	5786	0693	5615	2047
3134	1688	4071	3766	0570	2142	3492
0633	9002	1305	2256	5956	9256	8979
8771	6069	1598	4275	6017	5946	8189
2672	1304	2186	8279	2430	4896	3698
3136	1916	8886	8617	9312	5070	2720
6490	7491	6562	5355	3794	3555	7510
8628	0501	4618	3364	6709	1289	0543
9270	0504	5018	7013	4423	2147	4089
5723	3807	4997	4699	2231	3193	8130
6228	8874	7271	2621	5746	6333	0345
7645	3379	8376	3030	0351	8290	3640
6842	5836	6203	6171	2698	4086	5469
6126	7792	9337	7773	7286	4236	1788
4956	0215	3468	8038	6144	9753	3131
1327	4736	6229	8965	7215	6458	3937
9188	1516	5279	5433	2254	5768	8718
0271	9627	9442	9217	4656	7603	8826
2127	1847	1331	5122	8332	8195	3322
2102	9201	2911	7318	7670	6079	2676
1706	6011	5280	5552	5180	4630	4747
7501	7635	2301	0889	6955	8113	4364
5705	1900	7144	8707	9065	8163	9846
3234	2599	3295	9160	8441	0085	9317
5641	4935	7971	8917	1978	5649	5799
2127	1868	3664	9376	1984	6315	8396

TABLE X: Divisors for the Confidence Interval for σ of Minimum Length

Let $X_1, X_2, \ldots, X_n$ be a random sample of size n from $N(\mu, \sigma^2)$; let $\bar{x} = (1/n)\sum_{i=1}^{n} x_i$ and $s^2 = (1/[n-1])\sum_{i=1}^{n}(x_i - \bar{x})^2$. A $(100\gamma)\%$ confidence interval for σ which has minimum length is given by $[\sqrt{(n-1)s^2/b}, \sqrt{(n-1)s^2/a}]$. In the table, a is the upper number, b the lower number; $r = n - 1$, the number of degrees of freedom; and $\gamma = 1 - \alpha$, the confidence level.

r	$\gamma = 1 - \alpha$ 0.90	0.95	0.99	r	$\gamma = 1 - \alpha$ 0.90	0.95	0.99
2	0.206	0.101	0.020	16	8.774	7.604	5.649
	12.521	15.111	20.865		29.233	32.072	38.097
3	0.565	0.345	0.114	17	9.505	8.282	6.226
	13.153	15.589	20.973		30.480	33.362	39.469
4	1.020	0.692	0.294	18	10.242	8.969	6.814
	14.18	16.573	21.838		31.721	34.647	40.835
5	1.535	1.109	0.546	19	10.986	9.663	7.413
	15.350	17.743	22.985		32.959	35.927	42.195
6	2.093	1.578	0.837	20	11.736	10.365	8.021
	16.581	18.996	24.262		34.192	37.202	43.550
7	2.683	2.085	1.214	21	12.492	11.073	8.638
	17.839	20.286	25.602		35.420	38.472	44.899
8	3.298	2.623	1.611	22	13.253	11.788	9.264
	19.110	21.595	26.975		36.646	39.738	46.243
9	3.934	3.187	2.039	23	14.019	12.509	9.898
	20.385	22.912	28.364		37.867	41.000	47.586
10	4.588	3.773	2.496	24	14.790	13.236	10.539
	21.660	24.230	29.760		39.084	42.257	48.914
11	5.257	4.377	2.976	25	15.565	13.968	11.186
	22.933	25.547	31.158		40.299	43.510	50.243
12	5.940	4.997	3.477	26	16.344	14.704	11.841
	24.202	26.862	32.554		41.509	44.760	51.566
13	6.634	5.631	3.997	27	17.127	15.446	12.501
	25.467	28.172	33.947		42.717	46.006	52.886
14	7.338	6.278	4.533	28	17.914	16.192	13.168
	26.727	29.477	35.336		43.922	47.248	54.200
15	8.052	6.936	5.084	29	18.705	16.942	13.840
	27.982	30.777	36.719		45.123	48.487	55.511

TABLE XI: Distribution Function of the Correlation Coefficient R, $\rho = 0$

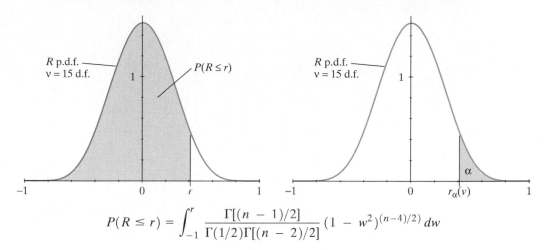

$$P(R \le r) = \int_{-1}^{r} \frac{\Gamma[(n-1)/2]}{\Gamma(1/2)\Gamma[(n-2)/2]} (1 - w^2)^{(n-4)/2} \, dw$$

$\nu = n - 2$ degrees of freedom	$P(R \le r)$			
	0.95	0.975	0.99	0.995
	$r_{0.05}(\nu)$	$r_{0.025}(\nu)$	$r_{0.01}(\nu)$	$r_{0.005}(\nu)$
1	0.9877	0.9969	0.9995	0.9999
2	0.9000	0.9500	0.9800	0.9900
3	0.8053	0.8783	0.9343	0.9587
4	0.7292	0.8113	0.8822	0.9172
5	0.6694	0.7544	0.8329	0.8745
6	0.6215	0.7067	0.7887	0.8343
7	0.5822	0.6664	0.7497	0.7977
8	0.5493	0.6319	0.7154	0.7646
9	0.5214	0.6020	0.6850	0.7348
10	0.4972	0.5759	0.6581	0.7079
11	0.4761	0.5529	0.6338	0.6835
12	0.4575	0.5323	0.6120	0.6613
13	0.4408	0.5139	0.5922	0.6411
14	0.4258	0.4973	0.5742	0.6226
15	0.4123	0.4821	0.5577	0.6054
16	0.4000	0.4683	0.5425	0.5897
17	0.3887	0.4555	0.5285	0.5750
18	0.3783	0.4437	0.5154	0.5614
19	0.3687	0.4328	0.5033	0.5487
20	0.3597	0.4226	0.4920	0.5367
25	0.3232	0.3808	0.4450	0.4869
30	0.2959	0.3494	0.4092	0.4487
35	0.2746	0.3246	0.3809	0.4182
40	0.2572	0.3044	0.3578	0.3931
45	0.2428	0.2875	0.3383	0.3721
50	0.2306	0.2732	0.3218	0.3541
60	0.2108	0.2500	0.2948	0.3248
70	0.1954	0.2318	0.2736	0.3017
80	0.1829	0.2172	0.2565	0.2829
90	0.1725	0.2049	0.2422	0.2673
100	0.1638	0.1946	0.2300	0.2540

TABLE XII: Discrete Distributions

Probability Distribution and Parameter Values	Probability Mass Function	Moment-Generating Function	Mean $E(X)$	Variance $Var(X)$	Examples
Bernoulli $0 < p < 1$ $q = 1 - p$	$p^x q^{1-x}, \quad x = 0, 1$	$q + pe^t$	p	pq	Experiment with two possible outcomes—say, success and failure; $p = P(\text{success})$
Binomial $n = 1, 2, 3, \ldots$ $0 < p < 1$	$\binom{n}{x} p^x q^{n-x}, \quad x = 0, 1, \ldots, n$	$(q + pe^t)^n$	np	npq	Number of successes in a sequence of n Bernoulli trials; $p = P(\text{success})$
Geometric $0 < p < 1$ $q = 1 - p$	$q^{x-1} p, \quad x = 1, 2, \ldots$	$\dfrac{pe^t}{1 - qe^t}$	$\dfrac{1}{p}$	$\dfrac{q}{p^2}$	The number of trials taken to obtain the first success in a sequence of Bernoulli trials
Hypergeometric $x \leq n, x \leq N_1$ $n - x \leq N_2$ $N = N_1 + N_2$ $N_1 > 0, \; N_2 > 0$	$\dfrac{\binom{N_1}{x}\binom{N_2}{n-x}}{\binom{N}{n}}$		$n\left(\dfrac{N_1}{N}\right)$	$n\left(\dfrac{N_1}{N}\right)\left(\dfrac{N_2}{N}\right)\left(\dfrac{N-n}{N-1}\right)$	Selecting r objects at random without replacement from a set composed of two types of objects
Negative Binomial $r = 1, 2, \ldots$ $0 < p < 1$	$\binom{x-1}{r-1} p^r q^{x-r}, \quad x = r, r+1, \ldots$	$\dfrac{(pe^t)^r}{(1 - qe^t)^r}$	$\dfrac{r}{p}$	$\dfrac{rq}{p^2}$	The number of trials taken to obtain the rth success in a sequence of Bernoulli trials
Poisson $0 < \lambda$	$\dfrac{\lambda^x e^{-\lambda}}{x!}, \quad x = 0, 1, \ldots$	$e^{\lambda(e^t - 1)}$	λ	λ	Number of events occurring in a unit interval, where events are occurring randomly at a mean rate of λ per unit interval
Uniform $m > 0$	$\dfrac{1}{m}, \quad x = 1, 2, \ldots, m$		$\dfrac{m+1}{2}$	$\dfrac{m^2 - 1}{12}$	Select an integer randomly from $1, 2, \ldots, m$

TABLE XIII: Continuous Distributions

Probability Distribution and Parameter Values	Probability Density Function	Moment-Generating Function	Mean $E(X)$	Variance $Var(X)$	Examples
Beta $0<\alpha$ $0<\beta$	$\frac{\Gamma(\alpha+\beta)}{\Gamma(\alpha)\Gamma(\beta)}x^{\alpha-1}(1-x)^{\beta-1}$, $0<x<1$		$\frac{\alpha}{\alpha+\beta}$	$\frac{\alpha\beta}{(\alpha+\beta+1)(\alpha+\beta)^2}$	$X=X_1/(X_1+X_2)$, where X_1 and X_2 have independent gamma distributions with same θ
Chi-square $r=1,2,\ldots$	$\frac{x^{r/2-1}e^{-x/2}}{\Gamma(r/2)2^{r/2}}$, $0<x<\infty$	$\frac{1}{(1-2t)^{r/2}}$, $t<\frac{1}{2}$	r	$2r$	Gamma distribution, $\theta=2$, $\alpha=r/2$; sum of squares of r independent $N(0,1)$ random variables
Exponential $0<\theta$	$\frac{1}{\theta}e^{-x/\theta}$, $0\le x<\infty$	$\frac{1}{1-\theta t}$, $t<\frac{1}{\theta}$	θ	θ^2	Waiting time to first arrival when observing a Poisson process with a mean rate of arrivals equal to $\lambda=1/\theta$
Gamma $0<\alpha$ $0<\theta$	$\frac{x^{\alpha-1}e^{-x/\theta}}{\Gamma(\alpha)\theta^\alpha}$, $0<x<\infty$	$\frac{1}{(1-\theta t)^\alpha}$, $t<\frac{1}{\theta}$	$\alpha\theta$	$\alpha\theta^2$	Waiting time to αth arrival when observing a Poisson process with a mean rate of arrivals equal to $\lambda=1/\theta$
Normal $-\infty<\mu<\infty$ $0<\sigma$	$\frac{e^{-(x-\mu)^2/2\sigma^2}}{\sigma\sqrt{2\pi}}$, $-\infty<x<\infty$	$e^{\mu t+\sigma^2 t^2/2}$	μ	σ^2	Errors in measurements; heights of children; breaking strengths of materials
Uniform $-\infty<a<b<\infty$	$\frac{1}{b-a}$, $a\le x\le b$	$\frac{e^{tb}-e^{ta}}{t(b-a)}$, $t\ne 0$ 1, $t=0$	$\frac{a+b}{2}$	$\frac{(b-a)^2}{12}$	Select a point at random from the interval $[a,b]$

TABLE XIV: Tests and Confidence Intervals

Distribution	θ: The parameter of interest	W: The variable used to test $H_0: \theta = \theta_0$	Two-sided $1-\alpha$ Confidence Interval for θ	Comments
$N(\mu,\sigma^2)$ or n large σ^2 known	μ	$\dfrac{\overline{X}-\theta_0}{\sigma/\sqrt{n}}$	$\overline{x} \pm z_{\alpha/2}\dfrac{\sigma}{\sqrt{n}}$	W is $N(0,1)$; $P(W \geq z_{\alpha/2}) = \alpha/2$
$N(\mu,\sigma^2)$ σ^2 unknown	μ	$\dfrac{\overline{X}-\theta_0}{S/\sqrt{n}}$	$\overline{x} \pm t_{\alpha/2}(n-1)\dfrac{s}{\sqrt{n}}$	W has a t distribution with $n-1$ degrees of freedom; $P[W \geq t_{\alpha/2}(n-1)] = \alpha/2$
Any distribution with known variance, σ^2	μ	$\dfrac{\overline{X}-\theta_0}{\sigma/\sqrt{n}}$	$\overline{x} \pm z_{\alpha/2}\dfrac{\sigma}{\sqrt{n}}$	W has an approximate $N(0,1)$ distribution for n sufficiently large
$N(\mu_X,\sigma_X^2)$ $N(\mu_Y,\sigma_Y^2)$ σ_X^2,σ_Y^2 known	$\mu_X - \mu_Y$	$\dfrac{\overline{X}-\overline{Y}-\theta_0}{\sqrt{\dfrac{\sigma_X^2}{n}+\dfrac{\sigma_Y^2}{m}}}$	$\overline{x}-\overline{y} \pm z_{\alpha/2}\sqrt{\dfrac{\sigma_X^2}{n}+\dfrac{\sigma_Y^2}{m}}$	W is $N(0,1)$
$N(\mu_X,\sigma_X^2)$ $N(\mu_Y,\sigma_Y^2)$ σ_X^2,σ_Y^2 unknown	$\mu_X - \mu_Y$	$\dfrac{\overline{X}-\overline{Y}-\theta_0}{\sqrt{\dfrac{S_X^2}{n}+\dfrac{S_Y^2}{m}}}$	$\overline{x}-\overline{y} \pm z_{\alpha/2}\sqrt{\dfrac{s_x^2}{n}+\dfrac{s_y^2}{m}}$	W is approximately $N(0,1)$ if sample sizes are large
$N(\mu_X,\sigma_X^2)$ $N(\mu_Y,\sigma_Y^2)$ $\sigma_X^2 = \sigma_Y^2$, unknown	$\mu_X - \mu_Y$	$\dfrac{\overline{X}-\overline{Y}-\theta_0}{\sqrt{\dfrac{(n-1)S_X^2+(m-1)S_Y^2}{n+m-2}\left(\dfrac{1}{n}+\dfrac{1}{m}\right)}}$	$\overline{x}-\overline{y} \pm t_{\alpha/2}(n+m-2)s_p\sqrt{\dfrac{1}{n}+\dfrac{1}{m}}$ $s_p=\sqrt{\dfrac{(n-1)s_x^2+(m-1)s_y^2}{n+m-2}}$	W has a t distribution with $r = n+m-2$ degrees of freedom
$D = X - Y$ is $N(\mu_X - \mu_Y, \sigma_D^2)$ X and Y dependent	$\mu_X - \mu_Y$	$\dfrac{\overline{D}-\theta_0}{S_D/\sqrt{n}}$	$\overline{d} \pm t_{\alpha/2}(n-1)\dfrac{s_d}{\sqrt{n}}$	W has a t distribution with $n-1$ degrees of freedom

TABLE XIV: *continued*

Distribution	θ: The parameter of interest	W: The variable used to test $H_0: \theta = \theta_0$	Two-sided $1-\alpha$ Confidence Interval for θ	Comments
$N(\mu,\sigma^2)$ μ unknown	σ^2	$\dfrac{(n-1)S^2}{\theta_0}$	$\dfrac{(n-1)s^2}{\chi^2_{\alpha/2}(n-1)},\ \dfrac{(n-1)s^2}{\chi^2_{1-\alpha/2}(n-1)}$	W is $\chi^2(n-1)$, $P[W \le \chi^2_{1-\alpha/2}(n-1)] = \alpha/2$, $P[W \ge \chi^2_{\alpha/2}(n-1)] = \alpha/2$
$N(\mu,\sigma^2)$ μ unknown	σ	$\dfrac{(n-1)S^2}{\theta_0^2}$	$\sqrt{\dfrac{(n-1)s^2}{b}},\ \sqrt{\dfrac{(n-1)s^2}{a}}$	W is $\chi^2(n-1)$. Select a and b from Table X in the appendix for interval of minimum length
$N(\mu_X,\sigma_X^2)$ $N(\mu_Y,\sigma_Y^2)$ μ_X,μ_Y unknown	$\dfrac{\sigma_X^2}{\sigma_Y^2}$	$\dfrac{S_Y^2}{S_X^2}\,\theta_0$	$\dfrac{s_x^2/s_y^2}{F_{\alpha/2}(n-1,m-1)},\ F_{\alpha/2}(m-1,n-1)\dfrac{s_x^2}{s_y^2}$	W has an F distribution with $m-1$ and $n-1$ degrees of freedom
$b(n,p)$	p	$\dfrac{\dfrac{Y}{n}-\theta_0}{\sqrt{\left(\dfrac{Y}{n}\right)\left(1-\dfrac{Y}{n}\right)\Big/ n}}$	$\dfrac{y}{n} \pm z_{\alpha/2}\sqrt{\left(\dfrac{y}{n}\right)\left(1-\dfrac{y}{n}\right)\Big/ n}$	W is approximately $N(0,1)$ for n sufficiently large
$b(n_1,p_1)$ $b(n_2,p_2)$	$p_1 - p_2$	$\dfrac{\dfrac{Y_1}{n_1}-\dfrac{Y_2}{n_2}-\theta_0}{\sqrt{\left(\dfrac{Y_1+Y_2}{n_1+n_2}\right)\left(1-\dfrac{Y_1+Y_2}{n_1+n_2}\right)\left(\dfrac{1}{n_1}-\dfrac{1}{n_2}\right)}}$	$\dfrac{y_1}{n_1}-\dfrac{y_2}{n_2} \pm$ $z_{\alpha/2}\sqrt{\dfrac{y_1}{n_1}\left(1-\dfrac{y_1}{n_1}\right)\Big/ n_1 + \dfrac{y_2}{n_2}\left(1-\dfrac{y_2}{n_2}\right)\Big/ n_2}$	W is approximately $N(0,1)$ when n_1 and n_2 are sufficiently large

Answers to Selected Odd-Numbered Exercises

Chapter 1

1.1-1 **(a)** $S = \{1, 2, 3, \ldots, 36\}$;

    (b) $S = \{w: 19 \leq w \leq 23\}$, where w is the weight in grams;

    (c) $S = \{HHH, HHT, HTH, THH, HTT, THT, TTH, TTT\}$.

1.1-3 **(a)** $f(x) = 1/10$, $x = 0, 1, 2, \ldots, 9$;

    (b) The respective frequencies are 21, 15, 16, 13, 6, 15, 13, 17, 17, 17;

    (d) The respective frequencies are 8, 13, 15, 10, 12, 10, 4, 9, 15, 0;

1.1-5 **(a)** Clutch size: 4   5   6   7   8   9   10

           Frequency: 4   5   6   4   6   7   1

    (c) No.

1.1-7 **(a)** Number: 0   1   2   3   4   5

    (b) Frequency: 4   16   27   35   14   4

1.1-11 **(a)** 0.261, 0.196;

    (b) 0.327, 0.315;

    (c) 0.270, 0.273;

    (d) Example of Simpson's paradox.

1.1-13 **(a)** $0.889 > 0.856, 0.948 > 0.921$,
           $0.914 > 0.855, 0.831 > 0.713$,
           $0.858 > 0.767$;

    (b) $0.867 < 0.891$;

    (c) Example of Simpson's paradox.

1.2-1 0.68.

1.2-3 **(a)** 12/52; **(b)** 2/52; **(c)** 16/52; **(d)** 1; **(e)** 0.

1.2-5 27/64.

1.2-7 **(a)** (i) 4/8, (ii) 2/8, (iii) 3/8, (iv) 4/8;

    (b) (i) 1/8, (ii) 0, (iii) 2/8;

    (c) (i) 5/8, (ii) 5/8, (iii) 5/8.

1.2-9 0.63.

1.2-11 **(a)** $3(1/3) - 3(1/3)^2 + (1/3)^3$;

    (b) $P(A_1 \cup A_2 \cup A_3) = 1 - [1 - 3(1/3) + 3(1/3)^2 - (1/3)^3] = 1 - (1 - 1/3)^3$.

1.2-13 **(a)** $1 - (1/2)^{10}$;   **(b)** $1 - (1/2)^{20}$;

    (c) $1 - (1/2)^{20}$;   **(d)** $1 - (1/2)^{10}$;

    (e) $(1/2)^{10} - (1/2)^{20}$;   **(f)** $(1/2)^{20}$.

1.2-15 **(a)** $S = \{00, 0, 1, 2, 3, \ldots, 36\}$;

    (b) $P(A) = 2/38$;

    (c) $P(B) = 4/38$;

    (d) $P(D) = 18/38$.

1.2-17 2/3.

1.3-1 4096.

1.3-3 **(a)** 6,760,000; **(b)** 17,576,000.

1.3-5 **(a)** 24; **(b)** 256.

1.3-7 **(a)** 0.0024; **(b)** 0.0012; **(c)** 0.0006; **(d)** 0.0004.

1.3-9 **(a)** 60; **(b)** 125.

1.3-11 **(a)** 2; **(b)** 8; **(c)** 20; **(d)** 40.

1.3-13 **(a)** 96; **(b)** 393,216; **(c)** 2,973,696.

1.3-15 **(a)** 362,880; **(b)** 84; **(c)** 512.

1.3-17 **(a)** 0.00024; **(b)** 0.00144; **(c)** 0.02113;

    (d) 0.04754; **(e)** 0.42257.

1.3-19 14,702,688.

1.3-21 **(a)** 560; **(b)** 3,688.

1.4-1 **(a)** 5000/1,000,000; **(b)** 78,515/1,000,000;

    (b) 73,630/995,000; **(d)** 4,885/78,515.

1.4-3 **(a)** 5/35; **(b)** 26/35; **(c)** 5/19; **(d)** 9/23; **(e)** left.

1.4-5 **(a)** 3/10; **(b)** 3/5; **(c)** 3/7.

1.4-7 **(a)** $S = \{(R, R), (R, W), (W, R), (W, W)\}$; **(b)** 1/3.

1.4-9 1/5.

1.4-11 **(a)** 1/220; **(b)** 3/110; **(c)** 9/55.

1.4-13 **(a)** 8/36; **(c)** 5/11;

(b) 8/36 + 2[(5/36)(5/11) + (4/36)(4/10) + (3/36)(3/9)] = 0.49293.

1.4-15 **(a)** 365^r; **(b)** $_{365}P_r$; **(c)** $1 - {}_{365}P_r/365^r$; **(d)** 23.

1.4-17 **(a)** 49/153; **(b)** 4/7.

1.4-19 **(f)** $1 - 1/e$.

1.4-21 11.

1.5-1 **(a)** 0.14; **(b)** 0.76; **(c)** 0.86.

1.5-3 **(a)** 1/6; **(b)** 1/12; **(c)** 1/4; **(d)** 1/4; **(e)** 1/2.

1.5-5 Yes; $0.9 = 0.8 + 0.5 - (0.8)(0.5)$.

1.5-7 **(a)** 0.29; **(b)** 0.44.

1.5-9 **(a)** 0.36; **(b)** 0.49; **(c)** 0.01.

1.5-11 **(a)** No.

(b) Only if $P(A) = 0$ or $P(B) = 1$.

1.5-13 $(2/3)^3(1/3)^2$; $(2/3)^3(1/3)^2$.

1.5-15 **(a)** $10(1/2)^6$;

(b) $[(120)(45)/15,504][7/15]$;

(c) Neither model is very good.

1.5-17 **(a)** $1 - (11/12)^{12}$; **(b)** $1 - (11/12)^{11}$.

1.5-19 **(b)** $1 - 1/e$.

1.6-1 **(a)** 21/32; **(b)** 16/21.

1.6-3 0.151.

1.6-5 $60/95 = 0.632$.

1.6-7 $50/95 = 0.526$.

1.6-9 **(a)** $495/30,480 = 0.016$; **(b)** $29,985/30,480 = 0.984$.

1.6-11 1/4.

1.6-13 0.54229.

1.6-15 0.8182.

Chapter 2

2.1-3 10; **(b)** 1/55; **(c)** 3; **(d)** 1/30; **(e)** $n(n + 1)/2$; **(f)** 1.

2.1-5 **(b)**

x	Frequency	Relative Frequency	$f(x)$
1	38	0.38	0.40
2	27	0.27	0.30
3	21	0.21	0.20
4	14	0.14	0.10

2.1-7 **(a)** $f(x) = \dfrac{13 - 2x}{36}$, $x = 1, 2, 3, 4, 5, 6$;

(b) $g(0) = \dfrac{6}{36}, g(y) = \dfrac{12 - 2y}{36}$, $y = 1, 2, 3, 4, 5$.

2.1-11 0.416.

2.1-13 2/5.

2.1-15 **(a)** 19/20; **(b)** 1/20; **(c)** 9/20.

2.1-17 **(c)**

x	Frequency	Relative Frequency	$f(x)$
0	13	0.325	0.253
1	16	0.400	0.422
2	9	0.225	0.251
3	2	0.050	0.066
4	0	0.000	0.008

2.1-19 78.

2.2-1 **(a)** 3; **(b)** 7; **(c)** 4/3; **(d)** 7/3; **(e)** $(2n + 1)/3$; **(f)** does not exist.

2.2-3 $-91/300$ or -30.33 cents.

2.2-5 $360.

2.2-7 $E(X) = -17/216 = -0.0787$.

2.2-9 $E(X) = \displaystyle\sum_{x=1}^{n} \frac{1}{x}\left(\frac{1}{n}\right)$; $\dfrac{137}{300} = 0.4567$;

$\dfrac{1}{100}\left[\dfrac{\ln(101) + 1 + \ln(100)}{2}\right] = 0.05110$

≈ 0.05187.

2.2-11 **(a)** $-1/19$; **(b)** $-1/37$.

2.2-13 $-\$0.01414$.

2.3-1 **(a)** 15; 50; **(b)** 5; 0; **(c)** 5/3; 5/9.

2.3-3 **(a)** 16; **(b)** 6; **(c)** 16.

2.3-5 7.

2.3-7 2.

2.3-9 **(a)** $\mu = (m + 1)/2$; $\sigma^2 = (m^2 - 1)/12$;

(b) $m = 7$.

2.3-11 **(a)** 4.473; 9.607; 3.937; 6.964;

(b) 938/117; 8,657/3,393;

(c) 226/33; 215/66;

(d) 251/100; 403/300.

2.3-13 **(a)** (ii) 127/64; (iii) $\sqrt{7359}/64$; (iv) 1.6635;

(b) (ii) 385/64; (iii) $\sqrt{7359}/64$; (iv) -1.6635.

2.3-15 **(a)** $25/8 = 3.125$, $55/64 = 0.8594$, $\sqrt{55}/8 = 0.9270$;

(b) $79/25 = 3.16$, $662/825 = 0.8024$, $\sqrt{21,846}/165 = 0.8958$.

2.3-19 $1809.80.

2.4-1 $f(x) = (7/18)^x(11/18)^{1-x}$, $x = 0, 1$; $\mu = 7/18$; $\sigma^2 = 77/324$.

2.4-3 **(a)** $(1/5)^2(4/5)^4 = 0.0164$;

(b) $\dfrac{6!}{2!4!}(1/5)^2(4/5)^4 = 0.2458$.

2.4-5 **(a)** 0.4207; **(b)** 0.5793; **(c)** 0.1633; **(d)** $\mu = 5, \sigma^2 = 4$, $\sigma = 2$.

2.4-7 **(a)** $b(2000, \pi/4)$; **(b)** 1570.796, 337.096, 18.360;
(c) π; **(f)** $V_n = \pi^{n/2}/\Gamma(n/2 + 1)$ is the volume of a ball of radius 1 in n-space.

2.4-9 **(a)** $b(20, 0.80)$; **(b)** $\mu = 16, \sigma^2 = 3.2, \sigma = 1.789$;
(c) (i) 0.1746, (ii) 0.6296, (iii) 0.3704.

2.4-11 0.1268.

2.4-13 **(a)** $\mu = 7.2, \sigma^2 = 1.44, \sigma = 1.2$;
(c) $\bar{x} = 7.32, s^2 = 1.4117, s = 1.1882$.

2.4-15 **(a)** 0.6513; **(b)** 0.7941.

2.4-17

x	$b(8, 0.2)$	$N_1 = 8,$ $N_2 = 32$	$N_1 = 16,$ $N_2 = 64$
0	0.1678	0.1368	0.1527
1	0.3355	0.3501	0.3429
2	0.2936	0.3299	0.3104
3	0.1468	0.1466	0.1473
4	0.0459	0.0327	0.0399
5	0.0092	0.0036	0.0063
6	0.0011	0.0002	0.0006
7	0.0001	0.0000	0.0000
8	0.0000	0.0000	0.0000

2.4-19 0.9841.

2.4-21 Take longer route.

2.4-23 **(a)** 0.0778; **(b)** 0.3456; **(c)** 0.9898.

2.5-1 **(a)** $b(1, 2/3)$; **(b)** $b(12, 0.75)$.

2.5-3 $\mu = 2, \sigma^2 = 4/5$,

$$f(x) = \begin{cases} 2/5, & x = 1, \\ 1/5, & x = 2, \\ 2/5, & x = 3. \end{cases}$$

2.5-5 $(4/5)^3(1/5)$.

2.5-7 **(a)** 0.4604; **(b)** 0.5580; **(c)** 0.0184.

2.5-9 **(a)** $0.9^{12} = 0.2824$; **(b)** 0.0236.

2.5-11 **(a)** $f(x) = (1/2)^{x-1}, x = 2, 3, 4, \ldots$;
(b) $M(t) = e^{2t}/(2 - e^t), t < \ln 2$;
(c) $\mu = 3, \sigma^2 = 2, \sigma = 1.414$;
(d) (i) 3/4, (ii) 1/8, (iii) 1/4.

2.5-13 **(a)** $f(x) = (x - 1)/2^x, x = 2, 3, \ldots$;
(c) $\mu = 4, \sigma^2 = 4, \sigma = 2$;
(d) (i) 1/2, (ii) 5/16, (iii) 1/4.

2.5-15 25/3.

2.5-17 **(a)** 29.29; **(b)** about 8 years; **(c)** 49.77.

2.5-21 **(a)** $\mu = 1, \sigma^2 = 1$; **(b)** 19/30.

2.6-1 **(a)** 0.693; **(b)** 0.762; **(c)** 0.433.

2.6-3 0.540.

2.6-5 0.558.

2.6-7 0.947.

2.6-9 **(a)** 2.681; **(b)** $n = 6$.

2.6-11 **(a)** $\bar{x} = 7.94, s^2 = 7.73$, yes.

2.6-13 **(a)** $\bar{x} = 4/3 = 1.333, s^2 = 88/69 = 1.275$, yes;
(c) yes.

2.6-15 **(a)** 2.967; **(b)** 2.792; **(c)** 1.671.

2.6-17 If $n = 80$, $Ac = 3$, $Oc(0.02) \approx 0.92$, $Oc(0.08) \approx 0.12$.

2.6-19 **(a)** 0.564 using binomial, 0.560 using Poisson approximation.
(b) \$598.56 using binomial, \$613.90 using Poisson approximation.

2.6-21 Using Minitab, **(a)** 0.0821; **(b)** 0.2138; **(c)** 0.4562; **(d)** 0.7576.

Chapter 3

3.1-1 **(a)** $\bar{x} - 1.1$; **(b)** $s^2 - 0.0350$; **(c)** $s - 0.1871$.

3.1-3 **(a)** Frequencies: [3, 15, 8, 1, 0, 5, 8, 2];
(b) $\bar{x} = 17.30; s_x = 2.34$.

3.1-5 **(a)** $\bar{x} - 16.706, s = 1.852$;
(b) Frequencies: [1, 1, 2, 14, 18, 16, 23, 10, 7, 2, 1, 1].

3.1-7 **(a)** $\bar{x} - 112.12; s - 231.3576$;
(d) Half of the observations are less than 48.

3.1-9 **(b)** $\bar{x} = 7.275, s = 1.967$.

3.1-11 **(a)** With class boundaries $90.5, 108.5, 126.5, \ldots, 306.5$:
Frequencies: [8, 11, 4, 1, 0, 0, 0, 1, 2, 12, 12, 3];
(b) $\bar{x} = 201$;
(c) With class boundaries $47.5, 52.5, \ldots, 107.5$:
Frequencies: [4, 4, 9, 4, 4, 0, 3, 9, 7, 5, 4, 1];
(d) $\bar{x} = 76.35$.

3.2-1 **(a)**

Stems	Leaves	Frequency	Depths
11	9	1	1
12	3	1	2
13	6 7	2	4
14	1 1 4 4 4 6 7 8 8 8 8 9 9 9	14	18
15	0 0 1 3 4 5 5 6 6 6 6 6 7 8 8 9 9 9	18	36
16	1 1 1 1 1 3 3 5 5 6 6 6 7 7 8 9	16	(16)
17	0 1 1 1 1 1 2 2 2 3 4 4 4 6 6 7 7 8 8 8 8 8 8	23	44
18	0 0 0 1 1 4 5 8 9 9	10	21
19	0 1 3 3 4 7 8	7	11
20	2 8	2	4
21	5	1	1
22	1	1	1

(Multiply numbers by 10^{-1}.)
(b) 11.9, 15.5, 16.65, 17.8, 22.1;
(c) There are three suspected outliers.

3.2-3 **(a)** Baby carrots: 1.02, 1.03, 1.04, 1.04, 1.06;
(b) Regular-size carrots: 1.00, 1.15, 1.21, 1.26, 1.43;
(c) Regular-size packages tend to be heavier.

3.2-5 **(a)** Frequencies for males: [1, 1, 3, 4, 20, 23, 16, 10, 3, 0, 1],

Frequencies for females: [5, 14, 32, 36, 13],

(c) Five-number summary for males: 1.4, 3.5, 4.0, 4.525, 6.5;

Five-number summary for females: 0.5, 1.325, 1.7, 2.0, 2.7.

3.2-7 **(b)** Five-number summary: 5, 35/2, 48, 173/2, 1,815;

(d) Inner fence at 190, outer fence at 293.5;

(e) The mean is greatly influenced by outliers.

3.2-9 **(a)**

Stems	Leaves	Frequency	Depths
127	8	1	1
128	8	1	2
129	5 8 9	3	5
130	8	1	6
131	2 3 4 4 5 5 7	7	(7)
132	2 7 7 8	4	7
133	7 9	2	3
134	8	1	1

(Multiply numbers by 10^{-1}.)

(b) 131.3, 7.0, 2.575, 131.45, 131.47, 3.034.

(c) Five-number summary: 127.8, 130.125, 131.45, 132.70, 134.8.

3.2-11 **(a)**

Stems	Leaves	Frequency	Depths
30f	5	1	1
30s		0	1
30•	8 8	2	3
31*	0 0 1	3	6
31t	2 3 3	3	9
31f	4 4 4	3	12
31s	6 6 7 7 7	5	17
31•	8 8 8 9 9	5	22
32*	0 0 0 0 0 1	6	(6)
32t	2 2 2 3 3	5	22
32f	4 4 4 4 5 5	6	17
32s	6 7 7	3	11
32•	8 8 9 9 9	5	8
33*	0 1	2	3
33t		0	1
33f	5	1	1

(b) Five-number summary: 305, 315.5, 320, 325, 335;

3.3-1 **(b)**

$$F(x) = \begin{cases} 0, & x < 0, \\ x(2 - x), & 0 \le x < 1, \\ 1, & 1 \le x. \end{cases}$$

(c) (i) 3/4, (ii) 1/2, (iii) 0, (iv) 1/16.

3.3-3 **(a)** (i) 3; (ii) $F(x) = x^4$, $0 \le x \le 1$;

(b) (i) 3/16; (ii) $F(x) = (1/8)x^{3/2}$, $0 \le x \le 4$;

(c) (i) 1/4; (ii) $F(x) = x^{1/4}$, $0 \le x \le 1$.

3.3-5 **(a)** $4/5$, $2/75$, $\sqrt{6}/15$;

(b) 12/5, 192/175; $8\sqrt{21}/35$;

(c) 1/5, 16/225, 4/15.

3.3-7 **(a)** $\pi/2$, $(\pi^2 - 8)/4$;

(c) $F(x) = (1 - \cos x)/2$, $0 \le x \le \pi$.

3.3-9 **(a)** d = 2; **(b)** $E(Y) = 2$; **(c)** $E(Y^2)$ is unbounded.

3.3-11 **(a)** $\mu = 0, \sigma^2 = 3/5$;

(b) $\mu = 0, \sigma^2 = 1/3$;

(c) $\mu = 0, \sigma^2 = 1/6$.

3.3-13 **(a)** $\mu = 0$; **(b)** $\sigma^2 = 1$.

3.3-15 $f(x) = \dfrac{e^{-x}}{(1 + e^{-x})^2} = \dfrac{e^{-x}}{(1 + e^{-x})^2} \dfrac{e^{2x}}{e^{2x}}$

$$= \dfrac{e^x}{(e^x + 1)^2} = f(-x).$$

3.3-17 **(a)** $1/e$; **(b)** $1/e^{19/8}$.

3.3-19 **(a)** $F_n(x) = nx$, $0 \le x \le 1/n$.

3.3-21 $\theta = 2.33$.

3.3-23 $740.74.

3.3-25 **(a)** $\mu = \$28{,}571.43$, $\sigma = \$15{,}971.91$; **(b)** 0.6554.

3.4-3 **(a)** $f(x) = 1/10$, $0 < x < 10$; **(b)** 0.2; **(c)** 0.6;

(b) $\mu = 5$; (e) $\sigma^2 = 25/3$.

3.4-5 **(a)** $G(w) = (w - a)/(b - a)$, $a \le w \le b$;

(b) $U(a, b)$.

3.4-7 **(b)** $\bar{x} = 20.486$, $s^2 = 20.344$;

(c) $0.472 \approx 0.467$;

(d) $0.472 \approx 0.462 = 12/26$.

3.4-9 **(a)** $f(x) = (1/3)e^{-x/3}$, $0 < x < \infty$; 3; 9;

(b) $f(x) = 3e^{-3x}$, $0 < x < \infty$; 1/3; 1/9.

3.4-11 $P(X > x + y \mid X > x) = \dfrac{P(X > x + y)}{P(X > x)}$

$$= \dfrac{e^{-(x+y)/\theta}}{e^{-x/\theta}} = P(X > y).$$

3.4-13 **(a)** $\lambda = 0.025$;

(b) exponential distribution, $\theta = 40$;

(c) $\mu = 40, \sigma^2 = 1600$; **(d)** (i) $1 - e^{-1/2}$, (ii) e^{-1}, (iii) e^{-1}.

3.4-15 **(a)** 10.524; 9.320; yes.

3.4-17 **(a)** $F(x) = 1 - e^{-(x-\delta)/\theta}$, $\delta \le x < \infty$;

(b) $\theta + \delta$; θ^2.

3.4-19 **(a)** $-\theta \ln 0.75 = 0.2877\,\theta$;

(b) $(\ln 0.75 + 1)\theta = 0.7123\,\theta$;

(c) $-\theta \ln 0.25 = 1.3863\,\theta$;

(d) $(-\ln 0.25 - 1)\theta = 0.3863\,\theta$.

3.4-21 **(a)** $e^{-1/8}$;

(b) $f(x) = ae^{bx}\exp[-(a/b)(e^{bx}-1)]$, $0 < x < \infty$;

(c) 0.24.

3.4-23 $e^{-1/2}$.

3.5-1 (a) $f(x) = \dfrac{1}{\Gamma(10)(3/2)^{10}} x^9 e^{-2x/3}, \ 0 \le x < \infty;$

(b) $M(t) = (1 - 3t/2)^{-10}, \ t < 2/3; \ \mu = 15;$ $\sigma^2 = 45/2.$

3.5-5 $f(x) = \dfrac{1}{\Gamma(20)7^{20}} x^{19} e^{-x/7}, \ 0 \le x < \infty; \ \mu = 140;$ $\sigma^2 = 980.$

3.5-7 (a) $\mu = 240/7 = 34.286, \ \bar{x} = 32.636;$

(b) $\sigma^2 = 28{,}800/49 = 587.755, \ s^2 = 548.338;$

(c) $0.605 \approx 0.591 = 13/22.$

3.5-9 (a) 0.025; (b) 0.05; (c) 0.94; (d) 8.672; (e) 30.19.

3.5-11 (a) 0.80; (b) $a = 11.69, b = 38.08;$

(c) $\mu = 23, \sigma^2 = 46;$ (d) 35.17, 13.09.

3.5-13 (a) $r - 2;$ (b) $x = r - 2 \pm \sqrt{2r - 4}, r \ge 4.$

3.5-15 0.9444.

3.5-17 (a) $e^{-(125/216)};$ (b) $120 * \Gamma(4/3) = 107.1575.$

3.5-19 1.96 yields an expected profit of \$660.99.

3.6-1 (a) 0.2784; (b) 0.7209; (c) 0.9616; (d) 0.0019; (e) 0.9500; (f) 0.6826; (g) 0.9544; (h) 0.9974.

3.6-3 (a) 2.326; (b) $-2.576;$ (c) 1.67; (d) $-2.17.$

3.6-5 (a) 1.96; (b) 1.96; (c) 1.645; (d) 1.645.

3.6-7 (a) 0.3849; (b) 0.5403; (c) 0.0603; (d) 0.0013; (e) 0.6826; (f) 0.9544; (g) 0.9974; (h) 0.99.

3.6-9 (a) 0.6326; (b) 50.

3.6-11 (a) Gamma $(\alpha = 1/2, \ \theta = 8);$ (b) Gamma $(\alpha = 1/2, \ \theta = 2\sigma^2).$

3.6-13 Both could be normal because of the linearity of the q–q plots.

3.6-15 (a)

Stems	Leaves	Frequency	Depths
7•	938	1	1
8*	032 089	2	3
8t	222 268 383	3	6
8f	442 490 528 572	4	(4)
8s	674 734 786	3	9
8•	850 873 920	3	6
9*	069 150	2	3
9t	243	1	1

(Multiply numbers by 10^{-4}.)

(b) Normal quantiles: $-1.645, -1.282, -1.036,$ $-0.842, -0.674, -0.524, -0.385, -0.253,$ $-0.126, 0, 0.126, 0.253, 0.385, 0.524, 0.674,$ $0.842, 1.036, 1.282, 1.645.$

(c) Yes, because of the linearity of the q–q plot.

3.6-17 0.025.

3.6-19 (a) The frequencies of 450, 451, 452, and so on are 2, 3, 3, 5, 6, 15, 19, 22, 11, 5, 5, 3, 0, 1; (b) yes.

3.6-21 (a) $N(0,1);$ (b) $N(-1,1);$ (c) $N(2,1).$

3.6-23 0.1437.

3.6-25 (a) $\sigma = 0.043;$ (b) $\mu = 12.116.$

3.6-27 (a) 0.0228; (b) 0.6844.

3.7-1 (a) $e^{-(5/10)^2} = e^{-1/4} = 0.7788.$

3.7-3 Weibull with parameters α and $\beta/3^{1/\alpha}.$

3.7-5 (a) 0.5; (b) 0; (c) 0.25; (d) 0.75; (e) 0.625; (f) 0.75.

3.7-7 (b) $\mu = 31/24, \sigma^2 = 167/567;$

(c) 15/64; 1/4; 0; 11/16.

3.7-9 (a)

$$F(x) = \begin{cases} 0, & x < 0, \\ x/2, & 0 \le x < 1, \\ 1/2, & 1 \le x < 2, \\ 4/6, & 2 \le x < 4, \\ 5/6, & 4 \le x < 6, \\ 1, & 6 \le x. \end{cases}$$

(b) \$2.25.

3.7-11 $3 + 5 e^{-3/5} = 5.744.$

3.7-13 226.21, 1,486.92.

3.7-15 $\mu = 345.54, \sigma = 780.97.$

3.7-17 $M = 83.38.$

3.7-19 $g(y) = \dfrac{c}{3y} \exp(-(1/3)\ln y), \ e^{0.12} < y < e^{0.24};$ $c = 1/0.0376730928.$

3.7-21 $a = 0.00023 \approx 1/4000, b = 0.07270 \approx 1/14,$ mode $= 79.155.$

3.7-23 0.4219.

Chapter 4

4.1-1 (a) $f_1(x) = (2x + 5)/16, \ x = 1,2;$

(b) $f_2(y) = (2y + 3)/32, \ y = 1,2,3,4;$

(c) 3/32; (d) 9/32; (e) 3/16; (f) 1/4; (g) dependent.

4.1-3 (b) $f(x,y) = 1/16, \ x = 1,2,3,4; \ y = x + 1,$ $x + 2, x + 3, x + 4;$

(c) $f_1(x) = 1/4, \ x = 1,2,3,4;$

(d) $f_2(y) = (4 - |y - 5|)/16, \ y = 2,3,4,5,6,7,8;$

(e) dependent because the space is not rectangular.

4.1-5 (b) $b(6,1/2), b(6,1/2).$

4.1-7 11/30.

4.1-9 (a) $f(x,y) = \dfrac{15!}{x! \, y! \, (15 - x - y)!}$ $\times \left(\dfrac{6}{10}\right)^x \left(\dfrac{3}{10}\right)^y \left(\dfrac{1}{10}\right)^{15-x-y},$ $0 \le x + y \le 15;$

(b) no, because the space is not rectangular;

(c) 0.0735;

(d) X is $b(15,0.6);$

(e) 0.9095.

4.1-11 $f_1(x) = 2e^{-2x}, \ 0 < x < \infty;$ $f_2(y) = 2e^{-y}(1 - e^{-y}), \ 0 < y < \infty;$ no.

4.1-13 $f_1(x) = x/2$, $0 \le x \le 2$; $f_2(y) = 3y^2/8$, $0 \le y \le 2$; yes.

4.1-15 **(b)** 1/3.

4.2-1 $\mu_X = 25/16$; $\mu_Y = 45/16$; $\sigma_X^2 = 63/256$; $\sigma_Y^2 = 295/256$; $\text{Cov}(X, Y) = -5/256$; $\rho = -\sqrt{2,065}/1,239 = -0.0367$; dependent.

4.2-3 **(a)** $f(x,y) = 1/16$, $x = 1,2,3,4$; $y = x + 1$, $x + 2, x + 3, x + 4$; $f_1(x) = 1/4$, $x = 1,2,3,4$; $f_2(y) = (4 - |y - 5|)/16$, $y = 2,3,\ldots,8$;
(b) $\mu_X = 5/2$; $\mu_Y = 5$; $\sigma_X^2 = 5/4$; $\sigma_Y^2 = 5/2$; $\text{Cov}(X, Y) = 5/4$; $\rho = \sqrt{2}/2$;
(c) $y = x + 5/2$.

4.2-5 $a = \mu_Y - \mu_X b$, $b = \text{Cov}(X, Y)/\sigma_X^2$.

4.2-7 **(a)** No; **(b)** $\text{Cov}(X, Y) = 0$, $\rho = 0$.

4.2-9 **(a)** $f_1(1) = 0.15$, $f_1(2) = 0.25$, $f_1(3) = 0.45$, $f_1(4) = 0.15$; $f_2(1) = 0.35$, $f_2(2) = 0.65$; $\mu_X = 2.60$; $\mu_Y = 1.65$; $\sigma_X^2 = 0.8400$; $\sigma_Y^2 = 0.2275$;
(b) $\text{Cov}(X, Y) = -0.0900$; $\rho = -0.2059$;
(c) $E(C) = 34.70$.

4.2-11 **(a)** $f_1(x) = x + 1/2$, $0 < x < 1$, $f_2(y) = y + 1/2$, $0 < y < 1$;
(b) $\mu_X = 7/12$; $\mu_Y = 7/12$; $\sigma_X^2 = 11/144$; $\sigma_Y^2 = 11/144$; $\text{Cov}(X, Y) = -1/144$; $\rho = -1/11$;
(c) $y = 7/11 - x/11$.

4.2-13 **(a)** $c = 1/154$;
(c) $f_1(0) = 6/77, f_1(1) = 21/77$, $f_1(2) = 30/77, f_1(3) = 20/77$; $f_2(0) = 30/77, f_2(1) = 32/77, f_2(2) = 15/77$;
(d) no;
(e) $\mu_X = 141/77, \sigma_X^2 = 4836/5929$;
(f) $\mu_Y = 62/77, \sigma_Y^2 = 3240/5929$;
(g) $\text{Cov}(X, Y) = 1422/5929$;
(h) $\rho = 79\sqrt{12090}/24180$;
(i) $y = 215/806 + 237/806x$.

4.3-1 **(d)** 9/14, 7/18, 5/9; **(e)** 20/7, 55/49.

4.3-3 **(a)** $f(x,y) = \dfrac{50!}{x!\, y!\, (50 - x - y)!}$
$\times (0.02)^x (0.90)^y (0.08)^{50-x-y}$, $0 \le x + y \le 50$;
(b) Y is $b(50, 0.90)$;
(c) $b(47, 0.90/0.98)$;
(d) 2115/49; **(e)** $\rho = -3/7$.

4.3-5 **(a)** $E(Y|x) = 2(2/3) - (2/3)x$, $x = 1, 2$; **(b)** yes.

4.3-7 $E(Y|x) = x + 5/2$, $x = 1, 2, 3, 4$; yes.

4.3-9 **(a)** $f_1(x) = 1/8$, $x = 0, 1, \ldots, 7$;
(b) $h(y|x) = 1/3$, $y = x, x + 1, x + 2$, for $x = 0, 1, \ldots, 7$;

(c) $E(Y|x) = x + 1$, $x + 0, 1, \ldots, 7$;
(d) $\sigma_Y^2 = 2/3$;
(e)
$$f_2(y) = \begin{cases} 1/24, & y = 0, 9, \\ 2/24, & y = 1, 8, \\ 3/24, & y = 2, 3, 4, 5, 6, 7. \end{cases}$$

4.3-11 $E(Y) = x$ and $E(X) = 0.700$; thus \$700.

4.3-13 **(b)** $f_1(x) = 1/10$, $0 \le x \le 10$;
(c) $h(y|x) = 1/4$, $10 - x \le y \le 14 - x$ for $0 \le x \le 10$;
(d) $E(Y|x) = 12 - x$.

4.3-15 **(a)** $f(x,y) = 1/(2x^2)$, $0 < x < 2$, $0 < y < x^2$;
(b) $f_2(y) = (2 - \sqrt{y})/(4\sqrt{y})$, $0 < y < 4$;
(c) $E(X|y) = [2\sqrt{y}\ln(2/\sqrt{y})]/[2 - \sqrt{y}]$;
(d) $E(Y|x) = x^2/2$.

4.3-17 **(a)** $h(y|x) = e^{-x}$, $0 < y < e^x$ for $0 < x < 1$;
(b) $E(Y|x) = e^x/2$, $0 < x < 1$;
(c) $f(x,y) = e^{-x}$, $0 < x < 1$, $0 < y < e^x$;
(d)
$$f_2(y) = \begin{cases} 1 - 1/e, & 0 < y < 1, \\ 1/y - 1/e, & 1 \le y < e. \end{cases}$$

4.3-19 **(a)** $c = 8$; **(b)** 29/93.

4.4-1 **(a)** 0.6006; **(b)** 0.7888; **(c)** 0.8185; **(d)** 0.9371.

4.4-3 **(a)** 0.5746; **(b)** 0.7357.

4.4-5 **(a)** $N(86.4, 40.96)$; **(b)** 0.4192.

4.4-7 **(a)** 0.8248;
(b) $E(Y|x) = 457.1735 - 0.2655x$;
(c) $\text{Var}(Y|x) = 645.9375$;
(d) 0.8079.

4.4-9 $a(x) = x - 11$, $b(x) = x + 5$.

4.4-11 **(a)** 0.2857; **(b)** $\mu_{Y|z} = -0.2x + 4.7$;
(c) $\sigma_{Y|z}^2 = 8.0784$; **(d)** 0.4230.

4.4-13 **(a)** $X + Y$ is $N(5, 7.4)$, 0.6434; **(b)** 0.5249; **(c)** 0.05.

4.4-15 **(a)** 0.3721; **(b)** 0.1084.

Chapter 5

5.1-1 $g(y) = 2y$, $0 < y < 1$.

5.1-3 $g(y) = (1/8)y^5 e^{-y^2/2}$, $0 < y < \infty$.

5.1-5 $M(t) = (1 - 2t)^{-1}$, $t < 1/2$; exponential distribution with $\theta = 2$.

5.1-7 **(a)** $F(r) = (x - 0.03)/0.04$, $0.03 < r < 0.07$; $f(r) = 1/0.04$, $0.03 < r < 0.07$;
(b) The interest for each of n equal parts is R/n. The amount at the end of the year is $50,000(1 + R/n)^n$; the limit as $n \to \infty$ is $50,000\, e^R$.

5.1-9 **(a)** Exponential; **(b)** in the *Solutions Manual*;
(c) $\exp(-e^{-2}) = 0.873$.

5.1-11 (a) $\dfrac{1}{2} - \dfrac{\arctan 1}{\pi} = 0.25;$

(b) $\dfrac{1}{2} - \dfrac{\arctan 5}{\pi} = 0.0628;$

(c) $\dfrac{1}{2} - \dfrac{\arctan 10}{\pi} = 0.0317.$

5.1-13 (b) (i) $\exp(\mu + \sigma^2/2)$, (ii) $\exp(2\mu + 2\sigma^2)$, (iii) $\exp(2\mu + 2\sigma^2) - \exp(2\mu + \sigma^2)$.

5.1-15 (a) $g(y) = \dfrac{1}{\sqrt{2\pi y}}\exp(-y/2), \qquad 0 < y < \infty;$

(b) $g(y) = \dfrac{3}{2}\sqrt{y}, \qquad 0 < y < 1.$

5.2-1 $g(y_1, y_2) = (1/4)e^{-y_2/2}, 0 < y_1 < y_2 < \infty;$
$g_1(y_1) = (1/2)e^{-y_1/2}, 0 < y_1 < \infty;$
$g_2(y_2) = (y_2/4)e^{-y_2/2}, 0 < y_2 < \infty;$ no.

5.2-3 $\mu = \dfrac{r_2}{r_2 - 2}, r_2 > 2;$

$\sigma^2 = \dfrac{2r_2^2(r_1 + r_2 - 2)}{r_1(r_2 - 2)^2(r_2 - 4)}, r_2 > 4.$

5.2-5 (a) 14.80; **(b)** $1/7.01 = 0.1427$; **(c)** 0.95.

5.2-9 840.

5.2-11 (a) 0.1792; **(b)** 0.1792.

5.2-13 (a)

$$G(y_1, y_2) = \int_0^{y_1}\int_u^{y_2} 2(1/1000^2)$$
$$\times \exp[-(u + v)/1000]\,dv\,du$$
$$= 2\exp[-(y_1 + y_2)/1000]$$
$$- \exp[-y_1/500]$$
$$- 2\exp[-y_2/1000]$$
$$+ 1, \quad 0 < y_1 < y_2 < \infty;$$

(b) $2e^{-6/5} - e^{-12/5} \approx 0.5117.$

5.2-15 Each has mean 1/3 (8 hours). The respective standard deviations are 0.0943, 0.0673, and 0.0552. In case (3), the probability that a person sleeps between 7 and 9 hours is 0.546 and between 6 and 10 hours is 0.869. If we let $\alpha = 48$ and $\beta = 96$, then $\sigma = 0.0391$ and the probability that a person sleeps between 7 and 9 hours is 0.7116.

5.3-1 (a) 0.0182; **(b)** 0.0337.

5.3-3 (a) 36/125; **(b)** 2/7.

5.3-5

$$g(y) = \begin{cases} 1/36, & y = 2, \\ 4/36, & y = 3, \\ 10/36, & y = 4, \\ 12/36, & y = 5, \\ 9/36, & y = 6; \end{cases}$$

$\mu = 14/3, \sigma^2 = 10/9.$

5.3-7 2/5.

5.3-9 (a) 729/4096; **(b)** $\mu = 3/2; \sigma^2 = 3/40.$

5.3-11 (a) 0.0035; **(b)** 8; **(c)** $\mu_Y = 8, \sigma_Y^2 = 4.$

5.3-13 (c) Using *Maple*, we obtain $\mu = 13{,}315{,}424/3{,}011{,}805 = 4.4211.$

(d) $E(Y) = 5.377$ with 16 coins, $E(Y) = 6.355$ with 32 coins.

5.3-15 (a) $f(x) = [1 - (1/2)^x]^3 - [1 - (1/2)^{x-1}]^3,$
$x = 1, 2, \dots .$

(b) $\mu = 22/7 = 3.1429, \sigma^2 = 430/147 = 2.9252;$

(e) Like flipping a number of coins equal to the dimension.

5.3-17 $1 - e^{-3/100} \approx 0.03.$

5.3-19 $G(y) = 1 - ([y/500]e^{-y/500} + e^{-y/500})^3, 0 < y < \infty; P(Y < 300) = 1 - ([8/5]e^{-3/5})^3 \approx 0.323.$

5.3-21 21,816.

5.3-23 5.

5.4-1 (a)

$$g(y) = \begin{cases} 1/64, & y = 3, 12, \\ 3/64, & y = 4, 11, \\ 6/64, & y = 5, 10, \\ 10/64, & y = 6, 9, \\ 12/64, & y = 7, 8. \end{cases}$$

5.4-3 (a) $M(t) = e^{7(e^t - 1)};$ **(b)** Poisson, $\lambda = 7$; **(c)** 0.800.

5.4-5 0.925.

5.4-7 (a) $M(t) = 1/(1 - 5t)^{21}, t < 1/5;$

(b) gamma distribution, $\alpha = 21, \theta = 5.$

5.4-9 (a) $(1/24)(e^{2t} + 2e^{3t} + 3e^{4t} + 4e^{5t} + 4e^{6t} + 4e^{7t} + 3e^{8t} + 2e^{9t} + e^{10t});$

(b)

$$g(w) = \begin{cases} 1/24, & w = 2, 10, \\ 2/24, & w = 3, 9, \\ 3/24, & w = 4, 8, \\ 4/24, & w = 5, 6, 7. \end{cases}$$

5.4-13 First die has 0 on four faces and 2 on four faces; second die has faces numbered 0, 1, 4, 5, 8, 8, 12, 13.

5.4-15 (a)

$$h_1(w_1) = \begin{cases} 1/36, & w_1 = 0, \\ 4/36, & w_1 = 1, \\ 10/36, & w_1 = 2, \\ 12/36, & w_1 = 3, \\ 9/36, & w_1 = 4. \end{cases}$$

(b) $h_2(w) = h_1(w);$

(c)

$$h(w) = \begin{cases} 1/1296, & w = 0 \\ 8/1296, & w = 1, \\ 36/1296, & w = 2, \\ 104/1296, & w = 3, \\ 214/1296, & w = 4, \\ 312/1296, & w = 5, \\ 324/1296, & w = 6, \\ 216/1296, & w = 7, \\ 81/1296, & w = 8. \end{cases}$$

(d) With denominators equal to $6^8 = 1,679,616$, the respective numerators of $0, 1, \ldots, 16$ are 1, 16, 136, 784, 3,388, 11,536, 31,864, 72, 592, 137,638, 217,776, 286,776, 311,472, 274, 428, 190,512, 99,144, 34,992, 6,561.

(e) They are becoming more symmetric as n increases.

5.4-17 **(b)** $\mu_Y = 25/3, \sigma_Y^2 = 130/9$;

(c)
$$P(Y = y) = \begin{cases} 96/1024, & y = 4, \\ 144/1024, & y = 5, \\ 150/1024, & y = 6, \\ 135/1024, & y = 7. \end{cases}$$

5.4-19 $Y - X + 25$ is $b(50, 1/2)$;
$$P(Y - X \geq 2) = \sum_{k=27}^{50} \binom{50}{k} \left(\frac{1}{2}\right)^{50} = 0.3359.$$

5.4-21 **(a)** $\displaystyle\sum_{k=3}^{10} \binom{10}{k} (1 - 2/e)^k (2/e)^{10-k} = 0.5167$;

(b) 0.792.

5.4-23 $1 - 17/2e^3 = 0.5678$.

5.5-1 **(a)** 0.4772; **(b)** 0.8561.

5.5-3 **(a)** 46.58, 2.56; **(b)** 0.8447.

5.5-5 **(a)** 0.90; **(b)** 0.90.

5.5-7 **(b)** 0.05466; 0.3102.

5.5-9 **(a)** 0.8962; **(b)** 0.8962; **(c)** 87.31; 229.58.

5.5-11 **(a)** 0.3085; **(b)** 0.2267.

5.5-13 25.

5.5-15 $0.8413 > 0.7734$, select X.

5.5-17 $E(Z) = 0; E(1/\sqrt{U}) = \Gamma[(r - 1)/2]/\{\Gamma[r/2]\sqrt{2}\}; E(Z^2) = 1$;
$E(1/U) = 1/(r - 2)$.

5.5-19 **(a)** 2.567; **(b)** -1.740; **(c)** 0.90.

5.5-21 **(a)** $t(2)$; **(c)** $\mu_V = 0$; **(d)** $\sigma_V = 1$;

(e) In part b, numerator and denominator are not independent.

5.6-1 0.4772.

5.6-3 0.8185.

5.6-5 **(a)** $\chi^2(18)$; **(b)** 0.0756, 0.9974.

5.6-7 0.6247.

5.6-9 0.95.

5.6-11 **(a)** The frequencies are 5, 6, 12, 18, 31, 31, 20, 17, 6, 4; the fifth and sixth classes; **(c)** yes.

5.6-13 $444,338.13.

5.6-15 0.9522.

5.6-17 **(a)** $\displaystyle\int_0^{25} \frac{1}{\Gamma(13)2^{13}} y^{13-1} e^{-y/2} \, dy = 0.4810$ (using *Maple*);

(b) 0.4449 using normal approximation.

5.7-1 **(a)** 0.2878, 0.2881; **(b)** 0.4428, 0.4435; **(c)** 0.1550, 0.1554.

5.7-3 0.9258 using normal approximation, 0.9258 using binomial.

5.7-5 0.6915 using normal approximation, 0.7030 using binomial.

5.7-7 0.3085.

5.7-9 0.6247 using normal approximation, 0.6148 using Poisson.

5.7-11 **(a)** 0.5548; **(b)** 0.3823.

5.7-13 0.6813 using normal approximation, 0.6788 using binomial.

5.7-15 **(a)** 0.3802; **(b)** 0.7571.

5.7-17 0.4734 using normal approximation; 0.4749 using Poisson approximation with $\lambda = 50$; 0.4769 using $b(5000, 0.01)$.

5.7-19 0.6455 using normal approximation, 0.6449 using Poisson.

5.7-21 **(a)** 0.8289 using normal approximation, 0.8294 using Poisson.

(b) 0.0261 using tables in book, 0.0218 using *Maple*.

Chapter 6

6.1-3 **(b)** $\bar{x} = 89/40 = 2.225$.

6.1-5 **(a)** $\widehat{\theta} = \bar{X}/2$; **(b)** $\widehat{\theta} = \bar{X}/3$; **(c)** $\widehat{\theta}$ equals the sample median

6.1-7 **(c)** (i) $\widehat{\theta} = 0.5493, \tilde{\theta} = 0.5975$,
(ii) $\widehat{\theta} = 2.2101, \tilde{\theta} = 2.4004$,
(iii) $\widehat{\theta} = 0.9588, \tilde{\theta} = 0.8646$.

6.1-9 **(c)** $\bar{x} = 3.48$.

6.1-11 $1/9.5 = 20/190 = 0.1053$.

6.1-15 **(a)** $\tilde{\theta} = \bar{X}$; **(b)** yes; **(c)** 7.382; **(d)** 7.485.

6.1-17 **(a)** $\tilde{\lambda} = \bar{X}$;

(b) $\bar{x} = 1.17$;

(c) $s^2 = 1.24$, yes.

6.1-19 7.2.

6.2-1 $[71.35, 76.25]$.

6.2-3 **(a)** $\bar{x} = 15.757$, **(b)** $s = 1.792$, **(c)** $[14.441, 17.073]$.

6.2-5 $[48.467, 72.266]$ or $[48.076, 72.657]$.

6.2-7 $[19.47, 22.33]$.

6.2-9 **(a)** $\bar{x} = 633/16 = 39.5625, s = 8.7949$;

(b) $[36.39, 42.73]$ using t, $[36.52, 42.61]$ using normal approximation.

6.2-11 **(a)** $\bar{x} = 16.706$;

(b) $s^2 = 3.4315, s = 1.852$;

(c) $[16.392, 17.020]$ with $t_{0.05}(95) = 1.661$, $[16.395, 17.017]$ using normal approximation.

6.2-13 $\bar{x} = 1.3347$, $s = 0.0630$, $[1.3151, 1.3542]$.

6.2-15 $[22.74, 25.48]$.

6.2-17 $[0.757, \infty)$.

6.2-19 **(a)** 29.49, 3.41; **(b)** $[0, 31.259]$;

      (c) yes, because of the linearity of the q–q plot of the data and the corresponding normal quantiles.

6.2-21 **(a)** $\bar{x} = 25.475$, $s = 2.4935$; **(b)** $[24.059, \infty)$.

6.3-1 $[-59.725, -43.275]$.

6.3-3 $[-5.845, 0.845]$.

6.3-5 $(-\infty, -1.828]$.

6.3-7 **(a)** Yes; **(b)** $[11.5, 13.7]$; **(d)** do not change.

6.3-9 **(a)** $\bar{d} = 0.450$; **(b)** $[-0.494, 1.394]$.

6.3-11 **(a)** $[-0.556, 1.450]$;

      (b) $[0.367, 1.863]$;

      (c) no for men, yes for women.

6.3-13 $[157.227, \infty)$.

6.3-15 $[-5.599, -1.373]$ assuming equal variances, otherwise $[-5.577, -1.394]$.

6.4-1 **(a)** $s = 6.144$; **(b)** $[4.406, 10.142]$ or $[4.107, 9.521]$.

6.4-3 **(a)** $\bar{x} = 273.04, s^2 = 3,155.54$;

      (b) $[2,079.43, 5,468.08]$;

      (c) $[45.60, 73.95]$; **(d)** $[44.02, 71.56]$; **(e)** yes.

6.4-5 $\left[\dfrac{\sum_{i=1}^{n}(X_i - \mu)^2}{\chi^2_{\alpha/2}(n)}, \dfrac{\sum_{i=1}^{n}(X_i - \mu)^2}{\chi^2_{1-\alpha/2}(n)} \right]$.

6.4-7 **(a)** $[142.72, 386.10]$; **(b)** $x = 220.69$, $s = 200.27$, yes.

6.4-9 **(b)** $\bar{x}$ or s; **(d)** $[148.11, 432.72]$ is an approximate 95% confidence interval.

6.4-11 **(a)** 0.5688; **(b)** $[0, 1.905]$, yes.

6.4-13 **(a)** 0.4987; **(b)** $[0, 1.835]$.

6.4-15 $g(a) = g(b)$ and $a = -b$; so $a = -t_{\alpha/2}(n-1) = -b$.

6.4-17 **(a)** $[0.97, 8.72]$; **(b)** $[0.214, 5.980]$; **(c)** $[0.155, 2.833]$.

6.4-19 **(a)** 8.9963, 0.7616; **(b)** $[2.67, 52.33]$;

      (c) both could be normal because of the linearity of the q–q plots.

6.5-1 **(a)** 0.0374; **(b)** $[0.0227, 0.0521]$; **(c)** $[0.0252, 0.0550]$; **(d)** $[0.0250, 0.0555]$; **(e)** $[0, 0.0497]$.

6.5-3 **(a)** 0.6900; **(b)** $[0.6613, 0.7187]$ or $[0.6607, 0.7179]$ or $[0.6606, 0.7179]$.

6.5-5 **(a)** 0.5061; **(b)** $[0.4608, 0.5513]$ or $[0.4609, 0.5511]$ or $[0.4607, 0.5513]$; **(c)** not necessarily.

6.5-7 **(a)** 0.1800; **(b)** $[0.0735, 0.2865]$; **(c)** $[0.0977, 0.3080]$; **(d)** $[0.0921, 0.3153]$.

6.5-9 **(a)** 0.6799; **(b)** $[0.6457, 0.7140]$ or $[0.6448, 0.7130]$ or $[0.6450, 0.7134]$.

6.5-11 **(a)** $[0.445, 0.515]$; **(b)** 91.04%.

6.5-13 $[0.207, 0.253]$.

6.5-15 **(a)** 0.2115; **(b)** $[0.1554, 0.2676]$.

6.5-17 **(a)** 0.2182; **(b)** $[0.1756, 1]$.

6.5-19 **(a)** 0.2726; **(b)** $[0.2254, 0.3198]$.

6.6-1 117.

6.6-3 **(a)** 1083; **(b)** $[6.047, 6.049]$; **(c)** \$58,800; **(d)** 0.0145.

6.6-5 **(a)** 257; **(b)** yes.

6.6-7 **(a)** 1068; **(b)** 2401; **(c)** 752.

6.6-9 2305.

6.6-11 451.

6.6-13 **(a)** 38; **(b)** $[0.621, 0.845]$.

6.6-15 601.

6.6-17 235.

6.6-19 144.

6.7-1 **(a)** $\hat{y} = 86.8 + (842/829)(x - 74.5)$;

      (c) $\widehat{\sigma^2} = 17.9998$.

6.7-3 **(b)** $\hat{y} = -7.877 + 0.0868x$.

6.7-5 **(a)** $\hat{y} = 10.6 - 0.015x$;

      (c) $\hat{y} = 5.47 + 0.0004x$;

      (e) horsepower.

6.7-7 Solve for α: $-t_{\alpha/2}(n-2) \leq \dfrac{\hat{\alpha} - \alpha}{\sqrt{\widehat{\sigma^2}/(n-2)}} \leq t_{\alpha/2}(n-2)$.

6.7-9 $[83.341, 90.259]$, $[0.478, 1.553]$, $[10.265, 82.578]$.

6.7-11 **(a)** $\hat{y} = 15.695 + 0.620x$;

      (c) $\widehat{\sigma^2} = 0.2595$;

      (d) $[0.4416, 0.7984]$.

6.7-13 **(a)** $\hat{y} = 1.896 + 0.538x$;

      (c) $\hat{\alpha} = 6.9667$, $\hat{\beta} = 0.5385$, $\widehat{\sigma^2} = 0.0549$;

      (d) $[6.8263, 7.1070]$, $[0.2081, 0.8688]$, $[0.0333, 0.1643]$.

6.7-15 **(a)** $\hat{y} = 3.575 + 1.225x$;

      (c) $[29.987, 31.285]$, $[0.923, 1.527]$, $[0.428, 3.018]$.

6.7-17 **(a)** $\hat{y} = 46.59 + 1.085x$.

6.8-1 **(a)** $[75.283, 85.113]$, $[83.838, 90.777]$, $[89.107, 99.728]$;

      (b) $[68.206, 92.190]$, $[75.833, 98.783]$, $[82.258, 106.577]$.

6.8-3 **(a)** $[2.246, 2.821]$, $[2.761, 3.174]$, $[3.133, 3.669]$;

      (b) $[1.821, 3.246]$, $[2.284, 3.651]$, $[2.696, 4.106]$.

6.8-5 **(a)** $[4.897, 8.444]$, $[9.464, 12.068]$, $[12.718, 17.004]$;

      (b) $[1.899, 11.442]$, $[6.149, 15.383]$, $[9.940, 19.782]$.

6.8-7 **(a)** $[19.669, 26.856]$, $[22.122, 27.441]$, $[24.048, 28.551]$, $[25.191, 30.445]$, $[25.791, 32.882]$;

(b) $[15.530, 30.996]$, $[17.306, 32.256]$, $[18.915, 33.684]$, $[20.351, 35.285]$, $[21.618, 37.055]$.

6.8-9 $\widehat{y} = 1.1037 + 2.0327x - 0.2974x^2 + 0.6204x^3$.

6.8-13 **(a)** $r = 0.143$;

(b) $\widehat{y} = 37.68 + 0.83x$;

(d) no;

(e) $\widehat{y} = 12.845 + 22.566x - 3.218x^2$;

(f) yes.

6.8-15 $\widehat{y} = 1.735042 - 0.000377x + 0.000124x^2$, yes.

Chapter 7

7.1-1 $\alpha = 7/27$; $\beta = 7/27$.

7.1-3 **(a)** 0.3032 using $b(100, 0.08)$, 0.313 using Poisson approximation, 0.2902 using normal approximation;

(b) 0.1064 using $b(100, 0.04)$, 0.111 using Poisson approximation, 0.1010 using normal approximation.

7.1-5 **(a)** $\alpha = 0.1056$; **(b)** $\beta = 0.3524$.

7.1-7 **(a)** $z = 2.269 > 1.645$, reject H_0;

(b) $z = 2.269 < 2.326$, do not reject H_0;

(c) p-value $= 0.0116$.

7.1-9 **(a)** $z = 1.758 > 1.645$, reject H_0;

(b) $z = 1.758 < 1.96$, do not reject H_0;

(c) p-value $= 0.0394$.

7.1-11 **(a)** $z = \dfrac{y/n - 0.40}{\sqrt{(0.40)(0.60)/n}} \geq 1.645$;

(b) $z = 2.215 > 1.645$, reject H_0.

7.1-13 **(a)** $H_0: p = 0.40$, $H_1: p > 0.40$, $z \geq 2.326$;

(b) $2.236 < 2.326$, do not reject H_0.

7.1-15 **(a)** $H_0: p = 0.037$, $H_1: p > 0.037$;

(b) $z \geq 2.326$;

(c) $z = 2.722 > 2.326$, reject H_0.

7.1-17 **(a)** $|z| \geq 1.960$;

(b) $1.726 < 1.960$, do not reject H_0.

7.1-19 $[0.007, 0.071]$, yes.

7.1-21 **(a)** $z = (\widehat{p} - 0.5)/\sqrt{(0.5)(0.5)/200} \leq -1.645$, reject H_0;

(b) $z = -1.697$; **(c)** p-value $= 0.045$.

7.1-23 $z = -0.473$, p-value $= 0.318$, does not support management's claim.

7.1-25 $z = -2.272$, p-value $= 0.012$, supports the manufacturer's claim.

7.2-1 **(a)** $1.4 < 1.645$, do not reject H_0;

(b) $1.4 > 1.282$, reject H_0.

(c) p-value $= 0.0808$.

7.2-3 **(a)** $z = (\overline{x} - 170)/2$, $z \geq 1.645$;

(b) $1.260 < 1.645$, do not reject H_0;

(c) 0.1038.

7.2-5 **(a)** $t = (\overline{x} - 3,315)/(s/\sqrt{30}) \leq -1.699$;

(b) $-1.414 > -1.699$, do not reject H_0;

(c) $0.05 < p$-value < 0.10 or p-value ≈ 0.08.

7.2-7 **(a)** $t = (\overline{x} - 47)/(s/\sqrt{20}) \leq -1.729$;

(b) $-1.789 < -1.729$, reject H_0;

(c) $0.025 < p$-value < 0.05, p-value ≈ 0.045.

7.2-9 **(a)** $t \geq 2.764$;

(b) $4.028 > 2.764$, reject H_0;

(c) p-value < 0.005.

7.2-11 **(a)** -4.60, p-value < 0.0001;

(b) clearly reject H_0;

(c) $[0, 14.573]$.

7.2-13 $1.477 < 1.833$, do not reject H_0.

7.2-15 $\overline{d} = 0.357$, $s_d = 0.561$, $t = 2.985$, p-value < 0.005.

7.3-1 **(a)** $t \leq -1.734$; **(b)** $t = -2.221 < -1.734$, reject H_0.

7.3-3 **(a)** $|t| = 0.374 < 2.086$, do not reject H_0 at $\alpha = 0.05$.

7.3-5 **(a)** $t < -1.706$; **(b)** $-1.714 < -1.706$, reject H_0;

(c) $0.025 < p$-value < 0.05.

7.3-7 **(a)** $t < -2.552$; **(b)** $t = -3.638 < -2.552$, reject H_0.

7.3-9 **(a)** $z > 1.645$; **(b)** $1.957 > 1.645$, reject H_0;

(c) p-value $= 0.0252$.

7.3-11 **(a)** $t = 1.20$, $0.20 < p$-value < 0.50, p-value $= 0.246$, do not reject H_0.

7.3-13 **(a)** $t = -1.67$, $0.05 < p$-value < 0.10, p-value $= 0.054$, fail to reject H_0.

7.3-15 **(a)** $z = 2.245 > 1.645$, reject H_0; **(b)** p-value $= 0.0124$.

7.3-17 **(a)** $r = 11$, $t = 3.177$, p-value < 0.005, reject H_0;

(b) $r = 14$, $t = 0.320$, p-value $= 0.754$, do not reject H_0;

(c) $r = 14$, $t = -0.112$, p-value $= 0.913$, do not reject H_0.

7.3-19 **(a)** $t = 3.440$, p-value < 0.005, reject H_0.

7.4-1 **(a)** $\chi^2 = \dfrac{10s^2}{525^2} \leq 3.940$; **(b)** $4.104 > 3.940$, do not reject H_0; **(c)** $0.05 < p$-value < 0.10, p-value $= 0.057$ using Minitab.

7.4-3 **(a)** $\chi^2 = 24s^2/140^2$, $\chi^2 \geq 36.42$;

(b) $29.18 < 36.42$, do not reject H_0.

7.4-5 **(a)** $59.53 < 64.28$, reject H_0;

(b) $0.025 < p$-value < 0.05, p-value $= 0.042$ using Minitab.

7.4-7 **(a)** $\chi^2 \geq 28.87$ or $s^2 \geq 48.117$;

(b) $\beta \approx 0.10$.

7.4-9 $F = 3.30 > 3.07$, reject $\sigma_X^2 = \sigma_Y^2$;
Welch's test: $t = -8.119 < -t_{0.05}(12) = -1.782$.

7.4-11 **(a)** $2.759 < 4.20$, do not reject H_0 at $\alpha = 0.05$;

(b) $s_x^2/s_y^2 = 0.818 < 2.96$, $s_y^2/s_x^2 = 1.222 < 3.18$;
do not reject H_0;

(c) $1.836 < 3.28$, do not reject H_0;

(d) $1.318 < 4.03$, $0.759 < 4.03$, do not reject H_0;

(e) $3.247 < 4.32$, $0.308 < 5.52$, do not reject H_0;

(f) $0.84 < F_{0.025}(24, 28)$, $1.19 < F_{0.025}(28, 24)$, do not reject H_0;

(g) $1.624 < 4.03$, do not reject H_0.

7.5-1 $7.875 > 4.26$, reject H_0.

7.5-3 $13.773 > 4.07$, reject H_0.

7.5-5 **(a)**

Source	SS	DF	MS	F	p-value
Treatment	31.112	2	15.556	22.33	0.000
Error	29.261	42	0.697		
Total	60.372	44			

(c) the respective means are 23.114, 22,556, and 21.120, with the eggs of the shortest lengths in the nests of the smallest bird.

7.5-7 $14.757 > 2.87$, reject H_0.

7.5-9 **(a)** $F \geq 4.07$;

(b) $4.106 > 4.07$, reject H_0;

(c) $4.106 < 5.42$, do not reject H_0;

(d) $0.025 < p\text{-value} < 0.05$, $p\text{-value} \approx 0.05$.

7.5-11 $10.224 > 4.26$, reject H_0.

7.5-13 **(a)** $F \geq 5.61$;

(b) $6.337 > 5.61$, reject H_0.

7.5-15 **(a)** $F = 12.47$, there seems to be a difference in feed supplements;

(b) yes, supplement B looks best and supplement C the poorest.

7.6-1 $18.00 > 5.14$, reject H_A.

7.6-3 **(a)** $7.624 > 4.46$, reject H_A;

(b) $15.538 > 3.84$, reject H_B.

7.6-5 **(a)** $1.723 < 2.90$, accept H_{AB};

(b) $5.533 > 4.15$, reject H_A;

(c) $28.645 > 2.90$, reject H_B.

7.6-7 **(a)** $1.727 < 2.37$, accept H_{AB};

(b) $2.238 < 3.27$, do not reject H_A;

(c) $2.063 < 2.87$, do not reject H_B.

7.6-9 **(a)** $0.111 < 3.89$, accept H_{AB};

(b) $9.692 > 4.75$, reject H_A;

(c) $2.606 < 3.89$, do not reject H_B.

7.6-11 **(a)**

Source	SS	DF	MS	F	p-value
Smoking History	84.899	2	42.449	12.90	0.000
Test	298.072	2	149.036	45.28	0.000
Interaction	2.815	4	0.704	0.21	0.927
Error	59.247	18	3.291		
Total	445.032	26			

7.7-1 $4.359 > 2.306$, reject H_0.

7.7-3 $5.373 > 2.896$, reject H_0.

7.7-5 $-0.45 < -0.3808$, reject H_0.

7.7-7 $[0.419, 0.802]$.

7.7-9 $|r| = 0.252 < 0.6613$, do not reject H_0.

7.7-13 $n = 9$.

(a) $\widehat{y} = 42.02 - 0.0060x$;

(b) $r = -0.656$, $0.01 < p\text{-value} < 0.025$, $p\text{-value} = 0.014$;

(c) $\widehat{y} = 55.67 - 0.0075x$, $r = -0.639$, $0.01 < p\text{ value} < 0.025$, $p\text{-value} = 0.017$;

(d) $r = 0.951$.

Chapter 8

8.1-1 $6.25 < 7.815$, do not reject if $\alpha = 0.05$; $p\text{-value} \approx 0.10$.

8.1-3 $7.60 < 16.92$, do not reject H_0.

8.1-5 **(a)** $q_3 > 7.815$;

(b) $q_3 = 1.744 < 7.815$; do not reject H_0.

8.1-7 $6.113 < 16.92$, do not reject H_0.

8.1-9 $2.010 < 11.07 = \chi_{0.05}^2(5)$, do not reject null hypothesis.

8.1-11 **(a)** Frequencies are 2, 4, 21, 33, 20, 14, 4, 1, 1;

(b) $q = 4.653 < 9.488$, do not reject H_0.

8.1-13 Grouping last two classes: $2.75 < 9.210$, not grouping: $3.46 < 11.34$; in either case, do not reject H_0.

8.1-15 $\bar{x} = 320.10$, $s^2 = 45.56$; using class boundaries $303.5, 307.5, \ldots, 335.5$, $q = 3.21 < 11.07 = \chi_{0.05}^2(5)$, do not reject.

8.2-1 $3.23 < 11.07$, do not reject H_0.

8.2-3 $2.40 < 5.991$, do not reject H_0.

8.2-5 $5.975 < \chi_{0.05}^2(2) = 5.991$, do not reject the null hypothesis; however, $p\text{-value} \approx 0.05$.

8.2-7 $8.449 < \chi_{0.05}^2(4) = 9.488$, do not reject the null hypothesis; $0.05 < p\text{-value} < 0.10$; $p\text{-value} = 0.076$.

8.2-9 $4.149 > \chi^2_{0.05}(1) = 3.841$, reject the null hypothesis; $0.025 < p\text{-value} < 0.05$; $p\text{-value} \approx 0.042$.

8.2-11 $23.78 > 21.03$, reject hypothesis of independence.

8.2-13 **(a)** $39.591 > 9.488$, reject hypothesis of independence;

  (b) $7.117 > 5.991$, reject hypothesis of independence;

  (c) $11.398 > 9.488$, reject hypothesis of independence.

  (d) $0, 0.03, 0.02$.

8.2-15 $3.603 < 12.59$, do not reject hypothesis of independent attributes.

8.3-1 **(b)** $\tilde{m} = 146, \tilde{\pi}_{0.80} = 270$;

  (c) $\tilde{q}_1 = 95, \tilde{q}_3 = 225$.

8.3-3 **(a)** $g_3(y) = 10(1 - e^{-y/3})^2 e^{-y}, 0 < y < \infty$;

  (b) $5(1 - e^{-5/3})^4 e^{-5/3} + (1 - e^{-5/3})^5 = 0.7599$;

  (c) $e^{-5/3} = 0.1889$.

8.3-5 **(a)** 0.2553; **(b)** 0.7483.

8.3-7 **(a)** $g_1(y) = 19(e^{-y/\theta})^{18} \dfrac{1}{\theta} e^{-y/\theta}, 0 < y < \infty$;

  (b) $1/20$.

8.3-9 **(a)** $L(\theta) = \exp[-\sum_{i=1}^{10}(x_i - \theta)]$ is a maximum when θ is as large as possible. Thus, $\hat{\theta} = Y_1$.

  (b) $g_1(y_1) = 10e^{-10(y_1 - \theta)}, \theta \le y_1 < \infty$.

  (c) $[y_1 - (1/10)\ln(20), y_1]$.

8.3-11 **(a)** $g_r(y) = \dfrac{n!}{(r-1)!(n-r)!}(1 - e^{-y})^{r-1}$
  $\times (e^{-y})^{n-r} e^{-y}, 0 < y < \infty$;

  (b) a beta p.d.f. with $\alpha = n - r + 1, \beta = r$.

8.3-13 **(a)** $g(y_1, y_n) = \dfrac{n!}{(n-2)!}(y_n - y_1)^{(n-2)}$,
  $0 < y_1 < y_n < 1$;

  (b) $h(w_1, w_2) = n(n-1)w^{n-1}(1 - w_1)^{n-2}$,
  $0 < w_1 < 1, 0 < w_2 < 1$;
  $h_1(w_1) = (n-1)(1 - w_1)^{n-2}, 0 < w_1 < 1$;
  $h_2(w_2) = nw_2^{n-1}, 0 < w_2 < 1$;

  (c) yes.

8.4-1 **(a)** 0.7812; **(b)** 0.7844; **(c)** 0.4528.

8.4-3 **(a)** $(6.31, 7.40)$; **(b)** $(6.58, 7.22), 0.8204$.

8.4-5 $(2.72, 2.84), 93.46\%$.

8.4-7 $(15.40, 17.05)$.

8.4-9 **(a)**

Stems	Leaves	Frequency	Depths
101	7	1	1
102	0 0 0	3	4
103		0	4
104		0	4
105	8 9	2	6
106	1 3 3 6 6 7 7 8 8	9	(9)
107	3 7 9	3	10
108	8	1	7
109	1 3 9	3	6
110	0 2 2	3	3

(b) $\tilde{\pi}_{0.25} = 106.0, \tilde{m} = 106.7, \tilde{\pi}_{0.75} = 108.95$;

(c) (i) $(102.0, 106.6), 89.66\%$; (ii) $(106.3, 107.7)$, 89.22%; (iii) $(107.3, 110.0), 89.66\%$.

(d) $[106.3, 107.7], 89.22\%$; $[105.87, 107.63], 90\%$.

8.4-11 **(a)** $\tilde{\pi}_{0.50} = \tilde{m} = 0.92$;

  (b) $(y_{41}, y_{60}) = (0.92, 0.93)$; 0.9426 using normal approximation, 0.9431 using binomial;

  (c) $\tilde{\pi}_{0.25} = 0.89$;

  (d) $(y_{17}, y_{34}) = (0.88, 0.90)$; 0.9504 using normal approximation, 0.9513 using binomial;

  (e) $\tilde{\pi}_{0.75} = 0.97$;

  (f) $(y_{67}, y_{84}) = (0.95, 0.98)$; 0.9504 using normal approximation, 0.9513 using binomial.

8.4-13 $y_4 = 5.08 < \pi_{0.25} < y_{15} = 5.27, y_{14} = 5.27 < \pi_{0.5} < y_{26} = 5.31, y_{24} = 5.30 < \pi_{0.75} < y_{35} = 5.35$.

8.4-15 **(a)** 14.75; **(b)** 0.8995.

8.5-1 **(a)** $-55 < -47.08$, reject H_0; **(b)** 0.0296;

  (c) $9 < 10$, do not reject H_0; **(d)** $p\text{-value} = 0.1334$.

8.5-3 **(a)** $y = 17, p\text{-value} = 0.0539$;

  (b) $w = 171, p\text{-value} = 0.0111$;

  (c) $t = 2.608, p\text{-value} = 0.0077$.

8.5-5 $w = 54, z = 1.533, p\text{-value} = 0.0661$, do not reject H_0.

8.5-7 **(a)** Let Y equal the number of weights less that 1.14 pounds. Then $C = \{y: y \le 4\}, \alpha = 0.0898$.

  (b) $y = 5$, fail to reject H_0;

  (c) $p\text{-value} = 0.2120$;

  (d) $z = 2.072 > 1.282, p\text{-value} = 0.0207$, reject H_0.

8.5-9 **(a)** $w = 145 > 126$ or $z = 3.024 > 1.645$, reject H_0; $p\text{-value} \approx 0.0014$.

8.5-11 **(a)** $C = \{w: w \le 79 \text{ or } w \ge 131\}, \alpha \approx 0.0539$, $w = 95$, do not reject H_0.

8.5-13 **(a)** $C = \{w: w \le 79 \text{ or } w \ge 131\}, \alpha \approx 0.0539$, $w = 107.5, p\text{-value} = 0.8798$, do not reject H_0.

8.5-15 $C = \{w: w \le 184 \text{ or } w \ge 280\}, \alpha \approx 0.0489$, $w = 241$, do not reject H_0.

8.5-17 **(a)** $w = 71$, reject $H_0, p\text{-value} = 0.0057$;

  (b) $w = 101$, do not reject $H_0, p\text{-value} = 0.7913$;

  (c) $w = 108$, do not reject $H_0, p\text{-value} = 0.8501$.

8.5-19 **(b)** $w = 223.5 < 224.25$, reject H_0;

  (c) $p\text{-value} \approx 0.01$;

  (d) reject H_0;

  (e) the p-values are approximately equal.

8.6-1 **(a)** $C = \{r: r \le 6\}, \alpha = 43/429, r = 6$, reject H_0;

  (b) $w = 95, p\text{-value} \approx 0.0027$, clearly reject H_0.

8.6-3 $(6!)(12!)\dbinom{11}{4} = 113,810,780,160,000$.

8.6-5 **(b)** $r = 11$, reject H_0; **(c)** $p\text{-value} = 0.0457$.

8.6-7 $r = 4$, p-value $= 43/1,716 = 0.025$, reject the hypothesis of randomness. There appears to be a trend effect.

8.6-9 $r = 7$, p-value $= 1,331/58,786 = 0.0226$, reject the hypothesis of randomness.

8.6-11 **(a)** $C = \{y: y \le 3 \text{ or } y \ge 11\}$, $y = 7$, do not reject H_0;

(b) $C = \{r: r \le 4 \text{ or } r \ge 12\}$, $\alpha = 43/858 = 0.05$, $r = 4$, reject the hypothesis of randomness. There appears to be a trend effect.

8.6-13 $C = \{r: r \le 10\}$, $\alpha = 919/22,287 = 0.041$, $r = 6$, p-value $= 1,133/1,002,915 = 0.0011$, reject H_0.

8.6-15 The number of runs, 10, is within one standard deviation of $\mu = 12$.

8.6-17 No, because the numbers of runs are close to the mean, 10.

8.6-19 Do not reject the null hypothesis at $\alpha = 0.10$.

8.7-1 $d_5 = 0.40 < 0.45$, do not reject H_0.

8.7-3 $d_{10} = 0.1643$ at $x = 49$, do not reject H_0.

8.7-5 **(a)** $\bar{x} = 1.3347$, $s^2 = 0.00397$;

(b) $d_{30} = 0.0738$ at $x = 1.28$, do not reject H_0.

8.7-7 $d_{40} = 0.04559$, do not reject H_0.

8.7-9 **(a)** $d_{70} = 0.1295 > 1.07/\sqrt{70} = 0.1279$, reject H_0;

(b) yes, the histogram is bimodal.

8.7-11 $d_{105} = 0.07$ at $x = 6$; fit is very good.

8.7-13 $\hat{\theta} = 0.28588$, $\hat{\delta} = -0.41588$, $d_{18} = 0.0873$ at $x = -0.31$.

Chapter 9

9.1-3 $3800(10/38) = 1000$; 307.79.

9.1-5 $2.67, 8, 16, 21.33, 64$.

9.2-1 **(a)** $k(\theta \mid y) \propto \theta^{\alpha+y-1} e^{-\theta(n+1/\beta)}$.

Thus, the posterior p.d.f. of θ is gamma with parameters $\alpha + y$ and $1/(n + 1/\beta)$.

(b) $w(y) = E(\theta \mid y) = (\alpha + y)/(n + 1/\beta)$.

(c) $w(y) = \left(\dfrac{y}{n}\right)\left(\dfrac{n}{n + 1/\beta}\right)$

$\qquad + (\alpha\beta)\left(\dfrac{1/\beta}{n + 1/\beta}\right)$.

9.2-3 **(a)** $E[\{w(Y) - \theta\}^2] = \{E[w(Y) - \theta]\}^2$

$\qquad + \mathrm{Var}[w(Y)]$

$\qquad = (74\theta^2 - 114\theta + 45)/500$;

(b) $\theta = 0.569$ to $\theta = 0.872$.

9.2-5 The median (or mean), because the posterior p.d.f. is symmetric.

9.2-7 $d = 2/n$.

9.3-5 **(c)** $2yz$ is $\chi^2(2n)$; $\left(\dfrac{2y}{\chi^2_{\alpha/2}(2n)}, \dfrac{2y}{\chi^2_{1-\alpha/2}(2n)}\right)$.

Chapter 10

10.1-1 **(a)** $\sum_{i=1}^{n} X_i^2$; **(b)** $\widehat{\sigma^2} = \left(\dfrac{1}{n}\right)\sum_{i=1}^{n} X_i^2$; **(c)** yes.

10.1-7 **(a)** $f(x; p) =$

$\qquad \exp\{x\ln(1 - p) + \ln[p/(1 - p)]\}$;

$\qquad K(x) = x$; $\sum_{i=1}^{n} X_i$ is sufficient.

(b) $\bar{X}$.

10.2-1 **(a)** $K(\mu) = \Phi\left(\dfrac{22.5 - \mu}{3/2}\right)$; $\alpha = 0.0478$;

(b) $\bar{x} = 24.1225 > 22.5$, do not reject H_0;

(c) 0.2793.

10.2-3 **(a)** $K(\mu) = \Phi\left(\dfrac{510.77 - \mu}{15}\right)$;

(b) $\alpha = 0.10$;

(c) 0.5000;

(e) (i) 0.0655, (ii) 0.0150.

10.2-5 $n = 25$, $c = 1.6$.

10.2-7 $n = 40$, $c = 678.38$.

10.2-9 **(a)** $K(p) = \sum_{y=14}^{25} \binom{25}{y} p^y (1 - p)^{25-y}$, $0.40 \le p \le 1.0$;

(b) $\alpha = 0.0778$;

(c) $0.1827, 0.3450, 0.7323, 0.9558, 0.9985, 1.0000$;

(e) yes;

(f) 0.0344.

10.2-11 With $n = 130$, $c = 8.5$, $\alpha \approx 0.033$, $\beta \approx 0.094$.

10.3-1 **(a)** $\dfrac{L(80)}{L(76)} = \exp\left[\dfrac{6}{128}\sum_{i=1}^{n} x_i - \dfrac{624n}{128}\right] \le k$ or $\bar{x} \le c$;

(b) $n = 43$, $c = 78$.

10.3-3 **(a)** $\dfrac{L(3)}{L(5)} \le k$ if and only if $\sum_{i=1}^{n} x_i \ge (-15/2)$ $[\ln(k) - \ln(5/3)^n] = c$;

(b) $\bar{x} \ge 4.15$;

(c) $\bar{x} > 4.15$;

(d) yes.

10.3-5 **(a)** $\dfrac{L(50)}{L(\mu_1)} \le k$ if and only if $\bar{x} \le$ $\dfrac{(-72)\ln(k)}{2n(\mu_1 - 50)} + \dfrac{50 + \mu_1}{2} = c$.

10.3-7 **(a)** $\dfrac{L(0.5)}{L(\mu)} \le k$ if and only if $\sum_{i=1}^{n} x_i \ge \dfrac{\ln(k) + n(0.05 - \mu)}{\ln(0.5/\mu)} = c$;

(b) $\sum_{i=1}^{10} x_i \ge 9$.

10.4-1 **(a)** $|-1.80| > 1.645$, reject H_0;

    (b) $|-1.80| < 1.96$, do not reject H_0;

    (c) p-value $= 0.0718$.

10.4-3 **(a)** $\bar{x} \geq 230 + 10z_\alpha/\sqrt{n}$ or $\dfrac{\bar{x} - 230}{10/\sqrt{n}} \geq z_\alpha$;

    (b) yes; **(c)** $1.04 < 1.282$, do not reject H_0;

    (d) p-value $= 0.1492$.

10.4-5 **(a)** $|2.10| < 2.306$, do not reject H_0;

    $0.05 < p$-value < 0.10.

10.4-7 $2.20 > 1.282$, reject H_0; p-value $= 0.0139$.

10.5-1 **(a)** 0.84; **(b)** 0.082.

10.5-3 $k = 1.464$; $8/15$.

10.5-5 **(a)** 0.25; **(b)** 0.85; **(c)** 0.925.

10.5-7 675.

10.6-1 **(a)** 0.9984; **(b)** 0.998.

10.6-3 $M(t) = \left[1 - \dfrac{2t\sigma^2}{n-1}\right]^{-(n-1)/2} \rightarrow e^{\sigma^2 t}$.

10.7-1 **(b)** σ^2/n; **(c)** $2/n$.

10.7-3 **(a)** $2\theta^2/n$; **(b)** $N(\theta, 2\theta^2/n)$; **(c)** $\chi^2(n)$.

Chapter 11

11.1-9 **(a)** There is a quadratic trend upwards.

11.2-1 **(a)** $158.97, 12.1525, 30.55$; **(f)** yes.

11.2-3 **(a)** $5.176(335.176), 0.5214, 1.294$; **(f)** no.

11.2-5 **(b)** $E(\sqrt{Y}) = \dfrac{\sqrt{2}\,\Gamma\left(\dfrac{n}{2}\right)}{\Gamma\left(\dfrac{n-1}{2}\right)}$;

    (c) $S = \dfrac{\sigma\sqrt{Y}}{\sqrt{n-1}}$ so

    $E(S) = \dfrac{\sqrt{2}\,\Gamma\left(\dfrac{n}{2}\right)}{\sqrt{n-1}\,\Gamma\left(\dfrac{n-1}{2}\right)}\,\sigma$.

11.2-7 LCL $= 0$, UCL $= 0.1024$;

    (a) 0.0378; **(b)** 0.1752.

11.2-9 LCL $= 0$, UCL $= 4.9496$;

    (a) 0.185; **(b)** 0.871.

11.2-11 (a) LCL $= 0$, UCL $= 13.99$; (b) yes.

11.3-1

2^2 Design			
Run	A	B	AB	Observations
1	$-$	$-$	$+$	X_1
2	$+$	$-$	$-$	X_2
3	$-$	$+$	$-$	X_3
4	$+$	$+$	$+$	X_4

    (a) $[A] = (-X_1 + X_2 - X_3 + X_4)/4$,
$[B] = (-X_1 - X_2 + X_3 + X_4)/4$,
$[AB] = (X_1 - X_2 - X_3 + X_4)/4$.

    (b) It is sufficient to compare the coefficients on both sides of the equations of X_1^2, X_1X_2, X_1X_3, and X_1X_4, which are $3/4, -1/2, -1/2$, and $-1/2$, respectively.

    (c) Each is $\chi^2(1)$.

11.3-3 $[A]$ is $N(0, \sigma^2/2)$ so $E[(X_2 - X_1)^2/4] = \sigma^2/2$ or $E[(X_2 - X_1)^2/2] = \sigma^2$.

11.3-5 **(a)** $[A] = -4, [B] = 12, [C] = -1.125$, $[D] = -2.75, [AB] = 0.5, [AC] = 0.375$, $[AD] = 0, [BC] = -0.625, [BD] = 2.25$, $[CD] = -0.125, [ABC] = -0.375$, $[ABD] = 0.25, [ACD] = -0.125$, $[BCD] = -0.375, [ABCD] = -0.125$.

    (b) There is clearly a temperature (B) effect. There is also a catalyst charge (A) effect and probably a concentration (D) and a temperature–concentration (BD) effect.

Index

Any entries referring to D1-D17 are found in Appendix D on the CD-ROM

Any entries referring to D1-D17 are found in Appendix D on the CD-ROM

Any entries referring to D1-D17 are found in Appendix D on the CD-ROM

Any entries referring to D1-D17 are found in Appendix D on the CD-ROM

Any entries referring to D1-D17 are found in Appendix D on the CD-ROM

Any entries referring to D1–D17 are found in Appendix D on the CD-ROM

Any entries referring to D1-D17 are found in Appendix D on the CD-ROM

Any entries referring to D1-D17 are found in Appendix D on the CD-ROM